THE EXCITEMENT AND FASCINATION OF SCIENCE:
Volume Two

Reflections by Eminent Scientists

THE EXCITEMENT AND FASCINATION OF SCIENCE:
Volume Two

Reflections by Eminent Scientists

Compiled by **WILLIAM C. GIBSON**
University of British Columbia

Reprinted from the *Annual Review of Anthropology*, the *Annual Review of Astronomy and Astrophysics*, the *Annual Review of Biochemistry*, the *Annual Review of Earth and Planetary Sciences*, the *Annual Review of Entomology*, the *Annual Review of Fluid Mechanics*, the *Annual Review of Genetics*, the *Annual Review of Microbiology*, the *Annual Review of Pharmacology and Toxicology*, the *Annual Review of Physical Chemistry*, the *Annual Review of Physiology*, the *Annual Review of Phytopathology*, and the *Annual Review of Plant Physiology*.

ANNUAL REVIEWS INC.   4139 EL CAMINO WAY   PALO ALTO, CALIFORNIA 94306

ANNUAL REVIEWS INC.
Palo Alto, California, USA

REPRINTS   The conspicuous number aligned in the margin with the title of each article in this volume is a key for use in ordering reprints. Available reprints are priced at the uniform rate of $1.00 each postpaid. The minimum acceptable reprint order is 5 reprints and/or $5.00 prepaid. A quantity discount is available.

International Standard Book Numbers: 0-8243-2601-6 (hard cover)
                                     0-8243-2602-4 (soft cover)
Library of Congress Catalog Card Number: 65-29005

Annual Reviews Inc. and the Editors of its publications assume no responsibility for the statements expressed by the contributors to this volume.

PRINTED AND BOUND IN THE UNITED STATES OF AMERICA

FOREWORD

Science without scientists would surely be an impossibility. Who, then, are those men and women who work so diligently and effectively, trying, as William Harvey said, "to read the book of Nature?"

Thirteen years ago Annual Reviews Inc. published its first volume of autobiographical chapters by scientists who had made great contributions and who could recount engagingly their development as scientists. Such chronicles are all too rare, alas, for they should be the heady diet of secondary school and college students searching for fact, rather than fancy, on which to base the selection of a field of study and possibly a career.

The initial appearance and growing popularity of these philosophical or historical chapters in the various Annual Reviews is entertainingly described in Dr. J. Murray Luck's Foreword to Volume I of *The Excitement and Fascination of Science*. That volume broke new ground in that it presented, for all to see, an intimate picture of many of our most distinguished workers in the vineyards of science, written while they were, in most cases, still contributing.

The style of each author is unique, and some of the writing simply brilliant. In some chapters the content carries the day. In any event, the fortunate reader who picks up these books will reap important benefits. The story of science is replete with instances of great personal decisions having been made on key, but minimal, evidence. Discoveries too may hang on the merest thread of chance. But, as Pasteur remarked, "fortune favors the prepared mind." These essays can be instrumental in preparing young minds.

Today the average age of scientific investigators is a great deal less than we tend to imagine. Because of this, these workers may face their successes and failures with limited philosophical resources. The present volume and its predecessor have between their covers enough spiritual leaven to carry boatloads of bewildered young people safely to a further stage in their evolution.

The chapters selected for this second volume of *The Excitement and Fascination of Science* embrace many fields of learning and have been gleaned from thirteen Annual Review series that have included autobiographical chapters over the past ten years. There was insufficient space to include in either volume all of the inspiring and enlightening autobiographical chapters published since the first such chapter appeared in 1950, but readers who wish to pursue their special subjects further may seek additional chapters of particular interest to them in their reference libraries, where they will find complete runs of each Annual Review series.

Warm acknowledgement is made of the unfailing assistance rendered by Dr. Joshua Lederberg and the staff of Annual Reviews Inc. in the publication of this volume.

Wm. C. Gibson

The Excitement and Fascination
of Science: Volume Two (published 1978)

CONTENTS

Ann. Rev. Physiol., Vol. 30

RESEARCH
PROVIDES SELF-EDUCATION

By E. F. Adolph

E. F. Adolph

Ann. Rev. Physiol., Vol. 30

RESEARCH
PROVIDES SELF-EDUCATION

By E. F. Adolph

The University of Rochester, Rochester, New York

I speak of research as a source of men's powers and not so much as a product of their endeavors. I speak of research as a mainspring of their educations, as a means of self-formation and self-renewal. I ignore men's specialties and their peak accomplishments. I suggest that the far reach of what men in all vocations do, depends considerably on their widespread experience in research modeled on the labors of scientists. It is their means of self-education.

Often education is conceived to be a formal pursuit by which a student gains knowledge and degrees. To me this view seems superficial and incidental. I think education comes mostly from within. The student, like the mature scholar, develops a project or idea. At first it seems his project poses problems that he cannot solve. In time, however, the forbidding becomes the familiar. He builds his own bridges between problem and answer; he no longer waits for bridges to be furnished by others. He becomes self-sufficient.

Is this student related to the scientist who spends his days in a laboratory clad in work-clothes? Yes, the scientist knows that manipulation and experiment get to the center of whatever problem he tackles. Notions that can be tested produce more satisfying answers than do vague ideas; testable ideas keep education itself footed in surety. Theory without continuous checking may fall to nothing. A built-in self-education treats part of every project as manipulative, whether or not it uses physical equipment.

The formal researches of those who spend most of their lives in organized scholarly effort represent the backbone of all learning. Meanwhile, the amateur researches of those who solve problems form part of their daily living, like breathing and eating. By these little researches many men live, while engaged in any occupation, even one that has no recognized place for research as a vocation. Such researches develop from everyday ideas and deepen the minds that work them out.

First I will talk about some aspects of research; then about its place in education.

MOTIVES IN RESEARCH

What impels scientists and others to do research? The usual motives mentioned are the advancement of knowledge, the public service, and the pursuit of a career. These are powerful urges but leave out the big benefit to the individual, namely his intellectual development.

1

THE EXCITEMENT OF SCIENCE

8243-2601/78/1127-0003$01.00 © 1978 ARI   3

Some ordinary problem faces the beginner; the problem has an available answer. He says, I will find an answer, and he succeeds in doing so. The second time he tries, he seems to know better than the first time how to seek an answer. Soon experience accumulates, and the reasoning processes, plus the searches in libraries and in laboratories, come more easily. Before he realizes it, he is launched in a type of endeavor that stays with him; the universe of self-education stands open to him. Later he finds his interests are cumulative, that one question leads to another. Already he is specializing, his efforts are channeled, and others look to him for certain types of answers. The exercise that started by accident may turn out to be intellectual.

Such prowess develops naturally in some beginning students. Maybe seeds were planted in preschool years, and sprouted into independent projects in days of unpressured adolescence. Individual effort becomes fun. Once such activity is tried, every success reinforces it. Soon the student can profit from the opportunity to live under his own initiative, free of the obligation to show accomplishments that are measured by conventional examinations.

Many instances of awakenings to the promise of self-education could be cited. The philosopher John Dewey, when a college student, first grasped the "excitement of the effort to understand the world" while he was attending a short course in textbook physiology. He suddenly saw what concrete scientific work could accomplish, and overnight changed from an indifferent student to a fiery one (1).

Research opportunity rarely came to college students, even in my day. Linked to a course given by G. H. Parker was an announced option of laboratory research. Though this option was meant for graduate students, I, a third-year undergraduate, applied for it and was accepted. Whether I wanted to test my choice of a career field, or to test my capacity to pursue an unknown kind of activity, I do not recall. In my research I managed to identify some sensory factors that led fruit flies to lay eggs. Later I wrote a manuscript describing my results, challenged to do so by Dr. Parker. Apparently he did not find my manuscript suitable for publication. Three years later he suggested that now was the time for me to write a definitive paper. This was published (2).

I believe the idea of research was already familiar to me through my associations with graduate students of varied interests, and earlier with teachers in school who delighted in guiding students into insect collecting, bird spying, and herbarium making. Those activities can arouse both questions and explorations.

Today opportunities for project-researches have become widely available to high school students. This movement toward early individual effort reflects a widespread realization that the young mind is often an inquiring mind, looking for an interest to latch onto. The opportunity to stir students who want to undertake their own projects also has brought new life to many teachers. Danger comes when students are pressed into an activity that they do not enjoy, for without enjoyment they shun involvement. The heart of the student is indispensable to a project, even as it proved to be in the days of "progressive" education that made a project out of every bit of instruction from kindergarten blocks onward. The notion that project education was the best education had great merit; fault lay in the attempt to press all education into a mold.

Experienced scholars will say: but research is intended for publication of results. I think publishable results do and will come out of only a small fraction of research activities. That fraction will be chiefly formal research. But much research is not of that kind; rather it is intended to benefit the doer.

However, formulation of results and implications is a part of every research. Only by careful analysis on paper will most decisions among possible interpretations be reached, and most intellectual values be realized. A news reporter gets his report straight by writing for an imagined reader. So the self-educating student writes for an imagined or real recipient and critic. The writing itself is an act of self-discipline; the words compel us to define our problem, arrange our evidence, and clarify our result. One's own censor is rarely a sufficient judge, hence one presents the results to a colleague or other willing listener, he serves as target of one's effort. He sees the work with a new eye, a fresh concept, and an independent judgment.

I hail the view that a specific research experience yields a general education, whether one ever again does formal research or not. A plausible basis for the view consists in the realization that no one can predict the type of problem or lack of knowledge that one will face a few years hence. To meet the unknown, one can be educated best by dealing with unknowns. When one gains self-confidence in solution of problems, a whole life stands to benefit.

At some stages, research may seem to be a self-centered way to obtain an education. But good stewardship will allow the self-educated person to become in turn an educator. He will help others to find a self-sufficiency, to stand alone.

So, while research can be a formal activity, perhaps designed in the public interest, it can also be designed as an individual activity. Sometimes individual activity develops into formal activity, little science into big science, amateur research into funded research. Hopefully the private rewards may be as great in one as in the other. For, once an individual is discovering that an answer gained to one question helps him to find the answer to another question, the individual needs no guidance, and perhaps advances best without guidance. He has become self-educating; he can enter any realm; he meets no authoritarian boundary to his explorations. Actually the amateur has one great advantage: formal research will usually limit him to "recognized" methods of thought and deed, while independence can lead him where there are no prescribed methods.

THE EXCITEMENT OF SCIENCE   5

I think these characteristics and rewards of research activities hold in all departments of learning and of practical life; I can find no differences between one field of endeavor and another. Everywhere I observe that the outstanding benefit from research accrues to the doer himself. Research is his means of self-stimulation and of self-reward. He needles himself into intellectual effort. He advances his education and his capacities.

PATHWAYS IN RESEARCH

What path does the budding researcher follow? I do not believe that careers of scientists are predestined. But hindsight can usually make them appear so.

By a combination of advice and fortune, I decided to work for a doctorate with L. J. Henderson. He was a self-made intellectual in a university that valued individuals (3). Even his departmental ties were tenuous; but his determination to do what he liked was of unlimited strength. His researches were of great variety, both inside and outside of laboratories. Though superficially he might appear to wander, his persistence was intense and his probing was deep. I sometimes think his monument might be titled like Galileo's book (4) "Two new sciences". Henderson's sciences encompassed the concepts of (a) multivariate equilibria and (b) functional interactions. He had an intense faith in man's capacity to comprehend empirical systems, both inorganic and living and social (Fitness, Order, Interrelations). To me he became a shining example of the man who continually educated himself through research.

My first faculty position was in a university and department where research did not constitute part of one's job. But teaching was plentiful and was one's own responsibility. My problem became: how meet the task of heavy teaching and at the same time enlarge my horizon? For I already had an intensive urge to "get ahead" derived during graduate work and during a privileged year abroad. From the research viewpoint my efforts were fruitful, for publishable results emerged. Concurrently I was helpfully welcomed on a part-time basis in a laboratory (gas toxicology) of the U. S. Bureau of Mines and a laboratory (hot atmospheres) of the Society of Heating and Ventilating Engineers. Also I spent summers at Woods Hole, as I will mention below. These connections aided my efforts; how much initiative of mine was needed to form the connections I cannot estimate. Brashness and accident no doubt combined to make livable the self-imposed years of heavy teaching and tenuous research. In my case, self-education by means of research soon led to solid opportunities in both teaching and research.

In those years I found summer activities especially important. Research became possible to hundreds of biologists at the Marine Biological Laboratory (Woods Hole) and similar "biological stations". Most stations were actually founded to furnish work places for those who were otherwise isolated in one-man departments of colleges. Even in the earliest days, almost a century ago, it became apparent that the stations did much more than furnish the modest facilities; they were places where those of similar interests gathered. Biologists encouraged one another (5). Only in later years did many biologists pursue research at their own universities and form departments large enough to be self-contained. Fortunately, even today a remarkable number of biologists foregather at various biological stations each summer who otherwise would be unable to cultivate research interests. In my own development, four consecutive summers at Woods Hole allowed me to experiment with a variety of marine organisms. I taught part of a course that included observation of living aquatic animals.

It may be appropriate here to state how I once came to formulate a new research interest, that in Physiological Regulations. I reached a juncture where I craved a period of exploration. Were my past researches, I wondered, chosen by expediencies and short sight? In order to explore, I spent a few months in another university. I would not do experiments but would "doodle" on paper as I examined some speculative notions. Most of the speculative notions soon reached frustration stages. But one of them, an "equilibration diagram", that related either the water gain or the water loss of an animal to its bodily water content, yielded a pattern that I could apply. In the months and years following, this one concrete relationship led into successive studies of other aspects of regulatory arrangements (6) and the end is not yet. Yes, many physiologists had enunciated concepts of regulation; the truth is that their statements did not sink into me until I had developed concepts for myself; thereupon I noted similarities to theirs. Here I am not concerned with the novelty of the relationships visualized; I am concerned with what developed in my own mind. I learned in general that multiple ways of regarding physiological phenomena exist, and that I need not stay in customary paths.

How did I know that my work on equilibrations and physiological regulations was not illusory? I was well aware that colleagues believed physiological researches should be planned to elucidate mechanisms. I think my courage derived from an acute realization that in my new endeavor I was learning more than in previous periods of concentrated research; that I was now planning broadly; that I gained much, just by virtue of the fact that there were few models to follow. I had to work harder to interpret my results, and had to present them in an elementary and definitive fashion. Evidently the new path challenged me. Whether there is always a gain when one forsakes the familiar I do not know. Somehow one avoids frustrations and resolves doubts. I incline to think a well-worn path leaves too little room for play. At any rate, in the end my colleagues recognized that my path led to fruitful concepts. And it happened that cybernetics and transfer functions came along in time to make the field of regulatory physiology respected. My chief point here is that I clearly reaped a personal benefit even before my new effort could be recognized to be a contribution to physiology.

THE EXCITEMENT OF SCIENCE 7

A life of research differs for each of its practitioners. Though no account of it can represent it adequately, one can view it with the same interest as one views biography. Little or big, research has its traits, sharply seen in the careers of others and eventually in the experiences of oneself.

ENJOYMENT OF RESEARCH

Research is for those who enjoy it. For a short while a student can be driven into research, for a longer while he can be led by a lure; but for enduring effort he thrives on what is within him. Where that inner urge came from can only be guessed; rather the urge (like love) seems to be special in each individual. Usually a moment of forceful conviction arrives and one says: research is what I want most to do.

What are some of the aspects of research that seem to furnish enjoyment? What are sources of this enjoyment? How does research, that seems to outsiders to be a detachment from life, hold warmth?

The kinds of satisfactions that scholars derive from their researches have been widely recognized. Some are: exercise of inquisitiveness, the chase, use of mind and hand, invention, exercise of skills, concentration of effort, framing of generalizations, competition in ideas, companionship in discussion, dispelling of superstitions, spicing of daily work, separation of fact from emotion, gain in reliability, and confidence in one's effort.

All these satisfactions tend to accumulate. They can be sources of pride so long as humility accompanies them. "Man needs to be convinced of the importance of the work he is doing" (Whitehead).

Of course, activities other than ostensible research can yield some of the same enjoyments. Fortunately, the researcher need not live without those other activities. However, research soon demands deep personal involvement; one may be warned in advance that by comparison many other activities will come to seem less enjoyable.

Here I do not propose to discuss each of these kinds of satisfaction. High quality is one's goal in every endeavor; slipshod work yields no satisfaction in research, big or little. Persistence of effort becomes indispensable; how can one rely on his observations unless he has repeated them often, and with variations of circumstances? How can one be confident that his concept will wear well unless he looks at it from many angles? Given solid quality in a report of research, only the fact that the same conclusions have been reached before will stand in the way of its publication. Therefore, to the individual, the quality of the work is the standard of his own satisfaction; to the fraternity of scholars, novelty of concept forms an additional criterion.

The human qualities of the researcher are revealed in his conversation. He has ideas to try out on his companions. He encourages his junior colleagues by shop talk, enchancing their devotion and cooperation. He heightens his own interest in how the results are coming out by arousing expectancy in others. He competes with colleagues for foresights and for vi-sions of possible significances. Such are the elements that make the difference between routine and enjoyment.

Ideas incite ideas, the self-educating scientist gains a habit of entertaining them. They do not come easily at first; he has to learn to be kind to them when they pop up. Further, ideas incite fresh ideas in others. There is something infectious about them. This is the scholar's highest function: to transmit to others the art of having thoughts based on realities, and to transmit the excitement that comes with them.

Satisfaction in one's efforts comes unawares. There is satisfaction in a meal, in a journey. A certain number of miles of travel satiates one. So it is with the activities of the scientist-scholar; a day's labor subconsciously demands a relaxation. But this metered satiety is only one of many satisfactions, and all of them together make up the measure of his believed accomplishment.

Perhaps some contrary mind would like to list the types of disappointment and frustration that come to researchers. Maybe the beginner can heed the difficulties he is likely to meet when he pursues elusive answers to his questions. For myself, the list of positive satisfactions in research here suffices.

What Is Self-Education?

I next mention some of the above characteristics of research activity, as they apply particularly to the education of university students. Everyone has aims in education, both for his own education and for the education of others.

I am unable to trace the various guidances, exposures, and frustrations that went into my early education. They, no doubt, determine my ideals of education, though their specific effects I do not recognize. What I do know is that somewhere and somehow I became self-educating. No longer did motives come from without but from within. In my own mind I link the attainment of self-education with the pursuit of research activities. Had it been otherwise, I suppose, this essay would not be written.

As a college student I was aware that I wanted to teach as well as research. I asked a faculty adviser whether I might learn about teaching in an education course. His reply was that when I had acquired something to teach I would be ready to present it to the students. This reply puzzled me, but years later it became plain that a teacher does his work by example and not by precept. He lives questioningly and transmits his questioning way of life.

Research is the greatest device for self-education invented by man, I judge. Research comes close to being a tool by which a man can lift himself intellectually by his own bootstraps. By it he cannot lose his present powers, but can increase them. He cannot stand still, he can only keep moving into new concepts and new understandings. He bares mistaken notions, amends hypotheses—and all by first-hand means. Bacon noted that "truth comes out of error more readily than out of confusion" (7). If one stands still he may only avoid errors; when one moves forward he tests his understanding. That process of education self-corrects itself.

The researcher becomes involved in what he does. Involvement is largely subconscious. The lures toward deed and thought take hold in both waking and sleeping hours. Stewing in a medium of notions and concepts seems indispensable to him who would reach a result. He feels excluded when involvement is prevented for a time, and comes back to commitment with pleasure. Hundreds of paths richly beckon. Some one of them seems satisfying. One says, this is what I was looking for. Actually one did not know what he was looking for, and when found, the new path satisfied an urge that one did not know existed.

Education and research both seem to say: You make the effort to frame answerable questions. You devise methods, such as experiments, suitable to answer them. If you take those steps, you will receive clear answers. This is a bargain that the universe offers. Few rules are known as to what kind of question is suitable—that is where experience counts most. But the scientist appears to have two advantages over the nonscientist: (a) He asks how, where, when, what; but not why. And (b) he can devise a decisive experiment.

Experience in extracting answers from an exacting universe creates attitudes in the student-researcher. He gains a faith, as did Empedocles, in the regularity of events. He embodies such faith in what some have called "scientific method". He acquires a feeling as to what "answerable" means. Sometimes he feels he can predict phenomena within narrow limits. He constructs hypotheses according to certain rules that he is almost unable to define. All this and more constitutes the research attitude.

Attitudes that develop in a research scientist are such as: He feels responsible for his work—thus, he cannot employ anyone to make decisions for him. He concentrates his effort, especially in the formation of concepts. He takes initiative in hard looking and hard thinking. He forces his portion of the universe to answer his questions wherever he thinks answers exist. Though he may flounder, and guess almost at random what to do next, he looks sharply for new directions.

Acquisition of such attitudes comes during the student's exposure to the thoughts and atmosphere of scholars at work. That exposure is today the basis of advanced education, and tomorrow perhaps of all education. The student "messes about" in experiments, books, and conversation. Something happens sooner or later to start him in pursuit of a notion. A special link latched his particular mind to a particular endeavor. Who could ever have prescribed that which would command his special motivations and capacities? From that moment of self-determination, self-education takes care of itself.

Occasionally one sees a junior colleague falteringly approach the boundary of self-confidence in his thinking. One wonders, is there some way in which I can encourage him to taste independence? Can he find an opportune release from conventional thinking? In my experience, the potential scholar can be drawn out by cautious questioning. He can be challenged to formulate inchoate questions that seethe in his mind. He will feel a need to dig deeper, and to found his inferences more accurately. Gradually he can define an area of thought where no one has previously penetrated. He forces himself into an independence he did not even seek. He makes his interest and effort visible to himself in the very act of trying to explain to me what he wants to do. What I describe here is the process that has gone on in "senior common rooms" and in "societies of fellows" for generations of scholars. I suggest that those doddering dons were not always purposeless when they made conversation.

The special quality of individual research lies in the motivation that results from raising one's own question. No question from another source will do so well. The limitation to any benefit in "problem-working" and in "teaching-machines" is that the student did not frame the problems.

The professor often thinks: the student will waste time arriving at a line of research endeavor and then it may not be a promising one. He won't know how to work at it, and I may not be interested in it. Let me save time by presenting to him one of my well-considered plans. Perhaps that represents a majority method of prodding the student into research. It may hasten one kind of productivity. But it may also deprive the student of exploration and decision-making at a critical stage in his career.

In my experience, each student or colleague with whom I associated in a research activity helped to educate me when he proposed a problem out of his own thinking. Not only did he learn as he drove himself to frame a question, but I learned a fresh approach, different from one I would have designed.

A widespread opinion is: the first research contributes most to one's education; after that, further research is largely repetition. But my personal experience indicates otherwise. The research I do today nets me just as many challenges as of old, not only because the immediate objectives are new, but because I look for more complex and comprehensive answers. I demand of myself a higher level of analysis and understanding. I have yet to enter a research that proceeded automatically to an anticipated conclusion. Always unexpected features and relationships appear; and they educate me.

My thought is that wherever someone raises a question, and follows through to an answer, that is research. And wherever research process and attitudes are learned by experience, that is education par excellence.

Education and Research

The importance of self-impelled education varies with one's view of the purposes of education. Anciently education was used as an instrument of suppression, saying "don't". More recently education was used to purvey information, like news media. Here I am describing education that encourates personal independence, like research.

The world needs naïve research as means of education. It needs sophisticated research as advancement of public achievement and welfare. Sometimes naïve research develops into sophisticated research and hence serves doubly.

Questions to be investigated arise in all sorts of ways. Even experienced investigators may raise naïve ones; therefore the student need not feel ashamed of the ones that come naturally to him. For example, Slama & Williams (8) tried to raise sexually mature bugs of the family Pyrrhocoridae, and failed. But in other laboratories the insects reproduced regularly. Evidently conditions differed. Fifteen environmental factors were tested; finally paper of another kind was used to line the insect cages. Now the bugs bred. What effect could the paper have? It turned out that when the larvae touched the paper originally used, they absorbed juvenile hormone, which induced extra molts instead of metamorphosis to adult. Only bugs of the one family were prevented from maturing; others were not prevented. The particular analog of juvenile hormone present in the paper came from the balsam-fir wood from which this paper was made. From a chance failure of reproduction, identification of a series of hormones and their specificities unfolded.

When I noticed a rat running along the wall of an old flour mill, I wondered whether the rat was subsisting on the flour spilled in the dirt nearby. So I asked myself, how much can food material be diluted by dirt or water or other nonnutrient? This led to laboratory experiments on quantitative intakes of nonnutrient mixtures that showed how limited were the alimentary loads tolerated by laboratory rats (9).

Students sometimes become animated by reading or hearing what others have done. They prefer to hear what ordinary people did with a problem that arose spontaneously. What did a student in our own school make out of his own resources? Such an account, especially in the first person, strikes sparks most readily. Only secondarily do students take in the famous deeds of Pasteur with his microbes (10) and of Darwin (11) on his five-year voyage. During that voyage Darwin educated himself in framing questions and recording observations designed to provide answers. Many men have described how they pursued down-to-earth projects, such as the insect studies engagingly related by J. H. Fabre (12). Such descriptions only rarely come to the ears of students at the right moment, for no one can predict the man, the hearer, and the juncture that are to meet.

Scientists may feel uneasy when the uninitiated spend their efforts on problems that have already been satisfactorily solved. I do not decry repetitions of previous researches, providing they start from a question raised in the mind of the repeater. He has the choice of finding the answer in published work or of investigating for himself. He may prefer the latter! Only when the results come up for publication or when public funds allocated to research are being used, is there need for restraint of this preference.

I often wonder whether the teacher who pumps information into students does so because he is directed to do it, or because he believes that more information will develop the student's mind. Shall one suppose that every student's mind is empty, and that the duty of a teacher is to fill it with accepted facts and concepts?

I question whether each page of predigested input may smother the flame of the student's own motivations, preventing his effort to deploy what he already knows. Can we find ways of helping him so that he will not be stifled, but encouraged to examine his own ideas?

I believe that instead of being told, the student can be aroused. If he can raise a question, then I can help him to search for its answer. But the answer without the question is something he will generally ignore. The question already in his mind, however wild, carries a desire to understand. He will exercise whatever resources he has to analyse and to reason. The teacher's job is to help him discover that he has resources and can use them to his own satisfaction.

Though the student's questioning is often aroused in an individual conversation, yet group or class education also works to that end. The aroused student is not always the one who is conversing at the moment. A rhetorical question or a difficulty in another student's concept, may be the trigger. Intense moments in class teaching come unannounced. Suddenly someone feels: this is ideal teaching and learning, a satisfying experience. Why cannot all group gatherings intrigue me like this one? If the teacher knew how, all classes would be built accordingly; but the teacher cannot always design such moments of excitement. He can, however, welcome them when they appear.

Further pursuing the above notions, I wonder that laboratory instruction is so often an exercise in following directions. Maybe minute directions will produce a complete set of data. But will the designer of those data be the only one to appreciate them? I say: occasionally, at least, let the student pose a question to which a laboratory procedure can yield an answer. Manipulations will then have a self-determined aim. And the aim will be consummated when the student draws a reasoned conclusion from the information he himself has designedly obtained. A follow-up conference will later reveal the significances of the whole exercise. The exercise has now been modeled as a research project, and built into it are the potential benefits of research itself.

Perhaps teachers need renewed faith in the capacities of students to initiate intellectual ventures. Students prove of specific value to teacher-researchers through the questions they ask. They may ask: What will you gain when you understand this problem? How do you know that is so? What consequence follows from this particular result? Teachers might agree with St. Benedict (535 A.D.) (13) "that God often reveals what is better to the younger".

Customarily we deny to students the adventure of participation in the search and probe by saying: You must first complete such and such courses. But, how do we know that a requirement does not kill the very excitement and initiative that we cherish? Ignorance can be corrected by the student once he is motivated, while blunted motivation can never be compensated, I fear.

Research also develops and sustains the teacher. It supplies him with exercise for the mind, forcing him to work in depth. New questions are his weekly fare. Research furnishes the impetus to develop new powers and prowess. It allows the teacher to exemplify the qualities that instruction is designed to encourage in the student—qualities of independence, logic, aptitude, and creativeness.

Self-education is today largely limited to advanced students because teachers who have ventured in research are almost exclusively found in universities. But whenever those who have ventured find themselves in other schools and homes they may exert their full capacities to hold the unexplored universe and the questioning attitude before themselves and before their students. We do not know what age may be the best for arousing mental excitement. Could it be the kindergarten age? If so, that is the age at which to encourage questions that can be explored.

Research by teachers promotes teaching when it strengthens the atmosphere around the student. Questions and projects need not be hidden from students but aired to them. One's best teaching can be along paths of thought where oneself has concentrated most, providing one is willing to show the students a beginning instead of a completed phase. Actually, only rarely would the same old paths of thought need to be retraced, alternatives will inevitably be taken, and at times will shake one's own notions to their foundations.

The teacher-researcher can unify what he does instead of splitting his life into two parts. He does well to make abundantly clear to students that his research activity benefits those he teaches, through the freshness and accuracy that pervade him. Whatever might seem hidden about research can be spread to benefit many.

The reader recognizes that much of my essay is a counsel of perfection. I too have done loose teaching, isolated myself from students, and camouflaged my lack of thought. One revises his life in the light of mistakes; "men are moulded out of faults" (14); and such revision itself is self-education.

In summary, I suggest that from research activities the student is likely to profit in his initiative, questioning, reasoning, self-confidence, and persistence; and the teacher can profit in his freshness, vividness, accuracy, timing, and skill. Research and teaching have this in common: they are ever-renewing challenges.

14   THE EXCITEMENT OF SCIENCE

Research Educates

I have pointed out that those who self-educate throughout life, may do so by some sort of research activity. Hence many men in many walks of life can profitably start their self-education early. Is not research as now practiced by scientists our one great contribution to educational method for all people? Students may be introduced to activity in, and feeling for, research in tender years. Also, teachers can share with students the benefits they derive from their own researches—especially the attitudes inculcated by them, and the modes of thinking and doing that they generate. Research activity may be encouraged wherever there is spontaneous curiosity. Its amateur uses may be as important as its sophisticated uses. Research can be a resource for all who capture its universal technique of question and answer.

LITERATURE CITED

1. Eastman, M. (J. Dewey) *Atlantic Monthly*, **168**, 671–85 (1941)
2. Adolph, E. F. *J. Exptl. Zool.*, **31**, 327–41 (1920)
3. Richards, D. W. (L. J. Henderson) *Physiologist*, **1**, ii, 32–7 (1958)
4. Galileo, G. *Dialogues Concerning Two New Sciences* (Crew, H., deSalvio, A., Transl., Macmillan, New York, 1914)
5. Lillie, F. R. *The Woods Hole Marine Biological Laboratory*, Chap. 2 (Univ. of Chicago Press, Chicago, 1944)
6. Adolph, E. F. *Physiological Regulations* (Cattell, Lancaster, Pa., 1943)
7. Bacon, F. *The New Organon*, **ii**, XX, first sentence (1620)
8. Slama, K., Williams, C. M. *Proc. Natl. Acad. Sci.*, **54**, 411–14 (1965)
9. Adolph, E. F. *Am. J. Physiol.*, **151**, 110–25 (1947)
10. Vallery-Radot, R. *The Life of Pasteur* (Devonshire, R. L., Transl., Doubleday, New York, 1902)
11. Darwin, C. *Journal of Researches during the voyage of H. M. S. Beagle* (Murray, London, 1845)
12. Fabre, J. H. *The Hunting Wasps*, Chap. 1 (Teixiera de Mattos, A., Transl., Dodd Mead, New York, 1915)
13. Benedictus. *The Rule of Saint Benedict* (Gasquet, C., Transl., Chatto and Windus, London, 1936)
14. Shakespeare, W. *Measure for Measure*, V, i. 435

16   THE EXCITEMENT OF SCIENCE

Reprinted from
ANNUAL REVIEW OF MICROBIOLOGY
Volume 27, 1973

FIFTY YEARS WITH VIRUSES

C. H. Andrewes

C.H. Andrewes.

Reprinted from
ANNUAL REVIEW OF MICROBIOLOGY
Volume 27, 1973

FIFTY YEARS WITH VIRUSES ❖ 1605

C. H. Andrewes
Overchalke, Coombe Bissett, Salisbury, Great Britain

CONTENTS

EARLY YEARS

The invitation to write an introductory paper for *Annual Review of Microbiology* came at an appropriate time: by the time this volume is published I shall have been working in the field of virus research for almost exactly fifty years.

I was once told by a publisher that I could not write a good autobiography because I was too modest a man: to be successful in this field one should project one's personality even to the point of blowing one's own trumpet. I shall now do my best to overcome my innate modesty, even including a bibliography consisting largely of my own publications. My work has been almost all in the field of viruses infecting vertebrates; plant virologists will, I hope, forgive me if I make little reference to their subject.

My first scientific knowledge of viruses came when I was a medical student in London at St. Bartholomew's hospital. My father was Professor of Pathology there and on the top floor of the pathology block was Mervyn Gordon, working with a grant from the Medical Research Council. His main work was on immunological aspects of vaccinia, but he was also one of the first to transmit mumps to an experimental animal. This was achieved by inoculating monkeys intracerebrally with filtrates of saliva from mumps patients. Gordon was a tremendous enthusiast; he would positively dance round his lab exclaiming: "We've got it, Andrewes, we've got it: they had it there, right under their noses and they missed it!" He also had a puckish sense of humor. The consultant staff, before their afternoon rounds, used to visit the P.M. room, which was next to Gordon's lab. They ascended by a lift,

1

the doors of which formed a sort of grille. One day the lift stuck on the top floor, imprisoning the very irascible senior surgeon. Gordon at once went to the room where he kept his monkeys and obtained a banana which he poked through an aperture in the grille.

In 1923 I had the opportunity of going to New York to work in the hospital of the Rockefeller Institute under Homer Swift. This was in the rheumatic fever service, and with the ignorance and confidence of youth I had little doubt that in the course of two years' work, that problem could be solved. At that time H. Noguchi had been successfully inoculating rabbits intratesticularly with vaccinia and other infectious agents, so we followed this technique, using material from patients with rheumatic fever. Serial passages were made and in due course a transmissible filterable agent came to light; it would produce pericarditis when injected into the rabbit's chest (38). Now Tom Rivers and W. S. Tillett had, in the previous year, obtained a virus from chicken-pox patients, using similar techniques; this even produced intranuclear inclusions similar to those of varicella (41, 42). The virus was, however, not neutralized by convalescent varicella sera, so they were cautious in their conclusions. It turned out, of course, that both groups were dealing with the same virus, a natural infection of the rabbit, called by Rivers Virus III (20). My recollection of how this became obvious differs somewhat from the account in Rivers' autobiography. At lunch one day, when we had found inclusions like those of Rivers' virus, Swift told him: "I've got something to show you under the microscope," to which Rivers replied: "I know only too well what I'm going to see." So far as I recall, Virus III was the first naturally occurring virus of a laboratory animal to turn up and lead workers astray. For Rivers and Tillett it produced intranuclear inclusions like those of varicella; for Miller, Swift, and myself a rheumatism-like pericarditis; and for J. B. Macartney, then in Australia, skin lesions recalling those of scarlet fever. Alas, in subsequent years, only too many unwanted viruses have been brought to light in experimental animals.

During the two years I spent at the Rockefeller hospital, I made many friendships, especially with Rivers and Peyton Rous, with whom I kept up a regular correspondence over many years.

In 1925 I came back to London to work for a time with W. E. Gye (30) and J. E. Barnard who were studying fowl tumors and had lately published papers claiming to have cultivated the viral cause of cancer. This was said to be cultivable in cell-free media and to be effective only in combination with a mysterious specific factor. Barnard had constructed microscopes which used ultraviolet light and gave higher resolution than did conventional instruments; with this he could see these cancer viruses. Barnard, a hatter by trade and a microscopist by hobby, had no biological training and always refused to examine specimens unless he knew their provenance. To ask him to do blind controls was considered an insult. In 1926 I spent many weary hours in operating theaters, waiting to collect tumor specimens for Gye to cultivate and Barnard to examine. However, one day Gye told me: "You know, Andrewes, you mixed up those specimens. The normal tissue control grew out a virus and the tumor one didn't." (I knew better.) It is now ancient history that there was altogether too much wishful thinking about this whole research and it is almost forgotten.

There was, however, good work, too. Rous was always very fond of Gye and grateful to him for saving his "tumor No. 1" from oblivion. "I was saved by an Englishman," he wrote. One paper published in 1926 has since proved of interest; Gye and I found that Rous sarcomata in the course of serial transplantation might for a while lose their filterability and later regain it (31). In retrospect, it is ironical that Gye, who believed passionately in "the virus theory of cancer" should have struggled so hard, yet without convincing success, to show that mammalian tumors could be transmitted by cell-free agents. Yet, only a few years after his death, viruses are being shown to cause cancers in one animal after another. The virus theory is today the basis of a great deal of fascinating research. Barnard's ultraviolet micros-copy was of very high quality; it was unfortunate for him that the advent of the electron microscope soon superseded it.

For about five years I had divided my interests between clinical medicine and laboratory research, but I was getting nowhere and in 1927 I joined the staff of the National Institute for Medical Research, then at Hampstead, as a full-time worker in the lab.

It was an exciting period in the early days of virology, with discoveries coming thick and fast, and my colleagues seem, in retrospect, to have been more colorful than those of today. (Doubtless today's leading workers will look equally glamorous to the next generation.) Laidlaw & Dunkin (34) had just published a series of papers on dog distemper as it affected dogs and ferrets. S. R. Douglas and Wilson Smith were working on vaccinia, Charles Todd on fowl plague, and J. R. Perdrau on herpes; soon after Macfarlane Burnet joined us for a year. I shared a small lab with Nicolau and Galloway who, as a result of their work, wrote a monograph on Borna disease (39). Nearly all the experimental laboratory work was carried out on rabbits; the usefulness of the mouse, oddly enough, did not become apparent for several years.

My first work at Hampstead was on the immunology of viruses; I worked chiefly with vaccinia and Virus III. I had found, as had several others at about the same time, that neutral mixtures of virus and antibody could be reactivated by dilution and in other ways (3); the nature of the reaction seemed then to be obscure. Later on, light was thrown when W. J. Elford and I described what we called a percentage law governing the neutralization of a bacteriophage by its antibody (17).

Elford had joined the staff soon after I did and soon devised his gradocol mem-branes which permitted reasonably accurate measurement of virus sizes (26). A series of papers was published by him in collaboration with myself and several others on the sizes of vaccinia, herpes, foot-and-mouth disease, Rous sarcoma (27), and all the other viruses which were available. We showed that bacteriophages were not all of the same size (16). In fact, it became apparent that the sizes of viruses generally covered a very wide range. Confirmation of the findings followed the use of inge-nious techniques involving centrifugation.

In the late 1920s tissue culture was beginning to come into its own in the study of viruses. In those days, suitable bacteriostatic agents were not available and very rigid precautions against bacterial contamination were necessary. Things were made easier, however, by the Maitlands' discovery that a virus could grow in tissues surviving in vitro; conditions for continual cell growth were not necessary (36).

THE EXCITEMENT OF SCIENCE   21

Using a medium such as theirs I was able to propagate Virus III in cultures of rabbit testis and to make two significant observations, both of them novel at that time. The virus produced specific changes (production of nuclear inclusions) in infected cultures; and these still appeared in cultures of testis of immune rabbits in the presence of normal sera, while they were not seen in normal testis cultures grown in immune serum (4). Similar results were obtained with herpes virus (5).

Perhaps the most exciting period of my life was that when Wilson Smith, P. P. Laidlaw, and I succeeded in transmitting influenza virus to ferrets, thus opening up the study of that infection to laboratory research (44). I have related elsewhere how the first virus to infect a ferret came from my own throat and how chance played a role in the successful outcome of the work (13). Our discoveries led us to communicate with Dick Shope who had shortly before discovered the virus of swine influenza (43). I visited the US in 1933 and met Shope, thus beginning a close friendship which continued until his death.

Before I deal more fully with the work on influenza, something must be said about some other work done in collaboration with Shope, arising out of his discovery in cottontail rabbits of the fibroma and papilloma viruses which often bear his name. Ever since my early association with Gye I had been fascinated by the possibility that viruses were concerned with cancer, a view very unpopular at that time, especially in America because of the influence of James Ewing. I had carried out immunological studies of several transmissible fowl tumors which were available (6). The Rous sarcoma was transmitted to pheasants, at that time an astonishing finding (7). We should, in those days, have been amazed at what has subsequently come to light in that field. A nonfilterable tar sarcoma of fowls was also transmitted to pheasants and there was a suggestion, though not conclusive evidence, that a virus might be concerned (8).

It was natural, therefore, that I should welcome an opportunity of doing some work with Shope's two new tumor viruses. A variant of the fibroma virus was found; this gave rise only to inflammatory lesions, not to the characteristic cell-proliferation (9). Most interesting was a finding, made in collaboration with C. G. Ahlström, now Professor of Pathology in Lund. Rabbits infected with the fibroma virus and also injected with tar developed generalized lesions and regression of tumors was much delayed (1); one transplantable sarcoma appeared.

The beginning of the Second World War put a stop to my work on the cancer problem. Influenza was of more practical importance, as too, were problems of controlling cross infections in air-raid shelters and elsewhere. Since the war my activities have been largely concerned with three subjects: influenza, the common cold, and virus classification. It will be convenient to consider these separately, though work in the three fields went on simultaneously.

POSTWAR YEARS

Influenza

In the early days of flu research, Laidlaw, Smith, and I made progress by reviewing the results of our daily rounds of the isolation cubicles of our ferret hospital. Life became much easier with the discovery that mice were susceptible (19), for cross infection among these hardly occurred. We confirmed the finding by Magill & Francis (37) that not all influenza viruses were serologically alike. This was done with the aid of the crude methods of neutralization tests in mice. Then came the techniques of infecting eggs devised by E. Goodpasture and improved by Burnet, and then, in 1942, G. K. Hirst's hemagglutination test. All these developments combined to make antigenic comparisons of flu viruses very much easier. It then very soon became apparent that the epidemiology of influenza had to be considered on a worldwide basis, for similar antigenic changes in viruses were happening in widely separate countries. A country-to-country spread of influenza B in 1946 made me feel even more that international collaboration was desirable (10). Accordingly, an unofficial meeting was held in Copenhagen in 1947 at the time of an international congress. R. Gautier, Assistant Director-General of the World Health Organization, was present. I recall that I had several ideas as to who might be responsible for organizing such collaboration, but, as happened on other occasions, the initial responsibility was thrown back at me; WHO asked the Medical Research Council to set up a World Influenza Centre at Hampstead. A center for the Americas was under T. P. Magill in New York, and there were corresponding laboratories in as many other countries as possible. By 1952 there were about 60 such laboratories and now there are about 80. Newly isolated viruses are sent to the central laboratories to be typed, and reagents are exchanged so that countries ideally may have the opportunity of making vaccines against a novel strain before it actually reaches them. If this ideal has been hard to achieve in practice, we have learned a great deal about the epidemiology of influenza as a result of the collaboration through WHO. So much so that when, in 1958, WHO decided to spend more money on virus research, they decided to organize regional laboratories to deal with other virus groups along the same lines as the influenza network.

Studies of antigenic changes in influenza A viruses have gradually revealed that these were of two types: antigenic drift is a gradual change which the viruses may undergo from year to year, perhaps in response to increasing immunity in populations; antigenic shifts of greater magnitude have occurred at roughly eleven-year intervals (1946, 1957, 1968) and the new families of viruses have been called A1, A2, A3 (40). It has seemed unlikely to many people that the major shifts could have appeared as a result of variation or mutation of a pre-existing strain; accordingly more and more attention is being given to the possibility that really novel strains have come from a nonhuman mammal or a bird. Influenza A viruses are known from pigs, horses, and birds of several species and doubtless will be found in other animals; recombinants between human and avian strains have been produced in the laboratory (46).

While these epidemiological studies have been in progress, there have been repeated attempts to produce effective prophylactic vaccines, beginning in 1938 with vaccines made from mouse lungs (45). Now, purified and concentrated vaccines can be made from virus grown in the allantoic cavities of fertile eggs and they have been used with and without adjuvants. They are undoubtedly of value when made from the appropriate strain and given at the right time, but their performance falls far short of the ideal and more and more attention is being devoted to producing an effective and safe, live attenuated vaccine.

The Common Cold

Interest in influenza led naturally to study of other upper respiratory infections and, first and foremost, to the common cold. In 1931 I had seen Dochez's work on colds in New York (25) and, on return to England, attempted to repeat it. Transmission to volunteers by means of filtrates was readily shown but attempts to repeat Dochez's cultivation of virus failed. After the war the problem was attacked once more and a common cold research unit was set up at Salisbury in 1946. No way of detecting virus was available except the production of colds by inoculation of volunteers held in quarantine. In the course of 25 years of work there was gradual progress from fumbling and stumbling in the dark towards something more rational. We talked then of *the* common cold virus and its properties, little supposing that a hundred or more were awaiting discovery. In one experiment we found that the common cold virus was inactivated by ether, in another that it was resistant. Which of the experiments had given the wrong answer? We now know, of course, that colds may be caused by ether-resistant rhinoviruses or by ether-sensitive coronaviruses (14). Our earliest efforts were directed to trying to grow a cold virus in eggs by one or another of the available techniques. We could not satisfy ourselves that we were successful and we spent an enormous amount of time and effort in testing the claims of success by several groups; such claims have not been generally confirmed.

Finally in 1952 some rather tenuous clues led to a concentrated effort to grow a virus in tissue culture. The work of Enders and his colleagues on polio (28) had made this a much more promising approach than it had seemed six years earlier. In 1953 we reported the successful propagation of a cold virus in tissue culture (15), and then promptly failed to confirm our own findings. After a lapse of some years this virus was successfully cultivated and possible reasons for the temporary failure became apparent (24). The virus was at first called DC, later rhinovirus type 9. H. G. Pereira and his colleagues spent several years in trying to get to the bottom of the mystery, and finally in 1960, D. A. J. Tyrrell and his colleagues discovered the trick of successfully cultivating what are now called rhinoviruses; they could also detect changes in the cultures, thus making available an in vitro test for virus (47).

There followed discoveries of the multiplicity of serotypes of rhinoviruses, elucidation of their properties, their epidemiological behavior, and then the discovery of another family of cold viruses, the coronaviruses (2).

Myxomatosis and Other Viruses

One virological event of major importance took place in the 1950s. Myxomatosis was successfully introduced into Australia in 1950 and into continental Europe in 1952; thence it spread to Britain in 1953. Australian workers made marvellous use of this unique opportunity to study the evolution of a host-parasitic relationship (29). Their studies provide a model which may well prove valuable in other situations in the future. I became involved in several studies of the virus, its behavior in the field in Britain and experimentally in the laboratory. One finding was, at the time, quite unexpected, though it would not necessarily cause surprise today; D. M. Chaproniere and I found no difficulty in cultivating the virus in tissues of nonsusceptible hosts, even in human tissues (22), and some study was made of underlying mechanisms (23).

Needless to say, in the course of the half century under review, important discoveries about all sorts of viruses were being made all over the world. These naturally had their impact on one's work and mode of thought, whatever virus one was concerned with, and I was marginally involved in not a few of them. Yellow fever and scores and scores of other arboviruses were cultivated or transmitted to animals. The work of Enders, Weller & Robbins (28) in growing poliovirus made quite a revolution in virology; not only did it lead directly to the production of highly successful polio vaccines, but use of the technique brought to light whole families of coxsackie and echoviruses, and later the adenoviruses. Cultivation of measles, varicella, and rubella viruses followed. Then, in 1957, Isaacs & Lindenmann, working in my virus division at Mill Hill, described interferon (33); knowledge about this changed many of our ideas about viral immunity, and research in the field is very active today, with hopes that it may lead to practical application in the fields of virus prophylaxis or even therapy.

Classification and Nomenclature of Viruses

In 1948 Dorothy Horstmann from Yale came to work with me for a year, and the work which we did together had a profound influence on my future activities (18). I knew that some viruses were readily inactivated by ether while others were not, so we proceeded by means of a standard test to examine all the viruses available to us. The results we obtained were placed alongside data about the sizes of viruses, as they were known from Elford's filtrations, and there then appeared the first glimmerings of a notion that there might be a logical way of classifying viruses. Francis Holmes, a plant virologist, had published in 1948 a supplement of *Bergey's Manual,* putting forward a scheme for classifying and naming all viruses (32). It was based almost wholly on tissue affinities. It was so full of anomalies and so contrary to emerging evidence about virus properties that I felt that some straight discussion of the subject was overdue. An approach to the International Committee for Bacteriological Nomenclature led to their appointing a virus subcommittee with myself as chairman. This met at the International Congress in Rio de Janeiro in 1950 and was there given official status. At these meetings we laid down eight criteria which, we felt, should guide us in virus classification. We placed first such things as morphology and chemical composition and last of all tissue affinities and symptomatology (11). This was a very important step forward.

It has been most interesting and encouraging to see how, during the twenty or so years since the 1950 Rio meeting, the whole subject has passed from a state of chaos to something fairly orderly. At Rio we saw only faint glimmerings of order. There was hardly any support for a Linnean system of binomial nomenclature, though we were being pressed to construct this by some bacterial taxonomists. Apart from proposing eight criteria as mentioned above, we did no more than appoint a number of small working parties to study and report about a few virus groups. At the following Congress, held in Rome in 1953, further progress was made. The subcommittee still felt that "the use of systems of classification for and the application of binomials to viruses as a whole are undesirable and should be discouraged". However, names for a few virus groups (since regarded as genera) were put forward— Poxvirus, Herpesvirus, Myxovirus, and Poliovirus. The psittacosis group of agents, then called Chlamydozoaceae, was considered to be more related to bacteria than to viruses and was rejected. I had proposed myxophilus as a name for the influenza-like viruses, on account of their affinity for mucins. However, we decided at Rome on a "–virus" suffix for all the virus groups, so myxophilovirus became shortened, at Albert Sabin's suggestion, to myxovirus. Poliovirus has, of course, ceased to be used in a wide generic sense. Names were also proposed for some insect viruses; those interested in these agents were with great difficulty persuaded to agree to the use of names with a –virus suffix and to abandon earlier ones based on names of virologists working in the field. The plant virologists were even more unwilling to work according to a common plan.

Meanwhile there began to appear more signs of order in the chaos. A number of us suggested that a classification might well be based on a few important characters (12); of these the chief were the nature of the nucleic acid (RNA or DNA), presence of an outer envelope (and associated sensitivity to ether), and number of capsomeres, where these were present. A year later Lwoff, Horne & Tournier proposed a system of viruses based on similar criteria with the useful addition of discrimination between those with cubical and those with helical symmetry (35). All this made it possible at the International Congress held in Montreal in 1962 to recognize officially four more genera: Picorna, Adeno, Reo, and Papova viruses. In the following year (1963) the International Committee of Bacteriological Nomenclature agreed that viruses were not bacteria and were entitled to have their nomenclature considered by a separate body. A provisional committee for nomenclature of viruses was set up as a preliminary to the creation of a virus committee (ICNV) equivalent to that dealing with bacteria.

The provisional committee, meeting in Paris in 1963, put forward for consideration far-reaching proposals, bringing together into one system viruses of animals, higher plants, and bacteria and arranging them in families and higher taxa. At the congress held in Moscow in 1966 a number of the provisional committee proposals were accepted but the designation of higher taxa was not (48).

The Montreal meeting had emphasized that the term arbovirus denoted a biological, not a taxonomic, character. It now appeared that the arthropod-borne viruses belonged to several genera, including not only the principal one, togaviruses, but also some of the rhabdoviruses, reoviruses, and other groups.

There now seemed a greater willingness to work towards a binomial nomenclature for viruses, together with a feeling that there was no need to hurry too rapidly toward that objective. Most workers recognize that in the virus field there are many difficulties in the way of too rigid a classification and nomenclature, and that sweeping changes will not be accepted, if there is no clear need for them. One particular difficulty remains: the ICNV has ruled that no new nonsense names or sigla can be accepted; but the plant virus subcommittee has proposed a whole series of names which do not conform to this rule and it proposes to continue to use them. As ever, the subject of taxonomy fails to stand out as a shining example of scientific harmony. Nevertheless, the progress made during the last 23 years has been highly satisfactory.

Steam and Dream Virology

In Britain, the term steam radio has been applied, perhaps a little derisively, to sound broadcasting as contrasted with television. At a meeting a few years ago a speaker dealing with a rather mundane aspect of virology apologized for it as being only steam virology. Whereupon the late Alick Isaacs suggested that the more high-flown branches of the subject, in which molecular biology, protein chemistry, and genetics are concerned, should be called dream virology. It will be apparent that my own interests during the 50 years covered by this review have been mainly in the field of steam virology. The interest in classification may be considered to be otherwise. At any rate there has been some balance between the two sides of the subject. The same applies to most virological work before the Second World War. In the early days most virologists were medically qualified and nearly all virological papers had an obvious relation to disease. There has gradually appeared a dichotomy in virus research. Viruses are of great importance from two points of view: they have to be studied because they cause diseases, often of economic importance, and their study has also served to throw light on fundamental biological processes. So, on the one hand, we need to know about the pathological changes they produce and the symptoms they cause, their epidemiology, and ecology; on the other hand are the questions of their structure and chemical composition, the biochemistry of their effects on cells, their genetics. Both sides of virology are of great importance, the steam and the dream, and it would be a pity to see a reduction in either. Of course one hopes that the dream virology will, in the long run, produce results of practical importance for prevention and cure of disease. One has only to think of the exciting work now in progress in the field of the relation of viruses to cancer and of the need for more knowledge before we can welcome really good antiviral chemotherapy.

Nevertheless, I feel that things are getting a little out of balance. During these 50 years the relative numbers of medical or veterinary graduates in virus research has been getting steadily fewer; the steam to dream ratio has been getting less and less. The trouble is that the dream virology has been getting too exciting, too glamorous. Young men and women coming into the field cannot avoid being fascinated, their imagination is stimulated, and perhaps they feel that steam virology has had all its days of triumph. When, in 1967, there appeared in addition to *Virology*, two new journals in the field, *The Journal of Virology* and the *Journal of General Virology*, I hoped that one of them would pay particular attention to the steam side of the subject. But not at all; today I read the lists of titles and it is rare that I find in any of the journals one which recognizes that viruses cause disease. For news in that field I have to look through the papers in a considerable number of other journals, dealing with all sorts of aspects of bacteriology, epidemiology, and pathology.

THE EXCITEMENT OF SCIENCE 27

Scientific meetings on virology attract more and more people. Twenty years ago, several of us working at the National Institute at Mill Hill began to organize such meetings from time to time. We thought we knew all the British workers in the field; we kept a list of them and wrote individually to each. We had no formal constitution, no officers, no committee, no published proceedings, in fact we used to call it "the non-existent society". But of course it grew and grew and became unmanageably large, till finally it took shape as the Virus Group of the Society of General Microbiology and many of its proceedings were published in the *Journal of General Microbiology*. Meetings attract more and more people and the discussions are much more formal and less rewarding than before. Another group, The South Wiltshire Virological Society, is based in Salisbury and it, too, grows steadily, though we have managed to retain most of its informality.

Those coming to such meetings belong mainly to two groups: the full-time workers in virus research (mostly dream) and the bread-and-butter virologists engaged in routine diagnostic work. Much valuable new knowledge comes to light through the activities of the latter group, but it mostly arises incidentally, as a spin-off from their main task. The gap, as I see it, is in purposeful intensive research in the steam field. As my interest has lain so largely in the field of respiratory infections, I feel particularly conscious of our lack of knowledge of epidemiology and ecology in this area. One must admire the application devoted to studies of the ecology of arbovirus infections. Nothing like the same attention is given to the ecology of respiratory infections, which are of far greater importance in most parts of the world. The subject is difficult and of fascinating complexity; why cannot its mysteries excite the rising generation of virologists?

In the late 1950s I had been working in the field of viruses for some 35 years and retirement from my position at the National Institute for Medical Research lay not far ahead. Since the 1920s, and particularly since I had been head of the Virus division at the Institute, I had tried to keep up with the literature of my subject and had accumulated a very extensive card index. The close relations between many viruses infecting man and those of other species were becoming obvious to many, yet medical and veterinary viruses were mostly dealt with in separate works. So I decided that, when I retired, I would try to distill some essence from my card index into a comprehensive textbook, to be called *Viruses of Vertebrates*. A few days before I retired I told my technician to throw away some thousands of cards containing all my records of animal experiments carried out over many years; any worthwhile results had been already included in published papers. I came in one morning to tell him to pack up my card index of the literature to send down to Salisbury, to which place I was retiring. "Oh," he said. "You told me to throw all the cards away." My horror and stupefaction can hardly be imagined, but by a fortunate chance the cards were found and could still be retrieved from a dustbin. The first month of my retirement was spent in sorting them out into a semblance of order. Keeping the records up to date and selecting what has seemed most worthwhile for the book has kept me busy; for the second and third editions (21), Pereira has been a valuable collaborator. Writing this and five other books has provided an ideal occupation for retirement; one can do most of one's writing in the winter and when the weather is unkind, and on pleasant days there is much to occupy one in the open air.

Literature Cited

1. Ahlström, C. G., Andrewes, C. H. 1938. *J. Pathol. Bacteriol.* 47:65–86
2. Almeida, J. D., Tyrrell, D. A. J. 1967. *J. Gen. Microbiol.* 1:175–78
3. Andrewes, C. H. 1928. *J. Pathol. Bacteriol.* 31:671–98
4. Andrewes, C. H. 1929. *Brit. J. Exp. Pathol.* 10:188–90, 273–80
5. Andrewes, C. H. 1930. *J. Pathol. Bacteriol.* 33:301–12
6. Andrewes, C. H. 1931. *J. Pathol. Bacteriol.* 34:91–107
7. Andrewes, C. H. 1932. *J. Pathol. Bacteriol.* 35:407–13
8. Andrewes, C. H. 1936. *J. Pathol. Bacteriol.* 43:23–33
9. Andrewes, C. H. 1936. *J. Exp. Med.* 63:157–72
10. Andrewes, C. H. 1946. *Aust. J. Sci.* 9:85–87
11. Andrewes, C. H. 1952. *Ann. Rev. Microbiol.* 6:119–38
12. Andrewes, C. H. et al 1962. *Virology* 15:52–55
13. Andrewes, C. H. 1969. *Int. Virol.* 1:220–23
14. Andrewes, C. H. 1973. *In Pursuit of the Common Cold.* London: Heinemann Medical Books
15. Andrewes, C. H., Chaproniere, D. M., Gompels, A. E. H., Pereira, H. G., Roden, A. T. 1953. *Lancet* 2:546–47
16. Andrewes, C. H., Elford, W. J. 1932. *Brit. J. Exp. Pathol.* 13:446–56
17. Andrewes, C. H., Elford, W. J. 1933. *Brit. J. Exp. Pathol.* 14:367–76
18. Andrewes, C. H., Horstmann, D. M. 1949. *J. Gen. Microbiol.* 3:290–97
19. Andrewes, C. H., Laidlaw, P. P., Smith, W. 1934. *Lancet* 2:859–62
20. Andrewes, C. H., Miller, C. P. 1924. *J. Exp. Med.* 40:789–96
21. Andrewes, C. H., Pereira, H. G. 1972. *Viruses of Vertebrates.* London: Bailliere, Tindal; Baltimore: Williams & Willkins, 3rd ed. 451 pp.
22. Chaproniere, D. M., Andrewes, C. H. 1957. *Virology* 4:351–65
23. Chaproniere, D. M., Andrewes, C. H. 1958. *Virology* 5:120–32
24. Conant, R. M., Hamparian, V. V., Stott, E. J., Tyrrell, D. A. J. 1968. *Nature* 217:1264
25. Dochez, A. R., Shibley, G. S., Mills, K. C. 1930. *J. Exp. Med.* 52:701–16
26. Elford, W. J. 1931. *J. Pathol. Bacteriol.* 34:505–21
27. Elford, W. J., Andrewes, C. H. 1935. *Brit. J. Exp. Pathol.* 46:61–66
28. Enders, J. F., Weller, T. H., Robbins, F. C. 1949. *Science* 109:85–87
29. Fenner, F., Ratcliffe, F. N. 1965. *Myxomatosis.* London: Cambridge Univ. Press. 379 pp.
30. Gye, W. E. 1925. *Lancet* 2:109–17
31. Gye, W. E., Andrewes, C. H. 1926. *Brit. J. Exp. Pathol.* 7:81–87
32. Holmes, F. O. 1948. *The Filterable Viruses. Supplement to Bergey's Manual of Determinative Bacteriology.* Baltimore: Williams & Wilkins, 6th 'ed.
33. Isaacs, A., Lindenmann, J. 1957. *Proc. Roy. Soc. London* Series B. 147:258–67
34. Laidlaw, P. P., Dunkin, G. W. 1926. *J. Comp. Pathol.* 39:222–30
35. Lwoff, A., Horne, R., Tournier, P. 1962. *Cold Spring Harbour Symp. Quant. Biol.* 27:51–55
36. Maitland, H. B., Maitland, M. C. 1928. *Lancet* 2:596–97
37. Magill, T. P., Francis, T. Jr. 1936. *Proc. Soc. Exp. Biol. Med.* 35:463–66
38. Miller, C. P., Andrewes, C. H., Swift, H. F. 1924. *J. Exp. Med.* 40:773–87
39. Nicolau, S., Galloway, I. A. 1928. *Spec. Rep. Sci. Med. Counc. London No. 121*
40. Pereira, H. G. 1969. *Progr. Med. Virol.* 11:46–79
41. Rivers, T. M., Tillett, W. S. 1923. *J. Exp. Med.* 38:673–90
42. Rivers, T. M., Tillett, W. S. 1924. *J. Exp. Med.* 39:777–802
43. Shope, R. C. 1931. *J. Exp. Med.* 54:373–85
44. Smith, W., Andrewes, C. H., Laidlaw, P. P. 1933. *Lancet* 2:66–68
45. Stuart-Harris, C. H., Andrewes, C. H., Smith, W. 1938. *Spec. Rep. Series Med. Res. Counc. London No. 228*
46. Tumová, B., Pereira, H. G. 1965. *Virology* 27:253–61
47. Tyrrell, D. A. J., et al 1960. *Lancet* 1:235–42
48. Wildy, P., Ginsberg, H. S., Brandes, J., Maurin, J. 1967. *Progr. Med. Virol.* 9:476–82

THE EXCITEMENT OF SCIENCE 29

Ann. Rev. Genet. 1976. 10:1–6
Copyright © 1976 by Annual Reviews Inc. All rights reserved

THEODOSIUS DOBZHANSKY:
THE MAN AND THE SCIENTIST

Francisco J. Ayala

Theodosius Dobzhansky
1900–1975

Ann. Rev. Genet. 1976. 10:1–6
Copyright © 1976 by Annual Reviews Inc. All rights reserved

THEODOSIUS DOBZHANSKY: ❖3097
THE MAN AND THE SCIENTIST

Francisco J. Ayala
Department of Genetics, University of California, Davis, California 95616

Theodosius Dobzhansky was born on January 25, 1900, in Nemirov, a small town in the Ukraine 200 kilometers southeast of Kiev. He was the only child of Grigory Dobrzhansky,[1] a teacher of high school mathematics, and Sophia Voinarsky. In 1910 the family moved to the outskirts of Kiev, where Dobzhansky lived through the tumultuous years of the Bolshevik revolution and the First World War. He graduated in biology from the University of Kiev in 1921, and from 1921 to 1924 taught zoology at the Polytechnic Institute. From 1924 to 1927 he was a lecturer in genetics at the University of Leningrad.

In 1927, Dobzhansky obtained a fellowship from the International Education Board (Rockefeller Foundation) and arrived in New York on December 27 in order to work with Thomas Hunt Morgan at Columbia University. In the summer of 1928 he followed Morgan to the California Institute of Technology where Dobzhansky was appointed assistant professor of genetics in 1929, and professor of genetics in 1936. Dobzhansky returned to New York in 1940 as professor of zoology at Columbia University, where he remained until 1962, when he became professor at the Rockefeller Institute (renamed Rockefeller University in 1965) also in New York City. On July 1, 1970, Dobzhansky became emeritus at Rockefeller University; in September 1971 he moved to the Department of Genetics at the University of California, Davis, where he was adjunct professor until his death in 1975.

On August 8, 1924, Dobzhansky married Natalia (Natasha) Sivertzev, a geneticist in her own right, who at the time was working with the famous Russian biologist I. I. Schmalhausen in Kiev. Natasha was Dobzhansky's faithful companion and occasional scientific collaborator until her death by coronary thrombosis on February 22, 1969. The Dobzhanskys had only one child, Sophie, who is married to Michael D. Coe, Professor of Anthropology at Yale University.

[1] Precise transliteration of the Russian family name includes the letter "r".

1

In a routine medical checkup on June 1, 1968, it was discovered that Dobzhansky suffered from chronic lymphatic leukemia, the least malignant form of leukemia. He was given a prognosis of a few months to a few years of life expectancy. Over the following seven years the progress of the leukemia was unexpectedly slow and, even more surprising to his physicians, it had little if any noticeable effect on his energy and work habits. However, the disease took a conspicuous turn for the worse in the summer of 1975. In mid-November Dobzhansky started to undergo chemotherapy, but continued living at home and working. He was convinced that the end of his life was near and dreaded that he might become unable to work and to care for himself. Mercifully, this never came to pass. He died of heart failure on the morning of December 18, 1975. The previous day, Dobzhansky had been working in the laboratory as usual.

Dobzhansky had incredible energy and very disciplined work habits. The list of his publications at the time of his death consists of 568 titles, including a dozen authored books plus several others that he edited. He was a world traveler and an accomplished linguist able to speak six languages fluently and to read several more. His interests covered a broad spectrum of human activities, including the plastic arts, music, history, cultural anthropology, philosophy, religion, and of course science. He was a good naturalist, and never lacked time for a hike whether in the California Sierras, in the New England forests, or in the Amazonian jungles. He loved horseback riding, but practiced no other sports. Throughout his academic career, Dobzhansky avoided administrative posts, and participated minimally in committee activities. He alleged, perhaps correctly, that he had neither the temperament nor the ability for management. Most certainly, he preferred to dedicate his working time to research and writing rather than to administration.

Personalities can hardly be described or even characterized in a few words. Dobzhansky's most obvious traits were, perhaps, magnanimity and expansiveness. He recognized and generously praised the achievements of other scientists; he admired the intellect of his colleagues, even when admiration was alloyed with disagreement. He made many long-lasting friendships, usually started by professional interactions. Many of Dobzhansky's friends were scientists younger than himself, who either had worked in his laboratory as students, postdoctorals, or visitors, or had met him on his trips. (Throughout his academic career Dobzhansky had about 30 graduate students and an even greater number of postdoctorals.) He liked to be called "Doby" by his friends even though these might be young scientists or students. To young biologists, he freely gave his time and encouragement. He had many friends abroad, and was largely responsible for the establishment or development of genetics and evolutionary biology in various countries, notably Brazil, Chile, and Egypt. He was conspicuously affectionate and loyal towards his friends; he expected affection and loyalty in return. Dobzhansky's exuberant personality was manifest in his friendships but also in his antipathies. When he disliked a person, this was obvious.

Dobzhansky contributed to evolutionary biology perhaps more than any other scientist since Darwin. Yet his prodigious scientific productivity includes milestone contributions to several areas of genetics. I shall mention a few.

Using translocations between the second and third chromosomes of *Drosophila melanogaster,* Dobzhansky demonstrated that the linear arrangement of genes based on linkage relationships corresponds to a linear arrangement of genes in chromosomes (*Genetics* 15:347–99, 1929). This linear correspondence had been postulated but proof was first provided by Dobzhansky (and independently by Muller & Painter also in 1929). In that same paper Dobzhansky presented the first *cytological* map of a chromosome—chromosome III of *D. melanogaster.* Comparing the linkage and the cytological maps of the chromosome, he showed that the relative distances between genes are different in the two maps; genes clustered around the center of the linkage map are spread throughout a larger portion of the cytological map. He correctly inferred that the frequency of crossing-over is not evenly distributed throughout the chromosome. Later he produced cytological maps of the chromosomes II (1930) and X (1932) of *D. melanogaster.* This work led to the hypothesis that the centromere (the "spindle fiber attachment" in the terminology of the time) was a permanent feature of chromosomes. He demonstrated that translocations decrease the frequency of crossing-over and advanced a hypothesis to account for this reduction (1931).

Dobzhansky first demonstrated that the determination of femaleness by the X chromosome is not due to a single gene or to a few genes, but to multiple factors distributed throughout the chromosome (*Proc. Natl. Acad. Sci. USA* 17:513–18, 1931). His publications on the genetic and environmental factors affecting sex determination started in 1928 and extended for more than a decade. These studies included work on *bobbed* mutants in the Y chromosome, and their role in male sterility (*Genetics* 18:173–92, 1933), as well as numerous publications on gynandromorphs and "superfemales." His publications on developmental genetics started in 1930 (*Biol. Bull.* 59:128–33).

Working with *D. melanogaster* in the laboratory headed by Y. F. Filipchenko at the University of Leningrad, he made the first systematic investigation of the pleiotropic, or manifold, effects of genes (*Z. Indukt. Abstamm. Vererbungsl.* 43:330–88, 1927), a phenomenon that maintained his interest for many years (e.g. *Genetics* 28:295–303, 1943).

Dobzhansky's important contributions to the study of position effects started in 1932 and continued for several years (a review in *Biol. Rev.* 11:364–84, 1936). In the last months of his life, Dobzhansky became intrigued with the hypothesis advanced by some molecular evolutionists that position effects prompted by chromosomal rearrangements may play a critical role in evolution.

Dobzhansky has been called the founder of experimental population genetics. He doubtless was its most eminent practitioner. He may also have been the first one to name and define the field: "The third subdivision of genetics has as its province the processes taking place in groups of individuals—in populations—and therefore is called the genetics of populations" (*Genetics and the Origin of Species,* Columbia Univ., New York, 1937, p. 11).

Dobzhansky's first contributions to population genetics appeared in 1924. (He had already published four papers between 1918 and 1923 on coccinelid beetles, three dealing with taxonomic problems and one with adult diapause.) The 1924 papers investigate local and geographic variation in the color and spot pattern of two *Coccinellidae* genera, *Harmonia* and *Adalia*. These ladybird beetles exhibit local polymorphisms, which in some species vary from one to another locality. Dobzhansky explained the genetic variation within and between populations as the consequence of the same fundamental evolutionary processes. Some cardinal themes of Dobzhansky's evolutionary theory are present in this work: the pervasiveness of genetic variation, geographic variation as an extension of local polymorphism, and as the first, but reversible, step towards species differentiation. Dobzhansky continued the study of natural populations of ladybird beetles until he left Russia in 1927. He occasionally worked with coccinelids in the United States (e.g. a 94 page monograph published in 1941).

The beginning of Dobzhansky's studies on the population genetics of *Drosophila* can be traced to 1933 when he published a paper on the sterility of hybrids between *D. pseudoobscura* and *D. persimilis* (then known as *D. pseudoobscura* race B). In a series of papers, he investigated the physiological, developmental, and genetic causes of hybrid sterility. This work developed from the convergence of two independent previous lines of investigation, the genetics of translocations and the study of sex determination. It led, in 1935, to the formulation of the concept of (sexually reproducing) species still accepted today: "that stage of the evolutionary process at which the once actually or potentially interbreeding array of forms becomes segregated in two or more separate arrays which are physiologically incapable of interbreeding." This notion establishes that reproductive isolation is what sets species apart. It is also an evolutionary definition that sees speciation as a dynamic process of gradual change. In 1937 Dobzhansky coined the term *isolating mechanisms* (*Am. Nat.* 71:404–20) to designate the phenomena that impede gene exchange between species. He identified, classified, and investigated the various kinds of isolating mechanisms. *Isolating mechanisms* is one example of the many useful terms coined by Dobzhansky that have become part of the language of evolutionary biology.

In a brief note it is possible to name only a few of the many problems of population genetics investigated by Dobzhansky: geographical, altitudinal, and temporal variation in the frequencies of chromosomal arrangements in natural populations; the occurrence of "concealed" variation, first of lethal genes and later of genes modifying fitness components such as viability, rate of development, and fertility; the origin and evolution of sexual and the other reproductive isolating mechanisms; mutation rates, their genetic control, and the equilibrium between mutation and selection in natural populations; migration and dispersion rates; effective population size; correlation between genetic polymorphism and ecological heterogeneity; causes of species diversity; heterosis; genetic load; coadaptation of gene pools; joint effects of selection, population size, and migration on traits with low heritability; and so on. Indeed, there may be no major problem of population genetics to which Dobzhansky did not make important experimental contributions.

Some characteristics of Dobzhansky's research strategy that contributed to his enormous success deserve mention. He worked both in the field and in the laboratory; whenever possible he combined both in the study of a problem, often using laboratory studies in order to ascertain or to confirm the causal processes involved in phenomena discovered in nature. He obtained the collaboration of mathematicians in order to design theoretical models for experimental testing and to analyze statistically his empirical observations. He was no inventor or gadgeteer, but he had an uncanny ability to exploit the possibilities of any new experimental apparatus or experimental method that suited him. He selected organisms that provided the best materials to investigate the problems that interested him: the biological particularities of *D. pseudoobscura* and its relatives, and of the *D. willistoni* group, made possible many of Dobzhansky's discoveries. He always worked at the utmost level of genetic resolution possible at any given time: he took advantage of the early methods of genetic analysis, then of various cytological tools, later of the giant polytene chromosomes, and of the techniques to produce chromosomal homozygotes. When gel electrophoresis came about, he immediately recognized its enormous potential as a tool to study population genetics problems; he felt that it was too late in his life for him to learn the technique but encouraged his students and collaborators to use it and collaborated in several projects using it.

Dobzhansky was an extremely successful experimental scientist. Yet his most significant contribution to biology was his establishment of the modern synthesis of evolutionary theory. By the early 1930s the work of Fisher, Wright, Haldane, and Chetverikov had provided a theoretical framework integrating Mendelian genetics with Darwin's theory of evolution. In *Genetics and the Origin of Species* (1937) Dobzhansky completed this integration in two ways. First he gathered the empirical evidence that supported and made acceptable the mathematico-theoretical framework. Second, he extended the integration of Mendelism and Darwinism much beyond the range provided by the mathematical models and produced a comprehensive evolutionary theory that carried population genetics up to the process of speciation and beyond. The modern theory of evolution would continue to develop in the future, but it was born with Dobzhansky's 1937 book in the same way as the original theory had come about with Darwin's *Origin of Species* in 1859.

The single-most important empirical fact established and argued by Dobzhansky is the ubiquity of genetic variation. He was quick to see the momentous implications of this fact for mankind. Dobzhansky set forth that the individual is not the embodiment of some ideal type or norm, but rather a unique and unrepeatable realization in the field of quasi-infinite possible genetic combinations. The pervasiveness of genetic variation provides the biological basis of human individuality. Dobzhansky evinced that it also leads to demystification of the much abused concept of race.

Dobzhansky saw that an adequate understanding of human nature is only possible in the light of evolution. In *Mankind Evolving* (1962) he expounds the implications of the theory of evolution for mankind. This book remains an unsurpassed synthesis of genetics, evolutionary theory, anthropology, and sociology. Dobzhansky's lasting interest in the relevance of biology to human affairs is revealed in the titles of some of his books, such as *Heredity, Race, and Society* (1946, with L. C. Dunn), *The Biological Basis of Human Freedom* (1956), *Heredity and the Nature of Man* (1964), *The Biology of Ultimate Concern* (1967), *Genetic Diversity and Human Equality* (1973).

Dobzhansky was a warm and compassionate man who had little patience with obscurantism, racial prejudice, or social injustice. In the 1940s and 1950s he published several articles criticizing Lysenko's biological quackery. He attacked antievolutionists and cogently rejected claims that Christian beliefs are incompatible with evolution. He relentlessly denounced what he called "bogus 'science' of race prejudice."

Dobzhansky was a religious man, although he apparently rejected fundamental beliefs of traditional religion, such as the existence of a personal God and of life beyond physical death. His religiosity was grounded on the belief that there is meaning in the universe. He saw that meaning in the fact that evolution has produced the stupendous diversity of the living world and has in fact progressed from primitive forms of life to man. Dobzhansky beheld that in man biological evolution has transcended itself into the realm of self-awareness and culture. A metaphysical optimist, he believed that somehow mankind would eventually evolve into higher levels of harmony and creativity.

Ann. Rev. Phytopathol., Vol. 8

MUSINGS OF AN ERSTWHILE PLANT PATHOLOGIST

F. C. BAWDEN

Ann. Rev. Phytopathol., Vol. 8

MUSINGS OF AN ERSTWHILE PLANT PATHOLOGIST 3500

F. C. BAWDEN

Rothamsted Experimental Station, Harpenden, Hertfordshire, England

Daunting as the prospect is of writing the prefatory chapter to a volume of the *Annual Review of Phytopathology,* the invitation to do so is such a compliment that it is irresistible. Similarly, although now used to seeing my name in papers on viruses and virus diseases only in sections dealing with history, it is not all pleasure to be called a "doyen of the subject" by the Editors. It is true their emphasis on my many years in plant pathology, combined with the generously long notice given to prepare the chapter, made acceptance all the easier by making my survival to do it seem improbable, but survive I have, and now face the task set by the Editors of being either autobiographical, philosophical, or both.

I find autobiography the simpler, so will begin by saying that, after taking the Natural Sciences Tripos and the Diploma in Plant Pathology at Cambridge University, I began research on July 1, 1930 as an assistant to R. N. Salaman, the director of the Potato Virus Research Station, Cambridge. It is an understatement to say that working conditions were primitive (the "laboratory" was a wooden hut and the most sophisticated piece of apparatus, a recalcitrant "Primus" stove), but I could not have had a better initiation to research than to work with Salaman. My debt to him is enormous, less for any formal training in techniques, than for the intangible benefits of simply associating closely with a man of vision, unusually gifted, and with exceptionally wide interests. Trained in medicine, he had to give up medical research in London because of ill health and he moved to the country where he turned his attention to, among many other things, the genetics of the potato. He soon became the authority on potato varieties and his interests in the crop steadily widened, not only to include the virus diseases that were ruining some of his most promising seedlings, but also to the origins of the crop and its great sociological and economic influence in Latin America and Western Europe. I caught his passion for the crop and have never lost it. Early influences have lasting effects, and the best advice I can give to anyone contemplating entering research is to choose carefully with whom you first work. To this I add, do well in your undergraduate work, less because what you learn will be technically useful later, but it will increase your chances of starting research with a good mentor.

The date I began was also propitious inasmuch as the International Botanical Congress was held in Cambridge in 1930, which gave me opportunity immediately to meet almost all of the few people in the world then working with plant viruses. The occasion I best recall and that influenced me most

1

THE EXCITEMENT OF SCIENCE

8243-2601/78/1127-0041$01.00 © 1978 ARI   41

was a discussion on nomenclature. It was becoming clear that the practice of assuming a virus to be "new" simply because it was found causing a previously undescribed syndrome, and naming it on the basis of the host in which it was found and the disease it caused, was leading to confusion and much synonymity. At this discussion James Johnson urged the need to use properties of the viruses, rather than symptomatology, as criteria for identifying them, and to stress the need for criteria other than host reactions, he suggested that viruses should be given numbers instead of names, for example, Tobacco virus 1 instead of tobacco mosaic virus. This suggestion was probably unfortunate, for it let the meeting, as many other similar ones since, get bogged down in discussing the type of nomenclature instead of dealing with the more important matter of virus identification. Neither then nor since could I see that it mattered whether viruses were given symptom names, numbers, or letters, or that the confusion would be any less by changing the type of nomenclature. What was needed to avoid synonymity and other troubles was not new names, but the better criteria for identification urged by Johnson. Or so it seemed to me, and it was in attempts to get methods that would relate clinically different strains, and separate clinically similar viruses, that I first gave attention to the in vitro properties of viruses and then to their serological behavior and the value of the plant-protection phenomenon, discovered independently by Thung and Salaman.

Salaman, with his medical training, had seen the possible diagnostic value of serology, and he brought me into collaboration with E. T. C. Spooner, who initiated me into the mysteries of serology. I still remember the excitement of the first successful complement-fixation test with potato virus "X" and of seeing the first specific precipitates with its antiserum, which led Spooner to the shrewd comment that "your viruses have tails." Of course, they did not but, as found later, viruses with filamentous or rod-shaped particles behave serologically exactly in the same ways as bacterial flagella, whereas viruses with isometric particles behave as somatic antigens. Indeed, it was work with plant viruses that was to explain the reasons for the differences between the behavior of these two types of antigen.

Salaman had the hope that serology would provide a quicker and simpler method of testing seemingly healthy potato plants for viruses than the grafting and inoculating to test species then being used in searching for virus-free plants that could be propagated to replace the infected stocks then in cultivation. It is one measure of progress since that such searches, always tedious and often fruitless, would no longer be needed because virus-free clones can now be reestablished from infected ones by heat therapy or apical-meristem culture. Although not applicable to all viruses, Salaman's hope was amply realized and serology has been used widely since as a rapid and specific screening test. However, its applications proved much wider than this; serology not only allowed clinically distinct strains of one virus to be related and gave some idea of the closeness of relationship but, most valuable to me, the precipitation test gave a simple assay, allowing the relative virus contents of different fractions separated in attempts at purification to be measured immediately. Also, in contrast to infectivity tests, assays done at different times could be related to one another, and the amount of virus assessed from the precipitation endpoint. I still remain puzzled to understand how it was that so many virus workers long remained reluctant to use these invaluable techniques. With hindsight, it is very evident they were even more valuable than those of us who used them appreciated. Thus, assays for infectivity and antigen content usually agreed, but not after some kinds of inactivation, as by radiations or nitrous acid, which left serological activity unimpaired while destroying infectivity. There, in the 1930s, was a pointer to infectivity residing in the nucleic acid, but no one appreciated its significance, and twenty years elapsed before the roles of the nucleic acid and protein were elucidated.

Serology had another very important consequence for me, by starting my collaboration with N. W. Pirie. He was working on the antigens of *Brucella* sp. with a colleague of Spooner's, and I asked him to look at some of the antigen preparations I was making of potato virus X. Now that so much is known about the structure of some viruses, and almost the only secret tobacco mosaic virus still holds is the sequence of nucleotides in its nucleic acid, it probably needs stressing that then they were wholly mysterious entities. It was known they caused diseases and how some were transmitted, but what they were was anyone's guess, though as most people working with them were renegade bacteriologists or mycologists, the commonest guess was similar to bacteria but smaller. It was reasonable to assume they contained protein, but the first convincing evidence of this came from our work showing that proteolytic enzymes inactivated potato virus X only in conditions at which the enzymes were proteolytically active.

Potato virus X is not the easiest virus to purify and when, in 1936, I moved to Rothamsted Experimental Station, Pirie and I turned our attention to tobacco mosaic virus. This was soon after Stanley had claimed its isolation as a crystalline globulin, a claim that did not convince us. Our reluctance to accept it was not the prejudice of the many biologists who were unwilling to accept that a virus could crystallize, but because of his description of the isolated material. Not only was 20% N unusual for a globulin, but some of the properties attributed to the material fitted ill with what was known about tobacco mosaic virus; also the chemical simplicity of a globulin conflicted with our ideas that potato virus X contained more than protein, and the serological activity reported seemed too little for a pure antigen.

Our suspicions were soon confirmed for, instead of a crystalline globulin, we obtained a liquid crystalline nucleoprotein, containing 0.5% ribose nucleic acid. When nucleic acid dominates biological thinking to such an extent that all specificities seem to be attributed to it, it may be difficult to imagine that it took some years to get agreement that nucleic acid was an essential component of tobacco mosaic virus. But science, as other activities, has its changing fashions, and in the 1930s proteins were fashionable and nucleic acids not.

There was never a dull moment in the work with tobacco mosaic virus (and with cucumber viruses 3 and 4, which despite their inability to infect tobacco proved to be distantly related to it serologically), for it was continually showing something novel. The many anomalous physical properties of virus solutions made it clear that previous attempts to estimate the particle size from sedimentation or filtration, based on the assumption that it was spherical, were wholly misleading. That it had anisometric particles was indicated by the intense sheen of solutions viewed in reflected light, but was made certain when concentrated solutions separated into two liquid layers, the bottom one liquid crystalline (spontaneously birefringent), and the upper not birefringent unless disturbed, when it showed intense anisotropy of flow. Only particles very much longer than they were wide could behave in this manner, but the phenomena did more than indicate the shape of the particles, for anisotropy of flow provided an even quicker test than serology for the virus. All that was needed to find what fraction of a preparation contained the virus, or whether a treatment had denatured it, was to shake a sample in a narrow tube between crossed polarizers. Also, with experience and a knowledge of the pH of the sample, the intensity of anisotropy reasonably indicated virus concentration. However, despite (or, perhaps, because of) its simplicity and value, very few virus workers adopted it as a test.

It was the crystallinity of viruses rather than their chemical constitution that attracted most attention in the 1930s, although the only feature that was to prove common to all viruses later studied was that they contain nucleic acid, either ribose or deoxy. However, although forming crystals per se had little biological significance, crystallinity had important consequences in allowing X-ray crystallography to be used to study the anatomy of viruses. The measurements Bernal and Fankuchen made on our liquid crystalline preparations not only gave the first indication of the width of the particles, but also showed that the virus differed from organisms in being composed of regularly arranged subunits of a uniform size. If not the foundation stone, these measurements were at least an important cornerstone in the now large edifice of Molecular Biology, though the name had not then been coined.

After finding that tobacco mosaic virus was a nucleoprotein, the obvious question was: Were other viruses?, because tobacco mosaic was by no means typical of the category. We therefore examined others and first returned to potato virus X, which also proved to contain 0.5% ribose nucleic acid and to be anisotropic, though it differed from tobacco mosaic virus in not forming paracrystals when precipitated with salt or acid. Next we turned to tomato bushy stunt virus, which proved interestingly different in having isometric particles, in forming true crystals, and in containing a larger proportion than tobacco mosaic virus of nucleic acid to protein. But none of these viruses was known to be insect-transmitted, so the generality of chemical composition was still far from established, and at least one insect-transmitted virus needed examining to know whether this feature was peculiar to a different kind of chemical composition. Hence, we tried potato virus Y and *Hyoscyamus* virus 3 (henbane mosaic virus), and tough nuts they proved. The paper in which we described our results seems a poor thing compared with those on tobacco mosaic or tomato bushy stunt viruses, but a lot more effort went into the work needed to get the evidence that these also were nucleoproteins containing ribose nucleic acid and had anisotropic particles. Whereas we could get 3 gm of tobacco mosaic virus from 1 litre of sap and did not need to hurry to do so, we got less than 1 mgm of these aphid-transmitted viruses, and if not got quickly we did not get it at all.

With the knowledge that these two viruses were chemically similar to the others, we were happy to be rid of them and were left with little desire to isolate and characterize still more. Instead, we turned our attention to virus inactivation, and I became increasingly interested in the interaction between viruses and their hosts, especially the factors that influence the susceptibility of plants to infection and the extent to which viruses multiply, and later to inhibitors of infection and multiplication. Also, with the outbreak of the second world war, it was necessary to do something more useful than purify viruses, and I returned to the practical task of improving the health of potato stocks. This time the problem was not to find virus-free clones to propagate in seed-growing areas, but to try to maintain the health of stocks in the ware-growing areas of England, which meant associating in field work on the epidemiology of leafroll and severe mosaic (potato virus Y), the main causes of stocks losing their cropping power in the south and east of England.

In 1940, I was appointed head of the Plant Pathology Department at Rothamsted and my horizons were further widened to include diseases caused by other pathogens than viruses. This was all very rewarding and educational, and soon led me to appreciate, better than I had before, the many general principles that unify the subject of plant pathology. This influence, I think, is amply shown by my little book *Plant Diseases,* first published in 1948, in which I tried to summarize these principles by describing how diseases originate, spread, and cause losses, what influences their development and their seasonal fluctuations, the methods whereby they can be combated, and to show their generality by using as illustrations diseases caused by bacteria, fungi, or viruses. Despite the contrary opinions of those who favor increased specialization and would separate bacteriology, mycology, and virology, my conviction that these are better kept together has become stronger rather than weaker. Why separate mycologists from virologists when their mutual interests should be increasing by the discoveries that fungi both suffer their virus diseases and are the vectors of some viruses that damage crops? Also, despite the great differences between bacteria, fungi, and viruses, the principles and practices of protecting crops from them do not differ. They rest in using varieties that best resist or tolerate infection, destroying sources of infection, planting uninfected stock in uninfested land and away from infected crops, and use of appropriate chemicals to protect a growing crop. A minor difference is that to protect against viruses, the chemicals will usually be aimed against the organisms that transmit them, whereas they will be aimed directly at bacteria or fungi.

Developing a control measure against an infectious disease in field crops often does not even demand knowledge of the cause, but only of the epidemiology of the disease, to know where the cause comes from, and how and when it spreads, so to know where it is most vulnerable to attack. It is fascinating to know that aster yellows is probably caused by a mycoplasm instead of, as long thought, a virus; this may allow an extra treatment by antibiotics, but it will not affect heat therapy or control by protecting plants against the vectors. Those working on aster yellows or similar types of disease were appropriately accommodated in departments of Plant Pathology, but would they have been in departments of Virology? And will there now be departments of Mycoplasmology? Possibly, but I hope not because if there were, I fear the workers would become increasingly concerned with minutia of the organisms and increasingly remote from pathology. Pathology needs specialists of many kinds, but will derive most benefit when these are working together with the common aims of understanding pathogenicity and improving plant health. Modern trends are tending towards fission into even narrower specialisms, and only one of the many reasons for welcoming the existence of the *Annual Review of Phytopathology,* and the founding of the International Congress of Plant Pathology, is the hope they will provide the strong cement needed to keep the subject from falling apart.

Forty years on, it is salutary to look back and contemplate the optimistic expectations that knowledge of the nature of viruses would help to control virus diseases. That it has not is vividly illustrated by the fact that the tobacco mosaic virus, whose structure and composition is most completely known, is still prevalent in tomato crops in the United Kingdom, whereas little more is known about potato leafroll virus than in 1930, yet it is now a rarity in our potato crops instead of the major cause of loss it was then. However, many pathologists seem still imbued with the faith I have lost, for how else to explain the increasing numbers being attracted to studying the detailed physical and chemical structure of virus particles? It surely cannot be only that the sophisticated and expensive equipment needed for the work has an irresistible glamour, although it is curious that taxonomy should be fashionable with viruses, whereas it seems to be languishing in mycology and in other parts of botany where it is more simply studied. For, of course, it is in taxonomy rather than pathology that the results of work on such things as size and shape of virus particles, number and arrangement of protein subunits, or position of nucleic acid and ratios of nucleotides, are likely to be useful. Taxonomy is a worthy subject, but I hope it will not attract too many virus workers from pathology, which is even worthier, especially as few pathologists will be likely to contribute as much new information as those already skilled in biochemical and biophysical techniques.

There is nothing easier than to put a virus through the current range of standard machines, some automatic or semiautomatic, that will purify it, photograph it, measure it, and analyze it, with a paper at the end containing the canonical measurements and pictures editors of journals readily accept, even though in essence it contains nothing new. It is much to ask someone to give up this easy approach to publication and tackle the more difficult problems in pathology, and I know it has been cynically said that "Scientific happiness is to have one experiment and continue doing it." Certainly, to add new knowledge to pathology will require more insight and ingenuity than to gain some information about the particles of another virus by switching machines on and off, and seeking it may be more frustrating. Nevertheless, I do urge some whose present interest in infected plants seems solely to produce virus to be examined in vitro, to turn to studying the diseased plant or the diseased crop, because these hold many secrets whose disclosure will probably be at least as interesting as demonstrating morphological similarities or differences between virus particles, and almost certainly more useful. If the machines must be used, why not put them to work seeking other components of infected plants than the virus particles? It is deplorable that, with so much known about the composition of different strains of tobacco mosaic virus, nothing should be known about the reasons for the strains differing in virulence or about the interactions with host systems that lead to characteristic symptoms. Current theories suggest the differences may reside in parts of the nucleic acid not concerned in specifying the virus protein, perhaps in other proteins these parts specify. Would it not be worth testing these theories by seeking such proteins, and, should they be found, then also seeking the host systems on which they work? But to anyone who does this, let me say, do not be prejudiced or blinded by current theories; they may be correct, but also they may have no more substance than the many past ones that held sway for a time but are now in limbo.

I should perhaps stress that nothing in the previous paragraph is meant as an adverse criticism of modern physical techniques or machines. Far from it, for they have been invaluable, not only for giving information about the structure of virus particles, but allowing a range of analyses and fractionations of systems previously intractable. They have, too, destroyed some concepts promulgated by those who were among the first to apply sophisticated machines in work with viruses. For instance, it is now ironic to compare some early results from high-speed centrifugation and electronmicroscopy that purported to show that virus multiplication led to a single endproduct, a "molecule" whose physical integrity was essential for infectivity, with a major current use, which is to separate and identify the various specific products, some infective, some not, some nucleoprotein, some not, that occur in infected plants. Inoculating plants with these components separately and after remixing is demonstrating various types of interaction between them, which can determine whether infection occurs and, when it does, what will be the products of virus multiplication; the accruing evidence that more than one type of particle of some viruses may be needed to establish infection almost introduces the concept of sex into virus behavior. This is a far cry from the concept of infection by single particles that exactly replicate themselves, and the discoveries could not have been made without the new machines and techniques. Used by skilled people in appropriate problems, they are wholly admirable. All that I protest is that increasingly they seem to be coming to be regarded as status symbols and to create the illusion that research on viruses demands expensive equipment. It does not; it demands only initiative, imagination, observation, and industry. Some types of work cannot be done without elaborate equipment, but there is no danger that these types will be neglected. It is the work that can be done more cheaply on epidemiology, transmission, and host susceptibility that now tends to languish, although the results from this will more probably have applications in improving the health of crops.

Plant pathology has had many successes to its credit during its brief history and methods have been developed to prevent the great losses some diseases used to cause. However, it would be vain to maintain that all is well. On the contrary, as has been borne in on me more and more by the visits I have been privileged to make to various countries during the past 25 years, ill health is the customary state of crops, most of which still go unprotected from diseases. Sometimes when I have been invited to see or report on some specific disease, I have been less impressed by this one, which has usually attracted attention to itself by being a killer, than by others, which were much more widespread and causing much more total loss, but neglected and largely unnoticed because that was how the crop always looked and some yield was harvested. Although in general the crops most studied by pathologists, and those growing in regions where plant pathologists are most plentiful, are least diseased, even where pathologists are thickest on the ground there are still important diseases for which control measures are unknown or inadequate; probably, too, if we take as analogies the late recognition of such prevalent diseases as sugar beet yellows in the United States of America or barley yellow dwarf in the United Kingdom, there are important ones everywhere yet to be diagnosed.

If, as I think it should be, the goal of plant pathology is to allow farmers to grow safely what crops they wish, where they wish, and as often as they wish, then it is still far off. For the control of some diseases, the plant pathologist can legitimately look to the plant breeder for help, but for others he must seek his own salvation. Any starry-eyed faith I might have had in immune varieties was early destroyed. When I began working with Salaman, he had just bred some of the first potato seedlings "immune" to blight (*Phytophthora infestans*) and it was very impressive to see how these stood firm when sprayed with spores while commercial varieties in the same test chambers rapidly decomposed. However, the next year, when there were some tubers from these seedlings to plant in the open, although the weather did not especially favor blight, the plants became attacked no less than the commercial varieties. After this demonstration of physiological strains of the fungus, I was puzzled that plant breeders should have continued their efforts to produce immune varieties, but continue they did, with the net result that we now have a large range of differential varieties that allow the different strains of the fungus to be distingished, but still have to rely on fungicides to protect potato crops from blight. Similarly, for diseases of other crops caused by airborne pathogens that spore profusely and can spread widely, we need additional control measures other than varieties whose immunity is likely to be only temporary.

In striking contrast to the way such potato varieties rapidly succumbed to blight, is the lasting hypersensitivity of such varieties as King Edward to potato virus X. Commercial stocks of this variety were free from this virus when I was a boy, and still are, without anyone having to do anything to keep them that way. This difference is not because viruses are any less variable than fungi; indeed the contrary may be true, but a new variant of a virus has much less chance of spreading than has one of a fungus, which can assure its dispersal by producing its spores and exposing them to the wind. The fact that every plant in some annual crops, such as lettuce, sugar beet or cucurbits, sometimes becomes virus-infected may give the impression that viruses spread as readily as do airborne fungi, but they do not. It is because viruses cause systemic infection that they are damaging; only one infection is necessary to affect the whole plant, whereas a single infection with such fungi as cause potato blight or mildew of cereals would be unimportant. It is odd that, with prospects of a lasting success seeming greater in breeding for resistance against viruses, so much more attention has been given to breeding for immunity to infection by airborne fungi. This is not to say the plantbreeder does not have an important part to play in lessening losses from such diseases as potato blight. He has, for there are valuable features short of immunity and that can be more lasting, as, for example, the continuing greater field resistance of the variety Majestic than of King Edward to blight. Also, there are more features than hypersensitivity that can be used to help in controlling virus diseases, as again evidenced by the slower spread of potato virus Y in the variety Majestic than in King Edward, a difference again that has persisted during the 50 or more years these varieties have been in cultivation.

Another thing that has always seemed curious to me is that, because I am a pathologist, people seem to think I will be pleased to see a diseased crop. Interested yes, but pleased no, and especially not when the disease is one for which control measures are known. Some moments of the greatest pleasure for me recently have been to contemplate the perfectly uniform stands of plants in crops of virus-free King Edward potatoes and compare them with memories of the uneven crop of 40 years ago, when not only every plant inevitably had paracrinkle virus, but many also had leafroll or virus Y, and some had all three. A trebling of yields is something in which it is legitimate to take pleasure and even, perhaps, pride. However, not too much pride, because magnificent as the tops look, and large as the yields may be, almost certainly the underground parts were infected by various damaging fungi, not all of which can yet be controlled.

While contemplating virus-free King Edward crops, it is impossible not to recount the cautionary tale of changes in knowledge about paracrinkle virus. This virus was discovered by Salaman during his search for virus-free plants, when he found that plants of the variety Arran Victory became crippled when grafted with scions from seemingly healthy King Edward plants. The disease differed from any seen in field crops, and attempts to transmit the cause by mechanical inoculation or aphids failed. The assumptions from these negative results then magnified into almost a belief that paracrinkle virus was unique to King Edward and was harmless to it, from which an eminent geneticist elaborated the thesis that it was not a virus, but a "plasmagene" that became a virus only when transmitted by grafting to other varieties. Ah well, it was fun while it lasted, and some amusement can also be derived from comparing these past assumptions with present knowledge, which is that paracrinkle virus decreases the yield of King Edward potatoes by at least 10%, it exists in strains some of which are aphid-transmitted and some not, it is readily transmitted mechanically and is related, not only to other viruses prevalent in potato crops, but to others found in plants as distant from the potato as the carnation.

Myths such as the plasmagene one mainly arise when speculation greatly outstrips facts, but what is factual is often far from sure. Lack of certainty is also often enhanced by sloppy phraseology and by reporting conclusions instead of describing results accurately. "Was not transmitted by *Myzus persicae*" is an unequivocal statement, but is not synonymous with "Not aphid-transmitted," although it often gets so translated in general descriptions of viruses. Similarly, "The only known method of transmission is by grafting" is different from "transmissible only by grafting," and although "not mechanically transmitted" sounds positive enough, it is something that cannot be established; all that can be said is "was not mechanically transmitted between such and such plants by such and such methods."

To have worked with viruses and been in plant pathology during the past 40 years has been both exciting and rewarding, even though in retrospect opportunities lost may stand out more prominently than those taken. Why were they lost? Mainly by lack of insight, inability to ask the most important questions at the right time, being too influenced by general opinion or, sometimes, by prejudices engendered by reacting too strongly against such opinions that seemed wrong. Thus, an example of the last may be our failure to uncover the phenomenon, discovered by Kassanis, of one virus depending on another for its ability to multiply in plants, from the results Pirie and I got with the Rothamsted culture of tobacco necrosis virus. Hindsight shows we had many pointers to the phenomenon, but we failed to read them, possibly because, in reacting against the general idea of a single endproduct of virus multiplication, when we were finding a range, we were conditioned to accept the seemingly noninfective component of our preparations as within the range that could be produced by infection with one virus.

At least one piece of advice can be distilled from applying hindsight to situations where insight was too little. It is: always keep an open and a questioning mind; query seeming facts almost as much as conclusions, and query your own no less than other people's. Never do an experiment "to prove something," a phrase that increasingly appears in papers. If you do you will probably succeed, and maybe miss something much more interesting. Hypotheses are necessary to design experiments but they are best forgotten when the experiment is done. Sift all the results and evidence carefully, and pay particular attention to anything that is anomalous. Question everything that is not convincing. Being sceptical may not always make you popular, and you will sometimes ask the wrong questions, but you will help rather than hinder the progress of the subject. I remember questioning the conclusion by that great pioneer L. O. Kunkel that his results proved that the cause of aster yellows multiplied in its leafhopper vector. I still think I was right to do so, because the evidence then was not conclusive and my questioning at least helped to stimulate work that proved the point. However, having had the temerity to question his conclusions, it is regrettable I did not also have the insight to ask the more important question: Is aster yellows caused by a virus?

However eminent a person may be, his results and conclusions need critical examination. Indeed, the more eminent, perhaps the more questioning is needed, because the statements and views of the eminent are more often taken as authoritative. Although it may be only cynical to say that the eminence of a scientist can be measured by the length of time his wrong ideas delay progress in his subject, it is probably true that imagination and receptivity to new concepts tend to dwindle as eminence increases. Let me be clear that what I am urging is not iconoclasm, but constructive criticism and healthy scepticism, questioning that will distinguish the good from the bad and the significant from the trivial. Also, let me emphasize that the history of plant pathology abounds in good and significant results, which provide a sound basis for the future development of the subject. I urge questioning only because, although we now seem to be in an age when mass protests and demonstrations are the order of the day, I sense that there is less of it by plant pathologists than there used to be. I hope I am wrong in this, as in another feeling I have, that there is a greater tendency for workers to engage in similar lines of work, to join already rolling bandwagons instead of starting one of their own. Plant pathology is rich in the diversity of problems it offers for study, and their solution will call, not only for an equal diversity of talents and techniques, but also for workers with inquiring and original minds. And they will be needed not only to operate equipment in laboratories, but at least equally to work with plants under glass or in the field, because it is there that the discoveries are most probably to be made that will improve the health of crops, something vital if farmers are to reap the full harvest of their labor and the rapidly increasing population of people is to be adequately fed.

Ann. Rev. Biochem., Vol. 43

RECOLLECTIONS

George W. Beadle

Ann. Rev. Biochem., Vol. 43

RECOLLECTIONS   ✗837

George W. Beadle
Professor of Biology, University of Chicago, Chicago, Illinois

In these exciting times when elementary and high schools teach modern biology, including many of the intricacies of biochemical genetics, the long slow process by which our present knowledge in this area was gained is not often fully appreciated. A third of a century elapsed before Mendel's work was "rediscovered" and properly appreciated. Archibald E. Garrod's (1) prophetic appreciation of the relation of genetics and biochemistry, beginning soon after the so-called rediscovery of Mendel, lay fallow for more than forty years despite the fact that he published widely and relatively voluminously. As late as a quarter of a century after the Mendel work came to light, Harvard's distinguished professor of biology, William Morton Wheeler (2), ridiculed genetics as a small bud on the great tree of biology, a bud so constricted at the base as to suggest its eventual abortion. Wheeler's colleague in paleobotany, Jeffrey (3), also expressed his disbelief in the work of the then flourishing school of *Drosophila* genetics. Fortunately, neither succeeded in significantly retarding the rapid advances then being made, many of them by two Harvard contemporaries, Edward M. East and W. E. Castle. It is of interest to note that Thomas Hunt Morgan (4) remained a skeptic about Mendelian interpretations for the first ten years after the rediscovery, that is until he established the sex-linked nature of the white eye trait in *Drosophila.*

Now that genetics is widely accepted as one of the most basic aspects of all biology, it is perhaps of interest that some of us old enough to have participated in or otherwise to know something of the history of present day genetics now record our recollections. I attempt to do so in the limited area of biochemical genetics of which I have had a small part.

Of the myriads of environmental influences large and small that have to do with the course of one's life, few are likely to be long remembered with any degree of clarity or confidence. Yet behavioral scientists are increasingly aware that what happens early in life can be of the greatest significance in later years. Unfortunately when one attempts to recall such thoughts and events as have influenced later attitudes and behavior, the uncertainties are many. Thus it is with a good deal of doubt and temerity that I attempt to record events influential in that part of my life that has had to do with biochemical genetics.

I was born in 1903 of parents who owned and operated a 40-acre farm near the small town of Wahoo, Nebraska. Both had grown up in similarly small

THE EXCITEMENT OF SCIENCE

8243-2601/78/1127-0055$01.00 © 1978 ARI

communities: father in Kendallville, Indiana and mother in Galva, Illinois. Both were inherently intelligent but limited to high school in formal education.

Because of its small size our farm was highly diversified, with field crops such as alfalfa, potatoes, and corn; truck crops including asparagus and strawberries for market; plus cattle, horses, hogs, and chickens. All of these were supplemented by retail selling of produce, including out-of-state apples and potatoes purchased in carload lots. In this and other ways I was intimately involved in matters of biological significance. We kept rabbits, ferrets, bees, cats, dogs, and for a time, a pet coyote. Hunting, fishing, and trapping were enjoyable pastimes. With these plus routine chores and farm work, life was never dull.

Mother died when I was four and a half. My older brother, a younger sister, and I were in part raised by a series of housekeepers, some very good, some poor, and one or two terrible.

My earliest years of formal school were in a genuine little red, one-teacher, wooden schoolhouse in town, which was a mile and a half from home. During my twelve years in this and other local schools I was exposed to perhaps a dozen teachers. Like our housekeepers, they were a thoroughly mixed lot.

With the accidental death of my older brother it was tacitly assumed I would eventually take over the family farm, a prospect I looked forward to with a certain amount of confidence and pleasure. But neither father nor I had reckoned with a young high school teacher of physics and chemistry, Bess MacDonald. She did not pretend to be, nor was she, a profound authority in either of the subjects she taught. But she did have a remarkable knack of interesting us, for example, in chemistry by challenging us with unknowns to identify by classical qualitative methods. But more than that, she took a personal interest in our aspirations and hopes.

I spent many nonschool hours with her at her home, during which she convinced me I should go on to college, even though I might eventually return to the farm. My psychological insight is not sufficient to describe our rather unusual relationship. Perhaps for me she was a kind of mother-substitute.

Father was not keen on the college idea, being convinced that a farmer did not need all that education. But determination won and I enrolled at the University of Nebraska College of Agriculture, fully intending to return to the farm. Had it not been tuition-free with an opportunity to work for living expenses, I doubt if I could have managed.

Again my plans were modified by teachers. In my first year I was so impressed by a required course in English that I thought to follow it up. In fact I was offered part-time employment reading student papers during my second year. Fortunately for English, the professor went off to Palestine to study the literature of the Bible. His successor did not pick up the commitment.

In rapid succession thereafter I became enamored of entomology, ecology, and genetics. I was happy with general and organic chemistry and did well in both, but was not carried away to the point of proposing to major in that general area.

After my second year I was given a summer job classifying genetic traits in a wheat hybrid population, this for Professor Keim of the Agronomy Department. In my spare time I read about genetics and found my interest increasing markedly.

I was given other assignments including laboratory instruction in an agricultural high school program given by the College at that time. I read student papers and examinations in the elementary genetics course. I had charge of a laboratory supply department set up to provide samples of crop plants and other materials for instruction in high schools that gave courses in agriculture. During summer periods I grew various exotic crop plants for this purpose, collected and mounted representative weed seeds, made up orders, mailed them out, and kept records. In my senior year I worked on a special problem on root development and survival of fall-seeded grasses of economic importance. I also devised a key for the identification by vegetative characters of local native grasses.

Keim was a remarkable person in many ways. He did not profess to be a great scholar. But he had an uncanny ability to size students up and encourage them, which he did with a kind of understanding I have never been able fully to fathom. Some he sent back to the farm, some to be county agricultural agents, others to teach high school, and a few to go on to graduate school. I've known half a dozen or more of the latter and have never known one to be a misfit.

The nearest he ever came to an error of judgment that I know of was in getting me a teaching assistantship at Cornell and admission to graduate school to work on the ecology of the pasture grasses of New York State. It might not have been a mistake if I had seen eye-to-eye with the professor who was to sponsor my thesis research. But I did not, and soon resigned my teaching assistantship to work in genetics and cytology with Professor R. A. Emerson. That was 1926. Shortly thereafter he gave me a part-time research assistantship.

This surely was one of the best things that ever happened to me. Emerson was the perfect employer, graduate advisor, and friend. He turned problems over to me. One of my special assignments was to complete a summary of all genetic linkage studies in maize up to that time (5). I had half time for course work and for my own thesis research.

These were indeed exciting times for all of us working with Emerson. He was the outstanding plant geneticist of his time and was a tremendously stimulating person to work with and under, and his group of graduate students at the time were outstanding. They included George F. Sprague, Marcus Rhoades, Barbara McClintock, H. W. Li, and perhaps a half dozen others.

Emerson's contributions to genetics came at a time when support for the new science was minimal and the doubters many. He moved from the University of Nebraska College of Agriculture to Cornell University in 1914, in part because he felt his work was judged by the Nebraska authorities to be too theoretical ever to be useful agriculturally. Thus it is of considerable interest to note that in addition to his remarkable work in basic genetics, which of course indirectly but significantly furthered the art and science of plant breeding, Emerson conscientiously assumed direct responsibility for more than his fair share of plant breeding. By genetically transferring resistance to the disease anthracnose to commercially desirable dry beans, he saved the important bean industry of New York State from utter collapse. He also succeeded in transferring disease resistance to commercially grown cantaloupes.

THE EXCITEMENT OF SCIENCE   57

Emerson was one of the first of the early American workers fully to appreciate the work of Mendel, this at a time when even T. H. Morgan was still a skeptic. As I pointed out in 1960 (6), Emerson never published until he had extracted the truth from his experimental material and verified it not once but many times in many ways. Predecessors had studied the inheritance of plant and aleurone colors in corn and had been distracted by incidental modifying factors and apparent inconsistencies with Mendelian principles to the point that some of them actually renounced those principles. It was Emerson's persistence, clear thinking, and hard-headed checking of facts that established the truth and showed beyond doubt that these apparently complex systems of inheritance in reality have an understandable genetic basis. His papers on kernel and plant color inheritance in maize are outstanding as solid experimental work, sound reasoning, and clear presentation. His early studies of quantitative characters, carried on in part through collaboration with E. M. East of Harvard, importantly influenced genetic thinking. His work on variegated pericarp led to the concept of unstable genes, another significant milestone in the history of genetics.

Important as were his own scientific contributions, in many ways it is Emerson the man most vividly remembered by those privileged to know him well. He was cordial in his relations with his friends and colleagues. The contagious enthusiasm and zest, so clearly displayed in his scientific work, were extended to other activities, bowling and hunting for example. During corn season he was first in the experimental garden and among the last to leave, an example that no doubt increased the productiveness of all who worked with him. Bag lunches eaten in the shade of the garden shed during these periods of intense field activity were of special interest to students and other associates. It was there that the unpublished lore of corn genetics and geneticists was most likely to be recalled. It was also a setting in which Emerson became best known to his students. It was also in such informal ways that he did much of his teaching. He was freely available to students but it was his policy that they come at their own instigation. At all times he was willing to be helpful but he did not direct student research in any formal manner.

Emerson's research materials were freely available, not only to his own colleagues and students but as well to investigators elsewhere. This generosity played an important part in making corn the best known of all higher plants from a genetic point of view and had the effect of interesting investigators throughout the world as well as significantly increasing general genetic understanding.

With the growth of the corn group the system of communicating unpublished information through conversation became inadequate. During 1932 at the International Genetics Congress at Ithaca a "corn meeting" was held where it was decided that a central clearinghouse of information and seed stocks would be established at Cornell. Out of this there evolved a series of mimeographed "corn news letter" edited by Marcus Rhoades and sent to all interested corn geneticists. Later this became the *Maize Genetics Cooperation News Letter,* a somewhat more formal organization for the dissemination of information not published in formal journals and for recording seed lines available for research.

One of the groups of mutant types being worked on from the earliest days of

Emerson's research were those affecting chlorophyll synthesis and function. I vividly recollect Emerson's attempts to interest plant physiologists and biochemists working on photosynthesis in making use of such mutants as tools in fathoming the physiology and biochemistry of chlorophyll structure and function. None responded, otherwise biochemical genetics might have moved forward more rapidly.

In my own attempts to improve my understanding of chemistry in relation to genetics, I audited courses in physical chemistry and biochemistry. The latter was given by James B. Sumner and it was during this period that he first crystallized the enzyme urease from the jack bean. Biochemists will recall the long lag between this accomplishment, and acceptance and confirmation of it as authentic and thus a significant forward step.

The time was clearly ripe for the new discipline of biochemical genetics. But few biochemists or geneticists were then intellectually or psychologically prepared, despite the fact that Archibald E. Garrod had a quarter of a century earlier clearly suggested a one-to-one relation between gene action and enzyme activity, and had published both repeatedly and voluminously (7).

My own graduate research in cytogenetics was both rewarding and significant. In part I worked on the genetic control of meiosis using corn lines in which chromosome behavior was markedly modified genetically. The asynaptic mutant was the first, polymitotic a second, and sticky chromosomes a third. I also worked closely with Emerson on the relation of corn to its nearest wild relative, a Mexican plant known as teosinte, this, incidentally, a relationship still not fully resolved and which I am now again actively investigating.

A significant turning point in my career came in 1931 with the completion of my graduate work. I had hoped to be awarded a National Research Council Fellowship to continue my corn cytogenetics work at Cornell, by far the best place to continue in terms of facilities and associates. But the wise chairman of the Fellowship Board, Charles E. Allen of the University of Wisconsin, intervened, pointing out that remaining for postdoctoral work in the same institution in which one took his PhD degree was in principle less desirable than moving to another institution where, other things being equal, new experiences and insights were more likely to be acquired. He said he would approve the award if I would accept my second choice as a place to continue. That was the California Institute of Technology where Thomas Hunt Morgan had recently moved from Columbia to establish a new Division of the Biological Sciences. Emerson approved and I concurred, little realizing at the time that this would be another best thing that ever happened to me.

Caltech biology was indeed tremendously stimulating. Among those who were there in genetics and related areas when I arrived as a research fellow were Morgan, Sturtevant, Bridges, Dobzhansky, Schultz, Anderson, Emerson (son of R. A. Emerson), Belar, and the Lindegrens. Darlington, Haldane, and Karpechenko spent time there as visiting scholars.

General enthusiasm was at a high level and persons in other fields were caught up in it. Linus Pauling took a personal interest in genetic crossing over. R. A. Millikan delighted in escorting visitors to Biology where he could give a masterly account of *Drosophila* investigations. Charles Lauritsen and associates were building a million volt X-ray tube which became available for medical and biological use.

At first I concentrated on my corn cytogenetics program but soon became actively interested in *Drosophila*, then by far the most favorable organism for genetic study. I worked with Dobzhansky, Emerson, and Sturtevant at various times on genetic recombination in the hope that this would tell us significantly more about the nature of the gene. It didn't do as much as we had hoped, though decades later it became clear that if we had really learned enough about recombination a good deal more about the nature of the gene could have been revealed.

An additional and significant turning point in my career began with the arrival in 1933–1934 of Boris Ephrussi from Paris as a Rockefeller Foundation Fellow. He was actively interested in tissue culture and tissue transplantation as a means of learning more about gene action.

We spent long hours discussing the curious situation that the two great bodies of biological knowledge, genetics and embryology, which were obviously intimately interrelated in development, had never been brought together in any revealing way. An obvious difficulty was that the most favorable organisms for genetics, *Drosophila* as a prime example, were not well suited for embryological study, and the classical objects of embryological study, sea urchins and frogs as examples, were not easily investigated genetically.

What might we do about it? There were two obvious approaches: one to learn more about the genetics of an embryologically favorable organism, the other to better understand the development of *Drosophila*. We resolved to gamble up to a year of our lives on the latter approach, this in Ephrussi's laboratory in Paris which was admirably equipped for tissue culture, tissue or organ transplantation, and related techniques.

Morgan arranged to continue my Caltech salary, then $1500 annually, which was 33% less than the previous year because of the great depression. Only years later did I find that this stipend was almost surely provided by Morgan personally. Caltech was in dire financial straits at that time and though Morgan was extremely frugal with Institute funds, he remained always generous in personally supporting causes he thought worthy. Leaving a wife and small son in Pasadena where living costs were unbelievably low at that time, I went to Paris to work with Ephrussi. Fortunately living costs were also very modest there, provided one could do with bare necessities. My daily subsistence expenses, room and food, were approximately two dollars.

In Ephrussi's laboratory we tried tissue culture without remarkable success or promise. We switched to *Drosophila* larval embryonic bud transplantation which turned out to be successful despite assurances from the Sorbonne's great authority on the metamorphosis of the blow fly that we could not succeed.

We knew from Sturtevant's work on naturally occurring mosaic flies that the character vermilion eye (absence of brown component of the two normal eye pigments) was nonautonomous in the sense that if one eye and a small part of the adjacent tissue were vermilion and the remainder wild type, the genetically vermilion eye would produce both pigment components. Obviously an essential part of the brown pigment system was produced outside the eye and could move to it during development. We confirmed this by transplanting genetically vermilion embryonic eye buds in the larval stage to wild-type host larvae. Although it was thought a priori by some to be extremely difficult if not impossible, a technique for doing this was devised. It involved two people working cooperatively through paired binocular dissecting microscopes focussed on one recipient larvae.

We confirmed the existence of a diffusable substance which we called vermilion-plus substance (8). A second mutant lacking brown eye pigment was found to behave similarly—the so-called cinnabar character. Reciprocal transplants between the two mutants lacking brown pigment showed that there were two substances involved, one a precursor of the second. We postulated that one gene was immediately concerned with the final chemical reaction in the formation of substance 1 and the second with its conversion to substance 2.

We investigated the twenty some other eye-color mutants then known in *Drosophila* and found just these two in direct control of the two postulated chemical reactions (9).

Since most biologically significant reactions are enzymatically catalyzed, we assumed the two eye-color genes, cinnabar and vermilion, directly controlled the two postulated enzymes. This was the origin in our minds of the one gene/one enzyme concept, although at that time we did not so designate it.

In formulating this interpretation we were much encouraged by the previous related work of Caspari and others (10) on related pigmentation in the meal moth *Ephestia* and also the work of Scott-Moncrieff (11) and earlier workers on the genetic control of anthocyanin pigments in higher plants.

It is of interest and I believe of some significance that Jaques Monod, then an instructor at the Sorbonne, took a keen interest in our work and spent a good share of his spare time in Ephrussi's laboratory following progress and discussing results with us. Later when Ephrussi returned to Caltech for a year where we continued our collaboration, Monod also came as a visiting investigator.

An obviously important next step was the identification of the two brown pigment precursors. Ephrussi and Khouvine worked on this aspect of the problem in Paris and I at Harvard with Kenneth Thimann and later at Stanford University with Tatum and Clarence Clancy. Tatum demonstrated a functional relation of one of the precursors to tryptophane and he and Haagen-Smit at Caltech came close to identifying it (12).

Butenandt, Weidel & Becker (13) in Germany took up the search and were able to identify the so-called vermilion-plus substance by trying then known relatives of tryptophane; it was kynurenine.

As an interesting sidelight, kynurenine had been isolated and identified years before by Clarence Berg of the University of Iowa, son of a Wahoo harness-maker who had lived only a few miles from the Beadle farm. Had we only known, we could have got kynurenine from him.

At about this stage in our work Tatum's father, then a pharmacologist at the University of Wisconsin, came to Stanford on a family visit. One day as he was visiting our laboratory he called me aside to tell me that he was concerned about the professional future of his son. "Here you have him in a position in which he is neither a pure biochemist nor a bona fide geneticist. I'm very much afraid he will find no appropriate opportunity in either area." I recall my response very clearly: "Professor Tatum, do not worry, it is going to be all right."

I recall another episode illustrating the doubts held as to the future of the new hybrid approach. We had tried earlier to interest in joining our group a young biochemist at Columbia University who had been recommended by Professor Hans Clarke. He declined because the three part-time positions he then held were financially somewhat more rewarding than the one for which we were responsible. On again meeting him more than two decades later he told me he had many times regretted not seeing more clearly the opportunities in biochemical genetics.

At about this time, 1940–1941, Tatum gave a course at Stanford on comparative biochemistry. Auditing his lecture one day it suddenly occurred to me that there was a much easier approach than we had been following for identifying genes with known chemical reactions. If, as we believed, all enzymatically catalyzed reactions were gene controlled in a one-to-one relation, it would obviously be much less time consuming to discover additional such relations by finding mutant organisms which had lost the ability to carry out specific chemical reactions already known or postulated. For two reasons the obvious organism to use for such an approach was the red bread mold *Neurospora*. First, its cytogenetics had already been worked out by the mycologist B. O. Dodge (14), whom I had earlier met at Cornell University, and the Carl Lindegrens whom I knew from my early years at Caltech. Second, we knew from the work of Nils Fries (15) in Sweden that many filamentous fungi not too distantly related to *Neurospora* could grow on chemically defined media containing a proper balance of inorganic salts, a source of carbon and energy such as a sugar, plus one or more known vitamins.

So why not determine the minimal nutritional requirements of *Neurospora*, produce mutant types by X or ultraviolet irradiation and then test these for loss of ability to synthesize one or more components of the minimal medium? We soon found the minimal medium to consist of simple inorganic compounds, a suitable carbon and energy source such as sucrose, plus the one vitamin biotin. That was 1941 and fortunately biotin had just become commercially available as a concentrate sufficiently free of amino acids and other vitamins to serve our purpose.

The 299th culture from a single ascospore, whose parent culture had been X rayed, proved not to grow on minimal medium but did so with added Vitamin B_6. It was then a simple matter to determine that a genetic unit, presumably a single gene, had been mutated by crossing the mutant strain grown on a supplemented culture medium with the original strain of the appropriate mating type and then testing cultures from the eight single spores derived from a single meiotic event. Our test showed that four such cultures required Vitamin B_6 while four did not, indicating change in a single genetic unit (16).

Could we produce more such mutant types with other requirements? The answer was yes, for other vitamins and for various essential amino acids. In sequences of biosynthetic reactions leading to a given endproduct we could identify genes for individual steps, in general one gene and one only for a specific biosynthetic step.

In addition to biochemical mutants, which for the most part are normal morphologically when grown on properly supplemented media, a variety of morphologically altered types were found, some quite bizarre in appearance. During the time we were accumulating these along with scores of nutritionally altered mutants Doctor Charles Thom, a widely recognized authority on fungi, especially of the genus *Penicillium* and related genera, paid us a visit. As we toured the laboratories he was obviously keenly interested but made few comments. After we had demonstrated a fair sample of the work under way and a number of morphologically diverse mutant types, Doctor Thom called me aside and said "You know what you need here?"

"What," I asked.

"A good mycologist," was the answer. "Those cultures you call mutants are not mutants at all. They are contaminants."

To the question of how, when crossed with the original type, they could segregate according to established Mendelian principles he had no answer. I'm sure he left convinced we were the most inept mycologists he had ever seen. He had never been an ardent admirer of genetics and we obviously failed to influence him in that regard.

At this stage of our investigations it was obvious that we could increase our rate of progress significantly by supplementing our research personnel. We were fortunate in obtaining the additional financial help needed for this from the Rockefeller Foundation which, through grants to the Stanford Biology group, had made the initial work possible. Herschel K. Mitchell, Norman H. Horowitz, David M. Bonner, Francis Ryan, Mary Houlahan, and others joined the team. Through C. Glen King of the Nutrition Foundation we received support for graduate students including Adrian M. Srb, August Doermann, David Regnery, Frank C. Hungate, Taine T. Bell, and Verna Coonradt.

Although the Research Corporation did not support our work financially, its officers gave us much appreciated encouragement in the following way: The Rockefeller Foundation had earlier made a $200,000 grant to the C. V. Taylor group of biologists at Stanford, of which Tatum and I were members. Knowing Taylor's persistence, persuasiveness, and ambition for his group, the officers of the Foundation had placed a condition on the grant, namely that he not apply for additional funds from the Rockefeller Foundation during a following ten year period. I of course knew of this, and thus inquired of Frank Blair Hanson of the Rockefeller Foundation if there was any objection to our applying to the Research Corporation for supplemental support of our special project. There was not, so on that same day I approached the Research Corporation and was told they would provide the needed $10,000. Just as the details of how formally to apply were being discussed a telephone call came to me from Hanson of the Rockefeller Foundation, saying that they had reconsidered our special situation and felt that since they had provided initial support they thought it appropriate to provide the requested supplement. On reporting this to the Research Corporation officers, I was immediately told it was right and proper that the Rockefeller Foundation should continue the support, but that if we would send them a carbon copy of our formal request they would agree to provide the $10,000 if the Rockefeller Foundation for any reason did not do so. That is the kind of confidence that really inspires a research team. The Rockefeller Foundation did make the grant, but it was only years later that I learned it had been Warren Weaver who had recommended that the exception be made. His record of judging projects that paid off, scientifically speaking, was one of remarkable success and I have always been grateful that we did it no serious damage.

During this visit to the Research Corporation R. E. Waterman, who had worked earlier with R. R. Williams in isolating and characterizing thiamine, pointed out to me that G. W. Kidder of Amherst was working with a protozoan and had obtained results very much like ours. He added that Doctor Williams knew the details, and that I would have a good chance of seeing him if I were to hurry over to the 42nd Street Airlines Terminal where he was waiting for an airport limousine. I did find him and was led to believe Kidder indeed had results very much like ours in *Neurospora*. We were of course anxious to learn more about it. On doing so we found that the work that had so understandably impressed Williams had to do with special cultural conditions under which *Tetrahymena vorax* could synthesize thiamin (17) and was not at all designed to answer the types of questions we were asking.

By 1942 we had gone a fair way in the process of identifying genes with specific chemical reactions. Then the classical work of Garrod (7) was rediscovered, or perhaps more correctly, properly appreciated, by J. B. S. Haldane and Sewall Wright (18, 19). Back in the early part of the century, very soon after the rediscovery of Mendel's paper and the confirmation of his principles, Garrod had demonstrated that the human disease alcaptonuria was a simple Mendelian recessive trait characterized by an inability to further degrade 2,5-dihydroxyphenyl acetic acid (alcapton or homogentisic acid), a metabolic derivative of phenylalanine. Unlike their normal counterparts who further degrade alcapton, alcaptonurics excrete it in the urine where, upon exposure to air, it oxidizes to a blackish compound. Not only did Garrod correctly deduce the relation of gene to enzyme and to chemical reaction, he also used alcaptonurics to identify intermediate compounds in the sequence of reactions between phenylalanine and alcapton. In a like manner he characterized several other genetically controlled metabolic reactions in man.

On learning of this long-neglected work it was immediately clear to us that in principle we had merely rediscovered what Garrod had so clearly shown forty years before. There were three differences of significance: First, we could produce many examples. Second, our experimental organism was far better suited to both chemical and genetic investigation. Third, ours was a time far more favorable for acceptance of the obvious conclusions.

Like Mendel, Garrod was far ahead of his time, but unlike Mendel, his work was not buried in a relatively obscure journal: Garrod published in standard journals and wrote a widely distributed book, *Inborn Errors of Metabolism,* first published in 1909 with a second edition in 1923 (7). His work was well known to Bateson, the early British enthusiastic advocate of Mendelism. Bateson and his associate Punnett advised Garrod on the genetic aspects of his studies of biochemical defects in man, and Bateson's classical 1909 book, *Mendel's Principles of Heredity* (20), referred to it in some detail. For reasons most difficult to understand it then dropped out of the genetic literature until revived in 1942.

In giving a seminar on biochemical genetics at the University of California at Berkeley, in the late 1940s I pointed out that among others, Goldschmidt's 1938 book *Physiological Genetics* (21) failed to mention Garrod's contribution. Professor Goldschmidt, who was in the audience, came up after the seminar and explained that he had known of Garrod's work and could not understand how he had omitted mention of it. Clearly, like many others, he failed to appreciate its full significance, else he could not have forgotten it.

In retrospect one wonders how such important findings could be so thoroughly unappreciated and disregarded for so many years. Obviously the time was not ready for their proper appreciation. Even in 1941 when Tatum and I first reported our induced genetic-biochemical lesions in *Neurospora* few people were ready to accept what seemed to us to be a compelling conclusion, namely that in general one gene specifies the sequence of one enzyme (or polypeptide chain). In 1945 I gave a series of some two dozen Sigma Xi lectures in as many colleges and universities of the country. The skeptics were many, the converts few. Even at the time of 1951 Cold Spring Harbor Symposium on Quantitative Biology the skeptics were still many. In fact the believers I knew at the time could be counted on the fingers of one hand, despite the eloquent and persuasive additional evidence presented at that meeting by Horowitz & Leupold (22).

In speculating on the long-continued reluctance of geneticists and others to accept the simple gene/enzyme concept so clearly implied in Garrod's early work, the anthocyanin studies, and the more recent microorganism studies, Horowitz (personal communication) tells me A. H. Sturtevant had once pointed out to him that this was because of a widespread belief in the so-called pleiotropic (many effects) action of genes. In the sense that the terminal results of a single gene mutation may appear multiple, this can be said to be correct. But in terms of the primary effect of such a mutation in replacing a single amino acid in a polypeptide chain for example, it is clearly not.

With the working out of the Watson-Crick double helix structure of DNA, its method of replication, and its role in protein synthesis, the difficulty in accepting the concept of one gene/one enzyme largely disappears, for it can now be stated as one functional DNA sequence/one primary polypeptide chain.

The work I have discussed was but a small part of a prelude to the magnificent new era of biology ushered in through the elucidation of the structure of DNA two decades ago. Our knowledge of living things at the molecular level has continued to increase exponentially. In a real sense genetics has come to be recognized as an integral and basic part of all biology, of biochemistry, biophysics, immunology, virology, physiology, the behavioral sciences, plant and animal breeding, and all the rest.

Largely as a result of its advances, the opportunities and challenges have never been greater in the areas of biology. Nor have the intellectual rewards to those adequately prepared and sufficiently motivated.

THE EXCITEMENT OF SCIENCE   65

In my own situation, I tried a quarter of a century ago what I thought of as an experiment in combining research in biochemical genetics with a substantial commitment to academic administration. I soon found that, unlike a number of my more versatile colleagues, I could not do justice to both. Finding it increasingly difficult to reverse the decision I had made, I saw the commitment to administration through as best I could, often wondering if I could have come near keeping up with the ever increasing demands of research had I taken the other route. My doubts increased with time.

As one bit of evidence that occasional satisfactions do accrue to academic administrators, I cite an example involving James D. Watson. On his return from the Cambridge Medical Research Council Unit shortly after he and Francis Crick had worked out the double helix structure of DNA, Watson continued research as a Senior Research Fellow at the California Institute of Technology, Division of Biology. His draft board address, however, remained Chicago, and at this time the board members concluded his deferment from military service had been sufficiently long and thereupon reclassified him 1A.

Being convinced his potential contributions to science would far outweigh anything he might do to promote the mission of the military, we set out to convince the authorities that his deferment should be continued. Successive appeals to higher and higher levels were consistently denied and the Watson file grew correspondingly thicker. Finally, through the help of the National Research Council, the appeal was carried to the highest level, the Presidential Review Board. At this level previous decisions were reversed and Watson assigned to what in Washington was facetiously referred to as "the rare bird category," a designation that seemed especially appropriate to Watson, a dedicated bird watcher.

Those of us involved were of course much pleased that our efforts had been successful. Personally I was never quite able to decide the appropriate sentiment to express to the military, condolences or congratulations.

Now, on retirement from administrative duties, I have returned to a relatively simple research project of four decades ago with Emerson, namely the origin of *Zea mays,* Indian corn. It involves a combination of genetics, ecology, archeology, biochemistry, and other related disciplines in ways I am glad to say I find intellectually and emotionally satisfying.

Literature Cited

1. Garrod, A. E. 1902. *Lancet* 2:1617
2. Wheeler, W. M. 1923. *Science* 57:61–71
3. Jeffrey, E. C. 1925. *Science* 62:3–5
4. Morgan, T. H. 1909. *Am. Breeders Assoc.* 5:365–68
5. Emerson, R. A., Beadle, G. W., Fraser, A. C. 1950. *Cornell Univ. Mem.* 180: 3–83
6. Beadle, G. W. 1950. *Genetics* 35:1–3
7. Garrod, A. E. 1923. *Inborn Errors of Metabolism.* London: Hodder & Stoughton. 2nd ed. 216 pp.
8. Ephrussi, B., Beadle, G. W. 1935. *C. R. Acad. Sci.* 201:98–101
9. Beadle, G. W., Ephrussi, B. 1936. *Genetics* 21:225–47
10. Caspari, E. 1933. *Arch. Entwichlungsmech. Organ.* 130:353–81
11. Scott-Moncrieff, R. 1936. *J. Genet.* 32: 117–70
12. Tatum, E. L., Haagen-Smit, A. J. 1941. *J. Biol. Chem.* 140:575–80
13. Butenandt, A., Weidel, W., Becker, E. 1942. *Naturwissenschaften* 28:63–64

14. Dodge, B. O. 1927. *J. Agr. Res.* 35: 289–305
15. Fries, N. 1938. *Symb. Bot. Upsal.* 3:1–188
16. Beadle, G. W., Tatum, E. L. 1941. *Proc. Nat. Acad. Sci. USA* 27:499–506
17. Kidder, G. W., Dewey, V. C. 1942. *Growth* 6:405–18
18. Haldane, J. B. S. 1942. *New Paths in Genetics.* New York/London: Harper. 206 pp.
19. Wright, S. 1941. *Physiol. Rev.* 21:487–527
20. Bateson, W. 1909. *Mendel's Principles of Heredity.* Cambridge. 369 pp.
21. Goldschmidt, R. 1938. *Physiological Genetics.* New York: McGraw. 325 pp.
22. Horowitz, N. H., Leupold, U. 1951. *Cold Spring Harbor Symp. Quant. Biol.* 16: 65–72

THE EXCITEMENT OF SCIENCE   67

Ann. Rev. Pharmacol., Vol. 9

ESSENTIAL PHARMACOLOGY

By J. Harold Burn

J. Harold Burn

ESSENTIAL PHARMACOLOGY

By J. Harold Burn

Emeritus Professor, University of Oxford, Oxford, England

The beginning.—When I went to Cambridge in 1909, I became a member of Emmanuel College, of which John Harvard had been a member some 300 years earlier. I had the good fortune to have as a tutor, whose duty it was to supervise my progress in a general way, a man called F. G. Hopkins! He was the pioneer biochemist who isolated tryptophan, who worked on what he called "accessory food factors," and discovered glutathione. He was one of the most modest, gracious and graceful men I have ever known.

At Cambridge I expected to make Chemistry my principal subject, but I found that for the first two years, I had to read Physics and two other subjects as well. With Hopkins' approval Physiology and Botany were chosen. I discovered that, compared with those who taught Physiology and Physics, the chemists were an unattractive lot, particularly those who demonstrated in the practical class. These seemed very undistinguished, had cheerless faces and wore rather shabby clothes. The physiologists and the physicists had far more personality. In Physics I heard lectures given by J. J. Thomson, who seemed to be not of this world, and one of the demonstrators was Aston who built the first mass spectrograph. In Physiology the Professor was Langley, always well-dressed, who rode to hounds (that is he hunted the fox) and spoke to no one so far as I could see. There was W. B. Hardy who, when not studying the behaviour of colloids, sailed a yacht in the rough seas of the English Channel. In the summer he pinned a notice on his door which said "Back tomorrow." Joseph Barcroft, an Ulsterman, also sailed boats, and Keith Lucas, who was in part an engineer, had a large motor vessel on the Cam. Hidden in his room was a young man called Adrian, who spent his spare time at one period painting pictures in the style of Marinetti for an exhibition. It is said that all the works were sold, mostly to eager, but unwary Faculty members.

Then there was the striking athletic figure of A. V. Hill, the very distinguished-looking W. M. Fletcher, and finally, helping Barcroft and Hopkins to fill the place with humanity, was H. K. Anderson.

To enter the Cambridge world after leaving the school world was like meeting sea breezes on the shore after being confined in a stuffy room, and for two years I was mainly interested in debates and discussions of politics, religion and literature, which at least taught me to form my own opinions. The result in the examinations at the end of the second year was, not surprisingly, a disaster, and this made it difficult to decide what subject I

1

THE EXCITEMENT OF SCIENCE

8243-2601/78/1127-0071$01.00 © 1978 ARI

should choose for my final year. The choice would determine my future life. I discussed it all with Hopkins, saying that Physiology was the subject I preferred. He thought that if I wanted to have some immediate security after my third year, I had better choose Chemistry, since there was always a market for chemists. But if I was prepared to leave the future to take care of itself, then I might choose Physiology, despite the fact that I was not a medical student. He advised me that I must not read text-books, but should study original papers, and should read them critically. He mentioned that I should look at the papers of H. H. Dale and P. P. Laidlaw. When I did so, I felt that they were of little use for the purpose of my final examination, because I could find nothing to criticize! At that time I was unable to see what a break into a new world they represented.

After my third year I began research with Barcroft on two small problems concerned with the oxygen capacity of haemoglobin. I learnt (to my surprise) that I could work accurately, but I was still only half-interested and was occupied with other things. I was secretary of a University Social Discussion Society, and of its Inner Circle which invited distinguished public figures to address them. One of these figures I remember was Beatrice Webb who had encyclopaedic knowledge on the reform of the Poor Law. I may say that she stayed, unforgettably, at the house of Mr. W. C. Dampier Whetham, and that he lived at Upwater Lodge, where I had dinner.

Barcroft, irritated no doubt by my wandering inclinations, told me one day in his ever-genial manner that what I needed was three years of a really hard time. When I was in the Army during the First War, I got them. Barcroft was easily the best lecturer in the department, and gave fascinating demonstrations (assisted by John). One, of which I have a photograph, was made on an anaesthetized cat in which a cannula had been inserted in Wharton's duct to collect the saliva. After an inch or two the cannula turned vertically upwards, and when the chorda tympani nerve was stimulated the saliva rose up the tube. There was also a cannula in the carotid artery, which turned vertically upwards as well, so that the blood (prevented from clotting by hirudin) ran high up this tube and oscillated between its systolic and diastolic height at a mean level of about 130 cm. The question which Barcroft then asked the students was whether continued stimulation of the chorda would drive the saliva above the height of the blood pressure. How many know the answer to that question?

So successful were Barcroft's demonstrations that when I went to Oxford in 1937, Edith Bülbring and I began a series of mammalian demonstrations there, and I believe they continue. I, for my part, still carry out about eight demonstrations each year in the Department of Pharmacology, Washington University, St. Louis, and there is a proposal that next year each group of four students be given a demonstration to practise until they can do it successfully, and then demonstrate it to the others, using closed-circuit television.

The main trouble in devising experiments (which are not biochemical) for a practical course in Pharmacology is that the student has to discover how to assemble his equipment and set up his preparation before making his observations. Too many fail to complete their work in time, and as a result feel frustrated. But if a small group remain with one experiment until they have mastered it, they learn much more from it, and can demonstrate it to others. They can even make a heart-lung preparation in the dog, and gain great satisfaction from doing so. The experiments must, however, be chosen with care.

Plunge into Pharmacology.—It was on January 1st 1914 that I went to work with H. H. Dale. To leave the strictly monastic life of Cambridge for the gayer life of London was exciting. The atmosphere did not conduce to single-minded devotion to the laboratory, and I can remember often feeling worn-out at the end of a morning's experiment, when Dale was obviously as fresh as at 10 a.m. But I was very happy to have arrived where mammalian experiments were done, and where, moreover, they were done in great style. Much has been said about the importance of Dale's discoveries, but not many have commented on their presentation. His paper "On some physiological actions of ergot" (1) is a good example, and Figure 26 is one that could be enlarged and chosen to adorn the drawing-room wall of an obstetrician as being a work of art. (It is the original demonstration of the action of pituitary extract on the uterus). All the kymograph tracings right up to 1936 were marked by their elegance, and few workers have been able to use the plethysmograph with such success, and to write with white ink on the varnished paper so beautifully as Dale.

All members of the staff of the Wellcome Physiological Research Laboratories had lunch together, and I soon realized that daily lunch is an essential part of laboratory life if workers are to profit by what they can learn from each other.

My work introduced me to the spinal cat and to the isolated organ bath, and these remained standard preparations for many investigations. Some workers forget that responses to autonomic nerve stimulation are greatly reduced in animals anaesthetized with pentobarbital.

Dale left the laboratories at the end of June 1914, and in August the First War began. I had been in Germany in 1913 and in Austria in 1914, with the result that during the first three months after war was declared I was very much anti-war, and had furious discussions, particularly with G. S. Walpole at the lunch table. The sudden development of a fierce hatred for Germans seemed entirely artificial. However, by October I began to feel that if everyone disagreed with me, I might be the one who was wrong, and that if I wanted to see a battlefield, I'd better get into it before it was all over. By good fortune a Cambridge Don invited me to join his Signal Company, and in July 1915 I found myself in France in charge of the Section of the 12th Division Signal Company which was attached to the 36th Infantry Brigade. The duty of the Section was to maintain telephone communication between the Brigadier-General and his four Battalion commanders. This meant plenty of shell-fire, but not taking part in an attack.

I then discovered all about courage, and how surprisingly right Bernard Shaw was in his play "Arms and the Man." In it an officer fighting for the Serbians against the Bulgarians runs away, and finds himself taking refuge in a lady's bedroom. She refuses to be frightened by him, and says "though I am only a woman, I think I am at heart as brave as you." He replies to her "I should think so. You haven't been under fire for three days as I have. I can stand two days without showing it much; but no man can stand three days: I'm as nervous as a mouse."

After being away from the trenches and the war for a few weeks, you could be quite indifferent to the bursts of shells and to machine gun bullets for 24 hours. But after that these things again had their effect in restricting your movements. The men who got the medals were those who could go fearlessly into action, capture a trench or destroy a machine gun, get wounded and then be out of the line for three months. Then they could return and repeat their brave deeds and again get wounded. The important thing, however, was not to be killed.

Before joining the army in October 1914, Dale suggested that I would be wise to take a medical degree, and I registered as a medical student who had done part of the course. In December 1917 I was able to leave the Army to do my Anatomy and clinical work at Guy's Hospital. Because the staff was depleted by the war, the opportunity for students was great. While I was a surgical dresser, I was told one afternoon that I must begin at once as an extern and visit patients in labour in their homes in the district around the Hospital. I was given a small bag and told to go to 2 Farthing Alley, Dockhead. I protested that I had had no instruction in Obstetrics, but I was told I must do my best.

At that time (December 1919) there were almost no qualified midwives, and when I arrived at the house, the patient was certainly in labour, attended by a Mrs. Gamp, straight from the pages of Charles Dickens. Before long a baby was born, and after cutting the umbilical cord, I put on a dressing. Then to my bewilderment, the mother delivered an object rather like a football. The only conclusion I could come to was that this was some kind of placenta, and I placed it in the chamber pot. About ten minutes later the mother delivered what was definitely a placenta, and I wondered what I had put in the pot. I took it out again and realized that it was a second baby born inside a caul. I removed the baby from the caul and all was well. Three days later both mother and twins were progressing satisfactorily. I did duty as an extern for five weeks, during which I attended 68 deliveries, including seven breech presentations, and one face presentation among other anomalies.

In September 1920 I rejoined Dale at the National Institute for Medical Research, where he was the head of the Department of Pharmacology and Biochemistry. My special duty was to study methods of biological standardization. The lunch-time conversations with the bacteriologists and patholo-gists were often interesting and informative. The bacteriologists were very much under the influence of Almroth Wright and had no thought of making a chemical approach to the killing of bacteria in the body. In 1911 the ethyl derivative of hydrocupreine had been found to be potent in killing them out-side the body, and indeed when given to mice it protected them against a gen-eral infection with pneumococci. But in doing so it caused blindness. I re-member Clifford Dobell arguing that any substance which killed bacteria would kill man. But probably the real difficulty was that the bacteriologists had no knowledge of Chemistry.

In addition to my special duty of working on biological standardization, in which subject I showed no sign of having ideas of my own, I learnt much from Dale by working with him both on the action of insulin and on the action of histamine. All these things helped me to acquire more knowl-edge of methods. On January 1st 1926 I went to take charge of a new labo-ratory at the Pharmaceutical Society in London which was set up mainly to study methods of biological standardization. There was an interesting devel-opment which concerned the alkaloids present in ergot. Very few today know the sequence of events which led to the discovery of ergometrine (also called ergonovine and ergobasine).

The ergot alkaloids.—In 1922 Dale (2) opened a discussion at the Royal Society of Medicine concerning Liquid Extract of Ergot, and the active substances in it. He recalled that Tanret in 1875 obtained a pure crystalline alkaloid from ergot which he called ergotinine; this was associated with an amorphous substance having the same chemical composition; it had a low specific rotation. The amorphous substance was thought to be ergotinine also. The crystalline alkaloid was naturally chosen for investigation, and since it was found to be inactive pharmacologically, the inference was drawn that it was also inactive therapeutically. However Barger and Carr found that the amorphous substance was not in fact similar in composition to the crystalline ergotinine and that it was an intensely active substance. They called it ergotoxine. Dale said that "from tracings of experiments conducted in the laboratory, ergotoxine might have been regarded as a stim-ulant of plain muscle responsible for the therapeutic action of ergot." The method described in the British Pharmacopoeia of 1914 involved the extrac-tion of ergot with water instead of alcohol, which was used in the previous Pharmacopoeia. Since ergotoxine was not soluble in water, it followed that the method "might almost have been designed to exclude ergotoxine." There were indeed other active substances in the watery extract—tyramine and histamine. But tyramine could be cheaply prepared from tyrosine and hista-mine from cheese. They were not substances peculiar to ergot. The charge brought against those responsible for the British Pharmacopoeia 1914, was that they did not understand what they were doing. When watery extracts were prepared, the specific active substances were thrown away and the ad-ventitious substances were retained.

The Secretary of the Pharmacopoeia Committee, Tirard, was present at this discussion, and said that 19 licensing authorities had been consulted, and not one of them had suggested omitting any preparation of ergot. A. W. Bourne, obstetrician at Queen Charlotte's Hospital, suggested that to resolve this question there should be clinical trials, using methods enabling measurements to be made, and the results might then replace clinical impressions.

Sometime in 1923 or 1924 I was asked by Dale to assist Bourne in carrying out some work of this kind, but for one reason or another the collaboration was brief, though long enough to show that a method for recording the intrauterine pressure in women in labour was practicable.

The Geneva Conference of 1925.—In 1925 the League of Nations Health Organization held a Conference in Geneva, at which Dale presided, to obtain international agreement on stable standards for insulin and other therapeutic substances "the purity and potency of which could not be determined by chemical means." Extract of the posterior lobe of the pituitary gland was one of these. For each substance a unit was chosen in terms of the activity present in a given weight of the powder which was to be the international standard. The standard powder for pituitary (posterior lobe) extract was declared by members of the Conference to contain 1 unit of activity in 0.5 mg.

When I went to the Pharmaceutical Society in 1926, it seemed to me that it would be fitting to attempt to determine what clinical effect would be produced by 1 unit. With Dale's agreement I approached Aleck Bourne again, and we proceeded further with observations on women in labour. A small sterile rubber bag attached to a sterile catheter was passed into the uterus when the os was sufficiently dilated. The bag and catheter were filled with water and connected by a long rubber tube to a mercury manometer, so that a rise of pressure within the uterus could be recorded on a drum. We discovered that in some patients the subcutaneous injection of two units produced as large and prolonged a contraction as was ever desirable, and were therefore able to suggest that this should be regarded as the maximum single dose. However, in some patients this dose had a very small effect. Our investigations therefore had little useful result, for it did not occur to us to try giving the extract by slow intravenous infusion. The intravenous drip was not invented until 1935, though we might have invented it ourselves.

Having made our observations on pituitary extract, we then decided to turn our attention to ergot, and to determine which was the important constituent. The task seemed simple. All we had to do was to examine the three substances, ergotoxine, tyramine, and histamine. This seemed preferable to spending time testing the effect of an extract itself. An extract would have to be given by mouth, and it would be difficult to know when its effect began because of the time required for absorption from the alimentary tract. The active constituents, however, could be given by injection, and any effect they had should be seen without delay. The only difficulty was about ergotoxine, which was not on sale. This was solved by using ergotamine instead, which had been isolated by Stoll in 1921 from ergot growing on Festuca grass. A careful comparison of ergotamine with ergotoxine had been carried out by Dale and Spiro. The result was that the two substances were indistinguishable by any test which could be applied. I myself made several comparisons by the method of Broom and Clark, using the reversal of the action of adrenaline on strips of the rabbit uterus. I found no certain difference between the two substances in any comparison.

Investigation showed that tyramine was without effect on the uterus of a patient in labour, and histamine, in the amount present in the full dose of an ergot extract, was also without effect. Ergotamine, however, when injected in a dose of 1 mg., produced a powerful contraction of the uterus lasting at least 16 hours. The results were taken to mean that a Liquid Extract of Ergot must be prepared so as to contain ergotamine or ergotoxine.

There was, however, one dissatisfied person. This was F. H. Carr, who together with G. Barger isolated ergotoxine. He was very anxious to know whether the effect of ergotoxine was actually the same in a patient as the effect of ergotamine. We did not feel able to make this comparison, because we had been warned by our experience with ergotamine that such trials on patients in labour were not without risk to the foetus.

However, Chassar Moir undertook the task under different conditions. He worked on patients in the puerperium after all the stages of labour were complete, and the foetus and placenta were delivered. He was then able to show that ergotamine and ergotoxine were indistinguishable, and thus he satisfied Carr. What was Chassar Moir to do then? It seemed as if he was left with nothing to do but dot the i's and cross the t's. He decided to have a watery extract prepared according to the 1914 British Pharmacopoeia, which would contain no ergotoxine or ergotamine, and then he gave it by mouth to one of his patients. I quote from his account (3).

> "Judged by all previous work, this preparation ought to have been inert, for analysis showed that it contained only a trace of alkaloid. The extract was used in 3 cases in doses of 4, 3, and 2 drachms. It was with the greatest surprise I found that far from being inert, this preparation surpassed by great measure the activity of any drug which I had previously used in the same manner. An equally surprising fact was that the effect appeared in a remarkably short time. In one case only 4 minutes elapsed between the swallowing of the extract and the onset of powerful uterine contractions."

Chassar Moir had discovered the presence of a new active principle in ergot which was soon realized to be the substance responsible for the usefulness of ergot in midwifery. That was in 1932. Chassar Moir and H. W. Dudley isolated the alkaloid in 1935, and gave it the name of ergometrine. Others were excited by Chassar Moir's discovery, and the alkaloid was also isolated in 1935 by Stoll who gave it the name of ergobasine, and by Dragstedt who gave it the name of ergonovine.

THE EXCITEMENT OF SCIENCE 77

The story was a splendid cautionary tale. It warned us (*a*) that the impressions of clinicians should be treated with respect; (*b*) that because one specific active principle has been found, it must not be assumed that there is not another; and (*c*) that control observations, however unnecessary they may seem, should never be omitted. If we had once given a dose of the 1914 B.P. Liquid Extract of Ergot to a patient by mouth, we would have discovered the new alkaloid ourselves.

The use of a perfusion pump.—Having given an account of a research which was a failure, and from which I could claim no credit whatever, I feel free to turn to a different investigation of which the outcome was more satisfactory. During the last year I worked with Dale a further study was made of the vasodilator action of histamine, following his classical paper published jointly with A. N. Richards in 1919. This work was done in the main by using a perfusion pump with the object of reproducing in the perfused leg of the cat effects of histamine which were regularly observed in the whole animal. The experiments were sometimes successful and sometimes not, and obviously there were factors in a perfusion experiment which were only discovered by perseverance.

When I went to the Pharmaceutical Society in 1926, I intended to do further work of this kind, and when Dale and Schuster produced a new pump in 1928, I obtained one and proceeded to investigate the action of tyramine. There was one reason for choosing tyramine, namely that Tainter & Chang (4) had shown that it differed from adrenaline in its relation to cocaine. Whereas Fröhlich & Loewi (5) observed that the rise of blood pressure produced by adrenaline was increased in height and duration in the presence of cocaine, Tainter found that the rise of pressure produced by tyramine was diminished or even abolished by cocaine. The difference suggested that adrenaline and tyramine did not produce a rise of blood pressure in the same way.

The Dale-Schuster apparatus was a double pump designed to deliver blood to the organ to be perfused, after which the blood was collected from the venous outflow and pumped through the lungs for re-oxygenation. The procedure was to prepare one hindleg of a dog for perfusion through the external iliac artery, then to bleed out the dog, and then to prepare the lungs. This meant that the leg to be perfused was without a circulation of blood for about 40 minutes, until the lungs were ready and the perfusion could begin.

The results obtained when adrenaline was injected into the cannula tied in the external iliac artery were exactly as expected. There was always vasoconstriction. But when tyramine was injected there was little or no response. Whereas in the spinal cat adrenaline was not more than 40 times stronger than tyramine in causing a given rise of blood pressure, in the perfused hindleg adrenaline was 1200 to 1800 times stronger than tyramine, and sometimes a larger dose of tyramine had a dilator action (6). Why was tyramine so weak in the perfused leg?

Experiments were then carried out to determine the relative constrictor effect of the two substances in spinal cats by observing the volume change produced in a limb enclosed in a plethysmograph; these cats were eviscerated, and the adrenals and kidneys were removed. A wide variation in the relative constrictor effect of the two substances was observed. In some cats tyramine had no constrictor action in the limb at all although an equipressor dose of adrenaline had a good one, while in other cats tyramine caused constriction equivalent to that caused by adrenaline in an amount only 40 times less. Thus the mystery still remained. There was an unknown factor responsible for tyramine having a very variable constrictor action in peripheral vessels.

A fresh attempt to identify this factor was made by setting up a hindleg perfusion in a dog. Beside it a heart-lung preparation was also set up, and when it was working well, the blood from the heart-lung was used to perfuse the hindleg. This was, so to speak, an attempt to synthesize a dog. During the early period in which the hindleg was connected to the heart-lung, there were repeated crises when, because of the persistent vasodilatation in the hindleg, the output of the heart-lung fell. These crises were overcome by injections of adrenaline. As the experiment continued these crises became less serious, and it was then observed that after the immediate effect of an injection of adrenaline had subsided, the response of the hindleg vessels to an injection of tyramine increased. Whereas at the beginning, injections of tyramine into the external iliac artery caused very little vasoconstriction, after the tone in the hindleg vessels had risen and been maintained at a level of about 100 mm Hg. for an hour, injections of tyramine had a much greater constrictor effect than before.

These observations led to the conclusion that the greatest constrictor effect of tyramine seen in the body could also be seen in the perfused hindleg, if there was a steady and continuous infusion of adrenaline into the perfused leg. Thus it appeared that the constrictor effect of tyramine was in some way dependent on the presence of adrenaline. The dependence was not, however, due to the increased tone produced by adrenaline, because the same increase of tone produced by infusing vasopressin did not increase the constrictor effect of tyramine (7).

These results with tyramine led to other experiments in which the effect of stimulation of the sympathetic fibres to the vessels of the perfused hindleg was studied. When a perfusion was started, the stimulation of the sympathetic fibres had little or no effect. When the tone was raised by adding adrenaline drop by drop to the circulating blood until the pressure reached a normal level, stimulation of the sympathetic fibres was then observed to cause vasodilatation. Finally, when the addition of adrenaline was continued for one hour or more, sympathetic stimulation caused vasoconstriction.

The interpretation of the changes was that during the period of preparation in which the hindleg was without a circulation, there was a loss of the transmitter from the sympathetic fibres due to anoxia. Therefore there was no response to sympathetic stimulation. (Anoxia was later shown to produce a large release of catecholamines from the adrenal medulla.) **When** the tone was first raised by the addition of adrenaline to the blood, the first change in the response to stimulation was a consequence of the stimulation of hitherto unsuspected vasodilator fibres, which produced a fall in peripheral resistance. When the tone was maintained for one hour or more by the addition of adrenaline, it appeared that adrenaline was taken up into the sympathetic fibres, with the result that when these were stimulated, the adrenaline was released and constriction followed (8).

This interpretation therefore explained the recovery of a normal response to sympathetic stimulation as due to a new phenomenon, namely that of uptake of adrenaline from the blood.

The explanation served to apply to the action of tyramine also, because Tainter and I (9) had found that tyramine, which acted like adrenaline in causing dilatation of the normal pupil of the cat's eye, did not dilate the pupil if the sympathetic fibres had degenerated. In this respect tyramine was completely unlike adrenaline, and the finding that the action of tyramine on the pupil was seen only when the sympathetic fibres were present, suggested that tyramine acted by releasing the transmitter substance from the fibres. The recovery of the constrictor action of tyramine in the perfused hindleg by a long-maintained infusion of adrenaline was thus consistent with the view that the adrenaline was taken up by the sympathetic fibres, and that the constriction produced by tyramine was due to the release of this adrenaline from the fibres. The outcome of this work, done in the early thirties, was unexpected when the work was begun, and the many experiments in which the dog hindleg was perfused eventually provided valuable information.

When, 25 years later, we learned that tyramine has no pressor action in an animal treated with reserpine, we showed in a few days that the pressor action was restored by an infusion of noradrenaline, and we showed also that stimulation of the sympathetic which caused vasodilatation in the perfused hindleg of the reserpine-treated dog, recovered its power to cause vasoconstriction after an infusion of noradrenaline (10). Thereafter the phenomenon of uptake was studied by many.

The training of Ph.D. students.—In the Department of Pharmacology in Oxford, where I went in 1937, I think that most of us might be said to have been under the influence of the discoveries made by Dale and his coworkers in the field of the chemical transmission of nerve impulses. It was these discoveries which enabled pharmacology to contribute so much to medicine as well as to physiology, and there are today many more similar discoveries still to be made. It is surely high time that the transmitter of sensory impulses should be identified, for there is good reason to think that it is liberated into the tissues and probably into the blood stream when stimulation is applied to the cut peripheral end of a sensory nerve (11). The impact of such a discovery on neurology would certainly be very great. It is possible that the release of the transmitter is a double event, and that the first step is the release of acetylcholine, which then in turn releases the specific agent. But while one would think that many pharmacologists would be trying to discover what this is, in fact it seems that very few of them are doing so, in spite of the problem seeming to be much simpler than the many problems of transmission in the brain.

Our work at Oxford was concerned at an early stage with chemical transmission, and in 1941 Edith Bülbring and I (12) showed that acetylcholine was a transmitter in the spinal cord. We made a preparation of half a dog, having two circuits of blood, one through the spinal cord and a quite separate circuit through the leg. Our observations also showed that the presence of adrenaline in the blood perfusing the cord increased many reflex effects. This action of adrenaline needs more investigation.

To workers who came to the Department to work for a Ph.D., or for a shorter time, we regularly gave problems which involved using isolated organs, preferably innervated. Edith Bülbring (13) devised the isolated rat diaphragm with the phrenic nerve. Dawes (14) prepared isolated rabbit atria so that they could be stimulated to contract at any given rate. He determined the maximum rate of stimulation at which the atria followed the stimuli, and found that this maximum rate was reduced in the presence of quinidine, or of substances with a similar action. The reduction depended on the concentration of quinidine.

Later on, other isolated preparations of the heart were made. McEwen (15) made a preparation of the isolated heart with the vagus nerves, and also one of the isolated atria with the vagus nerves. Huković (16) made a preparation of isolated rabbit atria with sympathetic fibres from the stellate ganglion. Then many learnt to use the preparation of rabbit ear vessels with the postganglionic sympathetic innervation described by Gaddum & Kwiatkowski (17). Those with sufficient skill learnt how to perfuse the superior cervical ganglion by the method of Kibjakow (18). Then there was the Finkleman (19) preparation in which stimulation is applied to the sympathetic fibres running to the small intestine, and finally the still better preparation of the rabbit colon described by Garry & Gillespie (20). In this the sympathetic and parasympathetic supplies can be stimulated separately.

When a chemist has completed his training for the Ph.D. degree, he has a wide experience of the standard chemical procedures. There seems to be much less agreement in pharmacology about what standard pharmacological procedures are. The methods which have been outlined above are among those in which the young pharmacologist can be trained. They are challenging and the skill acquired in learning to use them generally brings a feeling of satisfaction. Pharmacology should be a discipline of its own.

Giving communications.—Whenever anyone in the Department, whether he was a Ph.D. student, or a postdoctoral worker, or a senior member of the staff, or the Head of the Department, was proposing to give a communication to the British Pharmacological Society, or the Physiological Society, it was the regular custom for him to give his communication beforehand in the Department to as many of his colleagues as wished to come. This was to enable him to rehearse what he was going to say. He must not on any account read his communication, he must speak loud enough to be heard clearly at the back of the lecture theatre and he must say everything in ten minutes or little over. He was recommended to have not more than six slides, and if the slides contained tables of figures, these had to be in large type and to be few. It was common for a person to give two rehearsals, and not rare for him to give three. This often occurred when he did not present his matter clearly, and when his argument was not followed. The criticisms of the audience were a great help in improving presentation. As a result our communications were so well given that the Physiological Society asked us to draw up recommendations which they printed and sent to all those who gave notice that they wished to give a communication. These said that communications must not be read, should be rehearsed, etc. After all, if someone expects the members of a large Society to listen to him, it is common courtesy for him to practise giving it until the communication is given as well as possible. Moreover the value of a communication is often judged by the way in which it is presented.

Acetylcholine in the heart.—Among other studies which continued over some years was one concerning the action of acetylcholine (ACh) in the heart. In 1946 there were various suggestions that acetylcholine was synthesized in the heart, and had a function which was in some way excitatory. We demonstrated the formation of acetylcholine (21) by perfusing the isolated rabbit heart through the aorta with Locke's solution containing physostigmine, and recirculating the fluid through the heart for 40 minutes. At the end of this time the fluid was collected and freeze-dried, and the residue was extracted with ethanol. The extract, when tested on cat blood pressure, frog heart, and frog rectus, was found to contain acetylcholine and also (22) an adrenaline-like substance. The hearts when beating spontaneously beat at 56 beats per min, and produced 0.26 µg ACh per heart per 40 min. They were also driven electrically at 210 beats per min, and then produced 1.32 µg ACh per heart per 40 min. Thus the amount of ACh formed was proportional to the rate of beating. We had no clue to its function.

In 1949 John R. Vane (23) was examining a new antimalarial compound, proguanil, and found that when isolated rabbit atria were exposed to its action, the contractions declined and finally stopped. But when acetylcholine was added to the bath, the contractions started again with great vigour, and stopped again when the ACh was removed. Later, A.K. Armitage (24) made a similar observation when he exposed isolated atria to quinidine; again the contractions declined and stopped, and again ACh caused them to start again. However, he also found that when quinidine had caused the atria to stop, they started again when the amount of K+ in the bath fluid was reduced from 5.6 mM to 1.4 mM. This suggested that the arrest of the atria by quinidine might be due to a diminished permeability of the cell membrane to K+, so that in the presence of quinidine, the K+, which must escape through the cell membrane after a contraction in order to repolarize the membrane, could not escape. It could escape when the gradient of K+ concentration across the membrane was made steeper by reducing the external K+. This suggestion at once explained the action of acetylcholine in restarting the contractions stopped by quinidine, because as Harris & Hutter (25) showed, acetylcholine increases the permeability of the cell membrane for K+, so that the exit of K+ is facilitated and repolarisation is once more achieved.

The effect of acetylcholine in restoring atrial contractions was also demonstrated (26) in circumstances in which the atria were left in the bath until they ceased to contract, perhaps after 70 hours. The arrest of the atria in this case was shown by Goodford (27) to be due to a steady fall in the intracellular K+, so that the gradient from inside to outside again became insufficiently steep for repolarisation. But when acetylcholine was added, the permeability for K+ was increased and repolarisation was once more achieved. The atria then resumed their contractions.

The most exciting results, however, were obtained by Jean Marshall & Vaughan Williams (28) who found that atria ceased to beat when they were cooled to a temperature below 20° C, and that their contractions could then be started by acetylcholine. Jean Marshall (29), when working at Johns Hopkins later, was able to explain this by showing that on cooling below 20° C the transmembrane potential gradually fell, and when it was below 60 mV impulses were no longer propagated, and therefore contractions ceased. However when acetylcholine was added, the transmembrane potential rose and propagation of impulses began again. The effect of acetylcholine in raising the transmembrane potential was again probably due (at least in part) to its effect in increasing the permeability of the membrane for K+ and so facilitating repolarisation.

These results suggested that the transmembrane potential in rabbit atria could not be maintained below 20° C because it was normally maintained by a formation of acetylcholine which failed below 20° C. That this was so was demonstrated by observations on the activity of choline acetylase from rabbit atria. A. S. Milton (30) determined the Q_{10} for ranges of temperature between 37° C and 13° C. He found that from 37° C to 21° C the Q_{10} was approximately equal to 2.0 in every range of 4° C (i.e. from 37° C to 33° C, 33° to 29° C etc.) However between 21° C and 17° C it rose to 6.2, and between 17° C and 13° C it rose to 7.8. Thus synthesis of acetylcholine was greatly reduced below 20° C at the point where impulses were no longer propagated. This was a very satisfying conclusion, since it indicated that one purpose of the synthesis of acetylcholine in the atria was to enable the transmembrane potential to be maintained at a sufficiently high level. It is interesting to note that whether acetylcholine is restarting atria which have stopped at a temperature below 20° C, or whether it is stopping atrial beats at a normal temperature, it acts in both cases by raising the transmembrane potential. I have given an account of this evidence because it is not well known, and because it led to a conclusion both unexpected and interesting. Thus an investigation (31) of isolated atria with the right vagus nerve attached showed that when cooled to a temperature at which the spontaneous contractions stopped, one pulse applied to the vagus would not cause the atria to start again, but two pulses would do so. When the contractions had continued for 5 minutes, then without change of temperature one pulse would arrest the contractions. After 5 minutes arrest, two pulses would start them again, and so on. These observations were re-

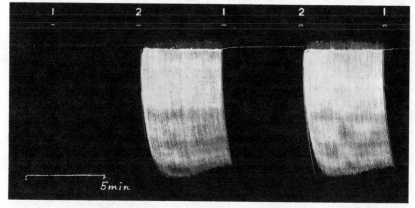

Fig. 1. Experiment showing the double action of the vagus on isolated rabbit atria. When the bath temperature fell to 17°C. the contractions stopped. At "1" a single pulse was applied to the vagus. It had no effect. Then at "2" two pulses at 1 sec interval were applied and the contractions then began. After 5 min, one pulse was applied to the vagus, and the contractions stopped. After 5 min more two pulses were applied and the contractions began again, to be arrested 5 min later by one pulse. These effects were repeated 30 times (31). (Reproduced by permission of the Editors of the *Journal of Physiology*.)

peated no less than 30 times in the course of an experiment. This fascinating situation, shown in Figure I, deserves further investigation.

The sympathetic transmitter and monoamine oxidase.—If a prefatory chapter such as this is to include confessions of failure, then it is essential that I should refer to papers which I published in 1952 and 1953 on the fate of the sympathetic transmitter, in which I put forward the view that the enzyme monoamine oxidase was the agent responsible for destroying noradrenaline. The evidence began with the finding that thyroid-feeding in rabbits diminished the amount of monoamine oxidase in the rabbit liver, and that thyroidectomy increased it. These observations were confirmed by others. But then it was demonstrated that the degeneration of sympathetic fibres in the nictitating membrane led to a fall in the amount of monoamine oxidase in the membrane, and before long I had reached the conclusion that this fall was responsible for the supersensitivity to noradrenaline which followed denervation. It was about two years before I faced the fact that the evidence from the catecholamine Corbasil (which has a —CH₃ group on the alpha carbon of the side chain) was sufficient in itself to show that the view was wrong. Denervation supersensitivity was present when Corbasil was injected, and yet it was not a substrate of monoamine oxidase. If a scientific worker can once make a mistake like that, and even give lectures about it, then any other ingenious ideas he may put forward must be eyed with the suspicion which is their due.

Work on auricular fibrillation.—The foregoing error was an error of insufficient thinking. But there was another error at that time made in relation to observations on isolated atria, and the effect of physostigmine upon them. I failed to realize that when atrial contractions were recorded by means of a lever working against a spring, the amplitude of the contractions on the drum depended on the rate at which the contractions occurred.

However, this work was the first stage of investigations of the action of physostigmine in the heart-lung preparation of the dog. These investigations proved to be very fruitful, and led us to a method of producing atrial fibrillation at will, and of stopping it at will. We applied electrodes to the tip of the right auricle. These were held in place by a spring, and did not pierce or damage the auricle. We also prepared a burette containing acetylcholine and could drive the solution into the superior vena cava at a slow uniform rate. When we stimulated the atria for 30 sec at a frequency of 14/sec, this caused fibrillation but the fibrillation returned to normal rhythm when the stimulation stopped. Similarly, when we infused the solution of acetylcholine into the blood at about 1 mg/min, the rate of the heart was slowed, but the rhythm was unaffected. However, when we applied the stimulation during the infusion of acetylcholine, then, when stimulation stopped, the atria continued to fibrillate as long as the infusion of acetylcholine continued. When the infusion was stopped, fibrillation gave place to flutter, and then to normal rhythm within one minute. We had hit on a means of producing fibrillation and of maintaining it for as long as we wished. We could allow it to continue for 90 minutes and then stop it by a turn of the burette tap. We could start it again ten times in one experiment (32).

We also studied ventricular fibrillation by perfusing isolated rabbit hearts through the aorta with Locke's solution. Stimulation of the ventricles at 25/sec for 3 minutes produced fibrillation which stopped when the stimulation stopped. This was at 32° C. But when the Locke's solution was modified in any way which would shorten the refractory period, then the fibrillation continued indefinitely after the stimulation stopped. Factors which had been shown to shorten the refractory period were oxygen lack, glucose lack, the presence of dinitrophenol or of sodium azide or of monoiodoacetate. That these factors shorten the refractory period is evidence that energy is required to maintain the normal refractory period which in cardiac muscle is very long (150 msec) compared with that in skeletal muscle (2 msec).

Why then should it be necessary to have a long refractory period in cardiac muscle? The answer is that a long refractory period protects the cardiac muscle against fibrillation in which the fibres are contracting asynchronously, and are therefore, like an asynchronous tug-of-war team, unable to exert force. The importance of the long refractory period is evident in the atria also, because acetylcholine greatly shortens the atrial refractory period. Fibrillation produced by initial stimulation of the atria during the infusion of acetylcholine is arrested when the infusion stops and the refractory period lengthens.

What then is the cause of fibrillation? If the fibres are stimulated at a high rate they fail to contract in synchrony, because the conduction of the impulse is slower and varies along different paths. When individual fibres do not contract synchronously, excitation spreads from one fibre which contracts to another which is resting. This excitation is effective when the refractory period is abnormally short and the fibre is excitable. Thus fibrillation occurs because the fibres stimulate one another (33).

Fibrillation may occur spontaneously in the perfused rabbit heart when glucose is absent for a long time, or if the K^+ in the perfusion fluid is reduced to 1.4 mM, which is one-quarter of normal. Both these changes reduce the refractory period. Dr. Carl Schmidt (34) in an Editorial made very favourable comments on these conclusions.

Ciliary movement and acetylcholine.—Because Gray (35) had shown that ciliary movement and cardiac contractions were similarly affected by a variety of factors, we began an investigation of the part played by acetylcholine in ciliary movement. The investigation was first carried out on the mucous membrane of the frog oesophagus, then on the rabbit trachea removed from the body (36), and finally on the gill plates of *Mytilus edulis* (37). In all three we measured the rate of particle transport and in the *Mytilus* gill plates we measured the rate of ciliary movement by a stroboflash as well. The particle transport was quickened in all three tissues by low concentrations of acetylcholine, and was depressed by high ones. We observed similar changes after applying physostigmine. The particle transport was depressed by *d*-tubocurarine. An extract of the mucous membrane of the rabbit trachea was found to contain acetylcholine, and the presence of both choline acetylase and of acetylcholinesterase was established. The gill plates of *Mytilus* were extracted, the extract purified and shown to contain acetylcholine by biological tests (parallel quantitative assays on three organs), and by chromatography. The presence of cholineacetylase and of acetylcholinesterase was also determined (38).

The importance of these results was that the gill plates do not contain nervous tissue and we had demonstrated the function of acetylcholine as a local hormone in a situation in which nerves were absent.

A worker at University College, London (39) was unable to repeat our observations on the effect of acetylcholine and of d-tubocurarine on the frog oesophagus. Most of her observations were made using a phosphate-buffered Ringer's solution. Milton (40) discovered that, when phosphate was present, the inhibitory action of d-tubocurarine was no longer seen; however, the action was restored in the same preparation when bicarbonate replaced the phosphate. Few are aware that phosphate can interfere with tissue respiration. According to Alt (41) even the inhibition of the respiration of kidney and liver slices by cyanide was reduced from 98 per cent to 11 per cent when bicarbonate buffer was replaced by a phosphate buffer. Our results with acetylcholine and d-tubocurarine were independently confirmed by D'Arcy, Grimshaw, and Pickering.

Pharmacology today.—In this prefatory chapter I have tried to express my view of pharmacology by a series of examples of the discoveries which can be made by what some would describe as old-fashioned methods. I have chosen these examples because the discoveries were of importance, and in my view they reveal the harvest still to be reaped by investigations of this kind. It is a great pity that so many are attracted to molecular pharmacology, because they are leaving fields unexplored which are highly relevant to the study of pharmacology proper.

What then is pharmacology proper? It is first of all the study of the mode of action of substances used in the treatment of disease, and of course the discovery of new substances for use in disease. Departments of Pharmacology were set up about the turn of the century as a step towards making the practice of medicine more scientific. Medical students were to be taught how drugs act. The success of the Departments was not always obvious because some teachers of pharmacology dwelt too much on actions of drugs which had no relation to treatment, with the result that 30 or 40 years ago there were physicians who thought that pharmacology as a discipline for medical students was not worth while. With the rapid increase of potent medicinal agents, the need for a knowledge of their mode of action became more and more necessary. Nevertheless there are those who think that this kind of pharmacology can be taught by the physicians. The practising physician is, however, rarely acquainted with the mode of action of drugs, because he has no time to study it. He is too much involved with diagnosis and the other aspects of patient care.

What is it that these members of electoral bodies would prefer to see? Some of them speak of molecular pharmacology as being a more appropriate study. One wonders in the first place how a knowledge of molecular pharmacology will help the doctor to treat his patient, and in the second place how much real understanding of molecular pharmacology the majority of students will gain from attending a course of lectures and practical classes. Molecular pharmacology is for the most part a subject only understood by those doing research in the field.

That simple methods can still break new ground and yield a rich harvest is well shown by the method of blood-bathed organs which has been lately introduced by J. R. Vane. He withdraws the blood from the carotid artery of an anaesthetized dog, and collects it in a reservoir from which it is pumped up above the dog to cascade over a series of isolated organs. The contractions of these organs are recorded on a series of drums. The blood is then returned to the dog. The organs over which the blood flows are chosen for their sensitivity to some particular substance. For example a strip of rat colon can be used to measure angiotensin in the blood, and the rectal caecum of a chicken can be used to measure adrenaline, distinguishing it from noradrenaline. They have used as many as 16 organs at once. This simple method has recently been used to demonstrate that angiotensin I is converted into the more active angiotensin II during passage through the lungs, and not, as previously thought, by the action of an enzyme in the blood. This method seems certain to yield much new information.

The release of noradrenaline.—Since 1959 when I reached the age limit of 67 for heads of departments in Oxford University, I have had the good fortune to spend my time in finding new tests for a hypothesis concerning the release of noradrenaline. The suggestion is that the sympathetic postganglionic fibre releases noradrenaline through the prior release of acetylcholine. Till now it has proved to be a wholly unacceptable view to many people on both sides of the Atlantic, largely, I think, because few have considered the details of the evidence since 1964. Yet, whether it is accepted or not, a hypothesis has a value, perhaps its main value, in stimulating investigations, and in challenging assumptions which have been accepted without question. Such an assumption is that the sympathetic impulse releases noradrenaline directly. Those who accept it must explain, for example, how a very low concentration of acetylcholine will abolish the response to stimulation of the postganglionic fibre, and prevent the release of noradrenaline. A large number of such observations have now been made which are not compatible with the view of a direct release. In time those who disagree with what has been proposed, will come to see that disagreement is not enough, and that they must, in their turn, explain these many observations. Observations concerning the transmission of nerve impulses should be one of the prime interests of pharmacologists. For me this continues to be an exciting subject which has filled the first nine years of retirement with activity.

The end.—Finally, would I choose a different career if I had my time again? After all it can be said that those who spend their lives in research leave little behind them. They are not sculptors or painters whose work may be preserved for hundreds of years. On the contrary, their research, whether of their earlier years, or even of ten years ago, is forgotten while they are still alive. Nevertheless they have their consolations. I can think of no way of spending life with more satisfaction than it was spent in the Oxford department, where we met for a brief break at 11 a.m., later for lunch in the library, and for tea at 4:15. At one period there was a grand piano, and there were two who played Mozart, Beethoven and Haydn as duets. We were a community like a College Common Room, and at the same time all (or nearly all) very interested in our research. There was usually a good sprinkling of workers from abroad so that we felt ourselves an international group. Moreover, our work was mostly of the kind that required hand skill, and gave an opportunity for some artistry in execution. I have a letter from a one-time statesman (Lionel Curtis) saying that the Vice-Chancellor (at that time Richard Livingstone) had told him we were the happiest family in Oxford. We used to have the Vice-Chancellors to lunch; they were all Arts men, and we wanted them to see what the scientific life was like. Thus I am quite certain that I would choose to have it all over again.

LITERATURE CITED

1. Dale, H. H., *J. Physiol. (Lond.)* **34,** 163–206 (1906)
2. Dale, H. H., *Lancet,* **ii,** 1275–78 (1922)
3. Moir, Chassar, *Brit. Med. J.,* **i,** 1119–22 (1932)
4. Tainter, M. L., Chang, D. K., *J. Pharmacol. Exp. Ther.,* **30,** 193–207 (1927)
5. Fröhlich, A., Loewi, O., *Arch. Exptl. Pathol. Pharmakol.,* **62,** 159–69 (1910)
6. Burn, J. H., *Quart. J. Pharm. Pharmacol.,* **3,** 187–204 (1930)
7. Burn, J. H., *J. Pharmacol. Exp. Ther.,* **46,** 75–95 (1932)
8. Burn, J. H., *Proc. Roy. Soc. Med.,* **27,** 31–46 (1933)
9. Burn, J. H., Tainter, M. L., *J. Physiol. (Lond.)* **71,** 169–93 (1931)
10. Burn, J. H., Rand, M. J., *J. Physiol. (Lond.)* **144,** 314–36 (1958)
11. Jancsó, N., Jancsó-Gabor, A., Szolcsanyi, J., *Brit. J. Pharmacol. Chemother.,* **33,** 32–41 (1968)
12. Bülbring, E., Burn, J. H., *J. Physiol.,* **100,** 337–68 (1941)
13. Bülbring, E., *Brit. J. Pharmacol. Chemother.,* **1,** 38–61 (1946)
14. Dawes, G. S., *Brit. J. Pharmacol. Chemother.,* **1,** 90–112 (1946)
15. McEwen, L. M., *J. Physiol. (Lond.)* **131,** 678–89 (1956)
16. Huković, S., *Brit. J. Pharmacol. Chemother.,* **14,** 372–76 (1959)
17. Gaddum, J. H., Kwiatkowski, H., *J. Physiol. (Lond.)* **94,** 87–100 (1938)
18. Kibjakow, A. W., *Pflügers Arch. Ges. Physiol.,* **232,** 432–48 (1933)
19. Finkleman, B., *J. Physiol. (Lond.)* **70,** 145–57 (1930)
20. Garry, R. C., Gillespie, J. S., *J. Physiol. (Lond.)* **128,** 557–76 (1955)
21. Briscoe, S., Burn, J. H., *J. Physiol. (Lond.)* **126,** 181–90 (1954)
22. Day, M., *J. Physiol. (Lond.)* **134,** 558–68 (1956)
23. Burn, J. H., Vane, J. R., *J. Physiol. (Lond.)* **108,** 104–15 (1949)
24. Armitage, A. K., *Brit. J. Pharmacol. Chemother.,* **12,** 74–78 (1957)
25. Harris, E. J., Hutter, O. F., *J. Physiol. (Lond.)* **133,** 58P (1956)
26. Bülbring, E., Burn, J. H., *J. Physiol. (Lond.)* **108,** 508–24 (1949)
27. Goodford, P., *J. Physiol. (Lond.)* **145,** 221–24 (1959)
28. Marshall, J. M., Vaughan Williams, E. M., *J. Physiol. (Lond.)* **131,** 186–99 (1956)
29. Marshall, J. M., *Circulation Res.,* **5,** 664–69 (1957)
30. Burn, J. H., Milton, A. S., *Brit. J. Pharmacol. Chemother.,* **14,** 493–96 (1959)
31. Burn, J. H., Rand, M. J., *J. Physiol. (Lond.)* **142,** 173–86 (1958)
32. Burn, J. H., Vaughan Williams, E.

M., Walker, J. M., *J. Physiol. (Lond.)* **128,** 277–93 (1955)

33. Burn, J. H., *Can. Med. Assoc. J.*, **84,** 625–27 (1961)

34. Schmidt, C., *Circulation Res.*, **9,** 1136–37 (1961)

35. Gray, J., *Proc. Roy. Soc. B.*, **96,** 95–114 (1924)

36. Kordik, P., Bülbring, E., Burn, J. H., *Brit. J. Pharmacol. Chemother.*, **7,** 67–79 (1952)

37. Bülbring, E., Burn, J. H., Shelley, H. J., *Proc. Roy. Soc. B.*, **141,** 445–66 (1953)

38. Milton, A. S., *Proc. Roy. Soc. B.*, **150,** 240–44 (1959)

39. Hill, J. R., *J. Physiol. (Lond.)* **139,** 157–66 (1957)

40. Milton, A. S., *Brit. J. Pharmacol. Chemother.*, **14,** 323–26

41. Alt, H. L., *Biochem. Z.*, **221,** 498–501 (1930)

Reprinted from
ANNUAL REVIEW OF PHYSIOLOGY
Volume 37, 1975

VARIETY—THE SPICE OF SCIENCE AS WELL AS OF LIFE
The Disadvantages of Specialization

Professor Alan C. Burton

Alan C Burton

Reprinted from
ANNUAL REVIEW OF PHYSIOLOGY
Volume 37, 1975

VARIETY—THE SPICE OF   ♦1121
SCIENCE AS WELL AS OF LIFE
The Disadvantages of Specialization

Professor Alan C. Burton
Department of Biophysics, University of Western Ontario, London, Ontario, Canada

Even the dogs may eat the crumbs which fall from the rich man's table; and in these days, when the rich in knowledge eat such specialized food at such separate tables, only the dogs have a chance of a balanced diet.

Sir Geoffrey Vickers[1]

I have been one of the fortunate dogs and I am sorry for the young physiologists of today, most of whom must sit at the separate tables, whether these are rich in knowledge or not. They must learn "more and more about less and less."

By the accident of being born into Edwardian middle class England, the vicissitudes of the economic state of my parents (my father was a dental surgeon, whose practice fluctuated with changes in suburban London), and the profound social changes of two World Wars, I have experienced the greatest variety of life styles; I have known family affluence (when I was eight my parents and six children had as many as five servants and lived in a 20-room house), as well as the insecurity of poverty and how a family must work together for survival, the New World as well as the Old, the prosperous twenties, and the Great Depression. I was educated under the extremes of education, from the stupidity of ultraclassical "private schools" (in the English sense) to modern grammar schools. My continuing education has been at London University and the Universities of Toronto, Rochester, New York, Pennsylvania, and my present University of Western Ontario. This variety has taught me that there is something to be learned from all kinds of people, that all types of education have some merits, and that there are many different ways of tackling problems in science.

I think this great variety contributed to my development. Perhaps it influenced my decision to pursue such a variety of disciplines in scholarship and research. My first research was an attempt to match the absorption bands of the major planets (Jupiter, Uranus, Neptune) in the laboratory; my latest has been an adventure in

---

[1]Preface to *The Art of Judgment.* Chapman & Hall, 1965.

1

THE EXCITEMENT OF SCIENCE
8243-2601/78/1127-0093$01.00 © 1978 ARI   93

theoretical biology in the control of cellular division. In between I have written papers in physics on atomic nuclear structure, the superconductivity of metals at liquid helium temperatures, the heating of electrolytes in high-frequency fields, and odd subjects like the floating of mercury droplets on water. My introduction to the field of biological research was more or less accidental, in that I was trying to discover why workmen making "shortwave" (then down to 5 m) machines developed fever. Shortwaves were then used as a tool for producing artificial fever, so I went to conferences on this to tell physiologists and physicans why the tissues became hotter.

A detailed list of the variety of my research problems in the more than 40 years that followed would be boring. They went from the study of animal heat exchanges, the mechanisms of temperature regulation, the control of peripheral blood flow, and the development of hemodynamic principles to the biophysics of the equilibrium of the wall of blood vessels and of the red blood cell, and to the biophysics of cellular membranes. The Second World War gave me the opportunity to learn of the fascination of applied physiology, of devising ways of obtaining objective and valid subjective data in the field, from design of life jackets and the flashing lights on them, to the best colors of life rafts for air-sea rescue. I enjoyed a chance to develop a "science of protective clothing" (which did not really exist before the war). I suppose there was a theme linking all this diversity; the desire to "put things in order" and my love of simple physical and mathematical analysis (what is this but the "shorthand of logic") to create "model systems" of living behavior, which is often so much more ingenious and much more successful than the devices of engineers.

Ever since a colleague introduced me to it, I have treasured the description written by Francis Bacon of the mind of the scientist. It is engraved on the wall of our Health Sciences building.

The Scientific Mind

A Mind, Nimble and Versatile enough to catch the Resemblances of things, which
is the chief point, and at the same time Steady enough to Fix and Discern their
Subtle Differences; endowed by Nature with the Desire to Seek, Patience to Doubt,
Fondness to Meditate, Slowness to Assert, Readiness to Reconsider, Carefulness to
Set in Order, and neither Affecting what is New nor Admiring what is Old, and
Hating every kind of Imposture.

Francis Bacon[2]

It has been my aspiration to live up to Bacon's description, which explains, perhaps, how anyone could so spread his efforts on such a variety of topics in science (resemblances of things).

In the first published history of the Royal Society (the first 62 years of it) Thomas Sprat[3] praised variety in science and academic goals in a delightfully quaint way

[2]Francis Bacon. 1955. *Selected writings on the interpretation of Nature,* p. 151. New York: Random House.

[3]Sprat's book is not so much a history of the Society as an apology explaining the need for existence of the Society. (Sprat, T. 1722. *The History of the Royal Society of London,* p. 245. London: Roy. Soc. 3rd. ed.)

(this quotation hangs in the hall outside the office of our Biophysics Department). It makes a plea for variety in experiments and in scholarly activity. I paid an art student to illuminate the words printed here in capitals in the manner of the medieval monks.

> "It is stranger that we are not able to inculcate into the minds of many men, the necessity of that DISTINCTION of my Lord BACON'S, that there ought to be EXPERIMENTS of LIGHT, as well as of FRUIT. It is their usual word, WHAT SOLID GOOD WILL COME FROM THENCE? They are indeed to be commended for being so severe EXAC-TORS of GOODNESS. And it were to be wish'd, that they would not only exercise this vigour, about EXPERIMENTS, but on their own LIVES, and ACTIONS: that they would still question with themselves, in all that they do; WHAT SOLID GOOD WILL COME FROM THENCE? But they are to know, that in so large, and so various an ART as this of EXPERIMENTS, there are many degrees of usefulness: some may serve for real, and plain BENEFIT, without much DELIGHT: some for TEACHING without apparent PROFIT: some for LIGHT now, and for USE hereafter; some only for ORNAMENT, and CURIOSITY. If they will persist in contemning all EXPERIMENTS, except those which bring with them immediate GAIN, and a present HARVEST: they may as well cavil at the Providence of God, that He has not made all the seasons of the year, to be times of MOWING, REAPING, and VINTAGE."

I am well aware of my susceptibility to the charge that I have been a "Jack-of-all trades, master of none." In my defense I would borrow from Edna St. Vincent Millay (forgive the alteration).

> But Oh! my Friends, and Ah! my Foes
> It's been a lot of Fun.

I suppose that I have been unusually fortunate that it was possible for me to work on all these problems, and yet find the generous support that enabled me to do so. I wonder if today any young scientists will be given such opportunities to be so general, rather than specialized, in their interest and researches. Most of them are doomed to stay in some restricted field of study. It is my contention that this is a great pity, that creativity in science (as in physiology) is likely to be discouraged, and that today's excessive specialization is likely to lead to impoverishment of ideas.

THE TREND TO ULTRASPECIALIZATION

Perhaps ultraspecialization, for example in physiological research, is an inevitable consequence in modern laboratories of the increasing complexity of technical methods and of analysis by computer. As someone has wisely said, if one uses relatively simple apparatus to answer a question, the answer is likely to be a very complicated one. To obtain a simple set of answers, very complicated methods usually have to be employed. With the very impressive technical methods at our disposal, we drift into spending a great deal of time and money on accumulating mountains of data, even where any underlying idea worth pursuing may have been lacking. I spent many years serving on committees of grant-giving bodies and on Study Sections of the National Institutes of Health of the US. Applicants almost always requested

12-channel recorders to report, simultaneously, 12 different physiological variables. I usually recommended that we give them instead a recorder with only, say, four channels, not because it cost less, but because the research would be likely to reach worthwhile new concepts much better with fewer channels. True, the 12-channeler might publish a more impressive article (to some readers). Probably the piles of data would be relegated to "data-storage" journal facilities, just as the data would be relegated to unused files in his own laboratory. But will any useful new idea emerge that would not better be supported by designing crucial experiments based on a preconceived idea, and settling the point by using only one or two channels? My views were not popular.

I have never forgotten a general lecture I heard many years ago on cancer research by the great Peyton Rous (of "Rous sarcoma" fame). He pointed out that while research required painstaking collation of reliable data, that was not enough to justify the effort to collect it. He illustrated this with the tale of a graduate student who was told by his supervising professor to decide for himself the problem in research that he (the graduate student) wished. After many months, and after taking up a great deal of the professor's time, the student had not made a decision. Finally the professor said, "All right, I will assign you a research problem. Cut off the tails of a thousand rats and analyze each of them for cholesterol." There is very little value in the most extensive and well-documented research, unless there is some idea behind it. With technical specialization this is apt to be forgotten.

A VIRTUE OF IGNORANCE

The temptation for the technical specialist is to continue to accumulate data with the apparatus and techniques he has developed long after they have served their purpose in opening new fields and prompting new basic questions. New ideas in research come from asking ourselves simple, often quite naive questions. The more "expert" we become in a narrow field of knowledge, the less likely we are to be willing to ask such questions, not only in other fields, but particularly in our specialized field. The diffidence of the "professional" is natural, since asking a "stupid question" might suggest that, with all his detailed knowledge, he lacks a grasp of fundamental principles.

A few years ago I was asked to address the Royal College of Physicians and Surgeons of Canada and the United Kingdom at their annual meeting in Toronto, on the role of biophysics in medicine. I chose the title "A Virtue of Ignorance," with the subheading "A Biophysicist Asks Simple Questions About Medicine and Medical Research," and gave several examples. One of these was that every medical student is taught the location of the auscultory areas on the chest where the stethoscope should be placed to best hear the various heart sounds originating in the four valves of the heart (aortic, pulmonary, mitral, and tricuspid) at the time of their closure in the cardiac cycle. Yet, if a transparency showing the anatomical position of these four valves is superimposed over the textbook picture of the position of these auscultory areas, it appears that they are nowhere near the points closest to the origin of the sounds, if it is assumed that these travel in straight lines through the tissues. My point is that if Dr. J. Faber, who was a graduate student with us at the time, and I had been "experts" in cardiology, it is very unlikely that we would have asked the question, "How do the heart sounds travel from their point of origin to the stethoscope of the physician?" The "experts" were sure they knew the answer. Dr. Faber provided the unexpected answer very conclusively by his research. After the heart sounds reach the stethoscope head they are travelling as sounds, but in most of the circuitous pathway from the heart valves these are not sounds at all! Instead they are transverse vibrations, travelling, with much less velocity than would sounds, over the surfaces of blood vessels or ventricular walls, to emerge where these walls are in contact with the thorax. From the auscultory areas the vibrations travel over the chest like ripples on a pond. Surely this fundamental knowledge should be the basis of understanding how the best places to hear specific heart sounds or murmurs (e.g. the murmur of patent ductus) in diseased states are often in unexpected locations. Variety in training and interest in physiology and biophysics can pay dividends in leading one to ask such simple questions.

SPECIALIZATION IN PHYSIOLOGY

I suppose it is because the majority of departments of physiology on this continent are in faculties of medicine, and draw their financial support from such faculties, that there has been undue emphasis on mammalian physiology in teaching and research. Physiologists should never forget that the normal behavior of cells and organisms, and its mechanisms, in the whole field of living things, is their province. While the application to the advance of medical science might justify a special interest in mammals, so much of the understanding we need may come from study of lower forms. *What does he know of mammals, who only mammals knows?*

It is not difficult to find departments of physiology in which nearly every member of the faculty is expected to join in a concerted research on some rather narrow branch of physiology, even though the teaching of the curriculum to medical students requires that many of them have to teach other topics in physiology. Not only is this likely to lead to poor teaching (in my opinion enthusiasm for the topic is about the most important ingredient in good teaching), but also to an eventual decline in the quality and originality of research in such departments, even in their own specialty.

"CENTERS OF EXCELLENCE?" THE RISE OF
ULTRASPECIALIZED RESEARCH INSTITUTES

Neglect of the virtues of variety is seen in its most extreme form in the proliferation in recent years of specialized research institutes, devoted to intensive pursuit of narrow goals. Even within a narrow field (e.g. cancer research) one finds restriction in the various institutes to pursuit of research exclusively based on one particular concept (e.g. virology, or disturbed biochemistry, or genetics). Some years ago in an attempt to place one of the PhD's of our department in a cancer institute, I tried to persuade the director, a friend of mine, that he should have someone in his institute that knew other ways than molecular biochemistry to attack the problem of cancer. I well remember his response: "Alan, I may be wrong, but I decided to put all my eggs in one basket." I did not think at the time of the apt reply, which would be to quote what I hope may become known as "Burton's Law." This is: *"However many hens sit on however many eggs for however long, nothing creative will result unless the eggs are fertilized."* Fertilization only occurs when there is interaction between different kinds of fowl.

We all resort to "defense mechanisms," as the psychologists put it. C. P. Snow, in his famous *Two Cultures,* pointed out one difference in attitudes between scientists and their colleagues in the humanities. Most scientists are a little ashamed that they do not know more history, philosophy, or sociology. In contrast, some humanists are actually proud that they know little about science. Ultraspecialization and compartmentalization of science, including physiology, has generated subcultures, and the same defense mechanisms are seen in some of the ultraspecialists. For example, I heard a public lecture from a very eminent molecular biologist (very eminent indeed; perhaps one should call him a biophysicist). At the end of the lecture he was asked what seemed to most of us a very relevant question. He replied with apparent pride: "I am interested only in proteins." I regret that this was not greeted with derisive laughter as it should have been, but with a reverent silence. Incidentally, his "ploy" was not as good "one-up-manship" as that of a very well-known physiologist, a specialist in nutrition, whose reply to a specific question at the end of his lecture was only to declare, "That is in my published work!" Why should he waste time on stupid people who had not studied his work?

ULTRASPECIALIZATION IN JOURNALS OF PHYSIOLOGY

Everyone knows of the explosive proliferation of journals in science, particularly in biology. Every month we receive announcements of new ones in one particular field. While, I am glad to see, a few of the new media of publication declare that they will coordinate and synthesize wide areas of knowledge and scientific enquiry, and some are new "interdisciplinary" journals, I think the majority of the new publications represent narrower and narrower subdivisions of science. I suppose that somewhere there must be provided a place for publication of the frightening mass of little "bits" of specialized information that leave the general reader saying "so what?" This is a problem we must solve somehow, like the Malthusian problem of excessive growth of human population; yet I have not heard of any proposed solution. My point is that the papers in such narrowly restricted journals tend to be written in more and more technical jargon, quite unintelligible to a generalist reader (even as well educated in a variety of science as I am!). Cross-fertilization of ideas is hardly served by this successive splitting of publication into narrow topics. The problem is timely, since members of the American Physiological Society are being polled (at the time of writing this) as to whether or not the *American Journal of Physiology* should be subdivided into several compartmentalized journals.

I have reason to be very grateful that a few journals still exist that have a declared editorial policy of providing a place for new ideas that synthesize concepts cutting across disciplines, and may be written by nonexperts with little expertise in topics on which they touch, i.e. those outside the "Establishment" of specialized knowledge. Three or four years ago I was excited by the discoveries of W. R. Loewenstein (physiologist and biophysicist) and his students of the existence and importance of intercellular communication in tissues where contiguous cells have "tight junctions," and of the possible role of this in controlling proliferation by "contact inhibition." After six months of reading the literature in control of growth in tissue culture and the many different theories of how it was mediated, I produced a general, highly speculative and simplistic, theory of "intercellular communication, cell clocks, and cancer." I think I would have found it difficult to have this accepted by most journals, even interdisciplinary ones, except the journal *Perspectives in Biology and Medicine* (I happened to know the editor). The result of publication of this set of "wild" ideas has been quite extraordinary, since I received over 1400 requests for reprints from all sorts of biologists all over the world (extraordinary in this day of xerography). Yet, after three years, I know of practically no references to this theory in the established "respectable" literature on growth and cancer. Indeed, in reading recent annual assessments by well-known experts of "where we stand in cancer research," I am struck with the complete absence of any reference to intercellular communication as a factor (the evidence of its importance in control of growth is most impressive).

My bold venture, uninhibited by my incompetence in the specialized fields on which it touched, certainly accomplished fertilization of ideas, even if to many the fertilizer used may smell excessively! I have succeeded since then in publishing a more detailed mathematic theory in another journal that is willing to explore new ideas, i.e. *The Journal of Theoretical Biology*. It is interesting that I am now having the greatest difficulty in publishing a paper that directly resulted from the theory. I spent a year studying the cancer statistics from different regions that were available to me, from the point of view that there might be a correlation with altitude. This was based on the idea that acid-base relations in the cell might be the key factor in my theory, and the knowledge that acclimatization to altitude involves chronic changes in acid-base relations. I discovered that there were quite remarkable correlations. The paper has been already rejected by two different general-interest journals. Rejection was on the advice of referees, who, judging from their criticisms, must have been ultraspecialized experts in a narrow subfield of cancer research. They were, apparently, not familiar with either the physiology of acclimatization to altitude, nor with statistical evaluation (I am of course highly prejudiced in the matter, and may be mistaken).

Somehow, while providing some way of storing all the detailed bits of unconnected scientific information that pour out of the laboratories, we must preserve the opportunities for publication of integrative concepts and generalizations, even if advanced by nonexperts.

THE EXCITEMENT OF SCIENCE   99

GRADUATE STUDENTS; DON'T FENCE THEM IN

The graduate student of today is given little opportunity to pursue any kind of research suggested by something that may have turned up in the course of his specific research project, which might well be of much more importance than what he was told to investigate. The restriction extends beyond his acquisition of the PhD, for it seems to be expected of him that his postdoctoral research, perhaps for the rest of his life, will be devoted to further investigation in depth of the narrow problem on which he was started. If his thesis was on, say, the physiology of some cells of the central nervous system, he is apt to think he must continue in this field. The impression of what is expected of him will be confirmed if he should be bold enough to seek postdoctoral experience in another field. He will find it difficult to find the support he needs or to be accepted into the specialized research laboratories or institutes that are eminent in the new field he may choose.

I had personal experience of this some years ago. One of my best PhD's, whose thesis was about the biophysics of capillary flow, sought postdoctoral experience in X-ray scattering work in molecular biology. Since he was very well trained in physics and had a flair for applied mathematics, I wrote to several eminent directors of research institutes where such work was done, highly recommending him for a place in their laboratories. I received the reply from one such that he never accepted anyone in his laboratory who was not already fully trained in X-ray scattering technique. The "rat race" of narrow specialization can easily lead to such deplorable parasitism on the body of science. Fortunately the reply from the United Kingdom was very different. The head of a laboratory was very willing to have my man, but had no place immediately because the laboratory was moving to new quarters. Then I received a letter, quite unsolicited, from another famous British scientist saying that he had heard of my student's desire. "Send him to me and I will look after him until there is a place in that laboratory for him." The contrast in attitudes was striking. The PhD in question is now back on this continent as a professor of biology, making significant contributions in the general field of relation of geometry of cells and tissues to their physiological function.

At the time of my retirement as head of our department a few years ago, my colleagues arranged with the editors of the *Canadian Journal of Physiology and Pharmacology* to consider, with the usual rules of criteria of acceptability, original papers by any of my former students, to be published in a single number of that journal.[4] (A delightful gesture with no grounds for the valid objections that might be raised to a special complimentary issue of a journal.) Ten papers were acceptable, i.e. this number of the former students happened to have acceptable contributions at the right time. It was remarkable how divergent were the topics of these papers from the particular topic of the PhD theses of the authors. At least I can say that I did not inhibit the curiosity and creativity of these students, and I think they have contributed more to science than if I had "fenced them in." Actually, one of the faculty of our present department obtained his PhD under my supervision some

[4]1970. *Can. J. Physiol. Pharmacol.* 48(6).

years ago. Though through the years he has made many fine contributions in research, he has yet to tackle the particular problem that I originally suggested as the topic for his thesis.

Our good students are "endowed by Nature with the desire to seek," as Bacon put it, and we must be careful to nurture that desire, not to frustrate it. Training to the level of the PhD should be regarded as preparing the student to tackle any problem in his general field (such as physiology or biophysics). It is training in scientific methods of how to reduce careful observations to quantitative data that can be used to test ideas about the models that may underlie biological behavior. Of course, to reach the level of the PhD the candidate must make some original contribution to new knowledge, adequately supported with evidence; but why should this be expected to remain his special interest for the rest of his career?

THE DEDICATED SPECIALIST

What I have said against ultraspecialization is not directed against a dedicated persistence in research in a narrow field by any individual. Some of the most important contributions to physiology have been made by those who refused to give up until every detail of the problem was elucidated, and this is likely to take a lifetime of research. There are admirable examples among the Nobel prize winners. What I am deploring is the segregation of some "perfectionists" in research into ultraspecialized departments or institutes (often they become directors of such institutes), where they are cut off from new ideas in other fields or disciplines. The most successful dedicated specialists that I have known did their work in environments where they were exposed to all sorts of ideas arising from the quite different research interests of their colleagues.

VARIETIES OF LOGIC

Just as the tools of experimental research have become more and more complicated and specialized, so have the tools of thinking (problem solving). We must not be trapped into forgetting that there are usually several quite different logical ways of solving a given problem. (I think someone has written a book about this, describing "horizontal" vs "vertical" thinking.) An intriguing simple example is the problem facing anyone arranging, say, a singles tennis tournament with 41 persons entered. The question asked is "How many matches are required to arrive at the champion?" To anyone trained in logic (simple mathematics), the problem is tackled in a routine manner; simple but a little tedious. A number, x, of entrants must be given a "bye" in the first round. The number can be proved to be $x = 2^n - N$ where 2^n is the power of 2 that is both greater than the total number of entrants N and closest to N. (2^n is 64 if N 41.) There will have to be 23 byes, and the remaining 18 will play a preliminary round. The 9 survivors will join the 23 byes, so the second round will consist of 32 players, 16 matches; the next round will have 8 matches, the next 4, the next 2, and then the final. The total number of matches turns out, in this numerical example, to be 40. Of course, one can, by knowing how to sum a geometric series, deduce algebraically that the total number of matches must be $(N - 1)$, but this is by no means very easily accomplished nor is the result obvious. However, if we want the answer only to the question of how many matches are needed in all, this can be given in a simple logical step, which perhaps would occur only to a nonmathematically trained person. This step is that every player, except the ultimate winner, may play in any number of matches, but eventually plays in only one match which he loses. Thus each match is uniquely associated with the loser of that match. Therefore, there must be $(N - 1)$ matches! We must, by constant communication with those who use other ways of thinking than ours, remain open to radical new methods of tackling specialized problems.

AN OCEAN OF UNDISCOVERED TRUTH

The young physiologist of today is apt to think that there remain very few questions of physiological function that could be investigated by the classic methods of physiology, such as denervation, blocking, or stimulation of nerves with observation of altered function in the relatively intact animal. Instead he may think the only type of research worth doing is with complicated techniques on isolated bits and pieces of animals. This point of view is quite unjustified. There remain, even in mammalian physiology (meaning of course dogs, cats, rabbits, rats, or, if there is enough money, monkeys), enormous areas of ignorance. Physiologists need to ask more simple questions. For example, Ulf Von Euler in his book on adrenergic innervation of various tissues, pointed out that the liver is very richly supplied with sympathetic nerves. Some of these, of course, may control the circulation of the liver, though vasomotor control in this region is not very great compared to others. What else are all these nerves controlling? Did any physiologists ever cut the nerves and see whether the many biochemical functions of the liver were affected; or stimulate these nerves to see what happens? Another naive question! In acclimatization to cold (certainly in the rat) the metabolism even of liver slices studied by the Warburg technique increases several times, yet the temperature of the liver was presumably never reduced in the chronic exposure to cold that elicited the changes seen. What triggered these changes in the liver cells?

To give an example in the biophysical field, the very devoted group of "sliding-modelists" (muscle contraction) find it very difficult to explain how muscles could shorten to less than half their resting length, yet nature abounds with muscles that contract to much smaller fractions than this. We do not even have to go outside the higher mammals to find them. The dilator pupillae (radial muscles of the iris) appear to shorten to ⅛ of their length! I know of no studies of the biochemical composition, tension development, ultrastructure, or anything else on these muscles. This is all the more astonishing since here is a thin sheet of muscle cells easily accessible and isolated. Is the apparent lack of curiosity in reality because these muscles are in a radial arrangement, not parallel fibers? or is it because in this specialized subculture of muscle physiologists, one does not ask embarrassingly simple questions, for which the experts have no answer? Instead they learn more and more about frog sartorius muscle, or in a few cases, insect flight muscles, and seem to think all contractile mechanisms must be basically related to what they find.

There seems to be an obligation on famous scientists, like other famous men, to utter memorable last words on their deathbed. Those of Sir Isaac Newton, certainly one of the greatest scientists of all time, are variously reported, so I have taken the liberty to paraphrase and to attempt to put them lyrically.

Sir Isaac Newton said it, just before he died:

> I know not what the world will say, or history teach,
> Of all my labors, less or more.
> To me, we have been children, playing on the shore.
> Picking up pretty pebbles from the beach.
> At times I've found one prettier than the rest
> And polished it, and boast with childish pride,
> While there beside us lay an Ocean, wide
> With undiscovered Truth.

That extraordinary combination of the self-serving, unscrupulous politican and mathematical genius, Louis LeCompt de Laplace, 50 years later felt he had to compete in such deathbed pronouncements; and said the same thing without poetic imagery.

"What we know is very little, what we do not know is immense." This is still true, particularly in physiology, in spite of the explosion of scientific knowledge.

Physiology is a lady of whom it can be said, "Age does not dim nor custom stale her infinite variety." Let us not spoil our appreciation of this by over-emphasis on narrow specialization. In doing so we risk stifling curiosity (as Bacon put it, our endowment with the desire to seek). The addiction of the creative scientist still is the thrill he finds in "adventures of the mind." Minor poets (very minor in my case) have so little opportunity to present their attempts, that I must be excused for trying verse to describe how solving a problem in physiology or biophysics seems to me.

DISCOVERY

Little rivulets of thought
Erode the broad surface of the problem posed,
Idle, wandering and aimless rills
Like garden freshets after heavy rain.

And now the streams have quickened, coalesced,
To eddy round the hillock of a doubt,
Find well-worn channels, ditches study-dug,
And flow with purpose in a common trend.

Ideas break surface with salmon splash,
While from the deep,
Wise intuition adds its hidden flow,
A rhythmic pulse is growing, surge on surge,
Insistent logic in bolero time.

At last the turgid waters will not stay—
Glide swiftly through the gorges of analogy,
Go leaping down the rapids of hypothesis, and break
Into a quiet flood of certainty.

If there remain any readers who have read this chapter to this point, and they have been influenced to read, in this Annual Review, not only the chapter reviewing their own specialized field of physiology, but also at least some of the other chapters, my purpose in this prefatory chapter will have been served.

Ann. Rev. Fluid Mech., Vol. 3

COMPRESSIBLE FLOW IN THE THIRTIES

ADOLF BUSEMANN

Adolf Busemann

Ann. Rev. Fluid Mech., Vol. 3

COMPRESSIBLE FLOW IN THE THIRTIES   8000

ADOLF BUSEMANN

Department of Aerospace Engineering Sciences
University of Colorado, Boulder, Colorado

1. INTRODUCTION

When the Wright brothers demonstrated on December 17, 1903, that human flight is possible, their range, height, and speed were just enough to prove this fact. From that date on, however, competition to increase these three parameters was a major challenge for both pilots and airplane manufacturers. All three of these parameters are somewhat limited on earth by the finite height of a useful atmosphere, by the limited distance between points on the earth's surface, and by the fact that incompressible flow experience helps a long way, but loses its validity somewhere after reaching half the speed of sound. Charles Lindbergh's flight across the Atlantic Ocean in 1927 brought the flying range into the proper order of magnitude; Anderson and Stevens reached an altitude of 72,400 feet or about three scale heights of the atmosphere in 1935; and the speed races in the famous Schneider Cup competitions from 1913 until 1931 approached half the speed of sound. All these accomplishments within the first thirty years of the birth of flight with heavier-than-air machines are still worth our admiration, even now when we are rather spoiled by the progress in space flight of the last decade going far beyond the limited atmosphere, the limited earth distances, and the speed of sound. The airplane wings did not quite come into the transonic range during the Schneider Cup races, but the propeller tips certainly did, and they brought home to the participants the change of flow behavior due to compressibility. This experience was a rather negative one for those propellers without variable pitch because of the need to compromise within a large range of tip speeds during the acceleration period.

The possibility of supersonic movement through the air was, of course, demonstrated by the ballistics of projectiles, and plenty of compressible-flow problems remained for the internal ballistics. To find, however, a more optimum shape for a spin-stabilized projectile, the nose of which already imitated the ship's bow, was not an urgent problem, while the larger part of the body had to serve as a piston in the barrel of the cannon and had to receive the spinning motion on the way out. Another field of progressing technology in compressible flow was that of turbines, starting with steam turbines late last century, and extending to gas turbines early this century by A. Stodola, of Zurich, who was a very prominent figure in this field. At first the experiments on simple efflux nozzles induced the misconception that a kind of sonic barrier exists in this type of flow, though it was broken by the convergent-divergent shape of the de Laval nozzle in 1889. A rather low efficiency of supersonic deflections as they were used in a few rotor and stator

1

8243-2601/78/1127-1107$01.00 © 1978 ARI   107

stages to overcome quickly the very high steam temperatures before the subsonic stages could do a better job was accepted with the excuse that energy
losses at high temperatures have a high rate of recovery in a Carnot cycle
within the later stages of the turbine, with admirable subsonic efficiency and
less extreme temperatures for the blades permitting a permanent exposure.

At this state of technology the research in compressible flow was more a
frontier attraction in science than an obvious practical necessity or even an
investment promising high return. Many scientists like L. Prandtl got interested in this field by writing corrections whenever a new publication about
the flow pattern in the minimum cross section of the Laval nozzle, or about
the shape of the plumes in the jet behind such nozzles, contained unjustifiable
assumptions or conclusions. During the very years when Prandtl introduced
the boundary-layer theory and applied it to the streamlining of bodies, which
brought us much closer to the d'Alembert paradox of zero pressure drag on
properly designed airships for incompressible flow, he also published a Bessel-
function treatment of small-disturbance plumes on axially symmetric jets of
supersonic speeds surrounded by air of constant pressure. He pictured also
in this paper larger disturbances of two-dimensional supersonic jets, which
exit through a slit or have side plates. These early pictures were so correct
that Prandtl could repeat them two years later in a paper where his famous
"flow around a corner" was actually singled out as a self-similar supersonic
flow phenomenon.

When Prandtl experimented with water to investigate vortices on the
incompressible boundary layers, he had a photographic camera to study the
details later at leisure and he made the velocity distribution visible by
putting powder of minerals (glimmering iron at first) into the water. Thus
Prandtl could obtain pictures of flow in the easy and modern way as the
famous Leonardo da Vinci—four centuries earlier—had to do the hard way.
Likewise in his study of compressible and especially supersonic flows the
photographic camera, in connection with the Toepler-Schlieren Method,
served as the tool to make permanent images of his experimental flows of
rather short duration. The trick to make the Mach-wave pattern visible was
to roughen up the walls of the flow by grooves perpendicular to the flow
direction. A similar utilization of modern optical and photographical
methods, including interferometers, helped the progress of understanding the
compressible flow around projectiles by the physicist E. Mach in Vienna and
the ballistician C. Crantz in Berlin. Mathematical treatments of supersonic
flow and its mapping into the hodograph plane were independent accomplishments by B. Riemann in 1860, P. Molenbroek in 1890, and S. A. Chaplygin
in 1902; but since they were made before the boundary-layer theory established the practical value of ideal mathematical flow theory, they did not
attract the immediate interest of large groups of scientists and engineers.

2. AFTER WORLD WAR I

Priorities in aviation changed quite a lot when, during World War I,
airplanes proved their value and even surpassed the airships. In times of
war the airship with its large lifting gas volume is, of course, an easy target

for anti-aircraft guns; but even its peacetime application for long-distance travel lost the popular belief that the only reasonable mode of transportation is by airship, while airplane flights are like risky circus stunts. The airplanes proved their safety while the dirigibles of rigid construction had to be filled with hydrogen—not yet helium—and were plagued by too many fires caused by the highly explosive mixtures of hydrogen and air.

Enthusiastic youth and war-time pilots made aviation a new field of practical interest, and in the new research centers hastily created during the war they continued their serious investigations. They approached their puzzling problems under a new perspective, regarding them in combination with results of other countries. In Germany, where under the Treaty of Versailles the engine in an airplane was limited to 50 horsepower, the new theory of lifting wings developed during the war, and its major result in M. Munk's doctoral thesis of 1918, which established the elliptical lift distribution on a single wing of given span as the optimum, supported progress toward larger lift-over-drag ratios on monoplane gliders of large span.

After the German runaway inflation was stopped in 1924, the prewar idea of surrounding Ludwig Prandtl with a Max Planck Institute (at that time still called Kaiser Wilhelm Institute), which had to be postponed during the war, became a reality in 1925.

The Max Planck Institute for Flow Research, side by side with the large wind tunnel built during the war, was the place where all those flow problems in Prandtl's past could be revived that had to step aside during the time when the Prandtl-type wind tunnel for the growing interest in aviation was designed, constructed, and improved. Among those overshadowed problems were compressible flow, cavitation in water, turbulent flow in pipes or in the atmosphere by unstable stratification, and other effects in the atmosphere influenced by the rotation of the earth. While Prandtl was very well supplied locally with mathematicians and physicists among his own students at the University of Goettingen, engineers for his wind tunnels and his institute were brought to him by his friends and relatives. Munich, his alma mater, furnished the lion's share (Max Munk, Albert Betz, Carl Wieselsberger); Aurel Stodola in Zurich recommended Jakob Ackeret; and Prandtl's brother-in-law, Otto Foeppl in Braunschweig, passed me on to Prandtl after my doctoral thesis in elastomechanics was finished. My original task was to complete the design and to supervise the construction of the rotating chamber for atmospheric studies according to Prandtl's sketches. This project proceeded rather well until the day arrived when the walls were installed all around. The afternoon before that day it was a great pleasure to try out the rotating platform at all speeds provided by the driving motor. As soon as the walls were closed and the internal lights turned on, the situation was quite different. It was nauseating in the true meaning of the word. The chamber was not the expected tool to study meteorology in an easy chair. You had to learn to keep your head still and to flick only your eyeballs. Instead of reproducing our known environment at higher rotational speed it became an introduction to bio-engineering or the art of living with our biological systems in a strange environment. Although before that time I was one of the first to become seasick in rough seas, after understanding the conflicting mechanics I am now one of the last ones to become seasick. Foreseeing speeds at which the centrifugal forces in the rotating chamber would not be comfortable for the observer inside, Prandtl had already prepared a large cylindrical opening in the upper bearing, to place an inversion prism with half the rotational speed into it for an equivalent observation from the outside.

In the Goettingen atmosphere around Prandtl I had learned plenty of fluid mechanics to be prepared for taking an active part in flow research. As an engineer it would have been appropriate for me to turn to the practical problems of aviation when my first project was finished. But it seemed that I entered our world too late; the earlier research assistants had pretty well divided up among them the urgent problems on the existing airplanes. Contrary to that situation, research work on compressible flow problems was wide open and I was very welcome to join Jakob Ackeret in this field of Prandtl's early loves. In line with my practice of not pretending that I can help where the stars of our profession, Th. von Kármán, J. Burgers, L. Prandtl, and others are already doing their very best, as in turbulence, I joined Ackeret in this frontier area of mechanics in measuring the lift and drag of profiles near and above the speed of sound. The new high-speed wind tunnel for this purpose had a cross section of 2 by 2 inches, which was much larger than the earlier equipment of 1904 to 1908 at which Prandtl worked with Th. Meyer, E. Magin, and A. Steichen.

3. Awakening of Interest in Compressible Flow

In spite of many far-reaching speculations about rocket flights, the general climate was not quite ready for such propositions to be taken seriously. Even when I was ready to present my first experimental results on lift and drag near sonic speeds, Prandtl himself suggested adding in parentheses "with regard to propellers" for the meeting in Danzig, 1928, to indicate some practical value of them. The meeting, however, proceeded in a quite unexpected manner, when Hermann Oberth had to defend his theory against Prof. Hans Lorentz, the host of the event at the Technical University of Danzig, who tried to prove that leaving the earth with a rocket-driven vehicle is impossible. His error was that he confused the "staging" of rockets with the "clustering" of smaller rockets for simultaneous combustion. It was not too hard for Oberth to straighten out this essential difference, at least for objective listeners. After that the younger generation, enthused by this discussion, was quite a pleasant audience for my paper, whether it was with or without regard to propellers.

The scientific world in general showed at least some interest in compressible flow as a special item. The two competing German handbooks of that time, the *Handbook of Physics* first and the *Handbook of Experimental Physics* soon after, included compressible flow under the title "Gasdynamics" to follow the many aspects of modern "Hydrodynamics" in a special chapter. The obvious choice to write this chapter was, of course, Ludwig Prandtl.

But Prandtl suggested J. Ackeret for the *Handbook of Physics*. Consequently he got another invitation from the *Handbook of Experimental Physics* and this time he passed it on to me. Undoubtedly it is a great honor to be selected to write such an article on a frontier science, but nevertheless, to write the same chapter as another young man from that same school—and after such a short interval—was not without danger. Since the most interested readers like Th. von Kármán, G. I. Taylor, and A. Stodola could be counted almost on the fingers of one hand, and they would certainly remember every word that Ackeret had said, the danger of ruining one's future by plagiarism was quite evident. So I had to sit down and try to stretch the natural difference between "physics" and "experimental physics" as a starting point. For an engineer of the steam-engine area, this was not too difficult a problem: If Ackeret uses simplified equations for perfect gases, experiments are made with real gases as we find them in the Mollier diagram for steam. Since the only visible progress in Goettingen since Ackeret was in making graphical constructions of perfect wind tunnel nozzles by applying the characteristics diagram, the shift to graphical representations for Ackeret's analytical equations may not only save the day but it may add another dimension for the perspective picture of some basic features in such a novel territory (deviation from local equilibrium was, of course, not considered). Otherwise the writing of that article was not too difficult for me after I had been used by Ackeret for trying out first reactions of the reader. (When he wrote his article I was sharing his office!) The only trouble was that the requested 60 pages of print were already doubled before nonsteady compressible flow, the oldest part of gas dynamics, was entered. The editor was willing to print those 120 pages, but told me not to go any further. This itself was quite a concession on his part, for this was at the time of the American bank crash when publishing in journals was next to impossible because of the flood of manuscripts by every young man who needed a break in his career.

I remember, when I gave in Prague in 1929 a short introduction in a ten-minute talk to the graphical integration of the supersonic flow around a conical tip with the intention of presenting the whole series of integrated "Apple Curves" at the International Congress of Applied Mechanics in Stockholm in 1930, that the German committee had to impose a strict rule on all speakers not to repeat any subject already discussed before. Under this rule I had to find a new subject. Conical flow was mentioned again at the Volta Meeting in 1935, but the complete result was published no sooner than in the C. Wieselsberger memorial issue of *Luftfahrtforschung* in 1942. At the German national meetings for applied mechanics there were some lonesome specialists like G. Weinblum for ship waves and Busemann for supersonic flow; at the international congresses of that time there was already quite a group of scientists interested in compressible flow who were collecting around G. I. Taylor in England, around von Kármán in the USA, around Ackeret and myself from the school of Prandtl, and around some French and Italian specialists. But none of us could really predict how soon our specialty would be introduced to the general society of aerodynamicists with more practical ambitions.

THE EXCITEMENT OF SCIENCE   111

4. The Fifth Volta Congress in Rome, 1935

No matter how optimistic a scientist could have been to defend the immediate value of his flow research, the invitations to the Fifth Volta Congress in Rome in 1935 under the title "High Velocities in Aviation" were still a surprise and a great challenge for every single one of the invited speakers. Words like lift and drag at supersonic speed were now the subtitles for special sessions to be discussed by internationally known theorists complementing the review of practical experiences during the Schneider Cup races and the progress since the final victory by the British participants. Annual Volta Congresses arranged by the Royal Academy of Science in Rome and supported by the Alessandro Volta Foundation began in 1931 with the subject "Nuclear Physics" and were of international interest combined with some special relevance for the Italian people. The Third Congress in 1933, for instance, was devoted to "Immunology" to discuss progress in fighting diseases in swamp areas that are uninhabitable, especially in warmer climates. But only the odd numbers were from the class of mathematics, physics, and natural sciences of the Academy, while the even numbers concerned humanities. The Second Volta Congress had the title "Europe" and the Fourth was about "The Dramatic Theater." This was quite a large variety of subjects, and the translators at these Congresses, who were extremely clever at comprising a whole paragraph of the speech into a few sentences in popular terms, had to change to a sentence-by-sentence translation for us, after Prandtl felt the need to correct them "it is not force but energy . . . etc." in our scientific expressions.

All invited scientists appreciated this unique opportunity to apply their collected experience fully to the progress in aviation, and each worked very hard on his specific subject between the invitation, in early January, and the delivery of the manuscript, the first of July, and even beyond toward the actual Congress from September 30 to October 6, 1935. Even von Kármán told me that he usually gave new publications to one of his assistants to report about in a seminar; but for this Congress he went over all the relevant papers himself and discovered a lot of ideas not revealed in those seminars.

Though most of the participants were known to each other, like members of one scientific family, two major political events of 1935 caused some difficulties. Hitler, after the murder of Ernst Roehm and the death of President Hindenburg in 1934, reestablished in a show of power the German Air Force, contrary to the Treaty of Versailles. Mussolini announced on October 3, at the time when my talk was originally scheduled, his intentions in Abyssinia, which were in strong contrast to the British policies. My troubles were not great in Germany. The fact that I was invited, together with the famous Prandtl, opened all of a sudden many doors for me. In the middle of March I was invited to see W. Dornberger and W. von Braun on their rocket research place near Berlin. One week later all three of us went to Munich to discuss the proposal of Paul Schmidt with regard to his buzzing thrust machine. In May I got my contract as division head for "Gasdynamics" in a new research center to be established near Braunschweig. On the trip to Rome I was asked to stop at Vienna and take a good look at Eugen Saenger, who worked on rocket flights with wings, and to decide whether I would like to have him as part of my Institute for Gasdynamics.

There was, however, a difficulty about my subject in Rome. It was originally "Supersonic Windtunnels" for the man who cleans up the tunnel flow according to the characteristics method, and it was "Supersonic Lift" for J. Ackeret, the man of the linearized lift theory. Because I had been four years away from Goettingen and working more theoretically in Dresden, while Ackeret was constructing wind tunnels even for the Italians at Guidonia near Rome, it was not too hard to arrange a switch of our subjects. Lift at supersonic speeds was obviously not connected to any sensitive development for the German Air Force, being too far out in speed. The swept wing design, however, derived originally to reactivate the high response of the air that tends toward zero at hypersonic speeds, could also be used to diminish the supersensitivity of the air approaching the speed of sound. This almost inverse application became a classified matter in 1936.

The British participants being invited as final winners of the Schneider Cup solved their political difficulties about Mussolini's actions by strictly avoiding any public appearances outside the Congress during the remaining three days. Except for these minor difficulties, the international relations at the Congress were as warm as one could hope. Only the Russian rocket expert, N. S. Rinin, did not appear in person after he mailed his paper on time, and it had to be read by the President of the Congress in the final session. The treatment of the foreign guests, who could even bring their wives with full payment of all expenses inside Italy, was almost like that of royalty. Only during lunch on Thursday, when many waiters of the Embassadore Hotel had to appear in their black uniforms in front of the Palazzo Venezia at Mussolini's balcony, they were in a great hurry serving us before their other duties started; but we could understand that this event had a higher priority for them. The President for the Fifth Volta Congress was General Arturo Crocco (the father of Luigi Crocco who worked in Princeton later), a very able chairman and aeronautical scientist in Italy, in both research and teaching. His comments after the delivery of the manuscripts and finally at the actual Congress added much to the spirit and vitality of the discussions. His career brought him in close contact since 1903 with the lighter-than-air and heavier-than-air aviation in Italy, and his latest field of interest was ramjets as an arrangement of negative drag, published in 1931.

5. PRACTICAL EXPERIENCE IN HIGH-SPEED AERODYNAMICS

Two British and three Italian experts discussed at the Congress experience accumulated during the Schneider Cup races and beyond. Their listing of areas of importance for future developments has not lost its validity in retrospect. I quote here from G. H. Stainforth of London: "More power, less weight, less frontal area, cleaner design with enclosed cockpit and smoother surfaces, lift- and drag-increasing devices, variable and reversible-pitch propellers, retractable undercarriage, retractable wings; to achieve the best results requires a compromise in altitude between the following four components: (1) power obtainable by supercharging, (2) lower drag due to lower density, (3) distance of journey and time for climbing, (4) comfort of the passengers with respect to pressure, oxygen, and temperature." Except for the missing jet propulsion and the heat created by friction at higher Mach numbers all these items are still up-to-date and some of them became standard equipment very soon after the conference.

6. Theory of High-Speed Aerodynamics

The theoretical aerodynamics of the Congress was divided by the President into two separate fields, general methods to master the involved mathematical and physical relations, as opposed to the treatment of particular questions; both fields were further subdivided into both subsonic and supersonic speed ranges.

(a) The general introduction to compressible flow was given by L. Prandtl, who illustrated his survey by many Schlieren pictures, especially of the transonic regime, which still was one of the greatest problem areas of interest to him. Geoffrey Ingram Taylor followed him in discussing "Well established problems in high speed flow," and his main concern was also with the understanding of transonic flow behavior to which he contributed the stepwise approximation of two-dimensional solutions by carving step by step new wax bottoms, according to the gas-density variations in an electrolytic tank of large extent with a shallow fluid layer. The lines of constant voltage may be used to represent either potential lines of the flow (in which case no lift is possible) or streamlines of the flow in this analogy. The simple method of carving the bottom after each test, fitting everywhere to the observed electric field, does not converge as soon as the flow field contains supersonic enclosures. A more sophisticated procedure of improving the bottom shape, which makes at least some distinction between the upstream and downstream directions within the supersonic portions, can be conjectured to imitate the nonsteady build-up in nature, but nothing like that has yet been found. The converging purely subsonic flows compare very well with the theoretical solutions including the increasing sensitivity when approaching the speed of sound locally; they also confirm wind-tunnel results. In supersonic flow fields G. I. Taylor was mostly interested in the conical flow field around a circular conical nose without angle of attack in which potential flow is preserved by equal shock losses on all streamlines. He compared exactly integrated results with the small-disturbance theory of von Kármán, which has quite an extent of validity for practical, though small, angles of cones.

(b) Treatments of particular questions encountered in aviation started with the paper of Th. von Kármán on "The problem of resistance in compressible fluids" in general, and wave drag in supersonic flow in particular. Drag in boundary layers exists at all speeds and is, of course, not independent of Mach number, especially when the friction paired with heat transfer causes density changes even for the simple constant-pressure case, because of temperature differences. While the boundary-layer friction itself may not vary too many orders of magnitude with Mach number, the boundary-layer separation is affected extremely by the supersensitive regime of Mach numbers near one and because of shock waves throughout the supersonic regime. At supersonic speeds the wave drag is added, and in the combination of both, a boundary-layer separation can sometimes be the lesser evil. Taking wave drag alone, Th. von Kármán and N. Moore started in 1932 the axially symmetric and small-disturbance theory for the body drag by assuming sources and sinks along the axis. The whole audience enjoyed it when von Kármán demonstrated how he now succeeded in finding the optimum shape for a body of revolution with given caliber and nose length. The fact appreciated most of all was the result that there is a simple analogy between the optimum nose and the optimum lift distribution for a wing of given span. The determination of forces on a source or sink in the field of an upstream arrangement of such singularities simplifies the function, valid off the axis, to an inverse second-power relation of distances along the axis, which can in turn be compared with the field energy of two-dimensional vortex-pair distributions along one axis. After this relation is established, the lift distribution integrated from one wing tip corresponds along the span to the cross-sectional area of the supersonic nose along its distance from the nose tip, while the total lift corresponds to the final cross section of the afterbody at the end of the nose length. Compared with the popular ogival nose of projectiles, the von Kármán nose, which is the integrated ellipse, has not constant curvature of the meridian line, but is somewhat blunt at the tip and curves into the afterbody rather sharply. Anybody who does not like this new shape may argue that the given length is not the most practical constraint for the actual problem; but, just as in the optimal lift distributions, other constraints can be used and the analogy between the two physically different problems very often reduces the new wave-drag problem to an already known lift-distribution result.

"Lift at subsonic speeds" was the title of a paper by Enrico Pistolesi of Pisa. The linearized Prandtl-Glauert relation prescribing how to reduce the thickness and the angle of attack for keeping the incompressible flow experience alive (by preserving its complete pressure field around the profile) may serve as a first impression of the changes due to compressibility; but higher-order calculations by Rayleigh and Poggi are also available and the G. I. Taylor electric tank values, as long as the flow stays completely subsonic, can be used to find the perfect flow field in two-dimensions. The final stall for any profile of finite thickness and angle of attack by boundary-layer separation was supplied from experimental data in wind tunnels.

Now the afternoon came on which I had to present my paper, though somewhat delayed because of Mussolini's declaration of war on Abyssinia. The title "Lift at supersonic speeds" meant the basic question: is there hope that supersonic flight is possible when already approaching the speed of sound the lift seems to vanish and the drag to increase vastly? My idea, to use a "razor-blade"-thin and straight wing to give even the most uncooperative flow no chance to find a pressure distribution to produce simultaneously low lift and high drag at small angles of attack was already mentioned by our president Crocco in his introductory survey. Finding a very high response, measured in dynamic pressure, just above the speed of sound but a diminishing response for high Mach numbers made me try for a lower "apparent" Mach number by turning the wing backward closer to the Mach cone. Such a swept wing configuration would keep the lifting force upward, would even turn the wave-drag force somewhat away from the flight direction, and might recover more flow response than is demanded by the increased friction drag, because of that useless velocity component along the span. President Crocco, after receiving my original manuscript, also suggested an appendix—which is therefore missing in the version of my talk given to the German authorities for publication in the October issue of *Luftfahrtforschung*. It was a more detailed explanation of my sentences: "One cannot find a single two-dimensional body of finite thickness and finite length without supersonic wave drag. For two such bodies one can construct solutions of perfect flow without drag." Whether such ideas far beyond the state of the art at that time did establish real hope for supersonic flight in the listeners is hard to say; but a supersonic airplane with swept wings and a propeller with swept blades in front of it was drawn as a cartoon on the head table during the banquet of the Congress.

7. WIND-TUNNEL RESEARCH AT HIGH SPEEDS

Following the theoretical papers of the Congress came wind-tunnel designs and wind-tunnel results from research laboratories. E. N. Jacobs presented the latest tests from NACA in Langley Field from the earlier 11-inch-diameter induction jet wind tunnel followed later by a larger one of 24-inch diameter. The results documented mainly the troubles with profiles when approaching the speed of sound and were shown in slides and movies of Schlieren pictures. M. Panetti from Torino reported on tests with moving bodies on rotating supports and he discussed also free-flight interferometer pictures on flying projectiles. Further tests also made on moving objects were from propeller blades described by G. P. Douglas at the National Physical Laboratory in Teddington; they required a correction from the S. Goldstein propeller theory for shed vortices in the case of lift. All these tests in the transonic flow regime were in reasonable agreement with each other and with the Prandtl–Glauert theory up to those forward speeds when stall sets in by supersensitivity of the near-sonic flow.

Wind-tunnel designs for all kinds of subsonic and supersonic speeds were presented by Jakob Ackeret of Zurich, some of them with a short time duration and others with permanent flow, the latter with multi-stage compressors and cooling equipment. Besides seeing these wind tunnels in sketches and photographs, we also had the opportunity during the Congress to inspect the new Italian aerodynamical research center in Guidonia near Rome, equipped with a variety of such high-velocity wind tunnels of Ackeret's design.

8. THERMODYNAMICS AND COMBUSTION

Like early flight, high-speed flight depends also on propulsion. Therefore, another chapter of the Congress was devoted to high-speed and high-altitude propulsion devices under the heading of thermodynamics and combustion. Not all of the engineering problems related to such engines may concern compressible flow, and in the discussion some participants of the Congress even said that they missed the evaluation of weight for a larger engine that works for shorter time with increased power. Nevertheless, there is much relation to compressible flow, just as I mentioned before. A. Stodola's steam and gas turbines run with good efficiency at the smallest number of stages both in the compressors as in the turbines and under endurable temperatures.

The first paper of G. C. Costanzi, of Rome, was an investigation of all altitude-related problems: "Stratospheric aviation." The next two speakers, H. R. Ricardo of London and A. Anastasi of Rome, subdivided the more specific problems of "High altitude engines: (a) thermodynamics and carburetion, (b) mechanics and cooling." Their concern was about superchargers, cooling, poppet valves that may be replaced by sleeve valves, and even the fact that the actual power generation from combustion may be surrendered to a vapor engine with any kind of a fluid between the combustion chamber and the condenser.

Less conventional problems were presented in the last two papers "Propulsion by jets with utilization of outside air," like ramjets, by Maurice Roy of Paris, and "Rocket propulsion with air admixing," prepared by N. A. Rinin of Leningrad. "Ramjets on the outer tips of helicopter blades" was one of the items presented by M. Roy. A survey of the work by many other rocket scientists besides his own—that of H. Oberth, F. Zander, E. Saenger, and K. Ziolkowski—was included in Rinin's contribution, and it was combined with ramjets and other means to involve the outside air in early stages of rocket flight.

These five papers are interesting historical documents with respect to the question of how we hit or miss the actual development of jet propulsion. H. E. Wimperis, London, finished a discussion remark with these words: " . . . Hence for such altitudes we must await the coming of the jet-propulsion engine. That, however, we have not nearly got. Some day, thanks no doubt in large part to the labours of this Volta Congress, we shall discover how to do it. And then our remaining task will merely be the discovery of the passengers who wish to fly at such altitudes!" That this "some day" was just five to ten years away, and that the jet age for passenger service was twenty years off, with so many passengers wishing to fly with them that air travel causes severe problems for trains and ocean liners was, of course, hard to foresee.

9. Conclusion

The Fifth Volta Congress in Rome was a great success in international cooperation and exchange of experience between the former Schneider Cup rivals. For high-speed flight and the compressible-flow sciences, it was the turning point from work in a grey area of our knowledge to an advancement in practical aviation. Whether such a success was accidental or could be duplicated in other parts of science may be hard to generalize. At any rate, the participants in the Congress went home with an enlarged view, and they found a better reception for their further research in their countries. Some immediate additions of high-speed research facilities were built not only in

Germany, at Braunschweig and Peenemunde, but also in other countries. On the other hand, when I once suggested a possible improvement of propeller blades, the responsible director of research explained to me that complete loss of efficiency experienced on the current propeller for high speeds enabled them to undertake jet-propulsion research without promising a very high efficiency in their first attempts. Believing, myself, in the future of a much larger speed range for which the propeller would require cowling, an increase in the number of blades, as well as a greater concern for the utilization of the engine exhaust, it was easy for me to realize that both developments would be ending in about the same product, the propeller changing toward higher speed increases in smaller diameters and jet propulsion getting there from the other direction.

Improving the rocket thrust by air-injection nozzles was a popular game in many countries especially to improve the first stage of rockets. For W. von Braun it was more the so-called heat barrier starting at a Mach number of two for aluminum skin that initiated his high-Mach-number research and his change to steel construction. In my own institute, Eugen Saenger's winged rocket vehicles demanded further studies not only in his laboratory for tests on high-breed fuels, improved combustion-chamber cooling for higher inside pressures and temperatures, and many other internal-flow and combustion problems, but also concerning the external flow around the vehicle itself, which led to his "flatiron" body for very high speeds. Such speculations required some checks of the assumptions by comparison with theory and wind-tunnel experience.

The existence of so many urgent problems made it possible to continue compressible-flow research and to exempt the personnel from military duties when the Second World War started. The international exchange of information, so much appreciated at the Volta Congress, froze during those years, and similar developments in different countries became visible only after the war. The optimum von Kármán nose was at first criticized for its restricted validity below the practical threshold of thin bodies of revolution and then it was improved upon in many ways. Jet-propulsion research was in progress in different places. The nonsteady expansion wave in three dimensions, being self-similar in time around an explosive of constant pressure, was treated independently by G. I. Taylor in England, by L. Sedov in Russia, and by me. Other features had a greater variety of appeal in different countries. The nonsteady ramjet-type compression of air for combustion at high vehicle speeds, enhanced by automatic valves and nonsteady mixing in the exhaust pipe with bypassing air, seems to have been a German specialty due to Paul Schmidt of Munich. The development of swept wings was also out of phase in the different countries. Nevertheless, the compressible-flow researchers remained an international family of mutual admiration; they tried hard, especially von Kármán, to improve international relations among scientists, while they did not blame one another for not having accomplished peace on earth prior to improving air transportation throughout the world.

Ann. Rev. Physiology 1977. 39:1–18

MY SCIENTIFIC ODYSSEY

John C. Eccles

Ann. Rev. Physiology 1977. 39:1–18
Copyright © 1977 by Annual Reviews Inc. All rights reserved

MY SCIENTIFIC ODYSSEY ◆1161

John C. Eccles
Contra (Ticino), CH-6611 Switzerland

I have been a wanderer over the world for more than 50 years of active scientific life. Except for my beginning at Oxford, I have not carried out even one scientific experiment as a guest in a laboratory. And after the first few months at Oxford I had, from 1928 onward, my own research room—later rooms. I have never had the good fortune in my wanderings to come to an institute with even primitive scientific equipment that could be used for my initial experiments. Like Odysseus, I have traveled the oceans carrying my own equipment with me like a snail with his house on his back. At each "port of call" I have had to set up and develop research rooms. I have left around the world a trail of elaborately designed shielded research rooms stripped of equipment! But I have been fortunate to discover expert technical and engineering assistance at each of my five ports of call after Oxford; otherwise my scientific life would have faltered. Electronics rapidly outstripped my understanding, but I always have maintained that the technical equipment must be the best; this criterion held right up to my last experiments in Buffalo in 1975. My indebtedness to my associates is immeasurable.

I recently had the opportunity to give an account of some aspects of my scientific life (8, 9). I have tried here to minimize repetition of my story, but necessarily some key events must be retold. Both of these earlier accounts ended with the years 1954–1955, which takes us no more than about halfway through the present story.

My first university training was in Melbourne, where I graduated in Medicine at the beginning of 1925, just after my twenty-second birthday. I was fortunate to be elected to a Rhodes Scholarship, and after some months as a house physician I left for Oxford, arriving at Magdalen College in October 1925 to study under Sir Charles Sherrington, the one man in the world whom I wished to have as my master. After two years in the Final Honour School of Natural Sciences (physiology and biochemistry) with a First Class and a Christopher Welch Scholarship, I was appointed, toward the end of 1927, to a Junior Research Fellowship of Exeter College, just as I was starting on my first research project. Apparently I was taken on promise. Exeter College was my delightful academic home for the next seven years. We were a small band of Fellows, but in their association I was able to mature culturally in

1

THE EXCITEMENT OF SCIENCE

8243-2601/78/1127-0121$01.00 © 1978 ARI   121

that natural manner characteristic of Oxford at its best. There were at Exeter famous scholars such as Farnell, Marrett, Madariaga, Dawkins, Soddy, Barber, Balfour, and Tolkien, but the closest associates were my contemporaries: Neville Coghill, John Wolfenden, Dacre Balsdon, Bill Kneale, and Jim Bessant who represented, respectively, the fields of English, chemistry, classics, philosophy, and theology!

Meanwhile I was inducted into research on the cerebellum by Denny–Brown and Liddell and on spinal inhibition by Creed. Thanks to the generosity of Creed, I inherited his modest research room. I was there joined by Ragnar Granit who arrived from Helsingfors at the beginning of 1928. At first the equipment was restricted to optical isometric myographs and a plate camera plus induction coils for single and repetitive nerve stimulation. We also had a massive antique pendulum for giving accurate timing for two induction shocks (to 0.1 msec). In 1928 Sherrington fitted up the research room with a string galvanometer—the height of luxury in those far-off days—and also with a Lucas spring pendulum. We began our collaboration studying first the motor unit and then the excitatory and inhibitory mechanisms of spinal reflexes from 1928 to 1931. That was for me a wonderful and intimate association, not restricted to neuroscience, for during our experiments the conversation ranged widely over literature, history, and art.

I have told elsewhere (7) of my first experimental success—demonstrating that the "angle" of the isometric muscle twitch was an artifact due to friction in the bearing of the recording myograph. The angle was an established dogma at Oxford. The isometric twitch was displayed with a rather flat summit and the angle occurred in the sharp onset of relaxation from this summit. I too believed in its authenticity. In fact the quality of an optical myograph was guaranteed by the sharpness with which it displayed the angle! In order to investigate the viscoelastic properties of the muscle before, during, and after the angle, I had constructed a myograph that would subject the muscle to a vibratory stretch at a fairly high frequency. For this purpose I had to redesign the support of the torsion rod, making it a friction-free knife edge instead of a V-shaped slot. This poor design had been accepted because the myographs delivered good angles! To my consternation, the new myograph failed to deliver angles, although with the same muscle the old myographs gave good angles. Superposition of the two mechanical traces revealed that the angle was a friction artifact. The summit of the twitch was recorded as flat because of the "hold" by friction which continued during the early stage of relaxation until it could "hold" no longer, the sudden "give" being registered as the angle. This discovery was the occasion of my first appearance (in December 1929) before the Physiological Society of which I had just become a member, and I now had won recognition.

One day in 1930 Sherrington announced that the Clarendon Press had persuaded him to write a book on the researches of the Oxford School. He accepted the invitation on the condition that it would be a conjoint effort, the chosen team of his associates being Creed, Denny Brown, Liddell, and I. I was much excited by the prospect of contributing to a book at so young an age and proceeded forthwith to write my three chapters. My associates generously criticized my amateurish efforts, but themselves wrote nothing for a year or so. I meanwhile was having the experience of writing and rewriting my chapters two or three times. Eventually all was completed; *The Reflex Activity of the Spinal Cord* (3) was published in 1932. I think the printing number was only 1000 and it did not sell out for many years—an indication of the small population of neuroscientists in those days.

England, in my time there (1925–1937), was a delightful and stimulating place for a young academic, although by present standards the laboratory facilities were primitive. There were almost no research grants and no secretarial assistance even for Sherrington. We had to type our papers and service and organize our equipment, which gradually became more complicated with string galvanometers giving place to oscilloscopes, cathode-ray oscilloscopes in 1933, and valve amplifiers. But in research the competition was not severe. The world literature was unbelievably small, so that one could easily survey the total publications, not only on the nervous system, central and peripheral, but also on all types of muscle and all types of sensory systems. Furthermore, one met personally almost all the great figures, for they came to visit Sherrington at Oxford. And we physiologists of Great Britain were united by our membership in the Physiological Society, which, I think, was then in one of its great periods. It was distinguished by the critical discussions that followed each paper. These criticisms were often severe, but it was an unwritten rule of the society that all criticism had to be accepted in a sporting manner. Under no account must it be taken personally. In the period of severe controversy of electrical versus chemical transmission for neuromuscular junctions and for sympathetic ganglia, it was a surprise to overseas visitors to find that I was throughout on the friendliest terms with Dale, Feldberg, Gaddum, and Brown. I have recently written at length on this theme in my lecture on the occasion of the Dale centenary (9). As a junior member of the Society I have the happiest memories of these meetings with their characteristic egalitarian style and of the warm friendliness of the senior members. I believe it to be the best of all scientific societies. And at that time it rejoiced in the award of three Nobel Prizes to its members: Sherrington and Adrian in 1932 and Dale in 1936.

In 1934 I had achieved a permanent position at Oxford with a Tutorial Fellowship at Magdalen College and a University Demonstrationship, but shortly afterwards Sherrington retired, and the Oxford Physiological Laboratory seemed to deteriorate after his departure. Then in the late 1930s there was the ominous rise of Hitler against the unprepared western alliance. So I decided—perhaps unwisely—to return to what seemed the security of Australia. There was the opportunity to create in Sydney a research institute matching the Hall Institute at Melbourne Hospital with Kellaway and Burnet. In 1937, on the advice of Kellaway, I accepted the Directorship of the Kanematsu Institute at Sydney Hospital. In retrospect I feel I should have stayed in England and weathered the storm, but instead I embarked on my Odyssean journeyings, never to return to my beloved England. It was a fateful choice.

Sydney was of course a lovely place to live, but the academic isolation was severe. The Sydney University Medical School was a very dim place, being little more than a teaching institution. Unbelievably, it was completely locked up by guards at 5 PM, even the professors had to scurry out to avoid imprisonment for the night! The Institute I was to direct was simply the routine pathology department of a large general hospital in the city some three miles from the University. Nevertheless, with good help from the Institute Committee and the Hospital Board, I was able to construct research laboratories on the top floor, utilizing for a start the equipment that Professor John Mellanby had kindly allowed me to bring from my two research laboratories at Oxford. I decided to study the electrophysiology of neuromuscular transmission in muscles of the cat hind limb because I thought that it could lead to results of clinical interest.

The academic wilderness soon blossomed. In 1938 Stephen Kuffler arrived in Sydney as a refugee from Austria, and by good fortune I heard of this young pathologist in search of a position. So he became a neurobiologist in the Kanematsu Institute, a novice with almost no background knowledge of the nervous system! In 1939 I managed to attract Bernard Katz from England on a Carnegie Fellowship. Thus in this way, through the machinations of Hitler, we three were sheltering securely in remote Australia and studying neuromuscular transmission in cats and frogs. There was criticism of our activities from some clinicians because this research on curarized muscles and anticholinesterases seemed so remote from clinical usefulness. Ironically it was soon to find an important application in the use of relaxants during surgical operations.

The security of our academic life in Sydney lasted until Japan entered the war. Then for two years I was deeply involved in various wartime projects, and the Kanematsu Institute became the Australian center for blood serum preparation and for applied research on such acoustic problems as noise protection and communication in the high noise levels of tanks and planes. Kuffler continued with his exquisite researches on the isolated single neuromuscular junction, while Katz was chosen to become a radar expert and rendered most distinguished service.

With all this scientific and war-oriented activity I thought my position in the Kanematsu Institute of Sydney Hospital was secure; however under new management Sydney Hospital proceeded in 1943 to make my position untenable. Unbeknown to me, living quarters for hospital residents were to be constructed on top of my Institute, preventing any postwar development, and there was no academic position for me in the Australian universities. I declined an invitation from Liverpool University to return to England, which seemed so bleak at that time; fortunately I was able to accept appointment as professor of physiology at the only medical school in New Zealand, which was an integral part of the University of Otago in Dunedin. So the next stage of my odyssey was to cross the Tasman Sea to New Zealand at the end of 1943, so that I should be ready to start the academic year in 1944. The happy and fruitful collaboration of the Kanematsu trio was broken; Kuffler was already in process of going to the United States at Ralph Gerard's invitation, and, on demobilization after the war, Katz returned to A. V. Hill's department at University College London.

The 6½ years at Sydney were notable for all three of us because of the lifelong friendships created. In these enduring friendships we were linked both on scientific and on personal grounds, even playing tennis together on my court almost every weekend! It is noteworthy that Katz and Kuffler have been closely associated ever since by the exchange of numerous graduate and postdoctoral students. For me that linkage has been less developed, but Paul Fatt came from Katz to me with the great successes I describe later, and Miledi was with me as a Rockefeller Fellow before he went on to develop his brilliant career with Katz. All through my 6½ years in Sydney, the academic isolation persisted, so that we three seemed to be huddled together in an alien world. Relieving features for me were the close wartime collaboration with Archie McIntyre, and the friendship with the distinguished biochemist, Rudolph Lemberg, who remained for 40 years in a hospital laboratory in Sydney with no official recognition by Sydney University. Katz, Kuffler, and I also had no official recognition, but we did function in an honorary capacity, giving lectures on neurobiology to the medical classes. There is an historic photograph of the three of us in 1942 determinedly walking in Sydney to catch the tram to the University. In 1972 we three were photographed again walking, but in a more relaxed manner, in the grounds of St. Catherine's College, Oxford. Mounted together, the two photographs provide a commentary on changed life styles!

In summary, our scientific discoveries were the end plate potential, both its pharmacology and its role in generating the discharge of impulses along the muscle fibers. This research was undertaken without knowledge of the work of Schäfer in Germany and Feng in China. Communications had become very bad. The action of anticholinesterases in increasing and prolonging the end plate potential finally convinced me in 1942 that acetylcholine was the sole transmitter. I wrote a letter of capitulation to Sir Henry Dale, which has recently been published along with Dale's reply (9). This period of collaboration in Sydney was the beginning of the magnificent contributions of Katz and Kuffler and their numerous distinguished associates on the biophysics of neuromuscular transmission.

And so my odyssey continued to the remote university in the south of the South Island of New Zealand, the closest university to the South Pole. Even there, I found stimulating challenges and the opportunity to develop scientifically. After my ten years of teaching at Oxford I was enthusiastic to attempt a similar program of lectures, practical classes, and discussion classes despite the greatly restricted facilities. In this I was fully supported by Norm Edson, an inspired teacher of biochemistry in the modern form that he had learned from Krebs. So the medical students were subjected to this intensive modernization of physiology and biochemistry. The hours available were adequate in the two years of the course that were entirely devoted to anatomy, physiology, and biochemistry, but our staff was most inadequate, no more than two or three in each subject. So in my first year in Dunedin I lectured to the second-year medical class in the whole of physiology—75 lectures in all, and I also did much of the first-year course. With the practical and discussion classes I found my total teaching time was 20 hours a week for 25 weeks. In addition, I had to spend many hours each week learning the whole of modern physiology, so that I could lecture on it with authority.

My research virtually came to an end during the first year when I was learning physiology for my lectures and also creating a completely new practical course with its specialized equipment. But this extreme operation has to be evaluated against the world situation; in 1944 and 1945 research had virtually came to an end for all except a few who were fortunately sheltered, as for example, Lloyd, Renshaw, and

Lorente de Nó at the Rockefeller Institute. Their publications at that time seemed to have come from another world. Meanwhile I was greatly encouraged by the excellent students in my classes. Many have become medical scientists and form the majority of the senior faculty of the newly founded medical school at Auckland University. In retrospect I feel I was almost fanatical in my zeal and ambition to develop model teaching methods for physiology, but it saved me from the narrowness that seems to be so common today, where a senior member of a teaching staff can for example lecture to medical students only on the biophysics of the node of Ranvier!

The year 1944 was important in my scientific life above all my post-Sherrington years because my intimate association with Karl Popper dates from that time. I had heard from Edson about the great stir that Popper was making among the scientists of Canterbury University College about 200 miles to the north in the city of Christchurch, so we invited him to give five University lectures on the philosophy of science. They were an enormous success among the staff and student body, and there were also two special seminars, one to physical scientists, the other to biological scientists. Many people, including myself, had our scientific lives changed by the inspiring new vision of science that Popper gave us. Our association has been intimate since that time and now we have collaborated in a book *The Self and Its Brain* (14).

Briefly the message we got in those memorable lectures was that science is not inductive, but deductive. A scientific project starts as a problem, for example with a theory that appears deficient or inadequate. New hypotheses are developed and tested experimentally, either to be falsified or corroborated, but the claim of verification should never be made. Thus there are two aspects of a scientific investigation: first, the development of a hypothesis using creative imagination; second, the rigorous experimental testing of this hypothesis in its most vulnerable aspects in an attempt at falsification. Thus the outcome may be rejection of the new hypothesis, or modification and further experimental testing, or, at best, corroboration and the possibility of further testing. Thus creative imagination is given the star role in scientific investigation. Even though an hypothesis is falsified, it can be counted as a scientific success in that it led to experimental testing with the discovery that the truth lies elsewhere. I was much encouraged, as I was concerned at the fate that seemed to be threatening my electrical hypothesis of synaptic transmission. Already I had given it up for neuromuscular transmission, and a similar fate seemed likely for ganglionic transmission. So the synapses of the central nervous system were the final haven for the hypothesis of electrical synaptic transmission. Anyway I was urged by Popper to formulate the electrical hypotheses of synaptic excitation and inhibition in models that invited experimental testing and falsification.

It was certainly a crisis in my life when the intracellular recording from motoneurons in the cat spinal cord was employed to test these two hypotheses. We were encouraged to attempt this intracellular recording by the successes of Nastuk & Hodgkin (13) and Fatt & Katz (12) in recording from skeletal muscle. With the expert assistance of Brock and Coombs we managed in my last year in Dunedin (1951) to discover the essential features of the postsynaptic electrical events produced by excitatory and inhibitory synaptic transmission (1, 2), and to falsify the electrical hypothesis of synaptic inhibition. Hence, belatedly, I was converted to the Dale hypothesis of chemical synaptic transmission even in the central nervous system. A detailed account of this story is given in my Dale Centenary Lecture (9).

During my latter years in Dunedin I had come to realize that the heavy teaching program seriously handicapped me in competition with the new wave of intensive neurobiology that was developing, particularly in America. There the teaching loads were much lighter and the financial support of research much greater. Dunedin was an acceptable home for me during the war and in the period of worldwide disorganization and reconstruction that followed, but I could foresee my failure in world class if I continued there. Fortunately a most ambitious project of a research university was being planned in Canberra, with Howard Florey as the advisor on the medical section. It was therefore with enthusiasm that in 1951 I accepted the Professorship of Physiology at the Australian National University in Canberra. At that time the University had a large grassy wooded tract of land on which there were two army huts for administrators. But there was good financial support by the Australian government and great enthusiasm!

So it was time for the next stage of my odyssey. However, it was to take 15 months before I could start experimenting in Canberra. Part of the time was filled in by travel to the United States, then to England, and finally back via the United States to Australia. After my departure from England in 1937 to the Antipodes I had only been briefly in America in early 1946 for two meetings of the New York Academy of Sciences (my first visit to the United States), so it was with great anticipation that I left the Antipodean isolation at the end of 1951 to return to England after 14½ years and to spend five months in residence at Magdalen College Oxford giving the Waynflete Lectures and preparing them for publication by the Clarendon Press, Oxford. The title both of the lectures and of the book was *The Neurophysiological Basis of Mind: Principles of Neurophysiology* (4).

A special feature of my return to England was that I came as a neophyte with my newfound enthusiasm for chemical transmission at both excitatory and inhibitory synapses in the central nervous system. In February 1952 there was a Ciba symposium on the spinal cord and also a Royal Society Symposium. At the Royal Society I had the pleasure of signing the Fellows Book some eleven years after my election in 1941—still the same book that Newton had signed. It was a strenuous period with my travel around England and much entertainment and good discussion. I had the feeling that England was at the beginning of a magnificent new postwar era. Certainly I was not misled in my estimate of the great scientific successes. Hodgkin, Huxley, and Keynes in Cambridge and Fatt and Katz in London were leading the world in neurobiology, and that lead has been maintained —as evidenced by the Nobel awards to Hodgkin, Huxley, and Katz. But, as a power in the world and as a great industrial country, England has declined in a way not anticipated in 1952.

I would have liked to return to England at that time, because the Antipodean prospects were as yet a matter of faith and hope. There were by then some centers of achievement in Australia, in particular in radio astronomy and in microbiology, but Australia was at a provincial level in other sciences. This had motivated Florey to work for a research university that was entirely postgraduate, giving only doctoral degrees. So in June 1952 I returned to Australia after a brief interlude in America attending the Cold Spring Harbor Symposium on the neuron. There was much discussion on intracellular recording. Woodbury and Patton had also succeeded with motoneurons, but Lloyd was severely critical of the whole project, which he thought could only lead to a vast and misleading literature on damaged and dying neurons! This attitude of Lloyd's was more than I could have hoped for, and aroused Bob Morison to compose two limericks which were regarded as unprintable in the published symposium! I realized that, with his superb technique and penetrating insight, Lloyd could have made great progress in intracellular recording in the study of excitatory and inhibitory synapses during the long latent period which necessarily would occur in the resumption of my research career. It was to take much time and travail before I could establish laboratories in the temporary buildings then being constructed in Canberra. Meanwhile, I had brought from New Zealand four magnificent electrical stimulating and recording units (ESRU), designed by Jack Coombs and built in New Zealand. At that time and for many years to come—in fact until the transistor era—they were the best general research instruments for electrophysiology in the world. Some were still in use in David Curtis's department until 1976, and I used the original ESRU until 1968. Without doubt the successes of our department in the Canberra period were dependent on the excellence of the ESRU's. For me it was very frustrating to have to wait month after month before I could resume the intracellular studies on motoneurons that had been interrupted in December 1951 on my departure from New Zealand. Paul Fatt had come to Canberra in the latter part of 1952 to continue with motorneurons the intracellular recording that had been such a success in the study with Bernard Katz on the neuromuscular junction. In the first months before the laboratories were ready he was intensely occupied in a critical appraisal of the literature on spinal motoneurons and came up with challenging ideas that helped to guide our research in those early Canberra years.

So in March 1953 the active Canberra phase of my life began and continued for over 13 years. Without doubt it was the high point of my research career. Koketsu from Japan joined Fatt and me in our study of Renshaw cells, which was one of my most satisfying research projects. The cholinergic excitation of these cells by collaterals from motoneuron axons was a striking vindication of Dale's principle (cf 9), namely that for the synapses formed by all axonal branches of a neuron there is the same chemical transmitter.

Other members of our research team during the first five years were Jack Coombs from New Zealand, Sven Landgren from Sweden, Bill Liley from New Zealand, my daughter Rose, Vernon Brooks from Canada, David Curtis from Melbourne, Anders Lundberg from Sweden, Ben Libet, Bob Young, and Kris Krnjevic from the United States, Ricardo Miledi from Mexico, and Arthur Buller from England. Already the department was taking on the international complexion that was to become so characteristic in later years. During these earlier years we worked in the rather limited facilities of the temporary hut. The grandiose permanent building was longer in coming than we had hoped. But somehow much research was accomplished in the three research rooms.

I felt very much the urgency of showing by our achievement that this extraordinary foundation by the Australian government and by the Prime Minister, Sir Robert Menzies, was delivering the academic goods. Menzies was one of the really great men I have been privileged to know. Canberra in those early days was a very small "city" for a capital, and very short on social amenities. Menzies advised that the first permanent building to be erected would be the residential college and faculty club, called University House. He rightly sensed that a university is a community of scholars and that there had to be good facilities for cross-cultural meeting in a faculty club. University House provided faculty club facilities rivaling any other such club in the world. I spent much time in those earlier Canberra years on the governing body of University House. Two Cambridge associates, Trendall and Oliphant, and I tried to give University House the style of Cambridge and Oxford, but appropriate for an entirely postgraduate college and in harmony with the international culture that was the hope of the postwar world. It was quite a challenge to mold the style and tradition of the past so that it would be assimilated by the young academics. Some were very raw material on arrival, but most became civilized without appearing to notice the transformation. We kept our satisfaction confidential.

I have dwelt long on this aspect of Canberra because I believe most American universities, particularly those financed by state governments, are miserly in their appropriations for cultural amenities for faculty. By contrast the student body is well catered for. The state university system of New York is a notable example of this absurd unbalance. As a consequence I spent the last seven years of my academic life in Buffalo in virtual isolation from my colleagues in other departments. The university was more like a trade school—and much of it was so oriented.

A remarkable feature of the Australian National University was its international orientation. No preference was made to Australians, and there were unrivaled facilities for overseas scholars. For example, if appointed to a research scholarship, the travel to Canberra of the scholar plus wife and children was fully paid and his emoluments dated from the time of departure. The emolument and the scholarship was about twice that for basic living and housing costs in Canberra, and furnished houses were provided for scholars with children. Married and single scholars lived in University House. There were no university or degree fees, and the return journey was also fully paid. The more senior staff had fellowships at appropriate levels, again with all travel costs paid and housing provided. These generous arrangements explain why there was such an international complex in this remote Australian university. This international generosity was encouraged by Menzies, who remarked that at last Australia was in the position to pay off some of the "academic debts" incurred during its growing and maturing stages, when Europe and America had been so extraordinarily generous in their help to young Australian scholars, as I for one well remembered from my years in England. Cynics may remark that it was an attempt to attract scholars from abroad to settle in Australia, but very few did outstay their appointments in Canberra. However, Australia did gain enormously in two respects. First, there was a transformation from the academic isolation of the prewar years that I had experienced on arrival in Sydney in 1937, and many Australian scholars returned to Australia, often after long sojourns overseas. Second, the new generation of young Australians had the great advantage of association in Australia with scholars from overseas. The academic renaissance spread through the whole university structure of Australia. Australian science is now in top world class, a remarkable achievement for a middle-sized country (13 million) so remote from the great centers of the world.

At this stage I should make reference to another of the activities to which I devoted great enthusiasm in those early Canberra years. Despite the existence of many scientific societies, Australia had been lacking a prestigious scientific body that could speak with authority to the government, to industry, and to the country. What we needed was a foundation playing a role in Australia equivalent to that of the Royal Society of London. Sir Mark Oliphant was the leading spirit in furthering this project. In order to give the new foundation adequate credentials, the Fellows of the Royal Society resident in Australia, together with a few senior scientists they nominated, petitioned the Crown for the foundation of the Australian Academy of Science, which would be modeled on the Royal Society of London. In this project we were most enthusiastically supported by the Royal Society. All went well, and in 1954 Her Majesty Queen Elizabeth II founded the Australian Academy of Science in Canberra. It was the second time in history that a British monarch had founded a scientific society, the first being the Royal Society of London by Charles II in 1660! Oliphant was the first president from 1954–1957, and I succeeded him in 1957–1961. Those were early creative years with many problems in the travail of birth pains. But the Academy has flourished and has contributed notably to Australian scientific advancement and to the influence that science plays in Australian affairs. A striking symbol of Australian science was the remarkable edifice of unique style, built by the Academy for its headquarters in Canberra. Oliphant initiated this project with great insight and courage and, in my period as president, the building was completed to the great joy of the Fellows and the amazement of the local inhabitants. It displayed an extremely simple geometrical form, an enormous copper-covered dome, a section of a sphere, broken only by the arches rising in scalloped form from some 16 "feet" immersed in a circular moat. I think the design was superb and the interior had excellent facilities, particularly the central meeting hall. I have most happy memories of the many great occasions associated with those early years of the Australian Academy of Science.

But I should return to the Australian National University with the department of physiology as an integral part of the John Curtin School of Medical Research. At last in 1957 the grandiose new building was completed. We had done so well in the temporary hut that I was somewhat overawed by the new magnificence and the greatly extended facilities. I had only one floor of one wing, but managed to plan the space so that there were six research laboratories and at least 12 studies. It was my belief that each research worker should have his own study, no matter how small, in order to have the privacy for working up his data, measuring, writing, and typing. The alternative of having several junior research workers sharing a larger space is almost universal. At Oxford I had this experience in my first year, 1928, where Denny Brown, Granit, Olmsted, Marcu, and I shared a large disused laboratory; but in 1929 I managed to acquire my own study, and since then I have almost always been able to provide separate studies for all of my associates. The increased accommodation of the new laboratories and the many applicants, particularly from overseas, resulted in a large expansion of research staff, particularly during the last half of my Canberra period—1959–1966. It was quite an organizational task to arrange for the coming and going so that there would be maximum occupancy and yet no overcrowding. In those years there were always more than 20 research workers and as many technical and support staff. Of course I did not personally supervise so many. I had by then a semi-autonomous section. Macfarlane had developed his own research group in endocrinology and climatology. Curtis had his group in neuropharmacology and neurochemistry, and Hubbard had a group in the biophysics and electron microscopy of neuromuscular transmission.

My own research interests had developed far from the initial intensive study of the biophysics of the motoneuron and the action of excitatory and inhibitory synapses thereon. There was so much that this new technique could be used for at the spinal level that I stayed almost a "prisoner" in the cat spinal cord for many years —in fact until 1962. As I look back on those years I can hardly imagine how we dared to attempt so much! The opportunity to present in an integrated form the initial stages of our work at Canberra was provided by the invitation to give the Herter Lectures at Johns Hopkins Medical School in 1955. These lectures were eventually published as *The Physiology of Nerve Cells* (5) by the Johns Hopkins Press.

Patterns of organization in the spinal cord provided the challenge for studies that Lundberg in part developed independently in 1956 and 1957. The inputs from muscle receptors by groups Ia, Ib, and II were studied in detail, particularly in relationship to the functions of the muscles supplied by the motoneurons—those that are homonymous or heteronymous and synergic, and those that are heteronymous and antagonistic. In this study we were carrying the pioneering work of Sherrington to the new level of enquiry made possible by intracellular recording.

In 1957 my colleagues, Ben Libet and Bob Young (11) from America and I utilized intracellular recording from axonotomized motoneurons to account for the enigmatic reflex responses of these chromatolyzed motoneurons that had been observed many years earlier. It was of special interest because of the finding that synaptic excitation resulted in dendritic spike potentials that were transitional to the generation of impulse discharges down the motor axon. There was also an investigation of the electrical properties of the chromatolyzed motoneurons and a full explanation of the earlier observations on reflex responses. I mention this paper because it has been undeservedly overlooked by later investigators and reviewers. These later papers have very little additional to report and the illustrations and measurements of our 1958 paper are superior to those of the later papers published in the 1970s. It seems that papers published more than a decade ago are ignored, regardless of content or merit!

Cross-union of various peripheral muscle nerves was attempted in 1957 in order to discover if it would result in some central reconstruction of connectivities by a kind of plastic response. Sperry had failed to find any significant changes many years earlier, but intracellular recording made it possible to do a more refined and quantitative study. Already we had amassed control studies on monosynaptic inputs using intracellular recording from many hundreds of identified motoneurons. However the first experiment completely changed our plans. I had cross-united in young kittens the nerve to the pale flexor, gracilis, with that to the red extensor, crureus. After a period of several weeks in order to allow time for regeneration, we proposed to see if there were changed connectivities to the respective motoneurons in the light of their transposed functions. For example, did crureus motoneurons now supplying gracilis muscle receive some connectivities appropriate to knee flexors? But, on exposing the muscles, there was an incredible display. The pale gracilis now innervated by crureus motoneurons had become bright red, the other gracilis showing the normal pallor. So we immediately set up mechanical recording and found to our delight that the red color of gracilis was matched by its much slower contraction. And, complementarily, the slow crureus innervated by gracilis motoneurons had become much faster relative to the control muscle on the other side. Arthur Buller had just arrived from England, so there was one year of intensive study on the influence of motoneurons in determining muscle contraction time. Not only did we study cross-union of various muscle nerves after a wide range of post-operative times, but we also studied the time courses of muscle contraction from birth onwards. Investigations in many laboratories have stemmed from these initial studies, not only mechanical (Buller, Close, and associates), but also electrophysiological, biochemical, histological, and pharmacological. The aborted study on central connectivities and cross-union was later taken up. Some changes were found, but even just after birth there was little evidence of the plasticity in the spinal cord that has now been shown to be so prominent at higher levels of the central nervous system.

Another good story began in 1959 when following up preliminary reports by Frank and Fuortes on presynaptic inhibition. The dorsal root potential and the dorsal root reflex had been known since the work of Barron and Matthews and Toennies in the 1930s and was studied by us in Dunedin in the 1940s, but the functional significance had remained an enigma. An intensive study for some two years in collaboration with Krnjevic, Schmidt, and Willis revealed the story that has been corroborated and enhanced by much subsequent work in other laboratories. It was postulated that there are special axon-axonic synapses on presynaptic terminals that act to depolarize these terminals and so to reduce the action potential and thus the emission of transmitter. Electromicroscopy first by Gray and later notably by Saito has displayed these axon-axonic synapses much as we diagrammed them, and recent work by Nishi, Nicoll, and associates has corroborated our suggestion that GABA is the transmitter. The main thrust of these investigations was a systematic topographic and modality study in the effort to define the way in which presynaptic inhibition was employed physiologically.

I was only partially associated with the refined biophysical studies that were carried out in 1959–1963 largely by the Japanese team of Ito, Araki, and Oshima. I was particularly happy at the finding that in an investigation of 34 species of anions the inhibitory transmitter was found to open up in the postsynaptic membrane gates that allow the passage of all anions regardless of species, provided that in the hydrated state their diameter is smaller than a critical size (2.9 Å). The only exception was formate that passed through despite being slightly larger than this critical size for the other 10 species of anions. It looked as if we had a simple model for the mode of action of the inhibitory transmitter. Unfortunately the investigations on cationic permeability of the inhibitory postsynaptic membrane were much more enigmatic, and the question of potassium permeability is still debated. Parenthetically, it may be noted that the ionic mechanisms of postsynaptic inhibition in hippocampal pyramidal cells was my last experimental study (until May 1975).

It was at this stage of my Canberra life that I received the Nobel award (1963) for the ionic mechanisms of synapses, and my Nobel Lecture was on the ionic mechanisms of postsynaptic inhibition. Earlier in that year I felt that the time was ripe for an extensive review of the whole field of synaptic mechanisms. This proved a heavier task than I had anticipated, but it appeared as *The Physiology of Synapses* (6) that was published just in time to be on display in Stockholm at the time of the Prize festivities in December 1963. The publishers (Springer Verlag) made an unprecedented effort in speed of publication in order to effect this felicitous timing. All was finally printed and in order for binding except for the subject index. The typescript reached the Heidelberg office on Monday, December 2. On Sunday December 8, I received in Stockholm some six bound copies by airmail special delivery, and, on Monday, the booksellers of Stockholm had display copies, and copies for sale were available on December 10, the day of the Prize award.

After some ten years of intracellular recording in the spinal cord I was happy to move into the much more complex and challenging problems presented by higher levels of the nervous system. The change occurred gradually. At first there were investigations led by Olov Oscarsson on the cells of origin of the spinocerebellar tracts. Cells of origin of another ascending pathway were also studied. Meanwhile Tom Sears was carrying out his refined studies on the control of respiratory movements by employing intracellular recording from motoneurons supplying intercostal muscles. At this time also stimulation of the motor cortex was shown to produce presynaptic inhibition in the spinal cord. However the decisive change occurred with studies on neurons of the brain stem under the leadership of Per Andersen.

Firstly, there was synaptic transmission in the cuneate nucleus with ascending actions from the spinal cord and descending from the cerebral cortex. The neuronal machinery involved in these actions was studied in detail. Next came the ventrobasal nucleus of the thalamus on the projection line to the cerebral cortex from the cuneate nucleus. There was study of the neuronal machinery and the role of inhibition in setting the rhythmic activity of the thalamocortical circuits, a theme that Andersen was later to develop so well.

The most important study was on the hippocampus, using the new techniques of intracellular recording and field potential analysis. Andersen already had extensive experience on the hippocampus, so good progress was assured. The most interesting discovery was that the basket cells of the hippocampus gave a very large and prolonged inhibitory postsynaptic potential of the hippocampal pyramids. It had been known since the time of Ramón y Cajal that the basket cells formed a dense terminal plexus (or basket) around the somata of pyramidal cells, which he believed to exert an intense excitatory action. The combination of depth profile and intracellular studies convincingly demonstrated that the action was an intense inhibition. So for the first time an inhibitory cell with its synaptic terminals had been identified histologically. At that time Renshaw cells had not yet been recognized histologically.

Having accomplished that identification so satisfactorily, I asked: where else are there basket cells? The answer being the cerebellum, we (Andersen, Voorhoeve, and myself) immediately in early 1963 changed our attention to the cerebellum. It was a more complex study than the hippocampus; nevertheless the clear answer came that the basket cells there are also inhibitory, this again being shown by depth profile and intracellular recording. Per Andersen had to return to Norway, so we made a pact. He was to have the hippocampus for his field, and I and my associates, the cerebellum. I regret to report that in 1975 I broke the pact by again working on the hippocampus, in a final electrophysiological study on the ionic mechanism of postsynaptic inhibition. It was an appropriate and very successful termination of my experimental life with my colleagues Allen, Nicoll, Oshima, and Rubia.

The beautifully organized structural pattern of the cerebellum was a great opportunity for an analytical study of the mode of operation of the two input lines, by mossy fibers and by climbing fibers, and of the five species of neurons. Of particular importance was Szentágothai's evidence for the origin of climbing fibers from the inferior olive. Remarkably clean results were obtained by stimulating through an electrode inserted into the inferior olive. The mossy fiber input gave a more complex picture, but, by utilizing various sites of stimulating together with depth profile recording and intracellular recording from Purkinje cells, a satisfactory picture emerged that enabled us to make models of the mode of operation of the neuronal machinery in the cerebellar cortex. With but minor variations this model still holds, so the comprehensive book published in 1967 with Ito and Szentágothai, *The Cerebellum as a Neuronal Machine* (10) still does not need extensive revision. This analysis of the neuronal operation in the cerebellum was greatly aided by a principle that I had proposed as early as 1954: that all the synapses formed by a neuron in the mammalian central nervous system have not only the same transmitter (Dale's Principle), but also the same action, either excitatory or inhibitory, there being no ambivalent neurons. So we could generalize from our analytical experiments and propose models of circuits that displayed the essential features of operation in all the complex interactions of the neuronal machinery.

Troubling me in my later years at Canberra was the early retirement age of 65 that was soon to overtake me. I had hoped to get this age extended to 68, but the administrators prevented this. I already knew the very impoverished conditions that would be my lot after 65—half salary renewable year by year and one laboratory with almost no support for staff or assistants. So I realized that soon my odyssey would take me from Australia across the Pacific. There were only two choices at that time: the University of British Columbia and the newly established Institute of Biomedical Research in Chicago. Both offered far more generous support than I would have in Australia, and there was provision for extension of my position for up to 70 years and even beyond. I was impressed by the grandiose plans for Chicago, and so, unwisely as it turned out, I accepted this position, which was the first I had ever been offered in the United States.

This decision relieved the Australian National University of what was clearly an embarassment—to have me there as an impoverished worker—and there was much ceremony on my departure, including an attractive portrait of me by Miss Judy Casab and a witty cartoon by Frith. Both were unveiled at the farewell banquet and there was a most generous speech by Lord Florey who was in Canberra on one of his numerous visits as Chancellor of the Australian National University. In my reply I alluded with approval to his frequent public statements that the value of the John Curtin School of Medical Research was not to be judged by the kilograms of publications, but by the value of its scientific discoveries for the people of Australia and of the whole world. Just to emphasize the point I displayed an official document prepared by my head technician, Lionel Davis, who was a Justice of the Peace, to the effect that the total publications of the department of physiology over 13 years weighed 10.8 Kg! The cost per Kg certainly would be discouragingly high for a business man.

Some more statistics for my 14 years at Canberra are that there were in the department of physiology 74 research workers from 20 different countries, and 411 scientific papers and 4 books were published. This is a tribute to the Australian government for supporting the Australian National University so generously, particularly in respect of overseas visitors. Another "human dividend" from this Canberra period were the excellent personal relationships that developed between the families of my visitors, no doubt fostered by the isolation in Canberra. After all these years they still feel members of a supranational society of Canberrans who had dared to adventure to the Antipodes!

The next stage of my wanderings was the briefest, the least successful, and the most unhappy of my research career. There were several reasons. Although at the start there was good material support in Chicago and a most prestigious governing body, it quickly became evident to me and to the Director of the Institute for Biomedical Research, that the American Medical Association was not enthusiastic —quite the contrary. When the retirement age was lowered to 68 years, contrary to our agreement, I realized it was time to look elsewhere. But there were severe internal problems within my own group, as well. It was time to go, and fortunately I had one chance. It was an invitation from the State University of New York at Buffalo with very generous financial support of a unit to be created for me as a Distinguished Professor of Physiology and Biophysics.

So the next stage in my journey was across land—not strictly Odyssean, across the ocean as heretofore. I have been often asked why I chose to go to Buffalo. The answer is very simple—I had nowhere else to go. My age of 65 was not encouraging to universities who may have been considering me as an associate. But in any case the Buffalo appointment was very generous so far as support and salary were concerned and President Meyerson and the University agreed to retirement at 70 to be revised upward from year to year thereafter. So everything started well, again in a temporary building, pending the construction of the grandiose new university.

I was happy with the temporary accommodation which was soon fitted with the best research facilities I had ever had. It was of course on a much smaller scale than Canberra, which was a good thing. The unhappy experience of Chicago had warned me of the problems of personality. So I had a small carefully selected group of research associates and some of my associates of former times came back to join me for periods in Buffalo with the happiest results: Robert Schmidt from Germany, Tomakazu Oshima (for two periods) from Japan.

After the debacle of Chicago, I had begun to wonder if I had lost the personal touch that had given me such good relationships with all my associates at all previous "ports of call." So it was very reassuring to find that at Buffalo I was again associated with a delightful group, not only the scientists and their families, but also the secretarial and technical staff. In all, counting my wife and myself, there were 30 scientists from 11 different countries. As this seven-year period came to an end with my voluntary retirement in 1975 at the age of 72, it was with great sadness that my wife and I said farewell. So the happiest group of my research career was dissolved, with tears from the ladies!

But of course outside our charmed circle we had had all the turmoil of the University revolt with threatened destruction by dissident students. Buffalo was one of the centers of the storm, and lacked leadership at a critical juncture. I felt that the only course of action was to continue actively in research, come what may. And that we did, being helped by the isolation of our temporary laboratory from the main campus.

At the end of the Canberra section of this story I told briefly of our success in being able to construct models for the mode of operation of the neuronal machinery of the cerebellar cortex. But such models have to be built into the wider picture of the input and output paths if they are to be used in providing explanations of the mode of operation of the cerebellum in the control of movement. Despite the emotional troubles we already had some good successes in this study when at Chicago. But at Buffalo we were much better equipped for this ambitious task, with on-line computers, digitimers, and, most importantly, a most versatile mechanical stimulator. This latter instrument was of the greatest importance in the study we made on the role of cutaneous sensing in cerebellar control. Robert Schmidt's experience with cutaneous receptors was of vital importance in this systematic study of cutaneous inputs onto cerebellar Purkinje cells. Hitherto it had been generally believed that muscle receptors were of more importance for cerebellar inputs than those of skin. In previous studies cutaneous stimulation had been crude: touching, brushing, squeezing, etc. With the instrumentation provided by our stimulator and an averaging computer, there was revealed the remarkable effectiveness of the cutaneous inputs particularly by the foot pads. Muscle receptors were much less effective, but the study of joint receptors awaits the development and application of good instrumentation. Suffice it to say that we were much impressed by the effectiveness of cutaneous inputs, and began somatotopic studies in the attempt to define better the way in which the cerebellar machinery was employed in controlling movement. This study on the vermis and pars intermedia of the anterior lobe of the cerebellum led on to sequential studies of the pathways from the cerebellum through the cerebellar nuclei and then by the next relay nuclei (the red nucleus and the medial reticular nucleus) on the pathway down the spinal cord to motoneurons. Our work was of course closely related to the comprehensive anatomical studies of the Norwegian school, but in conclusion it must be stated that as yet we are far from understanding the mode of operation even of the cerebellar anterior lobe onto spinal motoneurons.

In the latter two years I had encouraged Gary Allen to attempt the very difficult task of relating cerebral cortex to the cerebellar hemispheres. He had good success with the cat, and then proceeded to the primate, where these studies would be of the greatest importance attempting to understand how in man the cerebrum and cerebellum interact in the control of movement. About 88% of the human cerebellum is oriented exclusively to the contralateral cerebrum. Allen and his associates had made very good progress, and there was a good report of the whole project in *Physiological Reviews* (November 1974), yet his application for a National Institute of Health grant in 1975 was not funded. Since I was to continue as an adviser with frequent visits to Buffalo for this purpose, my plans for the continuance of my American association were thus terminated. This project of Allen's was almost unique in the world. Only at Kyoto is there the beginnings of a comparable study by Sasaki of the detailed topography of the cerebro-cerebellar connectivities in the primate, with a study also of the relay nuclei involved in this cerebro-cerebellar transaction.

So my active scientific life came to an end in 1975 and to my regret I feel that, after nine years of intense scientific effort in America I have left there no successors who would be continuing in the projects that we opened up. There was considerable scientific achievement as can be recognized from the more than 140 papers published from the Buffalo laboratory. But I was disappointed that so few young Americans came to work in Buffalo in the seven years I was there, altogether only five. Fortunately I had many co-workers from other countries: eight from Japan, four from Germany, two from Canada, two from Italy, two from England, and one each from Australia, Czechoslovakia, France, Lebanon, and Sweden. I have the feeling that the scientific fashion is for analytical work, and that there is far too little interest in synthesis, particularly when it involves the complex neuronal machinery of the brain. But, in biology, the findings of analysis achieve scientific meaning only when they are synthesized into principles of functional operation. In the final synthesis, models can be constructed that provide the basis for understanding some performance of the whole organism. For example the analytical success in disclosing the mode of operation of the neural machinery of the cerebellar cortex requires level after level of synthesis before it can provide a basis for understanding the cerebellar control of movement and posture.

The last journey of my odyssey is now ended, again across the ocean to Europe, where I live in Switzerland in idyllic mountain surroundings; I have here all my books and journals—many thousands of volumes and a large collection of reprints, so that I can continue my academic life, concentrating on the field that lured me into neurophysiology over 50 years ago—the mind-brain problem. I believe that the great successes of recent years in the study of the brain, and particularly of the human brain, have opened up exciting new prospects for limited successes in this problem that has perplexed mankind since the Greeks, and particularly since Descartes. I have had several attempts in this field since my Waynflete Lectures in 1952, but now realize their inadequacy. Surprisingly I was too timid! I have now developed a much stronger dualistic-interactionist philosophy and it is incorporated in a book that is being published conjointly with Sir Karl Popper, *The Self and Its Brain*. This is the first fruit of my life of retirement. I have much more planned because I realize that the present predicament of mankind results from the continuous process of denigration that has proceeded too far—far beyond the limits justified by our scientific understanding of the cosmos, of evolution, of genetics, and of the brain. Scientists and philosophers share the guilt of being dogmatic in promulgating claims to a knowledge and understanding that devolves from their inflated self-esteem. Mankind has been misled by these spurious claims. I see my task as twofold: to deflate this dogmatism, based not upon science, but on a this-worldly religion of materialist-monism often allied with Marxism; to help in building a new philosophy of man which recognizes that he is a creature that has transcended his animal origin through the building of culture and particularly of language, the World 3 of Popper. We academics have to be humble in our discussion of the nature of man, recognizing the ultimate mystery of the personal existence of a conscious self.

Literature Cited

1. Brock, L. G., Coombs, J. S., Eccles, J. C. 1951. Action potentials of motoneurones with intracellular electrode. *Proc. Univ. Otago Med. School* 29: 14–15
2. Brock, L. G., Coombs, J. S., Eccles, J. C. 1952. The recording of potentials from motoneurones with an intracellular electrode. *J. Physiol. London* 117:431–60
3. Creed, R. S., Denny-Brown, D., Eccles, J. C., Liddell, E. G. T., Sherrington, C. S. 1932. *Reflex Activity in the Spinal Cord.* London: Oxford Univ. Press
4. Eccles, J. C. 1953. *The Neurophysiological Basis of Mind: The Principles of Neurophysiology.* Oxford: Clarendon
5. Eccles, J. C. 1957. *The Physiology of Nerve Cells.* Baltimore, Maryland: Johns Hopkins Univ. Press
6. Eccles, J. C. 1964. *The Physiology of Synapses.* Berlin, Göttingen, Heidelberg: Springer
7. Eccles, J. C. 1970. Alexander Forbes and his achievement in Electrophysiology. *Persp. Biol. Med.* 13:388–404

8. Eccles, J. C. 1975. Under the spell of the synapse. *The Neurosciences: Paths of Discovery,* ed. F. G. Worden, J. P. Swazey, G. Adelman. pp. 158–79. Cambridge, Mass: MIT Press
9. Eccles, J. C. 1976. From electrical to chemical transmission in the central nervous system. *Notes and Records. R. Soc.* 30:219–230
10. Eccles, J. C., Ito, M., Szentágothai, J. 1967. *The Cerebellum as a Neuronal Machine.* Heidelberg, Berlin, Göttingen, New York: Springer
11. Eccles, J. C., Libet, B., Young, R. R. 1958. The behaviour of chromatolysed motoneurones studied by intracellular recording. *J. Physiol.* 143:11–40
12. Fatt, P., Katz, B. 1970. Membrane potentials at the motor end-plate. *J. Physiol.* 111:46–47P
13. Nastuk, W. L., Hodgkin, A. L. 1950. The electrical activity of single muscle fibres. *J. Cell. Comp. Physiol.* 35:39–74
14. Popper, K. R., Eccles, J. C. 1977. *The Self and Its Brain.* Heidelberg, New York, London: Springer. In press

Ann. Rev. Biochem., Vol. 40

SOME PERSONAL HISTORY AND REFLECTIONS FROM THE LIFE OF A BIOCHEMIST

JOHN T. EDSALL

John T. Edsall

Ann. Rev. Biochem., Vol. 40

SOME PERSONAL HISTORY AND REFLECTIONS   744
FROM THE LIFE OF A BIOCHEMIST

John T. Edsall

Department of Biochemistry and Molecular Biology, Harvard University
Cambridge, Massachusetts

CONTENTS

I was born in Philadelphia in 1902, the oldest of three sons. My father, David Linn Edsall (1), was then a young physician, practicing and caring for a considerable number of patients, but devoting much of his time to medical research in the Pepper Laboratory at the University of Pennsylvania Medical School. My mother, Margaret Tileston, was a New Englander, born in Salem, Massachusetts, who had decided at the age of twenty-five to go to college and had graduated from Radcliffe three years later with high honors. She and my father met after she had become a teacher in Philadelphia. On both sides of the family my ancestors had come to America early, most of them in the seventeenth century. On the maternal side nearly all of them had lived in the Boston area; on my father's side they had lived for several generations in the beautiful country of lakes and high rolling hills in the northern tip of New Jersey and the adjacent region of New York State, a region I later came to know well. My grandfather Edsall had been a lead-

1

ing citizen in the small town of Hamburg, New Jersey, where he ran a large general store; he served for some years as Sheriff of Sussex County and as a member of the State Senate of New Jersey. Of his seven sons, two—my father's older brother Frank and my father—went into medicine.

My father had an appointment at the Medical School of the University of Pennsylvania. He was already beginning to be recognized in some quarters as a man of promise, but his salary then was very small, and, even when it was supplemented by earnings from medical practice, the family had to live on a carefully planned budget. Father told me later that in those days he generally walked to the laboratory or hospital, a mile or two away, rather than pay a nickel for streetcar fare. At the same time we did have a part-time nurse to look after me, and we certainly had a cook soon afterwards, if not just at that time. From the point of view of most young married couples today, these represent almost unattainable luxuries, but in those days, to such families as those of my parents, they were virtual necessities.

We lived in central Philadelphia, at 1432 Pine Street. I was therefore a city child, with a fairly large fenced-in back yard to play in, and an outdoor balcony on the top floor, surrounded by wire netting. In the summers we went to Cataumet, on the Buzzard's Bay side of Cape Cod, where my grandmother Tileston had a large house overlooking a bay, and we rented a smaller one not far away. With two highly intellectual parents, I learned early to read, and devoured many books that most people would have considered beyond my age. When I was about six I read much of the textbook of astronomy that my father had used in an undergraduate course at Princeton; I skipped the more difficult parts, but became fascinated with the general accounts of stars, planets, and comets. My parents gave me a small hand telescope, with which I tried to observe the sky at night, but mostly with frustration, for I was strictly required to go to bed early, and in any case the artificial lights in Philadelphia made the stars dim.

Father worked exceedingly hard, but even so we saw him often. He played games with my brothers and me, and went on outings with us. Mother obviously adored us, and was constantly with us. Brought up in a New England family with a strong Puritan conscience, she was exacting in her demands upon herself, but gentle in judging others. Finding my mind responsive, she had read with me a good deal of poetry—I knew all of the "Rime of the Ancient Mariner" by heart, and some of Scott's novels and Shakespeare's plays, by the time I was ten. The love of poetry and imaginative literature has stayed with me ever since. At home, there was always lively talk, with the visitors who came to the house, of politics, literature, science, medicine, and travel, and even as a child I heard a good deal of it.

Father had become Professor of Pharmacology about 1907 and then, in 1910, Professor of Medicine, at the University of Pennsylvania. The latter appointment, however, was followed by a period of conflict within the School of Medicine which frustrated his hopes of seeing the school move rapidly toward becoming a great center of modern medicine as he envisaged it.[1] He therefore decided to accept what appeared to be a very attractive offer from Washington University in St. Louis, which was then undergoing drastic reorganization as a result of severe criticisms by Abraham Flexner in his famous report on American medical schools. Still he had misgivings about the wisdom of the move, and he left the family in Philadelphia for several months before deciding whether we should all move to St. Louis. During that time he found his misgivings more than confirmed; he decided that for him it would be a serious mistake to stay; and early in 1912 he accepted an invitation from Harvard Medical School to become Jackson Professor of Clinical Medicine at the Massachusetts General Hospital. Six years later he was to become Dean of Harvard Medical School, a post he held for seventeen years. So we moved to the Boston area which has, with some interludes, been the center of my life ever since.

This outcome made my mother immensely happy; it was a return to the region in which she had grown up, and my father's appointment was a signal honor—no one from outside of the Boston area had ever before been made head of the Medical Services at the Massachusetts General Hospital. The strain of the previous two years had been great, however, and she had worked unsparingly to keep the family budget in order, and pay off the debts they had incurred during this transition period of uncertainty. Shortly after we had moved to a house in the Back Bay, my youngest brother Geoffrey, aged four, developed a severe case of diphtheria. She wore herself out in nursing him and, although he fortunately survived, she developed pneumonia and died. After this terrible loss, we soon moved out of the house in the Back Bay to the suburb of Milton, where my grandmother owned ten acres of land and a large house standing vacant on the place.

I became a student at Milton Academy, in many ways an excellent school but one in which I did not feel much at home. I was shy, awkward, and more than a year younger than most of my classmates. My ineptitude in athletics, in a school that laid great stress on athletic achievement, gave me a strong sense of inferiority that was not much relieved by the fact that scholastically I always stood at or near the head of the class. These were on the whole lonely years except for the life within the family, where I shared a close bond with my father and brothers. I can well remember, however, the excitement that the elementary science course at school aroused in me, when I was about thirteen. I have a particularly vivid memory of our science teacher, Mr. Homer Le Sourd, demonstrating the decomposition of water by electrolysis into hydrogen and oxygen. Seeing the two gases evolving in a volume ratio of 2 to 1 in the two tubes above the electrodes, with

[1] The story is told in detail by George W. Corner (2) in his history of the Medical School of the University of Pennsylvania—see Chapter 12, "Revolution in the Faculty: and a Counter-Revolution"; and also in the biography of D. L. Edsall by Aub & Hapgood (1). At the time I was certainly quite unaware of the fact that my father's life was passing through a crisis, although my brothers and I must have felt the effects of it in subtle ways.

THE EXCITEMENT OF SCIENCE   143

the subsequent demonstration of the strikingly different properties of the
two gases, came to me almost like a revelation. Certainly the stimulus I received from that course had much to do with the direction of my later career. Growing up in a medical family, I had thought from early childhood
of becoming a doctor, but this idea now became combined with that of scientific work.

HARVARD COLLEGE AND MY TEACHERS THERE

In 1917, when I was fourteen, we moved to a house in Cambridge, and I
finished my preparation for college at the Browne and Nichols school there,
entering Harvard College in the fall of 1919, when I was just under seventeen. The college years brought a small group of intimate and lasting
friendships, of which the most important for my future career in science
was with Jeffries Wyman, with whom, in one way or another, I have been
associated ever since. My first college courses in chemistry and physics were
unfortunately not inspiring, and I almost gave up the thought of going into
a scientific career by way of medical school. However, I persisted, taking
chemistry as my major field, though bypassing the elaborate courses in analytical chemistry that then occupied two full-year courses for most chemistry students, in favor of a shorter and simpler course in the subject. My
grades were not very high, and I barely received a plain honors degree, but
in my last two undergraduate years I did experience inspiration from two
great science teachers—E. P. Kohler in advanced organic chemistry and
Lawrence J. Henderson in biochemistry. Kohler's lectures provided a magnificent example of sustained and searching thought, as the problems of organic chemistry, the tentative ideas for solving them, and the experimental
data and their interpretation gradually unfolded. His presentation, though
unhurried, demanded one's utmost attention; it was deeply scientific and at
the same time a work of art. Kohler's lectures were understandably famous,
but faculty members who asked permission to attend were invariably refused. Kohler said that he was speaking to students, and the presence of listeners with more advanced training might cause him to distort his presentation.

Henderson's influence, for me, went deeper; his lectures lacked the beautiful elegance of Kohler's, but the ideas and concepts he presented were of
the most fundamental importance in shaping my whole scientific outlook.
His book *The Fitness of the Environment* opened entirely new vistas on the
biological significance of the chemical elements and their compounds, especially water, carbon dioxide, and the carbon compounds in general. His
chapter on the ocean gave me an enduring sense of fascination with the
great waters of the world, and what they mean for life in general and for
man. The philosophical conclusions that he drew were in some respects either baffling or unconvincing to me, but that did not impair the inspiring
and original perspective that his presentation of fundamental facts imparted. Also, particularly in the years after I had graduated from college,

his work on blood as a highly organized physicochemical system, functionally adapted to its purpose, and involving a multiple set of interdependent variables, represented for me a glimpse of the organized complexity of biochemical systems that has illuminated my thinking ever since. Henderson and my father were close friends; just after my graduation from college in 1923, my father, my brother Richard, and I made an excursion to northern Vermont, where we visited Henderson at his summer camp in a beautiful location on Lake Seymour—the first of many such visits, continued later after we too had acquired a place in Vermont at Greensboro in the hills above Caspian Lake, some thirty miles to the southwest.

HARVARD MEDICAL SCHOOL: FIRST TASTE OF RESEARCH

In 1923 I went for a year to Harvard Medical School. After four months of anatomy came Otto Folin's biochemistry course, which was very different from Henderson's. I learned to do many kinds of analytical determinations with the old visual Duboscq colorimeter; an indispensable tool of the biochemist in those days, thanks largely to Folin himself, who was the great developer of colorimetric analytical techniques. Walter B. Cannon and his associates gave us an illuminating course in human physiology; but for me the most important experience of that year was with Alfred C. Redfield, then a member of the Physiology Department, later Professor of Biology at Harvard, and then Director of the Woods Hole Oceanographic Institute. Redfield took a few members of the class and set them to work on small research problems; they were then excused from much of the usual required laboratory work. I was fortunate to be one of them, and I studied the effects of pH change and of oxygen lack on the strength of contraction of the heart muscle of the tortoise. I began to get significant results almost immediately, largely because Redfield had solved the most difficult problems of the experimental technique before I ever started work with him. In one way, therefore, this experience was misleading; I did not suffer the periods of discouragement and confusion that are the normal accompaniment of original research. The work was interesting and exciting, and it gave me an abiding interest in the structure and function of muscle. Some of our findings were anticipated in a paper by A. V. Hill, which appeared just as Redfield and I were completing our experiments, but the rest of the work was eventually published, some eight years later, after further work had been done in Redfield's laboratory (3).

TWO YEARS IN EUROPE, AT CAMBRIDGE AND ELSEWHERE

In June 1924 Jeffries Wyman and I sailed for Europe to spend two years in Cambridge, England. First, however, we went to Austria for the summer, and settled down in Graz, with the primary aim of learning to speak and read German fluently. I had had four years of German in school, but still could neither read nor speak it readily. Jeffries lived with one family, and I with another not far away, and neither of us ever spoke anything but German except when he and I were alone together. The treatment worked; I have been able ever since to read German readily, and on later visits to Germany have lectured in German without having to read from a prepared manuscript, although knowing well my deficiencies in grammar and vocabulary. I should remind my younger colleagues, many of whom do not bother to learn much German nowadays, that in those days more important biochemical papers appeared in German than in any other language.

In Graz we had introductions to Otto Loewi, Fritz Pregl, the founder of microanalysis, and Fritz Reuter, the Professor of Legal Medicine, a man of exceptional charm and cultivation. Loewi had recently done the famous experiment, for which he later received the Nobel Prize, showing that stimulation of the vagus nerve of a frog produced a chemical substance—later shown to be acetylcholine—which passed out into the fluid surrounding the heart, a few drops of which would then inhibit the heartbeat of another frog. He demonstrated the "Vagusstoff" experiment to us, with delight and most convincingly. He also took us on a trip into the mountains of Tyrol, in the Oetztal; he himself did not attempt the higher climbs, and Wyman and I did them on our own. For the most difficult climb—which in fact a real mountain climber would have considered "Kinderspiel," as Professor Reuter put it—Loewi insisted that we must be roped, with two guides; it was his responsibility, he said, to see that we got back to our parents safe and sound. He seemed overfussy to us then, but I can appreciate his concern now.

Jeffries and I went to Cambridge at the beginning of the fall term, living in St. John's College and taking the Part II course in Biochemistry, which was then (1924) being given for the first time; previously biochemistry had been given as part of physiology. The new building of the Sir William Dunn Institute of Biochemistry had recently been completed, and was already crowded with workers attracted from all over the world by Sir Frederick Hopkins (Hoppy) and the group he had gathered around him. It was certainly due to his inspiration that biochemistry in Cambridge was taught and practiced with a breadth of outlook that I think was then unparalleled, with emphasis on the significance of the subject for biology in general, rather than its specific relations to medicine or agriculture, as in nearly all places in the United States at that time. The Reader, Hoppy's second in command in the Department, was J. B. S. Haldane. With his powerful, bulky figure, his thunderous booming voice, his vast learning which he was delighted to impart to others, and his dramatic experiments on himself—he was drinking considerable amounts of strontium salts just then and studying their metabolic effects—he was the most striking and picturesque figure in the department. His lectures on enzymes formed the basis of his famous book on the subject, which appeared a few years later. Joseph Needham was studying the chemistry of the developing egg, and his wife Dorothy was already known for her work in muscle biochemistry. Malcolm Dixon was studying oxidation-reduction systems, and Margery Stephenson was one of the then small group of biochemists in the world who devoted themselves to the biochemistry of bacteria. There was plenty of other research going on, and during afternoon tea staff and students gathered for half an hour or so, and exchanged ideas and jokes.

There were important influences outside of the Biochemistry Department. Sir William Hardy, who had started as a histologist and had become one of the great pioneers in the physical chemistry of proteins, a man of immense vitality and zest, was still very active, and I saw him both in the laboratory and on walks in the country. A. E. Mirsky and M. L. Anson, who had been at Harvard one class ahead of me, but whom I had never known until now, were doing important work on hemoglobin in Joseph Barcroft's laboratory, and I learned much from discussions with them on hemoglobin and protein denaturation. G. S. Adair, by his patient and beautiful work on osmotic pressure in the Low Temperature Research Station, had shown that the molecule of hemoglobin was four times as big as people had supposed it to be, and had set forth his famous equation for the binding of oxygen to hemoglobin. At the Molteno Institute, which was just a few steps from the biochemistry and physiology buildings, David Keilin, known chiefly up to that time as a distinguished parasitologist, with a profound knowledge of all sorts of bizarre parasitic organisms, had just rediscovered MacMunn's myohematin, which for good reasons he renamed cytochrome, and had started on his long and magnificent series of biochemical researches. Keilin was always kind, helpful, and inspiring to beginners like myself. I remember vividly his showing me the absorption bands of reduced cytochrome in muscles from a bee, on a microscope slide with a cover glass arranged to keep out oxygen, viewed through his microspectroscope; and the rapid disappearance of the bands when oxygen was admitted. He also gave me helpful guidance in experiments on phosphates in insect muscle. These were part of a small research on phosphates in muscle which I began, with some general advice from Hopkins, some six months after coming to Cambridge. They came to little, although they led to a short paper in the *Biochemical Journal* in 1926, but they involved one important experience, a three weeks' visit to the laboratory of Gustav Embden in Frankfurt-am-Main in the spring of 1925, to learn his methods of phosphate determination in muscle. At that time there was controversy between Embden and Meyerhof regarding the role of phosphates and lactic acid in muscle, with Embden postulating a "lactacidogen," perhaps a hexose phosphate, that found no place in Meyerhof's scheme. Embden was most kind in taking me into his laboratory for such a short time, and the younger people there taught me much in three weeks of hard work. I particularly remember Emil Lehnartz and F. Deuticke, whom I was to visit again after the Second World War when they had become professors in Münster and Göttingen. I am sure that no one ever bothered to refer to my little paper on phosphates in muscle; it became completely irrelevant within a year or two, with the discovery of phosphocreatine by Fiske and Subbarow, and of ATP not long after.

THE EXCITEMENT OF SCIENCE   147

Jeffries Wyman, who had started work in biochemistry in Cambridge, decided after one term that A. V. Hill's laboratory in London was the place where he really wanted to work. There he studied the physics of muscular contraction, and we remained in close touch. During the Christmas vacation of 1925, we returned to Austria to revisit friends in Vienna and Graz. In the long spring vacation of April 1926 we went off to Corsica with Robert Oppenheimer, whom we had known when he was a freshman at Harvard and we were seniors. Robert was spending that year at the Cavendish Laboratory in Cambridge, passionately eager to solve the problems of quantum physics. Heisenberg's first great paper on quantum mechanics had just appeared, Schrödinger's work was to appear only a few months later, and Dirac was a fellow graduate student of mine at St. John's College, though very few realized at that moment that he would make epoch-making contributions in the next few years. Robert Oppenheimer, unlike Dirac, was intensely articulate, and conveyed to me the deep excitement and promise of what was going on in quantum mechanics. My mind was far too slow to grasp what he could see and master swiftly, but the feeling that he gave me for the central importance of the subject stayed with me, and several years later I worked in Bright Wilson's seminar at Harvard to learn some of the basic essentials of quantum chemistry. Robert's interests, however, ranged far beyond science; he had studied philosophy quite deeply, he was devouring the novels of Dostoevsky and Proust, and he introduced me to several French poets that I had not read, notably Baudelaire and Hérédia. In the midst of all this he was passing through an intense psychological and spiritual crisis, the nature of which I would not attempt to diagnose, but which for a time (I believe) he felt threatened to destroy him. His capacity for work, even with such a handicap, was astonishing; but he was still deeply troubled in mind although he took part with enthusiasm in our long walks and climbs in the magnificent wild country of Corsica. There is no need to say now that he survived that crisis, since all the world knows his later history. We recognized at the time his extraordinary gifts, and expected him to make great discoveries, though none of us could have imagined him then as the director of a vast world-shaking enterprise in applied science, as he was to be in less than twenty years.

Those two years in Cambridge were of profound importance in my life. In science they immensely widened and deepened my vision of the scope of biochemistry. In my travels during vacations I came to know much of Europe well, with its enormous range of natural beauty, and the endless fascinations of its art and architecture. I had never been far from home for more than a short time; here I was on my own, and independent as never before.

HARVARD MEDICAL SCHOOL: PHYSICAL CHEMISTRY AND MUSCLE PROTEINS

Coming back to Boston in the summer of 1926, I started my first clinical year at Harvard Medical School, having had pathology and bacteriology in my second year at Cambridge. Much of that year seemed trivial and stupid, passing from a two weeks taste of one medical specialty to two weeks of another; and much of the time I was quite depressed. But, having gone so far, I decided to finish work for the medical degree. At the same time, however, I began the work that was to occupy me ever since. Medical students at Harvard did have some free afternoons, to do research or anything else they pleased, and I consulted Alfred Redfield about continuing my work on muscle. He remarked, "I think the most neglected part of the whole muscle problem lies in the muscle proteins. Edwin Cohn has started some work on muscle proteins. Why don't you go and work with him?" That was my introduction to Cohn's laboratory, and to the unique Department of Physical Chemistry at Harvard Medical School, tucked away on the fourth floor of the physiology building.

Cohn welcomed me in, and set me to work during my free afternoons on extracting what we then called simply muscle globulin, from beef muscle. We got it from a neighboring slaughterhouse, ground it, fresh from the killed animal, with a meat grinder, and stirred the ground meat rapidly into buffered potassium chloride solution. After prolonged stirring the resulting purification procedure was relatively simple. I diluted the filtered preparation to low ionic strength, centrifuged the precipitate, and redissolved it at higher ionic strength, repeating the process several times. The resulting protein preparation was a messy thing to handle, extremely viscous and showing no trace of any tendency to crystallize. Nevertheless it was fascinating, and I struggled with it month after month. In my last year as a medical student, thanks to the liberal policies that my father had initiated several years earlier as Dean, I had the chance, like a number of my classmates, to spend most of my time with the problem that interested me most, which for me was of course the muscle globulin. I learned much about the handling of proteins from Cohn's constant guidance and comments, and much also from Arda Alden Green (4), who was then in the midst of her beautiful studies on the solubility of horse hemoglobin as a function of pH, ionic strength, and temperature, and who had a natural talent for handling such complex materials. Jeffries Wyman came back in 1927, after receiving his PhD for his work on muscle in A. V. Hill's laboratory. He worked at an adjoining bench on the viscosity of proteins.

A turning point in my scientific life came with the arrival of Alexander von Muralt from Switzerland, to work on the double refraction of muscle, with a polarizing microscope and samples of muscle which he immersed in fluids of different refractive index, to reveal the orderly structure of the submicroscopic fibrillar elements within them. His work was going beautifully, but he received a sudden shock on looking one day into the *Chinese Journal of Physiology,* and discovering that the same research had already been done by a German physiologist named Stübel, working in China. This was a blow, but Alex recovered rapidly when Cohn suggested to him that he should examine my muscle globulin preparation for double refraction.

We set up a capillary tube for observation in the polarizing microscope, and forced the muscle globulin solution through it. The result was dramatic; there was no double refraction while the solution was at rest, but when it started flowing through the tube and we observed it between crossed Nicol prisms, the liquid became brilliantly luminous, showing strong double refraction. Alex had taken his PhD in physics at Zurich before going into physiology, and he perceived immediately the significance of what we saw. In spite of my background in physical chemistry, I was an ignoramus in this field; I had almost forgotten what little I had once known about double refraction. However, with some tutelage from Alex and much hard study, I learned fast, and realized that here we had evidence of long asymmetric protein molecules, oriented at random when the solution was at rest, but swinging into more or less parallel alignment when placed in a velocity gradient. Moreover these long thin protein molecules were presumably the very elements that gave rise to the double refraction of muscle, when oriented in the fiber, and by inference represented the essence of the contractile system. For the next two years we were busy working out the quantitative behavior of this fascinating protein, which we then called myosin, but which later has been recognized as actomyosin, thanks to the work of Szent-Gyorgyi and Straub. Alex designed an elegant concentric cylinder apparatus—the outer cylinder rotating, the inner one fixed—for producing a well-defined velocity gradient in the myosin solution, with a coupled pair of Nicol prisms above and below the liquid for observing the flow birefringence, which indicated the orientation of the myosin molecules. The two years that we spent working on myosin made me realize as never before the relations between the work of the morphologists and the biochemists for understanding the dynamics of muscle, and of life processes in general (5, 6).

One member of the laboratory at that time who should not be forgotten was our genial red-headed Italian dishwasher and general laboratory assistant, George Greco. Diligent, effervescent, ever conversational in somewhat broken English, George cherished a secret personal history which he revealed only to Alex von Muralt. Alex's medieval ancestors had ruled, from a strong castle, over one of the mountain passes between Switzerland and Italy, and levied tribute on all the merchants who passed through; and the von Muralts had always been one of the important Swiss families. George confided to Alex, while washing the laboratory glassware one day, that he was descended from a notable Italian Count, though in an illegitimate line. "You and I," he whispered to Alex, "are the only aristocrats in this laboratory."

Some Later Work on Myosin and on Flow Birefringence

Ten years later I returned to work on myosin (actomyosin). Jesse P. Greenstein, who had played a major role with us in the Department of Physical Chemistry for several years, had developed a quantitative method for determination of sulfhydryl groups in proteins by titration with porphyrindin. We applied this to the study of native and denatured myosin (7), demonstrating the presence of free –SH groups even in the native protein, with a marked rise in titratable –SH on denaturation with guanidine hydrochloride and other reagents. At the same time John W. Mehl and I studied the denaturation of myosin from rabbit and lobster muscle by observing the disappearance of the flow birefringence in the presence of a large variety of denaturing agents (8). Guanidine hydrochloride was effective even at concentrations of 0.2 to 0.3 M, and some other reagents at slightly higher concentrations.

At the time of the early work on myosin (1928–1930) by von Muralt and myself, there had been no good theory that related flow birefringence to the size and shape of the molecules producing it; but in the following decade the work of Paul Boeder, Werner Kuhn, Anton Peterlin, and H. A. Stuart provided a good quantitative theory for ellipsoidal molecules, relating the observed phenomena to the rotary diffusion coefficients of the molecules (9). This stimulated us to develop a more powerful instrument for studying the sizes and shapes of smaller protein molecules at high-velocity gradients. Joseph F. Foster in particular applied this method successfully in determining the molecular dimensions of fibrinogen (10) and even those of plasma gamma globulin and albumin (11). These latter molecules, considered for simplicity in calculation as ellipsoids of revolution, had axial ratios of about 6 to 1 for gamma globulin and about 3.5 to 1 for serum albumin. Even with the high-velocity gradient apparatus they could be oriented successfully only in solvents of high viscosity; generally we used glycerol-water mixtures. In spite of technical difficulties, the molecular dimensions that Foster obtained in these studies turned out to be in good agreement with those that other workers found by other methods, such as dielectric dispersion, viscosity, and low-angle X-ray scattering.

PHYSICAL CHEMISTRY OF AMINO ACIDS AND PEPTIDES; DIPOLAR IONS

About 1930 Edwin Cohn, recognizing how ignorant we were of the details of protein structure, initiated studies on the physical chemistry of amino acids and peptides, in order to approach the unknown by way of known structures. With Cohn's remarkable gift for drawing many people of varied talents and interests into the same orbit, this grew into a major enterprise which occupied a considerable group of us for a decade. I have already told most of the story in two biographical articles on Edwin Cohn (12, 13) and here I will elaborate only on a few points. We realized from the fundamental papers of E. Q. Adams and Niels Bjerrum that amino acids and peptides must be electrically charged molecules at all pH values— cations at low pH, anions at high pH, and dipolar ions (zwitterions) at some intermediate pH value, at which their net charge was zero. Dipolar ions, with a separation of some 3 Å between the positive and negative charge, even for α-amino acids, and much larger separations for peptides, must be studied in aqueous solution, or at any rate in quite polar solvents.

Jeffries Wyman, who by then had moved from the Medical School to the Biological Laboratories at Harvard College, developed a new and elegant method for studying dielectric constants in aqueous solutions. T. L. Mc-Meekin, who had joined the Physical Chemistry Department at the Medical School, was synthesizing a series of amino acids, peptides, and their derivatives, on which he did a long series of studies with Cohn to correlate structure with solubility in various media. Wyman's dielectric studies on Mc-Meekin's compounds demonstrated the enormous increase in dielectric constant produced by adding amino acids or peptides to water. The dielectric constant of pure water at 25°C was 78.5; it rose to just over 100 in a molar solution of any α-amino acid, to nearly 150 in a molar solution of glycylglycine, and to still larger values in the longer peptides. Indeed the molar increment in dielectric constant was a linear function of the number of atoms separating the positively charged amino group from the negatively charged carboxyl group. The very large dipole moments demonstrated by this work implied that there must be strong interactions between these dipoles and the ions of salts present in the surrounding medium; and McMeekin's solubility measurements showed this clearly. To explain such interactions obviously involved an extension, to dipolar ions, of the theory of interactions between ions developed by Debye and Hückel. Such an extension was beyond our powers, but fortunately we had the constant advice and help of George Scatchard and John G. Kirkwood at the Massachusetts Institute of Technology, who took the problem in hand and produced mathematical solutions for several more or less realistic models for dipolar ions. Our constant discussions with them were indispensable for understanding what we were doing and for suggesting new experiments.

My own contribution was in the study of Raman spectra, which revealed molecular vibrational frequencies from the shifts of frequency arising from the scattering of a monochromatic beam of light, incident on the solution. After a long struggle to learn the technique (it would have been easy enough for a real spectroscopist, but did not come so easily to me), results came pouring in. The data showed the characteristic vibrational frequencies of amino, carboxyl, and other groups in the ionized and un-ionized state, and demonstrated unequivocally the ionized state of both the amino and the carboxyl groups in isoelectric amino acids and peptides (14). The characteristic frequency changes produced by deuterium substitution threw further light on these structures (14, 15). The data showed much else besides—the structure of the guanidinium ion, for instance, which was significant for the understanding of arginine. I continued this work much later, after the Second World War, particularly to study histidine and other imidazole derivatives, and sulfhydryl compounds, with the more powerful spectroscopic techniques then available. I might add that the advantage of working with Raman rather than infrared spectra was that the Raman spectra could be recorded in great detail for substances in aqueous solution, with little interference by the frequencies arising from the water, whereas the infrared spectra of the solutes would have been almost completely blotted out by the intense absorption of the water. The limitation of the Raman technique, in those days, was its restriction to relatively small molecules; the intense background scattering in solutions of macromolecules fatally obscured the relatively weak Raman lines. Much later (1955) David Garfinkel, in the course of a long series of Raman studies in my laboratory, did obtain a Raman spectrum of lysozyme, but it was rather weak and not very informative,[2] although his studies on smaller molecules had yielded very important information (16).

We came to realize the powerful interactions between dipolar ions and the water molecules surrounding them. Cohn, McMeekin, and I discovered that the apparent molal volume of an amino acid in water—that is, the volume increment in the solution per mole of substance dissolved—was always substantially less than that of an isomeric uncharged substance. For glycine and its isomer glycolamide, for instance, the difference was about 13 cm³ per mole; and for molecules with a much larger separation between the charged groups the difference was about 20 cm³. The explanation was already at hand, from the ion studies of Drude and Nernst, and others: the intense electric fields surrounding the charged groups oriented the neighboring water molecules and packed them closely around the charges, with a resultant shrinkage in volume. This electrostriction effect was also well known to affect profoundly the heat capacities of ionic solutions; solutions of salts often show negative apparent molal heat capacities, the heat capacity of the solution being substantially less than that of an equal amount of salt-free water. Studying the scanty data in the literature, I found some evidence for similar effects in amino acid solutions also; but I found also a totally unexpected effect of nonpolar groups in *increasing* apparent molal heat capacities of organic solutes in water. In a homologous series, for instance, an added methylene group increased the heat capacity in water by about 20–25 calories per degree per mole; for the same compounds, measured as pure organic liquids, the increment was only about 5 cal deg⁻¹ mole⁻¹ (18).

This clearly pointed to some remarkable interaction between water and nonpolar groups, opposite in its effects on heat capacity to the electrostriction effect of charged groups. The work of J. A. V. Butler and others had also shown an evolution of heat and a puzzling decrease of entropy when substances containing nonpolar groups were introduced into water. I realized that some very important kind of interaction was involved, but did not have the wit to see what it was. I puzzled over it many times, but reached no clear conclusion. The state of our thinking then can be found in Chap-

[2] Two days after writing these words, I was delighted to discover the paper of Lord & Yu (17), in which, using a laser to excite the Raman spectrum, they report a spectrum of lysozyme containing an immense amount of detail, with quantitative intensity values. This far surpasses anything we were able to achieve, and should open up the whole field of Raman spectroscopy of proteins.

ters 7 and 8 of the book by Cohn and myself (19). The mystery began to disappear with the work of H. S. Frank, from 1945 on, on the structure of water and its interactions with hydrophobic groups, and with the penetrating analysis by Walter Kauzmann of the hydrophobic bond (20).

At any rate we did perceive the great importance of the nonpolar side chains in determining the free energy of transfer of amino acids and peptides between water and less polar solvents, such as ethanol, and formulated simple rules relating structure to solubility ratio in such solvents (see 19, Chapter 9).

PROTEIN CHEMISTRY IN THE 1930s

Protein chemists were a small fraternity in those days, and their laboratory equipment would look very simple and primitive to young biochemists today. Photoelectric spectrophotometers were still nonexistent; our principal analytical techniques were Kjeldahl nitrogen and dry weight determinations, supplemented sometimes by some of Folin's techniques with a visual colorimeter. We measured pH on the hydrogen electrode; the bubbling of hydrogen gas through protein solutions was not the most desirable thing in the world, but the quality of Cohn's titration curves for proteins was nevertheless very high. It was not until about 1935 that stable reliable glass electrodes became available. There was no ultracentrifuge in the laboratory until 1938. Up to about that time, if you wanted to do ultracentrifuge studies on a protein, you went to Uppsala and worked in Svedberg's laboratory. None of us in fact did this, though we followed each new publication from Svedberg's laboratory with the closest attention. In 1938 J. L. Oncley, who had joined us two years earlier and was doing outstanding work on the dielectric dispersion of proteins, installed an air-driven ultracentrifuge of the type developed by Beams and Pickels, under the guidance of Dr. Pickels, who was then at the Rockefeller Institute. In 1940 we obtained a Tiselius moving-boundary electrophoresis apparatus. During the war years, with our blood plasma fractionation work, both it and the ultracentrifuge were commonly running all day and well into the night.

Certainly our thinking was based on the belief that proteins were definite large molecules with a defined chemical structure, and not merely a miscellaneous collection of colloidal particles, as an earlier generation of colloid chemists had supposed. In this respect we were inheritors of the chemical point of view put forward by W. B. (Sir William) Hardy, T. B. Osborne, S. P. L. Sörensen, and Jacques Loeb. We held fast to the view that proteins were made up of amino acid residues linked in polypeptide chains, although this view came under attack in various quarters about 1925, and again in Dorothy Wrinch's famous cyclol theory in 1937–1940. The pioneer X-ray work of W. T. Astbury suggested how the polypeptide chains might be arranged in space: I well remember a visit he paid to us about 1936, with our discussion of the folding and unfolding of peptide chains, and the denaturation of globular proteins and their unfolding into fibrous structures. Even though most of Astbury's proposed structures were wrong, his findings were immensely important and his influence was inspiring. Likewise we felt that we had entered a new era when, in 1935, J. D. Bernal and Dorothy Crowfoot (Hodgkin) obtained the first X-ray diffraction photographs of crystalline proteins immersed in their mother liquor. This left no doubt in our minds that protein molecules were highly organized structures with a well-defined three-dimensional pattern; but we could not then realize what a long hard road remained to be traveled before those patterns were to be revealed in detail.

Of course we welcomed with enthusiasm the work of Sumner and Northrop, showing definitely that enzymes were proteins, and the work of Stanley on tobacco mosaic virus, which brought viruses into the realm of well-defined substances.

For all that, it is well to remember how ignorant we were. Amino acid analysis of proteins was an arduous enterprise, requiring large amounts of protein and yielding for the most part fairly inaccurate results after many weeks or months of labor. No complete amino acid analysis of any protein was available until the work of Erwin Brand and his collaborators on β-lactoglobulin in 1945 (21). Although we felt pretty sure that proteins were composed of polypeptide chains, we did not know the actual length of the chains in any protein, or how many subunits the protein contained. Svedberg's ultracentrifuge work on the hemocyanins and other proteins had indeed demonstrated that many proteins could be reversibly dissociated into subunits; and he had put forward the view that all proteins might be built up of subunits with a molecular weight of about 17,000. Cohn and I looked on this idea with extreme skepticism; and it was only many years later that I came to realize that Svedberg's idea, though wrong in detail, had far more truth in it than I had perceived earlier.

TEACHING AT HARVARD;
THE TUTORIAL SYSTEM IN BIOCHEMICAL SCIENCES

The Department of Physical Chemistry at the Medical School was primarily a research department. Edwin Cohn gave only an occasional lecture, at his own discretion. From the time I got my MD degree in 1928, however, I was involved in both the tutorial program at Harvard and in formal teaching.

The tutorial work was at Harvard College in Cambridge, some four miles away from the laboratory where I worked at the Medical School. It made exacting demands upon my time to be a member of two faculties, and in contact with two almost entirely different groups of people; but it was also a rewarding experience to be closely in touch with what was going on in both places. Fortunately Ronald Ferry was kind enough to invite both Jeffries Wyman and me to join the staff of John Winthrop House among its charter members, and friendships developed there with historians, economists, philosophers, political scientists, and others whom I might never have known but for the fortunate circumstance of belonging to the House.

The tutorial work was an immense intellectual stimulus, involving constant discussion and interchange of ideas with a small group of under-graduates each year, and the guidance of research for seniors who were candidates for the honors degree. Among these students who have continued in scientific careers are R. Gordon Gould, I. Herbert Scheinberg, Alton Meister, Alexander Rich, Gary Felsenfeld, Jared Diamond, W. French Anderson, Elliot L. Elson, Michael Chamberlin, David S. Eisenberg, Robert S. Eisenberg, and Joel Huberman. It has certainly been most valuable for my own outlook on the world to work with them, and with many other gifted students, at this early stage of their development.

By no means was all my teaching tutorial work. I gave a few lectures each year in L. J. Henderson's course; about 1940, when Henderson decided to give up his course altogether, Jeffries Wyman and I inaugurated a new course on the biophysical aspects of biochemistry within the Biology Department. Except for a break during some of the war years, we continued to teach it together until Jeffries resigned from Harvard in 1952 to become Science Attaché at the United States Embassy in Paris. Our thinking in the presentation of that course was the origin of our later book on *Biophysical Chemistry*, of which Volume I was published in 1958. The second volume, alas, has been long delayed, a delay for which I am chiefly to blame; but we still intend to complete Volume II.

I have continued to lecture at Harvard ever since, mostly on proteins, enzymes, and biophysical chemistry in general. Even during the strenuous years when I was Editor of the *Journal of Biological Chemistry* I continued to give a course of about thirty lectures in one term of each academic year, and the experience has been of great value to me in helping me to organize my thoughts and keep a general perspective on a broad area of biochemistry. My most strenuous teaching assignment came when I was a Fulbright Lecturer in the University of Tokyo in the spring of 1964. For three months, with an occasional week off, I lectured on biophysical chemistry and proteins three times a week, in English, to a class of about 35 advanced undergraduates and first year graduate students. I spoke as slowly and clearly as possible, writing a great deal on the blackboard and showing many slides, and taking about an hour and a half for each lecture. The students were clearly a highly superior group, and I was greatly impressed with the amount they apparently learned from me, considering the language problem involved. My friends Professor Haruhiko Noda and Dr. K. Maruyama attended all the lectures, made tape recordings, and afterwards worked up the subject matter into a book in Japanese, of which I am listed as a co-author, although in this case I cannot read my own work!

WORK AT HOME IN VERMONT; A YEAR IN PASADENA

In 1929 I married Margaret Dunham of New York. We lived in Cambridge, only a short walk from Harvard University; our three sons were born between 1930 and 1936. In the summers we occupied a cottage in Greensboro, Vermont, on the upper slopes of a high hill overlooking Caspian Lake, where my father had bought an old farm of about a hundred acres several years before. During those summers I would work several hours a day, learning advanced calculus and various parts of mathematical physics, so as to understand more deeply the physical chemistry of proteins and the theoretical basis of the Raman spectra that I was observing in the laboratory. Also it was an excellent place for thinking about the work I had done and writing it up. It was of course necessary to go back to Harvard at intervals to look up references and put papers into final shape, but I found that three or four hours of work in the undisturbed atmosphere of Vermont were often more productive than twice that amount of time at the University. With all this, we had the pleasure of living with our children in the beautiful surroundings of northern Vermont for about two months each summer, rather than in the city. From 1940 on, the war, and then the increasing load of responsibilities, made this kind of long working summer vacation impossible; but I am thankful to have had those quiet and delightful summers in the earlier years. We continue to come to Greensboro whenever we can, and it is there that I have written these recollections.

As the work in the laboratory developed, the idea of writing a comprehensive book on proteins, amino acids, and peptides became more and more compelling to Edwin Cohn and me. The deeper understanding of long-established facts, and the discovery of new facts in profusion, led to a vision of order replacing what had been chaos, which for me was thrilling and inspiring. I had the urge to portray that order in detail and in organized fashion. Cohn wrote much of the book, and George Scatchard was our constant advisor and helpful critic throughout. John G. Kirkwood and J. L. Oncley also contributed essential chapters. Most of the labor of writing, however, fell on me; and I would never have finished the job if I had not received a Guggenheim Fellowship in 1940–1941, which permitted the Edsall family to spend a year at the California Institute of Technology, where I could work almost uninterruptedly on the book. Margaret and I drove across the continent with our two younger children, David and Nicholas, taking three weeks to do so. This was a wonderful experience, for I had never even been as far west as Chicago before that. Pasadena was delightful, for the Los Angeles area was far less crowded with people than today, and smog was not yet a problem. Cal Tech was so small, compared to Harvard, that we soon came to know most of the faculty and felt very much at home. Linus Pauling and other members of the Chemistry Department, as well as those in Biology, furnished immense intellectual stimulus and an excellent atmosphere to work in. By the time we returned to Harvard in July 1941, the book was essentially complete, though it was not published until 1943, and other problems of terrible urgency awaited me.

THE WAR YEARS AND THE PLASMA FRACTIONATION PROGRAM

From the time that Hitler took over Germany in 1933 we had watched with alarm the spread of the power of the Nazi government, culminating in the outbreak of war in 1939. These grim events haunted us in the midst of all our work, and cast a shadow over everything. The fall of Norway and France in the spring of 1940 shook us profoundly; I and most of my friends were convinced that the United States must and should be involved in the war before very long. The Battle of Britain was proceeding as we made our way out to Pasadena in the late summer of 1940, and my year at Pasadena was the last for many years in which I could devote myself to quiet scholarly work.

By the time I returned, Edwin Cohn had already organized the laboratory on what was essentially a war footing, although the United States would not be officially at war for several months to come. We were already fractionating blood plasma, to obtain human serum albumin and other plasma proteins for clinical use by the Armed Forces. Albumin had great advantages over whole plasma as a plasma expander, particularly after we learned to pasteurize it by heating in the presence of a stabilizing agent, such as sodium caprylate, thereby killing the virus of serum hepatitis. Gamma globulin was used for temporary immunization against measles, and later against infective hepatitis also. Cohn had the vision to see what could be done by large-scale plasma fractionation, and the driving and aggressive energy to get the necessary Government support, to bring together large groups of scientists and clinicians working in a common cause, and to enlist seven major pharmaceutical firms in the large-scale production of plasma fractionation products by methods worked out in the Pilot Plant at Harvard Medical School. Cohn himself told the story in detail (22), I have already told it more briefly in two articles concerning him (12, 13), and have reviewed at length (23) the results that were achieved by the larger group of workers during those hectic years. My own share in this large enterprise chiefly involved the uses of fibrinogen and fibrin. We obtained two products, both of which were eagerly used by the neurosurgeons, fibrin foam with thrombin, and fibrin film. The former, in the development of which Edgar Bering played a major part, proved of great value in stopping bleeding during operations, especially in brain surgery. The latter, developed chiefly by the beautiful work of John D. Ferry and Peter R. Morrison, proved to be the first really safe and effective replacement for the dural membrane lining the brain, after some of the latter had been removed in an operation. During the war years both these products were made on a large scale by the methods developed in our laboratory, and from the testimony we received I would judge that they saved many lives. After the war equivalent products were developed elsewhere, from cheaper materials than blood plasma, and fibrin foam and film fell into disuse; but they did in any case serve as the models for these further developments.

So many people were involved in this enterprise that it would be impossible to mention all who played an important part. However, even in this brief, personal account I must speak of J. L. (Larry) Oncley, whose work on gamma globulins and lipoproteins, and in the total direction of the operation, was central; of Laurence E. Strong, fresh from his PhD in chemistry at Brown University, who directed the complicated operations of the pilot plant with scientific judgment and human wisdom; and of W. L. ("Pete") Hughes, whose insight into the interactions and crystallization of proteins was outstanding.

It was a big change in my life to work closely with clinicians and with industrial scientists and engineers, who were concerned with production problems. The pace of the work was terrific; as soon as one crisis was resolved, another arose to take its place. The feeling that we were contributing something vitally needed in the war sustained us and drove us on; and unlike many scientists engaged in war work we had the satisfaction of knowing that much of what we did would be of value in civilian medicine after the war also. Moreover the inherent scientific interest of the work was great; as the fractionation work proceeded, we came to realize how many and various were the proteins of blood plasma, and we were constantly identifying components that no one had clearly recognized before. The work of those strenuous years has certainly influenced my scientific life ever since. For a number of years after the death of Edwin Cohn in 1953 I served on the National Research Council Committee on Blood Plasma and Plasma Expanders, for part of that time as Chairman, and we grappled with many difficult problems, not always successfully. No one has yet found out how to eliminate the serum hepatitis virus from the whole plasma or whole blood, or even to assay reliably whether the virus is present or not. This represents a terrible gap in our knowledge, and it is of urgent importance for modern medicine to fill it. Fortunately recent work on the "Australian Antigen" has begun to offer what looks like a hopeful clue.

THE YEARS AFTER THE WAR:
RETURN TO BASIC RESEARCH AND MOVE TO HARVARD COLLEGE

In the postwar years the laboratory returned to basic work on proteins, but the center of our interests owed much to our work on blood plasma in the war. While Oucley was pushing forward with the study of lipoproteins and immunoglobulins, "Pete" Hughes discovered how to separate the mercaptalbumin fraction of plasma albumin, with one free sulfhydryl group per molecule, from the rest of the albumin with no free sulfhydryl. He crystallized mercaptalbumin as the mercury dimer, with one mercury atom linking two albumin molecules through their sulfhydryl groups (24). His fundamental discovery led several of us to quantitative studies of rates and equilibria in the dimerization process, which we could follow from moment to moment by light-scattering changes. Aroused by the work of Debye, I had already realized the power of the light-scattering method, and with Harold Edelhoch, Peter R. Morrison, and René Lontie, had carried out an extensive study of the interaction of albumin with other molecules and ions, as a function of the net charge on the molecule and the ionic strength (25). Walter B. Dandliker also contributed much to the light-scattering work. It was our good fortune that in 1951 Ephraim Katchalski came to the laboratory as a postdoctoral fellow, and he did a magnificent job in characterizing the thermodynamics and kinetics of the mercaptalbumin dimerization process (26). Rudolf Straessle then made use of a bifunctional organic mercurial to obtain another dimer in which the distance between the two albumin molecules was considerably greater; the dimerization process went far more rapidly in this case than in the simple mercury dimer. Robert H. Maybury and Richard B. Simpson worked out the relations in detail. Later Cyril M. Kay, in his PhD thesis, did a corresponding and very elegant series of studies on bovine mercaptalbumin, which had also been crystallized by Hughes (27).

Recognizing the importance of X-ray crystallography for proteins, we had persuaded Barbara W. Low to join the laboratory in 1948. There she embarked on the detailed studies of insulin which she has pursued since at Columbia, and did important research on albumin crystals, in which Frederic M. Richards, then a graduate student, obtained his first experience as a crystallographer. Another graduate student, Frank R. N. Gurd, was doing his thesis on lipoproteins with Oncley; he later became involved, with Philip E. Wilcox who had come from Wisconsin as a postdoctoral fellow, in studies of the interactions of amino acids, peptides, and proteins with metallic ions. Harold A. Scheraga and Geoffrey Gilbert worked with me on a cold-insoluble globulin from the fibrinogen fraction of blood plasma, and Charles Tanford began his career as a protein chemist with a searching and detailed study of the acid-base equilibria in albumin solutions. Ariel G. Loewy worked with me on the plasma factor that catalyzes the conversion of soluble into insoluble fibrin, and thereby began that excellent series of researches which he has pursued in conjunction with his teaching at Haverford College in subsequent years.

All these and other gifted young investigators provided an immense stimulus. In this incomplete list I have not attempted to mention the large number of workers who came from all over the world to learn Edwin Cohn's fractionation methods and develop them further. The size of the laboratory had now grown immensely beyond what it had been in prewar days. We used every bit of available space, although the space allotted to us had grown considerably. My own office was a very small room, tucked away in a corner; the shelves were crowded with books and journals, and there was room for perhaps two other people to squeeze in for discussions with me. The crowding was sometimes uncomfortable, but it also promoted the constant interchange of ideas, which was good for all of us.[3]

After Cohn's death in 1953, I moved from the Medical School to the

---

[3] I have given elsewhere (28) a brief history of the Department of Physical Chemistry at Harvard Medical School from 1920 to 1950; a more detailed version of this history was printed as one of the memoirs of the Laboratory of Physical Chemistry, but few copies of this survive.

University in Cambridge, where I could carry on laboratory work, tutorial teaching, and lecturing in close conjunction, and eliminate the four mile journey between the laboratory and my place of teaching. The Biology Department obligingly provided an office and laboratory space. I came just when an outstanding group of biochemists was assembling in Cambridge. Paul Doty was already in the Chemistry Department; Konrad Bloch and Frank H. Westheimer arrived there around the same time that I came to Biology. George Wald and Kenneth Thimann were long-established members of the Biology Department, and J. D. Watson joined the department soon thereafter. Soon the University established a Committee on Higher Degrees in Biochemistry, drawn from members of the Chemistry and Biology Departments, of which I was the first Chairman. That Committee has recently grown into a Department, with Jack L. Strominger as the present Chairman; the outstanding achievements of its members and its students are too well known to call for further comment here.

My research in the first years after moving to Cambridge centered chiefly on the ionization, and the interaction with metallic ions, of amino acids, peptides, and related compounds. Yasuhiko Nozaki, Masatami Takeda, Donald B. Wetlaufer, and R. Bruce Martin all played an important part in this work; and Susan Lowey contributed to it also, along with the studies on myosin which she had already started and continued in our laboratory. David Garfinkel did a major series of studies on Raman spectra (15). The pioneer work on ribosomes of *Escherichia coli* by J. D. Watson and Alfred Tissieres was getting under way before 1960, and they and their collaborators did much work in our laboratory, where the ultracentrifuge and other facilities were available, and where I followed with keen interest the development of this outstanding work. Several of their collaborators indeed had their home in my laboratory—notably P. F. Spahr, J. P. Waller, and for a shorter period J. I. Harris—and did important work on bacterial ribonucleases and on the complexity of ribosomal proteins.

From 1959 on, however, my own research interests centered primarily on the carbonic anhydrases of red blood cells. Since the time of my early studies with Henderson, I had always been concerned with blood as a system, and carbonic anhydrase was not only an essential part of that system but a fascinating enzyme in itself. It is among the most powerful of all catalysts for what is probably the simplest of all enzyme-catalyzed reactions; yet the mechanism of the process is still elusive. When we started work on it we had thought ourselves practically alone in the field, but we soon learned that B. G. Malmström, S. Lindskog, and P. O. Nyman—then in Uppsala, now in Göteborg—were already embarked on major researches on the subject; and not long afterwards we found that Professor Y. Derrien and Mme. G. Laurent in Marseille were also deeply involved. Fortunately there were plenty of problems for all of us to work on, and some overlapping in the researches was helpful to us all. As I write, the three-dimensional structure of human carbonic anhydrase C is now practically complete in the laboratory of B. Strandberg and A. Liljas in Uppsala, and the sequence work is progressing in the Göteborg laboratories, so one may hope that our own work on the physical chemistry and kinetics of these enzymes can soon be interpreted at a much deeper level. This is not the place to describe our work [I have given one review of it, written three years ago (29)], but I would note the names of those in our laboratory who have been responsible for it, and have initiated new approaches to difficult problems: Egon E. Rickli, Barbara H. Gibbons, S. A. S. Ghazanfar, Lynn M. Riddiford, Dirck V. Myers, J. McD. Armstrong, Jacob A. Verpoorte, Louis E. Henderson (now in the Göteborg laboratory, continuing the sequence work he started with us), Philip L. Whitney, Anna J. Furth, Shelby L. Bradbury, Allan J. Tobin, Julia F. Clark, Raja Khalifah, Friedrich Dorner, and Pierre Henkart. Guido Guidotti, whose laboratory has been so closely associated with mine for the last seven years, has also contributed important guidance and valuable suggestions to those involved in the work on carbonic anhydrase, in addition to his notable work on hemoglobin and on the membrane proteins of red blood cells.

EDITORIAL WORK

In the midst of the turmoil of the war years, my old friend M. L. (Tim) Anson persuaded me to join him in editing a new publication, to be called *Advances in Protein Chemistry*. It was indeed launched, in the midst of many other preoccupations, before the war ended. We picked the authors that we wanted to contribute, and I think that we generally picked well. I have told of the early days of that enterprise in a little article on Dr. Anson (30). With time it grew; we soon got Kenneth Bailey, who had worked beside me in Cohn's laboratory in 1939, to serve as our European editor, and he proved invaluable. In 1956 we were fortunate to enlist C. B. Anfinsen also; and after the untimely death of Kenneth Bailey in 1963 we persuaded F. M. Richards to join us. We have now published 24 volumes of that series, and I think it has made a real contribution to the advancement of protein chemistry. Editing it has involved some work, of course, but it has on the whole been fun, and I have learned a great deal from doing it.

For ten years (1948–1958) I served on the Editorial Board of the *Journal of the American Chemical Society*. I had to help decide some difficult problems regarding controversial papers, but on the whole it was not a strenuous job. My involvement with the *Journal of Biological Chemistry* was very different, and for ten years it became a central part of my life. It began innocently enough in 1950, when I was elected to the Editorial Committee, which had to keep a general oversight of the policies of the Journal. Since Rudolph Anderson was a wise and experienced Editor, with an excellent Board, this was at the time a light assignment. Four years later, however, under strong pressure from some of my senior colleagues, I was propelled from the Editorial Committee onto the Editorial Board, where Rudolph set me to work with a substantial load of papers each year. It was my job to reach an editorial judgment on each paper, and draft a letter to the author, including a decision on the paper. This went back to the Editorial Office in New Haven, and was usually the basis of the letter that Rudolph Anderson sent to the author. On a few occasions Rudolph wisely overruled my proposed decision, and always if I (or anyone else on the Board) favored rejecting a paper, he consulted at least one other Board member before reaching a final decision. All these procedures had been well worked out over the years, and they still remain in force, in the editing of the *Journal.*

Most unexpectedly, when Rudolph Anderson retired, I was asked to become his successor. I was staggered at first on contemplating the size of the job, but finally accepted, on the understanding that I would work half-time for Harvard and half-time for the *Journal,* so that I would teach in only one term. Even so, after I started work as Editor in Chief in 1958, it was for a time an overwhelming job. Fortunately I had a superb administrative assistant and head secretary, Elisabeth J. Cross, who supplied the organizational talent that I largely lacked. When she decided to leave us in 1960, Ada Wing, who had been her first assistant almost from the beginning, took her place and ran the office with superb efficiency and tact until I retired as Editor.

During my years as Editor, from 1958 to the end of 1967, the *Journal* doubled in size and in the number of members on the Editorial Board. It was the members of the Board who carried the heaviest load, some of them reviewing as many as 70 or more papers per year. They were free to consult referees as they saw fit, but it was still their responsibility for each paper they received to draft a decision letter, written as if addressed to the author, and to send it back to me for transmission to the author with such modifications as I might think necessary. In many cases I could use the drafted decision letter without change. It is the devoted and generally unrecognized work of all these members of the Board that enables the Editor-in-Chief to carry on and maintain the standards of the *Journal.*

Fortunately, before I had been long on the job, Robert A. Harte became Executive Secretary of the American Society of Biological Chemists and Manager of the *Journal.* I was thankful to turn over to him the financial management of the *Journal,* as well as a great many publication problems where his judgment was far better than mine. His expert knowledge of matters concerning scientific documentation and communication has also been of great value.

As the number of contributions grew, more help was necessary. Fortunately Konrad Bloch and Manfred L. Karnovsky agreed to become Associate Editors in 1961, and took over much of the load of responsibility that I had been carrying. After five years I asked for a year's leave of absence as a condition for my accepting a second five year term as Editor, and this was generously granted in 1963–1964. While my wife and I were away, first in Rome where I worked with Jeffries Wyman and Eraldo Antonini, and then in Japan as I have already told, Konrad Bloch, Manfred Karnovsky, and several of my other colleagues in the Boston area managed the editing of the *Journal*, and did it so well that I felt almost ready to take a permanent leave of absence. However I returned to the job in the fall of 1964, much refreshed by my year away. As the size of the *Journal* continued to grow, we realized that we needed more Associate Editors: in addition to Manfred Karnovsky, who fortunately continued to serve, we enlisted William H. Stein and Efraim Racker, and gave the Associate Editors far more independent authority than before.

SCIENCE, MANKIND, AND THE FUTURE

I have lived long enough to know as a child the relatively peaceful and stable world that existed before 1914. In fact it looks more secure in retrospect than it did to people at the time. However that may be, the world I have lived in since has been a world of wars, depressions, and great revolutionary upheavals. These terrible events have had a profound personal impact on the lives of many of my colleagues abroad, as one may see for instance by reading the autobiographical accounts by Karl Thomas, Hermann Fischer, and Albert Szent-Gyorgyi in earlier volumes of this series.

My own life, by contrast, has been much more sheltered and peaceful. I was too young to serve in the First World War, and was busy with war research in the laboratory during the Second. During the years of the depression I had a good position with adequate pay at Harvard. It was impossible, however, to be indifferent to the outside world. Our family was always deeply interested in politics, and we followed intently the course of world events. However, until the atomic bombs fell on Hiroshima and Nagasaki, my work in science and my concern with politics ran in different channels. After 1945, that was no longer possible.

Apart from the larger perils from the misuse of applied science of which many of us are so conscious today, new influences threatened the freedom and integrity of science. Demands for secrecy in research and security clearance for the researchers, justifiable as they were in certain sensitive areas, tended to spread into broader areas of scientific life and work, where they are poisonous and corrupting. Some of these issues came to a head at the meeting of the American Society of Biological Chemists in April 1954, when it became general knowledge that the US Public Health Service was tigators because of unevaluated adverse information in their security files. The investigators were not told what was going on, or given an opportunity to answer the alleged charges, which were in any case irrelevant to the criteria for awarding grants for unclassified research. This created a profound sense of outrage among the biochemists and other scientists gathered at the meeting. With Philip Handler, Wendell Stanley, and a few others, I helped to draft a resolution asking the National Academy of Sciences to investigate these alleged procedures of the Public Health Service, and the Society at its business meeting passed the resolution unanimously. President Bronk of the National Academy set up an excellent committee to investigate the subject and make recommendations. Its inquiry, however, necessarily consumed many months; and in the meantime I became aware of other cases in which excellent investigators lost their grants for reasons having nothing to do with merit. Scandalized by these events, I decided to speak out in protest as a private citizen. I wrote an article, published in *Science* (31), portraying and condemning what was going on, and declaring my own refusal to accept research support from the US Public Health Service as long as these practices continued. I was in a position to do this at the time, for I had just moved from Harvard Medical School to Harvard College in Cambridge; I was operating only a small laboratory, and had an adequate grant from the National Science Foundation to keep it going. By contrast an investigator who had been operating a big laboratory, with a large grant from the Public Health Service—for instance Edwin Cohn in his later years —could hardly have renounced the use of such funds without imperiling the livelihood and the future of the younger scientists working with him. The moral issues involved in taking such an individual stand are therefore complex; but I am glad that I was able to speak out as I did, and when I did. I cannot tell what influence my article in *Science* may have had; I know that a number of people, both inside and outside the US Public Health Service, expressed gratitude to me for writing it.

Several months later the report of the National Academy committee appeared; it was a strong and forthright document, and it firmly upheld the principle that grants for unclassified research should be awarded only on the basis of the scientific integrity and competence of the investigator. The Eisenhower Administration accepted the report and called upon all government agencies to put its principles into practice, and as far as my knowledge extends, they did so. Within another year I did apply to the Public Health Service for research support, which they have provided me most helpfully ever since.

I must add that although the Public Health Service ceased to use secret information in decisions concerning grants for unclassified research, it still maintained a blacklist excluding many highly qualified scientists from service on the committees that made recommendations on awards of research grants. I was not aware of this until 1968, when I learned of one particular case from a colleague. In the following year, an article by Bryce Nelson in *Science* (32) brought the matter to public attention. Vigorous protests from the National Academy of Sciences and from many other sources brought action from the government to eliminate this blacklist also (33), and I trust that these practices have now ceased. The whole story indicates, however, the need for unceasing vigilance and for outspokenness, if the free and open field of science is to be maintained.

In recent years my concerns regarding the larger implications of science and technology have broadened and deepened. Apart from the ever present threat of nuclear war, I have seen widespread deterioration in the world around me—the decay of our inner cities, and the outward spread of urban blight; the increasing contamination of once clean waters; the cluttering of the countryside with discarded automobiles and countless other forms of hideous junk; the proliferation of superhighways, designed without adequate concern for the total ecology of the region, uprooting people in great numbers in the cities, and devouring good agricultural, forest, or park land in the open country. I have seen new technological developments, such as the supersonic transport, promoted at vast expense without serious consideration of their adverse effects on man and the environment, and have helped to support the excellent work of my friend, Dr. William A. Shurcliff, in making the public aware of the hazards and drawbacks of this program.

I have also been deeply concerned with the threats to mankind of chemical and biological weapons, not primarily because I consider them uniquely horrible or inhumane—personally I would rather be killed by nerve gas than by napalm—but because they represent, next to nuclear weapons, the most dangerous potential agents for wholesale slaughter of great masses of people. Also there is an international agreement, the Geneva Protocol, renouncing their use, although the United States has not yet (September 1970) signed it. I would hold fast, whenever possible, to such agreements and work to strengthen and extend them. I have therefore been profoundly disturbed by the use of tear gases, defoliants, and herbicides by the United States in the war in Vietnam. Whatever the temporary tactical advantages of the use of such agents may be—and I think that many of the alleged advantages are highly questionable—I believe that the use of these agents in war is exceedingly dangerous, because it can lead to escalation and to the use of far more deadly chemical and biological weapons. This would be directly contrary to the vital interests of the United States and of mankind in general, and we should do our best to avoid the risk of such escalation.[4]

All these problems represent a part of the broader effort to adapt and control technology, and make it the servant of broader human values. A committee of the National Academy of Sciences has recently issued an important report bearing on this problem (34). In dealing with these issues we must learn to think in terms of the organized complexity of natural systems.

[4] In addition to these general considerations there is the profoundly disquieting evidence that herbicides such as 2,4,5-T—and probably 2,4-D also—are teratogenic in experimental animals, and are therefore likely to be so in man also. Thomas Whiteside, in a series of carefully researched articles in the *New Yorker* (1970), has brought these grave matters to public attention; but our government has lagged in imposing the restrictions on the use of these compounds that seem to be clearly called for by the experimental evidence.

A relatively simple example is L. J. Henderson's study of blood (35) as a system characterized in terms of seven major variables, a change in the value of any one variable necessarily involving changes in all the others. Even blood is of course in fact much more complex than this, and the natural systems that are modified by new technological developments involve a vast multiplicity of variables. It is natural for biologists to think in these terms, even though most of them do not think mathematically with ease, for they must deal from the start with the complexities of the living organism. With the aid of modern computers we should be able to deal with situations so complex that they were formerly intractable to human thought. To do this wisely, however, we must feed into the computer information on all the significant variables that are relevant to the system, and give adequate weight to each. Here the traditional economic cost-benefit analysis is likely to be deficient; it takes account of the more obvious economic variables, but is likely to leave out such vital matters as natural beauty, quiet and clean surroundings, and the many factors that make for a harmonious environment. There will be passionate disagreements over the relative weight to be given to these diverse and disparate factors, but we should be able to formulate the behavior of complex natural systems, and the effects of their modification by human action, in terms of a wide variety of assumptions about the relative weight to be given to different sets of values. Controversies regarding the policies to be followed will not abate, but I hope they will be based on a more critical evaluation of the evidence and of the possible choices of action.

We have been living through a period, unique in the world's history, of rapid growth of population, of material goods, and of energy supply. Soon, within a few moments of geological time, this growth must come to an end. Population will be stabilized at some reasonable level, or else population growth, after proceeding unchecked for another generation or two, will lead to catastrophe as mankind becomes more and more crowded and the earth's resources become increasingly exhausted. The liberation and utilization of energy by man must also be stabilized at a level that will avoid intolerable thermal pollution and other hazards. Waste products and valuable minerals must be recycled rather than discarded. We must aim to preserve the richness and variety of the world in its splendid diversity of landscape and of plant and animal life, if our descendents are to have at least as rich and full a life as the best lives that men can lead today. We should seek to maintain a world that will be a better place to live in than today, a thousand or a million years from now; for mankind will never find another home to compare with this ravaged but still magnificent planet.

ACKNOWLEDGMENT

I am indebted to the National Science Foundation for a grant (GS2723X) for studies in the history of biochemistry, during the period in which this article was written.

THE EXCITEMENT OF SCIENCE 167

LITERATURE CITED

1. Aub, J. C., Hapgood, R. 1970. *Pioneer in Modern Medicine: David Linn Edsall of Harvard.* Harvard Med. Alumni Assoc., Harvard Univ. Press
2. Corner, G. W. 1965. *Two Centuries of Medicine: A History of the School of Medicine, University of Pennsylvania.* Philadelphia: Lippincott
3. Edsall, J. T., Hunt, H. B., Read, W. P., Redfield, A. C. 1932. *J. Cell. Comp. Physiol.* 1:475–501
4. Colowick, S. P. 1958. *Science* 128:519–21
5. Edsall, J. T. 1930. *J. Biol. Chem.* 89:289–313
6. von Muralt, A. L., Edsall, J. T. 1930. *J. Biol. Chem.* 89:315–50, 351–86
7. Greenstein, J. P., Edsall, J. T. 1940. *J. Biol. Chem.* 133:397–408
8. Edsall, J. T., Mehl, J. W. 1940. *J. Biol. Chem.* 133:409–29
9. Edsall, J. T. 1942. *Advan. Colloid Sci.* 1:269–316
10. Edsall, J. T., Foster, J. F., Scheinberg, H. 1947. *J. Am. Chem. Soc.* 69:2731–38
11. Edsall, J. T., Foster, J. F. 1948. *J. Am. Chem. Soc.* 70:1860–66
12. Edsall, J. T. 1955. *Ergeb. Physiol. Biol. Chem. Exp. Pharmakol.* 48:23–48
13. Edsall, J. T. 1961. *Nat. Acad. Sci. Biogr. Mem.* 35:47–84
14. Edsall, J. T. 1936. *J. Chem. Phys.* 4:1–8; 1937. *J. Phys. Chem.* 41:133–41; 1937. *J. Chem. Phys.* 5:225–37, 508–17
15. Edsall, J. T., Scheinberg, H. 1940. *J. Chem. Phys.* 8:520–25
16. Garfinkel, D., Edsall, J. T. 1958. *J. Am. Chem. Soc.* 80:3318–23, 3823–26, and earlier papers in the same series
17. Lord, R. C. Jr., Yu, N.-T. 1970. *J. Mol. Biol.* 50:509–24
18. Edsall, J. T. 1935. *J. Am. Chem. Soc.* 57:1506–07
19. Cohn, E. J., Edsall, J. T. 1943. *Proteins, Amino Acids and Peptides.* New York: Reinhold. Reprinted 1965. New York: Hafner
20. Kauzmann, W. 1959. *Advan. Protein Chem.* 14:1–63
21. Brand, E., Saidel, L. J., Goldwater, W. H., Kassell, B., Ryan, F. J. 1945. *J. Am. Chem. Soc.* 67:1524–32
22. Cohn, E. J. 1948. In *Advances in Military Medicine.* ed. E. C. Andrus et al, I: Chap 28, 365–443. Boston: Little, Brown
23. Edsall, J. T. 1947. *Advan. Protein Chem.* 3:383; 1950. *Ergeb. Physiol.* 46:308–53, 354–78
24. Hughes, W. L. Jr. 1947. *J. Am. Chem. Soc.* 69:1836–37; 1949. *Cold Spring Harbor Symp. Quant. Biol.* 14:79–83
25. Edsall, J. T., Edelhoch, H., Lontie, R., Morrison, P. R. 1950. *J. Am. Chem. Soc.* 72:4641–56
26. Edelhoch, H., Katchalski, E., Maybury, R. H., Hughes, W. L. Jr., Edsall, J. T. 1953. *J. Am. Chem. Soc.* 75:5058–72
27. Kay, C. M., Edsall, J. T. 1956. *Arch. Biochem. Biophys.* 65:354–99
28. Edsall, J. T. 1950. *Am. Sci.* 38:580–93
29. Edsall, J. T. 1968. *Harvey Lect. Ser.* 62:191–230
30. Edsall, J. T. 1970. *Advan. Protein Chem.* 24:vii–x
31. Edsall, J. T. 1955. *Science* 121:615–19
32. Nelson, B. 1969. *Science* 164:1499–1504; 165:269–71
33. Nelson, B. 1970. *Science* 167:154–56
34. Committee on Science and Astronautics, US House of Representatives, July 1969. *Technology: Processes of Assessment and Choice.* Report Nat. Acad. Sci. 163 pp.
35. Henderson, L. J. 1928. *Blood. A Study in General Physiology.* New Haven, Conn.: Yale Univ. Press. 397 pp.

Reprinted from
ANNUAL REVIEW OF ANTHROPOLOGY
Volume 3, 1974

AMONG THE ANTHROPOLOGISTS

Fred Eggan

Photo by Joan Eggan

Reprinted from
ANNUAL REVIEW OF ANTHROPOLOGY
Volume 3, 1974

AMONG THE ANTHROPOLOGISTS

§ 9533

Fred Eggan
Department of Anthropology, University of Chicago, Chicago, Illinois 60637

ANTHROPOLOGY AND THE INFORMATION EXPLOSION

The increased tempo and changing structure of modern anthropology in the postwar period is nowhere better reflected than in the *Annual Review of Anthropology* and the predecessor volumes of the *Biennial Review of Anthropology*, which Bernard J. Siegel has been editing with distinction for over a decade. Henry Selby, in a recent review in *Science*, welcomes the new format and considers the *Annual Review of Anthropology* a great improvement over the earlier version: "Instead of long blunderbuss essays that attempt to make prose sense out of laundry lists of literature, we have smaller scale essays on the state of the art in those areas where something significant or interesting is going on." But I rather miss those essays—including Selby on social organization—as well as the periodic coverage of traditional topics and the relative leisure to read them. The authors were some of the best and brightest of our anthropological scholars, and their essays presented the trends of postwar anthropology with authority and distinction.

These trends included the shift from traditional physical anthropology to human biology, under the influence of modern genetics; the renewed emphasis on culture and culture change; the increasing concern with peasant society and urban life; the revival of psychological and philosophical anthropology; the revolutionary developments in modern linguistics; the shift in our conceptions of social structure; and the continuing search for new and better models. And once in a while a distinguished scholar like David Easton took the time to examine our efforts and provide us with some new ideas and directions in a field such as political anthropology.

While we can hope that the new series will improve on the high standards of its predecessors, information in most scientific fields is now doubling in volume every decade and anthropology is no exception. In fact, the rate in anthropology may be even higher if we think of our postwar expansion and the growing number of

1

papers at our various annual meetings. Even though the current *Annual Reviews* double the coverage and the number of topics, they may be overwhelmed in another decade, and we will be forced to depend on computerized bibliographies and possibly some access to data via central repositories and consoles.

The information explosion in anthropology likewise affects the audience. A decade ago the *Biennial Reviews* were aimed at three audiences: preprofessionals in anthropology, anthropologists with an interest in related fields, and members of other disciplines who wanted to know what was going on in anthropology. The current *Annual Reviews* will appeal more directly to professionals who are trying to keep up with their own fields. If you are not an ethnoscientist, you may be amazed at Oswald Werner's account of Ethnoscience 1972 (*Ann. Rev. Anthropol.* 1972. 1:271–308), despite his conclusion that "almost everything still remains to be done."

CURRENT STATUS OF ANTHROPOLOGY: SOME RECENT VIEWS

My predecessors in these introductory essays, Conrad Arensberg and Margaret Mead, have dealt with important aspects of the progress of anthropology. Arensberg, in his chapter on Culture as Behavior: Structure and Emergence (*Ann. Rev. Anthropol.* 1972. 1:1–26), notes that anthropology "remains the one science bridging the gap between the 'hard' and the 'soft' sciences, uniting the Two Cultures of Sir Charles Snow," but he is also concerned about the proliferation and sprawl which threatens its advance. He suggests that we need to pay renewed attention to the concept of culture: "Obviously it is culture which provides the theme which unifies, if tenuously, the great sprawl of the science. Culture, as man's special behavior, remains the theme still implicit in the background of all the subdisciplines. Perhaps it would be most useful to treat the nature, emergence, variability, and relationships of culture."

Arensberg goes on to outline what he terms "interaction theory"—a behavioral anthropology and sociology, based on the observation and precise measurement of interpersonal events, as developed in the pioneer work of Eliot Chapple and himself. The first full statement of the new method was Chapple and Coon's *Principles of Anthropology*, published in 1942, which attempted a recasting of anthropology through dispensing with the concept of culture. (I remember writing a favorable review for the *American Anthropologist* and being castigated by Ralph Linton for being too uncritical. I only learned much later that *The Study of Man* was never reviewed in the *Anthropologist!*)

Chapple's most recent book, *Culture and Biological Man* (1970), while continuing the systematic and quantitative observation. of interpersonal interaction, brings culture back into the picture, in a limited sense, at least. Arensberg calls attention to the existence within anthropology of "an operational generalization of social behavior into cultural forms," which may come as a surprise to many anthropologists. He thinks that "human ethology" can also account for formal and other "deep structures" of the mind that some observers believe to underlie cultural behavior.

While Arensberg and Chapple are finding their way back to culture and structure through interaction theory, G. P. Murdock has gone in the opposite direction. In the Huxley Memorial Lecture for 1971, which he gave under the title "Anthropology's Mythology," Murdock praises the corpus of descriptive ethnography but is very dissatisfied with our body of theory. "It now seems to me to be distressingly obvious that culture, social system, and all comparable supra-individual concepts . . . are illusory conceptual abstractions inferred from observations of the very real phenomena of individuals interacting with one another and with their natural environments." As reified abstractions they can be used for descriptive purposes but not as explanations of human behavior.

Murdock's solution is to jettison the whole body of anthropological theory —including much of his own work—which derives from such concepts, and to develop an increasingly intimate collaboration with psychology. He looks to psychology to provide an understanding of the underlying mechanisms of behavior, which then can be seen as producing differing forms of behavior under varying conditions of life.

Paul Bohannan, on the other hand, thinks culture is too important an idea to jettison and proposes that it be reexamined and strengthened. In his article on Rethinking Culture: A Project for Current Anthropologists (published in *Current Anthropology*, 1973, Vol. 14, No. 4), Bohannan suggests that culture traits are analogous to genes in certain respects, and that the concept of a "cultural pool" is a useful one. Culture is coded—in memory, behavior, language, art, writing, etc—but the most important aspect is that it is always encoded *twice*: once within the individual and once outside the human being in some other form. At a deeper level, he believes, genetics and culture are two parts of a single process, and culture can therefore play a vital role in evolution and in comparative studies.

These statements—so different in their treatment of culture—suggest that this central anthropological concept is going to be under active discussion during the coming decade. We can no longer take it for granted, and we must ultimately reach some consensus with regard to it. It is clear that anthropologists must pay greater attention to the individual in his physiological, genetic, and psychological aspects, and not merely begin with "social" or "cultural" behavior. Earl Count looks forward to a "human ethology"—a man-science which is yet unborn—and it is clear that we can learn from the ethologists.

My personal opinion is that Murdock's criticisms are too extreme. My understanding of the use of the concepts of culture, structure, and social system is not that they are used to *explain* behavior, but rather as the *field* within which behavior takes place. And if good fieldwork depends upon sound theory, a proposition which Murdock minimizes, the descriptive ethnography which he believes to be the "crowning glory" of our discipline must have some theoretical basis.

Of American anthropologists only Edward Sapir reached similar conclusions to those advanced by Murdock. In his earlier writings he took a conventional view of culture, and his *Time Perspective in Aboriginal American Culture: A Study in Method* is an early classic. But in *Cultural Anthropology and Psychiatry*, published in 1932, he wrote that: "The true locus of culture is in the interactions of specific individuals, and, on the subjective side, in the world of meanings which each one of these individuals may unconsciously abstract for himself from his participation in these interactions." He goes on to note that "the concept of culture, as it is handled by the cultural anthropologist, is necessarily something of a statistical fiction, and it is easy to see that the social psychologist and the psychiatrist must eventually induce him to carefully reconsider his terms. It is not the concept of culture which is subtly misleading but the metaphysical locus to which culture is generally assigned." Kroeber and Kluckhohn's comment in *Culture: a Critical Review of Concepts and Definitions* (1952) is to the effect that social scientists have taken surprisingly little notice of these drastic statements.

Margaret Mead, in her survey of Changing Styles of Anthropological Work (*Ann. Rev. Anthropol.* 1973. 2: 1–26), ranges over the post-World War II period and is particularly concerned with the way in which anthropology should relate to other disciplines and how it can overcome the process of fragmentation that has developed in so many fields. She asks whether there should be "more than one science concerned with the behavior of human beings, as individuals, in groups, as carriers of culture . . . Shouldn't we all be branches of one human science?"

I presented my own answer to this question in *One Hundred Years of Anthropology*, edited by J. O. Brew, a few years ago:

> We can begin to see a new organization for anthropology emerging, centering on man and his works and providing a spectrum of specialized fields which interlock with those of the social and behavioral sciences. Ultimately there will be no sharp boundaries—as there are none in the biological and physical sciences—but culture, society and personality will be the major foci of attention, both individually and in their relationships. What the role of anthropology will be in this larger field is not yet clear, but its broad comparative treatment of social and cultural phenomena should assure it of a central position.

This assessment—made less than a decade ago—was an optimistic one and seemed reasonable at the time. But as Mead notes, there has been "a kind of fragmentation, by areas, by schools, by instruments used, by approaches preferred, by style of work, into subfields which are as complex as whole cultures seen in their complete ecological settings." This is part of the proliferation and sprawl that Arensberg sees as threatening the progress of anthropology. And the current leveling off of support for anthropology will reduce cooperation and increase competition, thus enhancing the fragmentation that Mead describes.

Mead's chapter is a veritable *tour de force* in which she notes our various activities, relevant or not, and the changing contexts in which we do our work. Her comments on the new feminism, on the demands of minorities, and on ethical problems, are especially relevant, but she has significant things to say on almost everything. Her conclusions are worth stressing.

> Anthropology is entering a new era, the flesh pots are emptier, the difficulties of doing field work increase geometrically as the equipment grows more elaborate and the political situation in many parts of the world becomes more unsettled. In such meager times as these, anthropology can take several directions: an increased interest in professional careers that involve professional competence in related fields, like town

planning, health, nutrition, and political organization; an intensive reexamination of existing materials (where Lévi-Strauss has erected such a challenging theoretical structure); concentration on audio-visual recordings in an attempt to obtain the new kind of records of still living cultures with film and tape; a renewed dedication to the preservation of cultural diversity; and a greater involvement in an increasingly endangered planet. The problem remains of how to keep so many extraordinarily diverse and discrepant foci of interest and competence in active interrelationship. The very peculiarity of the task may be what will make it possible.

SOME PERSONAL REFLECTIONS

When I first entered the University of Chicago some 50 years ago I had never heard of anthropology. I had originally received a scholarship to Northwestern, but the valedictorian of our graduating class later decided she would rather go to Northwestern than to Chicago, so I arrived on the Midway in 1923. Frederick Starr had retired that June and there was an interregnum until Fay-Cooper Cole arrived the following year. I had taken a "scientific" course in high school, with an early interest in physics and chemistry, but I had no clear idea about what I wanted to do beyond a great curiosity about the world. I began in the newly established School of Business, where I learned some economics, but I soon discovered psychology and transferred my major interest to that field. I was also attracted by geography.

I first heard of anthropology while standing in line to register at the beginning of my junior year. Two students behind me were talking about courses, and Freddy Starr was still a famous name. He had come to the University in 1892 as part of the Department of Sociology and Anthropology, the first such department in this country, and after producing two outstanding PhDs in 1897, David P. Barrows and Merton L. Miller, looked forward to a separate department of anthropology. The following decades were not propitious for such an undertaking, and Starr settled for a career as a great undergraduate teacher and a campus character. When he retired in 1923 his former students raised $15,000 to provide him with a house in Seattle—such were the provisions for "retirement" in those days.

I had not heard of Starr, but the term "anthropology" attracted my attention so I looked it up in the University catalog while waiting my turn. Anthropology 280 was concerned with the study of peoples and races, and I decided on the spur of the moment to take it, thus beginning a lifelong involvement with the discipline. Cole was a dynamic and inspiring teacher whose enthusiasm for his subject was contagious. He had early joined the Field Museum of Natural History and had been sent first to the Philippines and later to Indonesia to make ethnological studies and secure collections. While in Borneo in 1923 he had received offers to develop anthropology at both Northwestern and Chicago, and on his return had taught at both institutions before deciding that Chicago was his first choice.

The University of Chicago in the 1920s was a typical college so far as the undergraduate program was concerned, but the graduate program was outstanding. The first comprehensive poll of scholarly opinion in 1925 showed Chicago in first place in overall standing so far as graduate departments were concerned.

Anthropological activities were still largely centered in museums, except for Boas at Columbia, but teaching and research were being developed at a graduate level in a few universities, and Cole came to Chicago with a plan to develop a graduate department of anthropology.

Cornelius Osgood and I began anthropology together, and we must have made some impression on Cole because we were invited to join a graduate seminar being organized on India. Cole had brought Edward Sapir to the University in 1925, and there were already a number of graduate students: Leslie White, Paul Martin, Robert Redfield, Charlotte Gower, Peter Roest, and others. We were excited at being allowed to join the seminar—until the topics were assigned. We protested that we were neophytes, with only two or three weeks of introductory anthropology, but the faculty decreed it to be a "working seminar." I was given the topic, "The Caste System of India," and disappeared into the stacks for a month where I read all the reports on caste in the census volumes and other tomes. I survived the experience and produced a paper, but I have been happy to leave the caste system to others ever since.

In psychology my initial interest had been aroused by William Sheldon, but I soon gravitated to L. L. Thurstone, who was adapting psychophysics to the measurement of attitudes and developing factor analysis. Combining the two fields, anthropology and psychology, I wrote a master's thesis on "Attitudes toward Races and Nationalities" in 1928, but I had already decided that I wanted to be an anthropologist. There was little support for graduate work in those days, particularly for anyone who was changing fields. Hence I took a job for 2 years teaching in a military academy and junior college in Missouri. I started on my 22nd birthday, and I taught psychology and sociology in the college and history in the academy. It was the hardest teaching I have ever done, but I learned on the job and saved enough money to return to graduate work in anthropology in the summer of 1930.

In the meantime Cole had developed plans for a separate department of anthropology and had added Robert Redfield, back from his pioneer study of Tepoztlán, Mexico, to the staff as an instructor. The new department was inaugurated October 1, 1929, with a promise of one million dollars for a museum and endowment, but the promised gift vanished with the stock market crash a month later.

During the 1920s anthropology began to expand in a number of centers both in the United States and elsewhere, and a remarkable group of graduate students assembled at Berkeley, Chicago, the London School of Economics, and Sydney. Columbia and Harvard had had an earlier start around the turn of the century, and between them had filled most of the university and museum posts then available.

Cole and Sapir made an excellent combination, and together they covered the various fields of anthropology as adequately as any combination existing at that time. Cole took responsibility for physical anthropology and archeology, though his training and research had been mainly in ethnology, and Sapir covered linguistics and ethnology, with excursions into personality and culture. With

Redfield's pioneer interests in folk culture and peasant society, a new facet was added which maintained continuity with the sociologists, particularly Wirth, Hughes, and Blumer, who were contemporaries.

Cole saw archeology as a means to achieve support for the proposed department, so he started an archeological survey of Illinois the first summer and proceeded to excavations the following years. He also took an active part in the Scopes' trial in 1925, and on his return from Tennessee learned that the President's office had been flooded with requests for his dismissal. The President called him in, but promoted him instead of firing him. His introductory classes soon doubled in size.

Anthropology in the 1920s at Chicago was a discipline with four subfields, and all graduate students were required to master the fundamentals of each before going on with their specialities. My first love was archeology, and I spent several summers excavating Indian mounds and village sites in Illinois and later at Awatovi, on the Hopi reservation in Arizona. We mainly learned archeology in the field and worked up our reports during the school year. My first experience with linguistics was a class in Navaho. Sapir was just back from a Laboratory of Anthropology field session on the Navaho reservation, and we started with the verb and worked in all directions from there. If I had had any native skills in language I would have become a linguist, but with graduate students such as Mary Haas, Stanley Newman, Morris Swadesh, Harry Hoijer, and Walter Dyk, the competition seemed too great. Physical anthropology in those days required a mastery of German—it was all in Martin's *Lehrbuch*. We learned to measure skulls—and each other—but most of our attention centered around evolution and race. The central field, in retrospect, was ethnology, and here we had the advantage of occasional courses from Leslie Spier and Paul Radin. Spier had a hard time with an elementary course but was magnificent in advanced courses, particularly on western North America. He also gave a memorable course on South America in which he did as much work as the rest of the students combined. Radin, in contrast, hated to teach and avoided the task whenever possible. But in the evenings, surrounded by admiring graduate students, he would talk for hours on anthropology and anthropologists.

The written examinations for admission to candidacy for the doctorate included all four fields and lasted a week. We were supposed to spend four hours on each set, but the "kindly" faculty usually allowed more time. Often two or three students reviewed together—John Provinse, Mark Watkins, and I were one such team—and we divided up the literature and educated each other. For one examination session we had 12 questions, and I remember that Watkins was still writing on the first one when we went to lunch. I only remember one question from the many: "Write a sketch of Navaho grammar," and I find it impresses even modern students.

Graduate students were encouraged to develop their own research projects, and the first PhDs exhibited the variety which continues to characterize the department. Leslie A. White wrote on "Medicine Societies of the Southwest," a gem of a study which has not yet been surpassed. Robert Redfield's "Tepoztlán, A Mexican

Village" needs no comment. Charlotte Gower continued the sociologist's interest in the ethnic groups of Chicago by studying "The Supernatural Patron in Sicilian Life," and then went to Milocca, a village in western Sicily, to check on the validity of her earlier findings. W. M. Krogman made a study of changes in the growth patterns of the skull and face of anthropoids, an interest he has continued with regard to man. Paul S. Martin analyzed the Kiva as a survival of an ancient house type. Continuing my interest in archeology, I had a thesis topic approved on "The Ethnological Interpretation of Archeological Cultures," and was admitted to candidacy for the doctorate in October 1931. For a variety of reasons I never completed that thesis.

Earlier that year Yale offered Sapir a Sterling professorship and the chairmanship of a newly established Department of Anthropology and Linguistics, so he left in July, taking most of the graduate students in linguistics along with him. In his place Cole arranged to bring A. R. Radcliffe-Brown from Sydney for a year as a visiting professor, and added Harry Hoijer to the staff to continue work in linguistics. We knew Radcliffe-Brown mainly through *The Andaman Islanders*, since despite several years in South Africa and in Australia, he was just beginning to publish on these areas.

I took his courses rather reluctantly, but soon found that he provided a new dimension to the anthropology then current in America. I had been impressed by the analytic studies of Lowie and Spier, but their conclusions provided little understanding of the cultures they studied. Radcliffe-Brown's analysis of the household structure, the institutions surrounding *lobola*, and the initiation rituals in South Africa provided a new and productive way of thinking about culture and society. I had already taken a course on "Primitive Society" from Spier, but "Family, Kin and Clan" clarified much of what had earlier puzzled me. And *The Social Organization of Australian Tribes*, which had just been published, made sense of aboriginal life for a whole continent.

Radcliffe-Brown's knowledge of the American Indian derived mainly· from a careful study of L. H. Morgan's *Systems of Consanguinity and Affinity of the Human Family*, published in 1871, and he had developed the idea that North American social organization could be analyzed on the model of Australia. On Cole's recommendation I became his research assistant and set out to summarize and organize the ethnographic data to that end.

The Laboratory of Anthropology had recently been established at Santa Fe, and their summer field school for graduate students in ethnology for 1932 was to be on the Hopi reservation under the direction of Leslie White. I decided to apply, and with my acceptance I was committed to social anthropology. In preparation for the summer I read most of the literature available on the Pueblos. I had visited the Southwest briefly in 1930 and had already come under its spell.

There were persistent rumors in the early 1930s that a great foundation was interested in providing funds for a large scale investigation of cultures in various parts of the world. Radcliffe-Brown told me that he had sent in a proposal for such a study before leaving Sydney, and apparently the foundation asked the American Anthropological Association to develop a parallel proposal.

Margaret Mead was then in New Guinea, and in *An Anthropologist At Work: Writings of Ruth Benedict* (1959), she published some of Benedict's letters which provide an account of the atmosphere of the confrontation. In her recent autobiography, *Blackberry Winter* (1972), she provides additional data:

A foundation, as Ruth wrote to me, had offered to give a million dollars for anthropological research. But there was a difficulty: Radcliffe-Brown was to be in charge. When I heard the news I made a beautiful plan. If we really got the money, it seemed to me the most useful thing we could do would be to offer fellowships to do field work to the most promising students in all the social sciences. This would mean that a large number of living cultures would be studied from the standpoint of economics, political science, and psychology and also that a whole group of specialists in the budding social sciences would each know at first hand what a culture was. It would have been a different world had this been accomplished.

But it soon became clear that intractability on both sides was going to defeat any plan for using the money. Radcliffe-Brown, for all his brilliance, was arrogant, dogmatic and dictatorial. And Boas and other leading American anthropologists decided that it was better to have no money than to have Radcliffe-Brown in charge of planning and allocating its use.

In retrospect it is clear that *both* structural-functional studies and historical studies needed to be done, but the efforts of Cole and others to reach a compromise were rejected. The development of sociocultural anthropology was held back for a decade, but the graduate students of the 1930s were the ones who suffered most. There were few jobs during the Depression years and only a handful of anthropologists were able to make studies abroad.

As early as 1901, Boas had outlined his program for comprehensive training in all branches of anthropology as a prerequisite for more adequate field work, and had stated his conviction that he knew exactly what was needed for American ethnology, and for this reason was concentrating control over ethnological work in his own hands. By 1925 he had succeeded beyond his expectations, since his students were in charge of most major departments and he had a strong voice in new appointments. But the view that ethnographic data could be gathered without theoretical considerations, and once enough data had been gathered the answers to theoretical questions would become clear, didn't work out.

It was beginning to be apparent that theory was needed, both to interpret the data and to gather it. Kwakiutl social organization defied easy analysis, and the thousands of pages of texts provided by native assistants only made it more confusing. By 1930 American anthropologists were looking for new and more satisfying paradigms, but they weren't about to have them forced on them by Radcliffe-Brown.

The establishment of the Laboratory of Anthropology at Santa Fe, New Mexico, in the late 1920s provided a research center for anthropologists, and their field training programs represented a great improvement over the older "sink or swim" procedures. Archeological students were accustomed to summer excavations, but even they profited by the presence of students from other institutions and the "field school" discussions. Linguists and physical anthropologists profited mainly from the presence of an experienced leader, but the program was perhaps best adapted to graduate students in ethnology. One result was to break down the departmental isolation and rivalry that had developed, but more important was the beginning of cooperative research and the opening of the field to all competent investigators. Margaret Mead notes that Boas had refused to let her and Reo Fortune work among the Navaho because they "belonged" to Gladys Reichard, and these "ethics of field work" lasted well into the 1930s.

Our 1932 field party consisted of Mischa Titiev (Harvard), Edward Kennard (Columbia), Jess Spirer (Yale), and myself, and we were joined for most of the summer by George Devereux, a French student en route to research in Indo-China. The Hopi Indians had been studied by many anthropologists, since they were more open than most of the Pueblos, in part due to their isolation which protected them from intensive acculturation and allowed them some choice with regard to the acceptance of innovations. Professor White and his wife Mary found us a Hopi house to live in in New Oraibi, and we set out—individually and collectively—to study Third Mesa social organization and what had happened to it when Oraibi split in two in 1906. That split had triggered other divisions, so by the time we arrived Old Oraibi was surrounded by Hotevilla, Bakavi, Moenkopi, and New Oraibi—all variants from a single ancestral pattern only a generation ago. Oraibi, in turn, was only one mesa among three. On First Mesa there were Walpi and Sichomovi, with the Tewa village of Hano, and the new settlement Polacca at the foot of the mesa. Second Mesa also had three communities: Shongopovi and its colony, Shipaulovi, and Mishongnovi. Each of the major villages had its own clan system, lands, and ceremonial system, and while the total population was only some 2500, it was clear that even a team of anthropologists could do no more than scratch the surface in a limited time.

What we learned about Third Mesa—and the Hopi more generally—has been reported in detail elsewhere, and most of our party have continued to work periodically on the reservation. Kennard learned to speak Hopi fluently and worked as a specialist in Indian languages for the Office of Indian Affairs for a period. He is currently completing a volume on the Hopi, with particular reference to modern changes. Titiev returned to Oraibi in 1933–34 for a longer stay, from which he wrote *Old Oraibi*, and more recently he has published his field diary under the title, *The Hopi Indians of Old Oraibi, Change and Continuity*, a day-by-day record of events and activities.

I utilized the data I had secured during the summer to make a comparative study of the social organization of the Western Pueblos, combining what I had learned from Radcliffe-Brown with my training from Cole, Sapir, and Spier, and it was accepted as a doctoral thesis in June 1933. E. C. Parsons borrowed it to read but didn't think much of it, since at that time she thought comparison was only legitimate if the items being compared were from the same historical source. Leslie Spier, on the other hand, not only approved of it but offered to publish the manuscript in his General Series in Anthropology. I was tempted but there were a few loose ends I wanted to investigate first. As it turned out the revised manuscript was not completed until after World War II and was finally published in 1950.

While writing my thesis I continued the analysis of ethnographic materials on the Southeast and Plains areas, along with taking a full program of courses. During the year, Kalervo Oberg, a graduate student with training in economics who had been sent by Sapir to study the Tlingit, returned from the field and we worked through his materials on social organization together. Though we urged him to publish his thesis, he never had the leisure to do the small amount of revising that he felt to be necessary until his recent retirement. *The Social Economy of the Tlingit Indians*, his "splendid study" of Tlingit life, as Wilson Duff characterizes it, wasn't published until 1973, a few months after his death.

The graduate students in the 1930s at Chicago had a variety of responses to Radcliffe-Brown, but I do not believe any of them thought of him as "arrogant, dogmatic, or dictatorial." Of those who contributed papers to *Social Anthropology of North American Tribes*, Sol Tax and Philleo Nash had been students of Ralph Linton at Wisconsin. Tax's contributions, in particular, are largely independent of Radcliffe-Brown, and Nash examined a nativistic religious revival which took place on the Klamath Indian reservation in 1871–78 in terms of hypotheses derived from psychoanalytic concepts. John Provinse, with earlier training in law, tested Radcliffe-Brown's categories for the classification of the underlying sanctions of social control against the ethnographic data from the Plains and found the categories in need of considerable modification. J. G. McAllister and W. H. Gilbert Jr. followed Radcliffe-Brown's teachings mainly in presenting a more adequate account of the social organization of the groups they studied, and Morris Opler was in opposition on a considerable number of points. And speaking for himself, Redfield notes that Radcliffe-Brown's associates in the department "have found him always an agreeable and efficient colleague, a generous contributor of time and effort, and a wise and open-minded counselor," a statement which I can support from my later experience.

There were no jobs immediately available in 1933, but I had run across some interesting problems in kinship among the Choctaw while surveying the literature. With a small grant from the Social Science Division and a borrowed car I set out for central Mississippi, where John Swanton had told me I might find the answers among the Choctaw who had hidden out in the Bogue Chito swamp when the tribe was removed to Oklahoma a century earlier. I was unsuccessful in my quest, though I later found the data I was interested in in some early documents, so I decided to spend the rest of the summer in Oklahoma among the Cheyenne and Arapaho. These tribes had been studied by a number of scholars, but their kinship systems were incomplete and the data in conflict. Here I outlined the terminological structure of each group and related this structure to the behavioral patterns and to the events of the life cycle. These were "cognatic" systems, though we hadn't conceptualized them as such in the 1930s. Noteworthy was the discovery of a systematic organization of joking and respect relationships which went well beyond the mother-in-law taboo. I also attempted a preliminary ordering of Plains kinship systems in terms of two major types and discovered that these kinship systems were particularly susceptible to change in terms of new ecological and social conditions. Later I was able to follow up many of these suggestions in greater detail, but in retrospect it was this early study which established the method of controlled comparison which I later conceptualized more clearly. It avoided many of the difficulties of the comparative method as generally practiced, and it was as interested in differences as in similarities. And having a historical orientation, it was not bound to synchronic formulations, though it did stress social integration. The influence of historical linguistics is apparent, but it wasn't a conscious model, so far as I can tell.

While I was writing a draft of *The Cheyenne and Arapaho kinship system*, Radcliffe-Brown made final arrangements for a postdoctoral fellowship which would enable me to spend 2 years in the Kimberley district in northwestern Australia under the auspices of the Australian National Research Council. Lauriston Sharp came by on his way to northern Australia under the same auspices and we promised to meet in Sydney. Later Raymond Firth came from Sydney on his way to the London School of Economics and confirmed the arrangements.

I still had a small sum left over from the summer's activities so I determined to return to the Hopi for a few months. Titiev was already at Oraibi so I joined him for awhile before moving to Second Mesa. I had very little money and no transportation, but I lived with the village chief of Shipaulovi and able to get an intimate view of the relations between the Second Mesa villages, as well as to see the ceremonial system in action. In February Titiev and I were invited to participate in the Powamu society ceremonies at Oraibi, where we danced as Powamu Katcinas.

John Collier had been appointed Indian Commissioner the year before and, with the Indian Reorganization Act under way, was planning to develop an anthropological unit in the Bureau of Indian Affairs. I was invited to attend a conference he was holding with the Navahos involving sheep reduction and soil conservation, both of which were to have momentous consequences for the future of the Navaho. I was impressed with Collier and his program, and when he asked me if I would be interested in a job in the Indian Service, I was sorry to have to tell him I was committed for the next 2 years.

Or so I thought. When I returned to Chicago in the spring to get ready to go to Australia I was in for a surprise. President Roosevelt had devalued the dollar in March 1934, and the Australian National Research Council got a considerable portion of their funds from the Rockefeller Foundation. In this situation they had no choice but cancel my fellowship. Cole, however, had come to my rescue. He had always wanted to send someone to the Philippines to see what had happened to the Tinguian, whom he and his wife had studied in 1907-8. With the assistance of Redfield, he had drafted a proposal and found the money. It was attractive but I would have preferred to go to Australia. I had been studying other people's tribes, and it would have been fun to have a tribe of my own.

The Philippines in 1934-35 was a colony on the verge of independence. They had been ceded to the United States by Spain in the Treaty of Paris in 1898, but the next year the Filipino leaders, who had organized an independent republic, started a rebellion against the Americans, which was not put down until 1901. Under American control they had built roads, constructed schools, improved health, and developed political skills, and were about to enter a 10-year Commonwealth period preparatory to full independence.

I left in May on the California Limited for San Francisco, and as we pulled out of the station I discovered that my neighbor in the observation car was a medical student from the University of Michigan and my future roommate on the President Lincoln. In Berkeley I visited Harold Driver, with whom I had spent one summer digging up Indian mounds in central Illinois, and he and his wife introduced me to artichokes. We boarded the ship in Los Angeles, where the scars of the recent Long Beach earthquake were still highly visible and the longshoresmen were on strike. I had earlier been a member of the Teamster's Union in Chicago and was apprehensive as to our reception, but we were loaded on a launch and managed to outflank the pickets. The old Dollar Line ran a monthly service around the world, with stopovers permitted anywhere, and we were headed for Japan, with perhaps a hundred passengers. The night before we arrived in Honolulu a fire broke out in a hold full of bales of cotton, and we spent the night on deck. With the fire under partial control the orchestra appeared, and we danced and sang "Smoke Gets in Your Eyes" until a destroyer showed up as an escort. We had had no fire drills so we hadn't been aware that most of the crew were novices and did not have the faintest idea of how to lower a boat.

A woman whom we had met on board invited several of us to join her for a tour of Honolulu. Her son, a captain in the Army, met us and took us around Oahu, ending up at the Outrigger Club on Waikiki where we ran the surf. The captain told us that an invasion by the Japanese was expected at any time, and he outlined the plans for the defense of the island against the invading forces. In the evening we bought a bottle of *okolehau* and went up by the Punchbowl where we watched the lights of the city and harbor and enjoyed the tropical night. We came down to find the burned bales of cotton on the docks and the boat about to sail.

The next two weeks were quite different. The captain, a retired Navy officer, set out to train the crew and we had drills almost every day. Most of the passengers were gone, but the remainder settled down to enjoy the trip. My roomate and I became friends with a Yale undergraduate who was spending his senior year on a trip around the world before joining the family firm, and the three of us decided to spend a month traveling around Japan.

We spent the first night in the old Imperial Hotel but soon gravitated to Japanese inns, learning enough Japanese to make our way. On a dare we climbed Mount Fuji, starting in the afternoon in early June, before the climbing season had opened. With the aid of a guide we got halfway by dark and made it to the top over snow, ice, and lava by the next noon. A team of Japanese meteorologists, who had been learning English from phonograph records, fed us and showed us around, and in turn we helped them with English syntax. In the afternoon we slid down the snow fields and then traversed the cinder fields with giant strides. We found a taxi at the base of the mountain and got to our inn, where we soaked in superhot water until our aching muscles relaxed.

We separated in Tokyo and I boarded the President Cleveland for the trip to Manila. We stopped at Shanghai and I thought longingly of the interior of China, but my conscience told me it was time to get to work. Manila before the war was a fascinating city, with the medieval Spanish city still intact behind its walls and moats, the modern Western residential areas along the bay, and the predominatly Chinese areas with their great markets. I went to see H. Otley Beyer, the one remaining anthropologist in the Philippines, who had already filled up two houses with his collections and library and was living in a third. He took me in charge and outfitted me in white cotton duck for Manila and brown cotton for the field. Manila has the worst climate in the Philippines, and soon after my arrival the first of several typhoons roared across Central Luzon, flooding the landscape and washing out bridges.

Beyer had come to the Islands in 1905 as a teacher and had been sent to Banaue to study the Ifugao and start an industrial school. When the University of the Philippines was organized he transferred from the National Museum and began a career which lasted until 1965. Except for a year at Harvard with Roland B. Dixon, he was largely self taught, but he had acquired an encyclopedic knowledge of the peoples and cultures of the Philippines, and he knew everybody of importance, whether Filipino or American.

Almost single-handed, Beyer had collected and preserved records, manuscripts, reports, and other documents, as well as books and pamphlets relating to the Philippines, and was in the process of collecting archeological materials which had recently been uncovered, along with tektites which were thought to have rained down from outer space in mid-Pleistocene times. He also had his classes in anthropology write essays on marriage and related practices in their home communities, but these accounts needed to be put in context before they were usable. Beyer doled out information as though it consisted of gold nuggets, but he was always interesting and informative and I spent as much time with him as I could manage.

Through Beyer I met a number of Americans and Filipinos, notably J. R. Hayden, a professor of political science on leave from the University of Michigan, who was Vice-Governor General, and who took time to introduce me to the political realities of the Philippines; also Dr. Paul Russell of the Rockefeller Institute, who had just worked out the vectors for malaria in Northern Luzon and told me enough about the habits of *anopheles minimus* so that I was able to escape malaria.

I was impatient to get to the field, and after the typhoon damage was repaired I set out for Abra Province by train and bus. Abra lies in the bend of the Abra river and includes most of its drainage area, with foothills and ranges on three sides. Up to the beginning of the 19th century Abra was inhabited primarily by pagan Tinguians. The more populous Ilocano coast had been pacified and christianized by the Spaniards in the sixteenth century, but the interior peoples had resisted acculturation until the mid-nineteenth century when missions were established and Abra was made a province. Much of the lowland Abra region was occupied by "Ilocanos," who had earlier been Tinguians, and the process was still continuing. In the foothills and mountains, however, the Tinguian still maintained much of their old life, and for the most part felt themselves superior to the Ilocanos in their midst.

I spent a year in Abra, learning some of the language, collecting data on every aspect of Tinguian life, and paying particular attention to social and cultural change. I resided for much of the time in a barrio of Patoc—renamed Peñarrubia by the Spaniards—where my principal mentor was Dumagat, the son of a headman whom Cole had brought to Chicago to help him with setting up exhibits at the Field Museum, and who had then stayed in America until the onset of the Depression. I later lived for a considerable period at Manabo, further up the Abra river, and visited practically every community in Abra.

The Tinguian, like other Philippine peoples, are not organized in "tribes" but rather in what we came later to call ethnolinguistic groups. The Tinguian communities were divided into barrios, grouped around a center, and they could grow by accretion as well as by population expansion. There were differences from community to community, but no sharp boundaries in either culture or language. Both locality and kinship were important, and kinship was cognatic and wide ranging, with marriage being arranged with distant cousins to maximize property holdings or establish alliances. Headhunting had been greatly reduced and had been partly replaced by a complex system of peace pacts on the model of marriage. Both irrigated and dry rice were grown, and there was an elaborate ceremonial system centering around wealth and prestige. Malaria was endemic and illnesses were believed to be caused by spirits who later were "called" by women mediums and made their demands known. The myths and legends the Coles had collected were still known, but new ritual myths were being added.

In Cole's time the mountain interior was little known and he had come to the conclusion that the Tinguian and Ilocano had a common origin, with the Tinguian representing the type of life before European influence. But I began to discover that there were also important similarities with the northern Kalinga. The Kalinga had spilled over the mountains in recent times and still intermarried with their relatives. It was possible that the lowland Tinguian were—in part at least—the result of an earlier migration from Kalinga, which had then been acculturated by long contact with the Ilocano of the coast. In southern Abra influences from the Bontoc region were also apparent, and I discovered one community that was half Tinguian and half Bontoc, with separate customs and institutions. It began to be clear that further research in the mountain province was essential to solve these basic problems.

In January I traveled over the Cordillera with a group of Tinguians. Beyond the range of kinship we were under the protection of peace pacts until we reached Salegseg, where we came onto the Mountain Trail. At this point the Tinguians sat down and sharpened their bolos—they were in the land of enemies and anything might happen. I left them at Lubuagan and went on by bus to Bontoc, Banaue, and Baguio, through some of the most magnificent scenery in the world. The rice terraces at Banaue are one of the few places where man has improved on nature, and I had seen enough of the mountain area to determine to return some day. When I finally did get back in 1949-50 I not only found answers to my earlier problems but a great deal more besides.

Beyer had invited me to the Second Congress of Prehistorians of the Far East, which brought together the European scholars of the various colonial countries in

Southeast Asia, and he immediately hauled me off to a Chinese tailor to be fitted for a dinner jacket—an item I wore just once. The Congress was interesting, but there was a carnival in progress and much social life in Manila, and the food at the French Restaurant almost obliterated the remembrance of rice three times a day.

I returned to Abra refreshed and ready to go back to work. I never felt lonely or alienated among the Tinguian, for they did their best to teach me their customs, putting me under the protection of their spirits and inviting me to their ceremonies, both public and private. I found Cole's accounts to be remarkably accurate, but the new road into Abra was providing an outlet for their rice surpluses and introducing more trade. Ilocano merchants provided a weekly market, though the Tinguian themselves continued to exchange goods in traditional ways. Their greatest problem was malaria, which flared up seasonally and kept the population reduced. The rituals and ceremonies were largely centered around the curing of illness, through the major ceremonies also involved the maintenance and enhancement of family prestige. I had expected the more isolated mountain villages to maintain the ceremonial cycle in greater purity, but they didn't have the surpluses of rice and animals necessary to carry them out. In fact, some were waiting for the arrival of missionaries which would give them an excuse to cut down on the more costly ceremonial activities.

I left the Tinguian, half convinced that I could be happy there indefinitely, and intending to return periodically to see how they were getting along. A few years later, however, the Japanese invaded the Philippines and many Tinguian fled to the mountains, to die of disease or exposure. When I finally returned after the war the survivors were reduced to subsistence activities, and much of the color had gone out of their lives.

While I was getting ready to pack up I received word that I was being offered a position as instructor at the University of Chicago, my time to be divided evenly between the Extension program and the department. I had the remains of a round-the-world ticket and I had looked forward to a leisurely journey home, but my apprenticeship was over and I was eager to get started teaching.

In the 5 years since I had returned to the university to work for my PhD I had shifted from archeology to social anthropology and had carried out field research among a variety of groups: Hopi, Choctaw, Cheyenne, Arapaho, and Tinguian. All of these had been studied by experts, but I discovered that it was still possible to work on new and interesting problems, and I never thereafter worried about anthropologists running out of peoples and subjects to study. Earlier studies, in fact, provided not only a larger field but data which might confirm or modify the conclusions that I was reaching and might provide evidence of possible change.

Teaching is hard work and I will soon have completed some 40 years of teaching—most of it at the University of Chicago. World War II shook us up and provided new experiences. I had been engaged in a pilot study of food and nutrition for the Department of the Interior when the war broke out and soon after went to Washington, D.C. to work for the Board of Economic Warfare. From there I went to the Philippine government-in-exile where I was engaged in research activities for President Quezon, working with several Filipinos I had known in happier times. After a period in the Army at the School for Military Government at Charlottesville, Virginia, I was sent to Chicago to develop a Civil Affairs Training School for the Far East, which ran for two years and provided me with more administrative experience than I ever wanted. I was briefly a Cultural Relations Officer in the Foreign Service assigned to the Philippines, but after I discovered I was not likely to get out of Washington for several months I resigned and returned to the University.

SOME FUTURE DIRECTIONS

Kierkegaard says somewhere that life can only be understood backwards but it must be lived forwards, a paradox that some of our existentialists have resolved by living only for the moment. But a discipline cannot live from day to day. It needs a philosophy and a direction. The postwar expansion of anthropology is so changed in almost all respects from the prewar period that it seems a different world, and it is for that reason that I have ventured to present some aspects of my own early training and experience.

In the 1930s anthropology was a way of life that enabled its practitioners to escape the worst features of American culture. For anthropologists in the Midwest the center of gravity was shifting from museums to the state universities and the unity that was anthropology was beginning to break up. With the increasing professionalization of anthropology the career lines continued to diverge, but became increasingly standardized as the production of PhDs went from a dozen in 1933 to over 300 in 1973.

The common purposes that characterized sociocultural anthropology in the early postwar period have now largely disappeared as well. A decade ago at the Conference on "New Approaches in Social Anthropology" sponsored by the Association of Social Anthropologists, held in Cambridge, England, it was clear that the younger scholars in both the United States and England were developing in similar directions, but this no longer appears to be so true, if we can judge by the papers for the Association's recent decennial conference on "New Directions in Social Anthropology."

The decade of change began with Edward Leach's *Rethinking Anthropology* (1961), his postwar essays which have done so much to reorient social anthropology in the directions of structuralism and Lévi-Strauss. He proposes to treat societies as man-made mechanisms rather than organisms, and he suggests: "Instead of comparison let us have generalization; instead of butterfly collecting let us have inspired guesswork." Since generalization "consists in conceiving possible general laws in the circumstances of special cases," he is back to Malinowski in part, though his generalization is more sophisticated as well as mathematical.

Lévi-Strauss, in his study of *Totemism*, published in 1962, called attention, however, to what he calls Radcliffe-Brown's second theory of totemism, presented a decade *earlier* in the Huxley Memorial Lecture for 1951, "The Comparative Method in Social Anthropology" (*Journal of the Royal Anthropological Institute*

LXXXI, 1952). Here, Lévi-Strauss finds that Radcliffe-Brown's formulation "brings about a reintegration of content with form, and thus opens the way to a genuine structural analysis, equally far removed from formalism and from functionalism. For it is indeed a structural analysis which Radcliffe-Brown undertakes, consolidating institutions with representations on the one hand, and interpreting in conjunction all the variants of the same myth on the other."

Lévi-Strauss was interested in whether Radcliffe-Brown was aware of the gap between the earlier "Sociological Theory of Totemism" and the second theory, and suggested that the answer might be found in his unpublished lecture notes or manuscripts. My colleague, Milton Singer, has become interested in this problem and has found clear evidence for the development of this structuralist position in Radcliffe-Brown's prewar publications where he utilized the logic of relations, as developed by Whitehead and Russell around the turn of the century.

The events of the 1960s, however, have been such that these seminal contributions have not been followed up as completely as they might have been, particularly in the United States. A decade of student unrest and discontent fired by the Vietnam War and related events has led many students of anthropology to seriously question what they are doing, and whether it is worthwhile.

Reinventing Anthropology (1972) a Pantheon antitextbook edited by Dell Hymes, brings it all together as various radical anthropologists and other scholars discuss the politics of power and the responsibilities of ethnographers. Dell Hymes suggests that if the study of man were being invented now, there would be no need for anthropology as it currently exists in the United States, and he goes on to critically examine our present organization and traditions. Many of the contributors are far more radical, however, and are quite willing to destroy anthropology in the belief that any new activity can only be an improvement.

Most anthropologists would rather repair anthropology than burn it down, however, and I would like to mention a few recent volumes which are both critical and constructive. Their variety indicates that anthropology is in little danger of falling into a rut, and their brilliance suggests that our best thinkers are still on the firing line.

One such volume is Robert Murphy's *The Dialectics of Social Life* (1971), which he subtitles *Alarms and Excursions in Anthropological Theory*. The "alarms" have grown out of the fact that he found orthodox theories inadequate to account for the data he secured from the Mundurucú and the Tuareg, and the "excursions" have been made in an attempt to find more satisfying answers to what is happening in culture—our own included. Murphy begins with an analysis and critique of functional theory, and suggests that the underlying causes for change in our theories are to be sought in the transformations of society. Thus he follows Talcott Parsons, rather than Thomas Kuhn, in postulating that functionalism arose in a time of relative stability, order, and complacency, and is a casualty of the profound disturbances of the recent period, a view that suggests a new role for "social ecology." This is an important and tightly argued volume that proposes a dialectical mode of analysis between functionalism and phenomenology, and emphasizes the contributions of Simmel and Freud, as well as those of Lévi-

Strauss. It also has many perceptive things to say about American anthropology as it is practised in our universities.

Robin Fox's *Encounter with Anthropology* (1973) is a more personal account of how he happened to become an anthropologist and a summary of his contributions to the study of kinship and marriage, Irish culture, and the pueblo of Cochiti. It is written with style and humor, but Fox is worried that anthropology is in danger of being shunted onto a siding as the evolutionary express whizzes by. He accepts the British verdict that anthropology is composed of odds and ends "held together by sentiment and dynastic interest," and he proposes that anthropologists who really care can be "mad amateurs foraging in the dawn of new discoveries." His own "foraging" has been in the directions of the biological basis for human behavior, and he sees the "anthropological ethologist" as a new emergent. And out of the study of the human condition and its evolution should come our moral philosophy. As our mentors for such study he lists Darwin, Marx, Freud, and Lévi-Strauss, each of whom has made an important contribution to the "understanding of the human species," and he suggests that anthropologists can go still further and start by restoring man to nature. After a century, he concludes, it is time to take Darwin seriously.

In contrast to these attempts to remodel anthropology in terms of other disciplines, Clifford Geertz' *The Interpretation of Culture* (1973), a collection of his essays written over two decades, is closer to the center of anthropology. His initial chapter, Thick Description, defines his present position with regard to culture as a system of shared symbols which give form and meaning to experience, and he has selected 14 of his essays which range over a wide variety of topics. Cultural anthropology for Geertz is not "an experimental science in search of law but an interpretive one in search of meaning," and he applies his insights with sophistication and style. My own favorite is Deep Play: Notes on the Balinese Cockfight, where he finds the whole Balinese social world condensed into the cockpit, but there is something for everybody.

Margaret Mead, in the Epilogue to *Blackberry Winter* (1972), asks the question: What is there for young anthropologists to do? Her answer is that the best work has not yet been done: "If I were twenty-one today, I would elect to join the communicating network of those young people, the world over, who recognize the urgency of life-supporting change—as an anthropologist."

The variety of recent answers to the question: What is anthropology and where should it go? suggests that the discipline is still wide-ranging and in a healthy state, and I think the reviews which follow will reinforce that conclusion. What is more uncertain is the support which anthropology can expect from the government and the public generally. Having experienced the Depression, I don't think this is a serious handicap. There is no necessary rate at which anthropology has to grow, and there is no evidence that our predecessors' ideas were less adequate because they had no federal support. There will be increased competition for fewer positions so that we will have to discover new ways to cooperate if anthropology is to continue to prosper. My strong impression is that anthropology is not only viable but is in good hands.

THE EXCITEMENT OF SCIENCE   189

Ann. Rev. Pharmacol., Vol. 12

SCIENCE AND FATE

Oskar Eichler

Ann. Rev. Pharmacol., Vol. 12

SCIENCE AND FATE

6525

OSKAR EICHLER

*Professor Emeritus, Pharmakologisches Institut der
Universität Heidelberg, West Germany*

When the life story of a German pharmacologist is related in conjunction with his scientific work, it also involves a narration of the difficulties which beset German Science during the decades following the First World War. The political events were often so overwhelming that one could hardly escape them.

Before the First War, everything ran smoothly. I myself, born in 1898, lived in a wealthy businessman's house in Gilgenburg, a small town in East Prussia, amid lovely surroundings and lying between lakes. In spite of its remote situation, I came in contact with scientific personalities early in life, because two close relatives (the physico- and photo-chemical Professor Luther, and the food scientist, Professor M. P. Neumann, known as Bread-Neumann) spent their holidays with us. It was thanks to them that I, at nine years of age, was able to go to the nearest "Humanistische Gymnasium" (High School) of the district town Osterode. The quiet pursuance of my education was completely disrupted by the beginning of the 1914 War. As my small home town lay only a few kilometers from the Russian border, we learned the horror of war from first-hand experience. Crossing the battlefield, where many dead lay, left a terrible impression on me, without my realizing when only 16 years old, that I myself would have to take part in these events as an active soldier two years later. In 1917, in one of the greatest battles in France, I came in contact with chemical warfare, something not without influence in my choosing the profession of Pharmacology and Toxicology, and an interest in this subject continued. After the end of the war, I began medical studies in Königsberg and Munich, with well-known and respected teachers. After the "Physikum" (pre-medical examinations), I wanted to change over to Chemistry. I took several semesters in the major chemistry practicals with success. But times were uncertain. As a doctor, one can always hope to be employed in the most desperate situations; not so as a chemist. The situation was difficult for us. There were days when we had practically nothing to eat. My father had completely lost his fortune as a result of the inflation.

I did my "Doktor-Dissertation" on perchlorate, under the supervision of Professor Hermann Wieland in Königsberg. This was a subject to which I always returned. In the so-called "Hofmeister's Reihe" (lyotropic series), I

1

THE EXCITEMENT OF SCIENCE

8243-2601/78/1127-0193$01.00 © 1978 ARI

classified the position of perchlorate beyond the thiocyanates, which until
then had occupied the last position.

In the Institute at the same time were Berend Behrens and Paul Pu-
lewka, with whom firm friendships were formed. During difficult times, that
happened more easily, as we found out yet again in the years after 1933.
During almost the whole of the year 1924, I had to work as a medical practitioner due to lack of money. House doctors could occasionally have lunch
in the hospital and even for that they were grateful. In contrast, I did well
in general practice. Nevertheless, at the end of 1924, I accepted Wieland's
invitation to return to his Institute, without pay. That time spent in the Institute was very stimulating. In 1923, Behrens began to work with isotopes
of lead, but no money was available and he had to abandon these experiments. And so Hevesy remained alone in this field for the time being. Frogs
and mice were practically the only experimental animals we could afford
and we had to build our own apparatus.

From April 1925 until the end of the year, I worked in the Medical Clinic
of Professor Krehl, in Heidelberg. This stimulating time was the nicest of
my years as an Assistant, not only because of the brilliant personality of
Krehl himself, but also because he would discuss anything, even his own
mistakes. He was great enough to admit his own mistakes or to seek the
advice of young assistants in the field of pharmacology. A few future
Chair-holders, Professors Reinwein, Bohnenkamp, Dennig, and Hoppe-Sey-
ler, who lived in the Assistant-wing of the hospital, guaranteed high levels
of discussion. At that time, Bohnenkamp had completed his work on heat
production by the frog heart. Hill's measurements were awarded the Nobel
Prize, showing how highly these results were esteemed. Unfortunately, I
later discovered that the method was not foolproof.

External conditions influenced my themes of work again and again. My
work deviated from a chemical theme when I went to Düsseldorf in 1926, to
Professor Hildebrandt's Pharmacological Institute. There, the money available was greater than I was ever to have again, though it was unfortunately
administered through dreadful bureaucracy. This favorable financial situation enabled me to carry out many costly experiments, using cats and dogs,
only for the sake of methodology (e.g. operations on the central nervous
system, Starling's preparations etc.). Much work was done on Cardiazol
(Metrazol), as well as on acetylene, which had been introduced by my first
teacher, Wieland, as an anesthetic but had not yet been investigated with
respect to its actions on the circulation. In spite of its minimal effects on the
circulation, acetylene could not be used as an anesthetic because of explosions. We tried, in vain, to find something that would make it nonexplosive.
At this time, I came in contact with the American Anaesthetists' Association.

During our diverse and many-sided experiments, the different characteristics of research became clear to me. One can explore a subject systematically, an approach favored today, if only on economic grounds. Research

workers who use this approach cannot be overlooked. Here, I am thinking of Feldberg or Koelle in acetylcholine research (of course, also Loewi and Dale), of Euler and Holtz in work on adrenaline, Lendle and Repke on digitalis, etc. One can work this way in a small field, for even in the smallest, one is still guided by all the laws of nature.

The other approach is that of the adventurer (*Adventures in Biochemistry*, a book published in 1931 in honour of Hopkins). In this way one may discover a great deal right away; important discoveries are frequently made but luck also plays a part. A particularly successful research worker using this approach is Selye. But danger lies therein: quisquis ubique habitat nusquam habitat (Martial).

Perhaps I myself had a tendency to work this way but benefited less by luck. However, in the initial stages of my work, purely theoretical ideas played a role. At this time, I reasoned out the spatial propagation of chemical reactions in a homogenous system [something like a model of nerve conduction (Meinecke 1)]. In a heterogeneous isothermic system, for example, in the super-cooling of molten substances, it is a frequent occurrence. Reactions that are spatially propagated are autocatalytic. In this way, the oxidation of oxalic acid by permanganate takes place. More easily understood is the oxidation of arsenious acid via bromate. If some potassium bromide is present in the solution, then the reaction proceeds as follows: $BrO_3^- + Br^- \rightarrow BrO^- + BrO_2^-$, which then converts the arsenious acid to arsenic acid. When a small amount of potassium bromide is added to the solution in a test tube the reaction begins immediately and propagates right through the tube, at first slowly and later at uniform speed, as long as substrate is available. The limiting factor is the diffusion of the newly-formed catalysts; thus a slower reaction at the beginning. One can formulate an approximate equation for the process: $v = a \sqrt{C.K.D.}$. For nerve conduction, this model equation gives much too slow a propagation speed. C cannot be increased and K is the reaction constant. The limiting factor is D, the diffusion coefficient. If one applies this to a heterogeneous system, as in a living organism, the diffusion on surfaces may be enhanced, as in the experiments of Scholander (2) and Ens (3) who showed that the diffusion of oxygen in the presence of hemoglobin, for instance, in membranes, could be increased eightfold, a result which could not be achieved by the partial differential equation alone. Also, if one observes the speed at which oleic acid propagates on the surface of pure mercury, what possibilities may exist? One would expect that the reaction speed could still be increased. That would be so, but for a further postulate—namely, that the reaction takes place continuously but the catalyst is constantly being removed as soon as it is formed. Only if an initial stimulus provides a greater amount of catalyst, which exceeds the capacity of the metabolic processes, can a reaction wave proceed over the nerves. At that time, Gerard, with Hill and Hartree, measured heat production in nerves and found continuous heat production, which was increased only slightly by stimulation. This semed to agree with these ideas.

Nevertheless, these reflections were abandoned and another example was sought for an autocatalytic reaction. I believed then that blood coagulation could be such a process. However, attempts to construct suitable apparatus to pursue this idea had to be abandoned because the time in Düsseldorf had come to an end. Everything remained at the theoretical level. "It lies not to me the truth to proclaim, but the truth to seek, and to reveal it for discussion" (according to Goethe).

Experiments were then begun in two different fields. We demonstrated tachyphylaxis of histamine, but further pursuit of work with histamine was forcibly interrupted again and again. Another line of thought led to the idea of teleology, which is a horror to the natural scientist after the philosophy of nature of Schelling and Hegel. But for me, the important statements by Kant in "der Kritik der Urteilskraft" were constantly relevant. Kant said that objective thoughts can represent only a regulative principle. Thus, to determine the real, ultimate objectives, one must have an overall understanding of the aims. Goethe called this one of the fundamental faults of the Greeks, e.g. Aristotelians, that they immediately jumped from an observed phenomenon to an explanation as psychoanalysts often do today. We judge only by the nearest observation, consequently we judge imperfectly and in only a temporary and therefore regulative way.

We tried to eradicate teleology from these thoughts using the Principle of le Chatelier. It is derived from an inanimate system and its statement denies any trace of teleology. Also, the conception of regulation should not appear, although elements of entirety are inevitable. Considering the analysis by Jendrassik (4), perhaps the idea of opposition is more readily applicable although it deals only with a descriptive connection. For example, in hypocalcemic tetany, calcium released from the muscles relieves the tetany. Indeed, some contracture of the muscles is connected with it, but this is a rapid reaction. Due to the tetany, lactic acid is formed which increases ionisation of blood calcium with consequent delayed calcium release from the bones which has a lasting effect. Such antagonisms are found in poisoning by sodium chloride and sodium iodide, both of which exhibit only the regulative principle (O. & L. Eichler, 11).

Still another trivial but well-known example: after an infusion of glucose, there follows a secondary release of insulin, which lowers blood sugar to levels below the starting point. Such effects are to be found with every action of a drug, being not fully congruous with the concentration of the drug, so that one should expect evidence of an opposing effect, as with the dose of glucose.

Following this principle, an action of adrenaline on release of histamine was sought and found. After a dose of adrenaline, a secondary hyperemia was observed not as a result of release of histamine from mast cells but according to Schayer, and often confirmed, by activation of histidine decarboxylase.

These thoughts were pursued further in Giessen where I went in October 1928 with my chief, Hildebrandt. There working conditions were very much more difficult and disclosed all the economic difficulties of the time. The budget was minimal. Without support from industry or from the "Notgemeinschaft der Deutschen Wissenschaft" (from which only limited funds were available) work would not have been possible. I was the only Assistant, and if research students applied, no technical assistance was available. It was quite impossible to pursue the expensive experimental projects of Düsseldorf. Also, that was not expedient considering the construction of the Institute. There was one large room directly above the lecture theatre, with a poorly supported floor. Every step caused swaying so that sensitive recording could not be used. Then there was a small room which served as a library (Naunyn-Schmiedeberg's Archiv and Ronas-Berichte were the only Journals), and balance room, and also housed a Leitz colorimeter. Another small room was suitable for chemical experiments. Therefore, there was a spontaneous turn towards chemical problems, with the exception of primitive toxicology studies using a combination of narcotics and analeptics.

Now the mathematics which had been formerly nurtured came to the fore. When I was with Krehl, I had already instructed some Assistants in differential and integral calculus. Now I studied the theory of functions, topology, elliptical integrals, partial differential equations, and calculations on variations and integral equations. One might rightly ask why I sacrificed so much time. I was attracted by some results of Hartridge and Roughton on the shape of erythrocytes, which is an interesting and relevant problem. In a rotating body, the oxygen molecule must attain the greatest possible saturation in the shortest possible time. Naturally this would be accomplished with the given amount of hemoglobin being in a flat body which, however, requires a greater surface area. This must also represent a minimum for economical reasons and therefore becomes a secondary condition; the problem of Dido. It seemed to me that a relationship between the structure and the dimensions of the pulmonary capillaries must emerge for each individual animal species. Many calculations concerning this were later lost and could not therefore be published.

This made me ask whether such an extensive pursuit of these problems is necessary for an individual research worker. To collaborate with a mathematician is always necessary but, in my experience, fruitful collaboration is difficult unless the researcher has some knowledge of mathematics. A mathematical exercise also serves as a training in the formulation of ideas, quite apart from the esthetic pleasure of the thoughts. I personally could not derive corresponding profit from the time spent on it, because after 1933, life simply did not allow uninterrupted work. But sometimes, one cannot otherwise encounter challenges. For example, with analytical functions of the theory of functions, if one knows the value of a function for the border of a surface or a space, one can then calculate any value for the inside. Could that also be possible one day for the structure of a cell? Anyway, this idea led me to investigate some of the chemical properties of the cell surface and to collect data, for example, from the surface of heart muscle cells where copper, magnesium, manganese, alkaline phosphatase, etc are bound. Using complexing agents, it was possible to release copper etc from the surface, or to inhibit the action of enzymes that require metals as coenzymes. By investigating some organic fluorine compounds, we found substances that formed complexes with manganese and magnesium and that inhibit alkaline phosphatase.

In Giessen during the summer semester of 1930 I qualified myself for teaching. The work was concerned with the anions in the lyotropic series. In this series, I added perchlorate, next to thiocyanate, as the final inorganic anion. In frogs, the symptoms of intoxication due to these two ions are practically indistinguishable. In general, there are close similarities as, for example, demonstrated by ion displacement in the thyroid gland. This also occurs in other organs that actively transport iodide, for example the bronchi.

I was specially interested in the mathematical formulation of the mechanism of poisoning, i.e. action depending upon concentration. As in the exact sciences, according to E. Mach, it is usual and easiest to express an effect proportional to an agent (diffusion, heat conduction). According to chemical kinetics, this assumption is often applicable to living objects as in the extensive application of the Michaelis-Menton equation in biology, especially as formulated by Lineweaver and Burk. These calculations are valid but only when one function is observed and for short observation times; whenever a longer period is chosen, the action is not proportional to the concentration in the whole animal. Also, after quantitatively-excreted substances, a slight action still remains. Today, it is usual to describe excretion according to half-life. Finally, one can express any curve as the sum of exponential functions. To explain it, the individual components are chosen arbitrarily. But if one reckons half the original concentration as half-life, then one is wrong because the next half-life expires in quite a different way. With the excretion of perchlorate in humans, I found practically only one component.

In accordance with physical nomenclature, e.g. with elastic bodies or magnetism, I proposed the expression hysteresis. Heubner proposed two expressions: pathobiose and allobiose. These would certainly be satisfactory for something like allergy or chronic poisoning, but for acute poisoning, hysteresis seems better to me. Thiocyanate and iodide were demonstrated as typical poisons with hysteresis (Eichler (5)). The expression presumes a proportional relationship between action and concentration. During the phase of excretion, a discrepancy occurs which is hysteresis. Here, there exists a link to the Ct-poisons which were introduced in chemical warfare and which are now used as carcinogens. Probably all poisons that cause mutations belong to this group (see Druckrey and Schmähl). However, this concept was rejected at a Colloquium in Geneva, 1969.

Also rejected, by Clark, was an idea that Straub had introduced following experiments with muscarine on hearts of "aplysia." The poisoning would then be visible only when the substance was entering or leaving the appropriate organ. Here also we can assume that action is proportional to the concentration, magnitude depending upon site and time. In local regulation, it would be helpful to know concentration changes with time at various sites in the tissue. If a regulation occurs quickly, then an effect will be visible only during absorption. If the drug in question acts after absorption and binding to tissue it cannot have an immediate effect. For example, we have demonstrated (with Sebening) the absorption of atebrine in thyroid tissue. Fluoride behaves similarly, being taken up in the apatite crystal of the bones. The question is whether Straub's idea is congruous with these findings. One can develop ideas in accordance with which an action will occur, if two stereoscopically adjacent elements of cells contain different concentrations of a drug. Here I would like to quote from Goethe: "Theory for its own sake is without use insofar as it makes us believe that connections between natural phenomena exist." This quotation applies also to the following.

In my studies at that time, I arrived at various conclusions derived from statistics. If one tests the sensitivity of a large number of isolated organs, then one can always record a spread of the effect. If these single effects on the organs were not interdependent, the spread must ultimately increase tremendously because the individual squares of the variations are added together. Since this contradicts experience there must be connections that counteract. One cannot be content with a statement and call it a general regulation or axiom. This brushes aside the problem in which regulations on basic structures play a role, as in cases of poisoning. So, analysis of a collective deals with the question of why some animals are more sensitive than others. Concerning this, in 1930, I began testing the toxicity of thiocyanate and iodide on frogs and found that poisoning with a two-molar solution was significantly greater than with half the concentration. The osmotic pressures alone are much less effective than those produced by the corresponding concentrations of sodium chloride solution. We concluded that the primary site of action in the muscle should be sought (naturally, the central nervous system might also be affected but these anions are taken up to a lesser extent there). In detoxication, excretion of iodide takes place via the kidneys and the water available is an important factor. With a lower concentration of solution, some decrease in toxicity might be expected but not to this extent. The findings could be explained if the following correlation holds: frogs that are sensitive, are poor excretors. Those that are insensitive require less fluid for excretion. Later, another finding indicated that muscle was responsible in part for the improved excretion. The muscle fibers increase in volume and the extracellular space decreases. The concentration in blood remains the same but the surface of the muscle fibers increases, resulting in a less dense occupation of the surface. And so animals reacting in such a way are poisoned less easily. This decrease in extracellular space will also encourage excretion.[1] Nevertheless, all possible factors must be considered. Better functioning of the kidneys is necessary. In a later publication, the above statement was confirmed by various means. But direct proof was lacking because we could not determine the excretion ability of uninjured frogs. Obviously, the problems investigated were extremely theoretical and for that reason I began to search for compensatory work with direct practical applications as was called for in science after 1933. To satisfy this need, I lectured on air raid protection and did some organizing work from autumn 1933 to summer 1934. For that I was aptly suited because of my previous lectures on chemical warfare. Further time was spent on planning the reconstruction of a hut as a Pharmacological Institute. I did most of the planning but gained no advantage from it.

In October 1934 I was called to Breslau and took over the Institute that had been formerly administered by Professor Riesser. I managed to keep Riesser almost two years in the Institute. I placed one floor at his disposal until he was appointed to a research post in Switzerland. His work was very different from mine so that the Institute had practically no apparatus of use to me. The research allowance was 8,000 DM per month, for the entire staff. What was saved on teachers and research was spent by the Government otherwise. If I had not obtained help by Straub, from Hoffmann-La Roche for the acquisition of apparatus, work would have been absolutely impossible. Also, I obtained further aid from "Kaffee-Handel" and consequently I worked on coffee problems. So I had to work for industry, since they provided the money, and I was, in a certain sense, bound. We could not use the so-called "Schwarze Kassen" which had formerly been for financing research and paying technical assistants. According to the original ordinance of the Kultusministerium, industrial donations to Institutes had to be given to the University administration, which could undertake their distribution arbitrarily. When that happened resources dried up because the donors wished to be certain that useful results would be obtained from those supported. This ordinance therefore had to be rescinded.

Our statistical problems mentioned previously were pursued further from another point of view (Eichler & Smiatek 6). I concluded that with similar substances with approximately the same site and mode of action, the position of each animal in the curve of the collectives must be the same. To test this hypothesis we used three anesthetics, chloroform, Avertin (tribomoethylalcohol), and Eunarkon (a short-acting barbiturate). The doses were so chosen that from a group of 400 rats, around 50% were anesthetized. Further, a correlation of the two alternatives was calculated between

---

[1] From these experiments, based on general experience, I concluded that (with very few exceptions) inorganic anions are to be found only extracellularly. This opinion was later disproved by Conway. I had the proofs at hand that anions were able to enter muscle cells reversibly, but I was concentrating on another problem and did not pursue this important fact.

the substances. Between chloroform and Avertin the correlation was une-quivocally positive. Both substances apparently act on the same structure as one would expect according to the theory of narcosis. Quite contrary to this, there was no correlation between Avertin and chloroform on one hand, and Eunarkon on the other. We concluded that they acted on different structures, which would comply with the conception of Pick on Eunarkon as a brain-stem anesthetic.

We then took up another question: whether the animals that were anes-thetized (lying on the side) with the smallest dose were also the first to die. At death, the respiratory center is depressed, therefore one can expect a smaller correlation or none at all. With chloroform, the correlation coeffi-cient amounted to only $+0.40$. With Avertin, no connection existed at all; a large number of the animals that were anesthetized with the chosen ED50 dose, did not die with the chosen higher dose, and vice versa. We see that the so-called therapeutic index itself shows a collective, i.e., there are ani-mals with a higher and others with a lower tolerance to anesthetics. Also, between Avertin and chloroform there was no correlation, which indicated, according to our way of thinking, that they act on different structures, to cause anesthesia and to depress the respiration center. This interpretation reveals the single mindedness of our thinking in which, using statistics, we could find no more than a comparison on significance. Today, one would say that various different receptors in the brain were attacked.

Sometimes it happens that earlier "sins" recur yet again. Thus, while still in Giessen, I once read in a newspaper about harmful effects of caffeine on the reproduction of rabbits, with misleading publicity about it. I saw at once that almost fatal doses were being used. It seemed clear to me that a correction was essential. Our investigation was made with a dose (0.1g/kg) (8) which was still high, but endurable over a longer time in rats and was repeated without interruption over four generations without an adverse effect. This type of problem has assumed great importance during the last decade. In 1936, I again began experiments on coffee. Because of these ex-periments, I was invited to give the main report at the meeting of the Phar-macological Society in Berlin, 1938, on this theme. All this led to my book "Kaffee und Koffein" (9) and now the second edition is being prepared. With the financial support of Darboven (Hamburg), my co-worker Profes-sor Vollmer began to investigate the action of diuresis on metabolism of minerals as a regulation opposing the increased excretion of sodium, etc. There was at that time no method for estimation of adrenocortical hormones. To obviate this difficulty, I again began to work on my earlier ideas regard-ing regulation. I concluded that if the caffeine dosage is suddenly stopped after a long pretreatment with caffeine, an induced counter-regulation should continue and become perceptible. Also, the plan was to introduce in Germany a method which was used in the U.S. with much success using a certain balanced mixture of inorganic salts from which many important results are still expected. With the much more effective diuretics available nowadays, this would be more effectively demonstrated.

In a concurrent set of experiments, we began to work on the metabolism of fats as influenced by caffeine. Everything was extraordinarily difficult because the Assistants were compelled to do military service. That could be done only during the term holidays, so that the most important time for work was lost. Ultimately, the two years' work by Dr. Hindemith, which had been giving promising results, was lost in the last phase of the war as we had to leave Breslau.

At the beginning of the war, I was called up and commissioned in the 6th Army for about six weeks in Poland, but from the end of November, was again able to continue my lecturing. To begin with, no importance was placed on scientific research. All the Assistants in the Institute were called up so that I had to give all the lectures. Furthermore, the enlightened Kultusministerium insisted that the students, granted leave of absence from the army to study, must be able to attend the lectures in any semester. This meant that lectures had to be repeated many times. Thus, I lectured 25 hours a week during both semesters; not practicals but lectures (Pharmacology I and II, 8 hours, for dental students I and II, 8 hours, medicinal plants, 4 hours, prescription writing, 2 hours. Pharmacology for pharmacists, 2 hours, industrial toxicology, 1 hour). Later I was granted an Assistant but then we had to take over work on war toxicology—the toxicology of explosives, mist, chemical warfare, and phosphorus combustion against which we had found an effective substance. The interesting research work was partly lost but in any case was not published. Nevertheless, the advancement of our other problems went ahead, even if slowly.

In 1934, Professor Wolfgang Heubner approached me about writing a book on the pharmacology of the inorganic anions (Hofmeister'sche Reihe). Earlier I had written a short article on chromium for the Handbook, which had pleased him greatly. By then I had worked for a long time on the Hofmeister'sche Reihe, so I agreed, particularly as I respected Heubner a great deal, and in the course of time I got to know him better. Finally, I wanted to see for myself what exactly had been done in this field. The literature was widely scattered and there was no generally accepted point of view. I worked intensively on it from 1934 to 1942, a hard time, since I did my work thoroughly (over 6000 references). There are certain laws that extend throughout the entire animal kingdom. The book dealt with different chemical properties—the precipitation of calcium by fluoride and phosphate, complex formation by thiocyanate, by pyrophosphate and distinct colloidal, chemical Hofmeister'sche effects, in order to gain an insight into pharmacology and into the inexhaustible theme of chemical and physico-chemical constitution and pharmacological action. Certain rules are to be found for colloids, enzymes, and inanimate membranes. But as soon as the whole cell is investigated with respect to the ions, all relationships cease, as though indicating how little we know about conditions within the cells. The Hofmeister'sche effect was found in whole animals in the same way as precipitation of calcium, but only with very large toxic doses. The action of thiocyanate and perchlorate on the thyroid had not been discovered at that time. Also, radioiodine was not yet in use as a more convenient and more sensitive test. But these two ions, whose actions appeared so easy to understand on this basis, showed differences on further inspection in spite of great similarity.

When one works for so many years on such a book, one learns a great deal. But more important are the consequences on further scientific research work. Many questions arose which I had not intended to pursue.

Investigations were begun on the frog heart with ferrocyanide and pyrophosphate, two anions of the Hofmeister'sche Reihe, which act identically on heart function. Here two different receptors in reaction constants, dissociation, exothermic heat, and chemical order were measured and calculated. It soon became apparent that the two anions both complex with copper and zinc (10).

We also became interested in fluoride complexes in consequence of the Handbook. We investigated some organic fluorides and found two complexing agents with magnesium and manganese, which inhibited alkaline phosphatase. By chance we found that histamine forms chelates with a series of heavy metals (copper, cobalt, nickel, and zinc), which we used immediately to investigate copper metabolism in the beating heart. Also, the previously mentioned storage of copper, manganese, and magnesium in the surface of heart muscle fibers was found. Soon we could give a complex constant for copper binding with tissue constituents. The results were lost in 1945 when my last luggage was burned during the great air attack on Dresden.

What interested me still more was the finding that Me-histamine formed two dissociation stages. That opened up the possibility of formation of ternary complexes. We had already observed the first sign of such a complex with protein, but in later research with Höbel it was proved conclusively although, as in the proteins investigated up till then, without firm bonding.

I have laid special emphasis on these experiments because the problem of fixation and release of pharmacologically active substances in the tissues is not yet satisfactorily solved; for example, the binding of histamine to heparin in the mast cells is not firm (Werle). With ternary complexes the following possibility exists. When a ternary complex exists in the cell for instance, as with copper, cobalt, zinc, or nickel, histamine is immobile. As soon as a substance arises in metabolism which forms a more stable complex with this heavy metal, then histamine will be set free and can fulfill a physiological function. But this is only one example and exactly the same can happen with thyroxin, adrenaline, insulin, etc. If one wishes to intrude into the mysterious realm of the structure of tissues, then one must piece together the structure in space out of an immense number of such small fragments, exactly like the analysis of a complicated organic molecule. Further research in Breslau had to be abandoned incomplete. With the experiments on the toxicology of explosives (the important results were lost), the question of whether one could protect the liver against poisoning became of vital interest. We found a suitable drug in *Carduus marianus* and wanted to proceed with its isolation. This was later accomplished by Vogel in the firm Madaus, Cologne, and a useful medicament was introduced. All the wearisome work of building up the Institute for ten years and all the maturing research results were abruptly annihilated in the last phase of the war.

We will omit the following very difficult years; the flight, the separation of the family in three different places and the loss of our entire property, including clothes. In the following years until 1948 I busied myself as far as possible with mathematics until on 3rd March 1948 I came to Heidelberg and in the Surgery Clinic, through the friendship of K. H. Bauer, had the use of two unequipped rooms as laboratories, at first without technical assistance. A revision of the volume on the inorganic anions and completion of the literature survey was made possible with the help of America House. During this time also I wrote a more natural philosophical book "Die Prinzipien des Lebendigen." I began with the single hit theory of radiation. It seemed to me that in all later deductions, the fact of regulation in this very place was continually neglected. In my cogitations, the idea of harmony on the various levels was introduced and the postulate formulated that disharmony in the cell system is necessary to enable development and adaptation. The idea of the possibility of two hits, which in a short time interval struck very close together, could not be confirmed by experience, but on one point the later development came near to my deduction, in which the size of the target changes with the preliminary conditions.

Then I used my time doing clinical pharmacological tests for which I sat at the hospital bed daily for several hours, making measurements. A series of publications, also a monograph on hydergin as a medicament to improve circulation in local pathological conditions, emerged from these experiments. At this time, in the neighboring Max-Planck Institute for Medical Research, under the direction of Professor Bothe, the cyclotron was again put into working order. It was then possible for us to produce radioactive isotopes ourselves, as far as the strength of the cyclotron went (15 to 20 Mev). I carried out the isolations and at the same time, in Bothe's Institute, I learned the methods for measuring radioactive substances from the beginning. All these projects led to a continuation of the Breslau research on the surface of heart muscle fibers with the aid of radioactive-labelled pyrophosphate and phosphate and also with dyes. The circumstances seemed reasonably good and much was planned. I had almost forgotten that I, exiled from home, had a wife and three children who lived in Hamburg under straitened circumstances. As a permanent civil servant, I could expect a salary of only 540 DM per month from the government. As a basis, that would perhaps have sufficed had it not been for a regulation that extra earnings had to be deducted from this salary. In addition, my laboratory was still completely empty, without even primitive apparatus and without a personal budget at my disposal. Therefore I was compelled, in order to proceed, to take over industrial work, renouncing any state salary so that the family could move to Heidelberg, as they did one and a half years later. We could then begin to procure our furniture, clothes, linen, and books. That I could not pursue purely theoretical projects is understandable and so my whole field of research was narrowed. One could not expect to be paid for working on mathematical formulae. Mathematical work makes sense for a pharmacologist only if he knows beforehand that a secure job can be guaranteed for a long time. When I began it in 1929, I could foresee nothing.

The themes that occupied me during the period from 1948 to 1958 are given in the following enumeration with their respective motivations.

(a) I received a small stipendium from the Strebel foundation for cancer and scarlet fever research. Therefore, I was compelled to work on cancer, which until then had not lain in my field. Indeed I had tried in vain to demonstrate a carcinogenic substance in roasted coffee. The early experiments were directed towards the isolation of nucleic acids from various vaccinated tumors under the action of cytostatics. But these experiments were not continued because finances ran out. In 1948, I obtained a series of substances from industry which were used in the treatment of wool fibers. First, we found the weak acting hexamethylolmelamin, later trimethylolmelamin which at the same time was discovered by Walpool and others in England. We made various investigations especially on the behavior of the cell nucleus in Ehrlich mouse ascites tumor, and also worked on its inactivity during longer treatment (research with Staib). When administered to humans in a series of experiments with several tumors they were found effective but very slow, although with only minor side effects. We could not continue these experiments further.

(b) While in Breslau, we had begun experiments with fluorine in collaboration with the dental clinician Professor Euler. In Germany, fluoride was known only as a rat poison. After a report at the dental congress in Wiesbaden in 1949, I suggested using fluoride in tablets for children. A memorandum was submitted to the "Hessischen" government but its publication was withheld by intrigues. However, I had at least patented a fluorine-containing toothpaste. The sale of the patent gave some money, part of which provided better equipment for the laboratories. After that we went ahead, together with Professor Ritter, to determine the distribution of fluorine in various enamel layers and found, by treatment with toothpaste, higher amounts of fluorine in the outer layers. These experiments were later continued and improved upon by Brudevold. We had spent much time on fluorine analysis and wanted to begin a survey of the fluorine content of the water sources, which in Germany had not then been done. Expenses were not granted because we were not in an official position, so the project had to be abandoned, as the analyses were too costly for us. We returned to our research in collaboration with Professor Ritter, using radioactive calcium as the test substance to investigate metabolism in the growth of teeth in vivo.

(c) In the clinic, since we were treating patients with thyroid disease, we constantly had radioiodide at hand. It was used not only for treatment but also to investigate other problems; testing of medicaments acting on the thyroid, preparations of spurge (not published) and others such as atebrin. On this point, I had previously been interested in the behavior of the trachea with regard to the excretion of iodide. It was concentrated in the bronchial secretions, by an active transport mechanism, which suggested that possibly a new method of testing expectorants could be developed with it. On this theme very little work had previously been done; the best was by Boyd, whose method we partially followed. That is a topic most pharmacologists avoid because the method is wearisome and unpleasant. However, that is a poor reason for not working on such a project. There are some who firmly believe that only mitochondria or microsomes and perhaps even receptors are worth working on, Contrary to this, I have taken advice from Goethe: "According to our advice everyone should keep to his original path and should not be impressed by authority, be harrassed by general conformity or be carried away by fashion." I am pleased that I had been encouraged in this way by Straub and Heubner. Thus, we continued this project in the important research by Höbel. A short time ago he developed yet another preparation from the trachea, which he has examined in minute detail. With it one can follow the transport mechanisms very well. Iodine transport in the trachea was clearly found to be different from that in the stomach and thyroid. So also are the actions of perchlorate and thiocyanate, a theme that ties up with older research work extending over several decades. That is what Hackental worked on.

In 1950, the editing of the well-known Handbooks of Experimental Pharmacology was assigned to me by Heubner. As Editor I came in contact with Alfred Farah, formerly of Syracuse, with whom a close friendship has been formed after many years of working together in harmony.

At that time Handbooks were frequently deprecated. The reviews were modern but some colleagues were of the opinion that the Handbooks grew obsolete too quickly. On that point it should be noted that the reviews are often very superficial and sometimes more of a literature survey. In spite of this, I would not wish these reviews to be dispensed with.

A Handbook of Pharmacology brings together, at a certain point in time, a whole theme from many points of view and sometimes one can even omit looking up the original papers. Concerning its going out-of-date, one should note that in pharmacology, at various times, different groups of compounds will be worked on and tested, and perhaps will again become interesting after several decades. Then one can always find a great deal of information in old Handbooks, for instance, derivatives of quinine, pyrazolone, cocaine, local anaesthetics. That becoming obsolete does not happen so quickly depends also on the fact that pharmacology has been using isolated organs continually for almost 100 years. In any case, the results give a lead and still today form a basis. Also, as Goethe asserted, in such encyclopedias, the mistakes will be remembered rather than forgotten as they deserve. Every Handbook article should be appreciated because it demands a great deal of self-denial. To do experiments is certainly more interesting, but such surveys are equally necessary because the work can be used in a thousand different ways. It stimulates new approaches and prevents the neglect of industrious work that would otherwise have to be repeated. Such literature surveys also belong to the great realm of the sciences. Writing reviews does not demand great knowledge but even the best research worker should write articles for the Handbooks. Such volumes in the Handbook of Experimental Pharmacology like those by Koelle, Bacq, Ussing, Berde, Jungmann, Erspamer, Graffi, Herken, and others seem to me to be of lasting value. I see a time coming when a prominent worker could be asked to revise such a Handbook because of his great experience and knowledge, for example, like the work undertaken by Brody. Because I am convinced of their worth and importance I have spent a great deal of time on them. Above all, the experimental production of disease in animals seems to me to be a necessary aid in research and so I have taken over the really laborious job of editing these books, even if reluctantly. I do not believe that one can achieve the same results with computers.

When I worked in the hospital, I also undertook the testing of various substances on patients (e.g. dibenamine, hydergin, polamidon, catechol amines, and arecoline). Although I finally held a Chair in Clinical Pharmacology, I could not be very effective because of a lack of co-workers, and I was really happy when finally, in 1958, I could take over the Direction of the Pharmacology Institute as successor to Eichholtz.

I came to an Institute in which most of the equipment necessary for my field of research was either lacking or obsolete. The structure was erected on the 500 year old foundations of an old monastery. The building itself was constructed in 1861 at the time of Helmholtz, who in those days also had his apartment there. A new building was planned for 1952 or 1953, during the time of Eichholtz and Fleckenstein. That the building was not erected was advantageous, for according to the plans, it would not have been sufficient. However, all applications for improvement were refused by the administration with the excuse that they were not worth while on account of the new building. I objected to this argument and by perseverance I succeeded in considerably augmenting the number of usable rooms. Slowly, it was being discovered that the prevalent neglect of science since 1914 should not be continued if we did not want to lose completely our former high prestige in science, with all the consequences for industry and education of students. But the investment of money in science was promoted very slowly and soon came to an end because of financial difficulties. For example, the animal house was in an indescribable condition. We did long-term animal experiments during which numerous epidemic infections were contracted, for instance, once in the course of three weeks about 500 operated guinea pigs were destroyed. There were the application for planning a new building, detailed plans, refusals because it was too big and too expensive, new plans by the administration that were a little cheaper—and so it continued for six long years. When it came to the actual building; it became the most expensive animal house that ever was built, if one adds on the time wasted by Assistants and myself in repeating infected animal experiments, loss of working time, etc.

In those years, money was granted to modernize the apparatus of the completely out-dated Institute. Of course, I paid special attention to the work with radioisotopes and an ultracentrifuge, freezing centrifuge, spectrophotometer, electrophoresis apparatus, and good balances were procured. I only mention these things to indicate that previously they were nonexistent.

Until then, my course of life had lacked continuity; always new situations arose, new themes were taken up—had to be taken up as money was obtained from industry—and finally they became interesting and challenged me to continue. Thus, in the Institute there were many lines of research in progress. A series of projects by Assistants who worked almost independently ran side by side. Thus, Priv. Doz. Dr. Ellen Weber's section worked on thrombocytes and their metabolism. That seemed to be a promising project with simple structures. Greatest importance was placed on measuring both the chemical and histological effects in parallel, as in Breslau.

Hackental worked on perchlorate metabolism with ^{36}Cl-labeled compounds. These experiments followed logically to a transposition to bacteria and expanded to nitrate (which was already presented basically in my book on the pharmacology of inorganic anions). Besides studying the metabolism of the trachea, work on binding of ternary complexes was also pursued by Höbel and Lippert. I was very disappointed that problems arising from a short publication by me after the war were taken up by others from various aspects without my being able to participate for many years. Since 1958, we have done many experiments, the results of which will shortly be published.

I have followed the principle of taking on industrial work occasionally for several reasons. Assistants could thus come to grips with practical problems and learn methods. Otherwise that would be done in a pharmacology course but until then there was none in Germany. Professor Schmier held such a course in my Institute for small groups of 10–15 students. There was not enough space, technical assistance, or money for all the students to take this course. Some money was provided by industry which permitted investigation of antipyretics and their combinations, flavinoids, etc. In addition, this money was greatly needed to supplement the scanty budget and the meager assistants' salaries. This is the way it was used in other Institutes also.

Later, on behalf of the Government, we worked on fog, its deposition in the respiratory tract and also its resorptive action. The experiments also forced us to design and build new apparatus employing the laws of the flow of gases. The themes were too diverse to conform to our earlier accomplishments, but we had the advantage that the staff of the Institute also learned about many other methods and problems. Every two to four weeks relevant divisions had to report on their results. At these meetings there was discussion and mutual stimulation so that the possibility of collaboration on a special subject was established. Three of my Assistants were able to qualify.

On April 1, 1968 I became an emeritus Professor. I was happy that now all the purely theoretical themes of work could be undertaken. Through the pressure of various obligations, lecturing, examinations, medical reports and meetings, these had been neglected. Michel de Montaigne believed that it represented an arrogance to maintain the world by one's self. There is nothing further from my mind. But many Germans have been stricken with such a destiny in life and so the continuation of their work was interrupted. And so to finish, I would like to add a quotation by the same author: "I cannot base my life's records in achievements. Fate has made them too inconspicuous. I base them on my thoughts and humour".

LITERATURE CITED

1. Meinicke, F. K. 1908. Räumliche Fortpflanzung chemischer Reaktionen. Dissertation, Leipzig
2. Scholander, F. P. 1961. Oxygen Transport through Hemoglobin Solutions. *Science* 131:585–90
3. Ens, T. 1964. Molecular collision exchange transport of oxygen by hemoglobin. *Proc. Nat. Acad. Sci.* 51:247–52
4. Jendrassik, L. 1964. Das Le Chatelier'sche Prinzip und die Gesetze der Störung der dynamischen Gleichgewichte. Verlag der Ungarischen Akademie der Wissenschaften, Budapest
5. Eichler, O. 1930. Zur Pharmakologie der Hofmeisterschen Reihe. *Arch. Exp. Pharmakol. Pathol.* 154:59–102
6. Eichler, O., Smiatek, A. 1937. Über die Beziehung der Empfindlichkeit für Chloroform, Avertin und Eunarkon untereinander. *Arch. Exp. Pharmakol. Pathol.* 186:702–20
7. Eichler, O. 1941. Versuche eine Gesetzmässigkeit in der Verteilung und Empfindlichkeit im Zusammenhang mit Veränderungen der extrazellulären Räume bei mit NaJ behandelten Fröschen aufzufinden. *Arch. Exp. Pharmakol. Pathol.* 198:442–71
8. Eichler, O., Mügge, H. 1932. Zur Frage der Schädlichkeit des Coffeins bei chronischer Zufuhr. *Arch. Exp. Pharmakol. Pathol.* 168:89–96
9. Eichler, O. 1938. Kaffee und Koffein. Springer-Verlag
10. Eichler, O., Wolff, E. 1944. Studie am Froschherzen mit Pyrophosphat und Ferrocyanid. *Arch. Exp. Pharmakol. Pathol.* 203:1–21
11. Eichler, O., Eichler, L. 1941. Wirkung von Na Cl und Na J auf die Ausscheidung verschiedener anorganischer Substanzen beim Frosch. *Arch. Exp. Pharmakol. Pathol.* 199:4–55

Ann. Rev. Earth Planet. Sci. 1977. 5 : 1–12

AMERICAN GEOLOGY SINCE 1910 – A PERSONAL APPRAISAL

James Gilluly

James Gilluly

Ann. Rev. Earth Planet. Sci. 1977. 5 : 1–12

AMERICAN GEOLOGY SINCE 1910 – A PERSONAL APPRAISAL

×10064

James Gilluly
975 Estes Street, Lakewood, Colorado 80215

The editors have asked me to submit a personal evaluation of the development of the geological sciences during the fifty-five years I have been active in the field. They have encouraged autobiography, but I hope to limit this to such aspects of personal history as throw light on the "sociology" of the science. A most striking aspect of the earth sciences is the tremendous expansion of the field in the last sixty years.

In 1910, while a freshman at Franklin High School, Seattle, I enrolled in a class in "Physiography"—really elementary geology. The instructor was J. J. Runner, a most inspiring and considerate teacher, later to be known to generations of students at the University of Iowa as "Uncle Joe." At the end of the term I had become so fascinated by the subject that I asked Professor Runner (in those days any high school teacher was ex officio a professor) how one could earn a living in geology. He replied that about the only employment for a geologist was as a teacher in high school or college or as a geologist with a state survey or the Federal Survey. He estimated that there were only about a thousand people in the United States who considered themselves professional geologists. It was a tight profession with very few openings.

Runner was probably right at the time. Only a handful of geologists were employed in those days by either mining or petroleum companies. The membership of the Geological Society of America that year was only 305. If there were indeed a thousand geologists in the country there was one in 92,000 of the population. Today there is at least one geologist in 7,000 people, 13 times as many proportionately. In 1910 I felt the profession was entirely too crowded to be counted on as a career; I dismissed it from my plans.

Three years later I entered the University of Washington in engineering. I tried this and two or three other majors without developing much enthusiasm for any of them. By the beginning of my senior year I had plenty of credits for graduation but no major. My close friend William T. Nightingale, later to become a highly successful petroleum geologist, was graduating in geology and was already employed by a major international oil company. He took me aside one day: "Since you have jumped around from one major to another, sampling them, why don't you jump once more and sample geology? You like the out-of-doors and you told me you liked the sample you had in high school. You have nothing to lose if you don't stay with it."

I decided it was worth the gamble; I dropped all the courses I had started and enrolled in elementary geology, mineralogy, and paleontology all at once. Under the flexible quarter system, I was able to graduate at the end of the year. My main instructor was that outstanding teacher, George E. Goodspeed, Jr.

Bill Nightingale was right; jobs were indeed "hanging on the bushes." Within two weeks of leaving the University I had employment with a small oil company mapping in the Judith Basin of Montana.

Some months before, at Goodspeed's suggestion, several of us had taken the two-day examination for Geologic Aid, the lowest rung on the U.S. Geological Survey's professional ladder. We were warned that Survey jobs were very few, but were urged to take the test as a measure of what we had learned.

Much to my surprise, less than two months after I had joined the oil company, I received an offer of employment from the Geological Survey. Although the salary offered was notably less than that of the company, a senior employee of the company advised me to take the Survey position: "The Survey training is much better than this company can give you." I followed his advice and accepted—a most fortunate choice.

I reported to Frank Reeves for a field season in central Montana, largely in the "Missouri Breaks" and adjoining country to the south. The work was hard, nearer seventy than forty hours a week. The "government sinecure," as my oil company boss called it with scorn as I resigned, was at least twice as demanding as the work with the oil company had been. But it was fascinating work and highly instructive.

Good fortune had it that a fellow neophyte, M. N. Bramlette, was also in the party. Bram is the keenest stratigrapher I have ever worked with, and though he had had little more training than I, working with him was a real education. He has been a close friend for fifty-five years, and during all that time I have never written a paper without thinking how Bram would appraise its logic and pertinence.

In 1921 the Geological Survey had about 60 field geologists, including such stars as Darton, Ransome, Keith, Paige, Spencer, Hewett, and Ferguson. There were eight or ten paleontologists, including Girty, White, Stanton, Woodring, Reeside, and Kirk. Of the four or five chemists, the leader was F. W. Clarke; the mineralogists were Schaller and Henderson; the petrologists Esper Larsen and Clarence Ross. Compare that small band with the present staff of nearly or quite a thousand geologists, more than a hundred chemists, and scores of physicists, seismologists, mineralogists, and geochronologists.

The scientific leaders of the "old Survey" included many outstanding men such as those just mentioned; no matter what the subject, there was nearly always a very knowledgeable scholar "just down the hall." Though there were a few exceptions, most of them were more than willing to discuss problems with the younger staff, loan specimens, give literature citations, and help in other ways. It was an outstanding school for a beginner; one wonders whether the present gigantic organization has retained this quality. (It is the prerogative of age to look back on the "good old days".)

During the spring semester of 1922, Bramlette, Rubey, and I used to commute Saturday mornings to Baltimore for the memorable morning-long course in stratigraphy taught by E. W. Berry at the Johns Hopkins. We had to catch the 6:00 AM interurban to make the 8:00 o'clock class but we all thought it worth the effort. Berry had a sardonic humor that kept one on his toes.

The Survey has always encouraged graduate study, so in the fall of 1922 Rubey and I took leave and entered the Yale graduate school. This was a most challenging experience. There is an old saying that one learns more from his fellow students than from even the most distinguished faculty. The faculty was indeed distinguished, Knopf, Schuchert, Gregory, Dunbar, Ford, and Longwell among the leaders. But my fellow students were of equal caliber and several became even more distinguished in later life. There were only about a dozen, but among them were T. B. Nolan, later to become Director of the Survey, William W. Rubey, George Gaylord Simpson, J. Frank Shairer, Carl H. Dane, and Ludlow J. Weeks. Of these, Nolan and Rubey were to become presidents of the Geological Society of America; Nolan, Rubey, Simpson, and Shairer all became members of the National Academy of Sciences; Rubey and Simpson became Penrose medalists; Shairer, a Roebling medalist and President of the Mineralogical Society; Dane, Chief of the Fuels Branch of the Geological Survey; and Weeks a group leader of the Geological Survey of Canada. Shall we say that the competition was keen? All these friends have continued to influence my work ever since; I owe much to them all, but especially to Nolan and Rubey, with whom it has been my good fortune to be closely associated throughout my professional life.

The most influential teachers in physical geology at Yale were Knopf and Gregory, with wholly contrasting styles. Knopf would generally give a single weekly assignment, perhaps only five or six pages long, to his graduate class in petrography. He would insist on our analyzing the article sentence by sentence until a clear consensus was arrived at: the author was either right, perhaps right but unproven, or definitely wrong. In the lab no mineral was satisfactorily identified with less than four diagnostic properties. (Simpson, who was to become the leading vertebrate paleontologist of his time, was far and away the best of us with the petrographic microscope.)

Gregory, on the other hand, with an innocent air, would assign each of us a handful of references on which to report in his seminar in physiography. When the class met we would find that the assignments were loaded. One of us would have read nothing but papers urging that glaciers protect the landscape from erosion; another only papers arguing that drastic glacial erosion occurs; one nothing but papers urging that erosional slopes all flatten with time; another that all slopes retreat parallel to the original slope. Obviously, when the reports were given, arguments were hot and heavy. Greg would say little, but he had a marvelous skill in prolonging the debates. One found oneself reading all the other assignments as well as one's own, simply in self-defense. When, after long debate—the "two-hour" seminar seldom lasted less than three or four hours—someone would try to get Greg to commit himself, his reply was always, "It doesn't matter what I think: I try to assign representative papers. What is important is, what do you fellows think?" Great teachers, both.

In the 1920s nearly every graduate student sat for his qualifying orals at the end of his second year of study, unless his undergraduate training was unusually weak. This was a far cry from the situation today, when many students delay for five or even eight years, thereby cutting off as much as twenty percent of their productive lives. One wonders whether many of the additional techniques—not theory—so acquired could not be picked up during, rather than before, a productive professional career. After all, few petrologists are expert in paleontology or vice versa. Why should every petrologist learn to operate high pressure furnaces? Of course he should learn to interpret the data gathered by the experimentalists, but he need not be an experimentalist himself.

Be that as it may, most of our group tried the oral hurdle after two years and nearly all passed creditably. The general expectation was to complete a thesis within the next few years, while earning a living, if one could find suitable employment. This was the road I took : although I had concentrated on petrology and mineralogy under Knopf and Ford, my thesis, presented two years later, was a stratigraphic and structural study of the San Rafaél Swell, Utah, done under the auspices of the U.S. Geological Survey.

In the 1920s the environment of the Geological Survey was as stimulating to its younger employees as that of any university. For the beginners there was the Geologues Club, which met monthly and consisted of the twenty-five most junior employees of the Geological Survey, Geophysical Laboratory, and the earth science division of the National Museum. When a new man joined one of these institutions he was invited to join the club. If he accepted, the currently senior member was ejected, with due ceremony. Generally, at each meeting two members would present formal papers. Criticism of both content and presentation was always searching and commonly effective. The quality of papers presented generally improved notably during the four or five years a tyro was a member of the club.

For the entire geological group in Washington there was the venerable Geological Society of Washington, which met, as it does today, on the second and fourth Wednesdays of each month between October and May. In the 1920s the Society was rather formal, with the President, speakers, and Secretaries in formal dress. The formality gradually disappeared during that decade and the meetings have been quite informal ever since. Any speaker before this Society may expect a highly skeptical audience and if he is not a foreigner the skepticism finds ready expression.

For those interested in "hard-rock" geology, there was the Petrologists Club, which met monthly at the Geophysical Laboratory or the Geological Survey during Prohibition, but exclusively at the Geophysical Laboratory after beer became legal. (Of course alcohol is never allowed on government premises.) The club consisted of such men as Cross, Washington, Bowen, Fenner, Zies, Shairer, Larsen, Schaller, Wright, Nolan, Greig, Ross, Burbank, Lovering, Tunell, Spencer, and others. It was a very lively place.

A somewhat smaller group, the Paleontological Society, commonly met at the National Museum; no mere field geologist could invade this arcane body.

Finally there was the Pick and Hammer Club, principally composed of Survey employees, though open to any geologist in the area. It met on no fixed schedule, but only when the program committee could persuade some member or visiting geologist to talk about his research. As most foreign geologists visiting the United States eventually reach Washington, it was a rare year when half a dozen international figures failed to appear on one or another of the programs.

Once a year the Pick and Hammer Club presents a show, pointing out and exaggerating the foibles of the Survey brass, in the tradition of the Gridiron Club of Washington. During the 1920s this was an elaborate occasion; everyone dressed formally for the show and the dance that followed it. Through the years the formality grew less and less until for the last thirty years it has disappeared completely.

This growing informality has been paralleled by forms of salutation. In the 1920s all the paleontologists were addressed as "Doctor," the rest of us simply by the surname: "Good morning, Hewett." Today everyone, with the possible exception of the Director, is hailed by his given name: "Hi, George." A remarkable change within a generation.

The Survey had very small funds for training in new techniques. In 1928 permission and a small field grant were given me to work for a month with Robert Balk in the Adirondacks in order to pick up the then-new techniques of the Cloos school of structural studies. Robert was most hospitable and cooperative; we became and remained close friends until his untimely death twenty years later.

A few years later, during the winter of 1931–1932, at the suggestion of Dr. T. W. Stanton, the Chief Geologist, I was permitted to utilize leave accumulated from past years and leave in advance of 1932 to spend a winter in Europe. This included a month in Innsbruck, studying petrofabrics with that most generous and cooperative scholar, Bruno Sander. Sander gave me two hours of his time every afternoon for a solid month, going over his very complex textbook, line by line.

As it happened, I was to utilize Sander's techniques in only one small project, but for forty years we have been friends and I still consider myself deeply in his debt. His writings are difficult, even for his Austrian compatriots, but this is chiefly because of his painstaking care to avoid going beyond the evidence. His logic seems to me impeccable.

In 1938, I accepted a professorship in the University of California, Los Angeles, where I remained until 1950, except for three years during World War II. I was able to retain "when actually employed" status on the Survey and thus continue field work during the summers.

During the war I returned to the Survey on a full-time basis, first as a roving inspector of the many war mineral projects necessarily manned by inexperienced geologists quickly assembled from all over the country.

Later assignment was to the Southwest Pacific Command as a military geologist. Our headquarters at first were in Brisbane, then Hollandia, New Guinea, studying maps and air photos in search of landing beaches, potential air strips, water supplies, and road metal in the Philippines. We landed on Leyte on D+1. I may say that we were successful with our first two assignments, but much less so in the others; deep weathering made good road metal hard to find in the Philippines.

Aside from the wartime interval I was associated with UCLA for nine years. Graduate teaching I found very rewarding but I was never happy with my undergraduate classes. I felt that I was either over the heads of the students or insulting their intelligence by laboring the obvious. I could never seem to find the happy medium. Finally, when the Regents of the University insulted the faculty by demanding new oaths of loyalty (though all of us had previously made such oaths) I realized that I had always been happier doing geology on the Survey than talking about it in the University. I resigned from the University and returned to the Survey, absorbing a pay-cut exactly equal to the increase I had received in the reverse transfer twelve years before. Many other Survey geologists have gone through similar cycles. Woodring, Lovering, Sims, Sheldon, Cloud, James, and Reynolds come readily to mind. Similar motivation?

From 1950 until my retirement in 1966, I was involved in the fascinating geology of north central Nevada. Here, with many able assistants, I was able to map four fifteen-minute quadrangles of highly complex, and therefore very interesting, geology. These were the most enjoyable years of my professional life; my one regret is that the hills are still rising at a rate faster than I can climb them. Since 1966 field work in rough terrane has become impossible for me.

MINERALOGY SINCE 1920

In the 1920s the best equipped mineralogical laboratories in the United States had, perhaps, a chemical bench with a research assistant, a goniometer, a petrographic microscope, immersion oils, and blowpipe equipment. Funds might be available for purchase of a few specimens and perhaps for a collecting trip every few years. Of course the Braggs had already gone far with their X-ray studies, but such equipment in the United States was confined to a few of the more advanced physics laboratories. No mineralogical department had X-ray equipment until late in the decade, most not until the 1940s. Even a universal stage was a rarity in America, though standard equipment in Europe. No one had ever dreamed of the electron probe, electron microscopes, or Mossbauer. Today the only goniometer one finds in a mineralogical laboratory is an X-ray machine. The last blowpipe expert in America, Foster Hewett, died in 1970. Were Palache or Schaller to enter a modern laboratory they wouldn't know where to begin. My guess, though, is that it wouldn't take them long to find out!

PETROLOGY

Fifty years ago petrology was mainly a field and microscope science, as, of course, much of it still is and always will be. Chemical analyses were so expensive that they were carefully rationed. H. S. Washington made an international reputation by compiling all the "superior analyses" in the world literature into a single volume.

The major figures were Sederholm, V. M. Goldschmidt, Harker, Escola, Niggli, La Croix, and Becke in Europe; Daly, Lindgren, Ransome, Larson, Knopf, Grout, and Buddington in the United States; and Coleman and Adams in Canada.

The only experimental work of consequence was being done at the Geophysical Laboratory of the Carnegie Institution of Washington. Here Bowen, Fenner, Merwin, Sosman, Greig, Goranson, Schairer, and a few others were engaged in a well-conceived program of experiments on simple systems of components of the rock-forming minerals. They were making considerable progress, though they were restricted to pressures of only one atmosphere, and hence to systems free from water and carbon dioxide, known to be important constituents of magmas.

Despite the limitations of these experiments as guides to igneous processes, Bowen, in 1928, wrote what is still a very significant contribution to petrology: *The Evolution of the Igneous Rocks.* The main concept is that crystallization differentiation dominates the evolution of magmas from a primary basalt through more and more salic differentiates to granite and aplite. Although Daly and Grout soon pointed out that the volumes of the successive differentiates are not consistent with dominance of Bowen's suggested processes, they may still be significant.

When experiments at high pressures became available twenty years later it was shown that the incongruent melting of olivine, a major element in Bowen's scheme, disappears at the pressures to be expected in the lower crust. The theory cannot then be adequate as presented—though some trends of rock evolution may conform to it.

Although some work on evaluating the solubility of water in molten silicates at high pressures and temperatures had been carried out by Goranson in the early 1930s it was not until new alloys had been developed that routine work with apparatus of Tuttle's design could be done on such systems. From the late 1940s, experimental petrology could go on under more realistic pressure conditions. It became possible to reach pressures and temperatures corresponding to those at the highest metamorphic grade.

No one, I suppose, would claim that experimental petrology has solved all the important problems in igneous and metamorphic petrology, but everyone must agree that tremendous advances have been made in the last decade by the brilliant work of Tuttle, Schairer, Ringwood, Green, O'Hara, Boyd, Yoder, J. V. Smith, and their confreres.

Many constraints on petrologic theory have been imposed by isotopic studies, which of course had to wait until the development of mass spectrometers in the late 1930s and early 1940s. Clearly many of the developments that have revolutionized both mineralogy and petrology have depended more on new instrumentation than upon new theory. After all, Gibbs and Van't Hoff are nineteenth century names.

PHYSIOGRAPHY—GEOMORPHOLOGY

For some unknown reason, the subscience concerned with the evolution of land forms, which had been known for at least a century as physiography, has become better known as geomorphology. Perhaps this is a more impressive name, as it has an extra syllable; certainly the words are synonyms. Whatever the name, the methods and content of the field have drastically changed for the better during the last fifty years.

In the 1920s most physiographic papers were almost entirely deductive—based on hypotheses largely developed by William Morris Davis. In his hands the results seemed quite reasonable, even though often incorrect, but most practitioners were far less able than he and few of the papers were based on sound stratigraphy. Peneplains were multiplied so that a single landscape might contain three or four. This is still advocated in Africa and Arkansas but fortunately it is going out of style. The measurements of erosion rates have generally made the recognition of Jurassic or even Oligocene landscapes somewhat less popular than it was.

The quantitative studies of stream regimens by Richardson, Simon, Rubey, Leopold, Langbein, Wolman, Schumm, and many others has revolutionized the study of stream erosion just as Bagnold and Hack have clarified the work of the wind. Nye, Carol, Seligman, and Sharp have elucidated many aspects of glacial action.

Fifty years ago the problems of pediment formation were just being noted by Bryan and Passarge. The studies by Denny, Bull, Sharp, and others have gone far to solve them.

Fifty years ago the Pleistocene Atlantic shorelines were thought to be horizontal from Florida to New Jersey; Hack has shown that they vary notably in elevation, even along this relatively stable shore.

STRUCTURAL GEOLOGY

Hans Cloos began his investigations of the internal structure of plutons during the depression in Germany following the first World War, when travel was difficult. He had to work on granite within walking distance, therefore the refined observations! His techniques became standard practice only in the late 1920s and early 1930s. Sander's petrofabric studies began in 1911, but it was well into the 1930s before his methods became widely applied. Under favorable conditions these methods permit extrapolation of surface structures to considerable depths.

In the early 1920s the only geophysical tool in routine usage was the magnetic dip needle, used in prospecting for iron ore. With the refinement of equipment during the antisubmarine war, a new tool—aeromagnetics—became available for seeking out magnetic anomalies. During the last 30 years most of the United States and Canada have been covered by aeromagnetic traverses and many patterns of anomalies worked out. While the interpretation of these anomalies is commonly controversial, many of them have disclosed structural trends at depth that would not otherwise be discovered.

Other geophysical techniques have been developed, mostly in the last fifty years. The first tool used in prospecting for oil was the torsion balance, followed by the gravimeter. The gravimeter was used in searching for salt domes in the Gulf coastal plains and in northern Germany. Seismic prospecting was about a decade behind; it is now standard practice throughout the petroleum industry, since virtually all favorable structures determinable from surface studies have been thoroughly tested.

Other geophysical tools in wide use in prospecting for radioactive minerals are the Geiger counter and the scintillometer. They are less useful in structural studies than are seismic methods, but are largely responsible for most of the uranium reserves thus far identified.

Geophysical methods have proven useful in many other ways as well. Simmons was able by gravity measurements to outline the shape of the Adirondack anorthosite mass at depth and to locate the feeding conduit. Mott has used gravity and seismic equipment to show that the Cornish and Devon plutons coalesce at depth and the combined mass extends to a nearly uniform depth of 12 km. Seismic studies of the Yellowstone Park area have shown that a siliceous pluton underlies much of the park and extends to a depth of 20 km. This salic intrusive is underlain by much denser rocks and quite surely does not overlie a "hot spot" extending deep into the mantle as has been suggested by several workers.

Of course the greatest development in structural geology has been the evolution of the idea of plate tectonics. This, too, depends largely upon geophysical data. Fifty years ago, or even twenty, few earth scientists in the northern hemisphere accepted the idea of continental drift; today few deny it. Several steps were involved in this reversal of opinion. First came Hugo Benioff's discovery of the alignment of intermediate and deep focus earthquake foci into zones that now take his name. Then came Mason's discovery of linear magnetic anomalies on the floor of the northeastern Pacific and the later finding that these anomalies parallel the oceanic ridges. Cox and Doell found a recognizable sequence of magnetic reversals and it was soon noted that the spacing of the sea floor anomalies was such as to strongly suggest that they record these reversals. Harry Hess first suggested that the sea floor is spreading and that the mid-ocean ridges are fault blocks bordering rift valleys. In short, the most convincing evidence for continental drift is found not on the continents but on the sea floor.

True, many reconstructions of former arrangements of continental plates are highly subjective and quite unsupported by geologic evidence, especially those for pre-Jurassic time. The most quoted such reconstruction, that by Bullard, Everett, and Smith, 1965, a computer-guided matching of the continental slopes bordering the Atlantic, produces an excellent fit between South America and Africa—a fit strongly supported by the match of geologic provinces—but a far less convincing fit of North America with Europe and North Africa. This reconstruction allows no room for the Paleozoic rocks of Central America. Nor does it take account of the post-Jurassic crustal shortening of the Betic and Atlas ranges. Before the narrowing of the Strait of Gibraltar by the counter-clockwise rotation of the Iberian peninsula, the strait was at least 300 km wide, as shown by paleomagnetic measurements in Spain and Morocco and the Betic-Atlas shortening. The poor fit is readily understood when one recalls that the present continental slope of North America was formed long after Jurassic time and the location of a pre-Jurassic slope is pure guess work. It may have been several hundred miles east or west of the present slope which was the line matched with Europe and Africa. In fact few pre-drift reconstructions are as convincing as that of Africa and South America. Yet the mutual relations of ridges, magnetic stripes, and sea floor ages as found by JOIDES drilling completely confirm sea floor spreading and continental drift.

The existence of large plates mutually displaced horizontally does not, however, eliminate important intra-plate orogeny. The southern Rocky Mountains lie, for example, several hundred kilometers from any plate boundary; the horizontal movement must have been passive with respect to adjoining crustal segments. But the vertical excursions of the mass are measured in kilometers. The Cretaceous strata of the Rockies from New Mexico to northern Montana are in places as much as 5 km thick and overlie continental Jurassic. The post-Cretaceous uplift has been such that the Jurassic has been eroded from a surface commonly 5 km or more above the level of the sea, a vertical excursion of more than 10 km over large areas, and in places even more. Plate tectonics obviously does not account for all orogeny. even though it does well with the Andes and Coast Ranges.

SEISMOLOGY

The Seismological Society of America was founded in 1910, at a time when there were probably fewer than thirty people in America who considered themselves primarily seismologists. Recently the Society admitted more than 180 new members in a single year! Here again the great advances have come from the improvement of instrumentation. The military objective of monitoring nuclear tests led, during the 1950s, to the great expansion of the world-wide instrumental network and to a very considerable improvement in the sensitivity of the instruments. The brilliant insight of Beno Gutenberg led him to the suggestion of a Low Velocity Zone in the upper mantle, but its existence could not be demonstrated until much better instruments, more advantageously located, could become available.

I have already mentioned the great improvement in structural interpretations made possible by the advances in seismology. Marine geology is largely dependent on seismic work, which has disclosed the layering in the sea floor.

STRATIGRAPHY AND SEDIMENTATION

Stratigraphy is the backbone of the science. Without sound stratigraphy structural studies are impossible, and all historical and most of economic geology depend upon it. In the last fifty years advances have been great. Although exceptions are becoming more abundant, the distinction between miogeosynclinal and eugeosynclinal suites of strata is very useful, both in historical and economic geology. Not all flysch is demonstrably antecedent to orogeny, but it is surely highly suggestive. Molasse deposition has long been recognized as following mountain building. Pebbly mudstones are now recognized as slide deposits, as are olistoliths and olistostromes. Stromatoliths are becoming useful as crude guides to age in the Precambrian.

Paleotectonic maps have become extremely valuable in petroleum exploration, and their compilation unravels much geologic history otherwise unsuspected. It is unfortunate that the Geological Survey has abandoned its project to compile maps for each of the Systems. I think it quite safe to say that with the approaching scarcity of economic mineral resources of all kinds, this useful tool must be ever more intensively pursued. Correlation charts, too, need frequent revision if they are to keep up with progress.

GEOCHRONOLOGY

Fifty years ago, a dozen or so crude measurements of lead-uranium ages had been made. K/Ar and Rb/Sr were decades in the future, and ^{14}C and fission track studies still further. Although there was only a handful of workers in the field, optimism was great. It wasn't until the 1960s that we learned about concordia, cooling ages, and other complexities. Here again improvements depended upon better instrumentation, notably Nier's improvements of mass spectrometers. Decay constants are still somewhat uncertain. To an outsider, lead-uranium ages on zircons seem to be least subject to "resetting." It is unfortunate that the Canadian Survey has classified Precambrian time on a scale based on K/Ar dates—the method most readily subject to resetting. It seems safe to guess that a revision will soon be in order.

Of course ^{14}C dates have revolutionized the study of the later Pleistocene and Holocene, even though we have been disappointed to find that radiocarbon dates are not calendar dates.

MAPPING

All must be delighted at the development of so many new tools in so many aspects of geology. But the working out of geologic history, structure, and resource potential basically still rests on field mapping. The laboratory is only a valuable supplement to working out the mutual relations of the various rock masses, which can only be done by field mapping. Unfortunately with the development of these ancillary techniques there has resulted not merely a relative but an actual lessening of field mapping. Stratigraphic sections are published without map control and are thus completely unreliable.

I don't know what the situation is today, but when the National Science Foundation began operations it would award no grants for field mapping, only for laboratory equipment. To an onlooker, this still appears to be the case. More surprising is the diminution of the mapping program of the Geological Survey, whose much increased staff carries on much less mapping than the smaller staff of twenty years ago. I think it is demonstrable that field study is infinitely more likely to locate an ore body than any number of Landsat views from 500 miles away. As with the human eye, the discrimination of a photo lens diminishes with the square of the distance.

The Landsat picture of the Front Range west of Colorado Springs covers the area of major mining districts: Cripple Creek, Leadville, and Climax. None of these districts is identifiable on the Landsat picture, but a group of rusty-weathering pegmatites are conspicuous. It is only a confirmed optimist who could think the money spent on Landsat can be justified by its value to metalliferous prospecting. A small fraction of the cost of the satellite would fund detailed mapping on a scale of 1/24,000.

METALLIFEROUS GEOLOGY

Geology of metals remains largely as it was fifty years ago: primarily structural geology. One major development has been the discovery of sensitive field tests for metals, thereby lessening the dependence on the placer pan and leading to many discoveries, some of major importance. The use of the Geiger counter and the scintillometer in uranium prospecting has been mentioned.

Much progress has been made in studies of mineralizing solutions, fluid inclusions, and alteration geochemistry, all of theoretical interest, if not of much economic importance. Isotopic studies of O and H yield values equivalent to ground water, as would be expected because of long cooling times; they throw no light on the original metalliferous solutions.

PALEONTOLOGY

Being innocent of paleontological skills, I cannot pretend to evaluate progress on a critical level. It is obvious, nevertheless, to even the most ill-informed, that foraminiferal research was in its infancy in the 1920s, and is now one of the most highly developed branches of the science. A few people were working on conodonts then; now they are among the most useful fossils. Diatoms were then beautiful curiosities; today, under the microscope of Lohman, they are good guide fossils to the Cenozoic. Coccoliths have become, under the critical eye of Bramlette, excellent guide fossils and perhaps the best of all groups for intercontinental correlations. Fossil spore and pollen assemblages have become highly useful in testing former continental assemblies. Thick sequences of marine rocks have been broken down into distinct formal zones that have been assigned absolute age spans by isotopic studies. These studies not only permit accurate correlations of individual zones, but also give valuable data on rates of deposition. My paleontological friends insist, too, that ecological interpretations have become much more reliable than of yore.

SUMMARY

With the relative explosion of workers in the earth sciences, it is not surprising to observe the great forward movement of the sciences concerned. There are obviously many more able scholars at work all through the field than ever in history. Clearly the next fifty years should see more advance than ever.

Ann. Rev. Physiol., Vol. 34

DISCOVERY AND UNDERSTANDING

RAGNAR GRANIT

Ragnar Granit

Ann. Rev. Physiol., Vol. 34

DISCOVERY AND UNDERSTANDING 1071

Ragnar Granit

*The Nobel Institute for Neurophysiology, Karolinska Institutet,
Stockholm, Sweden*

In asking me to write an introduction to these review articles the Editors
have generously given me a free hand. They may nevertheless have expected
me to write something about my research —its background in contemporary
work and in my own upbringing. If this was the idea, it did not appeal to me
in the least. The reason for this was that anyone interested in this kind of
information can have it in the books I have already written (see especially
6, 7, 9). To some extent these books not only report my work but also
illustrate my considerable delight in tracing the history of the ideas that
they have propounded. As to my personal attitude to our science, it emerges
in my book on Sherrington (8); nobody can write a book concerned with
values and evaluation without exposing his own values rather fully.

Grappling with the necessity of supplying something of general interest,
I remembered the frame of mind in which I had spent the early spring 1941
after a bicycle accident that crushed one knee. Reading could not then fill
all my time; besides, it compounded the constraint I felt, being confined to
intake alone, while all the time the creative urge demanded release in some
form of output. In this predicament I recalled an early lecture of mine to an
academic student body under the heading "Talented youngster looking for
a teacher," and this put me to writing a collection of essays, *Ung mans väg
till Minerva* (5) (Young man's way to Minerva) which was published that
autumn.

My book preceded Cannon's *The Way of an Investigator* (4) by a few
years. When his work appeared I read it eagerly and found a great deal of
overlap, both in its point of view and in its emphasis. Far more has since
been written on the same subject—more systematic, better documented
books covering the whole field (e.g. Beveridge 3). Thus, it was with feelings
of anxiety that I looked up my old work. Rereading it now and musing
over it, I found it, indeed, a book by a younger man than my present self,
written for young men fired by enthusiasm for a life devoted to science. The
tutor, slightly older than his listeners, speaks to them about the courting
of Minerva: he tells them of her apparent fickleness and real austerity, of
her views on ambition and success, and of much else, not forgetting to mention the radiance of her smile on the rare occasions when she bestows it.
There was about these essays an air of intimacy nurtured by convalescence.

1
THE EXCITEMENT OF SCIENCE

I have been asked by a publisher to translate them into English but do not trust myself to render into another language something that depended so much on its style of presentation.

Now, thirty years later, I return to such matters in a mood of detachment. Many people regard detachment as one of the great virtues. But it is probably not conducive to scientific creativity of the kind that was life itself to the young author of "Minerva." Passion is a better word for describing that attitude. Young people are out for themselves, to make discoveries, to see something that others have not seen. They may be satisfied with a modicum of analysis because there is always something round the corner to look at—perhaps something new and quite unexpected, exciting and important, at any rate a temptation hard to resist. Later in life one may feel it less compelling to discover something. Rather does one prefer to learn to understand a little of Nature's ways in a wider context. Then, detachment comes in handy. One realizes that it really is a great virtue: the virtue of those who have to weigh and judge. In this state of mind, I have decided to offer some comments on discovery and understanding. In the main I shall restrict myself to experimental biology.

By "discovery" we mean in the first instance an experimental result that is new. In a more trivial sense most results are new just as they always impart "knowledge" of some sort. For practical purposes I tend to ask myself when reading a paper: is this knowledge, or real knowledge? Similarly one may ask: is this result new, or really new? In the latter case it is a discovery, and a discovery tends to break the carapace of dogma around an established view, just as a bombardment with heavy particles tends to scatter the nucleus of an atom. In this type of discovery there is an element of unexpectedness. One of the best known examples is Röntgen's discovery of the rays that in many languages bear his name, a discovery that came as a surprise to him and to the rest of the scientific world.

There is a second and equally fundamental type of discovery: the delivery of experimental evidence for a view that is probable, yet not established, because such evidence as there is has not yet excluded alternative possibilities. An example of the latter type is the theory of chemical transmission at synapses, suggested by T. R. Elliot in 1905, but not proved until very much later (Loewi, Dale). This is the most common type of discovery: confirmation by evidence of one theory from a number of alternative hypotheses.

Either type of discovery, to deserve the term, must have far-reaching consequences, as the cases illustrated here indeed have had. Unless this criterion is satisfied, we are not willing to use a grand word like "discovery" instead of speaking modestly of a new result, more or less interesting, as the case may be.

The experimenter himself may not always understand what he has seen, though realizing that it is something quite new and probably very important. Thus, for instance, when Frithiof Holmgren in 1865 put one electrode on the cornea and another on the cut end of the optic nerve, he recorded a response to onset and cessation of illumination. This he held to be Du Bois-Reymond's "negative variation," that is, the action currents of the optic nerve fibres. These were what he had been looking for and therefore expected to find. Six years later, Holmgren started shifting his electrodes around the bulb and soon understood that the distribution of current he obtained required that the response had to originate in the retina itself. Dewar and M'Kendrick independently rediscovered the electroretinogram on the equally false supposition that a retina would display the photoelectric effect, at that time recently discovered by Willoughby Smith. In both cases the electroretinogram was the unexpected result of something expected. It was an important discovery, the first evidence for an electrochemical process generated by stimulation of a sense organ: evidence that something objective connected a physically defined stimulus to a sensory experience. Quite rightly Holmgren titled his first paper (in translation) "Method for objectivating the effect of an impression of light on the retina." It also satisfies the criterion that a discovery should have far-reaching consequences. I was myself concerned with three of them: the discovery of inhibition in the retina, the demonstration that an important component of light adaptation and dark adaptation was electrical in origin and not due to photopigments alone, and the development of the theory that generator potentials stimulate sensory nerves to discharge. Subsequent workers in this field could easily extend the list, if further proof of its importance should be required.

Quite interesting is the period of latency between the discovery of the electroretinogram and an elementary understanding of what it meant. In the present context the latency serves to emphasize that "discovery" and "understanding" really are different concepts and are not arbitrarily differentiated. There is in discovery a quality of uniqueness tied to a particular moment in time, while understanding goes on and on from level to level of penetration and insight and thus is a process that lasts for years, in many cases for the discoverer's lifetime.

The young scientist often seems to share with the layman the view that scientific progress can be looked upon as one long string of pearls made up of bright discoveries. This standpoint is reflected in the will of Alfred Nobel whose mind was that of an inventor, always loaded with good ideas for application. His great Awards in science presuppose definable discoveries. The following are his own formulations from his will: "The most important discovery or invention within the field of physics," "The most important discovery within the domain of physiology or medicine," "The most important chemical discovery or improvement." Only in chemistry, of which he had first-hand experience as an inventor of smokeless powder and dynamite, did he allow that a Nobel prize could also be given for an "improvement." It is well known that one of his major contributions to the invention of dynamite was in the nature of an improvement: he made the use of dynamite nearly foolproof by adding kieselguhr to the original "blasting oil" (nitroglycerol) that had proved so dangerous in practice. This finding may have made him realize that there are inventions and discoveries which have to be improved before their significance can be established. One should thread warily through these subtle distinctions. I can think of Nobel Prizes in Chemistry that have been given for "improvements" but do not remember ever in 27 years of Academy voting having heard any citation legitimized by this term.

It is easy to understand the emphasis, or rather overemphasis, on discovery as the real goal of scientific endeavour. By the definition used here, a discovery has important consequences and initiates a fresh line of development. It catches the eye and, in the present age, is pushed into the limelight by various journals devoted to the popularization of science—sometimes even by newspapers. In my youth we were much impressed by a philosopher at the University of Lund, Hans Larsson (who, I believe, wrote only in Swedish). I remember a thesis of his to the effect that in our thinking we try to reach points commanding a view. In science, discoveries often serve as viewing towers of this kind. The discoverer himself may not always climb to the top of his own tower. Others make haste to reach it, outpacing him. In the end many people are there, most of them trying to do much the same thing. The discoverer himself should be excused if he is possessed by a desire to find a peaceful retreat where he can do something else and quietly erect another lookout.

A systematic classification of types of discovery cannot be attempted here but some comments should be made. There are, for example, the discoveries that ride on the wave of a technical advance. At the time it became possible to stimulate nerves electrically, it became possible to discover any number of new and important mechanisms of nervous control. Small wonder that the great German physiologist Karl Ludwig could say to his pupils: "wer nur arbeitet, findet immer etwas."[1] Equally optimistic was Helmholtz when, as professor of physiology at Heidelberg, he said that it was merely necessary to take a deep dig with the spade in order to find something new and interesting. Transferring these amiable opinions of Ludwig and Helmholtz recorded by their pupil Frithiof Holmgren to the present age, one would, for example, expect every one of the large and busy brotherhood of neurophysiologists to turn out discoveries. But is this so? The question is rhetorical.

Today there is a much shorter period of skimming the cream off a new technique than there was in the 1860s. It is not uncommon to find that those workers who depend very largely on a specific technical innovation, soon become sterile even though they themselves may have had an honorable share in the development of the technique they are using.

Those who start with a problem and develop the technique for solving it

[1] "He who but works will always find something."

can in the long run look forward to better prospects. As an example one might take Erlanger and Gasser's use of the newborn cathode ray to measure conduction velocities of the component fibers of nerve trunks. On the basis of W. Thompson's formula for electrical cable conduction, Göthlin had made calculations in 1907 leading to the theory that conduction velocity in thick nerve fibers would be greater than in thin ones. Some fifteen years later Erlanger and Gasser, realizing that amplification made it possible to use the inertia-less cathode ray for tackling this question, took the trouble to overcome the deficiencies with which the early cathode ray tubes were afflicted and, as we all know, solved the problem of conduction velocity in nerve fibers of different diameter.

This is an interesting example of a rather common type of discovery, the one in which it is realized at the outset that something definite can be discovered, provided that the required technical solution can be managed. It presupposes that the experimenter knows how to formulate a well defined question and realizes what kind of obstacles prevented earlier workers from answering it. In the case of Erlanger and Gasser the basic result could hardly be called unexpected. Nevertheless most neurophysiologists are willing to classify their result as an important discovery, some perhaps merely because it had far-reaching consequences in physiological experimentation. I do so with a further motivation: many things can be predicted with a fair degree of probability, and in all good laboratories a number of such predictions, some passing fancies, others quite significant, are floating about. My respect and admiration goes to the people who reformulate such notions into experimental propositions and do the hard work required for testing them. These people are the real discoverers. The other day I saw in a student journal from the Royal Technical University of Stockholm my viewpoint expressed in a modern version: "It is easy enough to say Hallelujah, but go and do it!"

The sterilizing effect of a technique stabilized into a routine was briefly alluded to above. What then happens is that those adept in the routine easily turn into great producers of small things. Of course rejuvenation is possible. A good example is the technique of tissue culture which was for a long time in that particular state of aimless delivery but has since recovered its significance. In my own field of neurophysiology it seems that the technique of evoked mass potentials is balancing on a rather thin edge of functional relevance, all the time running the risk of becoming merely an accessory to anatomy. While this itself is a respectable science, physiology should have different aims in order to remain respectable in its own sphere. There should not be too many people within a field who care merely for the technically soluble and not for what is worth solving. However, this tempting subject will not be pursued now. Most workers, as they grow older, realize that some kind of borderline exists between those who are interested in a technique as an instrument for producing papers justifying grants, and those who see it as a possible way of furthering long-range projects.

"I have to admit," said Helmholtz in his *Vorträge und Reden,* "that those fields of study have steadily grown more pleasant for me in which one is not constrained to resort to happy coincidences and fancies" (my translation). With that basic attitude to a life in the field of science there is no alternative available than to try to realize some fundamental ideas about biological structures and their functions, that is, to promote understanding. Gradually understanding will ripen into insight. It cannot be denied that for some time "happy coincidences and fancies" may have a value that they otherwise would not possess, when fresh possibilities are opened up by a new technique. But will this inspiration last into one's old age? I daresay Helmholtz was right when he advocated working from a basis of understanding.

This attitude toward scientific work has the advantage of permitting the experimenters to devote themselves quietly to their labors without filling various journals with preliminary notes to obtain minor priorities. A disadvantage is of course the practical difficulty of persuading various foundations and research councils that their work is of some importance in a world such as ours is at present. The judgment required to appreciate the mode of progress I am advocating may not always be at hand. There is a well-known example in Fulton's biography of Harvey Cushing: After a visit to Sherrington in Liverpool in 1901 Cushing wrote in his diary: "As far as I can see, the reason why he is so much quoted is not that he has done especially big things but that his predecessors have done them all so poorly before." Sherrington, as we all know, had a good long-range program, and Cushing was no fool. One can only conclude that it can be very difficult to make others even understand the aims of long-range programs—much less support them.

There are so many instances of discoveries having led to major advances that one is compelled to ask whether it is at all possible to make a really important contribution to experimental biology without the support of a striking discovery. Sherrington's life and work throw light on this question. Most neurophysiologists would not hesitate to call him one of the leading pioneers in their field. Yet he never made any discovery. In a systematic and skillful way he made use of known reflex types to illustrate his ideas on synaptic action and spinal cord functions. Reciprocal innervation was known before Sherrington took it up, decerebrate rigidity had been described, many other reflexes were known, inhibition had been discovered, spinal shock was familiar—at least to the group around Goltz in Strasbourg, and the general problem of muscular reception had been formulated. What Sherrington did was to supply the necessary element of "understanding," not, of course, by sitting at his writing desk, but by active experimentation around a set of gradually ripening ideas which he corrected and improved in that manner. This went on for years—a life time, to be precise. Ultimately a degree of conceptual accuracy was reached in his definition of synaptic excitation and inhibition that could serve as a basis for the development that has taken place in the last thirty years. His concepts are still with us, now fully incorporated in our present approach to these problems.

The insight Sherrington ultimately reached can of course be called a "discovery," but to do so is contrary to usage. Within the experimental sciences the term "discovery" is not applied to theories acquired in this manner, even though the experimenter himself may feel that he has had his moments of insight coming like flashes of discovery after some time of experimentation.

Another example illustrating slow ripening of fundamental insight is provided by Darwin's life and labors. Back in England after the long cruise in the Beagle he went to work. "My first note-book," he said, "was opened in July 1837. I worked on true Baconian principles, and without any theory collected facts on a wholesale scale, more especially with respect to domesticated productions, by printed enquiries, by conversation with skillful breeders and gardeners, and by extensive reading.—I soon perceived that selection was the keystone of man's success in making useful races of animals and plants. But how selection could be applied to organisms living in a state of nature remained for some time a mystery to me" (1). Malthus' *Essay on Population*, a book still quite readable, gave him a "theory by which to work," because, he says, he was "well prepared to appreciate the struggle for existence" which would tend to preserve favorable variations, and tend to destroy unfavorable ones.

Darwin described flashes of insight in his work—as all scientists could do—but essentially it was twenty years of hard labor scrutinizing the evidence for his thoughts that in the end brought clarity. In 1858 he published a preliminary note together with Wallace who had independently arrived at similar conclusions; in 1859 appeared his *Origin of Species*. The idea of evolution was by no means new. His granddaughter Nora Barlow emphasizes that "to Charles Darwin it was the body of evidence supporting evolutionary theory that mattered, and that he knew was his own contribution" (2).

With some justification one can say that today the long, narrow and winding road to real knowledge has become harder to follow. In the face of innumerable distractions it has become increasingly difficult for the individual worker to preserve his identity. This, however, is necessary if he intends to grow and ripen within any branch of science. The point I want to make is that what we read, what we actively remember, and what we ourselves contribute to our fields of interests very gradually build up living and creative structures within us. We do not know how the brain does it, no more than we know how the world of sight gradually becomes upright again when for a while we have carried inverting spectacles. Our knowledge of the workings of our mind is of the scantiest. We simply have to admit that the brain is designed that way.

By "keeping track of one's identity" I mean cultivating the talents of listening to the workings of one's own mind, separating minor diversions from main lines of thought, and gratefully accepting what the secret process of automatic creation delivers. I can well understand that many people do not think much of this notion and prefer to regard it as one of my personal idosyncracies. Others who late in life look over their own activities, are sure to find at least something that looks like a main line of personal identity in the choice of their labors. Up to this point many colleagues are perhaps willing to agree. But a little more than that is meant when I maintain that an active brain is self-fertile in the manner described. I am convinced that if one can take care of one's identity, it, in turn, will take care of one's scientific development.

I am emphasizing all this so strongly because there are today so many distractions preventing scientists from enjoying the quietude and balance required for contact with their own creative life. The cities and the universities are becoming more restless, and the "organization men" with their meddlesome paper work of questionnaires and regulations tend to increase in number while the number of teachers relative to students decreases. This development tends to breed a clientele of anti-scientific undergraduates demanding more and more of the universities and less and less of themselves. The research workers withdraw into separate research institutes—furthering the deterioration of standards in teaching and in intellectual idealism in the faculties of our ancient sites of learning. Science does indeed need a number of pure research institutes, but university faculties left to themselves and engaged wholly in teaching can hardly be called universities; these should be capable of living up to the true "idea of a university" in the sense that it once was defined by Cardinal Newman in his well known book.

In all creative work there is need for a good deal of time for exercising the talent of listening to oneself, often more profitable than listening to others or, at any rate, an important supplement to the life of symposia and congresses. Perhaps this latter kind of life is also overdone in the present age. There are so many of these meetings nowadays that people can keep on drifting round the world and soon be pumped dry of what is easier to empty than to refill.

My plea for a measure of "self-contact" is really that of the poet and essayist Abraham Cowley (1618–1667) who said that the prime minister has not as much to attend to in the way of public affairs as a wise man has in his solitude. If there are those who experience nothing when trying to listen to themselves, this need not always indicate congenital defects. They may have been badly trained or may have been too lazy to absorb the knowledge and experiences that the brain needs for doing its part of the job.

Against this background one can raise the question of whether all creative originality in science is necessarily inborn or whether there also exists an acquired variety of this valuable property. I suppose most people share my view that great originality in creative work is part of a man's inheritance. But after half a century of scientific activity I have had the opportunity to observe the development of many contemporary as well as younger colleagues. I feel that a perusal of these experiences—without mentioning any names—might suggest an answer to this question or at least provide an opinion. It seems then that some of those who as young men did not show much promise of originality, although quite capable of the necessary intellectual effort, later have given original contributions to our science. How should this observation be interpreted? Obviously I may have been mistaken. On the other hand, one does not often make mistakes about real originality—quite apart from the fact that real originality often insists upon being recognized. I do not believe that the category of people of whom I am now speaking were wrongly assessed at an earlier date. Rather it is my conviction that these are the very people who without difficulties have managed to explore their own mental resources so as to make profitable use of them. They have had the capacity for listening quietly to their own minds and to the good advice of others, and in this way have grown, blossomed, and born fruit.

These conclusions will appear more evident if one considers progress within any individual branch of science. It is well known that in each phase of development the same ideas turn up in many laboratories of the scientifically active world. It is hardly necessary to add examples, as two good ones have already been provided: Holmgren with Dewar and M'Kendrick, and Darwin with Wallace. Even Newton himself said that he had stood on the shoulders of a giant. At the time when I regularly followed the scrutiny of proposals for Nobel prizes (requiring careful study of priorities), there were many opportunities to observe independent but overlapping discoveries as the bases of proposals from different sources. This is by no means surprising. Why should not well trained people who have read much the same lot of papers and monographs come to similar conclusions about the next step in a logical sequence? Since it is often difficult to foresee what each step implies for subsequent steps, parallel conclusions may in a number of instances lead to quite original contributions based on knowledge and perseverance.

In the last instance the front line of research is created by minds whose combined effort more or less perfectly represents the inner logic characteristic of a particular period. Many professional scientists have good intuitive contact with the broad lines of this development. This is expressed in the saying that something reflects the "characteristic note of the age." The original part of what is called "originality" is a capacity for understanding, intuitively as well as logically, what is an important step forward within any specific branch of science. A creative scientist has more numerous, better developed and more precise contacts with the characteristic note of his age, and can therefore, if health and perserverance do not fail him, make greater contributions than others.

I think I have said enough in defense of my thesis that acquired originality exists. Such acquisition requires intense work—preferably within one particular sphere of problems—and, obviously, enough talent to support a reasonable rate of intake. I have always believed that in most cases people have enough talent for handling research, at least when working with a team, and that failures should be accounted for by other factors which I need not enumerate in this connection.

All this implies that, in the present era of rapid communication by many channels, the individual scientist has but a share in the process of scientific discovery and understanding: even if he abandons his field of research, its development will continue, though perhaps in a slightly different way or at a slower pace.

From this standpoint it is profitable to contemplate the disturbing and often pathological quarrels concerned with the ownership of ideas. Ideas, notions, and suggestions are often thrown out in passing at meetings or in laboratory discussions and may sometimes fall on fertile ground. Who then is the owner? The one who made the suggestion may or may not have intended to make anything out of it. Again I maintain that the only definable ownership belongs to the man who develops the idea experimentally, or propounds it as a definite and well-formulated hypothesis capable of being tested.

Fights about priorities are never as violent as when a discovery is at stake. This is well known, and, if I mention such matters briefly, it is merely to point out the dangers of too much emphasis on the need for making a discovery, and to contrast it against the more peaceful life of development of understanding—without looking askance at "discovery" and what it may bring in its trail in the way of specific rewards.

I began by comparing the efforts of young men with those of men old enough not to be called young and by trying to show with the aid of two famous examples that it is by no means necessary to make any discoveries at all to do extremely well in science. It is not my intention to undervalue discoveries, but only to emphasize that it is really understanding that scientists are after, even when they are making discoveries. These are or can be of little interest as long as they are mere facts. They have to be understood, at least in a general way, and such understanding implies placing them into a structural whole where they illuminate a relevant step forward or solidify known ideas within it.

Since understanding or insight is the real goal of our labors, why make so much noise about discoveries? Why indeed? Perhaps because they provide instantaneous excitement—releasing the "eureka," whose echo we hear reflected across the centuries, and because they offer the immediate rewards found in the appreciation of colleagues, laymen, and donors. The alternative, the slow development of a world of conceptual understanding in the manner of a Darwin or a Sherrington is of course far more difficult to follow. If it is worth a great deal to have some good ideas when one wants to make a discovery, then it is an absolute necessity to have them if one intends to take that long road whose ultimate goal is to reveal fundamental principles guiding the development of knowledge in any field.

This second variant of scientific endeavour does not always suit the impatient passion of the young, ruled by an ambition which craves immediate satisfaction, but a little later in life it provides feelings of assurance and satisfaction in one's work. The pleasure of living to see a synthesis mature after years of labor helps the worker to maintain a more generous attitude toward the results of others and also to mention more freely the names of colleagues whose findings have contributed to the understanding ultimately achieved. Work becomes less competitive and the atmosphere of a laboratory friendlier. Such an attitude is particularly valuable in research institutes where people have to defend themselves by delivering results and have no chance of escaping into teaching or administration. The long-range program protects the individual worker and fosters insight of the kind that makes disputes about "intellectual ownership" meaningless.

LITERATURE CITED

1. Barlow, N., Ed. 1958. *The Autobiography of Charles Darwin*, London: Collins, p. 119–120.
2. Ibid. p. 157.
3. Beveridge, W. I. B. 1961. *The Art of Scientific Investigation*. London: Heinemann.
4. Cannon, W. B. 1945. *The Way of an Investigator*. New York: Norton.
5. Granit, R. 1941. *Ung Mans Väg till Minerva*. Stockholm: Nordstedt.
6. Granit, R. 1947. *Sensory Mechanisms of the Retina*. London: Oxford Univ. Press.
7. Granit, R. 1955. *Receptors and Sensory Perception*. New Haven: Yale Univ. Press.
8. Granit, R. 1966. *Sherrington. An Appraisal*. London: Nelson.
9. Granit, R. 1970. *The Basis of Motor Control*. London, New York: Academic.

Ann. Rev. Microbiol. 1977. 31:1–12
Copyright © 1977 by Annual Reviews Inc. All rights reserved

A "PURE" ORGANIC CHEMIST'S DOWNWARD PATH

Michael Heidelberger

Michael Heidelberger

Ann. Rev. Microbiol. 1977. 31:1–12
Copyright © 1977 by Annual Reviews Inc. All rights reserved

A "PURE" ORGANIC CHEMIST'S DOWNWARD PATH

❖1694

Michael Heidelberger

Emeritus Professor of Immunochemistry, Columbia University, New York,
New York 10032 and Adjunct Professor of Pathology (Immunology), New York
University School of Medicine, New York, New York 10016

We attribute to "pure chance" (automaton) all those events which are such as ordinarily admit to a telic explanation, but which happened on this occasion to have been produced without any reference to the actual result. The word "luck" (tyche), on the other hand, is restricted to that special type of chance events which (1) are possible objects of choice, and (2) affect persons capable of exercising choice.

Aristotle, as translated by Wheelwright (22)

I begin with this appearance of profundity because chance and luck have been frequent determinants along my way. And I add obstinacy, because I was known as an obstinate child and, having made up my mind at the age of eight that I wanted to be a chemist, though without knowing why, I stuck to it and became one.

I was born on East 127th Street in New York City on April 29, 1888. My grandparents were Jewish Germans who emigrated between 1840 and 1850, apparently because of the greater opportunities in the United States. I know little of my father's parents, except that his father, Michael, died in Silver City, Idaho. David, my father, was born in Philadelphia and left school early to earn a living. He became a partner in a firm that made carriage robes, but with the advent of automobiles, he went "on the road" almost six months of the year selling lace curtains for another manufacturer. My maternal grandparents settled in Norfolk, Virginia, and sent their oldest daughter, Fanny, my mother, to a young lady's finishing school, where she learned to play the piano. They also shipped her off to relatives in Nürnberg for a year. She and my father were married in 1884. Their first child died before I was born and I was followed in two years by a brother, Charles. My parents were devoted to each other, but the necessarily long absences of my father gave my mother much responsibility and resulted in a strictness that she herself characterized as "Spartan."

1

THE EXCITEMENT OF SCIENCE

As a consequence of her year abroad, Charles and I were required to speak German at the table and were accompanied to Central Park two afternoons a week by a governess who would tolerate nothing but French. We hated both burdens, and it was only years later, when these languages became essential, that I realized what a head start I had had.

Charles and I wandered in safety all over Central Park and knew several of its policemen by name. We ranged over many more distant parts of the city with the help of one or two friendly drivers of delivery wagons, this without the knowledge of our parents. We also gravitated often to the Metropolitan Museum of Art and the Museum of Natural History. All of these excursions were possible because my mother taught us all the primary school subjects in an hour or two each day, leaving the remainder for reading (I read *David Copperfield* and was taken to opera and concerts when I was eight years old), wanderings such as those above, and playing in the street with other children just out of school.

Later, when one of my teachers at public school, a Mr. Curtis, was invited to dinner, he was told that I wanted to become a chemist. "Buy him Cooley's 'Physics,' " he said, "he ought to know physics before chemistry, and if that doesn't discourage him, maybe he will really go on to be a chemist." I skimmed rather superficially through the book and was not frightened off. The real introduction to science followed in the eighth grade at the Workingman's School of the Ethical Culture Society, with a beautiful course in botany by Dr. Henry A. Kelly, followed in the Ethical Culture High School by his tour through zoology and well-taught courses in physics and chemistry by William E. Stark, with many hours in the laboratory. These led to my first research, an unpublished venture with Mr. Stark. Late in 1904, the Seventh Avenue-Broadway subway was about to open and the newspapers contained dire predictions of mass suffocation in the soon-to-be crowded trains. During the Christmas holiday, we lugged a five-gallon demijohn of water about a quarter of a mile to the nearest subway platform, emptied it there, replacing the water with air, and bore the stoppered demijohn back in triumph to the laboratory. There we exploded a small portion of the air with hydrogen in a eudiometer. Something went wrong, however, and we found only five percent of oxygen!

At the Ethical, I also learned to write good English from Percival Chubb and was taught higher mathematics, even trigonometry, by Matilda Auerbach, a superb but very strict teacher. As a freshman at Columbia, I started with an exciting course in qualitative chemical analysis, given for the first time according to the ionic theory by a young instructor, Hal T. Beans, against the wishes of the old-fashioned head of the department, a Professor Wells. He had Beans brought up on charges before the president, Nicholas Murray Butler, but Butler said Beans was right to teach in the most modern way possible. Another memorable course was Charles F. Chandler's Industrial Chemistry, which consisted mainly of personal anecdotes ranging far and wide. Quantitative analysis, the next year, with Floyd J. Metzger, led eventually to an offer of a fellowship if I would go on to a Ph.D. with him, but by that time I had tasted the joys of organic chemistry and decided on that for my doctoral work. I did get a master's degree, with Metzger, and we published two small papers together (17, 18). The second was rejected by W. A. Noyes, Senior, at that time the peppery editor and sole referee of the *Journal of the American Chemical Society,* on the ground that we did not know how to standardize potassium permanganate. Metzger, who had been watching me like a hawk, sent the paper back, saying that if Noyes didn't take it he would send it to the *Zeitschrift für analytische Chemie* with a note explaining why. Noyes took the paper.

Organic chemistry was taught by Marston Taylor Bogert and John M. Nelson (affectionately called Pop Nelson), both excellent teachers. After the preliminaries, Bogert would write two unrelated compounds on the board and ask you to synthesize one from the other. All around the room, a handful of us at each session would be racking our brains at the blackboard, while the rest of the class looked on in grim amusement, knowing their turn would come, too. I became interested in organophosphorus compounds as a subject for my thesis and Bogert wisely let me have a fling at the ones in the chemical museum. However, they were difficult to handle and sometimes caught fire in the air, and before long I was ready to join the other grads and work on quinazolines, with which one could figure out everything on paper and all compounds were nicely crystalline. An earlier student had made a quinazolone phthalone as a brown powder with hopefully useful properties as a yellow dye, and this topic was assigned to me. The more I purified the phthalone, the paler it became, so I added bromine to a derivative of it, hoping to intensify the color. Uncooperatively, the last traces of color disappeared. This intrigued me, and with Nelson's coaching I found that oxidative splitting had occurred (1). But this took time and I obstinately stuck to it, and at the end of two years I had only twenty new compounds instead of the forty traditionally required. Incidentally, one of my fellow students had his forty new compounds but was never able to do a C and H combustion correctly, also a requirement. Both of us were finally recommended for the Ph.D. degree, and guess to what eminence my colleague rose: a Professorship of Organic Analysis at a large, well-known university!

As for me, the question was "what next?" A part-time, older, German-American laboratory instructor named Hoffman had often pointed out the deficiencies of chemical, especially organic chemical, teaching and insisted that at least a year in a European laboratory under one of the recognized masters was a prerequisite to a successful career. My parents were willing to stake me to this, and I was selfish enough to accept what I knew was a sacrifice. But first they wished me to seek advice from old Dr. Samuel Meltzer, who had listened with his massive head against my chest as our family physician throughout my boyhood and had then become the distinguished head of Physiology at the Rockefeller Institute for Medical Research. He tried, wisely, I think, to discourage me, saying a scientific career was nothing for a poor man's son, but he gave it up and turned me over to the chemists, who would know the best person under whose direction to work. Thus, I had the luck to meet P. A. Levene, Walter A. Jacobs, and D. D. Van Slyke at tea that same afternoon. I had thought of going to Emil Fischer, with whom all three had worked, but they said he was aging and assigning his students to his assistants. On the other hand, Richard Willstätter was much younger, very able, and doing outstanding research on alkaloids and chlorophyll. Again, luck was with me, for Willstätter was willing to have me come.

That September, 1911, I boarded the French cabin liner "Chicago" (off-season rate, $50, for an outside cabin to myself, good food, and 11 days to Le Havre) and soon found myself fighting hard not to fall in love, for I had been brought up with the now strange notion that that was taboo until one could support a wife. My struggle was a result of a remarkable, attractive, and intelligent girl on her way to the Sorbonne for a Ph.D. (which she won) and who could read a whole page at a glance. Anyhow, we enjoyed being together and even renewed our friendship forty years later when we were both living alone and she was Professor of English Literature at a leading women's college. But I did fall in love with Paris during a week with as few hours as possible in a little back room on the top floor of a small, still existent hotel. Shortly after, Willstätter received me kindly and hopefully at the Federal Polytechnic Institute in Zurich. I became one of about twenty students to whose laboratory tables the Professor came twice a day. His first assistant was Arthur Stoll, who became the guiding spirit of the Sandoz Company of Basel, and his second assistant was Laszlò Zechmeister, with whom an abiding friendship developed and who came to the United States many years later as Professor of Organic Chemistry at the California Institute of Technology. I wanted to work on chlorophyll, but Willstätter said that would take two years—one to learn methods and another to penetrate further into its structure. But he had a one-year problem, a study of cycloöctatetraene, which had been synthesized in the laboratory by Ernst Waser, who then left for a job in industry. It would be an expensive problem, starting with a rare alkaloid, pseudopelletierine, and requiring much silver nitrate to remove nitrogen atoms by the Hofmann degradation (24). One had to pay for all materials, and when I doubted that my father, who was footing the bill, could afford it, Willstätter proposed that he and I should alternate in buying, whereupon he presented me with 500 grams of alkaloid and saw to it that it was always his turn when we had to buy silver nitrate and my turn for acetic acid or the like. I started with a practice degradation of a less-valuable alkaloidal derivative, and while I was watching the steam distillation, the mixture suddenly crystallized but fortunately remained fluid. Watching this unexpected happening intently, I did not hear Willstätter come up behind me. Suddenly, an arm with a pointing finger shot past my head and a voice hissed in my ear: "Was-s-s is-s-st das-s-s?" I was so startled I could not answer, and only later did I realize that sodium hydroxide had crystallized temporarily from the hot alcoholic solution. I am sure this made Willstätter realize that his new American student was very inexperienced in the handling of the unexpected or of a difficult problem, but that was precisely why I had come to him. Cycloöctatetraene, as made from the alkaloid, had the unpleasant habit of throwing bridges across the 8-carbon span, so that in addition to the final steps of preparation one had to do all necessary experiments with the product as rapidly as possible. This meant starting at 8 AM and working steadily until 3 the next morning. Not only did all details have to be thought out and all apparatus have to be readied for the appointed time, but there were difficulties unknown to the modern research worker. I mention one: the only way to obtain solid CO_2 was to invert a tank of the compressed gas on a diagonal wooden frame and blow the gas into a leather bag. Yields were low, and around midnight it was necessary to cross the street to the convenient pub, Café Friedegg (still there), and shoulder one of the tanks supplying the beer pumps. I had three of these nineteen-hour workdays, and the next day I was fit only for cleaning up.

There were also diversions. I played clarinet in the student orchestra, and in the middle of the overlong annual concert we played the Mozart quintet—the only time I have had a musical criticism in a daily paper, the "Züri Zietig," as the Neue Zürcher Zeitung was popularly called. Then there were fine operas and also concerts, for which one could borrow a regular student's card and be admitted for almost nothing. At these I heard Fritz Kreisler and Percy Grainger for the first time, and even the clarinetist, Mühlfeld, for whom Brahms wrote so superbly. I also had long Sunday walks with Hugh Clark, my English neighbor in the lab, a ten-day memorable trip at Easter to Italy, and, at the end of the academic year, the Professor's excursion up the lake to a fine restaurant, where, after dinner, I played a few selections and Zechmeister, with great seriousness, read a beautiful paper on "hippopotamuric acid."

Through all this I was worried about the future, and though offered an assistantship by Professor Eugen Bamberger at the Polytechnic, I wrote letters to a score of universities in the United States, proffering my services as instructor in organic chemistry. Not only was I turned down by all but the very last, but most of the answers bore only two cents postage instead of the then required five. The Swiss imposed a double penalty, so that being refused was expensive as well as discouraging. In the end, I received an appointment at the University of Illinois. After writing up the year's work with Willstätter for publication (23) (I asked him for a letter of recommendation and he said "Shall I write it the way I think, or the way you wish?"), I started for home, intending to see something of Germany and England on the way.

At Nürnberg, where I stayed a few days with relatives of my mother, a cable came from my father stating that I could have a job at the Rockefeller Institute if I returned at once and was approved by its director, Dr. Simon Flexner. As I had liked the Institute's chemists who had given me such good advice, I took an upper berth, second class, on the steamer that could get me back the fastest, the "Provence." Here, pure chance, as defined by Aristotle, was operative, or was it luck, since it aided in the choice? Getting into a compartment at random on the boat train from Paris to Le Havre, I found myself in the company of two professors from the University of Illinois and their wives, pleasant and interesting people, but with a lack of enthusiasm for their university that made me all the more inclined to accept the appointment at Rockefeller if I passed muster. This position was to assist Dr. W. A. Jacobs in synthesizing drugs for the cure of poliomyelitis, as part of the study of the disease by Dr. Flexner himself, and I started in September, 1912, as a Fellow of the Institute at a salary of $1200 a year.

One of the first things my father insisted upon after my return was for me to add up my expenses. I wondered at that, for he was not interested in details not mentioned in my letters, only in the total. I soon found out, for when I had calculated it as somewhat over $800, he immediately wrote out a check to my brother for the same amount. Alas, this benefitted Charles very little, for a sequel to an earlier attack of rheumatic fever, endocarditis, caused his death the following spring. Early in his illness, the Health Department of the City of New York immunized a horse with the strain of *Streptococcus viridans* isolated from Charles' blood. This first contact with immunology was not encouraging, for large injections of the horse's serum failed to help.

The Rockefeller Institute was rather small in 1912, with two buildings containing laboratories and another for the hospital. There were a few tables in a small lunch-room, a very necessary adjunct, as the nearest eating places were quite distant. Jacobs proved to be a very shy, kindly person and a keen organic chemist with original ideas. He made it clear that I was an essential part of the laboratory and that my name would be on every paper published, as my work would give him time for additional syntheses. He also taught me, when, for example, a reaction had to be refluxed for several hours, to start something else so that I could carry forward a number of syntheses simultaneously, keep working steadily, and not get things mixed up, all by his own example. He and his wife, Laura, were very hospitable and I was often at their home in Mt. Vernon, where Walter played the pianola very temperamentally. He had perforated rolls of all of Beethoven's piano works and of arrangements of the symphonies for the piano. At first we went to the lunch room of the Institute together, but as he always led me to the table at which the other chemists sat, I soon realized that I was missing much of the life of the Institute. After a month or so of this, I plucked up enough courage to ask if he would mind if I went to lunch at my own convenience. He did not, and then I began to meet men such as Jacques Loeb, Peyton Rous, Alexis Carrel, Hideyo Noguchi, and John Auer, all of whom would linger at the table and talk with young, unknown beginners like myself. I became an authority on Jacobs, often being asked questions about him, as he was scarcely known to the rest of the Institute because of his shyness and his habit of always lunching with the chemists.

For the chemotherapy of polio, Jacobs chose as his lead the slight therapeutic effect of hexamethylenetetramine, several quaternary salts of which had been made. He foresaw that a large number of such salts could be prepared from the almost limitless choice of chloro, bromo, and iodo compounds, both aliphatic and aromatic. We bought what we could, synthesized large amounts of others, and combined them with hexamethylenetetramine (11, 11a, 14, 15). Some of the new salts were highly bactericidal. One or two even seemed to delay the death of monkeys infected with the virus, but this was found to be the result of a loss of virulence of the virus itself.

At this juncture, with Dr. Flexner's encouragement, we turned to African sleeping sickness, or trypanosomiasis, which was making whole regions of that continent uninhabitable. Atoxyl, para-acetaminophenyl-arsonate, widely used by Thomas in its treatment (20), was dangerously toxic and not very effective. As Jacobs and I began to synthesize new and hopefully better arsenicals, we were joined by Drs. Wade H. Brown and Louise Pearce, who were to test our compounds for efficacy against trypanosomes in animals. Paul Ehrlich of Frankfort, the great pioneer of modern chemotherapy, had started on the same quest but it had led to "606" or Salvarsan, a cure for syphilis, instead. His most active synthetic against trypanosomiasis was the trivalent, highly toxic, and therefore unusable para-arseno-phenylglycine. Jacobs thought that the inactivity against trypanosomes of the less poisonous pentavalent para-phenylglycine arsonic acid might be a result of its highly reactive carboxyl group, which could combine with many components of tissues rather than with the parasites. He proposed masking the –COOH by converting it into –$CONH_2$, the amide, a grouping common to tissues, and this substance, sodium para-phenylglycinamide arsonate, our first and simplest arsenical, was more active than the many others that we synthesized in attempts to improve upon it. Dr. Flexner named it "Tryparsamide" and it was patented for purposes of control (10). A large batch was prepared by Powers, Weightman, and Rosengarten, a Philadelphia pharmaceutical concern, and when World War I was over, Louise Pearce conducted a field test in the Belgian Congo. She found Tryparsamide to be superior to the previously used drugs, and tests in this country showed it to be useful in the treatment of tertiary syphilis as well.

The value of Tryparsamide was eventually recognized by the Belgian government by bestowing upon us (and Thomas, who had introduced Atoxyl) the Order of Leopold II and monetary awards. Louise Pearce was given the lion's share deservedly, for she had risked her life and handled the field tests beautifully. I was angry, though, that Brown, who had merely tested our drugs on animals, received twice as much money as Jacobs, whose brilliant idea led to Tryparsamide, while I, the lowest in the group, was given as much as Jacobs. I immediately protested to the Belgian consul general and to the ambassador, but they claimed that nothing could be done after the king had signed the decree. Walter never complained, but my blood pressure still goes up as I write this.

Supplies of Salvarsan were very meager after the onset of World War I, and such difficulties were encountered in attempts to manufacture it here that Ehrlich was accused of deliberately omitting some essential step in his application for a patent. This was not true, however, for Jacobs and I, who were used to handling arsenicals, had no trouble in making the substance. By this time, I was a 1st lieutenant in the Sanitary Corps of the Army, assigned to U.S. Laboratory No. 1, the Rockefeller Institute, much of which was given over to training Army physicians in laboratory techniques. Jacobs and I thought we could improve upon Salvarsan and actually synthesized an analogue that was at least as active and less toxic. After exhaustive tests in animals by Brown and Pearce, it was tried with good results on about 100 human cases of syphilis. However, a second batch inexplicably resulted in several dangerous cases of dermatitis and we withdrew the drug.

In the early summer of 1915, when it appeared that we might soon be at war, I enrolled in a six-week officer's training course at Plattsburg, New York, to learn something of military life and soon found out how easy it was to acquire the art of killing. After that, armed with a Marksman's button and a commission as 1st lieutenant in a volunteer army that was quickly found to be utterly useless, I went cross-country to a summer camp at Center Lovell, on Lake Kezar in Maine, where I had my own canoe. The first evening, while I was playing Pergolese's "Nina" with my friend Stanley Ries at the piano, the door opened and a startlingly lovely girl walked in. Stanley stopped playing and said, "Meet Nina!" Would Aristotle have called this "chance" or "luck?" Companionship with Nina Tachau was a joy, and the following June we were married. Our son, Charles, was born in 1920; his mother died of cancer in 1946: we had been comrades for 31 years and had helped each other in our respective careers, for Nina was a writer and later became a frequent speaker for the American Association for the United Nations and chairman on foreign policy for the New York City League of Women Voters. While on this personal note, I might as well tell of living alone for the next nine years, bouyed by the excitement of the laboratory, the encouragement of warm friends, and visits to my son and his growing family. A good friend, Nellie Doogan, who had been in the family for years, came in daily to cook breakfast and clean; another, Anna Greene, cooked dinners two or three times a week. Then, one evening, I was invited to play the Mozart trio for clarinet, viola, and piano by Vally Weigl, widow of Karl Weigl, the composer. The violist was Charlotte Rosen, an outstanding musician and a cheerful, outgiving woman who had been a concert violinist in her native Germany and whose husband, a dermatologist, had died several years before. Though she lived in the same apartment house I did, I had never seen her, as she went to the second floor via the stairs while I used the elevator to the tenth. We walked home together, and again chance or luck and music brought companionship that ripened into marriage. With two happy marriages in one lifetime, I count myself among the most fortunate.

One of my duties as 1st Lieutenant at U.S. Laboratory No. 1 in 1917–18 was to march the visiting medical captains, majors, and lieutenant-colonels up and down while teaching them the Sanitary Detachment Drill, which I had learned at Plattsburg. This was also the year of a severe epidemic of influenza, at that time thought to be caused by an influenza bacillus. Dr. Martha Wollstein, of the Institute, prepared a vaccine from this bacillus and I persuaded all of my close relatives to receive injections of it. None of us caught the viral disease! Chance, once more!

At the close of the war, Jacobs and I decided that we had had enough of pure synthetic organic chemistry, but Dr. Flexner, an ardent believer in chemotherapy, insisted that we tackle bacterial infections, notably pneumococcal and streptococcal diseases. Lloyd D. Felton joined us for the testing in animals, and we started to synthesize increasingly active bactericides, including a number of cinchona alkaloidal derivatives (8–8b) more potent in vitro than Optochin (12), which had been used with some success against local pneumococcal infections. However, the combination "drug and bug" usually killed the test mice more quickly than the drug alone. One of the intermediates that we converted into such useless substances was para-aminophenyl sulphonamide, or Sulfanilamide, which the Tréfouels, Nitti & Bovet (21) found to be the active portion of the purple dye for which Domagk received the Nobel Prize in 1939. That so simple a substance could cure bacterial infections by a mechanism other than direct killing of the microorganisms never occurred to us. If it had, we might have saved hundreds of thousands of lives in the twenty years before Domagk, the Tréfouels, Nitti, and Bovet made their discoveries. I always told this story in lecturing on chemotherapy in the course on biochemistry at the College of Physicians and Surgeons of Columbia University and begged the students never to allow themselves to become slaves of an idea.

After nine and a half years of chemotherapy, Walter and I were still good friends but were eager to go on to something else, and, we agreed, independently. Dr. Flexner finally consented to our dropping out of chemotherapy, Felton went to the U.S. Public Health Service, and Walter went ahead with structural studies on cardiac-active glycosides, which we had started together (13), and related alkaloids. I wanted to find the active principles of several ancient Chinese drugs that had attracted notice in Western medicine, but Dr. Flexner did not consider that worth equipping a new laboratory and sent me over to Van Slyke, who was then chemist to the hospital of the Institute, to learn some biochemistry "so we can find you a job more easily somewhere else."

The hospital, devoted to research on a limited number of diseases and to the care and treatment of patients with these diseases, was directed by Dr. Rufus Cole, who also headed the team studying pneumonia. Van Slyke had a group actively probing the functions of the kidneys, and in 1921, when I started work with him, he had just begun a study of the equilibrium between oxygen and hemoglobin with Dr. A. Baird Hastings. They were having trouble obtaining enough purified oxyhemoglobin with intact oxygen-carrying power for their tonometric experiments. It devolved upon me to eliminate the trouble, and as this was primarily an organic chemical problem, I was soon making many grams at a time of crystalline equine oxyhemoglobin with virtually 100% oxygen-carrying power (4). This involved keeping the materials at low temperatures, so we put a No. 2 International centrifuge into a cold room. Eventually, my laboratory helper was threatened with a nervous breakdown as a result of the many colds caught from the necessary sudden changes in his working temperatures. Accordingly, I asked the president of the centrifuge company, then a frequent visitor to laboratories using his centrifuges, to build me one with an insulated brine coil around it, as the hospital had circulating brine to cool the banks of refrigerators and the cold rooms. The new instrument fulfilled its purpose and, thus, was the first refrigerated centrifuge invented and operated. Many of the visitors to the laboratory immediately ordered copies, as the wonder was how such a machine, so essential to the development of modern biochemistry, microbiology, and, indeed, molecular biology, had not been devised earlier. Thousands of refined, improved models are now in use throughout the world—my financial profit was $50 for writing the International Company's first descriptive booklet.

During the two years that I was making oxyhemoglobin and learning biochemistry, several more instances of good luck occurred. Dr. Karl Landsteiner, the famous Austrian pathologist and immunologist, became a Member of the Rockefeller Institute, and as he was interested in the immunological properties of hemoglobin, we were soon collaborating (9, 16) and I was learning my first immunological techniques from a great master. Then, Dr. Walter W. Palmer, head of the Department of Medicine at the Presbyterian Hospital, spent some months working with Van Slyke. Acquaintance with him was later to prove of crucial importance to me. And last, but not least, Dr. Oswald T. Avery, microbiologist of the pneumonia team, would come in from time to time with a small vial of brownish powder to say, "When can you work on this, Michael? The whole secret of bacterial specificity is in this vial." Eventually, the need for oxyhemoglobin ended and I was transferred to the pneumonia group, although "Van" generously let me continue to use the same laboratory. Thus began my career as a microbiologist.

In 1917, Dochez & Avery had published their discovery of the serologically type-specific "soluble specific substance" of pneumococcus (3), and five years had elapsed before a chemist undertook its characterization. In the meantime, Dochez had left the Institute, leaving Avery, or "the Fess," as he was affectionately called, to carry the study forward. At that time, there were only three "fixed types" of Pneumococcus, type I, Neufeld's "typical Pneumococcus," and types II and III, all other serological variants having been relegated to "group IV." Fess said we had better not start with type III, which formed the largest capsules, because some people called it *Streptococcus mucosus,* nor was type I a favorable prospect because its capsules were very small. That left only type II with its capsules of intermediate size. I needed type II antiserum with which to identify serologically active fractions, so Fess took me down to the cold room where there were shelves of large bottles of therapeutically used, type-specific horse sera. I spied a bottle of type II serum half filled with mold and, to Avery's horror, took it along for the chemical tests, sterility being unnecessary for rapid, qualitative precipitin tests. Another innovation that he was dubious about was centrifuging weak precipitin tests that did not settle rapidly by themselves, but he was soon convinced by a series of adequate control tubes.

In our initial studies, we concentrated batches of meat-infusion broth cultures on a large steam bath in a hood until we had accumulated the concentrate from 300 liters. Precipitation with alcohol and centrifugation gave a three-layer separation, with the active material in the gummy middle layer. The more we purified this, the less nitrogen it contained, which surprised us, as all immunologically active substances were supposed to be proteins. Finally, when it was virtually nitrogen free, Fess said, "Could it be a carbohydrate?" So I boiled a bit with acid and got a strong test for reducing sugars, one of which was shown to be glucose (5, 5a). An acidic component was recognized, but its identification as D-glucuronic acid and the finding of L-rhamnose were accomplished much later (2, 19). One must remember that our work was done before the introduction of paper chromatography, when one had to isolate the sugar itself or a characteristic derivative and the relatively large amounts of purified material required were not easy to obtain. The conclusive test that we had the "soluble specific substance" was to precipitate some with two liters of outdated antiserum and recover essentially the same polysaccharide from the washed, proteolytically digested precipitate. To test the resistance of the type II substance (SII) to carbohydrate-splitting enzymes, Fess suggested that we write to Peterson and Fred in Wisconsin for some of their thermophilic cellulose-splitting bacteria. These microbiologists obligingly sent us a culture, and we seeded it into a slurry of filter-paper pulp made up in a 1:20,000 nutrient broth solution of SII and put it into an incubator at 55°C to stew overnight. Next morning the filter paper was gone, but the culture was still 1:20,000 as to SII!

As soon as we started to purify our first batch of the "soluble specific substance" of type III, I knew it would be different chemically from SII, as it formed dense fibers when precipitated by alcohol. Moreover, SII was dextrorotatory whereas SIII was levorotatory. Both were almost nitrogen free and contained glucose, but obviously in a different combination. SII was a weaker acid than SIII.

From the products of hydrolysis of the latter, Dr. Walther F. Goebel, who had joined us, and I isolated a glucuronoglucose, which we termed an "aldobionic acid" (7, 7a) instead of the more correct "aldobiouronic acid." Such acidic bioses were later found to occur in enormous tonnage in the hemicelluloses and gummy exudates of trees and plants, but the first discovery of this combination of sugars was in the microscopic capsule of Pneumococcus type III.

Only after isolating SII and SIII did we turn our attention to type I, which immediately confirmed Fess's wisdom and unerring intuition in not working on it first, as its number, I, would have suggested. Not only were the yields small, but SI turned out to be an amphoteric polyelectrolyte and I threw the first batch down the sink, thinking it was nucleoprotein, while Zinsser and Mueller actually recorded their analogous "residue antigen" from type I as a polypeptide. It soon developed, however, that SI contained a large proportion of galacturonic acid, defining it as a third immunologically active, chemically different, type-specific polysaccharide (7b).

By this time, Avery's imagination, which always ranged far and wide, led him to the belief that if bacteria possessed immunologically reactive polysaccharides others should exist "free in nature." We accordingly bought the few plant gums then available, and lo!, gum arabic precipitated our type II antisera. While purifying a batch, I let it stand overnight in fairly strong hydrochloric acid. When I precipitated it next day with alcohol, pentose had been split off, exposing more chemical groups common to SII, so that the degraded substance reacted at a 150-fold higher titer than the native gum (6).

During all this period, I had to keep my eyes open for a job elsewhere, for although I was happy and satisfied, Dr. Flexner probably wisely insisted that as long as I stayed my work would be known as someone else's and that I should go elsewhere and stand on my own feet. Accordingly, when Dr. Samuel Bookman retired as Chemist to Mt. Sinai Hospital in New York after forty years of faithful service and the post was offered to me, I accepted it.

The year 1927–28 at Mt. Sinai was a very busy, interesting, and instructive one, with its close contact with medicine. It was necessary to reorganize the chemical laboratories, introduce new methods of analysis, and eliminate parallel determinations yielding the same information. This sometimes led to friction with the Medical Board. Although my associate, Dr. David J. Cohn, provided able assistance with the supervision of the vast amount of chemical routine, there was little sustained time for research. Accordingly, when Dr. Walter W. Palmer, Professor of Medicine at the College of Physicians and Surgeons of Columbia University, proposed that I join him as the first full-time chemist in a department of medicine, I was glad to accept. With me, from Mt. Sinai, went my technician, Check M. Soo Hoo, who was to serve ably and faithfully in that capacity for thirty-five years.

This brings me to the end of one epoch and to the start of others: twenty-seven years of ideal working conditions at P. and S. and "retirement," the story of which I hope to relate elsewhere.

THE EXCITEMENT OF SCIENCE   251

Editors note: The story of Michael Heidelberger's career will be continued in the prefatory chapter of Volume 48 of the Annual Review of Biochemistry, scheduled for publication in July 1979.

Literature Cited

1. Bogert, M. T., Heidelberger, M. 1912. *J. Am. Chem. Soc.* 34:183–201
2. Butler, K., Stacey, M. 1955. *J. Chem. Soc.,* pp. 1537–41
3. Dochez, A. R., Avery, O. T. 1917. *J. Exp. Med.* 26:477–93
4. Heidelberger, M. 1922. *J. Biol. Chem.* 53:31–40
5. Heidelberger, M., Avery, O. T. 1923. *J. Exp. Med.* 38:73–79
5a. Heidelberger, M., Avery, O. T. 1924. *J. Exp. Med.* 40:301–16
6. Heidelberger, M., Avery, O. T., Goebel, W. F. 1929. *J. Exp. Med.* 49: 847–57
7. Heidelberger, M., Goebel, W. F. 1926. *J. Biol. Chem.* 70:613–24
7a. Heidelberger, M., Goebel, W. F. 1927. *J. Biol. Chem.* 74:613–18
7b. Heidelberger, M., Goebel, W. F., Avery, O. T. 1925. *J. Exp. Med.* 42: 727–45
8. Heidelberger, M., Jacobs, W. A. 1919. *J. Am. Chem. Soc.* 41:817–33
8a. Heidelberger, M., Jacobs, W. A. 1920. *J. Am. Chem. Soc.* 42:2278–86
8b. Heidelberger, M., Jacobs, W. A. 1922. *J. Am. Chem. Soc.* 44:1098–107
9. Heidelberger, M., Landsteiner, K. 1923. *J. Exp. Med.* 38:561–71
10. Jacobs, W. A., Brown, W. H., Heidelberger, M., Pearce, L. 1918. *U.S. Patents No. 1280119–24 and No. 1280126*
11. Jacobs, W. A., Heidelberger, M. 1915. *J. Biol. Chem.* 20:659–84; 685–94
11a. Jacobs, W. A., Heidelberger, M. 1915. *J. Biol. Chem.* 21:103–43; 145–52; 403–37; 439–53; 455–64; 465–75
12. Jacobs, W. A., Heidelberger, M. 1919. *J. Am. Chem. Soc.* 41:1581–87; 1587–600; 1834–40
13. Jacobs, W. A., Heidelberger, M. 1922. *J. Biol. Chem.* 54:253–61
14. Jacobs, W. A., Heidelberger, M., Amoss, H. L. 1916. *J. Exp. Med.* 23:569–76
15. Jacobs, W. A., Heidelberger, M., Bull, C. G. 1916. *J. Exp. Med.* 23:577–99
16. Landsteiner, K., Heidelberger, M. 1923. *J. Gen. Physiol.* 6:131–35
17. Metzger, F. J., Heidelberger, M. 1909. *J. Am. Chem. Soc.* 31:1040–45
18. Metzger, F. J., Heidelberger, M. 1910. *J. Am. Chem. Soc.* 32:642–44
19. Record, B. R., Stacey, M. 1948. *J. Chem. Soc.,* pp. 1561–67
20. Thomas, H. W. 1905. *Br. J. Med.* 1:1140–43
21. Tréfouel, J., Tréfouel, T. J., Nitti, F., Bovet, D. 1935. *C. R. Soc. Biol.* 120: 756–58
22. Wheelwright, P. 1951. *Aristotle,* pp. 29–35. New York: Odyssey. 336 pp.
23. Willstätter, R., Heidelberger, M. 1913. *Ber. Dtsch. Chem. Ges.* 46:517–27
24. Willstätter, R., Waser, E. 1911. *Ber. Dtsch. Chem. Ges.* 43:3423–45

Ann. Rev. Physiol., Vol. 29

AN OLD PROFESSOR OF
ANIMAL HUSBANDRY RUMINATES

BY MAX KLEIBER

Max Kleiber

AN OLD PROFESSOR OF
ANIMAL HUSBANDRY RUMINATES

By Max Kleiber

Emeritus Professor of Animal Husbandry
University of California at Davis

When an old professor indulges in reminiscences some of his listeners may fear that he has reached the stage of rumination, and has therefore started to waste their time by rechewing old stuff. But a cow who indulges in chewing the cud assists her microbial co-workers by making material digestible which otherwise would be useless. I hope the readers of this preface will look at this more positive aspect of rumination and will bear in a friendly mood the frequent "I" and the seriousness with which I take myself and my cud.

Efficiency of Energy Utilization and Body Size

In the third volume of the *Annual Review of Biochemistry* (1934), Samuel Brody (1) discussed a German paper of mine on animal size and feed utilization [*Tiergrösse und Futterverwertung*, 1933 (2)] in which I deduced that the total efficiency (gain/food) of two animals is equal when, with equal partial efficiency (change in gain/change in food intake), their relative food intake (food intake/basal metabolic rate, or food intake per unit of the ¾ power of body weight) is equal. Since empirical data indicated that neither partial efficiency nor relative food capacity (maximum food intake per unit of the ¾ power of body weight) is correlated with body size, I concluded that in general the total efficiency of animal energy utilization is independent of body size.

J. Mayer (3) later confirmed this generalization and called it Kleiber's law. I keep this recognition by a colleague of Mayer's stature gratefully as a spiritual tonic for periods during which I am bothered by the awareness of my various obvious shortcomings. In 1961 my connection with law was officially cinched by the University of California with the degree Doctor of Law.

In the fourth volume of the *Annual Review of Biochemistry*, Brody (4) extended the discussion of body size and efficiency of energy utilization. At his request I sent him a copy of a paper which I had submitted to the *Journal of Nutrition*. It was mainly a critique of the Palmer-Kennedy efficiency quotient (5) (food consumed/gain in weight times weight), which obviously makes the efficiency (the reciprocal of the Palmer-Kennedy quotient) proportional to body weight, an assumption that makes no sense.

The *Journal of Nutrition* did not publish the article; my critique ap-

1

THE EXCITEMENT OF SCIENCE

8243-2601/78/1127-0255$01.00 © 1978 ARI   255

peared therefore only as a personal communication in Brody's review. Brody wrote me to Switzerland that Palmer was angry, would I please explain to him my criticism. This led to the exchange of several letters which I thought had clarified the matter. To my amazement I noticed later that Palmer et al. (6) in 1946 were still using their ill-conceived efficiency quotient, finding it satisfactory.

THE LILLIPUTIAN'S CALCULATION OF GULLIVER'S FOOD REQUIREMENT

An average Lilliputian was as long as Gulliver's middle finger, say 7 cm (7). If Gulliver stood 6 feet or 180 cm he was as tall as 26 Lilliputians piled up feet on heads. With the same density and isometric build Gulliver weighed as much as $26^3 = 17600$ Lilliputians. The Emperor of Lilliput decreed that "the said Man Mountain shall have a daily allowance of meat and drink sufficient for the support of 1724 of our subjects."

To find out how the Lilliputians calculated Gulliver's food requirement we can formulate for Gulliver (G) and Lilliputian (L) as folows:

$$\frac{\text{food } (G)}{\text{food } (L)} = \left(\frac{\text{weight } (G)}{\text{weight } (L)} \right)^{p} = 1724$$

therefore
$$p = \frac{\log 1724}{\log \dfrac{\text{weight } (G)}{\text{weight } (L)}} = \frac{\log 1724}{\log 17600} = 0.76$$

The Lilliputians calculated the food requirement as proportional to the ¾ power of body weight. This anticipates the results of Kleiber of 1932 (8), $p = \frac{3}{4}$, and those of Brody of a few months later in 1932 (9), namely $p = 0.734$, by at least 233 years since the Lilliputians applied this calculation at the time Lilliput was discovered, A.D. 1699.

If the Lilliputians in their remarkable anticipation of the "man mountain's" knowledge concerning body size metabolism and food requirement had stopped at A.D. 1839 when Sarrus & Rameaux (10) proclaimed the surface law of animal metabolic rate or even as late as 1883 when Rubner (11) or 1889 when Richet (12) rediscovered this law empirically and independently, they would have fed Gulliver only the equivalent of 675 Lilliputian rations since the square of the length ratio, 26^2, is 675.

IDEA AND ACTION

Two souls, alas! reside within my breast.
Goethe, *Faust*, Part I, Scene 2

Ulrich von Hutten confessed, "I am not a finely planned book, I am a man with his contradiction." Most people have splits in their souls. One of these splits is a discrepancy between their ideas and their actions. When they preach brotherly love but exploit or suppress or kill fellow human beings we call the split hypocrisy. People differ in degree of sensitivity to internal contradictions and in degree of tolerating or even liking such disharmonies. I seem to have a somewhat greater than average allergy against mental schisms. This means an enhanced tendency to express my ideas through my actions, to do as I preach, or one might turn this upside down and call it to think in harmony with what I am doing or preach as I do. Two incidents in my life seem to show this.

1. *An experiment in asocial isolation.*—Like many, perhaps most, young people I went through a phase of antisocial feeling during the first year of college. I assume that out of such a loathing of the social environment with its restrictions, but accepting it for practical reasons, Schiller wrote his "Robbers", and Goethe "Goetz von Berlichingen" and "Werther's Leiden". Most express their rebellion in less creative ways. In Zurich we wore sandals and went around without a hat and with soft "Schiller-collars", which at that time was just as much frowned upon by the solid citizens of the second decade of this century as long hair and beards are loathed today. But that protest in appearance was not enough. I said to myself: since I hate society I should get out and prove my independence as a hermit, so I searched for a place (accessible for my limited means) where the population density was especially low and emigrated with two younger friends to the West of Alberta, Canada. Each of us took up a homestead near the McLeod River about 100 miles west of Edmonton, less than 100 miles northeast of Jasper Park. This experiment was valuable. It made me aware of the degree to which *Homo sapiens* is a social animal. For a couple of months I worked on a farm near Red Deer to earn money for buying tools. Returning to my homestead I found a chance to teach a very intelligent twelve year-old girl who frequently rode on a huge black retired race horse to where I was building my blockhouse about a mile away from her home. Teaching arithmetic and geometry soon interested me more than digging out willowshrubs and cutting down poplar trees.

2. *Conscientious objector.*—After less than a year of homesteading and teaching in the Wild West, this experiment in asocialism was stopped by the first World War. I was ordered to return to Switzerland as a soldier for guarding the frontier. Partly supported by a generous contribution of my pupil's family, I traveled to New York and was sent to Marseille on an old ship called the *Germania,* loaded with French reservists.

Between periods of military service I returned to the College of Agriculture, having passed my experiment in radical individualism and being willing to prepare myself for useful service in human society. I was an ardent Swiss patriot and I hated to find in the army an undemocratic militarism which I thought was imported by the Swiss professional officers from their training in Prussia. I became an officer in order to fight against that un-Swiss undemocratic spirit. But more and more I realized that a democratic army may be an impossibility; military training, after all, must be training for war, for inhuman slaughter, for deception and therefore for unconditionally obeying orders rather than acting according to one's own judgment and following one's own conscience.

When a superior officer, obviously drunk at a party, ordered me to drink so much that I would have lost my judgment I refused and was later reprimanded. This incident was of course silly but it serves as an example of a serious problem. It must be admitted that once military discipline is broken, even at a point of no consequence, it may crack at another perhaps critical occasion. A lot of people recognize the schism between the freedom of the personal conscience and the dictate of military discipline but the degree to which one accepts this contradiction varies and here my allergy against a split between idea and action became especially active.

My internal struggle came to a crisis when, at a particularly dangerous turn of the war, two top-rank Swiss officers were caught spying for the German High Command. In Court they explained that they had acted as patriots because Switzerland was at the mercy of the Kaiser's *Wehrmacht*.

I began to suspect that the Swiss army under the command of officers trained in Prussia was not protecting Swiss freedom and democracy and Swiss neutrality, that to the contrary the army might become a threat. Freedom and democracy in Switzerland could only be secured by abolishing tyranny anywhere—by abolishing war and militarism.

I joined a youth movement whose slogan was "no more war" [that slogan is the title of a recent book by Linus Pauling (1958) (13)]. Regarding the means to achieve our goal our movement was split; some of us followed the advice of the Bolscheviki to go into the army in order to secure access to the weapons for the coming social revolution which the capitalists would try to suppress by force. Others, including myself, maintained we should fight against militarism, refuse participation in the armed services, and work for our cause by nonviolent means. Refusal of military service never was officially recommended by the Social Democrats, because it does not fit into disciplined party action; a conscientious objector always acted strictly as a person. I thus retained a considerable amount of individualism even after my experiment in the Wild West. I refused to follow the next order to active service and was sentenced to four months in prison which I regarded as a mild punishment for an officer. The prison term was even generously delayed to let me finish an exam. But the College of Agriculture kicked me out. Many students protested against this academic punishment of a conscientious objector which they felt was a political degradation of an academic institution.

Dr. Georg Wiegner, Professor of Agricultural Chemistry, sent Rubner's book *Die Gesetze des Energieverbrauchs bei der Ernährung* to me in prison and later remarked with a congratulation that I probably was one of a few students who had read that book through, because Rubner, a genius as an experimenter, was not a good writer. For some inmates, prison is a school for better crime, for me it helped my education in physiology, especially in animal energetics.

After release from prison I got a job in the Agricultural Department of the city of Zurich and married. Later I tried to direct a small production cooperative on a farm near Zurich. Professor Wiegner visited me there and told me I could render better service to mankind if I spent my effort in a way for which I was especially qualified, namely in scientific research. This sounded good to me since I painfully realized that despite training and good intentions I was not a good farmer. So I followed Professor Wiegner's advice gratefully. His prestige re-opened the door of my Alma Mater for me, and I became an assistant in Agricultural Chemistry.

The Swiss poet Kurt Guggenheim has written a fine historical novel, *Alles in Allem,* on Zurich's recent history (14). He makes the daughter of a German businessman in Zurich express sympathy for the conscientious objector, but also voice her disappointment that the man, after refusing military service, begs his Alma Mater to let him go on with his education. This seems to mar her picture of a proud hero. All I can say is that I did not want to create the picture of a proud hero. I just feared to live with a bad conscience. This is not the feeling of pride. A haughty refusal to follow my professor's friendly advice and offer might have satisfied the aesthetic sensibility of a rich young lady but for me it was more important to demonstrate that a conscientious objector does not have to be a holy man apart from others but can become a useful research worker and academic teacher.

Words and Meaning

(Semantic fuss)

In the beginning was the word
Gospel of John 1

In the beginning was the meaning
Goethe, *Faust,* Part 1, Scene 3

As ideas are preserved and communicated by means of words it necessarily follows that we cannot improve the language of any science without at the same time improving the science itself; neither can we, on the other hand, improve a science without improving the language or nomenclature which belongs to it. However certain the facts of any science may be, and however just the ideas we have formed of these facts, we can only communicate false impressions to others while we want words by which these may be properly expressed.
Lavoisier, *Elements of Chemistry,* Preface (15)

1. *Energy.*—The chapter on energy metabolism in the sixth volume of the *Annual Review of Physiology* (16) (1944) gave me an opportunity to fuss about bad language concerning energy. At the end of the 17th century, scientists distinguished between ponderable (with weight) and imponderable (without weight) substances. According to Lavoisier's law of conservation of matter (the basis for stoichiometric calculation in chemistry),

ponderable substance can be neither produced nor destroyed. It can only change the form in which different kinds (elements) are combined. According to Joseph Black's caloric theory, heat is an imponderable substance which also can be neither produced nor destroyed but can only change from sensible to latent heat. The caloric theory was disproved by Benjamin Thompson (who in 1792 became Count Rumford) who showed that indeed heat can be produced, for example by boring a cannon (17). Nuclear physics later revealed that matter is not conserved but can change to radiant energy. The Einstein equation $E = Mc^2$ should not be interpreted as indicating that energy can be produced. It indicates that mass, now to be regarded as a special form of energy, can be transformed to another form, namely radiant energy. Heat, work, chemical energy, or mass can be produced or destroyed by changing to other manifestations of energy; and energy itself remains the only generalizing concept for what, by definition, can be neither destroyed nor produced. Therefore expressions like "the energy production of cancer cells" by O. Warburg (18) or 'the energy-producing' mechanism" by H. Krebs (19) are self-contradictory.

2. *Turnover rate.*—In kinetic studies, especially investigations of intermediary metabolism in animals, which have been, and are, greatly helped through the use of isotopes as tracers, we define the *turnover time* as the duration of time during which a metabolic pool is renewed or turned over. It means the time interval during which as many molecules of which the pool consists have entered or left the pool, whose size remains constant, a condition which we presume when we speak of turnover in contrast to growth or decay. The *transfer rate,* or *flux rate,* indicates the number of molecules which enter or leave the pool per unit of time. This rate has the dimension: number of molecules per unit of time, or grams per unit of time.

Some authors use as an additional synonym for flux rate *turnover rate,* which other authors reserve for the reciprocal of turnover time. In a short note I argued that the use of turnover rate as a synonym for transfer rate was illogical because in tracer work we are not concerned with the rate at which molecules turn over (or rotate), which would be a problem of molecular physics. In tracer work, we are concerned with the rate at which a pool turns over. The letter in which an editor of *Science* declined to publish my note alleged that I was not sufficiently aware of Hévésy's work, but the note appeared in *Nature* (20). The year before (1954) I had made a trip to Stockholm to visit Hévèsy and his laboratory. That proved my admiration and respect for the major pioneer of tracer work.

Zilversmit (21) countered my critique of the use of turnover rate as a synonym of transfer rate by the suggestion that a good analogy for his use of turnover rate of pools was the turnover rate of patients in a hospital. But this is just an excellent analogy to make my point. The mean turnover rate of the patients (number of patients turning over per unit of time) would indicate how many times on the average each patient turns over (in his bed or otherwise) per unit of time.

The analogy of what we mean by the turnover rate of a metabolic pool is not the rate at which patients turn over but the rate at which the patient population (the analogy of the pool) turns over. Mawson (22) agreed with the use of turnover rate as the reciprocal of turnover time and suggested flux rate to express the number of molecules which enter or leave a pool. In a report of May 1, 1966 to the International Commission on Radiological Units the U. S. task group on tracer kinetics accepted the term turnover rate as the reciprocal of turnover time.

3. *Degrees of freedom.*—A mean of repeated measurements with a given variance has the smaller standard error the greater the number of measurements from which the mean is calculated. The smaller the error the more accurately the mean is determined, but in using statistics in biological research we say that with increasing number of measurements the degrees of freedom increases. The mathematician may have degrees of freedom when he constructs artificial series of figures but the experimental scientist has no such choice. He is bound by the results of his measurements. To call *degrees of freedom* what for us are more nearly *degrees of determination* rubs me the wrong way, perhaps all the more because freedom is so much and so disgustingly misused anyway.

4. *A billion.*—I am also bothered, in fact more so, when I have to call 10^9 a billion. The syllable "bi" means double. Some of us are bilingual when we speak two languages; we bisect, make bilateral arrangements, and ride bicycles. In Europe a billion means 10^{12} with twice the exponent of 10 for a million, $10^{6 \times 2} = 10^{12}$ is a billion and $10^{6 \times 3} = 10^{18}$ is a trillion; 10^9 is a milliard, 10^{18} a billiard which, as above, is also a trillion. That all makes sense but why is 10^9 a billion in America?

5. *Mol or mole?*—Wilhelm Ostwald called the quantity of a substance equal to its molecular weight expressed in grams a *mol*. He properly created a new word for a new concept. The spelling most common in this country, *mole,* spoils this good terminology. A mol stands specifically for one clearly established concept but mole can mean a lot of things. We might read that a mole digging in a mole of a castle consumed ⅕ of a mole of oxygen per day, or that the injection of a thousandths of a mole of a carcinogen produced a mole in the uterus.

I am grateful to the editors of John Wiley and Sons that they allowed me to use mol in my book, and I testified gladly that this was not an oversight on their part.

6. *Large and small calories?*—I was not so lucky with the editors of the *American Journal of Physiology* who changed my *kcal* to *Cal* which embarrasses me when I explain to my students that to call a kilocalorie a large calorie is just as silly as to call a mile a large foot, and they find Cal in one of my own and rather recent publications. I understand that a cer-

tain amount of uniformity in symbols used in journals is desirable but I feel there should be the possibility of progress in terminology and authors who give good reasons for a deviation from an accepted type of expression should not be suppressed by a rigid editorial adherence to the status quo. The author should be entrusted with a certain amount of semantic responsibility if he asks for it.

7. *Specific dynamic action.*—Specific dynamic action is an erroneous translation of Rubner's *spezifische dynamische Wirkung*. The proper translation would be specific dynamic effect but dynamic has to do with work or movement whereas the effect is an increase in heat production, that is a calorigenic effect.

LINEAR REGRESSION EQUATIONS

Writing usually means a great effort to me. When I have a sheet of paper in front of me and a pencil in my hand or when I sit in front of a typewriter, my mind seems to begin to labor like a motor with a lot of internal friction. Maybe this explains why I have a tendency to become belligerent when I read arrogant statements which seem to have come to paper without hesitations. Maybe I just envy the authors who seem so free of the type of writer's constipation which delays my own work.

When a belligerent mood is aroused, writing seems to go more freely. Sometimes a note thus produced is sarcastic and not friendly enough to be printed but at one such occasion a paper resulted which I still like and consider useful, since with the IBM machines taking over a lot of our routine calculations we will soon be even more swamped with empirical linear regression equations than we are already. The paper appeared in the *Journal of Applied Physiology* (23).

THE CALIFORNIA RESPIRATION APPARATUS FOR COWS
(Amateur engineering)

In 1929 I was imported to California from Switzerland to build a respiration apparatus for big animals, especially dairy cows.

Before I left Europe I visited Professor Møllgaard in Copenhagen whose respiration apparatus for dairy cows I had studied a few years before. He said that I was welcome as a person and a guest in his house as I had been earlier but he would not show me his laboratory since now I was representing the University of California.

Later I learned that Professor Møllgaard himself had earlier been approached but had demanded too high a salary for himself and an assistant whom he planned to bring along. Dr. Hart, then the head of the Department of Animal Husbandry, was not only a good scientist but also a shrewd businessman. He wondered if there were not perhaps cheaper European scientists on the market. He got together with C. B. Hutchison, about to become Dean of the College of Agriculture, but at that time still talent-hunting in Europe for an American Foundation. He knew my teacher Professor Wiegner at Zurich and had seen the respiration apparatuses for rabbits and one for sheep which I had built mainly to learn how to design one for cows. I was offered a temporary position as Associate in Animal Husbandry with a salary of $4500, I believe, per year for a period of three years. For me that was a wonderful opportunity and in May 1929 I arrived in Davis with my wife and our nine-year-old daughter.

To engage a beginner instead of Møllgaard, then at the height of his career, was of course a risk, but Dr. Hart took that risk and reassured me later at a time when one or the other of my attempts did not work out as I expected. "If one way does not work as you like go right ahead and try another", he would say. Dr. Hart could be a powerful and helpful friend.

To get our work started Dr. Hart introduced me to Professor Walker, the head of the Department of Agricultural Engineering where to my delight I saw excellent machine shops and met competent mechanics. We planned to build the respiration apparatus as a cooperative project. Unfortunately, this plan was nipped in the bud by a young man who was then business manager at Davis. This was my first great disappointment and I was shocked to witness how two internationally recognized professors were bossed rather crudely by a bureaucratic upstart who had no academic standing. The apparatus was to be built not by the obviously ideally suited Department of Engineering but by the administrative division, Construction and Repair, directed by an old carpenter. For metal installations, blueprints and specifications had to be worked out so that the Sacramento businessmen had a chance to bid for making this equipment. So, following my inclination to accept challenges, I learned to make drawings for blueprints and with Dr. Hart's help, worked out specifications. Since my skill as a draftsman remained rather limited I had to go over to machine shops and sheet-metal works in Sacramento and explain my blueprints. The Construction and Repair Department was mainly equipped for carpenter work. The double chamber of our apparatus is a 2×4 structure with wooden walls, later lined with galvanized iron. The chambers are ventilated by moving four copper pipettes, each holding 225 liters of air, up and down in a water trough. The pipettes were built at Sacramento but the driving mechanism was assembled in Davis, also the air sampling and air collecting devices, CO_2 absorbers, and methane combustors. By a special concession Dr. Hart was able to provide me with a vise, a drill press, and even a small lathe, and the glassblower of Dr. Lewis' department in Berkeley gave me an oxygen blow torch. I designed the ventilation, air sampling apparatus, and absorbing system so that they would work even if, with rather limited skill as a mechanic, I should construct the apparatus myself. I had a chance later, however, to get the help of a farmer who in his youth had worked as a gunsmith. After two and a half years the apparatus worked (24). We could burn up three liters of alcohol a day and get the proper result within less than 2 per cent error but the machine made an awful racket. When a highly trained engineer became my assistant, we changed water valves in the airline to ground in stopcocks and also changed from a switch which reversed the motor at the end of each aspirator motion to a rack-and-pinion drive.

I succeeded in coming to California because I was cheaper than Professor Møllgaard, and the California apparatus is, I believe, cheaper than most other similar apparatuses for big animals. Ventilation with a blower and testmeter may be cheaper, though, than my air-pipetting machine, and there are good reasons for using blower and testmeter for ventilation; I cannot so well understand a return to the old extremely expensive Blackslees mercury pump which makes an extra air-moistening tower necessary and has to be calibrated by a gas meter (25) (Møllgaard and Anderson, 1917). I also fail to see the justification for using closed chambers for respiration trials with large ruminants which not only use up daily the equivalent of 50 to 100 kg KOH but also make temperature, pressure, and airtightness especially critical. A closed chamber also causes methane to accumulate so that the experiments have to be interrupted every day. I suspect that some designers of apparatus assume that the most expensive must be the best.

TELEOLOGY

1. *In the past.*—Teleology is the doctrine that the world has been created according to a plan and that nature works toward a purpose (*telos*) or purposes. This theory plays a major role in the philosophy of Aristotle (384-322 B.C.). Through his great influence it remained important in the philosophy of the Christian era and its ghost still haunts the thinking of modern scientists and plays spooky tricks in many of the writings even of those who deny its justification. The doctrine was taught long before the time of Aristotle and had already been refuted by Empedokles (490-430 B.C.), a great thinker later worshiped by Galenos (130-200 A.D.) as the founder of the Italian medical school. Recently he was hailed by the followers of Mazzini as the "democrat *par excellence* of antiquity" who honored his democratic principles by refusing an offer to become king of Agrigentum. According to Empedokles the processes of nature occur by chance but among the products formed, especially plants and animals, only those survive which fit the conditions; misfits, such as horses with human heads, are weeded out. Thus survival of the fittest among offspring occurring by chance, Empedokles' alternate to classical teleology, anticipated by nearly two and a half millennia Darwin's *natural selection* as the alternate to the teleology of our time.

Galenos was more vague in his statements than Empedokles had been; according to Verworn (26) he suffered under the dualism of the natural acceptance of causality and a teleology inherited from Aristotelian philosophy.

Descartes (1596-1650) declared that we cannot know God's purposes but Boyle (1627-1691), of gas law fame, maintained that some of the divine ends are "readable", such as the marvelous adaptation of living creatures, and that "hence it is foolish to reject the teleological proofs for the existence of God".

To Galileo (1564-1642) the teleological terminology of the scholastics was, as Burt (27, p. 134) put it, no longer serviceable, but Newton (1642-1727) maintained a mystic credo and declared that the motions which the planets now have could not spring from any natural causes alone but were impressed by an intelligent agent (27, p. 289). As a faithful former student helps an old professor by keeping him active, so Newton tried to "keep God on duty searching the Universe for leaks to mend." This idea was repugnant to Huyghens and to Leibniz who postulated that God had created a world which did not need supervision and repairs (27, p. 299).

In 1742 a French medical doctor, Julien de la Mettrie, wrote a book on the natural history of the soul followed in 1748 by another, *Homme Machine*, man a machine. Both books got him into trouble with the clergy who accused him of atheism, and he had to escape to Holland. When he was no longer safe even in Holland, Frederick the Great of Prussia, the monarch who proclaimed that in his kingdom every man was free to save his soul in his own fashion, offered asylum to the persecuted atheist (28).

In an attempt to clear science of theology, the postulate that man is a machine is a rather tricky analogy because an essential characteristic of a machine is that it is planned for a purpose which implies a designer and that the best, or possibly the only, way to understand a machine is to understand the purpose which the designer had in mind. The study of man as a machine thus leads to teleology and that leads naturally to the question of the mind of the designer of man. This mind must work in a way similar to that of the human mind, if we are to understand its planning; we understand the planning of a machine because the designing engineer thinks as we think. So we are back at theology. Some atheistic teleologists of the period of enlightenment solved this problem by a switch from a moralizing stern biblical lord to a bright goddess, Nature; and as the priests claimed they learned God's purposes through supernatural revelation, so the atheistic scientists claimed that Nature revealed her purposes to those who were asking her properly through scientific research.

That type of naturalistic teleology, or nature theology, was clearly the frame of mind of Sarrus & Rameaux when they wrote (29) in 1839: "When nature can achieve an aim by various means she never uses one of the means exclusively to the limit, she makes these means compete so that each one of them produces an equal part of the total effect."

The attitude of a natural teleologist is not essentially different from Priestley's frankly theistic approach, not only to questions of ethics and religion in a narrower sense but to everything he did including his scientific research. He writes: "The most pleasing views of the unbounded power, wisdom and goodness of God are constantly present to his (the naturalist's) mind"—(30, p. 193).

In the preface to volume II of *Experiments and Observations on*

Different Kinds of Air (1774-1777) Priestley wrote as follows: There is nothing capital in this volume from which I can hope to derive any other kind of honor than that of being the instrument in the hands of Divine Providence which makes use of human industry to strike out and diffuse that knowledge of the system of nature, which seems, for some great purpose that we cannot as yet fully comprehend, to have been reserved for this age of the world . . . —(30, p. 246).

In Priestley's mind there were no different compartments, as for example in Faraday's where science and religion coexisted in complete separation (31, p. 110), for rational scientific thinking and for religion. Priestley was a consistent Unitarian also in this sense. The science of this great chemist and physiologist was integrated with his theology and his theology was permeated by the courage and freedom in the search for truth which characterizes a great scientist. One may question, though, whether this unity was not achieved at the cost of clarity wherefore it took Lavoisier, presumably an agnostic, to free Priestley's great scientific discovery from the confusion of the phlogiston theory, to show that Priestley had discovered an elementary gas, to name this gas oxygen, and to formulate the theory of combustion and metabolism as oxidations, a theory whose overall aspect is still valid today.

2. *Teleology today.*—Boyle's main argument for teleology was the marvelous adaptation of living creatures but Reichenbach wrote (32, p. 200), "Darwin's theory of natural selection is the tool by which the apparent teleology of evolution is reduced to causality. The need for teleology is eliminated by Darwin's principle."

Selection of the fittest among offspring whose characteristics result from gene inheritance and chance mutations makes the postulation of a plan superfluous. Bernatovics (33) describes the present situation as follows: "For most teachers of science, teleology and anthropomorphism are not issues to be debated but to be deplored."

Bernatovics presents a selection of samples of teleological language taken from modern textbooks of biology, chemistry, physics, astronomy or geology. Here is another example from Lehninger's excellent and lucid *Bioenergetics* (34, p. 159): "Just as the whole cell *had to evolve* active transport mechanisms located in its outer membrane *in order to* preserve the constancy of composition of the intracellular solutes . . . " [my emphasis]. A nonteleological formulation might read as follows: "Cells in whose membranes active transport systems operate may maintain intracellular solutes at a constant composition. If the latter is biologically advantageous, active transport becomes a criterion of natural selection and thus a factor in evolution."

(*a*) Teleological slip of my own.

"Let each man watch just where he stands and if he stands, beware of falling." This advice to Goethe comes to my mind when I find teleological slips in my own publications such as the following passage (35, p. 369):

"This increase in katabolic rate observed in magnesium deficiency may thus be related to an increase in the breakdown of tissue because of an inability to secure sufficient magnesium out of intact tissues after the stores have been depleted."

"Inability to secure" implies a frustrated will to obtain magnesium; it has a teleological flavor which was noticed by Blaxter & Rook (36). I should have cautioned the reader by "as if" to make him realize that my statement was meant metaphorically, for example as follows: "A lowering of the Mg level in the blood may lead to a lower Mg content in the tissues and this in turn may lead to an increased breakdown *as if* these tissues were sacrificed as a source of Mg for the rest of the body."

(*b*) An erroneous accusation.

A review (37) of a paper on metabolic rates of rats as a function of age by Kleiber, Smith & Chernikoff (38) is summarized by the statement, "The work is a valuable addition to the sum of information on the metabolic rate of rats but the teleological discussion, thinly disguised as physiology, is unhelpful."

This passage caused a good deal of hilarity among my students and co-workers familiar with my occasional exhibitions of a missionary zeal in preaching against teleological explanations of scientific observations (in fields other than sociology or psychology where of course human purposes are a legitimate object of research).

I wrote to the editor of *Nutrition Reviews:* "I do not claim that the criticized discussion is helpful, that depends on the reader, but I claim 1) That it is not disguised and 2) It is not teleological, being based on natural selection, the alternate to teleology." The article contains the statement: "*It is* as if an agent stimulating metabolic rate increased in old rats at a constant relative rate" and the sentence: "Our rats would then be *naturally selected* for high metabolic rate." True, the sentence "It [the agent involved] might be related to the condition that favors development of spontaneous tumors in ageing rats" contains "favors" which has a teleological flavor, but how could a condition "favor" in any other way but metaphorically?

3. *Don't eliminate metaphors!*—I think it would be a pity to eliminate from scientific language those poetical metaphors which are recognizable as such. I would feel sorry to miss such meaningful and stimulating statements as Fenn's (1924) that the muscle adjusts the extent of catabolic processes to the load which it "discovers" it must lift after the stimulus for contraction is over (39; 40, p. 16).

It would be deplorable if a statement like the following by J. B. S. Haldane (41) were banned from biological literature as teleological heresy. "If the insects had hit on a plan for driving air through their tissues instead of letting it soak in they might well have become as large as lobsters."

4. *Taking teleology seriously.*—The case is different when teleology is taken seriously as, for example, by H. Krebs (42) in his lecture at Johns Hopkins Hospital, 1954, under the title "An expansion into the borderland

of biochemistry and philosophy", when he states "that the wholesale dismissal of teleological consideration in biology is unjustified, that it is the primary purpose of the oxidative degradation of foodstuffs to generate phosphate bond energy."

This would make scientific research an effort to understand nature in terms of a creator's purposes, that would be theology, and for some scientists this may be a favorable condition for research as it was for Priestley; but others, and they are probably the majority of scientists, regard teleology with Reichenbach (32) as analogism and pseudoexplanation and see an open problem of causality where teleologists are satisfied with an answer based on analogy. The *Annual Review of Physiology* 18 (43) gave me an opportunity to voice disagreement with teleology.

5. *Function and purpose.*—Two years after Krebs' excursion into the borderland of biochemistry and philosophy, another Nobel laureate, A. V. Hill, joined Krebs' recommendation for a return to teleology. In an article "Why Biophysics?" (44) he states that "the idea of function of oragnization of design is an essential part of biology as it is of engineering . . . it is sensible for a physiologist to ask what the functional significance of an organ is . . . its relation to other parts of the machinery, its purpose in connection with behavior, survival, or inheritance . . . "

There is no argument that it is sensible for a physiologist to study the function of organs if we mean by function the relations of one organ to other organs as a part of an organism, that is, when we use function the way it is used in mathematics, expressing relation. What most of the physiologists object to is the use of function in the teleological sense as a synonym for purpose. Purpose has a proper meaning only in connection with a creator who designs organisms. A student of engineering properly speaks of the purpose of a part of an engine. He understands why the inventor of the machine had designed a part in a particular shape and position because the student of engineering has learned to think as the designer thinks. But can a biologist learn to understand what the inventor of a fish or a man had in mind when he designed these creatures?

A professor of engineering, in explaining a machine, properly starts with the purpose of the machine, but should a professor of biology likewise start with the purpose of an organism—and does a biologist whose conscience (or politics as Hill adds) forbids him to ask for purposes, who as a biologist avoids teleology and limit himself to causality, really "miss most of what is interesting"? Did Darwin make his successors miss most of what is interesting when he showed that evolution can be explained by natural selection, as an alternate to teleology?

To the contrary, instead of accepting an analogy between a creator of organisms and a designer of machines and hunting for divine blueprints, the darwinistically oriented physiologist is stimulated to search for causes and even if he does not completely succeed he usually finds a lot of what is interesting on his way.

THE DAVIS TRACER TEAM

1. *Peaceful use of atomic energy.*—The third epoch in the evolution of man started on December 2, 1942 when at the end of an abandoned football field of the University of Chicago "Man achieved the first self-sustaining chain reaction and thereby initiated the controlled release of nuclear energy"—(45, p. IX).

I wrote a jubilant letter to Switzerland, Science works! But my old friend, a well known educator, Fritz Wartenweiler, remarked I was crazy, man was not ready morally for that much power. I must admit he may be right. When warhawks direct the governments of nations, all increase in power increases evil, with or without nuclear bombs, though, Hiroshima . . . Vietnam. But I still hope and believe that the good in human nature will prevail over the evil, that free intelligence will defeat conformist stupidity, that reason will be victorious over nonsense, that generals will be subordinated to statesmen.

I was eager to participate in the peaceful uses of nuclear energy, in particular to use isotopes in metabolic research, especially research on dairy cows, the most important farm animal in California which can become an almost ideal tool for a physiologist who, supplied with isotopes as tracers, is eager to measure biochemical processes in a normally functioning animal.

The use of isotopes as tracers started before Fermi and his co-workers ushered in the nuclear age; it started A.D. 1911 when Hévèsy, working in Rutherford's laboratory, could not separate radium D from lead and, as a genius, turned failure into a success by using radium D as an isotopic tracer for lead. In Davis the use of isotopic tracers started in the Botany Division in 1942 when Robert N. Colvell used ^{32}P in translocation studies. For me it started with a Chemical Society lecture by Dr. E. A. Evans, Jr., at Davis and a following talk with him in our home in October 1944. Four years earlier Evans and Slotin had demonstrated carbonate fixation in liver slices incubated with ^{11}C.

In the fall of 1945 discussions among physicists, chemists, and biologists at Davis resulted in a project entitled "Metabolic research with isotopes". This project was activated by a proposal for cooperative research directed jointly by Drs. H. Young and M. Kleiber and accepted by Dean C. B. Hutchison in May 1946.

I wanted to supplement my earlier investigations on energy utilization involving respiration trials with dairy cows by research on the intermediary metabolism using a carbon isotope as a tracer. The isotope of carbon, ^{14}C, was not yet sufficiently available, though it was first discovered in 1940, arising from bombardment of carbon with deuterium in the cyclotron (46, p. 168). The ^{11}C was too short lived for extended trials with dairy cows; therefore, at that time the choice was the stable ^{13}C.

In 1946 I arrived in Chicago as visiting professor of applied biochemistry in order to learn tracer methods, especially work with ^{13}C, which usually involves mass spectrometry. During the period when I studied the mass spectrometer in Rittenberg's Laboratory at Columbia University in New York, the friendship and hospitality of Konrad and Lore Bloch made me feel at home in Cedarhurst.

By 1947, ^{14}C had become sufficiently available and the measurement of its soft radiation sensitive enough that we postponed the acquisition of a mass spectrometer for the Davis campus for the future when we would extend our tracer work to ^{15}N.

2. *A lucky accident.*—The work under our project began by an accident. ^{32}P was used in the Donner Laboratory at Berkeley to treat leukemia patients. A bottle with ^{32}P broke and the radioactive solution had to be recovered from the floor. The physicians declined to inject this sample into veins of human beings, so it became radioactive waste to be enclosed in concrete for dumping into the ocean. Fortunately, Dr. Garden, in charge of waste disposal, is according to unofficial information of Scotch ancestry, which involves characteristics similar to the Swiss. Anyway the idea of wasting so much of so valuable a material as that sample of ^{32}P hurt his soul. He called up Dr. Reiber at Davis, asking if the contaminated ^{32}P solution might possibly be of good use in agriculture. Again, fortunately, Dr. Reiber was one of the chemists who had helped to initiate our tracer discussion group. He called me up and asked, "Could we use ^{32}P?" I said, "Yes!" True, we had not planned to work with ^{32}P but with ^{13}C or ^{14}C. However, luck is a proud lady and should not be told to come at some other time or with something else on and I am glad that I said yes. Dr. Garden phoned "Come and get it."

My assistant A. H. Smith interrupted his Ph.D. thesis on microrespiration trials with mammalian egg cells, and we drove to Berkeley discussing en route what we might investigate with that ^{32}P and a cow. Our good friend Dr. H. B. Jones in the Donner Laboratory taught us how to handle radioisotopes. Fortunately Dr. Hewitt of Plant Pathology had a Geiger counter and Drs. Gardner and Patten of Physics knew how to operate it, and an undergraduate student with an unusually inquisitive and quickly grasping mind learned the tricks of that Geiger counter faster than the rest of us. That student, Arthur Black, was later endearingly called "Bones" because he prepared radioactive cow bones for radioassay in a self-constructed crusher with extraordinary devotion and corresponding noise.

On December 23, 1946, Dr. George Hart, Head of the Division of Animal Husbandry, thrust a bleeding needle into the jugular vein of the chosen cow and I, somewhat trembling I must confess, connected the needle to a bottle which contained half a liter of the awe-inspiring solution, and the infusion of the radioactive phosphate into the cow began. Dr. Loosli from Cornell University and Dr. Reiber from our own, as well as Drs. Patten and Gardner from our Physics Division, shared with us the joy and excitement of this first trial which included intensive laboratory work even over Christmas.

3. *Chicago meeting on isotopes in agriculture.*—Tracer work aroused the interest of progressive officers of the American Society for Animal Production early. The Investigating Committee (Wise Burrow, Chairman, George Davis, G. E. Dickerson, Max Kleiber, Paul B. Pearson) organized a panel discussion in the Sherman Hotel at Chicago, November 27, 1948, with the following program:

Isotopes in Animal Experimentation

Max Kleiber, University of California: "Phosphorus exchange in cows, an example of using radioactive isotopes for investigating turnover rates and or specific metabolic rates".

Konrad Bloch, University of Chicago: "Some applications of isotopic tracers to the study of intermediary metabolism in animals".

C. L. Comar, University of Tennessee: "Radioisotopes in animal nutrition research".

W. F. Libby, University of Chicago: "Availability of isotopes in the production of labeled compounds: Isotope farming".

James H. Jensen, Div. Biol. Med. United States Atomic Energy Commission: "Further considerations in using isotopes in animal experimentation".

4. *Atomic Energy Commission.*—Listening to Dr. Jensen's talk and getting acquainted with him may have saved the life of our newly born Davis Tracer Team when our project seemed doomed by withdrawal of financial support in the Division of Animal Husbandry. A desperate call for help was answered favorably by the U. S. Atomic Energy Commission (AEC). My assistant, who would have lost his job, was awarded a postdoctoral fellowship in 1948, and in 1949 an AEC contract "Studies on metabolism and biosynthesis in farm animals" was started. The reluctance of the administrators of the College of Agriculture to accept our AEC contract was fortunately overcome by the influence of Dr. Hardin B. Jones whose personal standing among his colleagues at Berkeley, together with the prestige of the Donner Laboratory and its chief Dr. John Lawrence, apparently outweighed the hostile attitude at Davis.

Of great importance for the development of our project was a fine cooperation with members of Professor Calvin's Bio-organic Group of the Radiation Laboratory where the labeled metabolites used in our experiments were synthesized under the direction of Dr. Bert M. Tolbert. Later, bio- as well as chemosyntheses were achieved in a laboratory organized for that purpose in the School of Veterinary Medicine at Davis by Dr. Georg Brubacher from Basel, Switzerland. Dr. Jerry Kaneko, an early member of our team and a technician and graduate student, acquired a good deal of skill in chemosynthesis and so did Dr. Black.

The youthful Tracer Team owed a great deal to the friendly and effective help of Dr. Donald Jasper in all functions requiring the skill and know-how of a veterinarian, from liver and bone biopsies to diagnoses of disease and treatment of sick cows.

In 1951 the Davis Tracer team suffered a major loss when Dr. N. P. Ralston left Davis to start his career as head of the Department of Dairy

Science at Michigan State University. He combined the grassroot interest
of an Animal Husbandman intimately acquainted with the practical problems of the farmer with the vision, skill, and curiosity of a progressive
physiologist. Fortunately some of the functions of Dr. Ralston on the team
were soon taken over by Jack Luick, a rather rare combination of a skillful
chemist and competent dairyman. His ability in dealing with people was
manifest in his research, undertaken together with a graduate student, G.
McLeod, on ketosis which required a close cooperation between dairy
farmers, farm advisors, practicing veterinarians, the School of Veterinary
Medicine, and the Tracer Team.

5. *The emblem of the team.*—After struggling through infancy with its
hazards, the Davis Tracer Team showed such a lusty growth that I have to
refrain here from even listing the names of its members, let alone discussing their achievements. It warms the old professor's heart to contemplate
how far the knowledge of his academic offspring exceeds his own without
making them strangers. Information on the activity of our family can be
found in the literature and it may be advantageous to mention in a prefatory chapter mainly what is not elsewhere found in print. Among these items
is the origin of the tracer team emblem. Mr. Chernikoff who had started
his career as a pharmacist in the old army of the Tsar, had escaped
through China and then earned his living as a technician in the Physiology
Department at Berkeley. When Professor Schock switched his field of action from pediatrics in California to Geriatrics in Pennsylvania, Mr. Chernikoff came to Davis as my technician. When we started to work with radi-
ophosphorus, Cherni appeared with a piece of sheet lead hung where sculptors usually put a fig leaf. From that time on a departing member of our
team is decorated with a fig leaf of lead.

END OF THIS RUMINATION

I am happy that I have chosen the right profession. A scientist is not
only free to seek the truth, it is his job. An academic teacher has not only
the right to tell the truth, it is his professional obligation. This fits especially human beings who are allergic to schisms in their soul. Scientific research and academic teaching saves them from the difficulty which was encountered by a prominent author who was said (47) to have written the strict
truth as a historian but had to tell lies as a White House aide. I am happy that
I found my place as a scientist and an academic teacher, a profession
which fits my charactersitics.

I enjoyed teaching basic ideas in my special field of animal energetics
and writing a book on that subject, *The Fire of Life* (48). I enjoyed lecturing also on subjects of general interest to students such as: "What you
should take with you from your University", and on subjects of general
human interest: "What science means to me" and "Scientists are human".
The latter may have stimulated a group of my friends to donate a beautiful
picture of Albert Einstein by Karsch for the entrance hall of the physics

building. I enjoyed a public debate with Dr. Teller on the question: "Should the U.S. resume testing?" I had my soul in sermons to Unitarians like: "The conscience of an agnostic" and "Freedom and truth". I am grateful for the opportunity to continue some of these activities, happy to discuss philosophical questions in monthly meetings at my home.

I feel satisfaction with the memories of some good fights during the time when the McCarthy spirit invaded our Campus at Davis during the period within which the University suffered through "the year of the oath" (49).

Just as a cow cannot chew her cud continuously, so can an old professor not put into a prefatory chapter all he has to ruminate. Some items left out may make a good cud to chew on at another occasion of contemplation.

LITERATURE CITED

1. Brody, S., *Ann. Rev. Biochem.*, 3, 295 (1934)
2. Kleiber, M., *Z. Tierernährung*, 5, 1 (1933)
3. Mayer, J., *Yale J. Biol. Med.*, 21, 415–19 (1948/49)
4. Brody, S., *Ann. Rev. Biochem.*, 4, 383 (1935)
5. Palmer, L. S., and Kennedy, C., *J. Biol. Chem.*, 90, 545 (1931)
6. Palmer, L. S., Kennedy, C., Calverley, C. E., Lohn, C., and Weswig, P. H. *Univ. of Minnesota Agr. Expt. Sta. Bull.*, 176 (1946)
7. Swift, J., *Gulliver's Travels* (1726) (Reprinted in Great Books, 36, p. 19, *Encyclopedia Britannica*, 1952)
8. Kleiber, M., *Hilgardia*, 6, 315–53 (1932)
9. Brody, S., *Missouri Res. Bull.*, 166 (1932)
10. Sarrus et Rameaux, *Bull. Acad. Roy. Med.*, 3, 1094 (1839)
11. Rubner, M., *Z. Biol.*, 19, 535–62 (1883)
12. Richet, Ch., *La Chaleur Animale* (Alcan, Paris, 1889)
13. Pauling, L., *No More War* (Dodd Mead & Co., New York, 1958)
14. Guggenheim, K., *Alles in Allem* (Artemis Verlag, Zürich, 1952)
15. Lavoisier, A., *Elements of Chemistry* (Great Books, 45, *Encyclopedia Britannica*, 1952)
16. Kleiber, M., *Ann. Rev. Physiol.*, 6, 123–43 (1944)
17. Brown, S. C., *Count Rumford* (Doubleday Anchor, New York, 1962)
18. Warburg, O., *Science*, 123, 321 (1956)
19. Krebs, H., *Bull. John Hopkins Hosp.*, 95, 19 (1954)
20. Kleiber, M. *Nature*, 175, 324 (1955)
21. Zilversmit, D. B., *Nature*, 175, 863 (1955)
22. Mawson, C. A., *Nature*, 176, 317 (1955)
23. Kleiber, M., *J. Appl. Physiol.*, 2, 417–23 (1950)
24. Kleiber, M., *Hilgardia*, 9, 1–70 (1935)
25. Møllgaard, H., and Anderson, A. C., *Forsøgslab. Vet. K. og Lands bohojskoles*, 94, 1–180 (1917)
26. Verworn, M., *Allgemeine Physiologie*, 10 (Fischer, Jena, 1922)
27. Burtt, E. A., *The Metaphysical Foundations of Modern Science* (Doubleday Anchor, Garden City, N.Y., 1932)
28. Lange, F. A., *Geschichte des Materialismus*, 1, 323 (Brandstetter, Leipzig, 1914)
29. Sarrus et Rameaux, *Bull. Acad. Roy. Med.*, 3, 1094–1100 (1839)
30. Brown, I., *Joseph Priestley* (Pennsylvania State Univ. Press, 1962)
31. Ostwald, W., *Grosse Männer* (Akad. Verlagsges., Leipzig, 1927)
32. Reichenbach, H., *The Rise of Scientific Philosophy* (Univ. of California Press, 1951)
33. Bernatovics, A. J., *ETC: A Review of General Semantics*, 17, 63–75 (1959)
34. Lehninger, A. L., *Bioenergetics* (Benjamin, New York, 1965)
35. Kleiber, M., Boelter, D. D., and Greenberg, D. M., *J. Nutr.*, 21, 363–72 (1941)
36. Blaxter, K. L., and Rook, J. A. F., *J. Physiol. (London)*, 121, 48 (1953)
37. Anonymous, *Nutr. Rev.*, 15, 77 (March 1957)
38. Kleiber, M., Smith, A. H., and Cher-

nikoff, Th., *Am. J. Physiol.*, **186**, 9–12 (1956)

39. Fenn, W. O., *J. Physiol. (London)*, **58**, 373–95 (1924)
40. Kleiber, M., and Rogers, T. A., *Ann. Rev. Physiol.*, **23**, 15–36 (1961)
41. Haldane, J. B. S., On Being the Right Size, *World of Mathematics*, 954 (Newman, J. R., Ed., Simon & Schuster, New York, 1956)
42. Krebs, H. A., *Bull. Johns Hopkins Hosp.*, **95**, 19–51 (1954)
43. Kleiber, M., *Ann. Rev. Physiol.*, **18**, 35 (1956)
44. Hill, A. V., *Science*, **124**, 1233–37 (1956)
45. Fermi, L., *Atoms in the Family* (Univ. of Chicago Press, 1954)
46. Kamen, M. D., *Radioactive Tracers in Biology* (Academic Press, New York, 1947)
47. Anonymous, *Progressive*, 9 (Jan. 1966)
48. Kleiber, M., *The Fire of Life* (Wiley, New York, 1961)
49. Stewart, G. R., *The Year of the Oath* (Doubleday, New York, 1950)

Ann. Rev. Pharmacol., Vol. 10

PHARMACOLOGY AND MEDICINE

HIROSHI KUMAGAI

Ann. Rev. Pharmacol., Vol. 10

PHARMACOLOGY AND MEDICINE

Hiroshi Kumagai

Department of Pharmacology, Faculty of Medicine, The
University of Tokyo, Tokyo, Japan

The establishment and independence of pharmacology in Japan dates back some 50 years, when Professor Juntaro Takahashi took the pharmacological chair in the University of Tokyo, Medical Faculty. This was in 1919, when he came back from Germany, where he had studied pharmacology under the guidance of Professor Schmiedeberg.

The social structure at that time was so arranged that the traditional way of living and thinking of the Japanese leaders had been switched from that of medieval Japan to western. In order to take part as a member of the modern countries of the world, the government made every effort for the modernization of Japan and its people.

The core of the policy incorporated in the advancement of Japanese civilization was the encouragement of the scientific way of thinking. This policy led to the systematic organization of the educational system of Japan as a whole, and primary school education became compulsory. This led to an abrupt elevation of the standard of education in Japan, and aroused the desire and willingness of the countrymen for higher education. The policy of the government for the realisation of scientific Japan and the willingness of the Japanese people culminated in the establishment of the Imperial University of Tokyo. The faculty of medicine of the University of Tokyo was established in 1879, and the medical educational program was formulated after the German system. The role played by German professors, who were the first professors in clinical as well as in basic medical departments, was decisive in the further development of medical education and the medical service system in Japan. Among German professors, Erwin Baelz (1876–1905) was the most outstanding figure both as medical teacher and as educator in general. He, as a far-sighted man, had a deep sympathy for Japanese history and culture and he advised the Japanese Government to develop medical service for the people of Japan, that is to say, medical service by the Japanese for the Japanese. But his advice was not accepted by the Government. This resulted in the development of German medicine in Japan, without the philosophy and way of thinking that have been a part of the German culture.

German medicine cultivated in Japanese soil grew up abruptly and covered the whole area of the country. The number of medical schools, for instance, now amounts to 46, including national, prefectural, and private.

1

The principle and pattern of Japanese medical education have completely followed that of Germany, and the most outstanding feature that governed in medical education was the idea of the superiority of academic scientific research over medical practice. This concept undoubtedly contributed greatly to the progress and development of modern medical science in Japan, and is the backbone of present Japanese medicine. This means that more weight was laid on medical research than on the welfare of the public. This pattern of medical education resulted in the dissociation of medical education and medical service.

As for the development of pharmacology in Japan, the department of pharmacology of the faculty of medicine, of the Imperial University of Tokyo, served as a source of training of professors of pharmacology for other medical colleges and medical faculties. The first of this type was the Kyoto Imperial University. Professor Haruo Hayashi, Professor Juntaro Takahashi (the first holder of the pharmacology chair in the Imperial University of Tokyo), and Professor Morishima, jointly established the first pharmacological chair in the Kyoto Imperial University. The professor of pharmacology in medical faculties and colleges in Japan to-day is a descendant or a successor of either the school of Hayashi, or the school of Morishima. Generally speaking, the eastern half of Japan, including Tokyo, belongs to Hayashi's school, and the western half of Japan belongs to Morishima's school. Among forty-six medical faculties each have a pharmacological department with one professor's chair, while Tokyo and Kyoto University medical faculties, have two professors' chairs. The leading policy of the professors at the early stage of the development of pharmacology was to follow the research pattern of the German system, and it has been a custom for professors in charge of pharmacology to go to Germany and study pharmacology. As the medicine that prevailed in Japan before the introduction of western medicine had been Chinese medicine, much emphasis was laid on the discovery of active principles from Chinese drugs, that is to say the scientific evaluation of Chinese drugs by means of pharmacological tools. It was fortunate that the Japanese pharmacologists had pharmaceutical scientists as intimate friends in the University, and the professors in departments of pharmaceutical science of the Tokyo and Kyoto Imperial Universities were specialists, as well, in organic chemistry. Much effort was directed toward the discovery of active principles in the natural herbs that had traditionally been utilized in Chinese medicine.

Nonetheless, the scientific interests of organic chemists in their special field, and those of the pharmacologists, do not necessarily coincide; rather, there sometimes arose discrepancies between them. This situation stimulated the establishment of a pharmacological department in the faculty of pharmaceutical science in Tokyo University. I, as a professor of pharmacology of the medical faculty of Tokyo University, cochaired the first chair in 1959, and therefore I have to take responsibility for the development of pharmacology in the pharmaceutical science field. Since that time, to have a pharmacological chair and department has become a requirement of every faculty and school of pharmaceutical science.

The Japanese Pharmacological Society was established in 1927, with membership composed solely of medical doctors. However, in 1959 membership was extended to scholars from pharmaceutical sciences. This led to a rapid increase in the number of members and size of the society. At an early stage of the establishment of pharmacological departments in faculties or colleges of pharmaceutical science, special efforts were directed to the teaching of biology to students who had specialized in chemistry rather than in biological sciences. Fortunately, because of the endeavours of the leaders concerned, the number of researchers in pharmacology from the field of pharmaceutical sciences increased steadily, and mutual understanding between medical pharmacologists and pharmaceutical pharmacologists become much improved. In my opinion, the research field covered by medical pharmacology may have to shift to a more physiological side, and that covered by pharmaceutical pharmacology more to structure-activity relationships of drugs or derivatives.

As the late Professor Gaddum has pointed out, "the pharmacologist borrows from physiology, biochemistry, pathology, microbiology and statistics, but he has developed one technique of his own, and this is the technique of bioassay." Bioassay without doubt is a technique of its own, however, technique must have a physiological meaning. In other words, we have to find out the biological mechanism that lies beneath the response of a tissue or an organ or a whole animal. This, as I believe, is the final goal that pharmacology has to aim at. Pharmacology by definition is the study of the responses of living matter to substances administered. Analysis and evaluation of responses to administered substances, functional as well as morphological, elicited by subcellular organelle, cell, tissue, organ, and especially the whole animal, chemical as well as physical, constitute the essential part of the study of the life phenomenon.

In my opinion, pharmacology is a science to elucidate the mechanism of the response of living matter to a given substance in its whole aspect. This, in turn, contributes to the elucidation of the physiology of living matter.

The Japanese Pharmacological Society, comprising medical, dental, veterinary and pharmaceutical pharmacologists, has now attained adulthood, nationally as well as internationally, both in size and in activity. Among driving forces that have caused the Japanese Pharmacological Society to develop rapidly, the introduction of American medical sciences after the end of World War II is decisive. Equipment and facilities came in torrents from the U.S.A. to Japan. Sophisticated young men visited the U.S.A. in enormous numbers spending one, two, or more years there, and they brought home the spirit of positiveness and independence from tradition, in every aspect of study and research.

Transfusion of American culture into postwar Japan brought progress in sciences together with confusion in social life. Among the changes and confusions we have encountered in social life as well as in academic study is the breakdown of the family system of Japan. For good or evil, the family system traditionally was the backbone of Japanese culture in its every aspect, in pre-war days. The breakdown of this family system encouraged a spirit of independence and a sense of freedom from all traditional, social, and family life, especially in the younger generation.

The energy of the younger generation culminated in a wonderful increase in gross national production. It also applied to the advancement of scientific activity as evidenced in the increase in number of member Societies of the Japanese Association of Medical Sciences (in which I have been active as the vice president since 1963) from 34 to 62 during the period of 1947 to 1969.

On the other hand, our medical education system and postgraduate education system have remained unchanged since the beginning of the Meiji era—that is to say, during some 100 years.

Social as well as legal aspects of our medical education system must be adapted to the drastic social changes and to the social needs of post-war Japan. Taking the intern system as an example, there is no objection by graduate students or instructors to the intern system for training physicians. However, in Japan, sufficient money for teachers, equipment, and arrangements was never provided by the government. In other words, we imported the seed of the system of internship in postgraduate education but we failed to provide sufficient care and fertilizer to make it grow.

The Japanese people by nature are very keen in modeling foreign patterns in every aspect of human life, in production as well as in consumption.

We imported Indian culture by way of China and created Japanese culture by modeling after it; however, in order to digest the foreign culture and to make it fit for the advancement and creation of our own culture, we have to provide every effort and means for the setting up of a nursery for growing foreign culture. If not, the foreign culture transplanted will be distorted or become the cause of confusion.

We Japanese have surely attained considerable economic development as viewed by our gross national production; however, we have many problems left to be solved, especially in medical education. To meet the ardent need for avoiding the confusion now raging in medical schools in our country, we have to work out the leading opinion as to medicine itself.

In this respect, I suggest that medicine is a science for the study of the laws of existence of the human race on this planet. In my opinion, to follow the natural law of survival or the struggle for existence is not sufficient for the ultimate welfare of the human race, we must aim at the co-existence and prosperity of all the human race.

For the realization of this principle, we have to work not only on human biology but also on human ecology. We have to keep in mind the fact that experimental biology and experimental medicine constitute only a part of medicine.

Recognition of this fact makes anyone who takes part in the study of medicine be more modest in drawing conclusions from his experimental data and makes him careful not to distort the truth.

Pharmacology by definition is the study of the response of living matter to substances (chemical or physical) administered, and thus it belongs to experimental biology and experimental medicine, and indicates the position of pharmacology in medicine.

We have, however, an important but difficult problem to be solved regarding pharmacology in medicine. The problem may be written as follows: Is it possible or permissible in medicine to extrapolate pharmacological findings obtained in experimental animals to human beings?

The only thing we pharmacologists can do both from a scientific as well as a humanistic point of view is to work out a precise and elaborate spectrum of actions of a given drug on subcellular structures, cells, tissues, organs and in the whole animal in terms of metabolism, excitation, excitation-contraction, and excitation-secretion, and apply the spectrum to the human body, and evaluate its effect. It is to be noted that pharmacotherapy itself is a kind of human experiment.

As pointed out by the late Professor Gaddum, we pharmacologists have to borrow every means possible from the advanced fields of sciences, i.e., physics, chemistry, molecular biology, genetic biology, and so on, to elucidate the mechanism of drug action, which in turn reveals the mechanism of living matter.

To accumulate facts or phenomena about the whole animal behavior, and consolidate it into a system, is surely an important process in elucidating life's phenomena; however, phenomena have to be based on physical or physicochemical facts, in other words on a material basis. This way of my thinking or philosophy is the deduction from my experiences as a pharmacologist during the past 37 years.

At the onset of my academic life in 1932 I was engaged in the study of uterine activity in an unanesthetized bitch by means of a chronic uterine fistula. I worked with Professor Azuma, who was a pupil of the late Sir Henry Dale, at the Imperial University of Tokyo. By this method, studies were carried out to analyse the effects of hormones on uterine activity and to find out the water soluble principle of domestic ergots.

In doing this kind of experiment my interest became directed to the activity of smooth muscle.

This led to the study of muscle in more basic aspects, and a first step was set up at the "Conference on the chemistry of muscle contraction" in 1957 in Tokyo under my chairmanship. Together with my pupil, Professor Setsuro Ebashi, I conducted physicochemical study of muscle energetically and found a muscle relaxing factor which was proved to be in the endoplasmic reticulum. This later was proved to be a protein that accumulates calcium, and thus the role of calcium in muscle contraction was established. The development of the study of muscle contraction may be traced in "Molecular Biology of Muscle Contraction, 1965"—Elsevier, Amsterdam.

As a result of these basic studies concerning muscle contraction, the role of Ca-ions as the basis of pharmacological action such as that of caffeine was clearly established. These findings are the outcome of contributions from colleagues from all over the world.

As I had a good successor, Professor Ebashi, in muscle study, I myself engaged in the study of the physiology and pharmacology of the central nervous system and especially the brain stem respiratory center and the unit discharge in inspiration and expiration, its localization, and its sensitivity to drugs.

Here also reigns my philosophy: phenomena must have as their basis a precise material background, that is to say, the respiratory center must have a neuro-cellular basis. In the study of the physiology and pharmacology of the respiratory center I am lucky to have had as my coworker Professor Fuminori Sakai, who succeeded to my chair in pharmacology at the University of Tokyo.

The scientific works in which I have been involved are limited and much is left to the hands of the worthy successors in my department.

In closing this manuscript, I emphasize the two important roles of the professor, one is to conduct academic study by himself, the other is to bring up competent and excellent successors. Scientific research must be carried out by cooperation not only on a global scale but also in the generational scale.

Reprinted from
ANNUAL REVIEW OF PHARMACOLOGY
Vol. 10, 1970

Reprinted from
ANNUAL REVIEW OF PHARMACOLOGY AND TOXICOLOGY
Volume 16, 1976

HOW I AM

Chauncey D. Leake

In 1959, when president of The American Society for Pharmacology and Experimental Therapeutics, and president-elect of The American Association for the Advancement of Science

Reprinted from
ANNUAL REVIEW OF PHARMACOLOGY AND TOXICOLOGY
Volume 16, 1976

HOW I AM   ♦6632

Chauncey D. Leake
University of California, San Francisco, California 94143

Although I am primarily a pharmacologist, I have a great diversity of related interests. This is natural. Pharmacology and toxicology are diverse disciplines, related to all the life sciences (including psychology and sociology); to the various clinical specialties of the expanding health professions; to agriculture, agronomy, ecology, forestry, oceanography, and even meteorology; to law, politics, and public policy, and to the arts and humanities, even to philosophy, especially the ethics, the logics, and the esthetics.

My entrance into pharmacology came by chance. During World War I, I was full of adolescent idealism to "save the world for democracy," and I was in service in March 1917. Universities were generous in those days: Princeton gave me a bachelor's degree (with a major in chemistry, biology, and philosophy), even though I missed half of the last semester of my senior year and did not turn in a thesis expected of me.

A top sergeant in a machine gun outfit training in Anniston, Alabama, I was transferred to the newly organized Chemical Warfare Service and sent to the Medical Defense Division, operating under Majors J. A. E. Eyster (1881–1955) and Walter J. Meek (1878–1963), in the physiology and pharmacology laboratories of the University of Wisconsin. The pharmacologist of the group was Arthur S. Loevenhart (1878–1929). He was in Washington. So was the biochemist, Harold C. Bradley (1878–1975).

Harold Bradley was a personnel officer for the Chemical Warfare Service, and it was he who spotted my chemistry-biology background. After the war, he taught me physiological chemistry, and we became close friends. His Madison home was a happy gathering place for his pupils. When he retired, he moved back to his father's home in Berkeley, California and became a leader in conservation. We enjoyed many pleasant evenings together at the Chit Chat Club meetings in San Francisco.

My job in the Chemical Warfare Service, in the basement of old Science Hall on the Madison Campus, was to study the effects of toxic war gases on the acid-base balance of blood. We used morphinized dogs. Samuel Amberg (1874–1966), who had come from the Mayo Clinic to work with us, and Walter Meek taught me well. We studied chlorine, chlorpicrin, mustard gas, and lewisite. When the war was over,

1

I was asked to stay on and run the necessary controls: what does morphine do to blood reaction in mammals?

In dogs, morphine causes first an increase in respiration, usually with vomiting, followed by respiratory depression, lassitude, and analgesia. I interpreted this as an initial decrease in cellular oxidative mechanisms in the medulla oblongata with subsequent increase of oxidative processes, in accordance with ideas expressed by Arthur Loevenhart and his pupil, Herbert Gasser (1888–1963). I showed that initial stimulation of the vomiting center by morphine is followed by depression, when vomiting cannot be induced either centrally by apomorphine or peripherally by stomach irritants. A ketosis develops after morphine, with a mild acidosis.

These findings (*J. Pharmacol.* 20:359–64, 1922; *Arch. Int. Pharmacodyn. Ther.* 27:221–27, 1922) suggested that it might be worthwhile to see what anesthetic agents do to blood reaction. So with my ever patient wife, Elizabeth, whom I was fortunate enough to marry in 1921, and with Alfred Koehler, a graduate student, we studied blood reaction in dogs under anesthesia with ether, chloroform, and nitrous oxide oxygen. We found an acidosis under ether and chloroform, with an initial alkalosis going into an acidosis under nitrous oxide and oxygen. The latter involves oxygen want if real anesthesia is to be obtained.

Meanwhile, Arno Luckhardt (1885–1957) in Chicago had introduced ethylene anesthesia. This has an advantage over nitrous oxide in that it can be successful with 15% oxygen (at sea level), instead of the 10% with nitrous oxide. Twelve percent is needed for satisfactory blood oxygenation. With Alrick Hertzman, I found that there is relatively little change in blood reaction with ethylene oxygen, and I thought this could be correlated with the generally superior clinical condition of patients under ethylene and oxygen.

This set me thinking about combining the chemical unsaturation of the carbon atoms in ethylene into the ether configuration. Long discussions were held on this, especially with my toxicology associate, Clarence Muehlberger (1896–1966). The compound in question, divinyl ether, was not in existence, but I determined to try to get it.

Meanwhile I had become engaged in many other search and research efforts. I reported a summary of our work on anesthesia and blood reaction at the Third Congress of Anesthetists in 1924 (*Br. J. Anaesth.* 2:1–20, 1924). With Frank G. Hall, I showed that alkalosis gives vascular constriction, while acidosis brings vascular relaxation. With Thomas K. Brown I found that infections, such as in experimental pneumonia, when there is respiratory involvement, are accompanied by an acidosis, due to an anoxic anoxemia.

With my wife, I became interested in blood regulation in anemias. Figuring that there might be some kind of a hormonal regulation, with red blood cell production geared to red cell destruction, we studied the effects of giving saline extracts of mammalian spleen and bone marrow to dogs. Stupidly we used healthy dogs, instead of trying to make them anemic. We found that spleen extracts give equivocal effects on erythrocyte counts, while bone marrow causes some increase. When we combined spleen and bone marrow we found a considerable rise in the red cell count. Liver, kidney, and heart extracts had no effect in our healthy dogs. So we introduced combined desiccated spleen and bone marrow for the treatment of secondary anemias. Clinically it seemed to be helpful. It had no effect in pernicious anemia. Years later, it seemed that spleen and bone marrow might give a kick to the reticuloendothelial system. My clinical balance was maintained by William S. Middleton, the sharp professor of medicine, who later became dean.

Those were busy years and happy ones, although we lived on a pittance. We canoed on Lake Mendota, picnicked on Picnic Point, and even got a secondhand auto for a trip to the East Coast. The university generously gave me a doctoral degree with publications in lieu of a thesis. My patient wife coached me in passing French in a qualifying exam. Walter Meek arranged for me to go to the Cleveland meeting of the Federated Societies in 1922, where Frederick Banting (1891–1941) thrilled us with his report on insulin. Meek got me into the American Physiology Society in 1923, but Torald Sollmann (1874–1962) kept me out of the Pharmacology Society until the following year, because as he said I was too young. Later he was a good friend.

I had become interested in historical and philosophical affairs, as a result of the stimulus of the William Snow Miller (1858–1939) seminar in medical history. One of my contributions was on the history of anesthesia (*Sci. Mon.* 20:304–28, 1925). This was later put into cadence, with a huge bibliographical chronology, and published by the University of Texas Press in 1947. I had written an account of Thomas Percival (1740–1804) and his misnamed "Medical Ethics" (*J. Am. Med. Assoc.* 81:366–71, 1923). This attracted the attention of Harvey Cushing (1869–1939), who asked me to come to Boston to present it there. He became one of my best friends and mentors. In 1927, this effort resulted in a book, *Percival's Medical Ethics,* which was published by Williams and Wilkins in Baltimore, under the direction of Charles C. Thomas, who later became his own distinguished publisher. This book fell flat as a mud pie. Amazingly, now, with much excitement over medical ethics, it is appearing in a second edition, nearly half a century after the first.

I became interested in William Harvey (1578–1657) and his great classic, *De Motu Cordis,* first issued in 1628. With coaxing from Charles Thomas, I undertook a new English translation to appear as a tercentennial tribute. Charles Thomas published it in fine format, and soon issued it in paperback, the first such in USA medical publishing. This has gone through five editions, the most recent in 1970. Harvey was probably the first to suggest giving drugs by injection into the blood circulation.

Meanwhile, Ralph Waters, the great anesthetist, came to Madison to develop anesthesia in the new Wisconsin General Hospital, the main clinical facility for the university. He came, in part, he said to have a chance to work with pharmacologists. He established a great residency training program in anesthesia, and we started a long work effort. In 1828, Henry Hill Hickman (1800–1830), in England, had used carbon dioxide as an anesthetic in dogs. No one had studied it since from that standpoint. Ralph Waters and I decided to try it out. We found that carbon dioxide, 30% with 70% oxygen, is indeed anesthetic, with no asphyxia involved. This we reported at the Seventh Congress of Anesthetists held at the University of Wisconsin in 1928, a century after Hickman.

Although I had started as a physiologist, Arthur Loevenhart persuaded me to move alone with him in pharmacology, and I soon was an associate professor, with major teaching responsibilities. In reality at Wisconsin, physiology, pharmacology, and physiological chemistry worked together. We had a joint weekly seminar in Arthur Loevenhart's pleasant laboratory; the large graduate student laboratory, under the calm, efficient eye of William Young, our English "diener," was a center of experimentation, with its big kymograph, and I had a convenient office-laboratory of 300 square feet, with a chemical bench and hood on one side, a microscope bench along the windows, and desks along the other side. I had graduate students with me: Peter K. Knoefel, later professor of pharmacology at the University of Louisville; George Wakerlin, later professor of physiology at Louisville and director of the American Heart Association, and Warren Stratman-Thomas, who worked so well on African sleeping sickness and its chemotherapy. The chemotherapy of trypanosomiasis was one of Arthur Loevenhart's major contributions.

From the way the four of us worked together in one office-lab came the idea for "The Student's Unit Medical Laboratory" (*J. Am. Med. Assoc.* 82:114–17, 1924). This attracted the interest of Abraham Flexner (1866–1959) and was the beginning of the now popular multidiscipline medical laboratories. Lathan Crandall, later director of research for Miles Laboratories, also worked with us on a broad study of the pharmacology of nitrites and nitrates, initiated by Arthur Loevenhart at the request of the DuPont interests to see whether or not it would be possible to get the headache out of dynamite. It is not, unless one has a nonnitroglycerine dynamite.

I was fortunate in having many keen colleagues at Wisconsin. K. K. Chen, who introduced ephedrine and later became research director for Eli Lilly, was one. Elmer Sevringhaus, who later was research director for Hoffman-LaRoche, was another. Another was Fred Jenner Hodges, later professor of radiology at the University of Michigan, and one of our greatest leaders in the field. We had some great graduate students: Samuel Lepkovsky, distinguished in vitamin nutrition and my colleague in Berkeley; Karl Link, who developed the coumarin anticoagulants, and Conrad Elvehjem (1901–1962), who isolated niacin, and later was president of the University of Wisconsin. These students came over from the College of Agriculture, where they had been trained by the vitamin pioneers, Edwin B. Hart (1874–1953) and Harry Steenbock (1886–1967).

When I arrived at the University of Wisconsin, I roomed with Edwin and Mrs. Fred, on Mendota Court, close to the lake with its swimming and boating. Edwin Fred was a kind mentor; he later became dean of the College of Agriculture and president of the university. He was a distinguished bacteriologist. My Wisconsin experiences have been a continuing inspiration to me.

In July 1927, I had a telegram from Carl L. A. Schmidt, professor of biochemistry at the University of California in Berkeley, asking me to come out in August for four months to teach pharmacology to the second-year medical students, then studying in Berkeley. Out we went, full of excitement, leaving the hot Midwest, and arriving in foggy, cool Berkeley, where the palms waved over fur-coated coeds. My office was in old Budd Hall, and the class met in the redwood Spreckles Laboratory where Jacques Loeb (1859–1924) had taught. The class was an alert one. I had time to do a bit of work on the effect of anesthetics on the osmotic resistance of erythrocytes. It was a stimulating atmosphere, with Karl F. Meyer (1884–1974) as professor of microbiology, and Herbert M. Evans (1882–1971), the neurotic discoverer of vitamin E, and pioneer on the estrus cycle and anterior pituitary hormones, as professor of anatomy.

The Medical School of the University of California was being reorganized, and President William Wallace Campbell (1862–1938), the great astronomer, asked me to come back in July 1928 as professor of pharmacology. We did so, but found that a laboratory had been provided on the top floor of the old medical school building in San Francisco. I had an office, and arranged a small laboratory under a mezzanine which was built for graduate students at the front of the large high general laboratory. It was a great place. With William Gilmore as an efficient "diener," we soon had the place humming.

Peter Knoefel came as a National Research Council fellow; Eric Reynolds (who later was president of the California Medical Society) lent us clinical guidance, and Hamilton H. Anderson, with Norman A. David and Anderson Peoples, worked with us while getting their medical degrees. Peter Knoefel and I began work on divinyl ether, which had been prepared for us, along with other unsaturated ethers, by Randolph Major at Princeton. Major later became director of research for Merck. With Mei-Yu Chen, a remarkably able worker, whom I had met at the Boston Physiology Congress, we found that divinyl ether has the anesthetic properties we had predicted for it. We had the good, practical help of Arthur E. Guedel of Los Angeles, who had been an associate of Ralph Waters. Our experimental study was summarized in 1933 (*J. Pharmacol.* 47:5–16), and the agent soon went into clinical use. But it is flammable, very powerful, and likely to injure livers if anesthesia with it is too long maintained.

We experimented with various halogenated hydrocarbons and both saturated and unsaturated ethers, but found nothing safe or useful. It remained for John Krantz, the keen pharmacologist at the University of Maryland, to study fluorinized compounds, which became available after World War II. These led to halothane, now a popular anesthetic, developed by my friend, Yule Bogue, of Imperial Chemical Industries.

K. F. Meyer, director of the Hooper Foundation for Medical Research, was interested in tropical medicine. He suggested that we study the chemotheraphy of amebiasis. This, we found, has about a 10% incidence in the USA, mostly as carriers. We got a dozen or so organic arsenical compounds from Lilly, and an equal number of halogenated hydroxyquinolines from Ciba, and started screening them. Hamilton Anderson handled the arsenicals, and Norman David the hydroxyquinolines.

First, Hamilton Anderson showed the unsatisfactory toxicity of emetine, the standard remedy for amebiasis. It injures heart muscle. Then, using natural amebic infestations in macaques, we found that 4-carbaminophenyl arsonic acid, called *carbarsone* for short, is only mildly effective, but safe. It also seems to have a general "tonic" effect. So we started using it clinically, after taking it ourselves, and finding that most of it is excreted within a day. Alfred Reed, a skilled clinician, directed clinical trials. Later, Hamilton Anderson, with his wife Jeanette, studied amebiasis and treated it successfully in many parts of the tropics in Africa, Latin America, and Asia. Herbert Johnstone (1903–1956) aided in much of this.

Quite as easy to show carbarsone better than stovarsol for amebiasis was it to show iodochlorhydroxyquinoline superior to iodohydroxyquinoline. Indeed, we found iodochlorhydroxyquinoline (Vioform ®) to be a useful intestinal antiseptic generally. Soon after Norman David introduced it for use in amebiasis, it was in wide use throughout the world for intestinal infections, and could be purchased by travelers over-the-counter, except in the USA. We noted no neurotoxic symptoms, as have been described, because our evidence showed it is not absorbed from the gut. A general review of our studies on the chemotherapy of amebiasis was prepared by me in 1932 (*J. Am. Med. Assoc.* 98:195–98). Norman David later became professor of pharmacology at the University of Oregon, Portland.

Our strategy in the amebiasis effort was simple: we eliminated many proposed remedies as ineffective, and concentrated on types of chemicals showing promise. We tried the same strategy in tackling the chemotherapy of leprosy, but soon abandoned the effort because we found the disease so hedged by politics and vested interests that clearly we were not welcome. We did, however, carefully study the characteristics of the organism, and used murine leprosy in test animals. George Emerson, a brilliant chemical biologist who came to us from Berkeley, devised the water-soluble chaulmoogryl-glycerophosphate for intravenous medication.

Meanwhile, Peter Knoefel studied acetals and aldehydes and added much stimulus to our efforts as he joined us each summer. Through K. F. Meyer, we brought over Myron Prinzmetal, later the distinguished cardiologist from Los Angeles, to work on "mussel poison." He persuaded us to offer a place for animal experimentation to Gordon Alles (1901–1963), a keen organic chemist from Los Angeles. Alles was trying to find a synthetic substitute for ephedrine, the price of which had gone sky-high, to use in treating asthma. Alles was supported by George Piness, the leading West Coast allergist from Los Angeles.

Alles was a most meticulous experimenter. He properly used molal solutions of the drugs he had made, thus being able to compare them in millimole doses on a strict molecular basis. He developed the amphetamines. I have told of his important work in the summary entitled *The Amphetamines* (Thomas, Springfield, Ill., 1958, 167 pp.).

Many graduate students came to work with us: Carroll Handley (1911–1958), later professor of pharmacology at Baylor Medical College in Houston; Benedict Abreu (1913–1965), later in charge of pharmacological research for Pitman-Moore in Indianapolis and professor of pharmacology at the University of Texas Medical Branch in Galveston; Nilkanth Phatek (1898–1971), a political refugee from Bombay and later professor of pharmacology at the University of Oregon Dental School; Michael Shimkin, keen administrator from Tomsk, who later became professor of medicine at Temple and professor of community medicine and oncology at the University of California at San Diego; James Morrison, later at Emory University, Atlanta; David Marsh (1919–1961), later director of research at McNeil Laboratories, Philadelphia; Jack Ferguson (1918–1959), later at the Medical College of Virginia, and E. Leong Way, later professor of pharmacology at the University of California in San Francisco, and famed for his studies on morphine derivatives.

Some of these students came from Berkeley as a result of a general course on pharmacology I offered there at the instigation of Carl L. A. Schmidt, others as a result of our cooperation with the School of Pharmacy in San Francisco. This had been reorganized by Carl Schmidt into a fine teaching and research institution, with a brilliant faculty, including the physical-chemist Troy Daniels, who later became dean; Robertson Pratt, microbiologist; John Eiler, biochemist; Warren Kumler, physical-chemist; and Louis Straight, spectrometrist. Michael Hrenoff helped develop mass spectrometry, and his brother, Arseny, was a devoted worker in our laboratory. We all worked rather closely together.

We had regular seminar sessions, which overflowed into the Crummer Room, and often for Sunday meetings in the Santa Cruz redwoods in a spot on the San Lorenzo River, which my wife and I developed for happy weekends. We had a large redwood circle with rustic benches around and a blackboard on a big tree. There, with our two sons, Chaunc and Wilson, we would entertain our friends and their families, and have vigorous scientific discussions. Arthur Guedel often came from Los Angeles, with friends, to debate metabolic effects under anesthesia. We tried out our proposed publications here, and then often reported them before the quarterly meetings of the West Coast section of the Society for Experimental Biology and Medicine. In these meetings we usually had sharp but pleasant debate with our colleagues from Stanford. Maurice Tainter, later the distinguished director of the Winthrop Laboratories, often joined us.

Maurice Tainter studied dinitrophenol as a stimulant of biological oxidation. We aided in studies on its toxicity. This suggested that we might use its respiratory-stimulating power to counteract the respiratory depression of morphine. So George Emerson easily made dinitrophenylmorphine, and it was well studied by Benedict Abreu, George Emerson, and Nilkanth Phatek. It is pain relieving without causing respiratory depression, and we had evidence that it might be less addictive than morphine. But bureaucratic red tape kept us from trying it clinically, except on ourselves.

By this time we had an arrangement with the Shell Development Company, under Clifford Williams, to test the toxicity of new organic chemicals and solvents, not only to protect the public but the workmen in the plant as well.

We had a heavy teaching load. We enjoyed it. I lectured to medical, dental, and pharmacy students jointly, but we split laboratory assignments. We offered a special course for nurses, and for collegiate students in Berkeley. We had postgraduate sessions. We made up our reference sources, and did not use texts. We tried to get our students to contribute. We were quite informal, and had the laboratory hung with pictures of the leading pharmacology contributors. Otto Guttentag, Charles Gurchot, and Salvatore Lucia added philosophical spice. Milton Silverman gave us helpful journalistic aid in writing. He got his degree from Stanford, but did the work for it (sugar synthesis under ultraviolet light) in our laboratory. He later became one of our best science writers, the author of *Magic in a Bottle* (Macmillan, New York, 1941, 332 pp.); *Alcoholic Beverages in Clinical Medicine* (Yearbook Medical Publishers, Chicago, 1966, 160 pp.); and with Philip Lee, *Pills, Profits and Politics* (University of California Press, Berkeley, 1974, 403 pp.).

THE EXCITEMENT OF SCIENCE   291

My wife and I finally made the Grand Tour in 1938. We visited A. J. Clark (1885–1941) in Edinburgh, with his great collection of poisons brought together by T. R. Fraser (1841–1920). We stopped in Oxford for the meeting of the British Pharmacology Society, where I reported on dinitrophenylmorphine, and where we watched J. H. Burn and Edith Bulbring do some of their careful experimentation. We enjoyed meetings with A. V. Hill and Sir Henry Dale. In Ghent we had a marvelous visit with Corneel Heymans (1892–1968), with his fine institute, and in neat 18th Century Darmstadt, with the Mercks. We went to the Physiology Congress in Zurich, where we enjoyed a visit with Hans Fischer out on the lake, and where we met Arthur Stoll (1887–1971) contemplating one of the huge military murals of Ferdinand Hodler (1853–1918). We later became good friends with the Stolls, who had one of the greatest art collections anywhere. Stoll, a pupil of Richard Willstätter (1872–1942), the great chlorophyll chemist, worked brilliantly on digitalis glycosides and ergot alkaloids. He directed the great Sandoz drug company.

The Zurich Congress was exciting. We had an extemporaneous discussion on the status of pharmacology as a science. This became the subject of my address as retiring president of the American Association for the Advancement of Science in 1961. I try to get double duty out of ideas, if I can!

So we worked cheerfully along. Ross Hart and Elton McCawley came as graduate students. Taking up our observation that the allyl radical stimulates respiration, they made N-allylnormorphine, in an effort to get the pain-relieving properties of morphine without its respiratory depression. But it turned out to be a morphine antagonist. As "nalorphine" it was widely used to detect morphine or heroin addiction and to treat narcotic overdose, as suggested by Eddie Way, another brilliant graduate student.

In order not to use too much space in the professional journals, we started *University of California Publications in Pharmacology* in 1938. This assured us prompt publication through the University of California Press, and we could distribute as we wished. Here we published many of our toxicity studies on new solvents made by the Shell Development Company. Here Benedict Abreu and Nilkanth Phatak published their important studies on nitrofuran antiseptics. These later came to wide clinical use. We also made furan local anesthetics, but these had no particular advantage. We even tried a furan aspirin, but this came to grief: animal experiments showed its value and apparent low toxicity. But when I took it, it deposited on my bladder wall, and I had to be hospitalized to be cleaned out. We had not looked at the bladders of our animals in our routine postmortem examinations of sacrificed animals in our toxicity studies. Now we do.

We had contact with the great cyclotron radiation laboratory in Berkeley, through Joseph Hamilton (1907–1957), one of our clinical associates, who was interested in radioactive iodine for thyroid studies. This he prepared himself in the Berkeley laboratory, and somehow got exposed so that he developed leukemia. He succeeded in correlating the deposition of iodine in the thyroid with the histology of the gland. This publication of 1940 was reproduced by Lloyd Roth at the University of Chicago Symposium on *Autoradiography of Diffusible Substances* (Academic Press, New York, 1969). Another worker with us was Charles Pecher, a Belgian, who studied whole body autoradiography in mice with radioactive calcium and strontium. He used these substances clinically in osteoblastic bone tumors. When we were blown into World War II, he enlisted and was killed within a year.

With war upon us, we turned our attention to war gases, the use of which was threatened by the Japanese. I went up and down the West Coast talking to high school students (who were sensible, and could tell their elders) about simple protection against war gases: if exposure is suspected, breathe through a wet handkerchief or rag, using urine if no water is available; get out of the area, shed outer clothing, and as soon as possible get a thorough scrubbing with soap and water. I was not popular with official stuffed shirts who issued elaborate instructions for identification of the gases and the different ways to handle each one—as if the enemy would oblige by shelling over one gas alone. David Marsh and I prepared an article for clinicians, based on our direct experiments with mustard gas on our own arms, showing that sodium hypochlorite solutions or soap and water would protect one, if applied within ten minutes of exposure.

In the midst of all this, when I was getting a little respite at the glorious Bohemian Grove on the Russian River, in the summer of 1942, I got an urgent phone call asking me to come to Texas to clear up a messy administrative situation at the University of Texas Medical Branch in Galveston. I went down, met the regents, and we liked each other, especially Lutcher Stark of Orange, who noted calluses on my hands and said I'd do. It was easy enough to get the place in order, with an open door to my office, and with Sunday afternoon receptions which Elizabeth, my always efficient wife, arranged at our home for faculty, students, and townspeople. But this all left no time for pharmacology. Actually, when we admitted two classes a year, and increased the size of each to 120, as part of the war effort, I continued to teach.

The pharmacologist at Galveston, Wilfred Dawson, had killed himself in frustration over the administrative situation, and I had to try to take over. Pharmacology had its own building, a wooden one, but quite adequate, and we soon had a seminar going, as well as classes. Both students and clinicians seemed interested in a new point of view. We used no text but asked the students to submit a couple of term papers properly documented, and graded from excellent to poor. We started publication of a quarterly *Texas Reports on Biology and Medicine,* which soon attracted contributions from all over, since we sent it without charge to the libraries of medical institutions throughout the world. The response in exchange journals made our library the best in the Southwest. It was especially gratifying to get a generous response from Russia.

As soon as the war was over, we planned and built new hospital and laboratory facilities. We promoted Galveston as a health resort, and I traveled over the South and Midwest spreading the gospel of the rational use of drugs, and the ever widening scope of pharmacology. I had helped that energetic gynecologist, William Bertner, organize the great medical center in Houston, with wise leadership from Fred Elliott, dean of the University of Texas Dental Branch, and Lee Clark, the efficient director of the M. D. Anderson Tumor Clinic and Hospital. I helped Baylor Medical School get started in Houston, getting Anderson Peoples and Carroll Handley to take over pharmacology. We gave library and laboratory equipment and books to these new developments, and even lent personnel to help get them going. Then they got the money, and soon had a huge and flourishing center. We aided in the establishment of the University of Texas Medical School in Dallas and tried to promote one in San Antonio.

I was fortunate in persuading Charles Marc Pomerat (1905–1964) to join our staff in anatomy, which had been so well maintained by Donald Duncan. Pomerat was a brilliant cytologist and teacher, a fine artist and bon vivant. He got our students interested in etching and lithography and established as fine a tissue culture laboratory as could be found. His great contribution was to use time-lapse cinematography to record changes in tissue culture under different conditions. I worked with him on drug effects on cells in tissue culture, and reported with him on this work at a tissue culture symposium he arranged with the New York Academy of Sciences. His studies even inspired me to write a half dozen "Tissue Culture Cadences." His biobibliography appears in the June 1965, issue of *Texas Reports on Biology and Medicine,* with many of his drawings and watercolors.

Meanwhile I began serving on many national committees dealing with problems of alcohol and with various aspects of medical education. These were tiresome. Frank Fremont-Smith (1895–1973), director of the Macy Foundation, made a great success of arranging series of conference discussions. I was invited to join one on neuropharmacology, chaired by Harold Abramson. This met pleasantly in Princeton and gave me a chance to get back to that delightful place. Elizabeth and I spent several months on three occasions in Princeton, when I worked at the Institute for Advanced Study, under Robert Oppenheimer (1904–1967) whom I had known at Berkeley. I was preparing a translation and annotation of the Hearst Medical Papyrus, a drug formulary from Egypt about 1550 BC, which was a prized possession of the Anthropology Museum of the University of California, which was in San Francisco when I first went there. It had attracted my interest then, and it still does. My work resulted in the Logan Clendenning Lecture at the University of Kansas (*The Old Egyptian Medical Papyri,* University of Kansas Press, 1952, 111 pp.).

With George Emerson in charge of pharmacology, my effort therein turned to writing reviews. I prepared one on drugs acting on the central nervous system, including the hallucinogenic agents, and then a more pertinent clinically oriented one on an analysis of drugs used in allergy (*Texas Reports on Biology and Medicine* 9:322–40, 1951). Ever interested in training programs, I wrote on the training of professional pharmacologists, and also on the training of physicians for general practice. Disturbed by the McCarthy witch-hunting, I wrote on the ideals of science in relation to national security (*Texas Rep. Biol. Med.* 13:434–45, 1955), and was denied clearance to go to Ecuador to survey medical education there. But I also wrote on ideals for a community health library.

Political machinations began again in Texas. I should have been alerted when a new bunch of regents fired James Hart, a distinguished barrister, as president of the University after a year of helpful leadership. The big corporations, anxious to avoid severance taxes on natural resources, set up the Texas Research League (nonprofit, and thus tax-exempt) to advise state agencies on operating with efficiency and economy. The university was hard hit. Most of the administrators resigned, as did I. When Charles Doan, the brilliant hematologist, and dean of the Ohio State University Medical School, asked me to join his program, we packed up and left Galveston to settle in Columbus. It took a decade for the University of Texas to recover. Now under the inspiring leadership of Truman Blocker, the University of Texas Medical Branch is flourishing, with an Oceanographic institute and an institute for the Humanities in Medicine, both of which Charles Pomerat and I had dreamed about. Galveston also has a superb seaorama, with a great aquarium, and performing seals, dolphins, and whales. We had tried to get an aquarium going years before, but it was a flop.

At Columbus my main job was to try to organize a laboratory for pharmacology. This was not easy in a predominately clinical school. Fortunately, I had the interest of Bernard Marks, and he took hold in admirable style, building a splendid group of teachers and graduate students. He worked on cardiovascular drug problems, while I went back to the stimulant action of spleen-marrow on the reticular endothelial system. I thought that stimulation of red blood cell production might aid in counteracting the fatal anemia following whole-body radiation. I used hundreds of mice, but had only slight indication that spleen-marrow fed ad lib before radiation might be helpful. The evidence was too slight to publish.

Now there were too many meetings: Macy Foundation Conferences on Central Nervous System and Behavior, National Research Council on Problems of Alcohol, The National Medical Library, Physiology Congress in Buenos Aires, History of Medicine Congress in Athens and Cos, and meetings of pharmacologists and the too big sessions of the federated societies at Atlantic City. I was up to my neck with travel. But my wife topped it off with two trips around the world.

With me home in bed, I was nominated from the floor and made president of the American Society for Pharmacology and Experimental Therapeutics. Ben Abreu told me I'd better get over to Chicago. I did not get a warm reception from the "establishment," but we soon were humming along with news letters and *Pharmacological Reviews,* and I started *The Pharmacologist.* I became involved with the American Association for the Advancement of Science, and presently was its president. My historical interests appeared in my address on retiring: "The Status of Pharmacology as a Science" (*Science,* Dec. 29, 1961).

But even with growing preclinical strength at Columbus, there were difficulties. We had a fine seminar, cooperating with Arthur Tye in pharmacy, and with Eric Ogden (1903–1973) who had built a great department of physiology. A new administration began cutting. With Bernard Marks running pharmacology well, I thought it was time to quit. We enjoyed Columbus. We helped get the symphony rolling under Evan and Jean Whalon, and the Kit-Kat Club was a stimulus to me.

Robert Featherstone had come to San Francisco from Iowa to take charge of pharmacology after Hamilton Anderson retired. John Saunders had built a big laboratory building, and pharmacology had new quarters, with much stimulus for growth. Robert Featherstone (1914–1974) was interested in mechanisms of anesthesia, promoted student research and asked me to guide it. So out we came again to San Francisco, where my brother Russell, found an apartment with a view for us near the school. I was given an office in the dining room of an old house below the school, and assured that I'd soon have fine quarters. I'm still there.

Actually I'm having a good time and enjoying it. Occasionally I'm asked to lecture in pharmacology to nurses or pharmacists, and I get to pharmacology seminars once in a while. Mostly I'm busy in my book-crowded office, trying to write what I think is worthwhile. The ethical problems of medical practice have long interested me; now they are hot subjects of debate. I'm interested in how theories of ethics apply. There is vast confusion in ethical theory in organ transplants, for example, and human experimentation raises many questions. Even euthanasia is pharmacologically related.

Since I can do quite as I wish, I offer a no-credit course on the ethics, the logics, and the esthetics. It is fairly well attended, but not by the graduate students for whom it is designed. I think that persons holding the degree, doctor of philosophy, should know something about the subject. I also teach a no-credit course on the history and philosophy of the health professions, quite as I have done yearly for over fifty years. This course is well attended. I offer coffee and cookies as bait.

Thanks to Henry Elliott, my former colleague, and now handling both pharmacology and anesthesiology well at our Irvine campus, I was asked to write a review of reviews for *Annual Review of Pharmacology,* when Murray Luck's great series was expanded in 1960 to include pharmacology. Thanks to Henry Elliott again, I have written such a review each year since. It helps to keep me abreast of the rapidly expanding field of pharmacology and toxicology. My colleagues, Harold Hodge and Charles Hine, help with toxicology, and Eddie Way always helps, as does Vi Sutherland, with details of pharmacology.

James Dille, developing a keen pharmacology program for the University of Washington in Seattle, persuaded the West Coast pharmacologists to organize the Western Pharmacology Society. He guided it well for many years. After a decade, he turned it to me as secretary-treasurer, and I enjoyed preparing the meetings and the proceedings, until I got in trouble with a president who wanted to run things his way, so I quit and let him do so.

Meanwhile I traveled a lot. Under the auspices of our State Department (having somehow been cleared), I went to the Brussels Physiology Congress, where I put on an exhibit entitled *Some Founders of Physiology,* and then via Helsinki, to Leningrad. There I enjoyed the high morale of the devoted staff of S. V. Anichkov, and of the keen workers at the Institute for Experimental Medicine, with their high regard for Charles Darwin. In Moscow it was a joy to know P. K. Anokhin (1898–1974), director of the Sechenov Institute for Physiology, who later visited and stayed with us.

My regard for Frank Berger was always high. He had developed meprobamate as a mild tranquilizer. Thanks to him, and Charles Hoyt, we went to Japan, and had a great time in the pharmacology laboratories in Tokyo, Osaka, and Kyoto. It was a special pleasure to meet Professor Hiroshi Kumagai, and his colleague, Setsuro Ebashi, and to learn of their fine work on muscle contraction. With Frank Berger again, I had the joy of meeting Silvio Garattini at his fine Instituto Mario Negri in Milan, at a conference on anti-inflammatory drugs.

Elizabeth and I enjoyed visiting Arthur and Martha Stoll at their lovely art gallery villa at Vevey on Lake Geneva, and going to Spain for an exciting visit with Francisco Guerra, who had done so well for pharmacology in Mexico. Francisco Guerra had explored Mexican hallucinogenic drugs, and had written widely on Mayan drugs. His baronial holdings in Santillana-sur-Mar include the famed Altamira Caves. Guerra had raised a company to protect his village in the Spanish

Civil War, and after losing a leg in the fighting, had fled to Mexico, where he became professor of pharmacology at the National University.

My interest in what goes on in pharmacological and medical fields led me to put out a monthly mimeographed sheet, "Calling Attention To," for the benefit of my friends. At Galveston and Columbus this grew to over 1500 copies a month. The postage was too much on my return to California, so I restricted the effort to books, a list of which appears monthly in *Current Contents—Life Sciences.* This is one of the helpful publications of the Institute for Scientific Information in Philadelphia, started on a shoestring by my long-time friend, Eugene Garfield, and now a large and vigorous enterprise. I had worked with Garfield while trying to convert what was then the Armed Forces Medical Library into a truly national library of medicine.

My pharmacological interests may sometimes seem to be far afield. I was once president of the Sex Education Society in San Francisco back in depression days. We ran a birth-control clinic, and our effort later grew into the Planned Parenthood program. In 1940, the students in Berkeley petitioned the faculty for a sex education course. I was put in charge. I made it a non-credit affair, with emphasis on human relations. If interpersonal relations are in order, sex relations will usually be so too. It was a shock to find 2600 students showing up for the weekly illustrated lectures. World War II stopped this undertaking.

The Hastings College of the Law in San Francisco was made by Sam Snodgrass, its informal dean, into one of the best in USA, simply by asking retired deans and professors from other law schools to join his faculty. His 65 Club included some of the greatest law teachers in USA. He asked me to join, in order to teach medical jurisprudence. I knew no law, but I had had much experience as an expert witness, and I did know something about toxicological analysis. The course went well for several years, until Dean Snodgrass died and a new, more formal and pedantic dean took over. He didn't like my informal grading system of good, bad, or indifferent, since law schools grade to decimal points. So I quit. But I enjoyed the experience.

Stanley Jacob, at the University of Oregon Medical School, had found dimethyl-sulfoxide (DMSO) to be an excellent solvent and skin absorbable. He used it to dissolve steroids for local application in arthritis. On trying DMSO alone, as a control, he found it was effective. It tends to dissolve collagen and fibrous tissue. So he began using it in a variety of clinical conditions for which it was indicated, and ran smack into the new regulations of the Food and Drug Administration, of which he knew nothing. He was accused of violating practically all of them. I persuaded James Goddard, FDA Director, that Jacob was not criminally minded, and we arranged a big symposium on DMSO under the auspices of the New York Academy of Sciences (*Ann. NY Acad. Sci.* 141:1–671, 1967). Schering AG, under the keen direction of Gerhard Laudahn, arranged for a notable clinical conference on DMSO in Vienna (*Dimethyl-sulfoxyd: DMSO,* Salabruck, Berlin, 1966, 219 pp). In spite of countless handicaps, Stanley Jacob has persisted in the careful, patient accumulation of data, so that DMSO now has a chance again of helping people.

My Princeton classmate, Harry Hoyt, who developed the Carter-Wallace Company with Frank Berger in charge of drug research, asked me to join his board of directors. This I did, and I have learned much of the tough problems continually faced by drug companies in trying to develop new drugs. It is nothing like the way we did it during the depression. But the drugs we introduced, spleen-marrow, divinyl ether, carbarsone, Vioform®, the amphetamines, and nalorphine, are still useful drugs, when used appropriately and properly. We never made a penny from our work, except Gordon Alles from amphetamine, but we were well content to have enjoyed the effort.

Actually, we had quite a close association at the big, open lab in San Francisco where we all worked together. We called ourselves the Blakians, in reference to James Blake (1814–1893), the brilliant London physician who by 1846 had shown clear relationships between chemical constitution and biological action using inorganic salts. He showed that their biological activities permitted their arrangement into families having similar actions, and there is the outline of the periodic table, 20 years ahead of Dmitri Mendeleev (1834–1907). Blake got into trouble and came to St. Louis in 1847, and then came to California in the Gold Rush. He was professor of midwifery and diseases of women and children at the University of California Medical School when it was organized in 1873, and became president of the California Academy of Sciences. He retired to Middletown where he returned to his earlier pharmacological studies. The Blakians made frequent pilgrimages to his grave in the Middletown cemetery. When I gave an account of his career at the Aberdeen meeting of the British Association for the Advancement of Science, I found that the old scandal, whatever it was, precluded the possibility of publishing my account in England.

When it comes to writing, scientists are often no better than their secretaries. I've been fortunate in having fine secretaries, keen, efficient, dedicated, often with us in family affairs, and always helpful in arranging laboratory parties. At Wisconsin there was Irene Blake, and at California, Marjorie Williams, both of whom died too young. At Texas there was Isobel Aiklin, now Mrs. Clemens, who visits my wife each year at Chautauqua, and Mary Jane Steding. At Columbus there was Jane Nolan, and now back in San Francisco there is Mia Lydecker, who really works for Milton Silverman, but who helps me when she can. Here also is Aline Steward, who actually is secretary for the group in environmental health, which shares the old house with me. Mrs. Steward keeps my mail in order.

My historical interest resulted in a large manuscript on the history of pharmacology. This was prepared for Pergamon Press *Encyclopedia of Pharmacology*. But I ran into editorial blocks, so now it appears from C. C. Thomas, who was my first publisher anyway. Peter Knoefel, working on Felice Fontana (1730–1804), pioneer student of the toxicity of venoms, comes over from Firenze once in a while to work with us, and to run to see what we can find on Blake. So we keep rolling along. Even at the Bohemian Grove, where each summer is "the greatest men's party on earth," I get a chance to talk at the lakeside on alcoholic beverages, and other drugs. My chief joy there, however, is lighting the Concert-at-the-Lake, throwing 250 foot beams of colored lights on the redwoods behind the orchestra in coordination with the music. This I've done for 40 years. So this is how I am: still at it for the excitement, the knowledge, and the fun of pharmacology, and related matters.

Ann. Rev. Microbiol., Vol. 25

FROM PROTOZOA TO BACTERIA AND VIRUSES
FIFTY YEARS WITH MICROBES

ANDRÉ LWOFF

André Lwoff

Ann. Rev. Microbiol., Vol. 25

FROM PROTOZOA TO BACTERIA AND VIRUSES   1560
FIFTY YEARS WITH MICROBES

ANDRÉ LWOFF

Institut de Recherches Scientifiques sur le Cancer
B. P. 8, 94—Villejuif (France)

To Marguerite

CONTENTS

INTRODUCTION

My father, born in Russia, in Sinferopol, came to France in 1880 at the age of 21. As a student, he had been involved in political activities and was sentenced to a few months imprisonment in the Peter and Paul prison in Saint-Petersburg, later Petrograd, then Leningrad. My father then left Russia, came to Paris, where he studied medicine and became a psychiatrist. My mother, too, was Russian. She had studied sculpture in Saint-Petersburg with Antokolski and, as a young girl, came to Paris as so many artists did.

My father soon became head physician in a psychiatric hospital and I

1

was born in the small village of Ainay-le-Château (Allier) in the center of France, on May the 8th, 1902. Then we went to Neuilly-sur-Marne, close to Paris, to another hospital. The psychiatrist had a house just at the outskirts of the hospital. At home, the servants, the cook, the chambermaid, and the gardener were inmates of the hospital. The cook sometimes answered voices which were speaking to her and which I could not hear. The chambermaid sometimes behaved strangely, and also the gardener. This was perfectly normal, of course, and things went all right. Maybe I should mention that one person was not an inmate of the hospital, namely, "Fraulein." She taught me German and at the time I became bilingual. However, Fraulein used to speak to herself . . . Have I been influenced in one way or another by this unusual environment? I am the last to be able to answer this question.

The house was surrounded by a garden with beautiful trees and flowers. There were also fruit trees and a vegetable garden. In the vicinity was a tennis court, and not far away a river where we used to swim and row. Moreover, I had been presented with a small rifle and used to kill rats and sparrows with bullets and even, once—a great event—a partridge in the field behind the garden. From time to time my father interrupted the games and I was summoned to his office to read Plato or Kant but, on the whole, life was marvelous.

Guests came to visit us. Relatives from Russia, of course, and from Italy and Austria, because the family had been scattered by emigration. There were also Russian socialists just out of jail or from Siberia, freed in one way or another. Also members of illustrious aristocratic Russian families living in Paris who had taken the wrong path and needed the care of a psychiatrist.

My father had decided that the study of mental illnesses should be part of my education, therefore at a tender age, I was taken to the wards. I do not know if this exposure has been useful but I am under the impression that the contacts, perhaps somewhat premature, with the manifestations of mental disorder considerably reinforced my inclination for scientific disciplines.

I was twelve when the first World War started. The German troops had reached a point twenty miles from our place. Later on during the battles, I could hear the roaring of the cannons from the battle line. From time to time at night German planes were trying to bomb Paris. The antiaircraft guns were close and the splitters were whistling and drumming on the roof. I listened to the strange music with curiosity, perfectly unaware of the danger. Later on, the psychiatric hospital was turned into a military hospital but I was not mature enough to realize the depth of the war tragedy.

I went to school, read an incredible number of books, and suddenly I decided to study biology in order to do research. How this idea germinated in my head, I do not know. Perhaps because of Elie Metchnikoff who was a friend of my father. When a patient had died of a disease which interested

Metchnikoff, he went to the hospital for the autopsy, and then came home for lunch. Cotton-plugged tubes showed outside the pocket of his coat and the cotton was soiled with blood. My poor mother was horrified. .

It was Elie Metchnikoff who showed me a microbe for the first time. It was in 1915 when I was thirteen. My father had taken me to the Pasteur Institute. Metchnikoff asked, "have you ever seen a microbe?" I had not. So, Metchnikoff took a glass slide, put it under the microscope and said "it is the typhoid bacillus. Look." I was very excited. What happened, I remember very well. I looked into the microscope and saw nothing. I was very impressed. Such was my first contact with bacteria.

Anyhow, at seventeen, I decided to study biology and to do research. My father explained to me that research is not always successful and as I would have to earn my living, I had better study medicine. At the time, before entering the medical school the student had to spend a year in the *Faculté des Sciences* studying biology, chemistry, and physics. This was not enough. As I was very enthusiastic I had spent two months in the Marine Biological Laboratory at Roscoff in Brittany during the summers of 1919 and 1920, and during my first years of medicine I took the botany and zoology examinations at the Sorbonne and again in 1921 went to Roscoff where a determining event took place.

Impressed by my youthful ardour, the assistant recommended me to Edouard Chatton. Thus, I had the privilege of becoming the pupil and the collaborator, later on the disciple and friend, of the most brilliant representative of the brilliant French school of protozoology and, in my judgment, the greatest protozoologist of all time. Could I note for the readers of the *Annual Review of Microbiology* that it is Chatton who, in 1928, separated the Eukaryotes from the Prokaryotes, and coined the two names. From 1921 on, we worked together two or three months each year in the marine laboratories at Roscoff, Banyuls, Villefranche-sur-Mer, Wimereux or Sète. Our collaboration was interrupted by the war and ended only with the death of Edouard Chatton in 1947.

In 1921, Chatton was professor in Strasburg. He recommended me to Felix Mesnil who was head of the Department of Protozoology at the Pasteur Institute. He had been the secretary of Louis Pasteur and the collaborator of Alphonse Laveran. In October 1921, I received a fellowship and worked part time whilst studying medicine. I earned 350 francs a month, that is, 70 U.S. dollars—and felt rich. I had, of course, a tendency to sacrifice my medical studies to my passion for ciliates. But still I learned to examine patients; I swallowed anatomy, histology, embryology, bacteriology, parasitology, physiology, and biochemistry, etc., and succeeded in passing the examinations for pathology, obstetrics, therapeutics, and the rest.

THE EYES OF COPEPODS

So, I passed the examinations. But in order to become an M.D. one had to defend a thesis. Whilst in the marine station in Roscoff I had worked on parasitic ciliates which evolved in copepods engulfed by coelenterates. Often they were red and soon it was clear that the eye of the copepod was the origin of the pigment. So I started working on copepod eyes and in 1927 I presented my thesis at the Medical School. The title was "Le cycle du pigment carotinoide chez *Idya furcata* Baird, Copépode Harpacticide." For the first time, a carotinoid pigment was described in an eye. The eyes of copepods are normally dark red. I raised copepods on a carotinoid-poor diet. The red layers became less important and a beautiful blue structure appeared which is normally masked. I showed, among other things, that the blue pigment is a combination of a carotinoid with a protid and proposed the hypothesis that it plays a role in the vision. The president of the "Jury" was the well-known histologist, André Prenant, but a gynecologist was a member of the examination board. He shrugged his shoulders in disgust. Who knows, perhaps he was not really interested in copepods. Anyhow, as a result of the work, my name is now on the list of world copepodologists!

However, there was a byproduct: whilst working at the Kaiser Wilhelm Institut in Heidelberg in 1932, I called to the attention of Richard Kuhn and Edgar Lederer the fact that carotinoids were present in crustaceans. They bought a few lobsters and discovered astacin.

MORPHOLOGY AND MORPHOGENESIS OF CILIATES

However beautiful a copepod with blue eyes might be, I was in love with ciliates. During the summer of 1921 in Roscoff I had been asked by Edouard Chatton to explore the gills of acephals. The first mollusc examined, *Dosinia exoleta,* showed a strange organism devoid of cilia which reproduces by ciliated buds. This was the type of a new genus, *Sphenophrya,* and of a new family, the *Sphenophryidae,* to which we later added two other genera, *Pelecyophrya* and *Gargarius.* The systematic investigation of the ciliates of the gills brought the discovery of many new organisms and the organization of new families into a new suborder, the *Thigmotricha.* I should make clear that the work on ciliates, with a few exceptions, was performed with Edouard Chatton and, from 1925 on, with Marguerite Lwoff.

The study of the morphology and morphogenesis of ciliates was greatly facilitated by the introduction of a new technique of silver staining. The Klein method, silver staining of dried specimens, was not applicable to marine ciliates owing to the formation of precipitates of silver chloride. But we noticed the absence of a precipitate and nice staining of encysted ciliates. So, the idea germinated that cysts could be replaced by gelatin. After proper fixation the ciliates were embedded in salted gelatin, then impregnated with silver nitrate which was reduced by ultraviolet light. The shape of the specimens was preserved and the kinetosomes were beautifully stained as also were the trichocysts and many fibers. Moreover, the whole organism was stained. So the ciliary system of many ciliates was described accurately for the first time, new structures were discovered as well as the intimate process of stomatogenesis. In *Tetrahymena piriformis,* before division, the kinetosomes of the stomatogenic kinety n° 1 start dividing toward their left. It is in or around the newly formed field of kinetosomes that the membranelles are organized and that the morphogenesis of the mouth takes place (with E. Chatton, M. Lwoff, and J. Monod).

One of the groups studied was *Apostomata*. Numerous new species were discovered and studied. We described the life cycle in the two hosts and the ciliary metamorphosis which take place during the cycle.

Some general features of the morphology of ciliates were disclosed: 1. the genetic continuity of kinetosomes and the concept of the "infracilia-ture"; 2. the origin of trichocysts; 3. the kinetodesma; and 4. the law of desmodexy. We shall examine them separately.

1. When one looks at a silver-stained ciliate one sees that new kineto-somes always appear in the immediate vicinity of a persisting one and that the two kinetosomes are united by a "desmose." Therefore, it was concluded that kinetosomes are endowed with genetic continuity and reproduce by di-vision. We know today that only a double-stranded nucleic acid made of two complementary parts can "divide" and reproduce its kind. No other mole-cule is reproduced by division. So, *sensu stricto,* a kinetosome cannot divide. How, then, does a kinetosome reproduce? The simplest approach is to as-sume that a kinetosome is located in a specific morphogenetic field or terri-tory. The molecules of this territory can increase in number. A new kineto-some can be organized only within the vicinity of a pre-existing one. The reproduction of a kinetosome is the result of the division of a specific mor-phogenetic territory.

Let us now consider a ciliate as a whole. Ciliates reproduce by binary fission or division. Before division, two new ciliates are organized within the framework of the parent. Division separates two new ciliates, one ante-rior, the other posterior, for which we proposed the name "proter" and "op-isthe," respectively. Whatever the case might be, if the reproduction of a ciliate is called division, then the reproduction of a kinetosome is also a division.

Some adult ciliates are deprived of cilia but are reproduced by ciliated buds. This is the case for most suctorians. The study of the suctorian *Po-dophrya* revealed that kinetosomes devoid of cilia are present in the adult and that they are the origin of the kinetosome of the bud. The ensemble of kinetosomes was called an "infraciliature."

2. Let us consider the origin of trichocysts. In the Apostomatous cili-ates, trichocysts are formed only during a short and unique definite period of the life cycle. One then sees all the kinetosomes "dividing" to their left and, from the daughter kinetosomes, one sees trichocysts growing. Needless to say, the electron microscopists studying apostomous ciliates were, up to now, unable to see this phenomenon which is very clearly visible with an ordinary microscope when one examines silver stained ciliates *at the right phase of the cycle.*

It is currently said that cilia are produced by kinetosomes. The kinetosome as such cannot synthesize the proteins of the cilium, but they have the property to organize and to orient the ciliary proteins, and also the proteins of the trichocysts and of the fibers. A kinetosome and its territory represent a remarkable morphogenetic machine.

3. The study of ciliates revealed the existence of a fiber which we called the "kinetodesma" and which always runs parallel to the rows of kineto-somes on their right. This is the law of desmodexy. The kinety is thus asymmetrical—one can recognize an anterior and a posterior end, a right and a left. With Chatton, we also saw that the kinetodesma was formed by fibers connected to the kinetosomes. This was rediscovered much later by the electron microscopists.

In some ciliates at a given phase of the life cycle, the kinetodesma of a given kinety disappears. As a result, the corresponding kinetosomes are scattered in disorderly fashion. This exemplifies the role of the cortex in the maintenance of the kinetodesma and the role of the kinetodesma in the or-ganization of the kinetosomal pattern.

Thigmotricha and Apostomata were our main concern. But, when work-ing in a marine laboratory, one comes across strange organisms which no-body yet has seen. We thus described a curious peritrichous ciliate, *Ello-biophrya,* which is appended to the gills of the acephalus *Donax vittatus* like an earring and reproduces by budding a strange and beautiful "embryo." We also described the *Conidophrys pilisuctor* a ciliate which, as embryo, empales itself on the secretory hairs of the amphipod *Corophium,* loses its cilia, encysts, and reproduces by budding.

Finally, I discovered an amazing ciliate whose nuclear apparatus is a protocaryon: *Stephanopogon mesnili* is homokaryotic. It was not possible to state any longer that ciliates are characterized by the differentiation of the nuclear apparatus into a macro- and a micronucleus. Maybe I should add that *Stephanopogon* is a true ciliate which possesses mouth characteristics both of Gymnostomes and of Hymenostomes, and is a most devilish animal. We may note here that Opalinids, which are homokaryotic, were generally and are sometimes still considered as ciliates despite the fact that Chatton and Brachon, a long time ago, showed that their division is longitudinal; it cuts between two sets of longitudinal kineties, whereas in ciliates the kine-ties are cut transversally by division.

I started playing with ciliates in 1921. The milestones of the work were monographs on *Stephanopogon, Conidophrys, Ellobiophrya, Apostomata,* and the two monographs on *Thigmotricha.* In 1948, I was invited to Har-vard to deliver the Dunham Lectures. The theme was the morphogenesis of ciliates and a book was published, *Problems of Morphogenesis in Ciliates,* which I will try to summarize.

The study of the development of ciliates shows that the activity of the kinetosomes, their organization into specific patterns, their division, and the expression of their potentialities, depend on their position and on the phase of the life cycle. As already said, before division of *Tetrahymena piriformis* occurs, the kinetosomes located around the middle of the "stomatogenic kinety" n° 1 start dividing actively toward their left; a dense field of kinetosomes is formed which is the base for the organization of the membranelles and the mouth. It is clear that kinetosomes are induced to reproduce by the properties of the underlying cortex. Hence, the conclusion that "if kinetosomes are necessary for morphogenesis, they seem not to command but to obey some mysterious force . . ." In certain ciliates, trichocysts, for example, are formed only at one phase of the cycle. The analysis of the movements of the cortical structures in various ciliates led to the hypothesis that the cortex differs according to its location and that cortical structures command morphogenesis. Morphogenesis is in part the result of the response of an apparently homogeneous population of kinetosomes to their environment, the cortex. An orderly and organized asymmetry like that of an egg or of a ciliate may be only the reflection of cortical properties. A constantly flowing or potentially flowing endoplasm cannot be asymmetrical. The organelles may be asymmetrical. But when the ciliate is considered as an organism the conclusion is reached that organized asymmetry, or simply organization, can belong only to a more or less rigid, a more or less permanent system, that is to say, the cortex. Finally, one finds that, in well-defined evolutionary series of ciliates the structure of the daughter ciliates just before division corresponds to that of the primitive type. Ontogeny repeats phylogeny.

It came as a surprise when Beisson and Sonneborn, in a paper which appeared in the *Proceedings of The National Academy of Science* in 1965, made me responsible for the statement that: ". . . kinetosomes . . . are directive or instrumental in morphogenesis" (this sentence is the only reference to my book). Impressed by this statement, Nanney wrote later: "Persistent reports of DNA associated with the ciliate cortex . . . lend plausibility to the suggestion (see Lwoff 1950) that gene-like elements imbedded in the cortex represent peripheral "nucleic" reservoirs responsible in some way for cortical characteristics." This, supposedly a view of mine, is stated by Nanney to be opposed to Sonneborn's ideas concerning the role of cortical structures in morphogenesis!

Of course, kinetosomes have a morphogenetic role in the sense that they are responsible for the production of cilia, trichocysts, and fibers. However, I have never thought nor said that kinetosomes were controlling the cortex or were directives in the morphogenesis. As a matter of fact, my conclusions as can be judged by the above quotations point to the opposite concept, namely, that the presence, location, organization, and activity of cortical organelles—including kinetosomes—are not determined by the kinetosomes but by the cortical structures and their "internal" environment as determined by the phase of development. Beisson and Sonneborn have reached the same conclusion. I am, however, unable to understand why my colleagues made me say *exactly the contrary* of what I clearly said. Owing to the authority of Sonneborn, his distorted interpretation of my concepts is now spreading. I regret to have been obliged to put things straight myself.

THE EXCITEMENT OF SCIENCE   307

Nutrition of Free-Living Protozoa

The work on morphology and morphogenesis of ciliates was performed each year at several marine biological laboratories during the months of July, August, and September and sometimes during the Easter holidays. At the Pasteur Institute I first took the course of microbiology. Later on, I became responsible for the classes in protozoology and parasitology and had the responsibility of maintaining pure cultures of trypanosomides, trichomonas, and amoebas. So, I gained some experience in handling protozoa and decided to work on the nutrition of ciliates. This was mad. So far as I can remember, I had read a review of an article dealing with the alleged pure culture of a ciliate in a medium in which ammonium sulfate was the nitrogen source and dextrose the sole organic substance. The author of the review said that these results shattered the doctrinal corpus concerning the power of synthesis of protozoa. The literature was searched for data concerning this power of synthesis. Nothing was found except concepts such as "heterotrophy" defined as the need for organic substances, and "autotrophy," the absence of need for organic substances which was considered as bound to photosynthesis. I could not discover anything about the nutritional requirements of protozoa. This is why I decided to investigate the nutrition of ciliates. In order to do so, "pure," that is bacteria-free, cultures were needed. From a wild culture of Tetrahymena piriformis, I obtained in 1923 a bacteria-free pure culture. A ciliate was put in a drop of sterile medium and transferred from drop to drop with a micropipette under a dissecting microscope. The ciliate grew in peptone solutions but refused to multiply in media with an ammonium salt and with dextrose as sole organic source. So it needed other "organic substances" which I tried bravely to identify. Knowing today the large number of amino acids and of vitamins necessary for this organism, it is retrospectively not surprising that my work failed, but not entirely as will be seen later. However, the first pure culture proved to be a useful tool for the solution of biological and morphological problems.

Flagellates seemed more promising. A number of groups are interesting because they comprise chlorophyll-bearing organisms as well as their chlorophyl-less counterparts. The nutrition of the "green" and "white" species was systematically investigated with the collaboration of Marguerite Lwoff first and, later on, Hisatake Dusi and Luigi Provasoli.

Some species existed in culture collections and I isolated a few strains by washing single individuals. The green flagellates utilize nitrates as nitrogen source. Most white organisms thrive in the presence of an ammonium salt but not with a nitrate. However, it was found that one species, Polytoma ocellatum, utilizes nitrates. This was the first example of a protozoan able to reduce nitrates.

It was known that Polytoma uvella utilizes acetic acid as a carbon source. The systematic study of white Chlamydomonadinas, Phytomonadinas, Cryptomonadinas, and Euglenidas revealed that some of them can thrive on acetic acid only, whereas others are satisfied with a number of lower fatty acids and some also lactic and pyruvic acids. The obvious idea was to study the green counterpart of the white flagellates in the absence of photosynthesis. *Chlamydomonas* grows beautifully in the dark if provided with acetic acid, as does *Haematococcus* also. It turned out that in the presence of acetic acid only, *Haematococcus* manufactures large amounts of carotenoids in the dark as well as in the light. Finally, we investigated a few Euglenidas. For *Astasia* (with H. Dusi) the fatty acids were the right carbon sources and this gave the key to the culture in the dark of green Euglenidas. They had been grown in the dark, but the cultures were always poor except when acetic acid was added. Acetic acid was supposed to act by lowering the pH, but we showed that, in fact, acetic acid acted as a carbon source. The green *Euglena* in the absence of photosynthesis behave as do the other flagellates (with H. Dusi).

All the investigated flagellates manufacture starch or paramylon. None of them, however—as already said—utilize glucose or any other sugar as an extrinsic carbon source. However, *Polytomella caeca* contains a phosophorylase which can transform starch into glucose-1-phosphate in the presence of phosphate (with Hélène Ionesco). It could be that the synthesis and utilization of starch in these organisms takes place without glucose as intermediate except in the phosphorylated form.

HEMATIN AS A GROWTH FACTOR

In 1932, I received a Rockefeller fellowship to work with Otto Meyerhof in Heidelberg at the "Kaiser Wilhelm Institut für Medizinische Forschung." I stayed there fifteen months, until the end of the year 1933. It was a very interesting year, not only from the scientific point of view. The government of Germany changed in the spring of 1933 and a tragic era began. It was clear that war would come sooner or later. The laboratory was very active and everyone behaved apparently as if science only was important. However, the numerous foreign workers attracted by the fame of Meyerhof did not fail to observe and to comment on the situation without any illusion for the future. Later on, Meyerhof, as a Jew, had to leave Germany with his family. His relatives who stayed disappeared in extermination camps together with millions of other human beings.

In the Pasteur Institute, Marguerite Lwoff had shown that hematin could replace blood for the growth of *Crithidia fasciculata*—at this time called *Strigomonas fasciculata*. I decided to investigate the role of hematin as a growth factor. The field of growth factors was rather confused. Microbiologists were convinced that a "trace" of a growth factor would induce an unlimited development of microorganisms. Growth factors were supposed to act on multiplication as catalysts. At the time, a few essential amino acids had been shown to be necessary for the development of bacteria, but no "vitamin" had been identified. There was only one exception, namely hematin. So, hematin as a growth factor for *Crithidia fasciculata* was investigated. It was first shown that blood and hematin acted quantitatively; the number of flagellates which developed was, within certain limits, proportional to the amount of hematin. Each flagellate needed 520,000 molecules of hematin. What was its function?

The respiration of flagellates grown in a medium in which the hematin concentration is the limiting factor is lower than the respiration of "normal" flagellates. When hematin is added the respiration increases, and the increase is proportional to the amount of hematin added. It was then easy to calculate the amount of hematin needed by one flagellate in order for its respiration to be normal: it was 720,000 molecules, in good agreement with the number 520,000 found by measuring the growth as a function of the growth factor. Thus, it was clear that hematin was not acting as a "catalyst" either on multiplication or on respiration. It entered into the constitution of the catalytic respiratory system and its action was quantitative.

Why is hematin necessary? Blood was known to be necessary for the growth of the bacteria *Hemophilus influenzae*. As it was found that hematin works, but that hematoporphyrin is inactive, the action of hematin had been ascribed to its iron atom. A number of "active" iron preparations had been proposed as a substitute for hematin and were supposed to work. However, hematoporphyrin does not differ from protohematin only in the absence of iron. Therefore, the specificity of the hematin molecule was investigated.

Protoporphyrin proved to be active. Thus, iron—the catalytically active atom of hematin—was not the growth factor. Moreover, a number of hematins were investigated, all of which were inactive. The activity of hematin was bound to the structure: tetramethyl 1,3,5,8-divinyl 2,4-dipropionic 6,7-porphyrin. This was the first study of the specificity of a growth factor. The need for hematin was the consequence of the inability to synthesize it.

The properties of hematin as a growth factor for the flagellate *Crithidia fasciculata* turned out to be the general feature of growth factors. Growth factors act quantitatively and not catalytically on growth. They enter into the constitution of catalytic systems. Their activity is bound to a specific structure.

Growth factors were for the first time defined as specific substances which the organism is unable to synthesize and which are necessary for its growth and multiplication. Trypanosomids are parasitic flagellates and the need for hematin is found only in parasitic organisms. It was concluded that the need for hematin was the result of the loss of the power to perform its synthesis. This concept was to be extended to all growth factors.

There is today a general consensus about growth factors. Let me tell what happened to me at the 2nd International Congress of Microbiology held in London in 1936. I had to open a session on growth factors. The title of the paper was "study of lost functions." I discussed especially the results concerning hematin. When I had finished, an eminent microbiologist and biochemist, the head of a brilliant school, stood and said, "I do not like substances which produce miracles." A beautiful execution—which I survived. The judge-executioner also survived. The miracle soon became the daily bread provided by text books but I still feel the rope around my neck, and how I was thrown into the emptiness.

Moraxella

The results concerning *Crithidia* were extended to *Hemophilus influenzae*. Moreover, with Ignacio Pirosky we showed that hematin was a growth factor for *Hemophilus ducreyi*. So it was decided to investigate systematically all the members of the *Hemophilus* group. Among them, in *Bergey's Manual*, the Morax bacillus was included. At the time, the group was defined essentially by the "need for body fluids." Officially, the Morax bacillus was unable to grow in the absence of serum. So I secured a culture of the bacillus and realized that it was widely different from *Hemophilus influenzae*, that it could belong neither to the genus nor to the family, and that it could not fit in any known bacterial genus. So I proposed the new name *Moraxella* given in the honor of Victor Morax who "invented" the organism known today as *Moraxella lacunata*. It grows in broth only if serum is added, and the question of the nature of the substance involved was posed. It turned out that the serum acts by neutralizing the toxic action of fatty acids present in broth. *Moraxella lacunata* grows in broth provided it is diluted with distilled water.

The systematic study of the various species of *Moraxella* was started and Alice Audureau discovered a new species (*Moraxella lwoffii*). We tried to grow it in synthetic media with an ammonium salt and glucose. It multiplied only when peptone was added. So I tried to identify the responsible growth factors. Thiamin was active, but was the organism really unable to synthesize thiamin from pyrimidine and thiazole? An astonishing phenomenon was observed. *Moraxella* grew in the presence of pyrimidine as well as thiazole. It was then realized that thiamine, pyrimidine, and thiazole were dissolved in ethanol. The addition of ethanol permitted growth. It was found that *Moraxella lwoffii* is unable to utilize any sugar as carbon and energy source but utilizes ethanol. It does not need any growth factor.

The wild strain of *Moraxella lwoffii* is unable to utilize malic acid as carbon source. I found a mutant able to do so and which possesses an enzyme converting malic acid directly into pyruvic acid without oxaloacetic acid as an intermediary step. The enzyme requires K^+ which could be replaced by rubidium or caesium but not by sodium.

Growth Factors for Free-Living Protozoa

Since 1932 the situation concerning vitamins had undergone considerable changes. The first vitamin, vitamin C, had been identified. This was followed by the identification of vitamin B_2 and B_1. During investigations on the carbon sources for flagellates it had been noticed that some flagellates would not grow in a medium containing acetic acid as sole organic substance. A "trace" of something was necessary. The growth factors for some of the flagellates were identified with the thiazole or pyrimidine moiety of thiamine.

Thiazole (methyl-2, β-hydroxyethyl-5, thiazol) is the only growth factor for *Polytoma obtusum, P. ocellatum,* and for *Chilomonas paramoecium.* Both thiazole and pyrimidine (methyl-2, amino-4, aminomethyl-6 pyrimidine) are necessary for *Polytomella caeca* (with Hisataka Dusi). The investigation was extended to an amoeba, *Acanthamoeba castellanii,* which was available in pure culture. In addition to numerous growth factors, it needed pyrimidine. All the investigated organisms are thus able to manufacture thiamine from pyrimidine and thiazole. This is not possible for the ciliate *Tetrahymena piriformis* which needs the complete molecule of thiamine (with M. Lwoff). Thiazole, pyrimidine, and thiamine were the first growth factors identified for free-living protozoa. The specificity of thiazole and pyrimidine was investigated.

The substitution of the -hydroxyethyl in the molecule of thiazole by one hydrogen or by methyl leads to an inactivation of the molecule, but the replacement of -hydroxyethyl by an acetoxyethyl is compatible with utilization as a growth factor by the flagellates (with H. Dusi).

The replacement of the aminomethyl in position 5 in the pyrimidine by thioformylaminomethyl or by a hydroxymethyl is compatible with the utilization. The substitution by a methyl inactivates the molecule as a growth factor. The substitution of the NH_2 in position 4 by hydroxyl or by $-OCH_3$ leads to inactivation. The transfer of the methyl from position 2 to position 6 also leads to inactivation.

The suppression of the hydroxyethyl of thiazole, the transfer of the methyl group of pyrimidine from position 2 to position 6, and the presence of a supplementary bond $-C-N=C-$ between pyrimidine and thiazole inactivates the molecule of thiamine as growth factor for the ciliate *Tetrahymena.*

GROWTH FACTOR V

In 1936, the Rockefeller Foundation gave me a second fellowship to work in Cambridge with David Keilin, then Director of the Molteno Institute.

Hemophilus influenzae can be grown in broth only if blood is added. It was known that blood provides two factors: one, the factor X, is hematin, the other one being the growth factor "V." "V" is often interpreted as a Roman figure for five, whereas it is a V as in victory. In fact, it stands for "vitamin-like" because it is known to be destroyed by heat; sensitivity to heat being long considered as a characteristic property of vitamins. At any rate, the growth factor V is not destroyed by "heat" in an acid medium. Nothing was known about its nature.

With Marguerite Lwoff, we decided to try to identify it. Very valuable help was received from David Ezra Green, who was then working in the Department of Biochemistry, and from Tadeusz Mann, David Keilin's collaborator. Thirty pounds of yeast were extracted and fractionated and the fractions tested for their "V" activity. The active substance was finally identified with coenzymes I or II, later on, phosphopyridino-nucleotides. Bacteria grown with factor V as limiting factor had a very low respiration rate and were unable to reduce methylene blue. The addition of coenzyme I or II restored the movement of hydrogen within 60 seconds. The need for growth factor V was found to be due to the lack of power to synthesize phosphopyridino-nucleotides. A new growth factor had been identified and its physiological role determined.

NICOTINAMIDE

After Paul Fildes had discovered that nicotinamide was the only growth factor for *Proteus vulgaris,* we investigated, together with Andriès Querido, the specificity and established the effects of various substitutions on its activity. Finally, a quantitative test for the estimation of nicotinamide was devised. We estimated nicotinamide in various organs and the first value for blood was found to be 0.75 mg per 100 ml. Vitamin PP had not yet been detected in milk and the hypothesis was put forward that the newborn synthesizes nicotinamide. With Madeleine Morel, we showed that nicotinamide was present in colostrum of human milk and that the nicotinamide content decreases for two to nine days and then increases up to a value of 15 to 34 mg/100 ml. The administration of nicotinamide was followed by a rapid but limited (0.5 mg/100 ml) increase. We also estimated vitamin PP of various tissues in various pathological conditions.

A SYSTEM OF NUTRITION

In the "good old days," autotrophy was defined as the ability to grow in the absence of any organic substance, and was considered to be correlated either with photosynthesis or with chemosynthesis. Heterotrophy was defined as the need for organic substances.

It was clear that organic substances could represent either energy and carbon sources or growth factors. Moreover, it was shown by Hisatake Dusi that some photosynthetic *Euglenas* need one or many growth factors not yet identified (later shown to be vitamin B_{12}). Thus, it was proposed in 1932 to consider separately the energy and carbon sources and the growth factors.

The problem was considered anew in 1946 in Cold Spring Harbor with C. B. van Niel, F. J. Ryan, and E. L. Tatum. We took into account the latest developments in microbial physiology. A nomenclature of nutritional types was proposed which was based upon energy source on the one hand, and the ability to synthesize essential metabolites on the other. Phototrophy, of course, corresponded to energy provided by photochemical reactions, of which there are two types: photolithotrophy and photo-organotrophy, depending on whether the exogenous hydrogen donor was inorganic or organic. Chemotrophy corresponded to the energy provided by dark chemical reactions with two types, chemolithotrophy and chemo-organotrophy, depending on whether growth depended on inorganic or organic substances. Autotrophy corresponded to the synthesis of essential metabolites—the term, prototrophy, is now commonly used; heterotrophy, to the need for growth factors.

For the definition of categories one considers separately the energy source, the hydrogen donor, and the power to synthesize essential metabolites. This principle is now widely accepted and the terms proposed are commonly used.

LYSOGENY

In 1949, I started working on lysogeny. At a time when genetic material had not been identified, a few bacteriologists had understood the strangeness of lysogenic bacteria, but knowledge concerning viruses and their reproduction was too cloudy. Moreover, a number of papers were obscured by useless polemics. Some textbooks of microbiology contained a paragraph about lysogeny but their reading did not bring much light; no adequate review or discussion was available.

My work had led to the following conclusions: (a) In lysogenic bacteria the phage is perpetuated in a noninfectious form which was called the prophage. (b) Bacteriophage is not secreted but is liberated by the lysis of a lysogenic bacterium. (c) The production of bacteriophage is the result of an "induction." Irradiation with ultraviolet light induces the quasi totality of the lysogenic *Bacillus megaterium* to produce bacteriophage (with Louis Siminovitch and Niels Kjelgaard). (d) Hydrogen peroxide is an inducer but only in organic media. Organic peroxides are inducers; the development of bacteria in an organic medium containing copper ends with the oxidation of thio compounds, the formation of hydrogen peroxide, and of organic peroxides which account in part for the "spontaneous" production of bacteriophage. (e) In *Salmonella typhimurium,* the fate of a bacterium infected with a temperate bacteriophage is decided within seven minutes (with Evelyne Ritz). (f) Lysogeny was defined as the perpetuation of the power to produce bacteriophage in the absence of infection.

Finally, the hypothesis was proposed in 1953 that the potential power of a cell to become malignant may be perpetuated in the form of a genelike structure—the genetic material of the oncogenic virus—and that carcinogenic agents induce the expression of the potentiality of this genetic material, which would culminate in the formation of virions. The history of lysogeny, together with the development and state of the new concepts, were discussed in the review "Lysogeny" which appeared in 1953.

In 1966, a collection of essays was dedicated to Max Delbrück on the occasion of his sixtieth birthday. The stories had been written by his friends, colleagues, and disciples. Max Delbrück had paradoxically played a role in the development of lysogeny. I say paradoxically because the founder of the "phage church" did not believe in the existence of lysogeny. Falling from the lips of Max Delbrück, the death sentence, "I do not believe" had been often heard by many of us. It was an excellent catalyst.

I contributed a paper entitled "The Prophage and I" which is the story of my own contribution—and I am not going to repeat myself. However, I would like to say that the lysogeny period had been something quite apart in my scientific life. In 1949, the "occupation" and its sequels were just over. After a long tragic period of isolation the outside world had flowed in with all its marvelous news, an awakening in flourish after a long sleep. Around us, groups were forming again, Louis Rapkine and Jacques Monod joined the Pasteur Institute. Young men, freed from the war, were starting their career, foreign scientists were visiting.

The 1946 Cold Spring Harbor Symposium had marked the rediscovery of freedom and the beginning of the new era. It was in this rather exceptional atmosphere that the experiments on lysogenic bacteria were started, that the work developed and that unexpected results cropped up.

THE POLIOVIRUS

In July 1953, I attended a Cold Spring Harbor Symposium on Viruses. It was sponsored by the National Foundation for Infantile Paralysis which gave considerable support to fundamental work on viruses, including bacteriophage, wisely considered as a good model.

After the meeting, I visited Harry Weaver, head of the Research Department of the Foundation. He invited me to lunch at the Banker's Club. The National Foundation for Infantile Paralysis was on Broadway, close to Wall Street. The Banker's Club was located on the top floor of a skyscraper. The weather was perfectly clear and the view of New York Bay extended for miles and miles. When I returned to earth, I suddenly became conscious of what had happened. The Foundation had invited me to spend a few months in the States, to visit a few laboratories in order to learn tissue culture and make contact with animal viruses. Thereafter, it would provide my laboratory with the necessary equipment and support my research for a few years. I would work on the poliovirus. I was abashed.

I have always been extremely sensitive to the charm, beauty, and personality of soul-inhabited cities, even if the soul is of stainless steel. Each one exerts its own specific influence and induces a given mood. The standing city was obviously not conducive to dreams but to action. I gave some thoughts to the wealth, power, and efficiency of American foundations. Of course, there was still time for reflection, but in a way it was too late; squashed on the bottom of the black Wall Street Canyon I was muddled by vertigo. Antivertigo would be more correct; and more fashionable too.

It was of course foolish to abandon the still quiet—although not for long —field of lysogeny in order to intrude into the jungle of animal virology.

THE EXCITEMENT OF SCIENCE   315

To enter a new field at the age of fifty-two is unwise anyhow; but unwise decisions debouch on the unexpected which is the salt of research. Beforehand, the unexpected is necessarily entirely hypothetical. The aposteriori nature of these remarks will not escape the perspicacity of the reader.

I forgot, or tried to forget, that I was worried about my future as an animal virologist and in March 1954 we, that is, my wife and myself, embarked courageously for a long trip in the United States. It started in Bethesda where we spent two weeks in the National Institutes of Health with Wilton Earle. Later on we visited, in succession, Joseph Melnick in New Haven, John Enders in Boston, Raymond Parker in Toronto, Jonas Salk in Pittsburg, Gerome Syverton in Minneapolis. I suppose there are a number of recipes for the rejuvenation of aging scientists. One of them is to become a student.

After a fascinating and beautiful drive we finally reached Pasadena. At the California Institute of Technology, we worked with Renato Dulbecco and Marguerite Vogt from July to December. How was the poliovirus released from the cell? Throughout the vegetative phase or at its end? In order to solve the problem it was necessary to study single cells, and a number of technical difficulties had to be overcome. Finally, the question was answered. Infectious particles are liberated all at the same time by the burst of the infected cell. On the way back I stopped in New York and delivered a Harvey Lecture, "Control and interrelations of metabolic and viral diseases of bacteria." The year had been busy.

Back in Paris, something unexpected happened. A colleague from the Pasteur Institute tried to persuade the Foundation that he, and not I, should receive a grant: he was defeated. He also made efforts that I should be forbidden by the Director to work on poliovirus but without success. I did what I had decided to do. Bad or good, I have always done.

It took some time to start the experiments. The Service de Physiologie Microbienne was really crowded. Jacques Monod had not yet completed the organization of his new laboratory and the density of scientists per square foot—I wish the United States had adopted the metric system—was high. As a matter of fact, the research on the poliovirus started only in the fall of 1955.

I had no idea, not the slightest idea, in what direction things would go. I knew only that I was expected to meet with success. A grant is more or less an investment. Of course, foundations are aware of the hazard of research and the National Foundation for Infantile Paralysis was very kind. It was only after two years when research was developing that Theodor Boyd, who had succeeded Harry Weaver, told me that the Foundation had been for a time worried because it felt responsible for throwing me into the adventure. The work had developed slowly. I had started to play with the virus, to study its multiplication under various conditions, and it was necessary to know the optimal temperature.

One-step growth cycles were performed at various temperatures. The curves of viral multiplication as a function of temperature showed that different strains exhibited different patterns. Beforehand, and even now, the sensitivity of development to temperature was expressed by a+ or a-. We proposed to express the sensitivity by the temperature at which the viral development is decreased by 90 percent. This was the *rt* (*r* for reproduction, *t* for temperature). To determine *rt* is a rather long procedure. First, one needs a series of water baths at different temperatures which have to be rigidly controlled. For certain critical values, a difference of 1° may modify the yield by a factor of 2. One needs to have a growth curve at each temperature, that is to say a number of estimations. It is probably the reason why the + +, +, ± and − are still popular.

The multiplication of the virulent strains was less sensitive to "high" temperatures—between 37 and 41°C—than the multiplication of the nonvirulent strain including the vaccine strains.

The vaccine strains of Hilary Koprowski and of Albert Sabin have a *rt* of 37.8. By growing the type I vaccine at 41°C a strain of *rt* 40°8 was obtained. An injection of 600,000 virions of the vaccine strain in the spinal cord of the monkey—the most sensitive route—does not produce lesions. With Albert Sabin, we injected the "hot" virions into the brain, the less sensitive route; three particles killed the monkey. A similar type of experiment was performed with the MEF strain of type II. The hot strain was much more virulent for the mouse than the normal one. The LD_{50} was correlated with the *rt*.

This type of experiment was extended to the virus of encephalomyocarditis. One particle of the wild strain kills the animal; the virulence is maximum. By growing the virus at lower temperature, strains of lower *rt* and of a higher LD_{50} were obtained (M. Lwoff, Y. Perol-Vauchez, and P. Tournier). This, a relation was established between virulence and the sensitivity to high temperatures. What does this mean?

FEVER AND THE FIGHT OF THE ORGANISM
AGAINST A PRIMARY VIRAL INFECTION

It had been known that by growing the poliovirus at low temperatures (23°C) strains devoid of virulence were obtained. That a strain, unable to grow, or growing poorly, at the temperature of the animal, would be devoid of virulence was not in the least surprising. The existence of a correlation between the ability to multiply at temperatures above the normal temperature of the animal and virulence posed a problem.

It had been known that an elevation of temperature can decrease the severity of a viral disease. Experiments showed that the value of the LD_{50} is increased when fever is induced within an animal. Suddenly, everything became clear. Fever is one of the mechanisms by which the organism fights against the primary viral infection. A virus is virulent when it can multiply despite fever. Fever is a byproduct of the inflammatory reaction. Moreover, in an inflammatory zone the pH can drop to values below 6.00 and the polio-virus is unable to multiply below 6.8. It thus appeared that the inflammatory reaction played an important role during a primary infection. Anyhow, prior to the production of antibodies only nonspecific reactions could be responsible for the fight of the organism against the virus.

During the past ten years many examples have been given of the relation between virulence and the resistance of viral development to temperature. The importance of fever in viral infection was recognized very slowly. Text books of virology are now often written by molecular virologists who are not interested in infectious diseases, and the books are strangely lacking in discussions concerning the fight of the organism against the viral infection. Is it so strange after all?

VIRUSES: DEFINITION, TERMINOLOGY, CLASSIFICATION

A virologist is necessarily bound to ask questions. One of them is, what is a virus? The question was asked and has been answered. He is also faced with problems of terminology; new terms were proposed. Finally, when one enters the field of virology rather late one can experience difficulties in recognizing the place of each virus, hence the need for a classification.

Definition.—Words should have a meaning. The "Concept of Virus" was discussed in a Marjory Stephenson Lecture delivered in 1957 before the Society for General Microbiology in London. It was proposed to define viruses as infectious particles possessing only one type of nucleic acid and which reproduced from their sole genetic material. A few other characteristics were sifted out: inability to grow and to divide, absence of metabolism, absence of the information for the enzymes of energy metabolism. We added later on, the absence of transfer RNA and of ribosomes and also of the corresponding information. Thus, by the virtue of a few discriminating traits, viruses were separated from nonviruses: the category, virus, was at last defined. An infectious particle could no longer belong to the group of viruses by the sole virtue of its size. A number of "small" bacteria were thus excluded from viruses and reinstated where they belong.

The concept of virus as it was proposed is now of universal acceptance. For historical purposes I should note that the concept had already been proposed in 1953 in the review "Lysogeny." At the time, however, nobody paid any attention to the proposals which were a few years in advance on the viral calendar.

Terminology.—Virologists interested in the structure of the infectious particle came across the inadequacy of terminology. With Thomas Anderson and François Jacob, we proposed three terms: *virion, capsid,* and *capsomere.* Later on, in Cold Spring Harbor, a group of virologists added *nucleocapsid.* All these terms are now part of the virological vocabulary.

Classification.—A synoptic table of viruses is certainly useful. Now, either you like order or not. If you like it, your love can be either active or platonic. If it is active then you are thrown into systematics. To classify is an amusing game—one tries to select characters and to define categories. However, there are drawbacks. First, categories do not exist in nature. They are creations of our mind; a category is the result of an arbitrary grouping. This does not matter as long as you are aware of the arbitrariness. If you were alone there would be no problem. But if you are not . . . The hell, according to Sartre, the hell, it is the others.

Moreover, when one builds categories, one has to provide them with names, hence nomenclature. A nomenclature has to be international. No wonder that there are conflicting views on nomenclature as well as on classification, hence discussions and even polemics. Before the war, I had been a member of the Judicial Committee of Bacterial Nomenclature. One day, I complained to one of my colleagues, an eminent biochemist, about the total lack of interest in the sessions. He said that if one would leave nomenclature to people interested in nomenclature the result would be a catastrophy.

This being said, with Robert Horne, Paul Tournier, and Peter Wildy, we discussed the problem of classification and finally succeeded in producing a system with the use of four characters. Once the work was brought to an end, Peter Wildy decided that he could not sign the paper because of hierarchy, the hierarchy of viral characteristics, of course. I have always regretted this decision. So the proposed system had to be issued without him. It became the L.H.T. system. It suffered the fate of all classifications, adopted by some, villipended by others. This system, however has a few advantages: 1. it exists, 2. it is the only one to exist, and 3. it allows us to classify viruses. If I were not a co-father I would be inclined to say that it is not such a bad system after all.

We had made use of four discriminating characteristics and I am still convinced that discriminating characteristics should be the basis for a classification. It is clear, for example, that the category, virus, can be defined only by the use of discriminating characteristics. What could be the use of nondiscriminating ones?

However, a number of virologists have not yet understood that the principles which apply to the category virus necessarily apply to categories of lower hierachial rank. They have escaped the difficulty by forgetting to provide their own definition of viruses. So the selective use of discriminating characteristics and the LHT system are not universally accepted and battles are raging. A taxonomical war, because it deals with categories which do not exist in nature and with opinions, is the equivalent of a religious war. There is, however, a difference. The heretics, that is the others, not being burned, the war cannot come to an end.

THE CANCER INSTITUTE

In October 1966, the director of the Centre National de la Recherche Scientifique asked me if I would within two years consider taking the directorship of the Institut de Recherches Scientifiques sur le Cancer in Villejuif (one mile south of Paris). This could only be a full time job. Acceptance would therefore mean abandoning the Pasteur Institute where I had worked for forty-five years and where I felt quite at home, and also the Sorbonne (Faculty of Sciences) where I had taught microbiology since 1959.

The decision was postponed. In 1967, I spent seven months with Renato Dulbecco at the Salk Institute and we worked together on the biology of the Simian virus 40. Back in Paris, I considered the situation.

The "Délégation Générale à la Recherche Scientifique et Technique" had, in 1961, offered the Pasteur Institute the funds necessary to build an Institute of Molecular Biology. However, the director and the board of trustees ruled that molecular biology held no interest whatsoever for the Pasteur Institute.

In 1965, the obstruction ceased. A new director took up the matter again and after a number of vicissitudes, the question was settled. But the edification would start in the spring of 1969 and the building be ready in the fall of 1971—only a year before my retirement. No opportunity was offered to me anywhere else. The attic still harboured various residues and I had no chance to develop what was since the adventure of lysogeny, my principal and enduring interest, namely, the cancer problem.

Good groups were at work at Villejuif. Why not spend a few years helping them to develop research on cancer? So I decided to move and on February the first, 1968, started my last scientific—or maybe parascientific—endeavour.

Remembrances

During the first part of my career at the Pasteur Institute, I had the fatherly and efficient support of Felix Mesnil who, just before his death in 1938, had obtained for me the creation of the Service de Physiologie Microbienne. So I organized an attic into a laboratory. Later on, I experienced some difficulties. For example, a director told me once that my work was devoid of any interest for the Pasteur Institute and I should throw out a few workers in order to save money! I paid no attention whatsoever to this preposterous command for I had decided long ago and once forever that the scientists transcended the director and the board of trustees, and that everything good for science was good for the Pasteur Institute.

In the Institute, salaries were low, promotions almost nonexistent, and the budget of the laboratories poor, but freedom, holy freedom, was provided with unlimited generosity. In passing, freedom, if provided without discernment, can be very costly. Yet, freedom is not enough. If the work could be pursued and developed it was, thanks to the help of the Centre National de la Recherche Scientifique, of the Institut National de la Santé et de la Recherche Médicale, of the Délégation Générale à la Recherche Scientifique et Technique, of the National Foundation for Infantile Paralysis, and of the National Institutes of Health. Thus, the Service de Physiologie Microbienne was amply provided with technicians, equipment, and a budget for daily life in such a way that money has never been the limiting factor for the work. May I be allowed here to express my deep gratitude to all those who have given a testimony of their confidence by generously supporting our work.

A scientist should be aware of the existence of competition and not be obnubilated by "the others," but I am now more and more conscious of the intensity of competition and of the pace of scientific development. Of course, in the past few decennaries, competition and pace have increased markedly. So, the alteration I observe in my mind might be a sign of the time as well as an evidence of maturation, or a symptom of aging, who knows? Whatever the case might be, science has always been competitive and it has certainly never been wise to enter, necessarily unprepared, widely different new fields of research. I have never given any thought to this aspect of scientific endeavour, and it is why, with perfect unconsciousness I have worked in succession on the morphology and morphogenesis of protists, on growth factors, and on various aspects of cell physiology, on lysogeny, on the virulence of viruses, and on the role and mechanism of nonspecific factors in the fight of the organism against viral infection.

In fact, during many years, research was performed simultaneously in different disciplines: the work on ciliates had started in 1921 and ended with the monograph on Apostomes in 1935, and with "Problems of Morphogenesis in Ciliates" in 1950. The work on nutrition of protozoa started in 1923 and was discussed in "Recherches biochimiques sur la Nutrition des Protozoaires."

The milestones of the work on growth factors were the papers on hematin (1933–34), on growth factor V (1936), and the book "La vitamine PP et les avitaminoses nicotiniques" (1942), l'Evolution physiologique, the editing of *Biochemistry and Physiology of Protozoa* (1951). The work on lysogeny extended from 1949 to 1953 and ended—almost—with the review "Lysogeny" (1953). The "Concept of Viruses" was published in 1957, the "System of Viruses" in 1962. In the meantime, "Biological Order" had seen the light (1962) and also the new concepts concerning the virulence of viruses and the fight of the organism against a primary viral infection (1959), and finally the mechanism of the action of fever on viral development (1969).

RETROSPECT

Biology in its widest sense has, since 1921, undergone extraordinary development: the structure and functions of vitamins, of growth factors and coenzymes, the steps of anabolism and catabolism, the activation and movements of oxygen and hydrogen, the cytochromes, the storing and utilization of energy, the antibiotics, the antimetabolites, the nature and structure of genetic material, one gene–one enzyme, the messenger, the operator, the repressor, the transcription and translation, the code, colinearity, the structure of proteins, allostery, the nature of mutations, the sexuality of bacteria, and, more widely speaking, molecular biology and also molecular virology.

It happens that I have been associated with, or known many, if not most, of the scientists responsible for these revolutions in our knowledge and in our thinking. In one way or another, by their achievements or their personality, they have influenced what I may describe as my evolution. A few men, however, played an especially important role in my scientific life. Edouard Chatton, my master, with whom I collaborated intensively for sixteen years; Otto Meyerhof who accepted me in the Kaiser Wilhelm Institut in Heidelberg; David Keilin who provided to me a kind hospitality in the Molteno Institute in Cambridge; Louis Rapkine, the friend too soon disappeared and, finally, the members of the Service de Physiologie Microbienne. I had been fortunate enough to attract a few exceptionally gifted scientists to the Pasteur Institute and to be able to provide everyone with everything needed for research. My collaborators have certainly influenced me at least as much as I might have influenced them. Their names and achievements are well known. Thanks to them, the attic has been for many years the theater of remarkable successes. The work was pleasurable despite its intensity, and the atmosphere festive. As research pertains to ludic[1] activity I should perhaps have described the attic as a playground. Anyhow, it has been for me a constant ravishment to see important problems solved, great discoveries blooming, and new concepts piling up day after day. I sometimes said to my friends that I never felt jealous . . . and that this was meritorious.

[1] From the latin ludus, meaning game.

BIBLIOGRAPHY

The prefatory chapter is not a review. The bibliography is therefore not organized according to the rules of *Annual Reviews*. It is not an exhaustive list of the author's publications but represents a selection of papers considered to be characteristic of the various scientific periods.

Sur une nouvelle famille d'Acinétiens, les Sphénophryidés, adaptés aux branchies des mollusques acéphales (avec E. Chatton).
C. R. Acad. Sci., 1921, *173*, 1495.

Sur la nutrition des Infusoires.
C. R. Acad. Sci., 1923, *176*, 928.

Reproduction d'um Hydraire gymnoblaste par poussées répétées de propagules.
Bull. Soc. Zool. France, 1925, *50,* 405.

Pottsia infusoriorum, n.g., n.sp., Acinétien parasite des Folliculines et des Cothurnies (avec E. Chatton).
Bull. Inst. océan. Monaco, 1927, *489,* 1–12.

Le cycle du pigment carotinoïde chez *Idya furcata* Baird, Copépode Harpactivide. Nature, origine, évolution du pigment et des réserves ovulaires au cours de la segmentation. Structure de l'oeil chez les Copépodes.
Bull. biol. France Belgique, 1927, *61,* 193–240.

Les infraciliatures et la continuité des systèmes ciliaires récessifs (avec E. Chatton et M. Lwoff).
C. R. Acad. Sci., 1929, *190,* 1190.

L'infraciliature et la continuité génétique des blépharoplastes chez l'Acinétien Podophrya fixa (O. F. Muller), (avec E. Chatton, M. Lwoff et L. Tellier).
C. R. Soc. Biol., 1929, *100,* 1191.

Contribution à l'étude de l'adaptation d'*Ellobiophrya donacis* CH. et LW., Péritriche vivant sur les branchies de l'Acéphale *Donax vittatus* da Costa (avec E. Chatton).
Bull. Biol. France Belgique, 1929, *63,* 321–349.

Imprégnation par diffusion argentique de l'infraciliature des Ciliés marins et d'eau douce après fixation cytologique et sans dessication (avec E. Chatton).
C. R. Soc. Biol., 1930, *104,* 834.

Détermination expérimentale de la synthèse massive de pigment carotinoïde par le Flagellé *Haematococcus pluvialis* Flot. (avec M. Lwoff).
C. R. Soc. Biol., 1930, *105,* 454.

L'apparition de groupements -SH avant la division chez les Foettingeriidae (Ciliés). (avec E. Chatton et L. Rapkine).
C. R. Soc. Biol. 1931, *106, 626.*

La formation de l'ébauche postérieure buccale chez les Ciliés en division et ses relations de continuité topographique et génétique avec la bouche antérieure (avec E. Chatton, M. Lwoff et J. Monod).
C. R. Soc. Biol., 1931, *107,* 540.

Recherches morphologiques sur *Leptomonas ctenocephali* Fanth. Remarques sur l'appareil parabasal (avec M. Lwoff).
Bull. Biol. France Belgique, 1931, 65, 170–215.

Recherches morphologiques sur *Leptomonas oncopelti* Noguchi et Tilden, et *Leptomonas fasciculata* Novy, Mac Neal et Torrey.
Arch. Zool. exp. et gén. (Protistologica), 1931, *71,* 21–37.

Bartonelloses et infections mixtes (avec M. Vaucel).
Ann. Inst. Pasteur, 1931, *46,* 258.

Rechesches biochimiques sur la nutrition des Protozoaires. Thèse de Doctorat ès-Sciences. Collections des Monographies de l'Institut Pasteur, Masson éd., Paris 1932.

Die Bedeutung des Blutfarbstoffes für die parasitischen Flagellaten.
Zbl. Bakt. I. Orig., 1934, *130,* 497–518.

L'appareil parabasal des Flagellés (avec M. Lwoff).
Arch. Zool. exp. et gén., 1934, *76,* 56.

Le pouvoir pathogène de *Trichomonas foetus* pour le système nerveux central (avec S. Nicolau).
Bull. Soc. Path. exot., 1935, *28,* 277.

Les Ciliés Apostomes. I. Aperçu historique et général; étude monographique des genres et des espèces (avec E. Chatton).
Arch. Zool. exp. et gén., 1935, *77,* 1–453.

Le cycle nucléaire de *Stephanopogon mesneli* Lw. *(Cilié homocaryote).*
Arch. Zool. exp. et gén., 1936, *78,* 117.

Etude sur les fonctions perdues. Rapport du 2e Congrès international de Microbiologie, Londres.
Ann. Fermentations, 1936, *2,* 419.

Les *Pilisuctoridae* CH. et LW. Ciliés parasites des poils sécréteurs des Crustacés Edriophthalmes. Polarité, orientation et desmodexie ches les Infusoires (avec E. Chatton).

Bull. Biol. France Belgique, 1936, 70, 86.
Les remaniements et la continuité du cinétome au cours de la scission chez les Thigmotriches Ancistrumidés (avec E. Chatton).
Arch. Zool. exp. et gén., 1936, 78, 84.
La pyrimidine et le thiazol, facteurs de croissance pour le Flagellé *Polytomella coeca* (avec H. Dusi).
C. R. Acad. Sci., 1937, 205, 630.
Le thiazol, facteur de croissance pour *Polytoma ocellatum* (Chlamydomonadiné).
Importance des constituants de l'aneurine pour les Flagellés leucophytes (avec H. Dusi).
C. R. Acad. Sci., 1937, 205, 882.
Le thiazol, facteur de croissance pour les Flagellés Polytoma caudatum et *Chilomonas paramaecium* (avec H. Dusi).
C. R. Acad. Sci., 1937, 205, 756.
Caractères physiologiques du Flagellé *Polytoma obtusum*. (avec L. Provasoli).
C. R. Soc. Biol., 1937, 126, 279.
Détermination du facteur de croissance pour *Haemophilus ducreyi* (avec I. Pirosky).
C. R. Soc. Biol., 1937, 126, 1169.
Studies on codehydrogenases. I. Nature of growth factor "V" (with M. Lwoff).
Proc. Roy. Soc. London, Series B, 1937, 122, 352.
Studies on codehydrogenases. II. Physiological function of growth factor "V" (with M. Lwoff).
Proc. Roy. Soc. London, Series B, 1937, 122, 360.
Rôle physiologique de l'hématine pour *Haemophilus influenza* Pfeiffer. (avec M. Lwoff).
Ann. Inst. Pasteur, 1937, 59, 129.
L'aneurine, facteur de croissance pour le Cilié *Glaucoma piriformis* (avec M. Lwoff).
C. R. Soc. Biol., 1937, 126, 644.
La spécificité de l'aneurine, facteur de croissance pour le Cilié *Glaucoma piriformis* (avec M. Lwoff).
C. R. Soc. Biol., 1938, 127, 1170.
Influence de diverses substitutions sur l'activité du thiazol considéré comme facteur de croissance pour quelques Flagellés leucophytes (avec H. Dusi).
C. R. Soc. Biol., 1938, 127, 238.
La synthèse de l'aneurine par le Protozoaire *Acanthamoeba castellanii*.
C. R. Soc. Biol., 1938, 128, 455.
L'activité de diverses pyrimidines considérées comme facteur de croissance pour les Flagellés Polytoma coeca et *Chilomonas paramaecium* (avec H. Dusi).
C. R. Soc. Biol., 1938, 127, 1408.

Dosage de l'amide de l'acide nicotinique au moyen du test *Proteus;* principe de la méthode (avec A. Quérido).
C. R. Soc. Biol., 1938, 129, 1039.
Révision et démembrement des *Hemophilae*. Le genre *Moraxella n. g.*
Ann. Inst. Pasteur, 1939, 62, 168.
La nutrition carbonée de *Moraxella Lwoffi* (avec A. Audureau).
Ann. Ins. Pasteur, 1941, 66, 417.
Recherches sur le sulfamide et les antisulfamides. I. Action du sulfamide sur le Flagellé *Polytomella coeca*. II. Action antisulfamide de l'acide aminobenzoïque en fonction du pH. (avec F. Nitti, Mme J. Tréfouël et Mlle V. Hamon).
Ann. Inst. Pasteur, 1941, 67, 9.
La nicotinamide dans les tissus du foetus humain (avec M. Morel et L. Digonnet).
C. R. Acad. Sci., 1941, 213, 1030.
Enrichissement du lait de la femme en **vitamine PP** après injection de nicotinamide (avec L. Digonnet et H. Dusi).
C. R. Acad. Sci., 1942, 214, 39.
L'évolution de la teneur en nicotinamide du lait de la femme et le besoin du nourrisson (avec M. Morel et M. Bilhaud).
C. R. Acad. Sci., 1942, 214, 244.
Conditions et mécanisme de l'action bactéricide de la vitamine C. Rôle de l'eau oxygénée (avec M. Morel).
Ann. Inst. Pasteur, 1942, 68, 323.
L'évolution de la teneur du lait de la femme en nicotinamide (avec M. Morel).
C. R. Soc. Biol., 1942, 136, 187.
Vitamine antipellagreuse et avitaminoses nicotiniques (avec L. Justin-Besançon).
1 volume in 8° de 284 pages. Masson édit. Paris, 1942.
L'agglutination réversible des Moraxella par les cations bi ou polyvalents (avec A. Audureau).
Ann. Inst. Pasteur, 1944, 70, 144.
L'évolution physiologique. Etude des pertes de fonctions chez les microorganismes.
Actualités scientifiques. Collection de microbiologie, Hermann éd. Paris, 1944, vol. in 8° de 308 p.
Un nouveau réactif biologique de l'acide p-aminobenzoïque le Trypanosomide *Strigomonas oncopelti* (avec M. Lwoff).
Ann. Inst. Pasteur, 1945, 71, 206.
Nomenclature of nutritional types of microorganisms (with C. B. van Niel, F. Ryan and E. L. Tatum).
Cold Spring Harbor Symp., 1946, 11, 302-303.
Essai d'analyse du rôle de l'anhydride carbonique dans la croissance microbienne (avec J. Monod).
Ann. Inst. Pasteur, 1947, 73, 323-347.

Production bactérienne directe d'acide pyruvique aux dépens de l'acide malique (avec R. Cailleau).
C. R. Acad. Sci., 1947, *224*, 678–679.
Nécessité de l'ion potassium pour la décarboxylation oxydative bactérienne de l'acide malique en acide pyruvique (avec H. Ionesco).
C. R. Acad. Sci., 1947, *224*, 1664–1666.
Sur le rôle du sérum dans de développement de *Moraxella lacunata* et de *Neisseria gonnorrhae*.
Ann. Inst. Pasteur, 1947, *73*, 735.
Nécessité de l'ion Mg pour la décarboxylation oxydative de l'acide malique et la croissance de la bactérie *Moraxella Lwoffi* (avec H. Ionesco).
Ann. Inst. Pasteur, 1948, *74*, 433.
Culture du Flagellé opalinide *Cepedea dimidiata* (avec S. Valentini).
Ann. Inst. Pasteur, 1948, *75*, 1.
Recherches sur les Ciliés Thigmotriches (avec E. Chatton).
Arch. Zool. exp., 1949, *86*, 169–253.
Recherches sur les Ciliés Thigmotriches. II. (avec E. Chatton).
Arch. Zool. exp., 1950, *86*, 393–485.
Induction de la lyse bactériophagique de la totalité d'une population microbienne lysogène (avec L. Siminovitch, et N. Kjeldgaard).
C. R. Acad. Sci., 1950, *231*, 190–191.
Problems of morphogenesis in ciliates. The kinetosomes in development, reproduction and evolution.
John Wiley & Sons Inc., New York, 1950.
Introduction to biochemistry of Protozoa.
In Biochemistry of Protozoa.
Academic Press, New York, 1951, 1–26.
Conditions de l'efficacité inductrice du rayonnement ultra-violet chez une bactérie lysogène.
Ann. Inst. Pasteur, 1951, *81*, 370–388.
Induction de la production de bactériophages et d'une colicine par les peroxydes, les éthylèneimines et les halogénoalcoylamines (avec F. Jacob).
C. R. Acad. Sci., 1952, *234*, 2308.
L'induction du développement du prophage par les substances réductrices (avec L. Siminovitch).
Ann. Inst. Pasteur, 1952, *82*, 676–690.
Définition de quelques termes relatifs à

la lysogénie (avec F. Jacob, L. Siminovitch et E. L. Wollman).
Ann. Inst. Pasteur, 1953, *84*, 222.
L'induction.
Ann. Inst. Pasteur, 1953, *84*, 225.
Lysogeny.
Bact. Rev., 1953, *17*, 269–337.
Recherches sur la lysogénisation de Salmonella typhi-murium (avec A. S. Kaplan et E. Ritz).
Ann. Inst. Pasteur, 1954, *86*, 127.
Kinetics of the release of poliomyelitis virus from single cells (with R. Dulbecco, M. Vogt and M. Lwoff).
Virology, 1955, *1*, 128–139.
Control and interrelations of metabolic and viral diseases of bacteria.
The Harvey Lectures, series L (1954–55), 92–111, Academic Press, New York.
The concept of virus.
J. Gen. Microb., 1957, *17*, 239–253.
The Mammalian Cell as an Independent Organism.
Spec. Pub. New York Acad. Sci., 1957, *V*, 300–302.
L'espèce bactérienne.
Ann. Inst. Pasteur, 1958, *94*, 137–140.
Factors influencing the evolution of viral diseases at the cellular level and in the organism.
Bact. Rev., 1959, *23*, 109–124.
Remarques sur les caractéristiques de la particule virale infectieuse (avec T. F. Anderson et F. Jacob).
Ann. Inst. Pasteur, 1959, *97*, 281–289.
Sur les facteurs du développement viral et leur rôle dans l'évolution de l'infection (avec M. Lwoff).
Ann. Inst. Pasteur, 1960, *98*, 173–203.
Tumor, viruses and the cancer problem: a summation of the conference.
Cancer Research, 1960, *20*, 820–829.
Les événements cycliques du cycle viral. I. Effets de la température (avec M. Lwoff).
Ann. Inst. Pasteur, 1961, *101*, 469–477.
Les événements cycliques du cycle viral. II. Les effets de l'eau lourde (avec M. Lwoff).
Ann. Inst. Pasteur, 1961, *101*, 478–489.
Les événements cycliques du cycle viral. III. Discussion (avec M. Lwoff).
Ann. Inst. Pasteur, 1961, *101*, 490–504.
Mutations affecting neurovirulence.

THE EXCITEMENT OF SCIENCE   325

In "Poliomyelitis." 5e Conférence Internationale sur la Poliomyélite. Lippincott éd., Philadelphia, 1961, 13–20.

Biological Order (Karl Taylor Compton Lectures).
M.I.T. Press, Massachusetts Institute of Technology, Cambridge, Mass., 1962.

The thermosensitive critical event of the viral cycle.
Cold Spring Harb. Symp. Quant. Biol., 1962, *27*, 159–174.

Proposals (with D. L. D. Caspar, R. Dulbecco, A. Klug, M. S. Stoker, P. Tournier and P. Wildy).
Cold Spring Harb. Symp. Quant. Biol., 1962, *27*, 49–50.

A system of viruses (with R. W. Horne and P. Tournier).
Cold Spring Harb. Symp. Quant. Biol., 1962, *27*, 51–55.

Un mutant du poliovirus insensible aux effets de la deutération (avec M. Lwoff).
C. R. Acad. Sci., 1964, *258*, 2702–2704.

The specific effectors of viral development. (The first Keilin Memorial Lecture).
Biochem. J., 1965, *96*, 289–301.

La synthèse du RNA chez le poliovirus. Effet de la guanidine (avec C. Burstein et E. Batchelder).
Ann. Inst. Pasteur, 1966, *111*, 1–13.

The classification of viruses (with P. Tournier).
Ann. Rev. Microb., 1966, *20*, 45–74.

Les effecteurs de l'infection virale primaire. (Conférence prononcée au Congrès de Microbiologie de Moscou, le 24 juillet 1966).
Extrait du Maroc-Médical, n° 500–47–67.

The Prophage and I.
In "Phage and the Origins of Molecular Biology." Edited by J. Cairns, G. S. Stent, J. D. Watson. Cold Spring Harbor Lab. of Quant. Biol., publisher, 1966.

Le rôle de la biologie moderne dans la médecine.
In "Scientia valemus," published by CIBA, Basel, 1967.

Death and Transfiguration of a Problem.
Bact. Rev., 1969, *33*, 390–403.

Ann. Rev. Pharmacol., Vol. 11

A PERSONAL BIOGRAPHY OF
ARTHUR ROBERTSON CUSHNY, 1866–1926

By His Daughter
Helen MacGillivray

ARTHUR ROBERTSON CUSHNY

A PERSONAL BIOGRAPHY OF
ARTHUR ROBERTSON CUSHNY, 1866–1926

BY HIS DAUGHTER
HELEN MACGILLIVRAY

Arthur Robertson Cushny would wish to be remembered by his work, which to the world is memorial enough. This, I am not qualified to assess. The many scientific papers and books he wrote, as well as the appreciation of his peers, are there to be read in scientific libraries. I have written this little account of my father as I knew him so that my grandchildren will be able to learn something of their remarkable great-grandfather. I have tried to follow the text he wrote in his scrap book, to "tell the truth with love."

In the family records, kept by my father with that enthusiasm for the remotest cousins shown by so many Scots, there appear many forceful characters, most of them worthy, but some less so as the lady who died "execrated by all her relations having blackmailed them for sixty-five years." One longs to know how. The Cushny (at first spelled Cushnie) family came from Morayshire and Aberdeenshire. The earliest ancestor mentioned in my father's records was James Cusnye (Cushnie), who was a reader at Aboyne in 1567. His salary was 20 lib. Of another forbear there was a ballad, one verse of which runs:

> Sandy Cushny's nae for me
> 'Cause I'm but a drover's dochter.
> So he's away to the parish of E'en
> To seek for Miss Gray and her tocher.
> He may go to Donside
> And wash his dun hide.

The rest is lost, perhaps fortunately. "Sandy" was Alexander Cushny, minister of Oyne, and Arthur Cushny's great-grandfather. He married Ann Gray. The "dun hide" refers to his dark complexion. Many Cushnys are very dark.

Like many gifted Scots, Arthur Robertson Cushny was a son, a grandson, and great-grandson of the manse. His father, John Cushny, was born at Rayne in 1826. He was ordained minister at Speymouth in 1848. In 1856 he married Catherine Ogilvie Brown, daughter of the Procurator Fiscal of Elgin. They had seven children, one daughter and six sons, all born at Speymouth. Arthur, born 1866, was the fourth.

John Cushny was a very earnest preacher, but his chief interests lay in public life in which he took a leading part in the locality. The living from this parish was very small for his large family, and this limited him a great deal. In 1871, he was presented by the Duke of Richmond to the much better living of Huntly in Aberdeenshire, and he was just beginning to get on his feet when he died at the age of 48. Arthur was only nine then and had rather faint memories of his father; to his mother, he was devoted.

THE EXCITEMENT OF SCIENCE

8243-2601/78/1127-0329$01.00 © 1978 ARI

Catherine Ogilvie Cushny, with her seven children, returned to Fochabers near her much loved Speymouth. The minister's widow must have had a hard struggle to make ends meet. Letters to her sister-in-law speak of "the plight we are in, and things go from bad to worse." She also mentioned the bitter cold of the winters in Fochabers. Probably there was not overmuch to keep them warm. Later, writing to Arthur of her pleasure at his winning a bursary for Aberdeen University, she states how every little is a great help. However, it must not be thought that she was in any way plaintive, for her letters were gay and full of interest in her children and her friends.

My father's stories of his boyhood were always happy ones. The boys roamed the lovely countryside, and my father was always a countryman at heart, with a good knowledge of and curiosity for birds, beasts, and plants. The Spey was a personality rather than a river, changing its course frequently, and as I was told when I visited there, demanding a life every year. There was a family dog, Sneeshun, the local word for snuff, which hunted squirrels and anything else that ran. The boys collected enough squirrel pelts to make a rug for their mother. She mentioned its return from the rugmaker in a letter, adding wryly that the making up had yet to be paid for. The redoubtable Sneeshun played a leading part in many of my father's stories. He came to a sad end, and I found a letter written by the youngest brother, Robin, to Arthur, who by then was away at the University. "My dear Arthur, You will be very sorry to hear of the death of poor old Sneesh. I was in the kitchen with Annie [the children's nurse whom I remember as a very alarming old lady] on Tuesday night and heard him give a scrape and a whine. So I went ben to the scullery and found him lying panting. I immediately thought of poison." There followed a graphic account of attempting to administer an emetic, and then "We did not keep any part of the old dog but buried him whole. The shepherd wanted to open him, but I said I would rather not and Tom said the same when he came home. Tom bids me tell you Gillice got your arm alright but has not got Bissett's leg. [Was this for anatomy? It almost smacks of Burke and Hare.] My leg is all right now [he had acute tibial osteomylitis, which later became chronic] and I am going to get a pony to ride over to school as Minnie will not be well for a long time. I must stop now, with best love, Yours truly, R. S. Cushny." It was signed with elaborate curlicues.

Arthur was said to have been a delicate child, but in what way is not specified. He grew to be a tall, fine looking man and, except for occasional migraines, he was never ill. An admiring lady told him in my hearing he was "a fine Highlander," a statement which annoyed him considerably, as he was quite out of tune with Celtic romanticism and often said that Morayshire was not the Highlands but a part of the Lowlands extended into the Highlands.

At school he was a good scholar and a very hard worker, and he remained so all his life. In October 1881, his mother wrote to her sister in Aberdeen to ask if Arthur could stay with her "till the competition is over—Of course you know he is *not* to remain during the winter, merely going up to prepare for the *grand* trial next year if spared to go forward. I hope he will get into the merit list poor laddie, for he has been working *real* hard. But at the same time he has a *real* pleasure in study. May God grant him health and strength for whatever his future course may be, and *direct* his *heart* to *good*." He was then fifteen. When he was a very small boy someone asked him what he was going to be when he grew up. He replied, "A professor." When asked why, he said, "Because they have such long holidays."

The year after the competition, he did get the hoped-for bursary and entered Kings College, Aberdeen, to study Arts. He won some prizes and medals in Arts subjects and graduated M.A. in 1886. He had begun to study medicine the year before, and quickly showed he had found his bent. Medals and prizes were won each year, and in 1889, he graduated with Highest Honours, won the Murray medal and scholarship for the best man of his year, and was the George Thompson Fellow for Pathology, Physiology, and Gynaecology.

He did not spend all his time over his books, but enjoyed the student life. I have the notes he made for a speech to an undergraduate society, moving that the General Practitioner's life was better than that of a doctor in the services. Characteristically, an advantage of general practice, in his view, was that a doctor could follow a disease "from its beginning to its close, and if he does his duty and has a reasonable amount of luck, need have no fear of it being snatched from his hand." He was speaking more than 70 years ago.

He made numerous friends in his student years, and kept in touch with many of them. Years later when we went to Scotland for summer vacations, he played golf with some of them. Once, at Aboyne, to make a foursome, my father and two surgeons from Aberdeen permitted a girl, said to be a fair player, to join them. Three crestfallen medical men came in at the end of the round, defeated by a "wee bit lassie," who turned out to be the Scottish Ladies Champion.

The family left Fochabers one by one. Cath was married; Alexander went to Shanghai, where his uncle had large business interests; John became an electrical engineer and went to India and later to South Africa, finally giving up engineering and becoming one of the earliest settlers in what is now Kenya. James, the least scholarly of the family, went to ranch in Mexico and later to Kansas. Tom died in Shanghai, in 1893 of dysentery. My father wrote of him, "He was the best fellow I ever met and my favourite brother." Robin, the youngest, began to study medicine at Aberdeen, but gave it up and went out to John in Africa, where he died in 1905. He was a charming man, and my father wrote, "had a personal magnetism which won for him hosts of friends, not always of the most judicious kind."

In 1887, Arthur's mother died at the age of 54. She did not live to see Arthur graduate with such distinction, but she had faith in his ability, and the day before she died, said to him, "Go on in your profession. Don't let anything discourage you."

Among Arthur Cushny's teachers at Aberdeen, J. T. Cash, Professor of Materia Medica, had a decisive influence. A colleague and fellow-student recalled years later that Arthur once said that Cash's teaching was an inspiration and made him feel he could give his life to pharmacology. So, when he was awarded the Thompson Fellowship, he went to Berne for a year to work in the laboratory of the physiologist, Hugo Kronecker, for further training in the methods of physiological research.

After the year in Berne, Cushny went to Strasbourg to work under the great Professor Schmiedeberg, the father of modern pharmacology. The following year, in 1892, when he was 26, he was appointed assistant to this man of genius.

While he was in Strasbourg, he met a 17-year-old English girl, Sarah Firbank. She was living with a French family and teaching the children English. I have always understood they met while skating. Having grown up in the North of Scotland, Arthur was quite proficient in the art, while Sarah was not. It was natural that the handsome young Scot should teach the pretty English girl to skate. A lighthearted friendship grew up. He was introduced to Monsieur and Madame Hatt, the French family with whom Sarah was living, and was presumably approved as a suitable escort for "Mees."

Subsequent to the transfer of Strasbourg to Germany, after the Franco-Prussian War, the German officers ruled the town arrogantly. Arthur's brothers, John and Robin, on a visit there, did not know that civilians had to step off the sidewalk for German officers. Robin was admiring some buildings and did not see the approaching officer. In a moment, a sword was being brandished over his head. The hot-tempered John immediately knocked the officer down. The brothers were led off to the police station, and they were released only because of Arthur's position with Schmiedeberg. In spite of international tensions, it was a pleasant town for young foreigners. There were expeditions to the Vosges and evenings at the opera. Public enthusiasm for Wagner was at its height; and Arthur and Sarah never lost their love of Wagner throughout their lives. They could sing all the "motifs," and Sarah played the scores on the piano. Years later, in London, although not strong at that time, Sarah would sit out the whole of the Ring when it came to Covent Garden, in far from expensive or comfortable seats. By that time, Arthur's enthusiasm must have waned slightly. I do not remember him going to the opera unless he could go in comfort.

During the year 1892, he returned to Aberdeen to receive his M.D., again with Highest Honours. He had already published several papers in English and German medical journals.

Under Schmiedeberg, the science of Materia Medica had undergone a change. Previously, it had been largely the study of the chemistry of substances used in medicine, with an alliance of the departments of chemistry and materia medica in medical schools, followed by clinical study. Schmiedeberg, however, was a trained physiologist, and showed the importance of studying the effects of drugs on living tissues under laboratory conditions. Materia medica became allied with physiology, and so developed the modern science of pharmacology. The American pharmacologist, J. J. Abel, who had also studied under Schmiedeberg, had built up from nothing a small modern department of pharmacology in the University of Michigan at Ann Arbor. He was later appointed to Johns Hopkins Medical School, and since he was traveling to Europe before taking up the appointment, he was asked by the Dean of Medicine to look around for a suitable successor. He asked young Cushny if he would consider the appointment if it were offered him. Cushny had intended to return to Britain to become a consulting physician, combining clinical work with pharmacological research. However, Schmiedeberg advised him to accept this post, and in October 1893, he was installed in the Chair of Pharmacology of the University of Michigan. He was 27 and younger than many of his students.

Since he was never precipitate in anything he did, and he considered Sarah Firbank much too young to know her own mind, (added to the fact that neither of them had any money), he did not ask her to go with him, nor was there any understanding between them before he left.

Unfortunately, there appear to be no letters remaining to tell of his early impressions of America, although he formed some friendships that lasted for many years. He was a member of a society of 12 young men, who called themselves the Apostles. They were all bachelors and at marriage ceased to be members. He seems to have been very gay, and was known to some as the "Butterfly," a name which seems remarkably inappropriate for one who is remembered as a large, bearded man.

To his class, men and women of his own age, he was an object of awe and affection. I have met very elderly men who were his students in the early days, who still spoke of him in this way. The brilliant originality of his approach to his subject, his dry humour, and his good looks impressed them. He could be very aloof. A colleague tells of when a girl, who had been "ploughed," came weeping to his room, he had only to say, "Hush! I hear Cushny coming," to make her flee. When he left America 12 years later, the Detroit Medical Journal wrote, "His admirable scientific poise of mind in teaching a subject which had scarcely emerged from the mists of quackery and empiricism and his dry Scotch humour have made Professor Cushny a favourite with Medical students in Ann Arbor."

While working with Schmiedeberg, he had already begun his famous work on the action of digitalis on the heart. He must have continued this as soon as he was settled in Ann Arbor, in spite of the rudimentary laboratory conditions, as in 1894 he published a paper on the subject in the *Transactions of the Michigan Medical Society*.

In 1895, he visited Strasbourg and became engaged to Sarah Firbank, who was still living with Monsieur and Madame Hatt. She was the youngest daughter of the large family of Ralph Firbank, a railway engineer, who had died when she was a child. Arthur and Sarah were married in London the following year.

When he took his bride to Ann Arbor, she was overwhelmed by kindness from all his colleagues and their wives. My mother often said that the nine years she lived in Ann Arbor were a particularly happy time, in what was an unusually happy life.

During 1896, Cushny was invited to collaborate in the *Journal of Experimental Medicine* in which he published three papers. He also received an invitation to be chairman of the Physiology and Pharmacology Section of the Twelfth International Congress of Medicine at Moscow in the following year. This, he evidently did not accept though we can only guess the reason. Probably, the distance was too great and he did not want to leave his young wife, and was not able to afford to take her with him. He was also elected to the Association of American Physicians.

In 1897, he published three more papers on the rhythm of the mammalian heart. My mother said that during these years he was possessed and driven by a passion for his work, and she did not see a great deal of him. His days were filled with his teaching and research; the evenings, taken up by writing. She had never before known a man to be so possessed by creative work, and said at first she felt resentful; but all her life she not only loved him, but admired him and moulded herself for him. In her eyes, he was always right. She quite sincerely regarded herself as a rather foolish person; but this was not so. She possessed gifts in her own right, her letters are witty and interesting; she was a pianist of more than average ability; and later, proved herself to be a queen amongst gardeners. She was much loved in the academic circles in which her life was spent.

In the summer of 1898, their only child, a daughter, was born. In terms of religion, my parents were, I believe, full of "honest doubt" a feeling prevalent among intellectuals of the period, in revolt against their fundamentalist upbringing, and they had not intended to have me christened. However, at this time, Arthur's brother, Alex, and his wife arrived on their way home from China, bringing with them the Cushny family christening robe and a silver mug. I think the robe must have brought out my father's strong Scottish family feeling, because I was christened after all. Characteristically, our stray mongrel dog, was observed to have followed us into the church and remained there quietly throughout the ceremony.

More papers were published in 1898, and at the same time, my father was working on his *Text Book of Pharmacology and Therapeutics* or the *Action of Drugs in Health and Disease*. The *British Medical Journal* said, in an appreciation after his death, "Pharmacology originated in Germany and the replacement of the old materia medica by the science of the mode of the action of drugs has been a slow process . . . . . but Cushny did more than any other person to bring about this change. His textbook . . . is recognized as the most trustworthy guide to the subject. His wide knowledge and exceptional powers of judgement made him one of the select number of persons who can write a great text book. The first edition was a pioneer piece of work for it was the first general text book of pharmacology in the English speaking world. Cushny treated the mode of action of drugs as an exact science, and his book contains only those facts that have been established by carefully controlled observations on animals or man. Naturally his attitude appeared to some to be unduly sceptical, for he had no hesitation in rejecting cherished traditions as unproven when they lacked definite objective evidence for their support. Now, however, it is generally recognized this is the only manner by which a science can be built upon secure foundations." This large book of 730 pages was published in 1899 and was immediately successful. It was required reading for most English-speaking medical students for many years. The eighth edition was published just before his death. In 1899 he was also invited to be collaborator in the *Archives Internationales de Pharmacodynamie et de Thérapie*. He published three more papers on the contractions of the mammalian heart, and the interpretation of pulse tracings, and another paper with G. B. Wallace in the *Pflügers Archiv für die Gesamte Physiologie des Menschen und der Tiere.*

In 1901, "the book," as his textbook of pharmacology become known in family circles, went into its second edition. This always involved much revising as new knowledge became available, and I remember the great sheets of galley proofs that lay on his desk, When the seventh edition was coming out, I was a student at University College and I felt tremendously important when I helped him read the proofs.

My earliest recollection of my father must have been when I was about three. We went for a walk beside what I suppose was the river, Huron. There were some cages, kept I think by the old man who collected the garbage, in which were some racoons and owls. My father made the owls snap their beaks at him. I suppose it was this performance and their piercing gaze and general reputation for wisdom that in some way ever after connected owls with my father in my mind. I can also remember, I suppose about the same time, being spanked for scribbling in a book and covering it with vaseline. Omar Khayyám, it was. I came across the disgraceful evidence years later when going through his books. My father had all the Scottish scholar's reverence for books and to maltreat one was a fearful crime. I have never since seen a book thrown face down or dog-eared without feeling distressed.

My parents were keen golfers, all the year round. In the winter they played on the packed snow with red balls. The winters were long and very cold, and the summers were equally hot. My father suffered from very bad migraine headaches, and I think the attacks came more often in the hot weather. We used to have rubber pillows filled with water from the ice box at night, and I can remember my mother refilling the rubber pillow with cold water and putting handkerchiefs wrung out in ice water on his forehead, while I crept about on my bare feet.

When I was two, we traveled to England and stayed with my father's Uncle Alex at Paines Hill. I remember nothing of this, but there is a splendid photograph of the whole party of relations, some 20 to 30 of them, from the aged great-aunts in bonnets and cloaks, down to me and my cousins in starched white frills or sailor suits. The young matrons, like my mother, wore long braided skirts, and leg of mutton sleeves and boaters. The voyage across the Atlantic was made in a cattle boat, as these were supposed to be very steady, and my father was a bad sailor. It must have been an unpleasant journey. My mother said the cattle never ceased moaning, and every night dead cattle were thrown overboard. The boat was very slow and infested by the largest cockroaches my mother had ever seen.

My father was a great admirer of Rudyard Kipling, and bought all his books as they came out. As soon as I could follow a story at all, he had read the jungle books to me, and elephants, as well as owls, were connected in my mind with my father. Elephants were wise, and very large. A big man, who habitually wore loose fitting clothes, might well have an "elephant look" to a small child whose first view of an adult is legs. Also, in the jungle books, the elephant protected the boy Toomai and showed him wonderful things such as the elephants dancing. When we went for walks in the woods and found an open space, it was the place where the elephants danced. My father's admiration for Kipling was an important ingredient of family life, and I think I thought he was actually a friend of Kiplings', although, in fact, he did not meet Kipling till many years later at some official dinner in London. Father was a great believer in the Empire and a fierce conservative, nowadays an unusual creed for an intellectual, but I think he would have hated that epithet for himself, and I am jumping too far ahead.

From the numbers of papers that were published, 1902 must have been a very productive year. In 1903, the third edition of "the book" appeared. Professor Starling, who was soon to be a valued colleague at University College London, wrote, "It is a boon to us in England to have a Pharmacology written from the experimental standpoint. . . . All our own men are amateurs." Professor Cash wrote after a later edition still, "It is satisfactory to know that your book is being read not only by advanced students and juniors who wish to advance in their knowledge of this important side of medical science, but that practitioners are making use of it, and therefore acquiring information not available in their student days which will greatly enlighten their work."

Cushny had a great gift for encouraging young scientists in the usefulness of research and would give them a great deal of his time. I still have letters from former students asking and thanking him for help in papers. A number of holders of chairs in pharmacology throughout the English-speaking world had been his assistants in the different universities in which he held chairs. In 1903 there were four important papers published by men working in his laboratory, as well as further papers by himself, including one on atropine and hyoscyamines for the *Journal of Physiology*. He later wrote a monograph on this subject.

In 1904, there was another paper on kidney secretion. Towards the end of his life, he published a monograph on kidney secretion, which aroused some controversy, but after his death he was proved to be right. In the same year, he was elected a member of the Society for Experimental Biology and Medicine of New York.

In 1905 came an offer of the new Chair of Pharmacology at University College London. I have the cable containing the offer and the salary of £500 a year. The pound sterling was worth a great deal more in those days, but even so, it was evidently not for the money that he accepted the offer. Letters came from all over the United States expressing dismay at his resignation and congratulations on the appointment .At that time in America, it was still felt that the true seats of learning were in Europe.

Before leaving, he put the mimeographed notes used for his laboratory course into more permanent form. In the same year, he and his then assistant, Charles E. W. Edmunds, who succeeded to the Chair two years later, published *A Laboratory Guide in Experimental Pharmacology*, which ran to many editions.

An important aspect of his work at Ann Arbor concerned the biological assay of drugs. It was Cushny who first suggested making use of animals to test the relative activity of different preparations of the same drug, such as digitalis. In the late 1890's the method was introduced into commercial practice, and the principle has since been widely extended. It is not generally known that Cushny was the father of the idea.

My parents were sad to leave their friends in Ann Arbor. They had been a popular couple. My mother was pretty and gay and unassuming; my father had brought distinction to the medical school, and was a loved and admired character. A colleague wrote at the time he left, "Personally, Cushny is characterised by a rare combination of natural and unconscious dignity and bonhomie. Like so many educated Scotsmen he is the best type of cosmopolite, appreciating the good qualities of a strange country and its people, and overlooking the minor differences that so strongly affect the provincial."

We sailed for England in the White Star liner, Cedric, after spending one night in New York, where I can still remember the noise of the traffic. In my father's scrap book, there is a "portrait" of one of the sailors, drawn by me. On arrival we stayed with my father's widowed sister in Richmond, Surrey. She had three sons, the youngest a little older than myself. We met him coming from school in cricket flannels with a school cap, a costume very strange to me. He gave me an acid drop dipped in sherbert. It was not always so easy, however, as I was teased a great deal because of my "Middle-West" accent, and was told I would be put in prison because I did not know "God Save the King."

My parents looked for a house somewhere within easy reach of the University and within their means, and eventually took Number 8, Upper Park Road, off Haverstock Hill in Hampstead. It was a street of tall Victorian houses, with stone steps to the front doors, which were adorned with stucco pillars. The street was lined with trees and pleasant enough. There were long narrow gardens behind the houses, but ours had been much neglected. I remember the walls were covered with ivy which was full of snails. I was put to work collecting these in a bucket. At the end of the garden was an old weeping red may tree, which was to be the setting for every sort of imaginative game. The neglected garden was soon transformed by my mother's green thumb.

My father naturally knew scientists from every part of the world, and these men and their wives frequently came to stay with us. In those days, even on a salary of £500 a year, it was not only possible to have two maids, but very necessary in that tall house with the kitchen in the basement. My parents were very unworldly. Ostentation or fashion really meant nothing to them. Once when my mother had been cajoled into buying quite a smart dress for some occasion, she proposed to wear her comfortable lace-up boots with it. After protesting vainly, a friend said, "Oh well Sarah, I suppose it doesn't matter, only a duchess would dare to wear boots like those." As I grew older and developed the adolescent's need to be exactly like everyone else, this unworldliness caused me much anxiety and heart-burning, but as a child it did not occur to me that our way of life was in any way different from that of my friends. Of course, amongst the families of professorial colleagues, it often was not different. My father belonged to the aristocracy of brains and was at ease with all men. My mother was naturally shy, but she was kind and gay, and because she sailed under his colours, she went everywhere with an easy manner, although she was always self-effacing. At a Buckingham Palace garden party, Queen Mary came across to speak to Father. "What did you do?" I asked Mother. "Oh I hid behind a tree." My father was not always perfect socially. He did not suffer fools with geniality. My mother, on the other hand, was so kind-hearted that she frequently invited the dull or unattractive to the house, because she thought no one else would. She tried to do this when Father would be away, but her kindness made them feel so at home that they tended to drop in after that, and so would find him there. Like other men, he liked women to be young and pretty, or very intelligent. When the silly or unattractive came to the house, he lapsed into complete silence. Often I have seen Mother making signals at him from the end of the table to exert himself. He would also sit in his armchair with a knee crossed, gazing intently at his foot which he waggled round and round. This was a bad sign, and very soon he would disappear into his study. The visitor would sometimes say, "Arthur seems very quiet," and Mother would reply, "Well, he's working *very* hard just now."

During those first days in Hampstead, while my mother tried to get the house into some order, my father was faced with a formidable task at the University. I quote from Professor Starling. "Cushny had already by his work at Ann Arbor achieved the leading position in his science . . . but at University College he was undertaking the creation of a department out of nothing. Previous to his appointment, materia medica and therapeutics were taught by a part-time lecturer, as was the custom in other medical schools.

The remuneration was meagre, and the department consisted simply of one ill-lit and badly furnished room. Nothing daunted, however, by these material disadvantages, Cushny set himself to the building up of the school in London with the calm optimism and the unfailing equanimity which characterised him in anything he undertook. His sane judgement of men and things made him at once a valued colleague, both in the college and in the laboratory; and there is no doubt that he fully appreciated the society into which he had fallen and the regard in which he was held by its members." This last sentence I think gives us the key. There was a fine medical faculty at University College—above all the two physiologists, Bayliss and Starling, who were well aware of his quality and who were almost certainly instrumental in getting him to accept the post. From the beginning the two departments worked in close collaboration. My father was skillful and ingenious at devising laboratory apparatus, which must have been particularly useful in these early days. He had to train his own "lab boys" and I remember a good deal of complaint, when he came home in the evenings, about those inept and careless people. After some time there arrived a young boy of intelligence called Condon. He became devoted to my father and showed great aptitude for laboratory work. When World War I began, he went into the army, but was soon seriously wounded and was discharged. He returned to work in the Pharmacology Department, and when my father went to Edinburgh in 1918, Condon went too. There, he became ruler over several "boys," and I suspect, of the graduate research workers too. He remained there with at least two of the professors who succeeded my father. No account of the later years in London or in Edinburgh would be complete without some mention of Condon, who was of such inestimable value to my father, and to my mother when father died so suddenly. Condon had at least two articles published in scientific journals concerning apparatus that he invented while working in the department. In 1956, Condon's fiftieth year of employment under Cushny and his successors was celebrated at the meeting of the Pharmacological Society in Edinburgh. He was introduced, I think, by Gunn, who told of Condon's first day of work under Cushny, as though he had been there. Gaddum was at that time Cushny's successor at Edinburgh.

On his appointment to London, my father was elected a member of the English branch of the Physiological Society. The annual conferences of the society were held abroad every two or three years, and my parents usually attended them. When the conferences were held in London, we always had some eminent foreign representative staying with us. The most beloved of these was Hans Meyer from Vienna. He not only was eminent, but rather unusually, looked eminent, with his bright eyes, acquiline nose, long pointed beard and thick hair enbrosse. I used to call him Herr Geheimrat, and he would shake his head and say, "Nein, nein, joost ze old Onkel." My parents were very fond of him, and were very anxious about him during and after the period of 1914 to 1918 when we heard the Viennese intellectuals were starving.

During these conferences there was a great deal of entertaining. I always enjoyed my mother's dinner parties from behind the scenes. I would go up to her bedroom to examine the ladies' evening cloaks, and then consume delicious left-over foods brought out to the kitchen. An extra waitress was engaged to help the rather inexperienced maids, who were all my parents could afford. On one occasion the cook knocked a box of matches into the creamed cauliflower which immediately turned bright red and doubtless poisonous. With great resource, I ran full-speed around the corner to where our great friends and distant relatives, the Wylies, lived. Fortunately, they were about to have stuffed tomatoes, and as they were a large family, there were plenty. I careered back with these in a silver entree dish. My mother said afterwards she was amazed when the hired waitress took this unexpected dish around the table, but etiquette, even in professorial homes, forbade the blinking of an eyelid.

My father retired to his study every evening at ten o'clock to work. He did his best writing, he said, from ten o'clock till about one in the morning. The study was a room on the first floor lined with shelves, which were filled not only with books but also with large quantities of scientific reprints, and unbound journals. The big desk was covered with papers in apparent confusion. Pipes and tobacco ash completed the air of cosy squalor. The housemaid was allowed in to clean the grate and light the fire, but otherwise nothing was touched. Once a year, however, my mother insisted on spring cleaning. She chose a time when my father would be away from home giving examinations or attending meetings. When he returned and found that this desecration had taken place, although he must have known it would happen, there was a tremendous outcry and declarations that hours of valuable time would be lost looking for misplaced books and papers. I found this clipping in his scrap book. I think it appeared in the Times.

> I hear the steady thumping on the carpet on the line,
> There are careless people dumping books and papers that are mine,
> They are tossing them and mixing them, so I shall never more
> Get them back in disorder as I had them fixed before.
> They have gone in force and taken firm possession of my den,
> They have swiped my scattered pamphlets and have burned them, Ne'er again
> Shall I find the tracts containing things I'd marked to read sometime,
> They are smoking in the alley, and the law permits the crime.
> They have robbed me of the cushion that was matted in my chair,
> They have put my pipe and ashtrays, well, I can't explain just where,
> They are rubbing, they are scrubbing there with all their might and main,
> And they shake their heads, assuming looks of sympathy and pain
> Showing that they think I'm crazy for presuming to complain.

In 1906 "the book" went into its fourth edition. He became a member of the committee on pharmacy of the British Medical Association and of the committee on proprietory medicines and an honorary member of the Association of American Physicians. He was also made a member of the Medical Research Club.

My parents always had a dog of some variety that was always much loved, and I was encouraged to keep pets of every kind, which I had to properly look after myself. I had at different times guinea pigs, a great speciality which went on for years, dormice, tortoises, hedgehogs, and almost every creature that could be kept in a London house or garden. In all of these, my parents took a benevolent interest.

All this seemed to some people incompatible with the experiments on animals that my father's work entailed. However, granted that it is important, if possible, to save human life and to deliver mankind from many disabling and painful diseases, experiments on animals seem unavoidable. The Medical Research Club tried to keep this point of view before the public, and to reassure it that unnecessary suffering was avoided. On one occasion, there was a meeting for public discussion with prominent anti-vivisectionists. Whether converts were made on either side is now unknown, but the affair was not without drama. Professor Starling, lean and ascetic with piercing blue eyes, standing on the platform pointed a long forefinger at his audience said, "Do you know that one in five of you is going to die of cancer?" A lady antivivisectionist rose with a cry of, "I can't bear it," and rushed from the hall. An old lady from the "anti" group came across and said cosily to my mother, "I think I will come and sit with you, you all look so nice." Certainly the pharmacologists and physiologists whom I knew were kindly and upright men, dedicated to the truth as they saw it in their science. Sometimes animals themselves were saved in this way, as when my father found by experiments on rats that a sort of madness which seized cattle in Africa was caused by eating the senecio weed in times of drought. Before this, it was thought to be due to an infectious disease and whole herds were slaughtered.

In 1907, my father was invited to accept the Chair of Pharmacology at the University of Pennsylvania at what was for those days a munificent salary of £1200 a year. He declined. This was the fifth chair offered to him without his having applied for any.

On May 9, 1907, when he was 41, he had the honour of being elected a Fellow of the Royal Society. That summer my parents must have been abroad in Switzerland, Heidelberg, and Strasbourg, probably combining a physiological congress with a holiday. My father kept a number of badly written postcards and letters written by me while staying with an aunt.

It must have been about this time that work began with Dr. (later Sir) James Mackenzie. Mackenzie had been a general practitioner in Burnley and came to London to be a heart specialist. Through Mackenzie's clinical work and Cushny's pharmacological work on the effect of digitalis on certain heart conditions, particularly auricular fibrillation, important foundations were laid for the modern knowledge of diseases of the heart, notably the therapeutic effect of digitalis.

Mackenzie was a great bear of a man, with a bushy grey beard and a broad Scottish accent. He was blunt in speech and said what he thought even to very exalted patients, a trait which they seemed to enjoy. Although he and my father were great friends, about politics, they could not agree; Father being a conservative and Mackenzie, a liberal. There were great arguments and, at the time of Lloyd George's Insurance Bill, I remember Mackenzie roaring and shaking his fist; but they remained good friends throughout.

Except when my parents went abroad, we always went to Scotland during the summer vacation, usually spending part of the time in the Highlands and then going to my father's boyhood home, Fochabers. I did not altogether enjoy these holidays, as they were usually spent at some fishing or climbing inn where there were no other children. I envied the more orthodox seaside holidays of my friends where there were lots of other children. My parents were great walkers, and at an early age, I tramped across the moors and climbed mountains.

When we were in Scotland we usually attended the "Red Kirk," the church at Speymouth where grandfather had been minister. Although as a descendant of a line of Scottish ministers my father knew a great deal about the church of Scotland and the Bible, particularly the Old Testament, his attitude to the church was one of affection for old associations and respect for the moral discipline of his Presbyterian upbringing. When I wrote to him from school asking him if I could give up Latin, I had this reply from him, "Latin is a good discipline to the mind. The people who learn Latin have generally better arranged minds than those who have not gone through that discipline . . . . They 'red up' their problems more. The same is true of those who have learned the Shorter Catechism. Those who survive this latter (and they are comparatively few) have an advantage over other men. They know that for them nothing is impossible in the way of learning, for have they not passed through the fiery furnace 'hetted seven times hetter' than ordinary souls can endure. You have not had the inestimable advantage of studying this terrible and wholly uninteresting compendium, but as a minor task have encountered the Latin declensions and I hope you are not going to show your back to the enemy." I did not. For the information of those of a softer age, the Shorter Catechism (it was not short) was the statement of faith and dogma which Scottish children used to learn.

My parents had decided that I should be taught no religion until I had reached a sufficiently mature age to decide for myself. In fact, this was impossible. When I went to a kindergarten presided over by two pious ladies, I was considerably embarrassed to find I knew nothing of these things so well-known to my companions. Later, Father began reading aloud to me the Old Testament, as literature with which I should be familiar. We did not get very far as it occurred to him that Noah must have allowed his aged relative, Methuselah, to drown. This involved so much research that the reading fell into abeyance, to my relief, and we continued later with Ivanhoe and Canterbury Tales, which were much more to my taste. Perhaps, nowadays he would be called a Humanist, I do not know. During World War I, I was much shaken when he said it was useless to pray for the safety of any person. His respect for absolute honesty forbade the consolation of prayer to a personal God whose existence he had been unable to prove. However, with the Christian moral virtues, he was endowed more fully than most men.

In 1908, Father was appointed a member of the Royal Commission on Whisky and other potable spirits. Two or three years before there had been police court proceedings in Islington as to the nature of whisky. The distillers and others were forced to conclude that there must be a definition of "whisky." The chairman was Lord James of Hereford. There were four Fellows of the Royal Society on the commission and senior government officials. As most of the witnesses were Scottish or Irish, some sly remarks were included in the evidence. Potstill and patent-still became topics of family conversation. The Times of August 10, 1909, had a leader on the final report: ". . . It is much better reading than we are accustomed to find inside the well-known blue covers. . . There are all sorts of theories about the secondary products which give the flavour, but the Royal Commission does not find that any of them rests on scientific basis . . . . if anybody finds himself the worse for whisky drinking, he had better face the plain truth that he is imbibing too much ethylic alcohol and not try to excuse his excess by throwing the blame on the secondary products." Naturally, the medical students found their professor's service on a Whisky Commission a ripe subject for various verses and jokes in their magazine.

My father was very abstemious. His one alcoholic indulgence was a glass of port after luncheon on Sundays. A favourite family quotation was Home's epigram:

> Firm and erect the Caledonian stood
> Old was his mutton, his claret good.
> Let him drink port the English statesman cried
> He drank the poison and his spirit died.

We had a tablemaid who always forgot to put out the port glass on Sunday. Father would recite the epigram, shouting fortissimo at the word "PORT". The tablemaid stood like a rock; never once did she catch the allusion.

The Royal Commission must have taken up a lot of time, but the research went on steadily, still concerned largely with the heart, and with optical isomers, though now there were some papers on alcohol as well.

In 1909, he was President of the Therapeutic Section of the Royal Society of Medicine, as well as a vice-president of the International Congress of Applied Chemistry and of the Physiological Section of the British Medical Association. In 1910 he was president of the Pharmacology and Therapeutics Section of the British Medical Association meeting in London. The fifth edition of "the book" was published. In that year, he received another call to the University of Pennsylvania, which now had a very fine pharmacology laboratory. In spite of the material disadvantages of University College, he declined the call, and also one to St. Louis University. He delivered the Presidential address to the Section of Therapeutics of the Royal Society of Medicine under the title, "A Plea for the Study of Therapeutics." He went to New York in 1911, where he delivered the Herter Lectures, and the Harvey Lecture, all on different aspects of irregularities of the heart. He went on to Philadelphia where he delivered the Weir Mitchell Lecture. He also lectured in Baltimore. From the reading of old letters, it is evident that he was away for several months. He visited various other places besides those where he lectured, such as Ann Arbor, Winnipeg, and Toronto. I presume his hosts saw that he caught his trains on time. He was one of those people who preferred to arrive on the station platform just as the whistle blew. My mother, on the other hand, was always in very good time indeed. In the early days of our Scottish holidays, when we had to get to Euston by horse cab, Mother was always so sure that the horse would fall, or we would break a trace, or experience some other disaster that we usually started at least two hours before the train was due to depart. When taxi-cabs replaced horse cabs, she still insisted on this early start. On one occasion we had more than an hour to wait. Father stalked off to the University, which is very near Euston, while we sat among the luggage. He reappeared careering down the platform as the guard was waving his flag. This was quite a common occurrence. When we transferred to the Highland Railway, there were frequent long stops at wayside stations. Mistrusting, with reason, the acumen of the guard, father would walk down to see that the luggage had not been put off before our destination. He always stayed away too long, until, with the whistle blowing and the green flag fluttering, we would lean from the window to cheer on a dignified figure in tweeds and knickerbockers loping along the platform. After he had actually been left behind at some Highland Station, where only one train a day passed, while Mother, a cousin, dog Fru, and myself rattled away without any tickets, each member of the party always carried his own ticket.

At the end of 1911, he went to Aberdeen to receive an honorary LL.D. from his old university. This gave him immense pleasure. For this degree, there was a splendid scarlet gown and a square black velvet cap, in which he looked very resplendent on special academic occasions.

In 1912, at last, through the generosity of the Carnegie Trustees, a new pharmacology department was built at University College incorporating the things Father wanted.

It must have been during the summer holidays of 1913 that there was a Physiology Congress at Groningen in Holland. It was decided that attendance at this meeting would be combined with a holiday in Holland, and that I should also go. There was a large German contingent at the Congress. My father spoke fluent German and often wrote papers in this language, and during his years in Strasbourg, he came to know the German character well. After this Congress, he said he was quite certain that Germany was determined on war soon. The attitude of the Germans at the meeting was markedly unfriendly and aggressive, and even those whom he had known well were uneasy in their attitude. Although he had many friends among German scientists, he greatly disliked the military type. After Groningen, we spent a few days on the windswept island of Ameland, one of the Frisian Islands. British visitors were very rare. In the evening when we sat in the glass-enclosed verandah of the small hotel, we saw a large part of the island's population with noses pressed against the glass studying the strange foreigners. The only other foreigners were a few German engineers. Perhaps when 1914 came, they thought we were spies.

That summer I went to St. Felix School. I realise now what financial sacrifices must have been made to send me there, although the meagre professorial salaries had by then been slightly increased. During my first term, my father was engaged on research concerned with senecio. Morbidly anxious to do the right thing and to remain inconspicuous, as a new girl should, I was horrified when he insisted on all of us gathering vast armfuls of ragwort and carrying it back to the hotel where we were staying, along with various parents of more exalted members of the school. I felt I was marked for the rest of my school life. In fact no comment was made, which shows perhaps how unnecessary it is for the young so often to feel shamed by their parents.

Collecting flora and fauna for scientific investigation often occupied a part of our holidays. Usually, it was for the purpose of extracting some special constituent, so large quantities had to be collected to yield a sufficient amount of the desired substance. Once a box of red spotted fungi was stored in the cupboard in the room where I was sleeping, and long before we went home the smell was horrible. Another time, the poison in a wasp's sting was the subject, and several nests were taken using chloroform. The stingers were removed with forceps, and partially anaesthetised wasps were later found crawling around in the most unexpected places.

The letters my father wrote to me at school had a turn of wit peculiarly his own, and usually ended with amusing messages to my friends, who still speak of him with affection and admiration. The following, written in reply to an idea I had of studying medicine, reveals something of his own view of his profession. ". . . Whether you have enough initiative to take up the research side of medicine I do not know and you do not know. It means in any case hard work and constant work, but there are moments of joy in it when an idea proves to be good and one feels that one has added a brick to the temple. But you can only find out your fitness by trying how far you have the necessary initiative, which is all in all . . . . I should be very joyful if you took up medicine and 'made good' as the Americans say, but I have no desire to see you make a bare pass into the profession and then lapse into a ruck. . . . . The first class there is always room for, the poor have to take what they can get and thank God. But the mediocre always expect more than they deserve and have to be repressed from treading on the skirts of the really first class. Ponder these things in your mind and we'll talk it over in the holidays."

Father's interests were far-flung, and 1914 rather surprisingly found him giving evidence before a Royal Commission on Sewage Disposal. During that year he received a nomination for Fellowship of the Royal College of Physicians of London but declined.

In the summer of 1914 we went as usual to Scotland, this time to Dinnet in Aberdeenshire. My parents, like most people, were anxious about the international situation, but nothing could be achieved by remaining in London and Father was, as usual, very tired after the summer term. Near us on Deeside were Professor (later Sir Herbert) Grierson and his family. Mrs. Grierson was the daughter of Sir Alexander Ogston, who also had a house on Deeside. He had been professor of surgery at Aberdeen when my father was a student. My father had a book bound in calf, stamped in gold with the University arms, and inscribed by Professor Ogston, which he had won as best student of surgery. Sir Alexander was an alarming old man, who only received a chosen few at his house. Father, of course, was one of these, while a flock of young Griersons and I waited on the hillside.

We first knew that the country was mobilising when an army nurse came into the hotel lounge and told her mother she had been called back. After that there was no news at all. There was no BBC then. On Sunday we went to the church, which was crowded. The minister announced from his pulpit that he had news that our Expeditionary force had been driven back with fearful losses; he used the word 'decimated.'

It is well known that at that time everybody seemed swayed by a tremendous demonstrative patriotism and a fierce hatred of the Germans, that belongs to a younger age than ours. My father for all his balanced intellect was not immune from this war fever. Perhaps because he knew the Germans so well, his hatred was the more intense. Like all teachers of young men, it was a bitter experience to see his students and assistants going off, so often never to return. He also had eight nephews, four already in the army. He himself was only 48 and it irked him not to be in it all the more directly. He hated the thought of young men dying while he lived.

By 1915 the Royal Society had formed a War Physiological Committee. Father served on this as well as on the Poison Gas Committee at the War Office, and when it was formed later, the Antigas Committee, a section of the Chemical Warfare Committee. When, by his special gifts and training, he could help the fighting men more directly, he felt easier. The research on poison gases had its dangers and it was impossible not to come in contact with the gases to a certain extent. He was what is called nowadays a 'Boffin'. He was pleased when he was allowed to go to the front for a few days to study the work at the scene of action.

Even the blackest days had moments of humour. A corps of Defence Volunteers was formed at University College and the professors as well as other members of the teaching and administrative staff joined. Some were old and none of them military types. In a letter to me at school he says, "We have begun target shooting with great zeal and some of us hit the target pretty often but some of us don't. And a good many of us cannot see the bull's eye at all and hope that a German is bigger, or that only the very fat ones will come over. Given a very fat one at twenty-five yards we could make great practice if he stood still long enough for the united corps to play upon him. . . Drill is getting fiercer and fiercer and to see the corps doubling up and down the playground would make you want some of them for your hockey-team." The corps went to Blakeney Point off the Norfolk coast over the 1915 Easter holiday. "The air is fresh and keen and makes us feel very fierce. This afternoon we 'advanced in rushes' it seemed for several miles before we got in for tea."

Zeppelin raids were taking place now and Father insisted on going outside to watch the thing caught in the searchlights, looking like an incandescent cigar. Mother was afraid that he would he hit by shell splinters, that rattled on the roof from guns on Hampstead Hill. On one of his journeys to Scotland to examine, he stayed with an old schoolfellow, a doctor near Stirling. He wrote, "A local manufacturer was here and received a telephone from the police to hurry off and put out his lights as a Zepp was near and this caused general alarm and discomfort in the household. I tried to reassure them and it never got this length. About eleven pm a hushed footstep came to the door and a very slight pull at the bell and on John going to the door a special constable whispered to him that there was 'a wee thing ower much licht in the kitchen.' John replied in similar tones and a whispered conversation went on for some minutes in case the Zeppelin should hear them. You may imagine my joy in listening to this from the study."

Father had a great interest in the history of drugs, and was an authority on William Withering, physician and botanist, early member of the Royal Society, and discoverer of digitalis. A paper on Withering was read to the History of Medicine Section of the Royal Society of Medicine in 1915. In that year, the sixth edition of "the book" came out.

In 1916, he gave the presidential address to his section of the British Association on the analysis of living matter through its reaction to poisons, and he was appointed to the Alcohol Committee of the Liquor Control Board. He was elected to the Council of the Royal Society, and also to its Sectional Committee of Physiology and the Government Grants Committee. There was at that time some concern over the increase of cocaine addiction in the country, probably due to war strain and a loosening of moral restraint. Father had a letter on the subject published in the Times, and gave evidence before the Parliamentary Committee on Cocaine. The following year, he published a book on the Secretion of Urine.

I was by then a student at University College. I observed my parent at the high table in the refectory and occasionally met him in his department where I had gone to seek Condon's advice about some chemical problem, which he understood a great deal better than I did. Otherwise, I naturally tried to avoid being marked down as the daughter of one of the professors.

Although my parents were quite unconcerned with the conventions involved with what, in present jargon, is called 'keeping up with the Joneses,' I felt that my father had very outmoded ideas about how I should behave, just at the time that war had exploded so many ideas about sheltered girl-hood. When I had tea in the extremely public restaurant of a large shop near the college with a young doctor from the hospital, I was in agony lest my father should walk in. There was trouble when he looked from the cloister windows and saw me playing tennis with some students of whom he did not approve. Lipstick he regarded as the mark of a fallen woman. Even scented soap was anathema to him, and only Pears unscented clear soap was allowed in the home. If one had had to use some other when not at home, immediately would come the command, "Go and wash off that stink." He was quite a formidable and almost Victorian parent for a rebellious adolescent.

Early in 1918, we learned that Sir Thomas Fraser, who for 41 years had held the Chair of Materia Medica in Edinburgh, was to retire, and that the post would certainly be offered to my father. He wrote to my mother that he was not certain he should accept it if it was offered and that they would have to talk it over carefully. For years almost no research had been done in the department, and it had fallen into a state in which research was hardly pos-sible. Until then the professor of Materia Medica had charge of beds in the Royal Infirmary. With private practice and hospital beds as well as teaching to occupy the holder of the chair, laboratory research had gradually died. There was now to be a new Chair of Therapeutics so that the professor of Materia Medica would be relieved of clinical duties. At the same time, an-other new Chair of Chemistry in relation to medicine was established.

A chair in the University of Edinburgh was looked upon as the climax of achievement for a Scotsman, even such a cosmopolitan one as my father, and eventually he accepted the offer. During the summer vacation of 1918, we left London for Edinburgh. Once again, he had the task of completely reor-ganizing and refitting his department. A large part of the course in Edinburgh had consisted of the recognition of the natural origins of therapeutic prin-ciples and of memorising the appearance of roots and leaves. The department was full of jars of these substances. Out they went to the dustbin. "Gone was the host of time-honoured drugs whose only qualification was founded on a hazy empiricism," as a writer in the university magazine put it. After this holocaust and the painting and cleaning, apparatus had to be set up. Condon did invaluable work. In October, term began and Father gave his inaugural address on "Progress in Materia Medica." The imposing figure in his red LL.D. robes, surrounded by members of the faculty, appeared on the plat-form. I had heard that his delivery was bad, but I had never before heard him lecture. He never learned to project his voice even after years of practice. I heard a girl behind me say, "Gosh, what a rotten voice." Those at the back of of the hall could have heard very little. I remember once in the large Mc-Ewan hall, whose acoustics were notoriously bad, I sat at the back when Father introduced the speaker. I saw his mouth open and shut, but heard no sound at all, let alone words. However, for those who sat in the front for his lectures, there was solid information salted with a dry humour that always pleased his audience. The following account of him taken from The Student, the undergraduate magazine, gives his students' view of him.

"It was mainly in his role as lecturer that the student came in contact with Cushny. His critical attitude will remain in the mind long after much of the solid matter of the orthodox lecture has faded into oblivion. Not that he was a brilliant lecturer in the ordinary sense. He was often inaudible and the ground he covered was never extensive. But he always illuminated any subject he touched, though following his book closely, he said many things that the dignity of print or perhaps the law of libel would not allow; made historical and literary references that were never pedantic; told tales of Dover who was buccaneer and physician, and of the Borgias, of worthy Germans who patented other men's discoveries, of osmotic pressure being a Mesopotamian word that should not be used if it could be avoided, of the whereabouts of the cockles of the heart and of the practice of hocus pocus. His lectures on alcohol will always be memorable; they were always humorous, but their humour never obscured their dignity.

"Occasionally he would adopt the Socratic method of teaching. Approaching a card box, he would lift a card, and looking at a name, enquire, 'Is Mr. So-and-So here?' If no answer were forthcoming, he would say, 'Perhaps it is Miss So-and-So, then?' If still no answer was forthcoming, he would say, with a twinkle in his eye, 'Ah, I see Mr. So-and-So has sent in his card by mistake', and try the next card, whose owner would be asked for his opinion on some subject. If no opinion were offered, the Mr. So-and-So in question would be told to think the matter over, and later on in the lecture would again be questioned. This process of questioning he never carried to extremes and he never allowed it to interfere seriously with a lecture. He always good-naturedly realised that a question and answer class is mainly question and little answer.

"In examinations his standard was high but an oral was always a friendly meeting and never an ordeal. It was not the appearance of an inquisitor that he presented as he sat on the other side of the table, glancing at one's paper, and then looking over his glasses and asking, 'Tell me, have you ever measured out ninety drops of anything?' It was rather the appearance of a benevolent uncle, still young in mind, half heartedly pretending to be a bogeyman.

"The writer has vivid memories of his journey home at the end of his first term of medicine. In the train there was an American whose extensive travels were indicated by the labels on his bags. In the course of conversation it appeared that he was a doctor of medicine, and much regretted having been able to make only a short stay in Edinburgh. His enquiries about professors in Edinburgh were met with much youthful enthusiasm and information, and at the name of Cushny his face lit up, and he eagerly asked about him. He said he had been one of his students when he was a professor in America, and had he known he was in Edinburgh, his one act would have been to go and see him once more. Such is typical of the place that Professor Cushny occupies in the feelings of his students, both here and in America."

In Edinburgh he gave nearly all the lectures, leaving his assistants free to carry on research in which he was always ready to give advice and help, and to suggest new lines. Young men from all over the world competed for the opportunity to work in his laboratory.

While my father was organizing his department, we were looking for a house. My parents were determined, to the surprise of their more conservative colleagues and their wives, to live outside Edinburgh and to have a large garden. At last, they found Peffermill House, a typical ancient Scottish mansion, said to have been the original Dumbie Dykes of Scott's "Heart of Midlothian." In one old edition of the Waverley novels, there is a picture of the house on the frontispiece. It had walls nearly three feet thick; two of the ground floor rooms had barrel-shaped ceilings. There was a spiral stone staircase, a source of great anxiety to my mother when the first grandchild came. At one corner of the house was a tower which made a little round closet to the adjoining rooms. The house stood in a large garden that had a stream running through it, and an uninterrupted view of Arthur's Seat. The garden was an absorbing interest and pleasure to my parents. Father had a water garden by the stream, and spent the weekends splashing round in rubber boots, heaving great stones about. There was a gardener already there. I think his life became considerably more busy, but also more interesting with mother working beside him, showing what he evidently regarded as an unladylike enthusiasm for digging. This, one must remember was only just after the World War I, and ladies were still supposed to be fairly ladylike, especially in Edinburgh. There were two cottages at the far side of the garden. The gardener lived in one and later, Condon and his Scottish wife were established in the other. Two geese, named Castor and Pollux, grazed on the green. In a field next to the garden, we installed two sheep to eat the grass. They had to be ewes so that there would be no agonizing over their eventual fate. Once, when Professor Barger, the biochemist, and his wife brought the author, E. M. Forster, to tea, the sheep escaped into the garden. Now whenever I see the name of that great writer I have a vision of him pursuing two sheep round the rose beds.

It must have been about 1921 when my mother began to have occasional but alarming heart attacks. She did not know how serious these were, nor did she know when they were impending, although we did. She, to the end of her life, talked of her "silly faints." My father with his knowledge of cardiac conditions must have known that any one of them could have been fatal, but he never told me this, and it was not till after his death that I was told what to expect. There was no treatment, and she would never have consented to live a restricted life, even if it would have done any good, which it probably would not have, so everything continued as usual. About that time too, through a chance observation in the classroom, Father found he had an abnormally high blood pressure. He must have known then that their time together could not be very long, but they were deeply happy at Peffermill. The house and garden seemed to fulfill everything they wanted in a home. There was a devotion between them that I cannot convey.

I was by then married to a former student of father's, the son of Dr. Charles Watson MacGillivray, who was one of the last surgeon physicians, and who had been a 'chief' at the Royal Infirmary of Edinburgh. Our son was christened Patrick Cushny by Dr. Peter Dunn, who had married my father and mother, and my husband and myself. The ceremony was performed in the Scottish custom at Peffermill House, and the baby wore the robe in which my father and his sister and brothers had been christened and which had travelled around the world for the christenings of my cousins and myself.

In Edinburgh, as in London, distinguished scientists were frequent visitors. In 1923 there was an International Congress of Physiology there. Since he held the view that science should know no frontiers, my father was strongly in favour of inviting members from former enemy countries. He persuaded the beloved Hans Meyer to come from Vienna. He came diffidently, fearing that some of his former friends would not welcome him, but my parents looked after him with affectionate care, and he found he still had many friends in England. My mother gave a garden party. The guests, especially those from the New World, were interested in the ancient house. Some of my friends, acted as guides, taking the visitors into the attics and telling them bloodcurdling ghost stories, made up on the spur of the moment.

In 1925, the University of Michigan conferred an honorary degree on their former professor of pharmacology. Both my parents went to America in the spring and saw many old friends, and my mother was able to visit her sister in Cincinnati. The citation for the degree conferred on the University Commencement Day was as follows:

"Dr. Arthur Robertson Cushny, Fellow of the Royal Society and Professor of Materia Medica in the University of Edinburgh. Coming to Michigan after a brilliant career abroad, he filled with distinction his professorship here from 1893 to 1905 and contributed greatly by his teachings and publications in the field of Pharmacology to the eminence of the Medical School. When he returned to the Old World he carried with him the affection of the New, which has followed with deepening satisfaction a reputation growing in fame from year to year."

The eighth edition of "the book" appeared in 1924 and the following year Longman Green & Company published *Action and Uses in Medicine of Digitalis and its Allies*, his second monograph on the subject. In the press was his *Biological Relations of Isomeric Substances*.

Professors are expected to take part in the administration of their universities in an advisory capacity. Father found time taken off from his proper work of research irksome, but he conscientiously attended all the Faculty meetings, where a colleague recalled he said little: "In controversy, his criticism was all the more effective because it was tempered by humour." He was interested in the rearrangements of the University Library and the administration of the Moray Research Fund. He also had to attend various meetings in London where he made the Athenaeum his headquarters, sleeping at the nearby Garlands Hotel, where they knew him well and gave him a ground floor room 'to save the stairs.' They would have been surprised to have seen him labouring in his water garden. The generally ordered atmosphere of an age gone by, which still prevailed at these places, pleased him. On one occasion he wrote to my mother: "I suppose you have rarely or never received a letter written with a quill. Now I am going to give you that pleasure." There followed a rather spluttery letter which ended: "Quills are not all they are cracked up to be."

As may be imagined from his revolutionary teaching of pharmacology, he had strong views on the revision of the British Pharmacopoeia. He went to London to attend a meeting about this in February 1926. Dr. Dale (now Sir Henry Dale) said of my father on this visit to London: "All who came in touch with him had remarked on his mental vigour and good spirits. I had a long talk with him on the Monday evening when he spoke with quiet optimism of plans for a well earned leisure after another five years of the work he loved and enjoyed." On February 24, he returned home and was talking to my mother when he suddenly had a cerebral haemorrhage and died early on the morning of the next day. He would have been 60 in nine days' time.

Perhaps no tribute is greater than one from a professor's own students. This is from The Student, the undergraduate magazine of the University of Edinburgh.

"Long after doses and actions have faded from the memory, will there remain in the mind the vision of a true scientist, of an experimenter and a lover of animals, of a sceptic and a healer, of one who founded the modern science of Pharmacology and filled his students with his own enthusiasm for it. Our seniors often talk of the 'giants' of their student days. We too have our giants and in days to come will look back upon Cushny as one of the greatest of them all."

Reprinted from
ANNUAL REVIEW OF MICROBIOLOGY
Volume 30, 1976

FROM THE PRECISE TO THE AMBIGUOUS: LIGHT, BONDING, AND ADMINISTRATION

W. D. McElroy
Chancellor, University of California at San Diego, La Jolla, California 92037

Ann. Rev. Microbiol. 1976. 30:1–20

FROM THE PRECISE TO THE AMBIGUOUS: LIGHT, BONDING, AND ADMINISTRATION

◆1670

W. D. McElroy

Chancellor, University of California at San Diego, La Jolla, California 92037

CONTENTS

I am not a true microbiologist, but I have used microorganisms as tools to study biological processes all my life. This interest can primarily be traced to the influence of three men. One was Professor L. R. Blinks, who supervised my undergraduate research at Stanford. He was a very patient man who allowed me to try things well beyond my comprehension and capabilities. Although during this period I was able to obtain an oxidation-reduction curve for phycoerythrin, the pigment from red algae, I am not sure I really understood the true meaning of that, unless it was related to the idea that accessory pigments could absorb light and transfer the energy to chlorophyll. Be that as it may, it was the rich research environment that was important to my future career. It has been an article of faith to me ever since that a strong research environment, and the earlier the better, is the critical factor in the education of young scientists.

The second major influence on my scientific career, Professor C. B. van Niel, brought adenosine triphosphate (ATP) to my attention. This was in the microbiology course he taught at the Hopkins Marine Station during the summer of 1941, just two months before I entered graduate school at Princeton University. This same course introduced me to the phenomenon of bioluminescence through observations on luminous bacteria. A paradox was evident to me even at that early stage. We were taught that energy transformation, to insure reversibility, involved small energy changes (7–8 kcal) in the form of phosphate bond energy; yet in luminous bacteria we were obviously observing an energetic change of the order of 70–80 kcal, a tenfold difference. This concept of phosphate bond energy, which only recently had been introduced into the United States from German and English laboratories, was to have a powerful and pervasive influence on my research and on the teaching of graduate and undergraduate students, for it soon became obvious that phosphate bond energy must be involved in all energy-consuming and -generating processes. However, the dilemma in energy transformation posed by the process of bioluminescence remained with me, and my concern for the problem is evident from some of the pages of Volume I of *Advances in Enzymology* which are soiled from my constant reading and rereading of Professor Fritz Lipmann's article on phosphate bond energy. The same can be said about Professor Herman Kalckar's article in *Chemical Reviews,* where opposed resonance was discussed. Neither article helped me to resolve this problem of energetics, but they kept the problem fresh in my mind for many years.

The third person to whom I am greatly indebted is the late Professor E. Newton Harvey, who introduced me to the many forms of luminescent organisms. His influence was primarily responsible for my continuing to work on the mechanism of bioluminescence even after leaving his laboratory at Princeton University.

ELEMENTARY AND HIGH SCHOOL TRAINING

I am sometimes asked how I became interested in science and a research career. I have no recollection that my early childhood headed me in this direction. I was born on a small cotton farm outside Rogers, Texas, a town of about 500 that has since grown to 1000. It was an unusually interesting and happy childhood. While growing up, I had the usual animals as pets, raised pigeons, and even had my own Shetland pony, which I rode to a one-room schoolhouse two miles away. I remember very little about the academic program because at that time I could not have cared less about the educational process. When I was in the third grade, my parents moved into the town of Rogers because that was the only place where my older sister could attend high school. Even here I did not find myself particularly academically inclined, although I did earn reasonably good grades. My interests at that time were primarily sports and other nonacademic activities, including camping, fishing, and hunting. I seem to remember that I missed more school days than I attended, although here again, probably thanks to the efforts of my family, I studied enough to get A and B grades through junior high school.

My parents moved to McAllen, Texas, in the Rio Grande Valley, where I finished my final year of junior high and attended high school. I was still primarily interested in sports and other extracurricular activities, although for reasons still not clear to me, I graduated with a high academic average. Even at that time, although I enjoyed mathematics tremendously and took all the advanced work that I could in biology and chemistry, the possibility of a career in science did not occur to me. However, I had an evening and summer job in a drug store, and I became fascinated with what the druggist did in filling prescriptions. I tended to spend more time behind the pharmacy counter than I did behind the soda fountain where I was supposed to be dispensing ice cream cones and Cokes. By the time I graduated from high school I was almost convinced that I wanted to be a pharmacist.

MIGRATION TO CALIFORNIA

Once again chance moved me toward a career in science. I had graduated from high school with excellent grades, but I also happened to be reasonably good at playing football. Trading on this latter ability, four of us from the McAllen High School team decided to migrate to California to play football at Pasadena Junior College. At that time Pasadena was the training ground for Howard Jones's football talent at the University of Southern California. It was at Pasadena Junior College that I first became seriously interested in an academic and scientific career, although even this was not too well defined. I did reasonably well in the athletic program at Pasadena and had the opportunity to go to USC, but I was intrigued by the stronger academic program at Stanford University. For this reason I accepted an athletic scholarship at Stanford to play football under the direction of Tiny Thornhill.

At Pasadena Community College I became acquainted with Professor Max Walter de Laubenfelds, who encouraged me to continue working in science. I took most of the biology and chemistry courses at Pasadena and became particularly interested in biology, which at the time translated into premedicine, my major when I transferred to Stanford in 1937. Professor L. R. Blinks's course in physiology introduced me to the excitement of doing original research. This was the major factor in my decision to do independent work with Professor Blinks and, in my senior year, caused me to change my mind about going on to medical school.

My two undergraduate years at Stanford were exciting ones. Not only did I have the opportunity to participate in original research with Professor Blinks, but I also took a fascinating genetics course taught by Professor George W. Beadle and his associates, who at that time were developing the newer concepts in microbial genetics, particularly the use of microorganisms as a tool in genetic studies. Professor Beadle's work had been primarily with maize and *Drosophila*, but he was convinced that if he was going to obtain the number of mutants and the precision that he needed, he had to deal with much larger numbers than were possible with the higher organisms. Therefore he looked for a microorganism that had a reasonably well-defined genetic system; as he explained at the time, *Neurospora* was one of the best organisms because of the chromosome number and the reasonably well-defined mechanism of growth and cell division. Not only was there a sexual cycle, but also there was a well-defined asexual process which allowed one to deal with a large number of spores. These decisive two years at Stanford dissuaded me from entering medicine and encouraged me to go to a graduate school instead.

In my senior year I changed from a premedical to a biology major. After graduation I was advised by Professor Charles V. Taylor, Chairman of the Biology Department, that it would be wise to go to Reed College for a year or two to study general biology under Professor L. E. Griffin before I proceeded to work on a research thesis. This proved to be a wise recommendation. At Reed I had an excellent opportunity to learn about general biology and natural history. I encouraged my physiology lab partner at Stanford, Mr. Howard Bliss, to join me at Reed the following year (1941), where we attempted—with the enthusiasm of youth—to revolutionize biology teaching at that institution. (At Stanford Bliss and I had taken Professor Beadle's outstanding genetics course, and later we were given the task of solving all the problems in the new book being written by Professors Sturtevant and Beadle. After reading the rough draft and galleys of this book, we thought that we were experts in genetics.) Because Dr. Griffin had no one on the staff who desired to teach such a course, Bliss and I offered Reed's first undergraduate course in genetics. In collaboration with Professor Demerest Davenport, we also rewrote the general biology laboratory manual and made a number of other contributions of perhaps dubious value to the biology curriculum at Reed. The environment at Reed encouraged one to experiment, to try new things, to be innovative, and to learn through mistakes. I daresay that in the process all of us—undergraduates, graduate students, and faculty—learned a great deal about how to use facts and make interpretations in a meaningful way.

My stay at Reed College was not entirely academic for reasons I suspect can very well be explained by my background. I was continually interested in things other than the academic program and at Reed my attention turned again to football. As most people know, Reed College is not noted for its high-powered intercollegiate athletics. However, Mr. Alfred W. Hubbard, who had just joined the Athletic Department, and I organized an informal football team. We played five games (the local CCC Camp team three times and another local college two times) and won all five—Reed's first undefeated football season. This feat received notoriety through *Time* magazine; that and other national coverage led Dr. Dexter Keezer, President of Reed, to announce the probability that neither Mr. Hubbard nor I would be reappointed if we were as successful the following season!

HOPKINS MARINE STATION

Following my first year at Reed, I returned to Stanford for a summer of special physiology courses, and at the end of my second year at Reed I enrolled in the Hopkins Marine Station (Pacific Grove, California) microbiology course, taught by Professor van Niel. For me this was an exciting period in the development of microbiology and biochemistry. It was approximately fifteen years since Kluyver and Donker had published their important papers on the unity in biochemistry. These papers were to go unnoticed for many years, only later appreciated when it became apparent that there was indeed a great deal of unity in this field. At that time we were talking about whether it was pyruvic acid or some C_2 fragment that condenses with oxaloacetic acid. Also at the time one could do ^{14}C experiments in photosynthesis and demonstrate, as predicted by van Niel's equations, that O_2 in the photosynthetic process comes from water. We also learned about the processes of oxidation, dehydrogenation, decarboxylation, the addition of water and taking off hydrogen, and ultimately the addition of phosphate instead of water to generate an energy-rich phosphate bond. These were all new concepts at the time and they were to have a powerful influence on a number of people taking the course, including Mike Doudoroff, Stan Carson, Bill Arnold, and many others.

The microbiology course was extremely intense, usually beginning at 7:30 A.M., with lectures for four hours, then a quick luncheon followed by another series of lectures continuing for four or five hours. Often after a brief dinner, we would work until midnight or 1 o'clock in the morning. No participant would ever forget this course because in addition to scientific knowledge it imparted a philosophy and method of thinking about the approach to biological research. We also took the invertebrate course (the microbiology course met three days a week) with excellent lectures and formal collecting by Professor Rolf Bolin. This required rising at 3:30 or 4:00 A.M., depending on the tides, to collect organisms from a defined niche in the environment and then returning to the laboratory to classify the organisms and to hear Professor Bolin lecture about what we had presumably learned from that morning's collection. That busy summer was the most intense science education I have ever experienced; van Niel's influence in stimulating my interests was critical to my future research career.

I remember some work that summer with Bill Arnold at the Hopkins Marine Station that was to have a major impact on my thinking about how to do biological experiments. This was when I first heard about the radioactive phosphate atom ^{32}P, which we could obtain in reasonable concentrations only by giving water containing the ^{32}P in dilute solutions to rats carrying cancerous tissue. It turned out that the cancerous tissue would accumulate the ^{32}P in much higher concentrations than one could obtain in any other way, and we used this tissue to feed the small jellyfish, *Polyorchis penicelata,* to determine whether the phosphate moved directly across the jelly or through the canals. It soon became obvious that the ^{32}P and the organic phosphate moved through the digestive route and through the canal system. We used Coca-Cola bottle caps to dry the samples and then measured the radioactivity with a gold leaf electroscope which Dr. Arnold had built himself. Here again by accident I found myself at an interesting time and place.

The invertebrate course at the Hopkins Marine Station was taught by Professor Rolf Bolin. He was a very good friend of John Steinbeck and Edward Ricketts. Those who have read *Cannery Row* will recall that Steinbeck developed a fictional character based on Ricketts. All of us were excited when Steinbeck invited the invertebrate class to an afternoon picnic and evening affair at his home. He was personally interested in marine biology and at every opportunity he encouraged the people at the Hopkins Marine Station.

GRADUATE SCHOOL AT PRINCETON

Perhaps the simplest way to describe my experience on going to Princeton in 1941 is to say that I was a country boy heading East. I moved into an environment that was quite foreign to my upbringing. For example, at that time graduate students were expected to live in the graduate college on the Princeton campus, where one had dinner in academic cap and gown. Except for graduation I had never worn this attire and never quite understood the idea behind the Princeton manner. Fortunately I was able to get an exception to the regular dining and living arrangements because by that time I was married and therefore was given permission by the Graduate Dean to live in an apartment off the campus.

Most of my first year at Princeton involved formal courses. In the middle of the year I met Mr. James Gregg, who was interested in developmental problems, and persuaded him that undoubtedly ATP had some major function in the control of development. I encouraged him to collaborate in an effort to look at the ATP concentration, particularly during the gastrulation process in frogs. I was convinced that the inward movement of the ectoderm during gastrulation must be controlled by the ATP concentration. Needless to say, neither of us had ever isolated ATP nor, as a matter of fact, had ever made an inorganic phosphate determination in a really quantitative way. This did not deter us from scrounging the only Klett-Summerson colorimeter at Princeton, which had just been acquired by Professor Ray Dawson. After several weeks we were able to assemble the necessary equipment and read the literature on how to isolate ATP. We initiated experiments on what I believed then to be an extremely important problem, but unfortunately we never continued it because of the beginning of the Second World War. We were able to isolate from the frog eggs several milligrams of ATP which I saved and was to use later at Johns Hopkins in some interesting experiments on fireflies. Fortunately, Professor Lester Barth at Columbia heard of the phosphate experiments and encouraged several of his students to continue them, and some very important developments about energy-rich phosphate and developmental processes resulted. Although there were a number of other interesting experiments, it was not until after the war and after my postdoctoral stay at Stanford University that I was able to continue my major research interest at the Johns Hopkins University.

At Princeton I had my first experience with Professor E. Newton Harvey, the father of bioluminescence in the United States. I had no idea that I would be working on bioluminescence when I went to Princeton. Then I was excited about a number of other biological and chemical problems. I was particularly interested in trying to enrich for photosynthetic organisms that could use different wavelengths of light for growth. I thought this could lead to the selection of organisms with unusual accessory pigments that would be capable of absorbing light at different wavelengths and subsequently transferring the excited state energy to the chlorophyll system. I never had an opportunity to really set up that selective experiment, even though I had learned in van Niel's course how to enrich for photosynthetic bacteria and other photosynthetic organisms.

I was diverted from this interest by experiments being conducted by Professors Frank Johnson and Henry Eyring on the effect of pressure on light emission from luminous bacteria. From these one could observe a powerful effect on the luminescent system, which suggested large volume changes occurring during certain physiological processes. This was particularly evident in the effect of pressure on the reversibility of the action of narcotics on luminescence. The observations made by Johnson and Eyring further suggested that the narcotics led to a major unfolding of the proteins, indicating a major volume increase; if so, the narcotic effect should be reversible by pressure. This immediately attracted my interest and I persuaded Professor Harvey to allow me to work with him on the mechanism of action of narcotics in the luminous bacteria. This use of luminous bacteria as a tool to study narcotics almost led me in the direction of pharmacology, but whether fortunately or unfortunately, events occurred that prevented me from doing this.

With the outbreak of the Second World War, I was quickly diverted to a war research project, under Professor Harvey, concerned with the bends and aviation physiology. However, when opportunity allowed, I was able to carry out an occasional experiment on luminous bacteria and to study the effect of narcotics on the light-emitting process. I was greatly impressed by some then-recent papers by Professors R. H. Burris and P. W. Wilson on oxidative assimilation and the effect of various inhibitors on the process. These led me to study the effect of various narcotics on the assimilatory system in luminous bacteria, which in turn led to major consideration of the mechanism of the action of narcotics. I returned to this interest after the war when I joined the faculty at Johns Hopkins.

One of the interesting aspects of the wartime work on bubble formation in fluids and in animals was the experiment using high-pressure techniques to eliminate gas nuclei in organisms and in liquids. During these experiments I used a Smith fermentation tube and yeast cells that had been pressurized at 10,000 pounds per square inch. In so doing I observed that when yeast cells were used to inoculate a medium that had also been pressurized in a Smith fermentation tube at the same pressure, no bubble formation occurred and no gas phase appeared. Yet we could demonstrate that CO_2 production was continuing at its normal rate and had built up to a pressure equivalent to several atmospheres. A small tap on the side of the vessel often gave rise to an explosive emission of CO_2 gas. This was an impressive demonstration of the importance of gas phases in the fermentation process, essential for gas production and the lowering of CO_2 tension. Although they were a diversion from my primary interest in biology and microbiology, these experiments gave me an opportunity to do some interesting theoretical and experimental work with Professors Harvey and Eyring on gases in solution, a subject that I undoubtedly would not have approached had it not been for the war research.

Learning to do research under the direction of Professor Harvey was a rewarding experience. He was tolerant of false starts due to inexperience and always took the time to point out how to do a particular experiment better. Most of all he was a storehouse of knowledge about luminescent organisms. We would spend hours discussing different types of organisms, where they could be found, and what was known about their luminescent characteristics. In later years we spent time together in Jamaica, Woods Hole, and other places doing joint research projects. It is fair to say that my lasting interest in luminous organisms can be traced entirely to Professor Harvey. Almost all scientists working on luminescence in organisms in the United States today were trained either by Professor Harvey or by his students.

POSTDOCTORAL FELLOW AT STANFORD

When the war was about over and the research projects were being terminated, I received a National Research Council postdoctoral fellowship to spend a year in Professor George W. Beadle's laboratory at Stanford University. By that time, Beadle had advanced considerably in his research on the genetic control of enzyme formation in *Neurospora*. I was anxious to learn about this tool because it seemed obvious that it would be an extremely powerful aid in analyzing any number of biological processes. At this time I was interested in trying to obtain mutants affecting the luminescent process in bacteria. In Beadle's laboratory I worked closely with Dr. Herschel Mitchell on a number of adenine mutants of *Neurospora*. We were particularly concerned with the pathway of adenine biosynthesis. Unfortunately, we made very little progress with this particular problem, but we did become interested in an enzyme which deaminated adenosine and adenylic acid.

Some of our experiments beautifully demonstrated the workings of the unprepared mind in scientific discovery. During our studies of adenylate deaminase, we observed that the enzyme from Takadiastase (a concentrated extract from *Aspergillus oryzae*) would deaminate approximately 50% of the adenylate added to the reaction vessel. We were convinced that the adenylic acid was impure. And yet, with repeated purifications and crystallizations of the adenylic acid, we still obtained only 50% deamination. We could never resolve this important problem. This of course was before the introduction of modern column chromatography. It is now obvious that we were dealing with a mixture of adenylic acids and that the two-prime compound was not deaminated. The five- and three-prime adenylate was readily deaminated by the enzyme. In retrospect this looks like a simple problem that we should have recognized. Unfortunately, Dr. Waldo Cohen had not yet made his important observations on the structural chemistry of adenylic acids. It is interesting to note that a colleague at the McCollum-Pratt Institute, Professor N. O. Kaplan, used this enzyme in collaboration with Dr. G. David Novelli and Professor Fritz Lipmann to determine the nature of the phosphate linkage in coenzyme A.

While working with Dr. Mitchell, we attempted to induce mutants in luminous bacteria. We had no knowledge concerning the biochemistry of the luminescent process, and we were not really in a position to know what to add to the medium components that might affect luminescence. We did know from earlier observations by Professor Michael Doudoroff that a flavine might be involved. However, we were not able to obtain dark mutants, nor, interestingly enough, any nutritional mutants such as J. Lederberg and E. Tatum had just demonstrated in *E. coli*. This mystery led me to do many experiments with the luminous bacteria in later years. Although we isolated a large number of nutritional mutants, we were never able to demonstrate that either X-ray or a variety of chemical mutagens could increase the muta-tion rate in luminous bacteria. As far as I am aware, no one has been able to increase the number of mutants in these organisms significantly above the normal mutation rate, except for certain dark mutants that require aldehyde for light emission.

JOHNS HOPKINS—PHOSPHATE BOND ENERGY

After my postdoctoral period in Professor Beadle's laboratory, I accepted an ap-pointment at Johns Hopkins in 1947 as an assistant professor. I remained there for the following 23 years as a faculty member in the Department of Biology. Those were my most exciting and productive years in science, made so particularly by the outstanding faculty throughout the university and the stimulating biology under-graduate and graduate students. I particularly remember as an assistant professor having the privilege of working with Professor B. H. Willier, then Chairman of the Department of Biology, Professor G. Wilson Shaffer, the Dean of the University, and Mr. P. Stuart Macauley, the Provost and primary financial officer of the uni-versity. All three gave me great encouragement and support, as well as the free-dom to pursue my research. Indeed, I had so many research interests then that it was almost impossible to contain my enthusiasm for starting many projects at once.

This was my first faculty appointment and I soon became involved in a wide variety of activities at the university. I was appointed to a host of committees, seldom missed the after-lunch pool game at the faculty club, and even found time in the early years to help coach the football team. The veterans returning to the campus after the Second World War created a stimulating teaching and research atmosphere; these students took learning seriously and were ready to get involved in all kinds of experiments and ideas.

My first major job on arriving at Hopkins was to establish the general course in comparative physiology and biochemistry. It should be clear from what I have already said that the course was greatly influenced by my training under Professors van Niel and Harvey. I spent almost the whole first semester making sure that I had set up my laboratory adequately to do a first-class teaching job for the undergradu-ate and graduate students. Professor Willier insisted that his students take my advanced undergraduate course in biochemistry because he was convinced that they should obtain a background in modern biochemistry and modern genetics if they were to tackle developmental problems.

Some of my very early research problems at Hopkins involved collaborative efforts with Professor Carl Swanson on the mechanisms of induction of mutations in *Neurospora* and other fungi. Some of us were very concerned with the molecular mechanisms of the mutation process. This led us to look at the effect of temperature and pressure on the mutation rate, and we were able to demonstrate that pressure applied even after the application of a mutagenic agent could reverse the mutation process. This exciting observation led us to speculate on the nature of the chemical process taking place during the induction of a mutation. Although we felt this was an important discovery, very few people then were interested in these ideas.

In the summer of 1947 we organized research on bioluminescence. Although we were growing luminous fungi and bacteria and other organisms at the time, it was clear that the firefly was the major source of luminous organisms during the summer in Baltimore. This work started when a student asked me what he might do with regard to his independent research project. I suggested that he collect some fireflies and see if we could repeat the experiment originally done by DuBois—namely, demonstrate the presence of a luciferin and a luciferase. A cold buffer extract was made of a few firefly lanterns and after a few minutes the light disappeared—the cold-water extract of DuBois. We extracted several other firefly lanterns with hot water, and after cooling this extract we added it to the cold-water extract. We observed immediately that light emission was restored in the cold-water extract. We attempted to isolate the factor that restored the light in the cold-water extract, assuming, of course, that it was luciferin as defined by DuBois. We had great difficulty trying to purify the luciferin, and finally I suggested that we try a mixture of various cofactors and ATP. (The ATP was that which Jim Gregg and I had isolated from the frog eggs while at Princeton.) It only took us a few minutes to observe that the mixture restored light in the cold-water extract, and through an elimination process we were able to show that unusually bright light could be obtained by the addition of ATP. By classical definition, therefore, ATP should be defined as luciferin. But this did not fit the concept of the oxidation of a substrate for light emission. This led us to suspect that there must be a true luciferin present that was not exhausted in the luminescent reaction when the cold-water extract ceased to emit light. In retrospect it is interesting to note that DuBois was the first to show a physiological function for ATP, even though it would be 40 years (through the work of Lohman and Fiske and Subbarow) before we would know that such a compound existed.

The following summer I organized a number of undergraduate and graduate students to collect fireflies in large quantities. Through a series of experiments, we were able to demonstrate the presence of a true luciferin in the firefly extract. I do not intend to review the following 20 years of research that led to the elucidation of the chemical mechanism of light emission in the firefly extract, but needless to say I had many colleagues, graduate students, postdoctoral students, and faculty members who aided me in isolating and eventually describing this system. It was an exciting time for all of us because it was the first time that a luciferin had been isolated, crystallized, identified, and eventually synthesized. In addition, it was the first time the enzyme luciferase had been isolated and crystallized. Today it is gratifying to realize that this esoteric biochemical system is being used in a number of research laboratories throughout the world, including clinical laboratories in many hospitals. Because the system is extremely sensitive and rapid for the detection of ATP, it has been adapted to a number of research problems. I remember once being asked whether this system would be of any use in either the production of "cold" chemical light for commercial use or in other research projects. Unfortunately, I said that I could not imagine what it could be used for other than assaying ATP and that we already knew how to do that reasonably well. Incidentally, while my colleagues used to chide me about my firefly-gathering techniques in Baltimore (we paid 25 cents for a hundred), today the research demand for fireflies is so great that they are supplied by a number of commercial firms, as well as our own major collection center at Oak Ridge, Tennessee.

We continued our work on the firefly system, but there were other luminous systems that also attracted us at the time. One of my students, Dr. B. L. Strehler, who went on to Oak Ridge to work with Dr. Edward Carter and later with Dr. William Arnold, started a research program on bacterial luminescence. At the same time, Dr. J. W. Hastings came to my laboratory as a postdoctoral student, and we started to look into the bacterial system from several different directions. Soon thereafter I received an invitation from Professor Perry Wilson to write a review paper for *Bacteriological Reviews* on bacterial luminescence. I accepted and invited Dr. Strehler to coauthor the manuscript with me. When he agreed, I indicated that I would send a rough outline of the proposed review with the hope that he would suggest the appropriate changes. In the outline, I indicated that we should discuss the nature of luciferin and luciferase in luminous bacteria. Although I didn't know it at the time, Dr. Strehler was busy attempting to isolate a cell-free preparation of the luminescent reaction. Subsequently he has told me that this outline stimulated him to move ahead even more rapidly to identify a factor in crude extract from luminous bacteria that stimulated light emission from a cold-water extract similar to that described for the firefly. As history has shown, Professor Strehler and his associates were able to demonstrate that this cold-water extract could be stimulated to emit light by adding reduced DPN. In other words, the addition of reduced DPN (DPNH) to a crude bacterial extract temporarily restored light. This was the real beginning of the unravelling of the bacterial system. Later Dr. Hastings and I demonstrated that reduced FMN was the main substrate for light emission in the bacteria and that the DPNH was needed to reduce the endogenous FMN present in the crude extract. Following these observations, Professors Milton Cormier and Strehler were able to show that a long-chain aliphatic aldehyde was also necessary for bright light emission. This was the second luminescent system that yielded to the isolation and identification of key substrate molecules necessary for light emission. Although Dr. Arda Green and I proposed that the aldehyde was consumed during the light reaction, only in recent years has it been demonstrated conclusively that the aldehyde is indeed used in the light emission.

SCHOOL BOARD AND OTHER EXTRACURRICULAR ACTIVITIES

I never seemed to learn the lesson that if one criticizes or comments on a particular subject, the result is often an appointment to a committee to do something about it. It so happened that I made the mistake of commenting on the Baltimore School systems relative to the quality of the entering students at Johns Hopkins. Within the year I was invited by the Mayor of Baltimore to join the Baltimore School Board; I accepted, of course, and served for 10 stressful years during the 1950s when the major social issue of integration was faced by many school boards, including that of Baltimore. Service on this board also gave me an opportunity to appreciate and understand the importance and the role of the superintendent of schools in major cities. Being a superintendent in a city the size of Baltimore is more harrowing than being the president of a large university. It is usually a no-win situation, as evidenced by the turnover in superintendents in the past two decades.

At the same time I also became involved in affairs in Washington, particularly at the Office of Naval Research (Panel on Microbiology chaired by Professor Roger Porter), the National Institutes of Health (NIH), and the National Science Foundation (NSF). Probably one of my most pleasant assignments was with the Office of Naval Research (ONR) right after the Second World War, where I worked with a number of competent administrators (Dr. Harve Carlson and Dr. Roger Reid) who understood the importance of supporting basic research. Most people do not realize the critical role the U.S. Navy played after the Second World War in supporting and advancing our cause of basic research in the universities. Had it not been for ONR, with its understanding of the importance of basic research for the good of the country as a whole and not just of the Department of Defense, we would not have the scientific leadership in this country we have today. In fact, in many ways ONR was the forerunner of NSF.

At the same time, the enlightened leadership at the NIH was also instrumental in advancing the broad scope of biomedical research in this country. I served for several years as chairman of the Biochemistry Panel and later of the Physiological Chemistry Panel, with its outstanding executive secretary, Dr. C. Donald Larsen. The excitement of new developments in biochemistry kept us all involved doing those things necessary to mount major research efforts in a number of areas; this produced, I believe, a revolution in biology, medicine, and agriculture. Later I was to assume membership on the Advisory Council of the National Institute of Arthritis and Metabolic Disease, where again I met many capable people and was educated in a number of areas in which I had been largely ignorant.

When people become critical of the peer review system, I reflect on my involvement in the biochemistry study sections and the Council on Arthritis and Metabolic Disease. At no time in my activities there did I see anything but the highest statesmanship in the peer review process. I have seen some of our very best scientists, including Nobel Laureates, turned down because of an inadequate description or research plan. I know of no occasion in which there was any suggestion of personal favoritism involved in the awarding of grants. And my experience has been that study section members tend to be even tougher on their best friends than they might be on people whom they do not know. When I served on the Woolridge Committee, charged with reviewing the total NIH programs, our detailed studies on this point entirely confirmed my personal observation. Of course, panels and councils do make mistakes in judgment, but overall I found NIH to be an extremely well-managed agency in the support of research. At that time I could not say the same about NSF. Although their peer review system worked well, I felt that the National Science Board was too isolated from its advisory committees. I served as a member of the divisional committee for Biology and Medicine at NSF for over two years, and I finally resigned because I did not feel we had access to the Director or the Board.

In 1962 I accepted President Kennedy's invitation to serve on the President's Science Advisory Committee (PSAC). This was a particularly exciting appointment, because it involved broad policy issues relating to the support of all science in the United States and the Committee had direct access to the President. One of my earliest assignments was as a member of the advisory committee on the support of oceanography. This gave me an opportunity to chair a subcommittee on biological oceanography and marine biology, an area which had become of great interest to me at Johns Hopkins. This assignment allowed me to visit most of the marine biological laboratories in the United States and to assess their productivity and needs.

This was the beginning of my major interest in using microorganisms to assess primary food productivity. In collaboration with Professors Howard Seliger, William Fastie, and James Carpenter and others from Hopkins, we established a temporary marine laboratory in Jamaica to study the phosphorescent bays that occur along that island's shores. We were able to develop underwater photometers and to demonstrate that the light intensity of the luminescent organisms were directly proportional to the cell concentration. Thus at nighttime it was possible—by towing the underwater photometer behind the boat—to determine the effective biomass of these dinoflagellates. This led us into a detailed study of the factors controlling the production of a high concentration of these organisms into what would normally be called a "bloom" in the open ocean. Again, it was soon clear that microorganisms were making a major contribution to the nutritional environment which stimulated these organisms to grow and to maintain a high, steady crop.

In 1962 Dr. Detlev Bronk asked me to chair a committee of the National Academy of Science to investigate the broad issues of what was then called the population problem. I did not quite realize what I was getting into at the time, but it developed into one of the most fascinating tasks I have ever faced. We assembled an extremely competent committee of biologists, social scientists, and others concerned with population issues to assess what the federal government should do, if anything, about the rapid population growth in the United States and the world. After a year of study, discussion, and probing, we published a small document which received unusual attention throughout the United States. It was not a detailed or scholarly publication, but it did point out the obvious in no uncertain terms: that the growth of the world population could not continue at the same rate without highly undesirable and almost predictable results. At that time population control was not a subject discussed publicly, at least not by politicians. However, Professor Jerome Weisner, then presidential science advisor, convinced President Kennedy to make a positive comment about the importance of understanding world population and doing something about it. With President Kennedy's recognition of the problem, the report had a major impact when it was published by the Academy. The committee's second publication concerned population growth in the United States. It too had a strong impact and seemed to spark a major national interest among many diverse groups. At that time it was difficult to learn from the various federal agencies how much money they were investing in research on human fertility, birth control techniques, and other research important to this worldwide problem. However, following these publications and with the efforts of a number of influential people and the Ford Foundation, it was possible to increase significantly the support of research in this area. It is gratifying to review what has happened in the last 15 years since that first Academy population report was released. We have seen major efforts by both public and private organizations encouraging individuals and nations to face up to the realities of population growth relative to available resources. Although the controversy still goes on in certain areas, most people, at least in the United States, now recognize the importance of balancing population with resources. During President Kennedy's administration and in the early years of President Johnson's, PSAC appeared to have some influence on the allocation of federal resources for the nation's science effort. Unfortunately, near the end of my term on PSAC, one could see a growing disenchantment of the Executive Branch with the science advisory apparatus. I presume this was because President Johnson was preoccupied with our unfortunate war in Vietnam. It was difficult for the academic science community to relate to that major political and military problem. In fact, a large portion of this community was opposed to the war, and I am sure this attitude was one of the frustrations faced by President Johnson. Although the President was pleasant and attentive, PSAC was obviously not one of his high-priority items in the final years of his administration.

ORIGINS OF THE McCOLLUM-PRATT INSTITUTE

In 1948, two years after accepting an appointment in the Department of Biology at Johns Hopkins, I was asked by Professor B. H. Willier, Chairman of the Biology Department, to meet with him and Professor E. V. McCollum, who had just retired as Head of Biochemistry at the Johns Hopkins School of Hygiene and Public Health. Over lunch Professor McCollum described to me the interest of Mr. John Lee Pratt in trace metals. Mr. Pratt, a former executive of General Motors, had long been interested in the role of cobalt in the nutrition of animals he was raising on his farm in Fredericksberg, Virginia, and he wanted to know what could be done about some of the problems he had encountered in trace metal metabolism. For several hours we discussed the general problem relative to our knowledge of trace metals, particularly from the agricultural standpoint, but more importantly, we agreed on our lack of understanding of the role of trace metals in biochemical functions. Professor McCollum asked me if I would put down on paper what I thought we might do at Johns Hopkins if Mr. Pratt were interested in supporting the research. None of my previous courses or research provided any more than an introduction to inorganic metal metabolism. I was greatly intrigued by the role of phosphate in metabolism from the standpoint of energy transformation (and I did understand the importance of iron in certain areas of metabolism and the cytochromes, hemoglobins, etc), but I had never really thought about the role of such metals as molybdenum, selenium, cobalt, copper, and other trace metals. However, after a few weeks of reading, I came to the conclusion that we should take an entirely different approach to the whole problem of trace metal metabolism. Up to then, it was clear that most of the work was of a general nutritional nature, and this was obviously why Professor McCollum was interested in the subject. (His earlier discoveries concerning the role of magnesium in nutrition occurred about the same time that Warburg and Christian discovered the role of magnesium in a number of enzyme systems, particularly those involving ATP.) Very little subsequent work had been done on the enzymology of trace metal metabolism, and most of that concerned the nutrition of plants and animals and not their mechanisms. I wrote a three-page memorandum to Professor McCollum indicating that a research program concerned with the mechanism of action of trace metal in enzyme systems was an exciting and possible approach. By unravelling their role one could hopefully arrive at a more rational basis for understanding both normal and abnormal metabolism. Professor McCollum was quite enthusiastic and recommended to Mr. Pratt that we start a center at Johns Hopkins to study trace metal metabolism, particularly as it relates to the role of metals in catalysis. Mr. Pratt agreed to the funding of an institute and I was asked to chair a committee to look for a director. Subsequently I recommended that we name the center the McCollum-Pratt Institute, and with the support of Professor Abel Wolman, chairman of our advisory committee, we obtained approval of the institute from the President of Johns Hopkins, Dr. Isiah Bowman, and the Trustees.

The McCollum-Pratt Institute was launched in 1949. At that time, right after the war, it was not easy to find a good biochemist-enzymologist interested in trace metals, but I did give President Bowman and Professor McCollum many names of good biochemists. Unfortunately, we were not able to attract any of these, and after six months' effort, the President—perhaps in desperation—asked me if I would be willing to be the Director. Being young apparently makes it easier for one to undertake things one knows nothing about, because I literally had no background nor knowledge about this subject. The only knowledge I could go on were the principles learned during my undergraduate and graduate years in science. But intuitively I was convinced that the way to proceed was to get good biochemists together and to encourage them to concern themselves with the basic problem of the function of trace metals in enzyme systems. After surveying the field, it became quite clear that there were few biochemists at that time really interested in this type of problem, so we developed a whole new strategy, namely, to hire outstanding biochemists interested in coming to Johns Hopkins, around whom we could infuse young people from the field of plant and animal nutrition who were concerned with trace element nutrition and metabolism. By combining good enzymologists-biochemists with students coming out of the nutritional laboratory, we hoped to stimulate a mutual interaction that would train the nutritionist more in the direction of the biochemist. Further, we anticipated that the problems would be of interest to the biochemists in the Institute.

As it turned out, we achieved a very happy union of these types. One of the first persons attracted to the McCollum-Pratt Institute was a young biochemist from Columbia, Professor Alvin Nason. I convinced Dr. Nason, who was working on the developmental aspects of corn embryos, to look at the nitrate-reduction problem in *Neurospora*. We had isolated a number of mutants that could possibly be of value in identifying the mechanism. Fortunately we were in contact with the people in

Long Ashton, England, where significant nutritional work was being done with the trace metal molybdenum. It had been known for some time that molybdenum was important in nitrogen fixation, but it had not been demonstrated how it might function in nitrate reduction, if indeed it really was necessary. We encouraged a young researcher from the Long Ashton laboratory, Dr. D.J.D. Nicholas, to join our laboratory to work with Dr. Nason on the problem of nitrate reductase. Dr. Nicholas was very knowledgeable about how to obtain a medium that was deficient in molybdenum, and he readily adapted this to *Neurospora,* where he was able to demonstrate the requirement for molybdenum when *Neurospora* used nitrate as the sole nitrogen source. Following this, attempts were made to obtain an active cell-free nitrate reductase.

At that time various coenzymes were fairly difficult to come by, although DPN was readily available. Only in the research laboratories did one have small quantities of TPN and FAD. By chance we were lucky enough to attract Professor Sidney Colowick and Professor N. O. Kaplan as a team to the McCollum-Pratt Institute. From their extensive enzymological research background, they had prepared and had available in the deep freeze a number of these coenzymes. Under these circumstances, Drs. Nason and Nicholas were able to demonstrate an active nitrate reductase when they used TPNH and FAD. As a matter of fact, a large number of *Neurospora* cell-free preparations, which we had frozen several years back, were quite active when these two cofactors were added. Under Dr. Nason's direction, the group purified the enzyme and demonstrated that the TPNH and FAD functioned to reduce molybdenum, which in turn reduced nitrate to nitrite. Again, having the right people in the right place at the right time seemed to be the critical factor. I need not comment further about the origins and functions of the McCollum-Pratt Institute except to indicate that this was an unusually productive time in my research career. I learned much more biochemistry and enzymology from the associates attracted to the Institute than I had learned previously in all of my graduate education. This experience demonstrated to me the critical importance of being associated with stimulating colleagues to discuss and share ideas and resources.

Probably the most educational aspects of the Institute—from a strictly academic standpoint—began when we started a series of major symposia to bring together, synthesize, and document the most current knowledge in specific aspects of biochemistry. Right after World War II there was a great dearth in the United States of major treatises or symposium volumes on various areas of biochemistry. We saw that void, and in 1950 we instituted the McCollum-Pratt Symposia to correct it. The first year we began modestly and thought it would be appropriate to hold a symposium involving trace metal. We chose copper metabolism and obtained some of the major experts from Australia, New Zealand, and England, where much of the nutritional work in animals and plants was being done. Interestingly enough, the enzymology—the role of copper in various proteins—was supplied by expertise from the United States. Following this symposium, it seemed obvious that the next symposium should be on phosphorus metabolism. Not only was it of great interest to me, but many of my colleagues, particularly Professors Sidney Colowick and N. O. Kaplan, were actively engaged in this type of research.

It was obvious from the beginning that we could not cover all areas in one meeting, and we decided that we would have two symposia. In 1951 to 1952, we held two major symposia on phosphorus metabolism. For the first time in the post-World War II era, the world leaders in phosphorus metabolism were brought together. It was exciting to hear from such people as Otto Meyerhoff, Fritz Lipmann, Herman Kalckar, Arthur Kornberg, Carl Cori, Luis Leloir, and other leaders in biochemistry. The formal meetings, the informal banquets, and the camaraderie that developed at that time did much to influence the function, role, and reputation of the McCollum-Pratt Institute for many years to come. An important product of the symposia was the resulting series of detailed, professional publications. In addition, we sensed a need for a general summary of the symposia for broader distribution, and we asked Professor Bentley Glass, also of Hopkins, if he would coedit these monographs and prepare a general summary. After the first summaries appeared, we realized that Professor Glass was an editorial genius. He would study the topic and somehow distill the essence into beautiful, logical prose. In my opinion, the two summaries published on phosphorus metabolism are classics in the synthesis of this subject. After 15,000 copies were printed and distributed, we had to let it go out of print. I am sure that many graduate students have read and reread Bentley Glass's summaries.

Following the two symposia on phosphorus metabolism, we had equally successful symposia on *Mechanism of Enzyme Action, Inorganic Nitrogen Metabolism, the Chemical Basis of Heredity* (where we first heard about cell-free protein synthesis and amino acid activation), *Amino Acid Metabolism,* and the *Chemical Basis of Development.* Naturally, one of the last ones was *Light and Life.*

All during the initial development of the McCollum-Pratt Institute, I was very anxious that it not evolve in isolation from chemistry, biology, and physics on the Johns Hopkins campus. During this period I worked closely with Professor Willier, Chairman of Biology, to make certain that the research and the educational programs in general were an integral part of a planned curriculum of the campus. As a functional unit, it was clear that the McCollum-Pratt Institute represented the biochemistry group in the biology department. Over time, the Institute basically became a major part of the Department of Biology, with little distinction between it and the rest of the department. In 1956, when I was asked to become chairman of the department upon Professor Willier's retirement, the McCollum-Pratt Institute fused even more solidly into biology and became an integral unit of the biology department. Mr. Pratt and Professor McCollum understood this opportunity and encouraged the McCollum-Pratt Institute to become more diversified in its interests.

AN EXCURSION TO WOODS HOLE

Although I had spent a few summers at Woods Hole with Professor Harvey doing physiological research on bubble formation, I had not been there often since the war years. In 1956 I was asked to assume leadership of the physiology course at the Marine Biology Laboratory (MBL). I replied that I was not a marine biologist, that my interests were in the more functional aspects of biochemistry. I was informed that possibly this was the direction that the physiology course should pursue. Naturally there was considerable controversy when I asked people concerned with ultracentrifugation and molecular weights of proteins, with enzymology, and those using isotopes in studying metabolic processes and others to join me. One senior MBL scholar observed that over the years he had expected the physiology course to become biochemistry and that the invertebrate zoology course would become the physiology course. We should not worry about names, he said, but rather whether the intellectual activity was suitable for the environment. I believe this comment eliminated most of the criticism of the physiology course and allowed us to develop a molecular biology–biochemistry curriculum. It was an exciting course and an exciting opportunity, with many visiting biochemists giving lectures in the course during the summer along with the regular staff.

At this time I also became interested in the role of microorganisms in the marine environment. Seeing the beautiful facilities at Woods Hole used only during the summer months gave some of us the idea that perhaps we should establish a marine microbiology institute at Woods Hole to function year round. The full-time people would work throughout the year, and during the summer they would be instructors in major courses for marine microbiology. This idea created considerable consternation among the MBL scientists and trustees, who worried that the year-round personnel would dominate the facilities and eliminate the summer program. Even the suggestion of a completely new building for such an activity, which had some sympathetic hearing from the microbiologists at ONR, did not alleviate criticism. Because this idea was rejected, in my opinion a great opportunity was missed to attract a large number of good microbiologists to Woods Hole on a permanent basis. I still believe this. There is today good microbiological work going on in the marine environment, but if we had started full-time efforts 20 years ago at Woods Hole, we might be further ahead in this important area of ocean research.

ACTIVITIES IN PROFESSIONAL SOCIETIES

I have long been interested in professional society activities. When I joined the Society of American Bacteriologists (now the American Society for Microbiology) in 1948, I recall that some of the most rewarding meetings were those in the physiology division of the Society. Then we had quite small meetings, with the exchange of ideas beginning at 9 o'clock in the morning and often going on until well past midnight. Growth of the Society has considerably interfered with this intimate relationship, and when I became president of the American Society of Biological Chemists, I argued for a large number of meetings. In my view it made no difference how many separate sections you had during the meeting; once you had conflict with a few you might as well add as many as possible to get the size of the individual meeting down to a level where intimate interchanges and vigorous discussions were possible. Unfortunately, even today this has not proven to be a viable approach to the interchange of scientific ideas because of the high degree of specialization in most areas. There are, I learned, problems relating to smallness as well as bigness.

Lately I have been concerned with a bigger aggregation of biological societies into an organization that could speak for all of biology at the national policy level. Occasionally this idea builds momentum, but then fades because it seems our best scientists often lack the interest to educate the public and, in particular, the Congress and the Executive Branch on science policy issues. In 1965, while President of the AIBS, I proposed that the Federation of American Societies for Experimental Biology and the American Institute of Biological Sciences should fuse their top leadership into an American Council of Biological Sciences to provide greater influence upon federal science funding. Although we eventually got this institution started, I still am dismayed at the scientific community's lack of activity in this direction.

TO THE NATIONAL SCIENCE FOUNDATION

In 1969 I was asked to become director of the National Science Foundation by Dr. Lee DuBridge, then Science Advisor to President Nixon. I had no desire to take on this position and earlier had indicated no wish to have my name on the candidate list. However, after an unfortunate incident involving Professor Frank Long of Cornell (which led to his withdrawal from the consideration for the directorship), Dr. Phillip Handler, President of the National Academy of Science, convinced me that I did have an obligation, that my acceptance might help restore some confidence in the scientific community vis à vis the federal government. This presented me with an extremely difficult decision. I had been on the Hopkins faculty for over 20 years and had derived great satisfaction from teaching and productive laboratory work. If I were to go to NSF, it would probably mean an irreversible decision to move out of science as far as laboratory research was concerned. I did, of course, accept the NSF directorship and subsequently found it was a broadening experience allowing me to influence science policy and administration in ways I could not have done elsewhere. But in some ways it was also an exercise in futility. It was my first experience as a full-time administrator, and I found trying to move a large bureaucracy is, to say the least, a very difficult and often frustrating experience.

At that time NSF had been mandated by Congress to reorganize and to add five additional Presidential appointees—a deputy director and four assistant directors. Recruiting these individuals was not easy. Many first-rate scientists simply had no desire to leave the laboratory and had little interest in the federal policy-making apparatus, which by its nature had to be sensitive to the politics of broader national issues. Fortunately, at the time, there were people on the presidential staff, including Dr. DuBridge, who understood this problem and were helpful to me in attracting to the Foundation several outstanding assistant directors.

In these recollections I will not comment in detail on the activities of particular interest to me at the National Science Foundation because I am probably still too close to those activities to be fully objective. I will mention, however, one activity in which I became increasingly interested and involved: the effort to develop a new NSF activity to apply research to societal problems. Before I went to NSF, I sensed that the country needed a new thrust in its science effort, one which would encourage our best scientists and engineers to investigate areas that were probably going to become national problems 5, 10, or 15 years hence. We needed scientists who, in my opinion, could look ahead and perceive the areas where new knowledge was needed to solve a foreseen national problem. However, it was clear to me that if this new concept was infused into the basic science division of NSF, the mixture of function could damage the basic research effort of that division. Both the Office of Management and Budget (OMB) and the Congress were sensitive to this concern and allowed us to establish a completely separate division, Research Applied to National Needs (RANN). (The RANN program, soon to reach 60–80 million annually, grew out of an existing pilot effort called IRRPOS.) At the same time we increased the support of the basic research division, because we successfully argued that basic, nondirected research was the fundamental underpinning of the entire scientific establishment. Contrary to the opinions of many, RANN was never intended to be concerned exclusively with applied research but rather to be the mechanism for identifying and studying national problems where both basic and applied research were necessary to identify alternate solutions. Founding RANN was an important NSF milestone, and I still believe the concept is viable and important to this country.

RANN was also planned to bring the nonacademic component of science and technology back in contact with the academic environment. The huge federal funding of research in the universities after the Second World War reduced the necessity for university scientists to look toward industry for support. For this reason the two groups were no longer in close communication. This was an unfortunate separation, because the free flow of ideas and knowledge between the universities and the industrial community is critically important if we expect to accelerate major technological advances in the solution of some of our societal problems. Today we have great difficulties melding together the universities and industries in a coherent way to meet such problems as energy, food production, transportation, land use, water use, and other major problems. I believe the National Science Foundation still has an opportunity for further leadership here.

After spending almost three years at NSF, I felt I had done about all I could do to move the broad policy of the Foundation into new directions. When the opportunity arose to return to an academic environment as Chancellor of a challenging young university of the highest quality (the University of California at San Diego), I accepted and became an academic administrator. This new position is primarily administrative, of course, but I hope to be able to keep in touch with the laboratory and to do some work on bioluminescence in microorganisms, on bacterial mutations, on the mechanism of enzyme action, or even on bacterial and mold metabolism.

Reprinted from
ANNUAL REVIEW OF ANTHROPOLOGY
Volume 2, 1973

CHANGING STYLES OF ANTHROPOLOGICAL WORK

Margaret Mead

Margaret Mead.

Reprinted from
ANNUAL REVIEW OF ANTHROPOLOGY
Volume 2, 1973

CHANGING STYLES OF ANTHROPOLOGICAL WORK

❖ 9516

Margaret Mead
The American Museum of Natural History, New York City, NY

The last 20 years have seen an enormous growth of institutions devoted to anthropological enterprises, membership within the discipline, and students, textbooks, and paraphernalia. From a tiny scholarly group that could easily be fitted into a couple of buses, and most of whom knew each other, we have grown into a group of tremendous, anonymous milling crowds, meeting at large hotels where there are so many sessions that people do well to find those of their colleagues who are interested in the same specialty. Today we look something like the other social science disciplines, suffering some of the same malaise, and becoming cynical about slave markets and worried when grants and jobs seem to be declining.

It has been a period of excessive growth; it is astonishing, looking back, to recount how many large enterprises have been undertaken. The year 1953 marked the end of the Korean war and the final exodus from Washington of almost all the remaining anthropologists who had lingered on to make the kind of contribution to national affairs that had developed in wartime. Most of the ventures that had been specifically influenced by the immediate post-World War II world drew to a close: Columbia University Research in Contemporary Cultures, Studies in Soviet Culture, American Museum of Natural History Research in Contemporary Cultures A and B, (189, 213), the Coordinated Investigations of Micronesian Anthropology, the period of affluence in the Foreign Service Institute, the intensive exploitation of the Human Relations Area Files, and the preparation of manuals and directives for participation in technical assistance and foreign aid (13, 191, 269). A few anthropologists stayed on for several more years, but activities of the House Un-American Activities Committee had disillusioned anthropologists with government, and as their participation in government shrank, so did the receptiveness of government agencies to anthropological contributions because there was no one to inaugurate them, receive them, or interpret them.

Anthropologists came out of the war years with several important new orientations. They had learned that their skills could be applied fruitfully to problems affecting modern societies and the deliberations of national governments and nation

1

states. They had learned to apply themselves to problems they had not themselves chosen, and to work with members of other disciplines. While this was most conspicuously true in the United States, the English style of operational research involved anthropologists in many unfamiliar fields, and some of the theoretical approaches of the French underground also meant a novel use of anthropology. A rationale for the kind of contribution that anthropologists could make to problems of national and worldwide scope was developed (171). But, at the same time, the experience of anthropologists during the war was summed up in the dictum, "you can't advise an adviser." If anthropologists were to participate in public activities, it meant some of them had to accept positions within various parts of the establishment, and this they became increasingly unwilling to do. The contradiction between a willingness during World War II to become involved and a disinclination to become involved later has not yet been resolved. This, in addition to the rejection of the Vietnam war, may account for the rather meager participation of anthropologists in the last 20 years in problems involving technical assistance, modernization, arms control, the prevention of nuclear warfare, peaceful uses of atomic energy, population control, and the environmental crisis (122).

In the earlier years, anthropology was so slightly established that the usual academic punishment for unorthodoxy was confined to excluding from it those who expressed very large heresies, such as the racist approach or an overemphasis on the dependence of American Indian culture on importations from Asia. After the war, although such major unorthodoxies persisted to a certain extent, the scene shifted to rewarding those who took part in currently popular minor theoretical discussions. Students were advised to concentrate on very recent polemics, and small, specific discussions of kinship, or variations in response to Lévi-Strauss (144, 160), became the road to academic advancement.

This exodus from any situation connected with national policy coincided with a tremendous growth in research opportunities and academic appointments. The National Institutes of Mental Health and National Science Foundation programs, and in the 1960s the development of programs vaguely conceived as foreign aid that would be relevant to political purposes abroad, such as the Ford Foundation Area Studies programs, all provided funding for academically phrased research. There were not enough funds for the tasks that needed doing, but there were more than enough for the strength and capabilities of those who were mature enough to lead and direct these programs.

An enormous number of new possibilities opened up in the United States through the establishment of new departments in old universities, transformation by upgrading old institutions, and a proliferation of new institutions. The establishment of new universities in the United Kingdom, new forms of cooperation in Paris (through coordination provided by the Sixieme Section de L'Ecole Pratique des Hautes Etudes and the Maison de la Science de l'Homme), and the new universities in Australia, India, and the new countries of Africa and Oceania also widened the field.

Movement from academic post to project to a different academic post was very rapid, and in the scramble for projects and posts there was little time for writing up the large amount of field work that was being funded. A great many young men and women wrote their dissertations, their three short papers, did a few reviews, presented a paper or so, and went on to a tenured appointment with very little opportunity to show their mettle. As the period of educational expansion slows down, it may well be that one legacy from those years will be a layer of middle management, the members of which reached their positions by a kind of gamesmanship that is no longer as relevant to a world where stringency, frugality, and specific capacities critically appraised are again in vogue.

Internationally, there have been a series of large enterprises: the Wenner-Gren Conference of 1952, which held us together for another decade (150, 280); *Current Anthropology,* with its network of associates and commentators; the big University of Chicago symposium on evolution (279); the Wenner-Gren Conferences at Burg Wartenstein and in the United States, the International Congresses of Anthropology and Ethnology, and a strong anthropological presence at the Pacific Science Congresses; the development of the Center for the Study of Man at the Smithsonian; the Laboratoire d'Anthropologie Sociale in Paris; the New Guinea Research Unit of the Australian National University; and the present (culminating) effort for the Ninth Congress of Anthropological and Ethnological Sciences to be held in Chicago in September 1973. These clusters of impressive activities owe a great deal to the imagination and energy of three people: Sol Tax, Editor of *Current Anthropology;* Lita Osmundsen, Director of the Wenner-Gren Foundation; and Clemens Heller of the Ecole Pratique des Hautes Etudes, in Paris. As available openings for research, personnel, and funds have been identified and deployed around the world, these people have enlisted the cooperation of many institutions and seized opportunities for amplifying organizational efforts.

This has also been a period of large research programs, varying from the Caribbean program of the Institute for the Study of Man, the continuing area-based projects, like the Chiapas project (296), Watson's New Guinea project (297), Goldschmidt's African project (114), the associated sets of fellowships and activities of the East-West Center in Hawaii, the New Guinea Research Unit of the Australian National University, the Arctic Institute of North America, Vayda's ecology project on the New Guinea Rainforest, the widespread studies emanating from the University of Manchester, the cluster of studies in French West Africa and French Oceania, integration of studies on political and social organization in the Netherlands at the Hague, and the Australian Institute of Aboriginal Studies. And there are many, many more programs in which there has been an attempt to bring together groups interested in an area with various related anthropological specialties. The large scale long term projects have had, I think, no higher rate of failure —failure to complete work, write it up, or integrate results—than other social science projects which had less complicated logistical problems (177).

The catholicity of anthropology, with its ideological insistence on the psychic unity of man, and its traditional disregard of race, ethnicity, age, and sex as criteria for academic posts or research capabilities, combined with the political instabilities of the post-World War II world, has resulted in another form of international cooperation and interpenetration of different nationally based anthropological traditions. When I was a graduate student, Jochelson, a refugee from the new Russia, was writing the last of his volumes on the Koryak (148) in the tower of the American Museum of Natural History, and an Ansa African student from the Gold Coast (187) was giving us lessons on how Africans viewed their own kinship systems. During these early years, a Nez Perce American Indian left Columbia to be welcomed in the Soviet Union as a member of a persecuted minority, and Boas and Manuel Gamio were planning joint studies in the United States and Mexico. In the early 1930s Radcliffe-Brown was teaching at the University of Chicago and—to the great disapproval of his American colleagues—was being considered for the directorship of a large international effort. Since World War II, the mix of students, students as informants, and faculty members drawn from all over the world, partly due to political upheavals, partly due to the great number of foreign fellowships available, has increased enormously. Americans teach in Japan, Frenchmen in India, Indians in the United States, Ceylonese in Australia, and the number of anthropologists from the new African, Asian, and Oceanic countries has increased. The vociferous group making political demands for more representation of Third World points of view has almost completely ignored this movement of scholars between countries, which has also included Third World political leaders who have studied anthropology in some Euro-American country. It is true that anthropology was born among the Euro-American metropolitan powers and flourished most vigorously among countries with colonial interests or identifiable minorities within their gates. But it is also true that this very anthropological emphasis has brought with it an insistence upon the comparable capacities for political and social development of all the members of *Homo sapiens,* and so has made a substantial intellectual contribution to the process of political democratization in different parts of the world. The current emphasis upon the symbolic processes found in primitive man (166), while it stems (at least within the French tradition) from Lévy-Bruhl, nevertheless reflects just as much interest in the existence of primary process thinking in modern man and the importance of symbolism and ritual in modern life as it does an emphasis on the more vivid and bizarre elements of the culture of preliterate peoples (28, 49–51, 73, 74, 101, 236, 288, 289). One of the belated effects of criticism of Freud's *Totem and Taboo* (88, 195) has been to establish that primitive man could think in ways as rational as modern man, as Boas maintained, and that primitive children might be more rational than primitive adults, and that civilized adults might display the kind of egocentric thinking of civilized children (190, 278). With this it has been established that it is necessary to rethink our whole overevaluation of rational linear thinking, with its dependence upon script and script-like processes (24, 28, 47, 178, 260). There has also been a new interest in cross-cultural studies of cognition and perception (61, 261, 290) and a return to a study of multimedia contexts.

At the same time, there are ways in which anthropologists have remained almost incredibly ethnocentric. Forty years ago Radcliffe-Brown suggested a system of kinship nomenclature that was not only cross-cultural but also a way in which kinship relationships could be read back as reciprocals (242). (I put this system on my typewriter in 1931.) Yet in 1972 Conrad Arensberg (12) can call for real "scientific models" and still express beautifully generated kinship relations as abbreviations of English terms *fa si da* or *mo br da*. We do little better in archeology, or physical anthropology, and only slightly better in linguistics. It is true that we need an agreed-upon terminology, but it is equally true that we need terminology that is not ethnocentric, favoring members of one linguistic group or world religion over another.

The last 25 years has also been a period of massive individual enterprises. These have included: Murdock's progressive exploitation of the Human Relations Area Files, first in *Social Organization* (218) and later in his *Ethnographic Atlas* (219); Lévi-Strauss' enormous and detailed study of the myths of the world (166); Yehudi Cohen's intensive use of cross-cultural comparisons and systematic attempts to grapple with the complexities of large areas (58–60), combining field work and integration of the literature; the Whitings' continued cross-cultural studies (304); Mandelbaum on India (182); Goldman's *Ancient Polynesian Society* (112); Leach's *Social Systems of Highland Burma* (159); and the Geertz's work on Indonesia and Morocco (97, 99, 102). These activities are comparable to the syntheses of an earlier day in which art, folklore, and material culture were organized within theoretical frameworks for the benefit of other students. They are notable for the way in which the individual integrating intelligence, handling vast amounts of material collected by hand, is still the prevalent style, only slightly helped by IBM cards or computers.

Traditionally, the sciences have advanced with the help of complementary interrelations between theory, instrumentation, and the stimulating or diverting effects of the climate of opinion within which their practitioners were working. There has been an increasing amount of critical and historical work on anthropological theory, such as Stocking's very distinguished work on Boas (273) combined with a series of trivial father-killing attacks (251, 302), reassessments of Malinowski to the point of boredom of all those participating, Marvin Harris's portenteous evaluation of everyone else (130), and the interesting experiment of using biography as a method of assessment of periods and national schools of anthropology in the *International Encyclopedia of the Social Sciences* (201). Encyclopedias have become one of the ways that the anthropological ideas which originated in the West have been incorporated into the thinking of other parts of the world, e.g. the *Educational Encyclopedia* published in Hebrew (77) and the Japanese version of the *Encyclopedia Britannica*. But the Wenner-Gren conference on The Nature and Function of Anthropological Tradition, held in 1968, which attempted to delineate the various national streams in anthropology failed both to draw together a sufficiently representative group and to integrate the results (162).

Biography, in a more personal sense than the theoretical evaluations, has also flourished: *Alfred Kroeber* by Theodora Kroeber (151), *Franz Boas* by Herskovits (135), the new series edited by Charles Wagley on Columbia anthropologists (173, 222), *An Anthropologist at Work, Writings of Ruth Benedict* (192) are examples. Autobiographies of various sorts include (166, 185, 271), *High Valley* (244), *Return to Laughter* (42), *Women in the Field* (111), and *Blackberry Winter* (206). There have also been passionate ethnologies like Jules Henry's *Culture Against Man* (132), and Colin Turnbull's *The Mountain People* (287). The furor over the publication of Malinowski's *Diary* (181) represents a low point in the discipline's degree of sophistication. In the inflamed political atmosphere of the 1960s, Malinowski was attacked because of his private diary, which records his tribulations and miseries as he did his magnificent field work. A Polish word, which he used for the Trobrianders when he was most emphatically fed up with them, was translated as "nigger." The increase in self-evaluation and puzzled, troubled exposure of difficulties in the field has not been accompanied, as it might have been, with greater charity or detachment. Anthropologists have continued to be highly personal, unskilled in separating their own affects from their material, polemic, given to *ad hominen* arguments and, as if they were all members of one giant extended family, personal rather than relevant nit-picking.

The question of whether anthropology should be regarded as a self-sufficient science, and as such be required to generate its own propositions and hypotheses without the use either of concepts or the findings of other human or social sciences, remains a subject of controversy (220). At the Wenner-Gren Conference in 1952, David Mandelbaum demanded that the field of culture and personality should make its own contributions and validate its own premises (280), and this same demand has been made by psychoanalysts, because they objected to the importation of physiological data into the validation of their discoveries, as in the famous study of Benedek & Rubenstein (35) in which the findings of the analysts are juxtaposed and validated by records of endocrinological occurrences. Importations from other sciences, whether social and human or biological, are treated by Arensberg (12) as analogies, and yet what he wished to identify as anthropology is "interaction theory," with a heavy dependence on physiological theory and measurement (55).

The argument has several facets. Should there be more than one science concerned with the behavior of human beings, as individuals, in groups, as carriers of culture? Are we not enormously hindered when sociologists, social psychologists, clinical psychologists, psychiatrists, or ethologists attack the same problems as anthropologists, and develop their own set of terminologies, methods, and literature? Shouldn't we all be branches of one human science, which would include human biology, evolution, history and prehistory, distinguished by our methods rather than by the areas or fields that we study? How can we compensate for the damage done by attempts to synthesize a science of human behavior that relies on secondary sources, as Freud did in *Totem and Taboo* (188, 195), as George Land is doing in *Transformation* (157), or on a smaller scale as Barkun (18) is doing in his study of millenarian movements? As long as each subdiscipline of the human sciences persists in its myopic, academically bounded contemplation of its own navel, we will have over-reliance on secondary sources, because access to all the small self-contained subdisciplines will be impossible for the synthesizer.

In addition to the objections that can be raised equally to any of the discussions within the human sciences, there are special conditions in anthropology which complicate the matter even further. One of these is the circumstance that we are dealing with vanishing materials; primitive cultures are swallowed up, remote, isolated human populations interbreed, rare languages vanish when the last two old women who speak them die, archeological remains are destroyed by road-building and dam construction. The data can never be re-collected in the light of later paradigms. Thomas Kuhn's illuminating discussion of the way paradigms are finally replaced in the natural sciences (153) simply does not apply to any branch of anthropology, and only to a limited degree to some of the other human sciences. True, a Bavelas experiment may be repeated on a later group of MIT students (30) or Bell Telephone employees, but won't the difference in period contaminate the experiment, as the later subjects have been reared in a different social milieu, eaten different food, breathed different air? The explicit demand of the natural sciences that an experiment be replicable is simply impossible in anthropology. The nearest approach we can make to it is to preserve observations in as complete a form as possible. Sound-synch film today is the closest we can come to the preservation of a complex event which will be subject to later analysis in the light of new hypotheses. With a 360° sound-synch camera we will come even closer. But replicability we cannot produce. The anthropologist must take earlier data into account; he cannot simply wipe the slate clean and begin all over again as the physical scientist can, and he must therefore continue to use the kind of tools and understandings that will enable him to work with data collected under very different conditions in the past (16).

Furthermore, anthropology shares with other field sciences, with ethology and geology, an extraordinary degree of interpenetration between particular sets of data and theory. Anthropology continues to have such special and peculiar characteristics which on the one hand provide the cement that holds us together as a discipline and on the other hand limit and define our work by the conditions under which it is done. In the laboratory sciences, one laboratory is very like another, better or worse equipped and endowed, but nevertheless laboratory scientists in Japan, Africa, Germany, and the Americas are linked by common methodologies, and scientists can move easily from one country to another. But in the field sciences, the actual conditions of work, bound in as they are with the geography, cultural areal style, politics, logistics, and state of equipment, are so intimately related to the discipline that while the processes of dealing with them provides a basic bond of sympathy between ethnologists a world apart in theory and national origin, they also preserve the extraordinarily idiosyncratic, apprentice style of the discipline.

But this close relationship between the nature of a particular area and the temperament, capabilities, and theoretical orientation of individual field workers has certain other consequences. As anthropology has expanded, the literature has proliferated to such an extent that it is almost impossible to keep up with publications in one's own area, so that an area of specialization shrinks accordingly from the Pacific, to Melanesia, to the Solomon Islands, for example; from New Guinea, to the Highlands, to a part of the Highlands; from Mesoamerica to Mexico; from South America to jungle dwellers only. We do not, on the whole, relinquish our close relationship to our own field materials, and most theoretical work that matters is tightly bound, although sometimes rather remotely, as in the case of Lévi-Strauss, or Radcliffe-Brown in his later years, to the individual's own field work. But the sense of scope which it was possible for Boas, or Kroeber, or Lowie, or Haddon to have is becoming lost under the torrent of publications, many of them unpredictably trivial or unidentifiably magnificent. It is not, I believe, the kind of loss that we anticipated a generation ago, when the 1952 Wenner-Gren conference (150) was designed to hold together the diverging classical fields of archeology, linguistics, ethnology, and physical anthropology. Instead of a split into these larger fields, as we feared, there has been a kind of fragmentation, by areas, by schools, by instruments used, by approaches preferred, by style of work, into subfields which are as complex as whole cultures seen in their complete ecological settings. This fragmentation may perhaps reflect the greater sense of holistic imperatives just as much as it does the sort of narrowing of approaches that one finds in biology—the development of embryology, the merging effect found in biochemistry, or the development of experimental ethology.

There has been an increased but still rather limited response to general systems theory, as variously reflected in the work of Bateson (24), Vayda (293), Rappaport (243), Adams (3), and an interest in the use of computers, programming, matrices, etc (105). But the interaction between general systems theory [as represented, for example, by the theoretical work of Von Bertalanffy (33, 39)] has been compromised, partly by the state of field data, extraordinarily incomparable as it inevitably is, as well as historical anthropological methods of dealing with wholes. General systems theory has taken its impetus from the excitement of discovering larger and larger contexts (163), on the one hand, and a kind of microprobing into fine detail within a system, on the other (41, 258). Both of these activities are intrinsic to anthropology to the extent that field work in living societies has been the basic disciplinary method. It is no revelation to any field-experienced anthropologist that everything is related to everything else, or that whether the entire sociocultural setting can be studied in detail or not, it has to be known in general outline. General systems theory, in a sense, is no news at all, as Von Foerster found out when he attempted to organize a conference of general systems people and anthropologists (162). In a sense, the situation is comparable to that found by the Committee for the Study of Mankind, in which a committee that included Robert Redfield tried to get each discipline to consider its relationship to the concept of Mankind. Anthropologists replied, "we are related already," and so they were. Something similar may be said of attempts to date in mathematical anthropology (107, 149). The kind of information that a computer program can finally provide, on a level of a particular culture, is simply a reflection of how detailed field work has been done, and to the careful field worker, on kinship, for example, it provides no illumination. This is, however, in strong contrast to the uses to which computer programming can be put, as in the work of Alan Lomax (174, 175), where it has been possible to map world styles in song and dance, combining Murdock's technologically defined areas (219) with analysis of films, records, and tapes made by many field workers in many parts of the world.

Without the kind of unification that might be provided by a recourse to some sort of agreed-upon complex mathematical analysis, which would include information retrieval, prediction, and genuine standardization of data collection, the discipline fragments in the way in which Bali, for example, has been studied. Anthropologists have taken samples of one village in detail and studied trance in selected spots. Orchestral music has been recorded and defined for the whole island. Temple ceremonial has been studied at different levels of complexity, and there have been restudies of specific spots. The construction of one ceremonial object found in many ceremonies has been detailed, photographic records of various aspects of the culture have been made, along with cinematographic studies of artistic behavior, etc. Such methods of unsystematically interrelated probes, some taxonomic, some following older categories of analysis, some searching for large chunks of unanalyzed materials, relate each field worker to a different network of interdisciplinary fellow scholars and scientists. One has only to mention the names of Bateson & Mead (25, 26), Kunst (154), Jane Belo (33, 34), Colin McPhee (179), H. J. Franken (86), C. Holt (142), C. and H. Geertz (96, 100, 103), to illustrate this.

So we have had such incomparable, cross-cutting developments as the Society for the Study of Oceania and the Society for Visual Anthropology, Urban Anthropology, and Structural Anthropology. There are those who have developed *festschrifts* for Boas, Herskovits, Radin, Bateson, and the group in applied anthropology who worked on interaction theory (12). And there are those archeologists who combine archeology and ethnology; those who combine primatology and the study of hunters and gatherers; anthropologists who are using video tape; those who are interested in the study of child development, or millenary movements, or ethnohistory, or blood types, ethnoscience, latent structures, and somatotyping. Any given anthropologist of any experience today will be found to have his interests anchored and flourishing in half a dozen fields. So, for example, the late Oscar Lewis was interested in: Mexico (163, 169, 170), Puerto Rico (171), and Cuba; restudies (168), poverty (172), urbanism; the use of tape, multiple familial interviewing, collection, transcription, translation, and integration by one field worker (169), integration of interviews collected and transcribed by many field workers (171); relationship of recorded ethnographic materials and those whose lives are recorded; photography, projective tests, child-rearing, and effects of social revolution. Journals read and articles written span such a variety of fields that it is not surprising that when I asked 80 colleagues, variously selected from many areas and lines of association, to name the five most important books of the last 5 years, only four books were mentioned more than twice. Each respondent revealed his deep involvement in one or more sets of overlapping fields. The interpretation that one informant placed on it, that nothing of very much importance has really happened in the last 5 years, is simply inaccurate. A great deal has happened, but no consensus can be reached, because of the extraordinary diversity within the subject.

This diversity is somewhat paralleled by the increasingly rapid oscillation between the search for universals and the emphasis on diversity. Kluckhohn signaled the search for universals, and the whole series of value studies, eagerly grasped at by other disciplines, represents attempts to find cross-cultural units of analysis which lose the depth of the cultures from which they come. The earlier period of *etic* research, in which etic or cross-culturally viable units that disregarded the specific cultural unities of the cultures within which they were found, has given place to a much more sophisticated use of comparisons. We contrast, for example, the earlier correlation studies from the Human Relations Area Files, their criticism by Norbeck (228), and Mead & Newton (215), and Cohen (58), the substitution of very detailed studies of clusters of cultures, Vizedom (295), and Textor's type of "more" or "less" associations (281). This trend is beautifully summed up in Levi-Strauss' Gildersleeve Lecture (167) in 1972, in which he says:

> Should we insist on sticking to the "etic"/"emic" distinction, this can only be done by reversing the acceptances currently given to those terms. It is the "etic" level, too long taken for granted by mechanistic materialism and sensualist philosophy, which we should consider as an artefact. On the contrary, the "emic" level is the one where the material operation of the senses and the more intellectual activities of the mind can meet, and altogether match with the inner nature of reality itself. Structural arrangements are not a mere product of mental operations; the sense organs also function structurally, and outside us, there are structures in atoms, molecules, cells, and organisms. . . . .When the mind processes the empirical data which it receives previously processed by the sense organs, it goes on working out structurally what at the outset was already structural. And it can only do so inasmuch as the mind, the body to which the mind belongs, and the things which body and mind perceive, are part and parcel of one and the same reality.

At the same time, anthropological interest in diversity has been given a boost by the worldwide responses to such homogenizing trends as the green revolution, which sums up both tendencies. At the same time that there is a drawing together of a huge genetic pool of human experiments in the domestication of a particular grain and the development of synthetic types appropriate for the assumed average growing conditions, there is the rediscovery of the dangers of monocrops and a return to the previous uses of diversity in the horticulture of such areas of poor soil as East Africa (46) and the ecologically varied slopes of the mountainsides of Peru or Guatemala. This renewed enthusiasm for diversity—which is after all our particular concern and heritage—was demonstrated by the symposium held at Brown in the spring of 1971 (240).

Today there is a growing movement towards ecological synthesis and systematization of world materials (8, 284). This is evidenced by the considerations of the planetary environment, the simulations of the Club of Rome (217), the search for an Index for the Quality of Life, the development of a Law of the Seas, the search for appropriate regulation and provision of cross-culturally usable soft-ware in the satellite program, the design of new towns and regions, the development of new energy sources, the invention of new life styles, the revision of education to fit a world of rapid cultural change, and the extreme divergence that exists at present between the experience of the generation in power and the young people under 30 (204).

The more that anthropologists respond to these movements, the more they are almost inevitably forced into narrower specializations, as they try to keep up with the specialized vocabulary of the area of worldwide problems with which they are dealing, the places within anthropological literature where the particular area is being tackled, and those anthropologists—who may be at the moment anywhere in the world—who are concentrating on one aspect of the whole. Who would have expected one of the most brilliant pieces of research relevant to planetary political organization to be developed by a field worker in the Congo (307) and published initially in an African newsletter, or that we would be able to tie it in easily with

Morton Fried's paper at the NYAS on tribalism (89) and Adams' discussion (3) of Central American national political structures? Would we have expected that in the field of urbanization it would be difficult for a student of social integration in the city, who follows work at the Athens Ekistical Institute with graduate work in Sydney, to be at once conversant with legislative proposals for utility corridors and attempts to think about the present status of unemployed youth; or that in the field of child development, it is equally difficult to integrate Lawrence Malcolm's studies of protein deficiencies in New Guinea (180) to the Society for Social Anthropology's gingerly approach to problems of childhood by way of studies of socialization (186)?

There is another set of these clusters which centers about the use of specific instruments, either psychological or technical, and still another of those who are interested in particular theoretical approaches. One of the great advances of recent decades has come in the whole field of semiotics, ushered in by the conference at Indiana University (260). We now have the well-developed fields of Kinesics (41), Proxemics (126–128), Choreometrics and Cantometrics (174), and Paralinguistics (260), all dependent upon a fine scale analysis based on film and tape, and more recently on video tape. These specialities overlap with the interaction studies of Arensberg (12), Bateson (23, 26), Oliver (231), and the interaction chronograph of Eliot Chapple (55), which in turn overlap with cybernetically oriented field studies like Rappaport's (243), studies in Psychiatry (24) and Primatology (71), and studies of conference techniques and reporting (203, 208).

Other groups are formed around the use of psychological instruments, a sophisticated use of Human Relations Area Files (172, 219, 224, 281), the Whitings' continued studies of judgments of selected anecdotes (156, 304), use of questionnaires on mother-child behavior (5), Rorschachs, Raven Matrices, Mosaics, etc. Each user has to have at least a working knowledge of what is being done by others using the same instruments, and these uses are likely to cross every other subdivision, geographical or subdisciplinary.

Another consequence of the proliferation of data is that each theorist became so tied in with the particular field experience within which his theoretical stance became clarified for him—but not for others—that there has developed a complementary tendency to ignore other people's work and start from scratch. It is perhaps not accidental that in *The Nature of Cultural Things* (129), Marvin Harris only uses the work of a social psychologist, Roger Barker (17), who himself insisted on starting from scratch, ignoring all previous work that had been done in the same field. So we have insights, some of them developed decades ago, which prove periodically illuminating to a new generation, such as Bateson's *Naven* (23) or Chapple's *Interaction* chronograph (12). And finally we have books about books about books, like Murphy's *The Dialectics of Social Life* (221), and those by Barnes (19) and Jarvie (147). However, one distinctive feature of the present day set of attack and counterattack is the willingness to discuss, analyze, dissect, propound, and expound the findings of Lévi-Strauss during the course of his work, where in previous periods, except for book reviews, very little of this was done until a master was dead (160).

It is not surprising that there has been such a proliferation of field manuals and

THE EXCITEMENT OF SCIENCE   387

field work reports, and quasi-autobiographical discussion of field problems, such as *In the Company of Man* (48), *Crossing Cultural Boundaries* (298), Hilger's story of an Araucanian (138), Golde's *Women in the Field* (111), Williams' *Field Methods in the Study of Culture (305)*, Freilich's *Marginal Natives* (87), and including the mammoth and outrageously delayed handbook by Naroll & Cohen (225). But the writers of most of these manuals have to rely on methods that are not very much of an improvement over those of the 1920s and 1930s. They expose a student to intimate and detailed accounts of the troubles and struggles that other anthropologists have had, just as we expose a student to exercises in a variety of languages and accounts of a variety of kinship systems, with the hope that somehow they will be able to incorporate a sense of how to do their field work with different equipment and under quite different conditions. The traditional tendency to avoid teaching concrete methods of field research has been exacerbated by the extremely rapid changes in technology of taping, filming, photographing, preserving, developing, viewing, retrieving, and preparing materials for suitable forms of publication and exposition. When to this is added the complexity of preparing for a field trip, in terms of selecting and testing equipment, and the length of time it takes to process grant applications, it is perhaps not surprising that efforts to do any systematic teaching in the use of both kinds of instruments break down. But when we add to this a lack of training, and the requirement that predoctoral students elect a narrow problem that often precludes their making the absolutely essential study of the culture first, it is not surprising that a great deal of incomplete work has come out of the areas which have been popular for predoctoral study. It is even possible that the present financial stringency may keep a certain number of graduate students at home, doing book theses and learning how to organize materials before they plunge prematurely into an area which they may only learn to dislike.

There have been curious discrepancies in the application of anthropology also. Ecology became fashionable over a decade ago, and a whole school has grown around the meticulous reporting of terrain, crop, land ownership, ethnobotanical knowledge, soil fertility, and relative shares of food allotted to families of men of different rank. One might have expected that as the environmental crisis deepened there would be contributions from the field of anthropological ecology. Equally, it might have been expected that a field so deeply concerned with the study of kinship and related problems might have contributed to the whole question of population control. In fact, there are only a handful of workers in either field.

It also might have been expected that those who had clamored most loudly for a scientific and objective approach would have eagerly availed themselves of the new exactness of recording provided by film and tape (1, 2, 14, 16, 26, 94, 189). Actually, it was not until video tape appeared that this kind of instrumentation received much approval, and this, I believe, is because video tape will permit the anthropologist to join the sociologist and social psychologist in distancing himself from his data. Someone else, several someone elses, can code the hours and hours of video tape, and the traditional close tie between observation and recording is broken. Those who demand that anthropology be objective and scientific for the most part have been uninterested in improving upon the pencil as a recorder of anecdotes, subsequently given a rank order by three trained observers. So, in spite of the very greatly increased number of anthropologists, both the funds and the personnel needed to make records of existing primitive peoples are missing, and those who emphasize the use of film are likely to be told they aren't doing real anthropology.

The post-World War II period has been characterized by two kinds of turning towards high cultures: intensive work in countries like India and Japan; and attention to subgroups within our own society, such as transvestites, drug addicts, those living in communes, and ethnic minorities. However, work on the white majority is still somewhat suspect; the demand that anthropology should be comparative seems often to be translated into the demand that there must be something strange and other worldly about the people whom the anthropologist studies. This provision was originally not only a way of fulfilling our responsibility to record vanishing cultures, but was also believed by Boas to be a way of attaining a limited degree of objectivity. Boas did not believe that objectivity was possible or even desirable within one's own culture, where the responsible anthropologist, like any responsible citizen, had to take sides on matters of social justice. He believed it was possible to learn that a member of another culture, far away physically and in technological level, might smack his lips aloud as a sign that he had eaten well, and that this was good manners, and still be critical of the manners of ill-bred persons in our midst. But we would, he used to tell his students, be a little more tolerant as we came to understand that manners were learned cultural behavior and not matters of absolute right and wrong.

However, this achievement of a scientific objectivity, and even the achievement of the ethically desirable stance of tolerance, looks very different today from the way it did 50 years ago. Those who are studied, whether they be members of other races, other ethnic groups, the poor, the oppressed, the imprisoned, feel that to use their lives to obtain a kind of objectivity is to treat them as objects, not as subjects (65, 292). And all over the world the previously dispossessed and ignored are actively demanding an identity which the rest of mankind must respect. For many anthropologists, the recognition of these new demands has coincided with new situations which they welcomed, such as the greater ability of previously nonliterate people to participate in research, to write about their own cultures (20), to become ethnologists themselves, and to engage in a mutual interchange, instead of an exchange in which one side was at least partially ignorant of the motivations of the other (234). But such recognition has not by any means been universal; in many divisions of the human sciences, human beings are spoken of as digits, as middle-aged white ethnic males, or black unemployed females; nameless, faceless, they appear as statistics, or as individuals described in terms that the reporter has taken no pains to make bearable. The protests of English-speaking "objects of study" became merged with various forms of political protest and partisanship for the oppressed (291). The possessive "my people" or "my village" appears arrogant where once it appeared affectionate and personal. In fact, the application of anthropological methods to our own society, especially when they are applied to groups other than our own, contaminates the study of other peoples, who become, not the primitive peoples whom we fully respected as representatives of whole cultures, but instead members of disadvantaged groups contending for a place in the sun (291, 292). And the anthropologist, who has developed the idea of culture, in the name of which they are pleading their case, is simultaneously attacked for having somehow been responsible for their primitive state, which they now wish to redefine or repudiate (64). When this is combined with the identification of applied anthropology in fields like technical assistance or community development, as colonial, neocolonial, or imperialistic maneuvers, the whole ethic of research and applied research becomes ambiguous (123, 124).

Another of the current ambiguities in anthropology is the question of the new feminism. Anthropology has traditionally been very receptive to the participation of women. In England, in France, in the United States, and more recently in Japan, women students have been given opportunities to do research, in spite of the objections of introducing women into parts of the world where there are many physical dangers. Yet today there is a lively movement among young women anthropologists against their own departments, against the paucity of data on the primitive societies that have been studied on matters of interest to women: pregnancy, childbirth, women's health, the menopause, etc. There is also a movement against theories of the early division of labor between men and women, which they assert neglect the contributions that prehistoric women made to the development of culture. Too often none of these claims and accusations are analyzed with the degree of cultural sophistication which anthropology should provide. Scarcity of data on women in primitive cultures is due primarily to the fact that most women anthropologists were more interested in doing the same kind of work that men did, rather than studying women and children. Early theories of the consequences of the division of labor, which assigned disproportionate roles to man the hunter, had to wait for correction until studies made on hunters and gatherers provided data on the relative contributions of each sex to subsistence, now estimated at something like 80 percent produced by women. (There is, it is true, one major discovery which had to have been made by women, and that is the discovery of the role of paternity; only women were in a position to make it.) But the revival of discredited speculations about previous matriarchies does no credit to historical perspective and threatens to cloud discussions which should proceed on a different level. University establishments do discriminate against women in all departments. We have been short of data on the female contribution to food supply, and we know far less than we would like to about many aspects of women's lives in primitive societies. "Male" speculations about matriarchies may make attractive daydreams, just as the widely spread myth of the island of women—a male nightmare fear that women could get along without them—is at present being turned into a feminine daydream by some feminist extremists. But as in the question of race, it is a great mistake to let contemporary partisan politics distort a disciplined look at the facts as we now know them.

I wish to return again to the subject of applied anthropology. Applied anthropology involves working with interdisciplinary teams and administrators and politicians. This field, so highly and promisingly developed during and immediately after World War II, has languished during the last two decades (214). A number of conditions have contributed to this decline: loss of interest in psychoanalysis—

which mediated between the field of mental health and anthropology (66, 68, 193, 267, 268); disillusionment with government where many such projects originated; easily found, well-paying jobs in academia and the realization that anthropology is becoming an academic-based discipline with rather rigid hierarchical relationships, from the prestigious universities, through a series of lower echelons, all academic, and the awareness that the ambitious young anthropologist is likely to be penalized for working outside these frozen hierarchies. Anthropology was a vocation until World War II; when it became a field that men and women entered instead with high sensitivity to their career hopes and problems and posts which would fulfill their professional ambitions, something was lost. It may be that the contrast between the relations of pre-World War II field workers and the people they studied or attempted to help in a variety of ways, and the postwar inflation of the field by those who thought of it as just one way of making a living, may also be somewhat responsible for some of the extravagant political accusations of the last decade.

Anthropology, like all the social sciences, has been subjected to intentional politicizing, by the demands of minorities, including students and women, by a questioning about the relationship of scientific work to the state of the world, the inequities of the establishment, the hope of revolution, the mysticism about the people themselves. For anthropology, the intense polarization about these issues has been intensified by a number of special problems: (a) almost all of our field work has been done on the cultures of those who are now subsumed under the term Third World; (b) almost all of our field work involves complicated relationships with some form of officialdom—foreign offices, district officers, Indian agents, or customs officers—and we are also likely to encounter movements and activities which are illegal, from methods of burying the dead, the illegal cultivation of opium, and militant nativistic cults, to politicized rebellions and conspiracies. Furthermore, we have been the discipline concerned with race, with the comparable abilities of all members of the human species, with records of past glory or past primitiveness, with problems of language—dialects, developments of national languages, new orthographies. In a world that is teeming with the rising expectations of minorities, with new nations, and revolutionized or modernized old nations, almost every anthropologist stands somewhere in a crossfire position, knowing too much about the renascence of old customs, or providing information that may be used by some agency of modernization, suppression, or militarization, etc.

The original response to anthropologists participating in national activities during World War II culminated in the code of ethics developed by the Society for Applied Anthropology in 1953 (209), with its insistence on anthropologists taking responsibility for all foreseeable effects. Since then, there have been a succession of crises. There were objections to participation in secret research which was to be guarded against by open publication as represented in the Beals report (31). This was followed by the seemingly contradictory demand that information be hidden from government agencies that might misuse it, which resulted in the Thailand investigation in 1971 (10a, 10b, 11a, 11b) and the recognition that anthropologists not only had to protect their informants and the cultures they studied, but also identifiable communities which might become targets. They are also called upon to become more active protectors of threatened minority primitive peoples around the world, such as the aboriginal peoples in the Amazon valley who are being threatened at present.

The question of the participation by a people in studies of their culture has been raised in many ways: in demands that the Indians receive a percentage of research grants, on the grounds that someone, anthropologist or society, was exploiting them for gain; the assertion that only an ethnic group was equipped to study itself; and proposals, notably by Alan Lomax, that the material collected by anthropologists —especially on music and dance and folklore—be fed back to the people themselves as an element in their cultural renewal. These demands parallel the discussions in medicine which ranged through a suggestion of giving experimental subjects the status of coinvestigator in a relationship of collegiality (123), to a recognition that the aims of the investigator and experimental subject might be so different that only organization on the part of the proposed subjects would meet the situation. The classical position of trust and cooperation between an anthropologist and his informants, no matter how disparate their education, in which both were devoted to recording a vanishing culture and assuring the safety of its artifacts, has now been replaced by a relationship in which the anthropologist must sometimes either espouse the cause of some ethnic group within a revolutionary formula, or be forced to acknowledge that there are no longer such shared values.

It has been a period of minimal detachment and capacity for cultural self-consciousness and loyalty. We could pride ourselves that no anthropologist denounced a fellow anthropologist in the various anticommunist witch hunts of the late 1930s and early 1950s. But the 1960s have involved us in a mass of denunciations and counter-denunciations, in the failure of seniors to recognize the implications of the generation gap, or in a refusal to consider any discussions whatsoever of racial differences, some of which, like the campaign against Carleton Coon's book (62) and the extreme views expressed in the symposium on Science and Race (210), do us little credit.

Intellectual ferment has taken as many and as diverse forms as the formation of clusters of co-workers and recognition of subgoals. Evolution, formerly a battle cry which assumed lines drawn up on both sides, has become a respectable central object of discussion (6, 196, 230, 264, 303), illuminated by studies of the behavior of primates in the wild (115) and attempts at teaching chimpanzees to communicate (238), studies of the brain (43, 239, 241, 300), and by the claims of the various types of structuralists for a basic brain-based grammar (56, 164) and dialectics of opposites (166). While the study of evolution has been enriched by the new kinds of archeology—which united thousands of years of selective adjustment to the same terrain, or by the combinations of studies of living hunters and gatherers and associated primates and ungulates—it has also been given extra urgency by considerations of the present technological crisis. A serious consideration of man-made crises and the need for a middle technology has revived interest in material culture, for example, and the role of museums, and attempts to understand earlier artifacts by making them. It also brings into focus the role of conscious purpose in our increasingly man-made, interdependent world (28).

Ethology has seriously entered into the theoretical considerations of anthropologists only during the last two decades (255), and comparative studies of man and other creatures, the arguments over aggression and war (91, 119, 176, 203, 264) and a reconsideration of the degree of patterning of instinctive behavior have proceeded in parallel with discussions of the structure of the brain and its products (248). Here again, political ideology has clouded the issue and the clarity of the arguments; new findings (235) call for drastic revisions in over-elaborate schemata of human development. The specter of behavior modification, of the loss of autonomy and freedom, not only haunts any discussion of biological engineering, but also hinders investigation into the functioning of the brain, the effects of psychedelic drugs, and the interpretation of insights provided by natural and laboratory experiments with animals (165). There seems little doubt that cross-disciplinary research in the wild, taking in wider and wider considerations of the total environment (78, 238), is a more promising field for anthropological cooperation with other sciences than patching together results of isolated laboratory procedures. The very circumstance that recent studies of the brain involve the whole brain (300) reinforces the traditional anthropological preference for the study of whole cultures and whole societies, with the integrative capacity of our major scientific resource, single human minds (197).

The peculiar history of anthropological field work has introduced a new dimension into the discipline, as field workers have been able to make restudies of earlier work, especially their own. The limitations of purely synchronic records of a people at a given moment in time, and subjects who had the same speed of movement and life-span as the investigator, has been mitigated by the rapidity of culture change. When people studied at 20 or 30-year intervals are changing within a world scene where everything else is changing as well, new opportunities for research have been automatically introduced, as new tools, new concepts, and new conditions enable the field worker to study quite new problems. Boas inaugurated this kind of thing when he took recording and film equipment to the Kwakiutl when he was in his sixties (192), and since then we have had a long series of restudies: by Oscar Lewis (170), Redfield (246, 247), Firth (81), Mead (199), to mention only a few. The possibility of studying fully identified groups over a long period has enormously increased the capacity of anthropology to include individual differences and continuities of personality within statements of cultural regularities. This in turn has made it possible to distinguish levels of analysis more sharply. Taking Lévi-Strauss' categories into the field may illuminate field research, but importing far more detailed studies of myth-making or myth-telling into Lévi-Strauss' work would only be disruptive.

Anthropology is entering a new era, the flesh pots are emptier, the difficulties of doing field work increase geometrically as the equipment grows more elaborate and the political situation in many parts of the world becomes more unsettled. In such meager times as these, anthropology can take several directions: an increased interest in professional careers that involve professional competence in related fields, like town planning, health, nutrition, and political organization; an intensive reexamination of existing materials (where Lévi-Strauss has erected such a challenging theoretical structure); concentration on audio-visual recordings in an attempt to obtain the new kind of records of still living cultures with film and tape (14, 16, 94, 95); a renewed dedication to the preservation of cultural diversity; and a greater involvement in an increasingly endangered planet. The problem remains of how to keep so many extraordinarily diverse and discrepant foci of interest and competence in active interrelationship. The very peculiarity of the task may be what will make it possible.

Literature Cited[1]

1. Adair, J., Worth, S. 1967. The Navajo as filmmaker: A brief report of research in the cross-cultural aspects of film communication. *Am. Anthropol.* 69: 76–78
2. Adair, P., Boyd, B. 1967. *Holy Ghost People.* 16 mm black & white film, sound (60 min). New York: McGraw-Hill
3. Adams, R. N. 1970. *Crucifixion by Power: Essays on Guatemalan National Social Structure, 1944–1966.* Austin: Univ. Texas Press
4. Adams, R. N., Preiss, J. J., Eds. 1960. *Human Organization Research.* Homewood: Dorsey
5. Ainsworth, M. D. 1967. *Infancy in Uganda.* Baltimore: Johns Hopkins Univ. Press
6. Alland, A. 1967. *Evolution and Human Behavior.* New York: Natural History Press
7. Alland, A. 1971. *Human Diversity.* New York: Columbia Univ. Press
8.*Albertson, P., Barnett, M., Eds. 1971. *Environment and Society in Transition.* NY Acad. Sci.
9.*Albisetti, C., Venturelli, A. J. 1968–1969. *Enciclopedia Bororo.* Compo Grande, Mato Grosso, Brasil: Museu Regional Dom Bosco (first volume publ. 1962)
10a. Am. Anthropol. Assoc., M. Mead, chairman, 1971. Charge to the ad hoc committee to evaluate the controversy concerning anthropological activities in relation to Thailand. *Newsletter Am. Anthropol. Assoc.* 12, No. 3
10b. Am. Anthropol. Assoc. 1971. Report of the ad hoc committee to evaluate the controversy concerning anthropological activities in relation to Thailand, Sept. 27, 1971. Part I: Anthropological Activities in Thailand; Part II: Guidelines on Future Policy
11a. Am. Antrhopol. Assoc. 1972. Council rejects Thai controversy committee's report. *Newsletter Am. Anthropol. Assoc.* 13, No.1:1,9
11b. Am. Anthropol. Assoc., M. Mead, 1972. Thailand controversy response to the board's response to the discussion. *Newsletter Am. Anthropol. Assoc.* 13, No.2:1,6
12. Arensberg, C. M. 1972. Culture as behavior: Structure and emergence. *Ann. Rev. Anthropol.* 1:1–26
13. Arensberg, C. M., Niehoff, A. H. 1964. *Introducing Social Change: A Manual for Americans Overseas.* Chicago: Aldine
14. Asch, T., Chagnon, N., Neel, J. V. 1971. *Yanomama.* 16 mm color film, sound (43 min). Center for Documentary Anthropology, Brandeis Univ.
15.*Balandier, G. 1971. *Political Anthropology.* New York: Pantheon
16. Balikci, A. 1969. *Netsilik Eskimos of the Pelly Bay Region of Canada.* 9 films, 16 mm color, sound (approx. 30 min each). New York: Universal Education and Visual Arts
17. Barker, R. G., Wright, H. F. 1971. *Midwest and Its Children.* Hamden: Shoe String Press (first publ. 1954)
18. Barkun, M. *Disaster and the Millenium.* To be published
19.*Barnes, J. A. 1971. *Three Styles—The Study of Kinship.* Berkeley: Univ. California Press
20.*Barnett, D. L., Mjama, K. 1966. *Mau Mau from Within.* New York: Monthly Review Press
21.*Barth, F., Ed. 1969. *Ethnic Groups and Boundaries.* Boston: Little, Brown

[1]The following kinds of references are included: (*a*) literature cited in text; (*b*) publications which were suggested by one or more of the author's colleagues (starred); (*c*) books the author considers especially interesting; (*d*) references to the author's own work, particularly articles which have extensive bibliographies.

22. Bateson, G. 1956. The message "this is play." In *Group Processes*, ed. B. Schaffner, 2:145–242. New York: Macy Found.
23. Bateson, G. 1958. *Naven*. Stanford Univ. Press. 2nd ed.
24. Bateson, G. 1972. *Steps to an Ecology of Mind*. San Francisco: Chandler
25. Bateson, G., Mead, M. 1962. *Balinese Character: A Photographic Analysis*. N.Y. Acad. Sci. (first publ. 1942)
26. Bateson, B., Mead, M. 1952. *Character Formation in Different Cultures* series. 6 films, 16 mm, black & white, sound. New York Univ. Film Library
   a. *A Balinese Family* (17 min)
   b. *Bathing Babies in Three Cultures* (9 min)
   c. *Childhood Rivalry in Bali and New Guinea* (20 min)
   d. *First Days in the Life of a New Guinea Baby* (19 min)
   e. *Karba's First Years* (20 min)
   f. *Trance and Dance in Bali* (20 min)
27. Bateson, M. C. 1970. *Structural continuity in poetry*. PhD thesis. Harvard Univ., Cambridge
28. Bateson, M. C. 1972. *Our Own Metaphor*. New York: Knopf
29. Bateson, M. C. 1973. Ritualization: A study in texture and texture change. In *Pragmatic Religions*, ed. I. Zaretsky, M. Leone. Princeton Univ. Press
30. Bavelas, A. 1951. Communication patterns in task-oriented groups. In *The Policy Sciences*, ed. H. D. Laswell, D. Lerner, 193–202. Stanford Univ. Press
31. Beals, R. L. 1969. *Politics of Social Research: An Inquiry into the Ethics and Responsibilities of Social Scientists*. Chicago: Aldine-Atherton
32. *Behavioral Science* 1956 to date. Mental Health Res. Inst. Univ. Michigan, Ann Arbor
33. Belo, J. 1960. *Trance in Bali*. New York: Columbia Univ. Press
34. Belo, J. 1970. *Traditional Balinese Culture*. New York: Columbia Univ. Press
35. Benedek, T., Rubenstein, B. 1939. Correlations between ovarian activity and psychodynamic processes: I. The ovulative phase; II. The menstrual phase. *Psychosom. Med.* 1:245 ff., 461 ff.
36.*Bennett, J. W. 1969. *The Northern Plainsmen*. Chicago: Aldine
37.*Berger, P., Luckmann, T. 1966. *Social Construction of Reality*. Garden City: Doubleday
38.*Berlin, B. 1970. A universalist-evolutionary approach in ethnographic semantics. In *Current Directions in Anthropology*, ed. A. Fischer. *Bull. Am. Anthropol. Assoc.* 3:3–18
39. Bertalanffy, L. von 1969. *General System Theory: Essays on Its Foundation and Development*. New York: Braziller
40.*Binford, L. R., Binford, S. R., Eds. 1968. *New Perspectives in Archeology*. Chicago: Aldine-Atherton
41. Birdwhistell, R. L. 1970. *Kinesics and Context*. Philadelphia: Univ. Pennsylvania Press
42. Bowen, E. S. 1964. *Return to Laughter*. Natural History Library 36. Garden City: Doubleday (first publ. 1954)
43. Braud, L. W., Braud, W. G. 1972. Biochemical transfer of relational responding (transposition). *Science* 176:942–44
44. Brown, D. 1971. *Bury My Heart at Wounded Knee*. New York: Holt, Rinehart & Winston
45.*Butzer, K. 1971. *Environment and Archeology*. Chicago: Aldine-Atherton
46. Campbell, J. 1971. *Agricultural development in East Africa: A problem in cultural ecology*. PhD thesis. Columbia Univ., New York
47. Carpenter, E. 1970. *They Became What They Beheld*. New York: Outerbridge & Dienstfrey
48. Casagrande, J. B., Ed. 1960. *In the Company of Man*. New York: Harper
49. Castaneda, C. 1968. *Teaching of Don Juan: Yaki Way of Knowledge*. Berkeley: Univ. California Press
50. Castaneda, C. 1971. *Separate Reality*. New York: Simon & Schuster
51. Castaneda, C. 1972. *Journey to Ixtlan: The Lessons of Don Juan*. New York: Simon & Schuster
52. Caudill, W., Lin, Tsung-Yi, Eds. 1969. *Mental Health Research in Asia and the Pacific*. Honolulu: Univ. Hawaii Press
53. Caudill, W., Weinstein, H. 1969. Maternal care and infant behavior in Japan and America. *Psychiatry* 32:12–43
54. Chagnon, N. A. 1968. *Yanomamo: The Fierce People*. New York: Holt, Rinehart & Winston
55. Chapple, E. D. 1970. *Cultural and Biological Man: Explorations in Behavioral Anthropology*. New York: Holt, Rinehart & Winston
56. Chomsky, N. 1972. *Language and Mind*, enl. ed. New York: Harcourt, Brace, Jovanovich
57.*Clastres, P. 1972. *Chronique des Indiens Guayaki*. Paris: Plon
58. Cohen, Y. A. 1964. *The Transition from Childhood to Adolescence: Cross-*

Cultural Studies of Initiation Ceremonies, Legal Systems, and Incest Taboos. Chicago: Aldine

59. Cohen, Y. A. 1968. *Man in Adaptation: The Cultural Present.* Chicago: Aldine

60. Cohen, Y. A. 1971. *Man in Adaptation: The Institutional Framework.* Chicago: Aldine

61. Cole, M., Gay, J., Glick, J. A., Sharp, D. W. 1971. *The Cultural Contexts of Learning and Thinking: An Exploration of Experimental Anthropolgy.* New York: Basic Books

62. Coon, C. S. 1962. *The Origin of Races.* New York: Knopf

63.*De Laguna, F. 1960. *The Story of a Tlingit Community: A Problem in the Relationship Between Archeological, Ethnological, and Historical Methods.* Washington, D.C.: GPO

64. Deloria, V. Jr. 1969. *Custer Died for Your Sins.* New York: Macmillan

65. Deloria, V. Jr. 1970. *We Talk, You Listen.* New York: Macmillan

66. De Reuck, A. V. S., Porter, R., Eds. 1965. *Transcultural Psychiatry: A Ciba Foundation Symposium.* London: Churchill

67. Deutsch, K. W. 1966. *Nationalism and Social Communication.* Cambridge: MIT. 2nd ed.

68. Devereux, G. 1967. *From Anxiety to Method in the Behavioral Sciences.* The Hague:Mouton

69. Devereux, G. 1968. *Reality and Dream.* New York: Doubleday (first publ. 1951.) 2nd ed.

70. Devereux, G. *Dreams in Greek Tragedy.* Oxford: Blackwell. In press

71. DeVere, I., Ed. 1965. *Primate Behavior.* New York: Holt, Rinehart & Winston

72. Dillon, W. S. 1968. *Gifts and Nations: The Obligation to Give, Receive and Repay.* New York: Humanities Press

73. Douglas, M. 1966. *Purity and Danger.* New York: Praeger

74. Douglas, M. 1970. *Natural Symbols.* New York: Pantheon

75.*Duchet, M. 1971. *Anthropologie et Histoire au Siecle des Lumières: Buffon, Voltaire, Rousseau, Helvetius, Diderot.* Paris: Maspero

76.*Dumont, L. 1970. *Homo Hierarchicus: The Caste System and Its Implications.* Univ. Chicago Press

77. Educational Encyclopedia (Hebrew) 1959–61. Thesaurus of Jewish and general education, ed. M. M. Buber, Vol. 1, 2. Jerusalem: Ministry of Education and Culture, and the Bialik Institute

78. Eisenberg, J. F., Dillon, W. S., Eds. 1971. *Man and Beast: Comparative Social Behavior.* Washington: Smithsonian Inst. Press

79.*Epstein, T. S. 1962. *Economic Development and Social Change in South India.* New York: Humanities Press

80. Erikson, E. H. 1969. *Gandhi's Truth: On the Origins of Militant Nonviolence.* New York: Norton

81. Firth, R. 1967. *Tikopia: Ritual and Belief.* London: Allen & Unwin

82. Firth, R., Ed. 1967. *Themes in Economic Anthropology.* New York: Barnes & Noble

83. Foerster, H. von, Ed. 1950–1956. *Cybernetics.* New York: Macy Found. 5 vols.

84. Fortes, M. 1969. *Kinship and the Social Order.* Chicago: Aldine

85.*Fox, R. 1968. *Kinship and Marriage: An Anthropological Perspective.* New York: Penguin

86. Franken, H. J. et al 1960. *Bali: Studies in Life, Thought and Ritual.* The Hague: van Hoeve

87. Freilich, M., Ed. 1970. *Marginal Natives: Anthropologists at Work.* New York: Harper & Row

88. Freud, S. 1960. *Totem and Ţaboo.* Transl. A. A. Brill. New York: Random House (first publ. 1918)

89. Fried, M. H. 1966. On the concepts of 'Tribe' and 'Tribal Society.' *Trans. N.Y. Acad. Sci.* 28:527–40

90. Fried, M. H. 1972. *The Study of Anthropology.* New York: Crowell

91. Fried, M., Harris, M., Murphy, R., Eds. 1968. *War: The Anthropology of Armed Conflict and Aggression.* Garden City: Natural History Press

92. Fromm, E., Maccoby, M. 1970. *Social Character in a Mexican Village.* Englewood Cliffs: Prentice Hall

93. Gans, H. J. 1962. *Urban Villagers.* New York: Free Press

94. Gardner, R. 1964. *Dead Birds.* 16 mm color film, sound (83 min). Cambridge: Peabody Museum, Harvard Univ. Distributed by Contemporary Films, New York

95. Gardner, R., Heider, K. G. 1968. *Gardens of War.* New York: Random House

96. Geertz, C. 1959. Form and variation in Balinese village structure. *Am. Anthropol.* 61:991–1012

97. Geertz, C. 1963. *Agricultural Involution: the Process of Ecological Change in Indonesia.* Berkeley: Univ. California Press

98. Geertz, C. 1964. *The Religion of Java.* Glencoe: Free Press (first publ. 1960)
99. Geertz, C. 1968. *Islam Observed: Religious Development in Morocco and Indonesia.* New Haven: Yale Univ. Press
100. Geertz, C. 1972. Deep Play: Notes on the Balinese cockfight. *Daedalus* Winter: 1–37
101. Geertz, C., Ed. 1972. *Myth, Symbol and Culture.* New York: Norton
102. Geertz, H. 1961. *Javanese Family.* Glencoe: Free Press
103. Geertz, H., Geertz, C. 1964. Teknonymy in Bali: Parenthood, age grading and genealogical amnesia. *J. Roy. Anthropol. Inst. Gt. Brit. Ireland* 94: 94–108
104.*Gellner, E. A. 1969. *Saints of the Atlas.* Univ. Chicago Press
105. General Systems Yearbook of the Society for General Systems Research 1956 to date, ed. L. von Bertalanffy, A. Rapoport. Washington: Soc. Gen. Syst. Res.
106. Gilbert, C. 1968. *Margaret Mead's New Guinea Journal.* 16 mm color film, sound (90 min). New York: Nat. Educ. Telev.
107.*Gillespie, J. V., Nesvold, B. Eds. 1970. *Macro-Quantitative Analysis.* Beverly Hills: Sage Publ.
108. Gluckman, M. 1964. *Custom and Conflict in Africa.* New York: Barnes & Noble (first publ. 1955)
109. Gluckman, M., Ed. 1964. *Closed Systems and Open Minds: The Limits of Naivety in Social Anthropology.* Chicago: Aldine
110. Gluckman, M. 1965. *Politics, Law and Ritual in Tribal Society.* Chicago: Aldine
111. Golde, P., Ed. 1970. *Women in the Field.* Chicago: Aldine
112. Goldman, I. 1970. *Ancient Polynesian Society.* Univ. Chicago Press
113. Goldschmidt, W. 1971. *Exploring the Ways of Mankind.* New York: Holt Rinehart & Winston
114. Goldschmidt, W., et al 1965. Variation and adaptability of culture. *Am. Anthropol.* 67:400–47
115. Goodall, J. V. L. 1971. *In the Shadow of Man.* Boston: Houghton Mifflin
116.*Goodenough, W. H. 1970. *Description and Comparison in Cultural Anthropology.* Chicago: Aldine
117. Gorer, G. 1965. *Death, Grief and Mourning.* Garden City: Doubleday
118. Gorer, G. 1966. *The Danger of Equality.* London: Cressett
119. Gorer, G. 1966. Man has no killer instinct. *New York Times Magazine,* Nov. 27:47ff.
120. Gorer, G. 1971. *Sex and Marriage in England Today.* London: Nelson
121. Gould, R. A. 1969. *The Yiwara: Foragers of the Australian Desert.* New York: Scribner
122. Graubard, S. R., Ed. 1965. Science and culture. *Daedalus* Winter issue
123. Graubard, S. R., Ed. 1969. Ethical aspects of experimentation with human subjects. *Daedalus* Spring issue
124. Gwaltney, J. L. 1970. *Thrice Shy: Cultural Accommodations to Blindness and Other Disasters in a Mexican Community.* New York: Columbia Univ. Press
125. Hagen, E. E. 1962. *On the Theory of Social Change.* Homewood: Dorsey
126. Hall, E. T. 1961. *The Silent Language.* New York: Fawcett (first publ. 1959)
127. Hall, E. T. 1963. A system for the notation of proxemic behavior. *Am. Anthropol.* 65:1003–26
128. Hall, E. T. 1966. *The Hidden Dimension.* Garden City: Doubleday
129. Harris, M. 1964. *The Nature of Cultural Things.* New York: Random House
130.*Harris, M. 1968. *The Rise of Anthropological Theory: A History of Theories of Culture.* New York: Crowell
131. Heider, K. G. 1970. *Dugun Dani.* Chicago: Aldine
132. Henry, J. 1963. *Culture Against Man.* New York: Random House
133. Henry, J. 1972. *Pathways to Madness.* New York: Random House
134. Henry, N. B., Ed. 1959. *Community Education.* Nat. Soc. Study Educ. 58th Yearb., part 1. Univ. Chicago Press
135. Herskovits, M. J. 1953. *Franz Boas.* New York: Scribner
136.*Heusch, L. de 1972. *Le Roi Ivre ou L'Origine de L'Etat* (Les Essais CLXXIII). Paris: Gallimard
137. Hilger, M. I. 1954. An ethnographic field method. In *Method and Perspective in Anthropology,* ed. F. Spencer. Minneapolis: Univ. Minnesota Press
138. Hilger, M. I. 1966. *Huenun Namku: an Araucanian Indian of the Andes Remembers the Past.* Norman: Univ. Oklahoma Press
139. Hill, R. B. 1972. *The Strengths of Black Families.* New York: Emerson Hall
140.*Hodgen, M. T. 1964. *Early Anthropology in the Sixteenth and Seventeenth Centuries.* Philadelphia: Univ. Pennsylvania Press

141. Hogbin, I. 1970. *The Island of Menstruating Men.* Scranton: Chandler
142. Holt, C. 1967. *Art in Indonesia.* Ithaca: Cornell Univ. Press
143. Holt, C. 1972. *Culture and Politics in Indonesia.* Cornell Univ. Press
144. Homans, G. C., Schneider, D. M., 1955. *Marriage, Authority, and Final Causes.* Glencoe: Free Press
145.*Horton, R. 1960. A definition of religion and its uses. *J. Roy. Anthropol. Inst. Gt. Brit. Ireland* 90:201–26
146. Ianni, F. A. 1972. *A Family Business: Kinship and Social Control in Organized Crime.* New York: Sage Found.
147.*Jarvie, I. C. 1964. *The Revolution in Anthropology.* New York: Humanities Press
148. Jochelson, W. 1908. *Material Culture and Social Organization of the Koryak.* Mem. Am. Mus. Natur. Hist. 10, part 2. Leiden: Brill
149.*Kay, P., Ed. 1971. *Explorations in Mathematical Anthropology.* Cambridge: MIT Press
150. Kroeber, A. L., Ed. 1953. *Anthropology Today.* Univ. Chicago Press
151. Kroeber, T. 1970. *Alfred Kroeber.* Berkeley: Univ. California Press
152.*Kronenberg, A. 1972. *Logik und Leben.* Wiesbaden: Steiner Verlag
153. Kuhn, T. S. 1968. *The Structure of Scientific Revolution.* Univ. Chicago Press
154. Kunst, J. 1949. *The Cultural Background of Indonesian Music.* Amsterdam: Indish Inst.
155. Kunz, R. M., Fehr, H., Eds. 1972. *The Challenge of Life.* Basel, Stuttgart: Birkhäuser Verlag
156. Lambert, W. M., Minturn, L. 1964. *Mothers of Six Cultures: Antecedence of Child Rearing.* New York, London: Wiley
157. Land, G. *Transformation.* New York: Random House. In press
158. Lawrence, P. 1964. *Road Belong Cargo.* New York: Humanities Press
159. Leach, E. R. 1965. *Political Systems of Highland Burma.* New York: Beacon (first publ. 1954)
160. Leach, E. R. 1970. *Claude Lévi-Strauss.* New York: Viking
161. Lee, R. B., De Vore, I., Eds. 1968. *Man the Hunter.* Chicago: Aldine
162. Leeds, A., von Foerster, H., Eds. 1965. *The Potentiality of Systems Theory for Anthropological Inquiry.* New York: Wenner-Gren Found. Anthropol. Res.
163. Lehman, F. K. 1959. *Some anthropological parameters of a civilization: The ecology and evolution of India's high culture.* PhD thesis. Columbia Univ., New York. 2 vols.
164.*Lenneberg, E. H. 1967. *The Biological Foundations of Language.* New York: Wiley
165. Lévi-Strauss, C. 1961. *Tristes Tropiques: An Anthropological Study of Primitive Societies In Brazil.* New York: Atheneum (first publ. 1961 as *A World on the Wane*)
166. Lévi-Strauss, C. 1964–1971. *Mythologiques.* 4 vols. Paris: Plon.
    Vol.1: *Le Cru et le Cuit*
    Vol.2: *Du Miel aux Cendres*
    Vol.3: *L'Origine des Manières de Table*
    Vol.4: *L'Homme Nu*
167. Lévi-Strauss, C. 1972. Structuralism and ecology. *Barnard Alumnae* Spring issue: 6–14
168. Lewis, O. 1960. *Tepoztlan: Village in Mexico.* New York: Holt, Rinehart & Winston
169. Lewis, O. 1961. *The Children of Sanchez.* New York: Random House
170. Lewis, O. 1963. *Life in a Mexican Village: Tepoztlan Restudied.* Urbana: Univ. Illinois Press (first publ. 1951)
171. Lewis, O. 1966. *La Vida.* New York: Random House
172. Lewis, O. 1966. The culture of poverty. *Sci. Am.* 215:19–25
173. Linton, A., Wagley, C. 1971. *Ralph Linton.* New York: Columbia Univ. Press
174. Lomax, A., Ed. 1968. *Folksong Style and Culture* (Symp. Vol. 88). Washington: Am. Assoc. Advan. Sci.
175. Lomax, A. 1972. The evolutionary taxonomy of culture. *Science* 177:228–39
176. Lorenz, K. 1971. *Studies in Animal and Human Behavior,* vol. 2. Cambridge: Harvard Univ. Press
177. Luszki, M. B. 1958. *Interdisciplinary Team Research: Methods and Problems.* New York: University Press
178. McLuhan, M. 1962. *The Gutenberg Galaxy.* Univ. Toronto Press
179. McPhee, C. 1966. *Music in Bali.* New Haven, London: Yale Univ. Press
180. Malcolm, L. A. 1969. Growth and development of the New Guinea child. *Papua and New Guinea J.* 6:23–32
181. Malinowski, B. 1967. *A Diary in the Strict Sense of the Term.* New York: Harcourt, Brace
182. Mandelbaum, D. G. 1970. *Society in India.* Berkeley: Univ. California Press. 2 vols.
183.*Marshack, A. 1972. *The Roots of Civilization.* New York: McGraw-Hill

184. Marshall, J., Gardner, R. 1958. *The Hunters.* 16 mm color film, sound (72 min). Cambridge: Film Study Center Peabody Mus., Harvard Univ.
185. Maybury-Lewis, D. 1968. *The Savage and the Innocent.* Boston: Beacon Press (first publ. 1965)
186. Mayer, P., Ed. 1970. *Socialization.* Scranton: Barnes & Noble
187. Mead, M. 1937. A Twi relationship system. *J. Roy. Anthropol. Inst. Gt. Brit. Ireland* 67:297–304
188. Mead, M. 1950. The comparative study of cultures and the purposive cultivation of democratic values, 1941–1949. In *Perspectives on a Troubled Decade: Science, Philosophy, and Religion, 1939–1949,* ed. L. Bryson, L. Finkelstein, R. M. McIver, 87–108. New York: Harper & Row
189. Mead, M. 1951. *Soviet Attitudes Toward Authority.* New York: McGraw-Hill
190. Mead, M. 1954. Research on primitive children. In *Manual of Child Psychology,* ed. L. Carmichael, 735–80. New York: Wiley. 2nd ed.
191. Mead, M., Ed. 1955. *Cultural Patterns and Technical Change.* Mentor Books. New York: New Am. Libr. (first publ. 1953)
192. Mead, M. 1959. *An Anthropologist at Work: Writings of Ruth Benedict.* Boston: Houghton Mifflin
193. Mead, M. 1961. Psychiatry and ethnology. In *Psychiatrie der Gegenwart: Forschung und Praxis, III:Soziale und Angewandte Psychiatrie,* ed. H. W. Gruhle et al, 452–70. Berlin: Springer
194. Mead, M. 1963. Anthropology and the camera. In *The Encyclopedia of Photography,* ed. W. D. Morgan, 166–84. New York: Greystone
195. Mead, M. 1963. *Totem and Taboo* reconsidered with respect. *Bull. Menninger Clin.* 27:185–99
196. Mead, M. 1964. *Continuities in Cultural Evolution.* New Haven: Yale Univ. Press
197. Mead, M. 1964. Vicissitudes of the study of the total communication process. In *Approaches to Semiotics,* ed. T. A. Sebeok, A. S. Hayes, M. C. Bateson, 277–87. The Hague: Mouton
198. Mead, M. 1966. *The Changing Culture of an Indian Tribe.* Cap Giant 266, New York: Putnam (first publ. 1932)
199. Mead, M. 1966. *New Lives for Old: Cultural Transformation—Manus 1928–1953,* with new preface. Apollo Editions. New York: Morrow (first publ. 1956)
200. Mead, M. 1968. Cybernetics of cybernetics. In *Purposive Systems: Processes of the First Annual Symposium of the American Society for Cybernetics,* ed. H. von Foerster et al, 1–11. New York, Washington: Spartan Books
201. Mead, M. 1968. Incest. In *International Encyclopedia of the Social Sciences,* ed. D. L. Sills, 7: 115–22. New York: Macmillan. 17 vols.
202. Mead, M. 1969. Crossing boundaries in social science communication. *Soc. Sci. Inform.* 8:7–15
203. Mead, M. 1969. From intuition to analysis in communication research. *Semiotica* 1:13–25
204. Mead, M. 1970. *Culture and Commitment: A Study of the Generation Gap.* Garden City: Natural History Press & Doubleday
205. Mead, M. 1970. *Ethnicity and Anthropology in America.* Wenner-Gren Conf. Ethnic-Identity, Burg Wartenstein Symp. 51. To be published
206. Mead, M. 1972. *Blackberry Winter: My Earlier Years.* New York: Morrow
207. Mead, M. Systems analysis and metacommunication. In *The World System,* ed. E. Laszlo. New York: Braziller. In press
208. Mead, M., Byers, P. 1968. *The Small Conference: An Innovation in Communication.* Paris, The Hague: Mouton
209. Mead, M., Chapple, E. D., Brown, G. G. 1949. Report of the committee on ethics. *Hum. Org.* 8:20–21
210. Mead, M., Dobzhansky, T., Tobach, E., Light, R. E., Eds. 1968. *Science and the Concept of Race.* New York, London: Columbia Univ. Press
211. Mead, M., Heyman, K. 1965. *Family.* New York: Macmillan
212. Mead, M., Macgregor, F. C. 1951. *Growth and Culture: A Photographic Study of Balinese Childhood.* New York: Putnam
213. Mead, M., Metraux, R., Eds. 1953. *The Study of Culture at a Distance.* Univ. Chicago Press
214. Mead, M., Metraux, R. 1965. The anthropology of human conflict. In *The Nature of Human Conflict,* ed. E. B. McNeil, 116–38. Englewood Cliffs: Prentice Hall
215. Mead, M., Newton, N. 1967. Cultural patterning of perinatal behavior. In *Childbearing—Its Social and Psychological Aspects,* ed. S. A. Richardson, A. F. Guttmacher, 142–244. Baltimore: Williams & Wilkins
216. Mead, M., Schwartz, T. 1960. The cult as a condensed social process. In *Group*

THE EXCITEMENT OF SCIENCE   399

Processes: Transactions of the Fifth Conference, October 12–15, 1958, ed. Bertram Schaffner, 85–187. New York: Macy Found.

217. Meadows, D. et al 1972. The Limits to Growth. Washington: Potomac Assoc.

218. Murdock, G. P. 1949. Social Structure. New York: Macmillan

219. Murdock, G. P. 1967. Ethnographic Atlas. Univ. Pittsburgh Press

220. Murdock, G. P. 1972. Anthropology's mythology. Proc. Roy. Anthropol. Inst. Grt. Brit. Ireland for 1971, 17–24

221. Murphy, R. F. 1971. The Dialectics of Social Life: Alarms and Excursions in Anthropological Theory. New York: Basic Books

222. Murphy, R. F. 1972. Robert Lowie. New York: Columbia Univ. Press

223.*Nader, L., Ed. 1969. Law in Culture and Society. Chicago: Aldine-Atherton.

224. Naroll, R. 1970. What we have learned from cross-cultural surveys. Am. Anthropol. 72:1227–88

225. Naroll, R., Cohen, R., Eds. 1970. A Handbook of Method in Cultural Anthropology. Garden City: Natural History Press

226. Norbeck, E. 1970. Religion and Society in Modern Japan: Continuity and Change. Houston: Rice Univ.

227. Norbeck, E. 1971. Man at play. Natur. Hist. Mag. 80:48–53

228. Norbeck, E., Walker, D. G., Cohen, M. 1962. The interpretation of Data: Puberty rites. Am. Anthropol. 64:463–85

229. Norbeck, E. et al, Eds. 1968. The Study of Personality. New York: Holt, Rinehart & Winston

230.*Nurge, E., Ed. 1970. Modern Sioux. Lincoln: Univ. Nebraska Press

231. Oliver, D. L. 1955. A Solomon Island Society. Cambridge: Harvard Univ. Press

232.*Ortiz, A. 1969. The Tewa World. Univ. Chicago Press

233.*Otterbein, K. F. 1970. The Evolution of War: A Cross-Cultural Study. New Haven: Human Relations Area Files Press

234.*Owusu, M. 1970. Uses and Abuses of Political Power. Univ. Chicago Press

235. Payne, M. M. 1973. The Leakey tradition lives on. Nat. Geogr. Mag. 143:143–44

236.*Peacock, J. L. 1968. Rites of Modernization. Univ. Chicago Press

237.*Pelto, P. J. 1970. Anthropological Research: The Structure of Inquiry. New York: Harper & Row

238. Pfeiffer, J. E. 1972. The Emergence of Man. New York: Harper & Row (first publ. 1969)

239. Pietsch, P., Schneider, C. W. 1969. Brain transplantation in salamanders: An approach to memory transfer. Brain Res. 14:707–15

240. Poggie, J., Lynch, R., Eds. Modernization: Anthropological Approaches to Contemporary Socio-Cultural Change. Westport: Greenwood. In press

241. Pribram, K. H. 1969. The neurophysiology of remembering. Sci. Am. 220:73–86

242. Radcliffe-Brown, A. R. 1930. A system of notation for relationships. Man 30:121–22

243. Rappaport, R. A. 1968. Pigs for the Ancestors. New Haven: Yale Univ. Press

244. Read, K. E. 1965. The High Valley. New York: Scribner

245. Redfield, R. 1930. Tepoztlan: A Mexican Village. Univ. Chicago Press

246. Redfield, R. 1962. A Village that Chose Progress: Chan Kom Revisited. Univ. Chicago Press (first publ. 1950)

247. Redfield, R., Villa Rojas, A. 1962. Chan Kom: A Maya Village, abr. ed. Univ. Chicago Press

248. Richardson, F. L. W., Ed. Allegience and Hostility: Man's Mammalian Heritage. To be published

249.*Rigby, P. J. 1969. Cattle and Kinship Among the Gogo. Cornell Univ. Press

250.*Rivière, P. 1969. Marriage Among the Trio. London: Clarendon

251. Rohner, R. P., Ed. 1969. The Ethnography of Franz Boas. Univ. Chicago Press

252. Romanucci, L. Violence, Morality and Conflict. Palo Alto: National Press Books. In press

253.*Rosman, A., Rubel, P. G. 1971. Feasting With Mine Enemy. New York: Columbia Univ. Press

254.*Sahlins, M. 1972. Stone Age Economics. Chicago: Aldine-Atherton

255. Schaffner, B., Ed. 1955–1960. Group Processes. New York: Macy Found. 5 vols.

256.*Schlegel, A. 1972. Male Dominance and Female Autonomy. New Haven: Human Relations Area Files Press

257. Schneider, D. M. 1968. The American Kinship. Englewood Cliffs: Prentice Hall

258. Schwartz, T. 1962. The Paliau Movement in the Admiralty Islands, 1946–1954. Anthropol. Papers Am. Mus. Natur. Hist. 49, Part 2

259. Schwartz, T. 1963. Systems of aerial integration: Some considerations based

on the Admiralty Islands of Northern Melanesia. *Anthropol. Forum* 1:26–97

260. Sebeok, T. A., Hayes, A. S., Bateson, M. C., Eds. 1964. *Approaches to Semiotics.* The Hague: Mouton

261. Segall, M. H., Campbell, D. T., Herskovits, M. J. 1966. *The Influence of Culture on Visual Perception.* Indianapolis, New York: Bobbs-Merrill

262.* Selby, H. A. 1970. Continuities and prospects in anthropological studies. In *Current Directions in Anthropology,* ed. A. Fischer. *Bull. Am. Anthropol. Assoc.* 3:35–53

263.* Silverman, M. G. 1971. *Disconcerting Issue.* Univ. Chicago Press

264.* Simons, E. L. 1972. *Primate Evolution: An Introduction to Man's Place in Nature.* New York: Macmillan

265.* Singer, M. B. 1972. *When a Great Tradition Modernizes: An Anthropological Approach to Indian Civilization.* New York: Praeger

266.* Sinha, S. 1970. *Science, Technology, and Culture: A Study of the Cultural Traditions and Institutions of India and Ceylon in Relation to Science and Technology.* New Delhi: Res. Counc. Cult. Stud., Munshiram Manoharlal

267. Soddy, K., Ed. 1961. *Identity; Mental Health and Value Systems.* London: Tavistock

268. Soddy, K., Ahrenfeldt, R. H. 1965. *Mental Health in a Changing World.* London: Tavistock

269. Spicer, E. H. 1952. *Human Problems in Technological Change: A Casebook.* New York: Sage Found.

270. Spindler, G. D., Ed. 1955. *Education and Anthropology.* Stanford Univ. Press

271. Spindler, G. D. 1970. *Being an Anthropologist: Fieldwork in Eleven Cultures.* New York: Holt, Rinehart & Winston

272. Steiner, S. 1968. *The New Indians.* New York: Harper & Row

273. Stocking, G. W. Jr. 1968. *Race, Culture and Evolution.* New York: Free Press

274.* Strathern, M. 1972. *Women in Between: Female Roles in a Male World, Mount Hagen, New Guinea.* New York: Seminar Press (Academic)

275.* Sturtevant, W. C. 1964. Studies in ethnoscience. *Am. Anthropol.* 66:99–131

276.* Sudnow, D., Ed. 1972. *Studies in Social Interaction.* New York: Free Press

277. Swidler, W. W. 1972. Some demographic factors regulating the formation of flocks and camps among the Brahui of Baluchistan. In *Perspectives on Nomadism,* ed. W. Irons, N. Dyson-Hudson. Leiden: Brill

278. Tanner, J. M., Inhelder, B., Eds. 1956–1960. *Discussions on Child Development.* London: Tavistock. 4 vols.

279. Tax, S., Ed. 1960. *Evolution After Darwin.* Univ. Chicago Press. 3 vols.

280. Tax, S. et al, Eds. 1953. *An Appraisal of Anthropology Today.* Univ. Chicago Press

281. Texter, R. B., Ed. 1967. *Cross-Cultural Summary.* New Haven: Human Relations Area Files Press

282.* Thomas, K. 1971. *Religion and the Decline of Magic.* New York: Scribner

283.* Tiger, L., Fox, R. 1971. *The Imperial Animal.* New York: Holt, Rinehart & Winston

284. Tiselius, A., Nilsson, S., Eds. 1970. *The Place of Value in a World of Facts.* Nobel Symp. 14. New York: Wiley

285.* Tuden, A., Plotnicov, L. 1970. *Social Stratification in Africa.* New York: Free Press

286. Turnbull, C. M. 1961. *The Forest People.* New York: Simon & Schuster

287. Turnbull, C. M. 1972. *The Mountain People.* New York: Simon & Schuster

288. Turner, V. W. 1967. *Forest of Symbols: Aspects of Ndembu Ritual.* Cornell Univ. Press

289. Turner, V. W. 1969. *The Ritual Process: Structure and Anti-Structure.* Chicago: Aldine

290.* Tyler, S. A. 1969. *Cognitive Anthropology.* New York: Holt, Rinehart & Winston

291. Valentine, C. A. 1968. *Culture and Poverty: Critique and Counter-Proposals.* Univ. Chicago Press

292. Valentine, C. A. 1972. *Black Studies and Anthropology: Scholarly and Political Interests in Afro-American Culture.* Reading, Mass.: Modular Publ.

293. Vayda, A. P. 1969. An ecologist in cultural anthropology. *Bucknell Rev.* March issue

294. Vincent, J. 1971. *African Elite.* New York: Columbia Univ. Press

295. Vizedom, M. B. 1963. *The concept of rites of passage in the light of data from fifteen cultures.* PhD thesis. Columbia Univ., New York

296. Vogt, E. 1961. *A model for the study of ceremonial organization in highland Chiapas.* Presented at 60th Ann. Meet. Am. Anthropol. Assoc., Philadelphia

297. Watson, J. B. 1963. A micro-evolution study in New Guinea. *J. Polynesian Soc.* 72:188–92

THE EXCITEMENT OF SCIENCE 401

298. Watson, J. B., Kimball, S. T., Eds. 1971. *Crossing Cultural Boundaries.* San Francisco: Chandler
299. Wax, M. L., Diamond, S., Gearing, F., Eds. 1971. *Anthropological Perspectives on Education.* New York, London: Basic Books
300. Westlake, P. R. 1970. The possibilities of neural holographic processes within the brain. *Kybernetik* 7:129–53
301. White, L. A. 1959. *Evolution of Culture.* New York: McGraw Hill
302. White, L. A. 1963. *The Ethnography and Ethnology of Franz Boas.* Austin: Memorial Mus. Univ. Texas
303. White, L. A. 1966. *The Science of Culture: A Study of Man and Civilization.*

New York: Grove Press (first publ. 1949)
304. Whiting, B. B., Ed. 1963. *Six Cultures: Studies of Child Rearing.* New York, London: Wiley
305. Williams, R. R. 1967. *Field Methods in the Study of Culture.* New York: Holt, Rinehart & Winston
306. Williams, T. R. 1972. *Introduction to Socialization: Human Culture Transmitted.* St. Louis: Mosby
307. Wolfe, A. W. 1963. The African mineral industry: Evolution of a supranational level of integration. *Soc. Probl.* 11:153–64
308.* Young, M. W. 1972. *Fighting with Food.* Cambridge Univ. Press

Ann. Rev. Entomol., Vol. 18

JOHN RAY: INDEFATIGABLE STUDENT OF NATURE

CLARENCE E. MICKEL

John Ray 1627–1705

Ann. Rev. Entomol., Vol. 18

JOHN RAY: INDEFATIGABLE STUDENT OF NATURE[1] 6038

CLARENCE E. MICKEL[2]

Department of Entomology, University of Minnesota
St. Paul, Minnesota

> *His Latin gave him a great reputation as not only the most eminent of contemporary naturalists, but in the eyes of Cuvier and Haller a principal founder of scientific zoology, ornithology, ichthyology and the greatest botanist in the memory of man.*
> *His works are the basis of all modern zoology.*
>
> Memorials of John Ray, Ray Society,
> London, 1846. pp. 65, 104–6

The last half of the seventeenth century is remarkable for the number of biologists who contributed such epoch-making and revolutionary ideas to biological thought that a new era in the history of biology may be said to have dated from that period. During the previous century encyclopedic works on natural history had been published by Gesner, Aldrovandi, Jonston, and Wotton, but these were all based on the works of Aristotle and, except for Gesner, included no original observations or new concepts. The seventeenth century saw the development of knowledge based on observation and experimentation, and the beginnings of the development of the scientific method. Among the naturalists who applied these principles were William Harvey (1578–1657), the discoverer of the circulation of the blood; Anton van Leeuwenhoek (1632–1723), whose portrayal of bacteria in 1683 and of spermatozoa in 1677 are triumphs of observation with the optical means at his disposal; Jan Jacob Swammerdam (1637–1680), the first to deal with the subject of insect metamorphosis in his *General History of Insects;* Robert Hooke (1635–1703), who figured the boundaries of the cell walls of cork, and gave remarkable delineations of the insect compound eye and the larvae of a gnat; Marcello Malpighi (1628–1694), who was the first to observe the capillary circulation of the blood (inferred but not seen by Harvey) and who figured and described the structure and metamorphosis of the silkworm; Francesco Redi (1626–1697), who introduced the use of the experimental

[1] Paper No. 1429, Misc. Journal Series, Minnesota Agricultural Experiment Station, St. Paul, Minnesota.

[2] The author is indebted to Canon Charles E. Raven who has given a definitive understanding and unusually detailed account of John Ray's life in *John Ray, Naturalist, His Life and Works,* Cambridge, 1942.

1

THE EXCITEMENT OF SCIENCE

8243-2601/78/1127-0405$01.00 © 1978 ARI   405

method in discrediting spontaneous generation; and John Ray (1627–1705), who, using precise methods of observation and description, developed the first fundamental principles of the taxonomy of plants and animals.

John Ray was born November 29, 1627 at the village smith in the hamlet of Black Notley, one and a half miles south of Braintree, in Essex. He was the son of Roger Ray, the blacksmith, and his wife Elizabeth. Two older children preceded John, a brother Roger, born in 1624 (who died in childhood in 1632), and a sister Elizabeth, born in 1625. The question immediately arises: how could the son of the village blacksmith in a remote rural area obtain the necessary elementary education to qualify him for entrance to Cambridge University at the age of sixteen and a half, at a time when the boundaries between social classes were extremely rigid? The answer would seem to be that Joseph Plume, the local rector at Black Notley, who lived only a few hundred yards from the smith, recognized the talents of the young boy John and encouraged his father to send him to the grammar school in Braintree. There he acquired an excellent grounding in Latin, a beautiful and very legible handwriting, a trained memory, and an orderly mind. Samuel Collins, the Vicar of Braintree and a man of some importance, was probably influential in Ray's admittance to Trinity College, Cambridge and in his appointment to a sizarship (a Cambridge University assistantship requiring menial service) in 1644. Later, when this appointment was not forthcoming, plans had to be changed and he was finally admitted to Catherine Hall.

Ray entered Cambridge at a time when the country was in the midst of a civil war, and Cambridge had been a center of military activity since the war's beginning. Cambridge was also a center of Puritanism and was in opposition to the established Church; and in Ray's time many of the faculty were opponents of science. The curricula at the time were directed exclusively to grammar, logic, and rhetoric. Mathematics, science, and even philosophy and theology in any profound sense, were wholly lacking. Natural science was not recognized as a legitimate or profitable field of study and no laboratory of any kind existed in the University. The whole objective of education was to prepare students for the clergy and to give a classical education to the nobility.

After two years at Catherine Hall, Ray transferred to Trinity College where he was awarded a sizarship. Languages were certainly the chief study during his undergraduate years at Cambridge and he became as fluent in Latin as in English, a fluency he retained to the end of his life. Later, Ray chose Latin as the language of his publications, and his mastery of the language brought him attention not only in England but on the Continent, and gave worldwide coverage to his work.

It must have been a real struggle for the young Ray with his provincialisms of speech and rural background to acquire the language and manners of the aristocracy, and to gain the poise, dignity, and modesty which enabled him to hold an honored place in a society in which class distinctions were still almost feudally rigid.

He received his BA degree in 1647–1648 and his MA degree in 1651. An appointment to a minor fellowship in 1649 gave him the freedom to study natural history, although his time was limited by college duties. He was appointed Greek lecturer in 1651 and again in 1656, mathematics lecturer in 1653, and humanities lecturer in 1655. By this time he was already observing and collecting plants in the vicinity of Cambridge, and we find that in the autumn of 1654 he was present at the dissection of four birds: a bittern, a curlew, a yardhelp, and a duck-like bird.

It was early in this period that Francis Willughby, a young man whose name has always been most intimately connected with Ray, came to Cambridge. He was the son of Sir Francis Willughby, of Middleton Hall in Warwickshire, and Lady Cassandra, daughter of Thomas Ridgeway, first Earl of Londonderry. He was a man of delicate physique, ardent and restless temperament, and great ability and industry. Not a great deal is known regarding him, except what is written in Ray's books. His generosity in financing the travels of Ray, in giving him a home during his life, and an annuity in his will, gave Ray the courage and the means to devote himself to science. Early in their association it was agreed between the two of them that Ray would assume responsibility for the plants, while Willughby would undertake the birds, beasts, fishes, and insects.

After 1657 Ray was free of the lectureships, but served in a number of college offices: Praelector Primarius, 1657–1658, Junior Dean, 1658–1659, and Steward, in 1660 and 1661.

At the time Ray was elected a fellow in 1649, it was required in Trinity College that all fellows must be in priest's orders seven years after completing the MA. This would mean that Ray must be ordained not later than 1658. By 1660 he had not satisfied this requirement. He had a deep dislike for ritual and much sympathy with Puritanism. When he was urged to continue his place in Trinity College in 1660 he was forced to make a choice and finally decided that the invitation of the College must be accepted and that his life's work was with the University. He was therefore ordained in London on December 23, 1660; but his security with the University was short lived. During the reign of Charles II, the reactionary Parliament passed the Act of Uniformity in 1662, a law drawn up with the view of suppressing liberty of conscience. The Act decreed that all clergymen and all who held any office in the Universities must take an oath in accordance with the act, or forfeit their office. To Ray an oath was an oath and to accept the Act was to subscribe to a lie. As a man of honor, a teacher concerned with truth, he could not sacrifice truth to expediency. He could not retain his fellowship or hold any position in the Church on those terms. He refused to take the oath and resigned his offices in the College. August 24, 1662 found Ray free and without employment, a teacher with no pupils, a cleric without an assignment, prohibited by his profession from secular employment, and prohibited by law from his profession. The blacksmith's son returned to his village, a seeming failure by his own choice.

It is not clear at what point in Ray's life he became interested in the observation of plants. His interest may have been aroused while still in grammar school at Braintree, or more likely while he was an undergraduate at Cambridge. At any rate by 1654 his explorations of the plants in the fens and woods in the vicinity of Cambridge was familiar to his friends. He made an extended trip alone on horseback in 1658, visiting Derbyshire and North Wales to observe plants and items of zoological interest. By 1660 he had written and published his first important botanical work: *Catalogus Plantarum circa Cantabrigiam nascentium.* It is this publication that gives us our first view of Ray as an entomologist. The work was not only an important contribution to botany but also includes the first published descriptions of the life histories of several insects, and demonstrates that Ray's interest in observing insects and recording in detail what he saw at first hand, antedates the agreement between Willughby and himself, whereby the former was to be responsible for the study and publication with respect to insects.

These first contributions to entomology include brief notes on the caterpillar, chrysalis, and imago of the small tortoiseshell *Vanessa urticae,* and an account of the irritability of the caterpillar of the privet hawk *Sphinx ligustri* when touched, its change to a pupa with a sheath for the projecting proboscis, and finally the moth that emerges from it. He calls attention to the mistake that Mouffet and others make in stating that the head of the caterpillar is changed into the tail of the butterfly: "in every caterpillar I have seen the exact opposite is the fact." There is an account of Mouffet's hazel-caterpillar cocoon (constructed of hair), of its pupa, and the moth, probably the pale tussock *Dasychira puribunda,* which he compares with the silkworm moth. There is a note on the luminous myriapod, probably *Geophilus electricus,* called Julus from the resemblance to hazel catkins (juli) which he says is a Scolopendra.

The subject of parasites of caterpillars occupies one of his longest notes. He begins by stating that the caterpillars of the large white *Pieris brassicae,* reared on cabbage, readily eat rape, though refusing other plants. He then states,

> I shut up ten or so of these in a wooden box at the end of August 1658. They fed for a few days, and fixed themselves to the sides or lid of the box. Seven of them proved to be viviparous or vermiparous: from their backs and sides very many, from thirty to sixty apiece, wormlike animalcules broke out; they were white, glabrous, footless, and under the microscope transparent. As soon as they were born they began to spin silken cocoons, finished them in a couple of hours, and in early October came out as flies, black all over with reddish legs and long antennae, and about the size of a small ant. The three or four caterpillars which did not produce maggots after a long interval changed into angular and humped chrysalids which came out in mid-April as white butterflies.

He finally solved this problem in his *Historia Insectorum,* 1710, (p. 114):

> Whence these maggots arise is a great problem. I think that the ichneumon

wasps prick these caterpillars with the hollow tube of their ovipositor and insert eggs into their bodies: the maggots are hatched by the warmth of them, and feed there until full grown; then they gnaw through the skin, come out, and spin their cocoons.

Note that in the above account he describes the appearance of the animalcules under the microscope as transparent. Ray called any lens a microscope, but never possessed an achromatic lens since these were not developed until more than a century after his death.

A note by Ray concerning the rose gall bedeguar or spangiola *Rhodites rosae* refutes the theory, mentioned by Spigel, Mouffet, and Aristotle, that beetles are born of the small maggots in the galls. Ray says: "This is incorrect. I saved some of these galls; and the maggots hidden in them through the winter came out in the following May as flies."

And finally he records his observations of the spittle-bugs:

> Of the cause and origin of this foam the common crowd of philosophers is under a gross illusion; some call it star's spit and believe it rains down from heaven, like Manna and honey dew—which are found in abundance on very few plants; others that it is an exhalation from the earth; others an exudation of the plant itself. I have discovered that it is vomited from the mouth of an insect, a tiny creature that always lurks in the middle of the spittle. For if you wipe off the froth of foam, you will see for yourself the same foam very soon again poured out from the creature's mouth in such abundance that it will soon enwrap and conceal itself in it: so it can lie hid there in safety from all harm from frogs, small birds and other enemies while it is still feeble and cannot save itself by jumping or flight. This insect is almost like a louse in shape but shorter for the size of its body, yellow green in color with large protruding eyes: its hind legs are shaped for jumping, whence we are ready to agree with the learned men who have asserted by experiment that it turns into a locust such as we call Grasshopper. We, to confess the truth, have not yet investigated its origin or its final state.

This problem was also finally solved and the answer published in *Historia Insectorum* (p. 67):

> The hind legs are scarcely longer than the rest: rudiments of wings appear on the shoulders: a long proboscis is bent back on the belly between the feet: they often change their skin which is found next the spittle. While they are still enwrapped in the spittle, they crawl; after leaving it, they always move by jumping, the hind feet being now stronger and longer. They resemble cicales more than locusts: they fly higher than locusts: their wings conceal their whole body.

The above observations on the life histories, metamorphosis, parasitism, and insect secretions must be regarded as the first scientific observations and precisely written descriptions of insects and their ways and means of living. As Charles E. Raven has said so well:

> Ray stands out for the extent of his learning and soundness of his judgment, for his independence in rejecting traditional beliefs and demanding

proof, for the thoroughness of his observations and the accuracy of his state-ments. In Entomology the field was unexplored: Thomas Mouffet had left a large book based upon Edward Wotton, Conrad Gesner and his friend Thomas Penny, but is a very incomplete, ill arranged and unreliable compilation. Ray, whose concern with insects has been often ascribed to the influence of Willughby, reveals in these early notes not only a knowledge of all that had been written but the acute insight, the power of exact objective description, and the indefinable flair for a correct interpretation which are the marks of a great scientist. In a pre-scientific age, when speculation was limited by no experience of what could or could not happen, Ray's power of discarding legendary lore and fanciful explanations and of fastening upon the right line of investigation establishes for him a strong claim to be one of the fathers of modern science.

After the publication of the Cambridge Catalogue, Ray made another exploratory journey, this time with Willughby as companion. Their itinerary took them to North England and the Isle of Man. This was apparently the beginning of the partnership between Ray and Willughby that resulted in the expansion of Ray's interest to zoology as well as botany and ultimately led to his editing and publishing Willughby's notes on ornithology and ichthyology and his own works on mammals and reptiles and insects.

A third exploratory trip was made to Scotland from July 26 to September 7, 1661 with a former pupil, Philip Skippon. The record of this trip includes not only the plants, but also a list of fish seen, and a description of the soland geese, the scout, the cattiwake, the scart, and the turtle dove. A comment on the mores of the time gives us a glimpse of the general superstition that existed, "At the time we were in Scotland divers women were burnt for witches, to the number of about one hundred and twenty."

A fourth journey, around Wales, was made May 8 to July 24, 1662. His companions were Willughby and Skippon. A previous journey had involved visits to famous churches and castles, but Willughby's enthusiasm for field observations of animals resulted in most of the time being spent in exploring little known islands and bits of coast where sea birds and rare plants were the prime objects of interest. The two friends were in agreement that for the naturalist, the study of museum specimens and the literature must be subordinated to personal observation and knowledge of the organism in its native habitat. Ray held to this principle until his death. At the end of the journey Ray returned to Cambridge, refused to take the oath required by the Act of Uniformity, resigned his fellowship at Cambridge, and returned to Black Notley.

The winter of 1662–1663 he spent at Saxmundham in Suffolk as a tutor in the home of Mr. Thomas Bacon. This was only a stop gap, since Willughby had proposed an extended journey on the Continent that began on April 18, 1663 when Ray, Willughby, Skippon, and Nathaniel Bacon (probably not the son of Thomas Bacon) met at Dover and proceeded to Calais. The record of this three year sojourn in Europe was published in 1673 with the title "Observations in the Low Countries." They visited Belgium, the

Netherlands, Germany, Switzerland, and Italy. By March 1664 they had visited many points in Italy and had arrived at Naples. Here the party divided, Willughby and Bacon staying in Naples, Ray and Skippon going on. Willughby spent the summer of 1664 in Spain and then returned to England. Ray and Skippon remained in Italy until March 1665 when they went to Switzerland and France and did not return to England until the spring of 1666. Ray devoted a great deal of time during these two years studying and collecting plants, birds, fishes, and stones. The winter of 1664 he attended lectures on anatomy given at the University of Padua by Pietro Marchetti. He visited the Museum of Aldrovandi and attempted to see Malpighi, but he was absent from the University of Messina. He also made observations on a comet and visited natural historians in the universities of the Low Countries, Germany, Italy, and France. All this was to serve him well in the years ahead, for during the last years of his life after 1679, he rarely left Black Notley.

In the years following the European tour, Ray devoted much of his time with Willughby to arranging the material they had gathered. In 1667 he and Willughby made a journey to the west of England. He was in London on November 7 when he was elected to the Royal Society. The years 1668 and 1669 were devoted to the writing of the *Catalogus Plantarum Angliae*, published in 1670 and dedicated to Francis Willughby. It was at this time that Ray changed the spelling of his name. While at Cambridge in Trinity College and thereafter, he spelled it "Wray," but in a letter to Martin Lister August 22, 1670 he announced that he was returning to the spelling "Ray," used by his forefathers.

His last journey for collecting materials began in July 1671 when he traveled to the north for the purpose of observing plants in the field. In late May 1672 he was planning a further journey to the west country when his plans were abruptly ended by the death of Willughby on July 3 in his 37th year. This was an even greater blow to Ray than the loss of his fellowship at Cambridge ten years before. Willughby was a man of high ideals, sensitive, alert, and energetic. By birth and training he belonged to the world of politics and society, and had an interest in government and education, all outside of Ray's concern. His chief interest in natural history was the insects, but he, together with Ray, did much field work and dissection of birds, mammals, and fishes. Ray left the insects largely to Willughby, and shared with him the others, dissecting a bird and mammal at Padua, a porpoise at Chester, and fishes at various times and places.

Ray had been a member of the Willughby household since their return from the Continental tour. His position there had been easy. It was the custom of the day for every large land owner and most of the smaller squires to have a resident chaplain. Willughby was conforming to custom when he invited Ray into his home. After Willughby's death, his mother, Lady Cassandra, apparently saved Ray from considerable unpleasantness, but Mrs. Willughby seemingly had no scruples about putting him in his place. Willughby's will made Ray one of five executors, and in addition he was given the responsibility for the education of Willughby's two sons, and an annuity of 60 pounds a year. The responsibility of publishing Willughby's work was an even greater tie. "I am," he wrote in a letter from Middleton, "like now to set up my staff here, at least so long as my old lady (Lady Cassandra, Willughby's mother) lives."

Following Willughby's death Ray assumed the responsibilities laid upon him by Willughby's will: the duties of an executor and the education of the children. Within a year he married, Margaret Oakley, with Mrs. Willughby's permission, a member of the Willughby household, on June 5, 1673. His life in the Willughby household was less than pleasant. The widow Willughby had no interest in the scientific interests of her husband nor his scientific friends. When Lady Cassandra died on July 25, 1675 a change was made at once; Mrs. Willughby removed the children from Ray's responsibility and made it impossible for Ray and his wife to continue living at Middleton. After a brief stay at two different locations, and following the death of his mother on March 15, 1679, the Rays returned to Black Notley where they remained until Ray's death.

After Willughby's death Ray set himself the goal of producing books on birds, fishes, mammals and reptiles, and insects, as well as botany. As early as 1674 he was already working on the birds. Willughby's notes formed the basis of this work, but Ray had shared with Willughby in the field work and enlarged on Willughby's notes by including his own personal observations and the writings of other students of birds, and he proposed a classification of outstanding merit. It was customary until 1844 to regard Ray as merely the editor of the book. It is now known that Ray spent an immense amount of time organizing the undigested notes of Willughby, adding knowledge of his own, substituting his own classification for Willughby's, and skillfully editing the whole. It is characteristic of Ray's modesty and his loyalty to his friend that he presented it for publication to the Royal Society with Willughby's name as author, and that during his lifetime he spoke of it as Willughby's *Ornithologia*. It was published early in 1676.

The move from Middleton to Black Notley enabled him to devote twenty years to study and writing. The *Methodus Plantarum* was published in 1682 and Willughby's *Historia Piscium* in 1686. The latter was begun immediately upon the publication of the *Ornithologia*. It may be that during the last year at Middleton he organized Willughby's notes and wrote some of the manuscript. After the break with Mrs. Willughby it was difficult for him to get Willughby's papers, and the evidence at present indicates that almost all of his book is Ray's work. His loyalty to Willughby and the memory of his stimulating and energetic cooperation compelled him to publish the *Historia Piscium* as a posthumous work. Ray was evidently a man who deeply appreciated any kindness or consideration and was perhaps overzealous in showing his gratitude and acknowledging obligation. Willughby had given him a home when Cambridge and the Church had rejected him and Willughby had pro-

vided in his will the means to carry on his research and to devote the rest
of his days to science. He felt bound to honor his friend and co-worker with
the authorship of the *Ornithologia* and the *Historia Piscium*. He must have
felt that this was the least he could do after their agreement that Ray would
be responsible for plants and Willughby for the birds, mammals, fish, and
insects. The book has its defects, due to the condition of knowledge at the
time, but where it is based upon first-hand knowledge it is regarded as remarkably accurate. The descriptions are a tremendous improvement over
those of his predecessors; this was acknowledged by Cuvier when he wrote
"they are often more accurate and intelligible than those of Linnaeus." It
may be said that, considering Ray's lack of time and material, the ignorance
of anatomy, habits, and classification of his day, the difficulty of procuring
material, and the pressure of his work on plants, the *Historia Piscium* is an
outstanding achievement.

Four daughters were born to the Rays: the twins, Margaret and Mary,
August 12, 1684; Catherine, April 3, 1687; and Jane, February 10, 1689. In
later years these children were active cooperators in the gathering and rearing
of insects that provided much of the information for the *Historia Insectorum*.

The last great works on plants were the *Methodus Plantarum* in 1682, the
three volumes of *Historia Plantarum* in 1686, 1688, and 1704, and the *Synopsis Britannicarum* in 1690.

Having completed his great works on plants, and having honored his responsibilities to Willughby with the *Ornithologia* and the *Historia Piscium*,
Ray now turned to the mammals and reptiles. The *Synopsis Animalium
Quadrupedum et Serpentini Generis* was started in 1690 and published in
1693. Very little is known of the circumstances which encouraged Ray to
undertake this work, but since the book reveals his extensive knowledge of
the literature of physiology and comparative anatomy, and a long experience
of observation and dissection of the mammals, including the dissection of the
human body, he must have long contemplated producing a work in this field
of zoology. The book includes a classification that marks a great advance; he
discusses spontaneous generation (which he emphatically rejects) and
makes extensive use of comparative anatomy, which prompted Cuvier to
claim that he was the first zoologist to make use of this field of science. He
discusses the problem of defining "what is an animal" and deals with the
question of whether individual animals were each created in the beginning or
are constantly produced by fresh generation, and are animalcules which are
increased and perfected by generation situated in the ovum of the female or
the sperm of the male. The two latter problems are discussed at length but he
readily admits, "To be frank, many doubtful points can be mooted which I
confess myself unable to solve, not because they do not have definite natural
causes but because I am ignorant of them." His discussion of classification
rejects the distinction of animals into viviparous and oviparous because it is
inadequate, stating that all animals come from eggs and describing the differences between the eggs of the two groups.

In the early days of his fellowship at Cambridge one of his duties was to address the chapel at stated intervals. These addresses or sermons were known as "commonplaces." In these he began to develop the principles and philosophies which finally resulted in the published work *The Wisdom of God Manifested in the Works of Creation*. This was probably the most popular and influential of Ray's works. The first edition of 500 copies was a small volume of 249 pages. It was reprinted in a second edition of 382 pages in 1692, a third edition of 414 pages in 1701, and a fourth edition of 464 pages in 1704. A French edition came out in 1714 and a German edition in 1717. Altogether 16 editions appeared before the end of the eighteenth century and two more appeared during the nineteenth.

Ray had discussed the structure and functions of the living organism in the introductory essays of *Historia Plantarum* and *Synopsis Quadrupedum*, going from comparative anatomy to physiology and behavior, but he had not gathered together and organized the results of his studies, nor had he attempted to explain life in the terms of science and philosophy. *The Wisdom of God* abounds with references to and discussions of problems ranging from the nature of atoms or the influence of the moon upon tides to those of the shape of bee's cells, the movements of birds and fishes, the structure of the eye, and the growth of the fetus in the womb. With each edition he enlarged the book with additional observations and inductions.

He mentions the importance of air for living things: that fishes die in a sealed vessel, that insects breathe through many orifices on each side of their bodies (spiracles) so that if these are stopped with oil or honey the insect dies, and that plants have a kind of respiration, as discovered by Malpighi.

Of entomological interest is his full account of the comb and cells of the honey bee and of the ant, "which as all naturalists agree hoards up grains of corn for the winter and is reported by some to bite off the germen of them lest they sprout: which I look upon as mere fiction; neither should I credit the former relation, were it not the authority of scripture, because I could never observe any such storing up by our country ants."

It is impossible in this brief treatment to list all the scientific statements and arguments that Ray brings up in his book when dealing with animal and plant physiology, the usefulness of animals and plants to man, the character and properties of the earth, planets, sun and stars, his discussion of the human body, and numerous other topics. We would, however, call attention to a statement by Ray concerning the principles that should guide one in scientific investigations:

> It may be (for aught I know and as some divines have thought) part of our business and employment in eternity is to contemplate the Works of God. . . . I am sure it is part of the business of a Sabbath Day. . . . Let it not suffice to be book learned, to read what others have written and to take upon trust more falsehood than truth, but let ourselves examine things as we have opportunity, and converse with Nature as well as with books. . . . Let us not think that the bounds of science are fixed like Hercules his pillars and in-

scribed *Ne plus ultra*. The treasures of nature are inexhaustible. . . . I know that a new study at first seems very vast, intricate and difficult; but after a little resolution and progress, after a man becomes a little acquainted, as I may say so, with it, his understanding is wonderfully cleared up and enlarged, the difficulties vanish, and the thing grows easy and familiar . . . . Some reproach methinks it is to learned men that there should be so many animals still in the world whose outward shape is not yet taken notice of and described, much less their way of generation, food, uses, manners, observed. If man ought to reflect upon his Creator the glory of all his works, then ought he to take notice of them all and not to think anything unworthy of his cognizance.

It has been said that the *Wisdom of God* more than any other single book initiated the true adventure of modern science and is the ancestor of the *Origin of Species* and *L'Evolution Creatrice*.

One last project to which Ray had dedicated himself on leaving Cambridge and one which his friend Dr. Tancred Robinson was continually urging him to complete was the history of insects. In the beginning he had left the insects to Willughby, but nevertheless he made notes of observations of his own, and so in 1690 he turned to the work on entomology. He was now 63 years old, in poor health, made miserable by constant pain and sleeplessness due to ulcerations of his legs. While he collected and noted a considerable amount of material himself, he had no access to collections or manuscripts after he left the Willughby home and settled in Black Notley. He was not lent Willughby's notes on insects until the last year of his life, but he incorporated them into his manuscript in full. The only literature available on the subject was Jan Jacob Swammerdam's *General History of Insects* (1669), which Ray regarded as the best book ever written on the subject; the work of Johan Goedart, whose notes and plates had been edited and published by Ray's friend, Martin Lister; and Mouffet's *Insectorum Theatrum*. We might say that he had to start from scratch. Here then is Ray, the entomologist, who in the last fourteen years of his life, with the help of four little daughters, reared and described the life cycle of nearly three hundred local Lepidoptera.

We find him during these years collecting specimens, rearing countless insects, and recording his observations. All of this was accomplished in his own home, where he encouraged his children to participate in gathering and rearing material, and with the help of Thomas Simpson, who apparently was in Ray's employ. The exploits of the children are frequently recorded in the published work, and several of the reared insects are named for them, "Katherine's Eruca, Jane's Chickweed Caterpillar, etc." His wife is connected with a notable observation describing a moth (*Pachys betularia*). He writes: "it emerged out of a stick-shaped geometer caterpillar; it was a female and came out from its chrysalis shut up in my cage; the windows were open in the room or closet where it was kept, and two male moths were caught by my wife who by a lucky chance went into the room in the night; they were attracted, as it seems to me, by the scent of the female and came from the outside." This was first known record of assembling, and Ray hit on the correct explanation of it the first time. Insect scents with the power of attraction have now been chemically identified and termed "pheromones," a popular topic for entomological symposia.

Ray must have had an unusual ability to arouse the enthusiasm of his family, his gardner Simpson, and also the neighbors for his project and he managed to maintain their interest and cooperation in contributing material over the years. It is to be regretted that Ray himself, more or less bedridden, had to depend on others to such a great extent. We miss the talent for observation in the field that characterized his active, earlier days. Two examples will illustrate this point, both quoted from his published work.

On the 22nd June, 1667 I saw a Wasp, one of the largest of his tribe—I do not now recollect the species—dragging a green caterpillar three times longer than itself . . . Before my very eyes it carried it almost the full length of a measuring rod, that is some fifteen and one-half feet; and then deposited it at the mouth of a burrow which it had previously dug for itself. Then it removed a ball of earth with which it had sealed up the entrance; went down itself into the hole; after a brief stay there came up again; and seizing the caterpillar which it had left near the opening carried it down with it into the burrow. Soon, leaving it behind there, it returned alone; gathered pellets of earth and rolled them one by one into the burrow; and at intervals scratching with its fore feet, as rabbits and dogs do, flung dust backwards into the hole. It kept repeating the same operation with dust and pellets alternately until the burrow was completely covered up; sometimes it descended in order, as it seemed to me, to press down and solidify the soil; once and again it flew off to a fir tree nearby perhaps to look for resin to stick the soil together and consolidate the work. When the opening was filled and leveled with the surface of the ground so that the approach to it was no longer visible, it picked up two pine-needles lying near and laid them by the burrow's mouth, to mark, as is probable, the exact spot. Who would not wonder in amazement at this? Who would ascribe work of this kind to a mere machine.

This was originally published in Philosophical Transactions [6(76):2280–81] October 26, 1671, two hundred years before Fabre!

Another example from about the same period, signed F. W. but observed by both Willughby and Ray and written by Ray, concerns the leaf-cutter bee (*Megachile willughbiella*).

These bees fashion sections of rose leaves carefully rolled up and stuck together into cylindrical chambers: they might be called "Cartrages" in English from the exact resemblance they bear to the paper wrappings filled with gunpowder used for the larger guns; in the trunks of willows soft and decaying they dig cylindrical burrows exactly the size of these capsules; these burrows run up or down following the grain or fibre of the wood, never across it; the capsule is placed at the bottom of the burrow, or at the top if the burrow runs upward, and exactly fills the space, the round end of the capsule touching the bottom of the burrow; at the concave end of the capsule the round end of another capsule is tightly fixed; and so, five, six, or seven capsules, one upon another, are found in a single burrow.

By 1690 when he began the preparation of his work on the insects, personal field observations were no longer possible because of his poor health. Thus he turned to description and classification of the material he had collected over the years and the specimens that his friends and correspondents sent him. He was fully aware of the difficulty that he faced with respect to classification. The only proposal for insect classification existing when he started his work was that of Swammerdam, who had studied and described metamorphosis and suggested it as a basis for classification. Ray was therefore breaking new ground; he was aware of Swammerdam's work, but even so he writes of the uncertainties and lack of knowledge that made the problem so difficult. He had gained considerable experience in his work on the *Historia Plantarum*, the *Ornithologia*, the *Historia Piscium*, and the *Synopsis Quadrupedum;* now during the last years of his life he would attempt a history of insects. That he realized the difficulties is evident from a letter to Edward Lhwyd, July 7, 1690:

Reviewing my notes concerning insects and considering the things themselves I find it a thing of infinite difficulty to draw up any tolerable epitome of the History of such as are found with us; they being almost innumerable. The two great heads or Summa genera that I would divide them into are 1. *Quae nullam subeunt mutationem.* (Those without metamorphosis). 2. *Quae metamorphosis aliquam petiuntur.* (Those undergoing some metamorphosis). The second genus I may divide either as they appear at their first hatching from the egg, and so they will be either *Polypoda, erucae;* or *Hexapoda;* or *Apoda, eulae.* Of the first sort come Butterflies; of the second Beetles, of the third Flies of all sorts, Bees, etc. Or 2, as they appear after they have undergone their last change, and so they may be subdivided into *Coleoptera,* Beetles, etc. or *Anelytra* which are either *alis farinaceis, Papiliones,* or *alis membranaceis pellucidis,* and those either *Tetraptera,* Bees, Wasps, Hornets, etc. or *Diptera,* Flies, Gnats, etc. I am yet doubtful about the process of *Locustae forficulae* and *Cimex sylv.* which though winged insects, yet I suspect undergo no metamorphosis. Howbeit they are not at this first exclusion winged, but their wings grow out afterward.—I am doubtful concerning ants whether the flying may not be the males, the creeping the females; for they are found together in the same hills. Neither am I yet full satisfied concerning the flying and creeping Glow-worms. The number of *Erucae* alone in this island is incredible, some plants having three or four sorts feeding upon them; and if we should make the *Papiliones* a distinct genus from them, as all that write the History of Insects have done, we should double the number of species. The Beetle-tribe I hold to be no less numerous than they; and the Flies perchance more. So that I know not but that the species of insects may not be equal to or exceed those of Plants.

Here is the first recognition of the enormous number of existing insects and the complications issuing therefrom in bringing order to the knowledge about them. Here, also, are the beginnings of the classification that was published in 1705 in *Methodus Insectorum* and later reprinted in the *Historia Insectorum.*

It appears that from the beginning Ray planned to include Willughby's notes on insects in writing the *Historia Insectorum,* but he had long ago sent

Willughby's papers to his son, Sir Thomas Willughby, and he now requested the loan of the notes on insects to make the book on insects more complete. At first he was unsuccessful, but they finally arrived in 1704 at about the time he finished the *Methodus Insectorum*. So in August 1704 he set for himself the task of completing the history of insects. He realized that he might not live to finish it, but he determined to leave it in such a state that an editor would not find it difficult to arrange for publication.

In the five month period after he received Willughby's notes, Ray had produced a manuscript which, while incomplete, was in such a state that it could be edited for publication. Unfortunately, this was not to be. He died on January 17, 1705. A few days before his death the collections and manuscript were sent to Samuel Dale of Braintree, who had been an assistant and cooperator in Ray's studies of insects. Dale was urged by Hans Sloan to undertake the editing, but he declined. The script was then taken by William Derham. He kept the papers until the summer of 1708, having done nothing with them, and then sent them to the Royal Society. After more than a year's delay the *Historia Insectorum* was published in 1710, with no sort of preface or explanatory note.

A critique and description of the contents of *Historia Insectorum* has been given by Charles E. Raven in his book *John Ray, Naturalist, His Life and Works*. Raven comments that "The book is a medley, an unedited collection of material, such as any author undertaking a piece of research amasses before the final drafting of his work. The *Methodus*, or scheme of classification, published as a sixteen page pamphlet in 1705, was reprinted as an introduction; but in the book itself this arrangement is very loosely followed." Sections on certain groups contain only notes, sections on other groups contain lists of species with short descriptions, and some sections contain attempts at classification. Ray's treatment of the butterflies and moths illustrates the principle which guided him in the study of insects: to consider the biology and structure of the whole insect from egg to adult. Raven has pointed this out with respect to the moths:

> Here Ray evidently felt that he was breaking new ground; for in fact, apart from a few large and brightly colored species, they were then almost entirely unknown. No one hitherto had even collected them seriously; no one had ever realized the importance of studying their metamorphoses and working out full descriptions of each stage. He seems to have grasped what Lepidopterists for one hundred and fifty years after him were slow to appreciate that, if a true understanding of these insects or a correct classification were to be obtained, it was not enough to amass collections of the imagines. From Linnaeus down at least to Haworth the perfect insect alone was regarded as furnishing characters for the division and subdivision of groups, and a study of the earlier stages was virtually ignored. Here as elsewhere Ray reveals a scientific endowment, a sense of the wholeness and continuity of life and a flair for the right method of approach, which was centuries ahead of his time. Where he has completed his task and brought together descriptions of larva, pupa and imago, his records can easily be identified; if he had been able to

carry forward his work or even to arrange the material actually gathered, he might have saved entomology from a vast amount of superficial and mistaken classifications. He would have done for insects what he did for plants—start future students upon the study of the organism as a whole, open up and provide material for a thorough appreciation of taxonomy, and save us from being afflicted with arbitrary and fanciful systems of classification and nomenclature. If only the Royal Society or some of his wealthy friends had realized the mass of material that he had assembled, and recovered from Dr. Dale the boxes of insects to which his notes so carefully refer, and had plates made from them so that each description might have its appropriate illustration, a century of pioneering could have been saved and Ray's own inimitable work been made available. As it is the History remains a locked museum, known indeed and studied by Linnaeus, but replaced in public esteem and general utility by Petiver's very inferior junk-shops, and Eleazar Albin's picture-books.

From the standpoint of entomology it is unfortunate that Ray did not undertake the *Historia Insectorum* until the last years of his life when he was chronically ill, confined to his house a large part of the time, and dependent upon others for aid in his observations of the insects which he so carefully reared. However, considering his devotion to the study of plants which was his first love, and that his major work in that field was published in three volumes over a period of years from 1686 to 1704; his loyalty and devotion to Willughby which compelled him to give his attention to the *Ornithologia* and the *Historia Piscium* immediately after Willughby's death; the fact that he had to begin on an entirely new basis since Willughby's notes on insects were not made available to him until the year before his death, it is apparent that he could not have done otherwise. In fact, his illness and confinement to his home are probably responsible for the devotion of his last years to the insects, since these he could rear and observe in his own home, and his family and friends could, without great effort, supply him with material. We can only speculate that if entomology had been a major interest in his early years, his insistence on observing the whole animal in its natural environment, his genius in interpreting what he saw, and the ability to describe it clearly and succinctly, would have resulted in an epoch making *Historia Insectorum* that would have wielded tremendous influence on later investigators.

THE PERSONAL LIBRARY OF JOHN RAY

There is no evidence that Ray was himself a book collector, though he necessarily possessed a considerable working library. His books were for use and they sometimes suffered in consequence, even when they were borrowed copies. In the year 1699 he had to write to Dr. Hans Sloan: "Your Herman's 'Parad. Bat.' and Boccone's two books I intend to remit by next week's carrier, with thanks for the use of them. I must beg your pardon for having in some measure defaced them . . . by sullying them myself, being forced to use them by the fireside, and partly by a childs unluckily scattering ink upon them." Conditions at Black Notley were evidently not always favorable to the care of books. Nevertheless, at his death, Ray's widow found herself in possession of a library of some value, with which she was able to relieve her straightened circumstances. In 1704 she wrote to Dr. Hans Sloan: "I do intend to dispose of Mr. Ray's books and will get Mr. Dale to make a catalogue of them which shall be sent to you, as likewise an account of what papers Mr. Ray left and doubt not your assistance therein."

The books were sold in March 1708 by Ballard, and a copy of the catalogue is preserved in the British Museum [S-C. 326.(6)]:

> Bibliotheca Rayana: or, a Catalogue of the Library of Mr. John Ray, Late Fellow of the Royal Society. Consisting of very valuable Greek, Latin, French, English, etc. Books, in Divinity, Physick, Philosophy, Philology, History, etc. in all Volumes, and of the most Celebrated Editions of Aldus, Stephanus, Elzivir, etc., several on large Paper, gilt Back, etc. Which will be sold by Auction, at the Black-Bay (alias Latin) Coffee House in Ave-Mary-Lane, near Ludgate Street, on Thursday, the 11th of this Instant March, 170-⅞, beginning every evening at Five-o-clock, till the sale is ended. By Thomas Ballard, Bookseller. Catalogues may be had. . . .

Each of Ray's books is listed separately in the catalogue, and the width of his interests and reading is fully demonstrated. The total number of volumes in Latin, French, Italian, and English is 1350, including 284 in folio. Ray's mind, however, was serious and did not allow him much recreation with the English poets, the only representatives of lighter literature being Chaucer (1602), Donne (1650), Cowley (1656), and Quarles (Argilas and Parthenia).

LITERATURE CITED

Boulger, G. S. 1896. *Dictionary of National Biography*, 47:339–44. London

Boulger, G. S. 1917. Unpublished Material Relating to John Ray. *The Essex Review*, Vol. 26

Dale, S. The Life, by a worthy friend, printed in *A Compleat History of Europe for the Year 1706* under the heading "Additions to the Remarkables of the year 1705." Reprinted by R. W. T. Gunther in *Further Correspondence of John Ray*.

Derham, W. 1718. *Philosophical Letters Between the Late Learned Mr. Ray and Several of his Ingenious Correspondents, Natives and Foreigners*. 198 leaves. Royal Society of London

Derham, W. 1760. *Select Remains of the Learned John Ray, with his Life*. London: George Scott. 336 pp.

Gunther, R. T. 1928. *Further Correspondence of John Ray*, pp. xxiv, 6, 332. London: Ray Society

Gunther, R. T. 1934. Letters from John Ray to Peter Courthope. *J. Bot.* 72: 217–23

Keynes, G. L. 1951. *John Ray, A Bibliography* (limited edition of 650 copies). London: Faber & Faber. 163 pp. (Lists all of Ray's publications with facsimiles of title pages, comments on the texts, detailed information concerning the format of the various editions of each work, and the present availability and location of copies.)

Lankester, E. 1848. *The Correspondence of John Ray*, pp. xvi, 502. London: Ray Society

Lankester, E. 1846. *Memorials of John Ray*, pp. xii, 11, 220. London: Ray Society

Raven, C. E. 1942. *John Ray, Naturalist, His Life and Works*. Cambridge, England: University Press. 502 pp. (An exhaustive and most important study of sources relating to Ray's life and work; an illuminating account of his life, critiques of his publications, and evaluations of his contributions to the knowledge and philosophy of natural history.)

Ann. Rev. Microbiol., Vol. 23

SEQUENCES IN MEDICAL MICROBIOLOGY.
SOME OBSERVATIONS OVER FIFTY YEARS

BY STUART MUDD

Stuart Mudd

Ann. Rev. Microbiol., Vol. 23

SEQUENCES IN MEDICAL MICROBIOLOGY.
SOME OBSERVATIONS OVER FIFTY YEARS   1520

By Stuart Mudd

Veterans Administration Hospital, Philadelphia, Pennsylvania

Contents

PREFATORY CHAPTER

A given field of science, such as microbiology, may be thought of as a vast tapestry hung in a dimly lit gallery. The focus of research interest and teaching emphasis at any particular time may be likened to a spotlight scanning specific areas of the whole. The point of emphasis is that the total tapestry continues to exist and to have relevance, even though the preoccupation of investigators and teachers under given circumstances may comprehend only a fraction of the whole.

Large areas of microbial physiology (1), the basis of photosynthesis (2) and the immunology of parasitic infections (3) have been illumined by those who have preceded me in preparing Prefatory Chapters for the *Annual Review of Microbiology*. These authors have shared with us also the events which have been critical in orienting them to scientific investigation and in guiding them along the paths of discovery. This is the human and personal element in science which is all too rarely revealed in the published literature.

My task will be to scan a few areas of medical microbiology and immunology, and to tell a little of how one thing has led to another. This I would not attempt in a complete and systematic way; such a record would be far too tedious to write, and certainly to read. Rather, I hope to mention in an episodic way events along the trails of progress which have proved to be critical in respect to orientation and motivation; *forsan et haec olim meminisse iuvabit* (4).

I was born in St. Louis in 1893, the only child of Dr. Harvey Gilmer Mudd and Margaret de la Plaux Clark Mudd. My father (1857–1933) (5, 6), his older brother, Dr. Henry Hodgen Mudd (1844–1899) (7), and their double uncle, Dr. John Thompson Hodgen (1826–1882) (8), had successively been leading practitioners and teachers of surgery from about 1850 until almost 1930. These men "left behind them footsteps in the sands of time." Small wonder, then, that friends of the family expected me as a matter of course to "follow in my father's footsteps." This I was strongly in-

THE EXCITEMENT OF SCIENCE

8243-2601/78/1127-0423$01.00 © 1978 ARI   423

clined to do, because of love for my father and because of the surgical tradition in the family. Nevertheless, as will appear as we go on, several crosscurrents proved more cogent than this ambition.

One of my earliest memories is of my distaste for writing the little essays that were required of us in school. But my mother, who was a lady of strong character and a disciplinarian, took a dim view of my attitude in this matter. She insisted that I write and write well, until I finally developed a certain skill and actually enjoyed exercising it; in consequence, I was on the editorial board in each school which published a student paper, including Lawrenceville, and was an editor of the Daily Princetonian. So I entered Princeton with an inclination to become a writer. But what to write about?

During my Freshman year at Princeton, Professor Hugo De Vries of Amsterdam came to lecture on "Evolution by Mutation." As a candidate for the Daily Princetonian I went to the laboratory of Professor Edwin Grant Conklin so that I could prepare a knowledgeable write-up of De Vries. That great teacher and distinguished student of evolution took an hour and a half of his valuable time to explain to the Freshman "heeler" of the Princetonian, the mysteries of meiosis in the maturation of egg and sperm. I left his laboratory in a state of wonder and exhilaration. Here was my future field! It may be of some comfort to younger readers to know that this wonder and exhilaration of a far-off afternoon in 1912 are still alive in 1969, as continuing motive and reward for investigation of life processes.

Thereafter I had the privilege of sitting in the crowded amphitheatre of Guyot Hall while Professor Conklin looked over our heads and with awed voice told of some of the unfathomed mysteries of life. It was my privilege also to write up Professor Conklin's classic "Heredity and Environment in the Development of Man," for the Daily Princetonian (and incidentally for the Smith College Weekly.) The last time I saw him was very many years later, when I was commissioned by Haverford College to invite him there as a Phillips Lecturer. He said, "Dr. Mudd (he was a courtly man), I don't think I should do this. You know I am in my ninetieth year. I still climb the stairs to my laboratory in Woods Hole, make my observations with the microscope, and do my drawings, but I don't believe I should come to Haverford." He ended our long, and to me, inspiring association by telling me I was one of the pupils he was proud of.

The person other than Professor Conklin who most deeply influenced my years at Princeton was Professor Henry Norris Russell, the great astronomer. I well remember one day when he was lecturing on the data he had collected on the dimensions and the temperatures of the stars. I became so completely fascinated and absorbed in his data that I suddenly blurted out, interrupting his lecture, "Why then the stars must do so and so. . . ." The interpretation I suggested was essentially the same as Professor Russell himself had already proposed in a paper I had not seen. His theory of stellar evolution subsequently won him the Gold Medal of the Royal Astronomical Society.

Professor Russell honored me with his friendship, and for many years thereafter whenever I returned to Princeton I called on him to learn how the Cosmos was doing. As I now recall, the prognosis for the Cosmos was surprisingly variable. In the early years Professor Russell conveyed to me the then prevailing idea that the days (or light-years) of the Cosmos were numbered, because it would always be impossible to derive more energy from break-up of the atom than was required to break it up. The Universe was running down from loss of free energy. Later, after the discovery of atomic fission and atomic fusion, Professor Russell initiated me in the mysteries of hydrogen fusion and the incredible stores of energy thus provided. I am afraid that this experience made me a little skeptical of the finality of hypotheses of cosmogeny, and even now I venture to predict that the prevailing "Big Bang" hypothesis of the beginning of the Universe as we know it, may eventually be succeeded by a more easily credible hypothesis.

My first experience in personal research was in the summer of 1916 at Woods Hole. A young comparative anatomist, Dr. Franklin P. Reagan, wanted to investigate certain controversial questions regarding the ontogeny of the blood-vascular system. For this, a continuous monitoring of the development of the fertilized eggs of the little fish, *Fundulus heteroclitus,* was required. He asked me to be one of the monitors. About 2 A.M. one morning I first saw the expected "blood islands," and one small controversy in this field of perennial controversy was on its way to resolution (9).

One should not mention Woods Hole without paying tribute to the great biologists of that era who carried on their work during the summers at Woods Hole: Professor Conklin, Professors Frank and Ralph Lillie, Jacques Loeb, W. J. V. Osterhout, McClung, Thomas Hunt Morgan, are the ones I remember best. In my desk I still have a picture of my "class" in life saving, including children of Professors Osterhout, Morgan, and McClung. Jacques Loeb's children were older, but Robert Loeb and Leslie Webster served well as "drowning persons" to be rescued by my students at the annual water sports carnival. We had fun at Woods Hole.

I would express particular gratitude to Professor Ralph Lillie and his course at Woods Hole on Comparative Physiology. This course placed great emphasis on the critical importance of cell surfaces in all sorts of biologic phenomena. Cell surfaces have fascinated me ever since, in particular in reference to immunity, antibodies, and phagocytosis. It is not without amusement that I realize that some fifty years and two hundred and twenty odd publications after this course, my present preoccupation is still with cell surfaces. For it seems that the specific recognition factors on the surfaces of lymphocytes, in their interaction with foreign cells, initiate the whole complex of transformations which are the subject matter of the newer cell-mediated immunology (10–12, 98a).

THE EXCITEMENT OF SCIENCE   425

I entered Washington University Medical School, St. Louis, in the Fall of 1916 with enough advance credits from Princeton so that I was excused from certain routines, and could work toward an M.A. as well as an M.D. Here, I made a mistake I have regretted the rest of my life. I was still under the illusion that I was preparing for surgery. I was poorly advised and my intuitive perception of where the action was going to be (usually not too bad), on this occasion deserted me. I elected to work for an M.A. in human anatomy, with the blood supply of the human pituitary gland as my thesis subject. Had I been more foresighted I could have laid a firmer foundation in chemistry, which would have been far more useful than what I did. However, I found a certain macabre satisfaction at being on call at the morgue, injecting the internal carotid arteries of new arrivals there with colored gelatin, and subsequently dissecting the arterioles which entered the pituitary gland from the Circle of Willis (13).

My appetite for research was further whetted at Washington University. Professor Montrose Burrows, in a lecture in pathology, had expressed some ideas which seemed to me improbable, not to say "for the birds." Student protest followed a milder pattern then than it does now, and was met in kind by the faculty. I asked Professor Burrows if we might have the use of the pathology cold-room to test the ideas experimentally. Upon his agreement, the participation of two fellow students, Samuel B. Grant and Alfred Goldman, was secured.

The question at issue was what effect did chilling of the body surface have upon the blood supply of the skin and mucous membranes of the upper respiratory tract. We assembled a nondescript collection of electric fans, thermos jugs and galvanometers, and I fabricated in my father's garage thermocouples and wire applicators to hold the thermocouple terminals in place on skin, tonsils, nasopharynx or nasal turbinates. The idea was that the surface temperature must be determined by the balance between heat supplied by the local circulation and heat loss to the ambient atmosphere or respired air. Like most ideas, this proved to be somewhat more simple than the actual operational situation. For we soon learned that when the body was chilled by an electric fan in the cold room, hyperventilation could not be prevented, and this upset the postulated equilibrium. So we had to add a metronome, smoked drums, and tambours to our collection. Now the subject could intentionally hyperventilate, monitored by the tracing of respiration on the smoked drum, and breathe in synchrony with the metronome; as chilling proceeded the steady-state of his respiration could thus be maintained.

A series of careful studies showed beyond question that chilling of the body surface caused reflex vasoconstriction of the skin and the mucous membranes in question. Little gestures of refinement were added, such as having the subject breathe amyl nitrite, which caused beautiful spikes on the temperature curves. We were well pleased with these clear demonstrations (14). In retrospect, however, it seems that the significant point of the work slipped through our fingers. Did the chilling and the ischemia of the mucous membranes really lower resistance to the opportunist flora involved in upper respiratory infections (15)? I am not sure that a really convincing answer to this significant question has been found to this day. Certainly we "catch colds" primarily by inhalation of infected droplets and air-borne droplet residues (16). Any effects of chilling must be secondary or insignificant.

After two years at Washington University I transferred to the third year at Harvard Medical School with an M.A. in Anatomy and a good preparation in the preclinical subjects. I had particularly enjoyed physiology under Professor Joseph Erlanger. He withstood the usual pressure from students to allow them to imbibe facts and later regurgitate them on demand, with very little digested. Professor Erlanger insisted that the facts should at least be assimilated and metabolized, a requirement which I found entertaining, however unfair many students regard it.

In the fall of 1918, when I transferred to Harvard Medical School, Boston was overwhelmed by the great influenza pandemic. Hospitals were flooded with patients; there were not enough nurses to go around. The third year class was assembled in an amphitheatre, and the situation was explained to us. Volunteers to act as nurses were called for. I had had a bad history of recurrent respiratory infection, and I was thoroughly scared. However, Tracy Putnam stood up immediately. I thought that if Tracy could volunteer, I guessed I could, too. So I was given Pfeiffer bacillus shots and assigned to a young member of the football squad who was a patient. There was really no treatment at that time except good nursing care. I did the best I could for him until he died.

Ideas of etiology of influenza at that time were derived from observations in the preceding pandemic, so the medical students were vaccinated with Pfeiffer's bacillus, a mistake which is perpetuated in the name *Hemophilus influenzae*. It was not until after the pandemic that it was learned that the initial cause was a virus which rarely killed of itself, and that the huge mortality was associated with secondary infections by whatever opportunist bacteria were prevalent in the area in question: *H. influenzae*, pneumococcus, *Streptococcus*, *Klebsiella pneumoniae*, or *Staphylococcus*, usually.

This was the most devastating pandemic which the western world had experienced since the waves of the Black Death had swept over Europe. Official estimates of world mortality were in excess of twenty million deaths. Those old enough to remember may recall the disastrous defeat of the German Navy in the Battle of Jutland in the First World War. It has been said that had German intelligence known the state of the British Navy during this pandemic they could have made their assault then and been victorious. There were not enough well men in the British Navy at one time to operate the warships effectively.

THE EXCITEMENT OF SCIENCE   427

The following summer (1919) I spent at one of Dr. Grenfell's hospitals at Indian Harbour, Labrador. There, I heard tales of how influenza was carried by ship from port to port along the Labrador, and utterly devastated the Eskimo villages it attacked. One story was of a village whose inhabitants had been wiped out except for one little girl. The hungry dogs were eating the bodies of victims, but this child had a pet husky which drove off the other dogs and saved her.

Here is a problem for medical ecologists and epidemiologists. Why such singular devastation on first contact with an infectious disease: influenza on the Labrador, measles in the South Pacific, syphilis around 1500 A.D. when first widely prevalent in Europe as "the great pox," tuberculosis among primitive peoples? What are the immunological mechanisms involved in protection by *Durchseuchung?* Personally, I do not believe that the facile explanation of persistent circulating antibodies or enhanced capacity to elaborate circulating antibodies is adequate to explain all the facts, nor do I believe that selection of resistant genotypes is adequate either. Certainly, at the very least the cellular immunology which is now under such intensive study by the younger immunologists must be brought into the picture.

I was so horrified by this pandemic that I later tried to persuade Dr. Wilbur Sawyer of the Rockefeller Foundation that a worldwide surveillance of influenza should be established. Whether or not our conversation had anything to do with it I do not know, but fortunately such a surveillance network has been established and is now maintained by the World Health Organization. My former associates, Drs. Werner and Gertrude Henle, and former student, Dr. Florence Lief, have had principal roles in this surveillance. Of course, vaccination and antibiotic therapy of the secondary infections has now made the occurrence of any great mortality very unlikely.

At Harvard I came under the tutelage of several great physicians, if not always great teachers: Drs. Harvey Cushing, Henry Christian, S. Burt Wolbach, Professor Walter Cannon, I particularly remember. One occasion I recall helped undermine my orientation toward surgery. Professor Cushing used to stand at the blackboard and draw a cross-section of the brain at any desired level, simultaneously using both his hands. The beautifully symmetrical result impressed itself upon me as a feat I could never even hope to rival. However, in one clinico-pathological conference, in which Professor Wolbach loved to catch his eminent clinical confreres off base, he presented the history of a man who had died suddenly, and asked for diagnoses. Professors Christian, Cushing, and others made educated guesses. Dr. Cushing's was "I don't know, but it must have been something in the brain or something in the belly." Dr. Wolbach then turned to the students and asked for their guesses. I said "ruptured ventricle," at which Dr. Wolbach looked sharply around and the students laughed at such an obvious blooper. When it turned out that this was indeed a case of ventricular infarction and rupture, the message that had been gradually growing in my mind for some time became clearer. Obviously, I could do better in the realm of ideas than in such fine muscular coordination as Dr. Cushing, or, indeed, as my father possessed. This skill, which had been in my family for so long, obviously skipped me, although it has fortunately manifested itself again in my son, Dr. Stuart Harvey Mudd, a biochemist now at the National Institutes of Health in Bethesda.

When I graduated from Harvard Medical School in 1920, I was offered positions in Professor Milton J. Rosenau's Department of Hygiene, and in Harvard's young and small Department of Biophysics. I accepted the latter because by this time I was firmly convinced that fundamental progress in biology required that biological phenomena be understood in terms of the chemistry and physics which underlie them. This was not a popular idea in the Society of American Bacteriologists at this time, but very soon true believers such as C. E. Clifton and C. B. van Niel became influential in the Society. Times have changed. Now, active investigators of molecular biology are so highly placed in teaching departments in our schools of medicine as to make one feel some concern lest the complex phenomena of infectious disease become lost to sight among the operons and microsomes.

While I was in the Department of Biophysics I had an ancient Ford nicknamed "Pegasus" because of the way its loose side curtains flapped like wings in the wind. Pegasus and I increasingly often found ourselves headed for Groton, where a certain Emily Borie Hartshorne was a student in the Lowthorpe School of Landscape Architecture. Before her graduation in 1922, on a sunny day during the preceding winter when the roads were sheathed in ice, Pegasus took Emily and me for a long drive in the country. My nervousness must have communicated itself to Pegasus for he tried to leap over a stone wall. His anatomy was in such poor condition on our return that we had to climb Beacon Hill in reverse gear. However, I was happy, for Emily had accepted me.

She has the inquiring and critical mind of a born scientist. She realized that I needed her help, so she gave up her position in a landscape architect's office and joined forces with me as a volunteer in the Biophysics laboratory —a happy collaboration which lasted for ten years.

Emily and I were preoccupied at this time with the movement of particles such as bacteria along interfaces and through capillary spaces (17–20), which we hoped would throw light on the early stages of bacterial invasion. Moreover, at this time in the development of virus technology, those infectious agents were known as "filterable viruses." It seemed important, therefore, to investigate the physical-chemical factors involved in filtration (21).

At about this time my friend Jacques Bronfenbrenner, I believe, told Dr. Simon Flexner that I was a promising young investigator, and Dr. Flexner invited us to join his staff at the Rockefeller Institute. I came to the Institute ardently believing that physical chemistry was the avenue of approach to a deeper understanding of biological phenomena. Dr. Flexner told me later that he had regarded this as mere youthful enthusiasm, but that later he had become convinced. Dr. Flexner at a banquet in his honor, I believe in 1924, recalled the advice given him by the Director of the Naples Biological Station: "Unless you give your young men enough latitude to make fools of themselves, your Institute will never be any good." I daresay Dr. Flexner regarded me as a case in point.

Thomas M. Rivers at this time was elaborating methods of manipulation and analysis of viruses and viral infections which were of significance to the enormous proliferation of knowledge of virus diseases which followed. It is interesting to recall, lacking this later knowledge, how investigators, even of the first rank, went astray in their conceptions of etiology. Dr. Flexner himself believed that "globoid bodies" were the cause of poliomyelitis. The great Hideyo Noguchi, justly famous for establishing the presence of *Treponema pallidum* in the lesions of syphilis of the central nervous system, thought that a *Leptospira* was the cause of yellow fever, a mistake for which he atoned with his life when he later succumbed to the virus of yellow fever. And Peter Olitzky described *Bacterium pneumosintes* as the cause of influenza.

Dr. Flexner, however he regarded me, allowed me to concentrate on the electroendosmotic transport of fluid across mammalian serous membranes, work, however, which attracted little attention then or later (22).

Those were exciting years at the Rockefeller institute. Under the general aegis of Dr. Rufus I. Cole at the Rockefeller Hospital, Oswald T. Avery[1] and Michael Heldelberger and Rebecca Lancefield were well advanced in the investigation of the bacterial and immunochemistry, respectively, of the pneumococci and the hemolytic streptococci. These studies were destined to become classic paradigms for the study of extracellular bacteria as infectious agents. They were primarily concerned with the antigenic architecture of the various serologic groups and types of these infec-

---

[1] Dr. Avery was affectionately known to his many friends as "Fes," because of his professed regret at never having been a professor. He told me once how he lured Michael Heidelberger into the study of pneumococcal capsular carbohydrates, work which, of course, has been lifelong and is one of the monumental achievements in the chemistry of cells. As nearly as I can recall his words, Dr. Avery said:

"I was doing some kitchen chemistry on pneumococci, and I had some gummy precipitate in a test-tube. I shook it under Dr. Heidelberger's nose and said: 'See this, Michael? If you can find out what this is, we will learn a great deal about pneumonia.' Michael said: 'I am busy with Dr. Jacob's cardiac alkaloids—go away; I am not interested.'" Later, Dr. Avery shook another tube under Michael's nose and repeated the invitation. Avery and Heidelberger, of course, became the foremost investigating team in all of bacterial chemistry.

tious agents and the circulating antibodies elicited by them in their human and animal hosts.

At about the same time at the Rockefeller Institute, Dr. P. A. Levene was elucidating the structure of nucleosides and nucleotides, substances which subsequently were to become basic to genetic theory.

Dr. Karl Landsteiner arrived from Austria at the Rockefeller Institute to begin his fundamental studies on the specificities of antigens and antibodies. His English was not yet fluent and he asked me to help with translation of his papers. Also, he was still preoccupied with the specificities of blood, and he used to appear at our laboratory door, saying to Emily: "I am a vampire. I want your blood" (which is AB).

In the stimulating atmosphere of the Rockefeller Institute at that time it was not difficult to come to realize that the primary events in parasitic invasion are encounters between infectious agents and host cells; that the chemistry of the surfaces of infectious agents and host cells are definitive in these encounters; that the surfaces of the disease agents can be modified by antibodies. Host cells of obvious importance are the cells of the blood, and these are relatively easily accessible to study. Such considerations as these were to orient my work for many years toward the study of the surface chemistry of bacteria, the modification of these surfaces by specific antibodies, and the behavior of leukocytes in phagocytosis.

A number of clues should have alerted me to the fact that these considerations, correct as far as they went, nevertheless overlooked other equally important considerations. The most obvious of these is that in all cases of facultative or obligate intracellular parasitism, events subsequent to the entrance of the parasite into host cells are as critical as the entrance itself. This is true of some viral infections, even though it is strangely overlooked by so many virologists, who might seem to believe that effective immunity in a virus infection is merely a matter of having the proper circulating antibody on hand in the necessary concentration. Certainly, in tuberculosis (23), in brucellosis (24, 25) in listeriosis (26), and in salmonellosis (27, 28) it does not suffice to have the infectious agent taken up by granulocytes and macrophages. To eradicate the infection the macrophages must be activated to destroy the intracellular parasite.

However, the above paragraph is hindsight, and at the time Emily and I worked at the Rockefeller Institute and later at the Henry Phipps Institute under Professor Eugene Opie, we studied, contentedly enough, the surface properties of blood cells (29), of bacteria and of their corresponding antibodies (30, 31), and the mechanism of phagocytosis. In 1927, we were most fortunate to have the pathologists Balduin Lucké and Morton McCutcheon, and the hematologist Max Strumia join us as collaborators in the phagocytosis studies. Some references for those interested are (29–42). What they showed in brief is the following:

Phagocytosis involves a spreading of the surface membrane of granulocyte or monocyte over the surface of the bacterium or other particle undergoing ingestion. The granulocyte is a much less viscous cell and phagocytosis is a more rapid process than with the monocyte; the latter is, however, capable also of pinocytosis, or envelopment by petaloid pseudopods. In later years, when the electron microscope became available, it was shown that the ingested particle is completely enveloped in the plasma membrane of the ingesting cell (43). The phagosomes move centripetally in the cell, where they encounter, at least in the granulocyte, lysosomal granules which fuse with the phagosomes and discharge digestive enzymes into them (44).

Virulent strains of pathogenic bacteria usually have surface components on which the membranes of phagocytes do not readily spread. Examples are the capsular carbohydrates of the pneumococci and Smith-type staphylococci, the M substances of hemolytic streptococci, and the somatic carbohydrates of the Salmonellae, the O and Vi antigens. Phagocytosis in immune persons or animals is usually facilitated by circulating antibodies specific for these surface components. Combination with such antibodies makes the surface properties of the bacteria approach those of a globulin film, which is, in Bernard Shaw's words, "palatable to leukocytes."

After some years of these studies, Emily urged that we observe and photograph the actual process of phagocytosis. Accordingly, we bought the necessary equipment and carried out the observations recorded in her last study in my laboratory: "The process of phagocytosis. The agreement between direct observation and deductions from theory (38)." This ended a most happy interlude of collaboration. However, the wisdom of her practicing and training for a profession for which she was gifted by nature, namely, marriage counseling, has been borne out by her distinguished success and international recognition as Professor of Family Study in Psychiatry at the School of Medicine, University of Pennsylvania.

One amusing episode resulted from this last paper. We were attending a reception in London on the occasion of the Second International Congress of Microbiology. Emily was wearing a pretty and very feminine summer dress when I introduced her to Professor Neufeld, then head of the Robert Koch Institute in Berlin. He exclaimed: "Mudd und Mudd—Mechanismus der Phagocytose—Es ist unglaublich"! Evidently she did not closely resemble the young ladies whom he was accustomed to see at the Robert Koch Institute.

Another incident which reflects an atmosphere which is presently changing is the opposition we encountered from our (nonscientific) administrative superior at the Henry Phipps Institute at that time, with reference to our work in relation to birth control. At all events, we successfully demonstrated the species specificity of mammalian spermatozoa (44a).

The Department of Bacteriology at the University of Pennsylvania became almost nonexistent with the retirement of Professor David H. Bergey. Professor Eugene Opie was made chairman of a committee to select an acting head to recreate a strong department, and he recommended me to undertake the job. My appointment with Dr. Opie had been as an experimental pathologist. I spent the summer of 1931 reading a standard textbook of bacteriology from cover to cover—an exceedingly dull task as I recall it. Topley and Wilson's splendid *Principles of Bacteriology and Immunology* had become available in 1929, and I took myself and the students through this year after year, despite the protests of the Vice-President for Medical Affairs and even of Professor G. S. Wilson himself that I was "so wrong" to ask so much reading of the students. Anyway, they learned a good bit about infectious diseases. Approximately 120 students a year went through our course about thirty times in almost as many years. (We gave the course an extra time or two during the exhausting period of continuous instruction during the Second World War.)

Dr. Harry E. Morton, who is now Chief of the Microbiologic Division of the William Pepper Laboratory of the Hospital of the University of Pennsylvania, and who was given the 1968 Kimble Award at the Conference of Public Health Laboratory Directors, carried a large share of the teaching burden through those earlier years. When the course was later revised Dr. John A. Flick, now Associate Professor of Microbiology in the Division of Graduate Medicine, and Dr. Joseph S. Gots, a brilliant investigator and teacher now with my successor, Professor Harold Ginsberg, were principally responsible, and were assisted by Drs. Warren Stinebring and Frank A. Kapral. When the principal importance of virology became apparent, Professor Werner Henle established lectures and laboratory work in viral diseases.

The passing scene in medical microbiology underwent dramatic changes during those thirty years. I may illustrate by our teaching on the therapy of pneumococcal pneumonia. At first there was no specific therapy for pneumococcal pneumonia except for antisera against the capsular polysaccharides. At the Second International Congress of Microbiology held in London in 1936 (45), a considerable amount of time was given to papers establishing that antipneumococcal horse serum was, in fact, efficacious against Type I infection, and perhaps also against Type II. At the same time, however, rabbit sera against many more types had been introduced, and I recall acceptance by the Council on Pharmacy and Chemistry of the American Medical Association, on which I then served, of a long series of types of antipneumococcal rabbit sera. It was indeed said that the Lederle Company had let a contract for a half million dollar rabbitry in which to produce antipneumococcal sera.

At the same Congress in London, however, Colebrook (46) reported spectacular success in treating puerperal sepsis with Prontosil Soluble. This dye was soon shown to owe its efficacy to sulfanilamide. Dr. Leonard Colebrook brought the news of this breakthrough in chemotherapy to America in 1936–37. Sulfanilamide was soon followed by sulfapyridine and other drugs of the sulfa family. Before the huge Lederle rabbit hutch could be built, therapeutic antipneumococcal rabbit serum gave way to sulfa therapy which, in turn, was displaced by penicillin.

Now, history is completing the circle. The miracle drugs are found to be something less than miraculous, and pneumococcal pneumonia is now recognized as a serious problem. The U.S. Public Health Service is setting up a very large field trial under Dr. Robert Austrian of the prophylactic value of active immunization against a selected mixture of pneumococcal polysaccharides (47).

Up to the time I became department head I had been extremely active at the laboratory bench. With the responsibility of building up a department through whose course 120 medical students passed each year, selecting and guiding graduate students, not to mention various administrative chores, I became a sort of nonplaying captain of a team in which post-doctoral Fellows and graduate students carried out most of the work at the bench. I do feel, however, that a somewhat broad outlook on what is worth researching, possible procedures, and the significance of the experiments when performed, is not without its place in the laboratory scheme of things.

We had a very considerable roster of graduate students in medical microbiology. Some one hundred Ph.D.'s were awarded during these thirty years, and thirty-odd Masters in Medical Microbiology as well. Each student, according to his background, research interests, and capabilities was matched with a member of the faculty for this research. Teacher and student collaborated, and when material was ready for publication it was published like any other collaborative work. At the end of his course of study, the student could bring these publications together with the necessary "connective tissue writing," as his thesis.

This system has been criticized by traditionalists in certain fields who seem to believe that a doctoral dissertation should be something quite apart and slightly remote. My contention, however, is that the primary reason for a student undertaking a doctoral course in natural science is to learn to become a scientist, capable, within himself or herself, of making significant contributions to science. What comes out of the laboratory, we hope may be science, but what goes into making that science is certainly a judicious blend of knowledge, intuition, imagination and skill; in brief, its essential dimensions are those of an art. Skills have been acquired and intuitions sharpened since time immemorial by personal contact with more experienced and skilled craftsmen. Why not give the student the benefit of such association, and let him learn to contribute to science by contributing to it in fact under wise guidance? A similar philosophy of learning is, of course, regularly followed in such helping professions as clinical medicine and psychiatry, psychology, and social work.

One of the freedoms I have particularly enjoyed since retirement from my Professorship is the freedom to decide things simply on what I believe to be their merits, without bothering about all the hedges and ditches of the academic steeplechase. May I illustrate some of the obstacles from the experiences of two students, each a child of a scientist whom I myself revered.

Isabel Morgan is a daughter of Professor Thomas Hunt Morgan, the great geneticist. At the end of her highly contributory work in our streptococcal program, her dissertation was questioned because it was "too short." I told the Dean that I had recently participated in an international Congress in the Royal Danish Academy of Science; that the principal adornment in that Academy was a marble statue of Christian Gram; that, insofar as my knowledge went, Gram's sole contribution to science was in introducing the Gram stain. But what a contribution! A bronze bust of Dr. Isabel Morgan Mountain, in recognition of her subsequent fundamental contributions, at the Rockefeller Institute, to the understanding of immunity in poliomyelitis, now adorns the Warm Springs Foundation.

Philip E. Hartman is the son of Professor Carl Hartman. Philip had broad social interests in college and did not always attend as he should have to his chemistry. In his first year of graduate study he failed his course in medical biochemistry. The Professor said, "Mr. Hartman has had his chance in our course, and he has failed. He is out." I said, "But you cannot do things like that! Phil will take organic chemistry this summer, he will take your course and pass it next year and he will proceed with his graduate work." He did precisely that, and he is now a Professor at Johns Hopkins and one of the foremost students of biochemical genetics in the country. I am exceedingly proud of the fact that his son is named Paul Stuart Hartman, "Paul for my twin brother and Stuart for my professor."

Our graduate students, with few exceptions, have taken their places in universities, research institutes, public health laboratories, and industry. It is indeed a pleasure to meet them again at the American Society for Microbiology.

In consideration of the obligations of my years as department head, it is perhaps not surprising that in research I responded to promising events and opportunities as they presented themselves. Had I been a microbe (which is not quite the case), I should doubtless have been classified as an opportunist.

Three hypotheses, independently arrived at by Jerome Alexander, Felix Haurowitz (48), and myself (49), were proposed in 1930–32 to explain antibody formation. These hypotheses all postulated the synthesis of antibody by successive additions of amino acids at positions of stereochemical fit on the surface of antigen as template. These so-called instructional hypotheses and modifications of them were later challenged by the "clonal" hypotheses, in various modifications, as proposed especially by Burnet. Voluminous discussion has followed. Some form of clonal hypothesis of antibody formation is now more acceptable to most authors (10). I think it can be said, however, that the template idea offers a fair model of the synthesis of proteins in general, with messenger RNA as template and the proper amino acids attached by microsomal sRNA at positions coded for in the DNA genetic blueprint.

A problem which preoccupied my department during the 1930's was the preservation of serum, plasma, and other labile biological products by drying *in vacuo* from the frozen state (50–57). This story is long, involved and tortuous and need not be recounted here. Freeze-drying techniques were developed through successive stages until they became practicable procedures for a variety of purposes and on several scales of operation.

In addition to the technological development there was a rugged struggle to keep this useful art from crippling commercial entanglements. That this struggle was successful is attested by the following letter (58):

Lyophile-Cryochem Company
Incorporated
Tabor Road—East of Adams Avenue
Philadelphia, Pa.

May 26, 1941

Dr. Stuart Mudd, President
American Human Serum Association
Haverford, Pa.

Dear Doctor Mudd:

As President of your Association, we should like you to have information concerning any possible bearing the patent situation surrounding desiccation from the frozen state might have with regard to application to blood substitutes, either for hospitals or military purposes.

Commercial investment of well over $500,000 has had a large part in making possible the tremendous advance in the art of desiccation from the frozen state and has made practicable the large scale production of plasma at a much reduced cost for this emergency. As a result of that commercial investment, 25 patents in the United States and others in foreign countries, as well as additional applications for patents pending, have come to exist in this field. However, great conflict in competing commercial institutions existed and this hampered the general use of the process. In order to correct this condition, the Lyophile-Cryochem Company was formed for granting licenses under all the inventions and this has resulted in making generally available to everyone all the improvements in the field of vacuum desiccation from the frozen state, including the inventions in process, products, apparatus and containers.

Apparently some confusion exists concerning administration by this new company of the many patents. From the onset, the company has had as its policy the granting of royalty-free licenses to public institutions including hospitals, universities, and all governmental agencies for research and experimentation and for production on a non-profit basis of products derived from human beings, such as serum and plasma. Recently the company has announced its decision also to grant royalty-free licenses to commercial institutions which may be selected by the War Department for production of plasma collected by the Red Cross during the present emergency for the military forces of our country.

We should welcome any suggestions you or your Associates may have as to how we may assist in the important work of your Association.

Very truly yours,

JAS/EWF/B

Lyophile-Cryochem Company
(signed) John A. Silver
President.

Suffice it to say that freeze-dried serum or plasma was available for treatment of shock and hemorrhage in the Battle of Britain and to American, Canadian, and British forces in the Great War. Many other applications have, of course, been developed since.

Another opportunity which knocked on my door was afforded by William Firth Wells. While still at Harvard, Wells and his wife, Dr. Mildred Weeks Wells, presented a very fine demonstration of their early work on air-borne infection at a meeting of the American Public Health Association. I was greatly impressed with this, and when I learned that these investigators were available I arranged for them to come to the University of Pennsylvania. During the ensuing years the Wells continued their work in air hygiene and air-borne contagion at this University. A number of eminent investigators played supporting roles in these developments: Professor Joseph Stokes, Jr., Professor Esmond R. Long, Drs. Max B. Lurie and Werner Henle and, eventually, Vice-President for Medical Affairs, Alfred Newton Richards. My own very modest contributions are recorded in the references (59–63). The total effort of the Wells' work is integrated in a volume, *Airborne Contagion and Air Hygiene* (64) published in 1955.

Since the work of the Wells and others has established beyond question the reality and fundamental importance of confined air as a vehicle of contagion of respiratory illness, the question may well be asked why we should today be facing a pandemic of influenza, and every winter we should suffer misery and economic waste in the form of nondescript respiratory illness, as well as tuberculosis and various childhood contagions. Similarly, it might well be asked, when water and food transmission are so well understood, why cholera and dysentery should still have endemic prevalence, and why the world traveler should regard "Montezuma's revenge," "Delhi belly," or similar affections as only to be expected. Obviously, the application of medical knowledge is lagging.

In 1939, I believe it was, the report came to my attention that physicists at the Siemens-Halske Werke in Berlin had been able to take electron pictures at incredibly high useful magnifications. I happened to visit the California Institute of Technology that summer, and I asked a young physicist about the reports. Oh, you can forget about them, he said, in effect. An electron beam strong enough to produce an image would incinerate any biological specimen. I have never ceased to wonder how this physicist's judgment could have been so nearly correct in theory and so totally wrong in operational terms.

Later, in 1950, Dr. James Hillier and our group reported (65–66) critical experiments on the fixation of biological specimens by the electron beam. Specimens on stainless steel mounting grids were subjected to electron micrography. They were then exposed to an acid bath and again micrographed. The areas of specimen protected by the wire were hydrolyzed and cleared by the acid bath; the areas which had been exposed to the electron beam appeared morphologically unchanged. Clearly, the electron beam had acted like a very special "fixing agent," changing the chemistry but not obviously the appearance of the exposed specimen. A similar resistance to high temperatures following exposure to the electron beam had previously been recorded by von Borries & Glaser (67) and by Winkler & König (68).

In 1938, I believe, Dr. V. K. Zworykin, head of the RCA Research Laboratories, had invited Dr. L. Marton from Brussels to set up his pioneer electron microscope in the RCA Laboratories. The high-powered physicists and engineers at RCA were as innocent about biological specimens as I was about precision circuits and electron lenses. I will never forget my first visits to the RCA Laboratories. Pictures were to be seen of objects resolved at several scores of times the light microscope is capable of. I felt much as Alice must have felt when she dropped down the rabbit's hole. However, with Dr. Marton's aid (69), David Lackman and I (69a) published, in 1941, micrographs which were among the first electron pictures of biological specimens to appear in the United States.

Dr. Zworykin, in 1940, secured the brilliant services of Dr. James Hillier from Professor Burton's Laboratory in Toronto, and the first of the series of RCA electron microscopes took shape. Meanwhile, the laboratories of the Siemens-Halske Werke in Germany had been largely bombed out of existence, and it became obvious that the task of exploring the practical potentialities of the electron microscope had fallen to the lot of investigators on this side of the Atlantic.

Dr. Zworykin asked me to form a committee of the National Research Council in biological applications of the electron microscope, and provided the necessary stipend for an RCA Fellow. Of the considerable number of applicants for this Fellowship, I selected Dr. Thomas F. Anderson, a well-trained young physical chemist. This choice was against the advice of a number of persons who felt that this Fellow should be primarily trained as a biologist. Dr. Anderson's creative leadership in applications of electron microscopy ever since certainly has more than justified his selection.

We scanned the fields of biology to find areas and investigators who could profit most from ultra-high resolution micrography. The investigators selected were invited for an initial week of intensive work with Dr. Anderson. If their results during the initial week were sufficiently promising they were scheduled for an additional week later. Dr. Anderson displayed great versatility and ingenuity in cooperating with the specialists to make this work scientifically significant. At the end of the two years of the RCA Fellowship (1940–1942), thirty-odd studies had been prepared for publication and a sound basis for application of electron microscopy to biology had been laid.

It became evident during this period that the electron microscope was peculiarly fitted to fill in a "blind area" in the study of minute structure, the area between light microscopy on the one hand, and X-ray diffraction on the other. In this area obviously lay viruses and the fine structures of bacteria. It also became evident that the vast riches of the ultrastructure of cells and tissues could not at once be exploited, pending the development of better techniques of fixing, embedding, and ultrathin sectioning. I talked persistently with Dr. Hillier to the effect that if tissues and organs could only be sliced thin enough that practically every laboratory concerned with research in histology or pathology would have to have an electron microscope. This highly trained physicist finally became sufficiently persuaded so that he took a furlough to work at the Sloan-Kettering Institute, where he devised one of the earliest ultramicrotomes and prepared beautiful electron micrographs of liver sections.

The early electron microscopy of bacteria and viruses concentrated on such structures as could be resolved without sectioning: flagella, capsules, cell walls, nuclear sites, metachromatic granules, spores, and antibodies in combination with surface structures. These studies were summarized in 1944 in articles in the *Journal of the American Medical Association* with the following rather prophetic introduction by Dr. Austin E. Smith, Secretary of the Council on Pharmacy and Chemistry (70, 71):

> The electron microscope is giving us a new understanding of the physical structure of the microscopic and submicroscopic agents of disease, the bacteria, rickettsias and viruses. It appears that these microparasites may be arranged along a scale of diminishing size and complexity of organization which tends to parallel the scale of their decreasing metabolic independence and increasingly obligatory parasitism.
>
> Simultaneously new insight into the fundamental host-parasite relationships is currently laying the foundation on which a more useful understanding of susceptibility and resistance to these disease agents will be based. Finally the principles and mechanisms underlying the spectacular recent successes in chemotherapy are beginning to assume a clarity which would seem to point the way toward eventual extension of chemotherapy into areas in which it has not as yet been successful.
>
> These rather wide prospects are presented by the chairman of the National

Research Council Committee on Applications of the Electron Microscope, who is also a member of this Council.
COUNCIL ON PHARMACY AND CHEMISTRY
Austin E. Smith, M.D., Secretary.
J.A.M.A., *126,* 561-571 (1944).

These and other structures have, of course, been studied in the intervening twenty-five years with varied and more refined techniques, and with brilliant analyses of ultrastructure, particularly of the viruses. However, I should like to mention three areas in which our pioneer and continuing efforts have borne fruit. Electron micrographs by Anderson & Stanley (72) visualized the specific combinations of tobacco mosaic and bushy stunt viruses with their specific antibodies. Mudd & Anderson visualized antibodies in combination with the flagella of typhoid bacilli (73). The art has now progressed to the point where individual antibody molecules may be shown binding virus capsids together, or, under appropriate conditions, forming loops with the two binding sites of the antibody attached to specific antigen (74).

Internal granules, whose nature was then not known, shown in the cytoplasm of many bacteria, have since been recognized as metaphosphate, important in the storage and transformation of energy in the bacterial cell (75-77).

One other controversial point may be mentioned which refined electron microscopic work has clarified. In 1951 we described sites of oxidation-reduction in bacteria which were suggestively like mitochondria of yeast cells (78, 79). After my report of these sites, Professor C.-E. A. Winslow, whom I revered, asked what I thought they were. I replied that I was not sure but that they seemed to have properties suggesting mitochondria. Professor Winslow remarked that Adam and Eve, walking in the Garden of Eden, had seen a curious little green animal by a pond. Eve asked what it was. Adam said he was not sure, but it behaved like a frog. Eve said, "Well, if it looks like a frog, jumps like a frog and croaks like a frog, why not call it a frog"?

Few others have gone as far as Eve and myself in calling these sites bacterial mitochondria. They are known as "chondrioids," "mesosomes," and by a variety of other terms. Advanced electron microscopy, however, has shown that they have oxidative-reductive functions equivalent to those of mitochondria (80, 81).

Emily and I were each invited to give a paper at the 14th Japanese Medical Congress in Kyoto in 1955. This well-organized Congress in one of the truly beautiful cities of the world was indeed a notable affair. The Japanese, realizing that the foreign visitors would get little from the Japanese language part of the program, arranged delightful excursions to the gardens, temples and shrines of Kyoto, the Cherry Blossom Festival, and even down the Hosugawa rapids by boat. The time was early April and the cherry blossoms and Japanese maples were at their height.

On the occasion of my talk I found myself placed beside a former classmate of the class of 1920 at the Washington University Medical School in St. Louis. I was to talk chiefly from lantern slides about the intracellular induction of coliphage maturation (82), so I asked my former classmate what to say if the pictures should get out of focus. Later, I interrupted a talk in reasonably good English to say "Sho ten, dozo." It has never been wholly clear why the thousands of Japanese in the audience laughed so heartily. The focus was sharpened anyway.

During Emily's paper, Professor Tomio Ogata, Vice President of the Congress and our particular friend, presided. As she finished, with a whimsical exaggeration of the Japanese idiom, Dr. Ogata bowed to her and said: "This is a poor thing, but please accept it," and handed her a pearl necklace.

Following the Kyoto Congress, a succession of Japanese post-doctoral Fellows came to work in our Department, and every one of them contributed to our research program. Over a period of busy years, the light microscope, the electron microscope, and analytical chemical procedures were coordinated in studying the internal organization of a variety of pathogenic bacteria, and of course the experimenter's faithful friend, *Escherichia coli.*[2] Our Japanese friends contributed notably in this work (74, 76, 77, 83–87). Our most recent Japanese colleague, Dr. Kazuhito Hisatsune (88, 89), is a carbohydrate chemist from the Institute of our first Japanese post-doctoral Fellow, Dr. Kenji Takeya, who is now Professor at Kyushu University. I am pleased to learn that at Kyushu I am known as "Muddosan."

The International Congresses of Microbiology have been a source of pleasure, of broadening horizons, and of valued friendships. I have participated in them all save for the first one: Paris (1930), London (1936), New York (1939), Copenhagen (1947), Rio de Janeiro (1950), Rome (1953), Stockholm (1958), Montreal (1962), and Moscow (1966). In Brazil and Rome I was representative of the Society of American Bacteriologists; in Stockholm, Vice-President; in Montreal, President; and in Moscow, Past-President of the International Association. It is unfortunate that the average member of the National Societies has no way to realize how much effort goes into the preparation of the Congresses, nor how much work is done by the Association Presidium and Committees between Congresses in the broader interests of microbiology. It might help if the Association Presidium would adopt a policy of regular annual reports to the national societies and if these were circulated to memberships.

The forthcoming Congress in Mexico should be a good one. Emily and I

[2] Even the beloved Microbiologist, Leonard Colebrook, was known to his many friends in Britain as "Coli."

THE EXCITEMENT OF SCIENCE   441

inspected the facilities in Mexico City and they are very fine. The preparative committees in Mexico started preparations years ahead of time (which is important). Our warm good wishes for the success of the Tenth Congress in Mexico City!

As my retirement as Professor of Microbiology in the School of Medicine, University of Pennsylvania, approached (65 was then the age of compulsory retirement), I became very much concerned, for I could not easily imagine my life without work. Moreover, a creative set of younger investigators of ultrastructure, Professors Edouard Kellenberger in Geneva, Woutera van Iterson in Amsterdam, R. G. E. Murray in Canada, and Thomas F. Anderson in Philadelphia, had developed sophisticated techniques and skills in electron microscopy which I could not match. It was, therefore, a matter of the utmost good fortune for me that the United States Veterans Administration offered me a position as special investigator in 1958.

There were many excitements and alarms (90–93) at that time with regard to the frequency of staphylococcal infections, particularly in hospitals. Useful as these conferences were in calling attention to staphylococcal infection and its epidemiology and in insisting on more meticulous attention to hygiene, they were but little concerned with the fundamental biology of the staphylococcus itself and its intimate interaction with host cells and tissues. Indeed, interest in the biology of the staphylococcus had waned with the success of penicillin and other antibiotics. No such determined and sustained investigation as the classical studies of Avery and Heidelberger on the pneumococcus, and Rebecca Lancefield on the hemolytic streptococcus had been made on staphylococcus. It was, therefore, natural enough that I should gather a small team of younger scientists who actually had been my former students, and go to bat against the staphylococcus. It did not take us long to learn that *Staphylococcus aureus* has more on the ball than the pneumococcus, streptococcus, or any other known bacterial pathogen.

Having been imprinted at the Rockefeller Institute with the immunology of the pneumococci and streptococci I felt, half unconsciously, that another pus-forming organism, *S. aureus,* would be found to follow a similar pattern. We were due for many surprises.

In studies relative to exotoxins of staphylococcus we have been most fortunate in enjoying the collaboration of Dr. Gareth Gladstone of Oxford, a fine investigator and a fine gentleman (94). One of our surprises was in finding that patients with deep-seated and chronic staphylococcal infections, even though draining staphylococci from their lesions, were usually not making optimal antibody response to α-toxin and Panton-Valentine leucocidin. Titers of antibodies to both could be significantly augmented by injection of the corresponding toxoids (95).

Surprises in regard to the invasive characteristics of staphylococcus were in store for Nancy Lenhart, who performed the agglutinin titrations, and for Drs. Isabel Li and Mehdi Shayegani, who studied phagocytosis and intracellular killing, and in particular for me. We found no significant differences in the agglutinin titers against a given staphylococcus in human populations, whether these were normal, endemically infected populations or clinical cases (96). Moreover, phagocytosis and intracellular killing of ordinary staphylococci (we except the rare encapsulated Smith-type strains) are opsonized by normal human serum, slightly after heat inactivation, efficiently by fresh human serum (97, 98). The meticulous investigations of Dr. Shayegani are indicating that these opsonins can be adsorbed also by *Staphylococcus albus* and by *Bacillus subtilis*. This study is in progress. Cell wall preparations of *S. aureus* and chemical fractions for Dr. Shayegani's work on phagocytosis and intracellular killing are being prepared by Dr. Hisatsune, who, for two years, has had the privilege of working in association with my very distinguished colleague and friend, Professor Carl-Göran Hedén, at Karolinska Institutet, Stockholm.

An aspect of our program which is exciting us very much now concerns "the angry macrophage," to borrow George Mackaness's term. It is well established that the macrophages of animals in which delayed hypersensitivity has been induced can be activated by challenge with homologous antigen administered in appropriate amount and manner. Such "turned on" macrophages are greatly augmented in their capacity to destroy intracellular parasites, and this capacity is not limited to the parasite which induced the hypersensitivity, but is nonspecific. Professor Robert Good has series of patients who illustrate these principles in human subjects (98a).

Can we now activate the macrophages in such a way as to afford protection in significant situations? With respect to mice repeatedly infected with *Staphylococcus,* Dr. James Taubler has demonstrated delayed hypersensitivity (99, 100), and is showing that their macrophages can be mobilized in such a way as to afford significant protection against challenge with viable staphylococci. We have also shown that mice made hypersensitive to tubercle bacilli by infection with strain H37Ra and activated with Old Tuberculin, are significantly protected against vaccinia virus (100a).

Dr. Arthur G. Baker is demonstrating that a high proportion of his patients with chronic upper respiratory infections, usually superinfected with staphylococci, show delayed hypersensitivity to a lysate [Staphylococcal phage lysate (SPL), Delmont Laboratories, Swarthmore, Pa.] of staphylococci, and this hypersensitivity decreases during the course of weekly injections with SPL (101). We are trying to establish a solid experimental basis for these interesting clinical observations.

Our investigations on staphylococci are recorded in a considerable number of experimental papers and critical reviews and are fully summarized in chapters on "Host-Parasite Interaction in Staphylococcal Infection" (102), and in a chapter on "A Successful Parasite" (103). I shall not bore the patient reader with further details beyond saying that the trail of this ubiquitous but elusive pathogen is quite the most intriguing one I have ever followed, and, indeed, the end is not yet.

The late Dr. M. G. Sevag was a creative biochemical investigator and teacher of graduate students in our department for about a quarter of a century. Unfortunately, Dr. Sevag's fine work never received the recognition it deserved for a number of reasons, among which are the facts that his papers were very hard to read, and that in his early work at least he had a distinctly Lamarkian orientation at a time when the currents of genetics were running strongly in the opposite direction. With the recognition that mutational frequencies can be altered by many chemicals as well as physical factors and that such frequencies can be diminished as well as increased by chemical substances, the situation has changed.

Almost as soon as it became evident that the development of drug resistance would become a severely limiting factor in the usefulness of antibiotics and other antimicrobial agents, Dr. Sevag dedicated himself to exploring the nature of drug resistance, and to the search for means of preventing drug resistance. A long series of studies with a succession of graduate students followed, in which the metabolic pathways of strains of *S. aureus* and *E. coli* were contrasted with the metabolic pathways of resistant mutant strains.

Sevag and co-workers demonstrated that antibiotics, in addition to acting as inhibitors of mechanisms of bacterial metabolism, are mutagenic. In the continuing presence of the antibiotic, clones emerge which exhibit heritable alterations in metabolic characteristics, including specific resistance to the antibiotic in question (104). Sevag reasoned that substances which form stable complexes with DNA might reduce the frequency of the requisite mutational events. Subsequently, he found that quinacrine hydrochloride [Atabrine] and more recently phenothiazines and dibenzocycloheptenes exhibit such antimutagenic effects and, in combination with the antimicrobial agents, under appropriate conditions prevent the emergence of resistant clones.

Obviously, clarification of these principles and refinement for their clinical application required meticulous quantitative study of the relative concentrations of antimicrobial and antimutagenic agents, the specificities of resistance and other matters. These studies have been and are being conducted by Samuel J. De Courcy, Jr. (105, 106).

I have ventured to present the following modification of the interpretation of the comprehensive work of Dr. M. G. Sevag and his collaborators (103):

1. That contact with sublethal concentration of sulfonamide or antibiotic induces changes in the DNA which are expressed phenotypically as altered patterns of metabolism. This is a fact attested to by a large body of experimental studies by Sevag and co-workers over a period of 15 years.

2. That in the "sublethal" concentrations of the primary exposure to anti-infective agent there is selection of such induced metabolic patterns as are compatible with growth in the inducing agent.

3. That subsequent exposure of the clones so produced to lethal concentrations of the homologous anti-infective plus quinacrine hydrochloride permits survival because no further mutations are required.

4. That exposure of the above induced clones to lethal concentrations of another anti-infective plus quinacrine hydrochloride yields no survival because the latter drug, acting as antimutagen, prevents the requisite induced mutations.

A therapeutic trial of the above principles has been reported from the service of Michie at the Children's Hospital in Philadelphia (107). A combination of quinacrine hydrochloride and an antibiotic was used in a series of urinary tract infections. The combined therapy was reported as successful in eight of ten patients in whom previous treatment had failed. A more extensive study of the efficacy of anti-infective agents, with and without quinacrine hydrochloride, in cases of nonobstructive uropathy is in progress (108).

To observers of the pageant of medical microbiology it has become apparent in recent years that foci of research interest and teaching emphasis are increasingly becoming events at the molecular level. These are obviously of cardinal importance in deepening the understanding of microbiology, as of others fields of biology. However, infectious diseases as holistic phenomena are still and will continue to be among the facts of life, and oncoming generations of scientists and physicians must be prepared to deal with them understandingly. It has been my great good fortune to have known personally many of those whose professional lives have been given to the study of infectious diseases in depth. I have, therefore, asked such representative men and women to contribute chapters on their respective areas of professional experience to a volume on "Infectious Agents and Host Reactions." This is not to be another textbook, of which there are already excellent examples available, but an expression of such experiences, insights, and judgments as the authors themselves believe need to be recorded. The response has been most heartening. Twenty-five of the twenty-six chapters are already in the hands of the W. B. Saunders Co., the publisher (103).

Editing this volume has indeed been a learning experience for me. By far my most significant reward has been to come to see resistance to infectious agents as a very broad integration of the classical immunology mediated by antibodies, and the "newer immunology" involving transformed cells of the lymphocytic-macrophage system. Too many of us have been content with circulating antibodies, to combine with infectious agents and effect their phagocytosis, or with the cellular immunity of some chronic, obligate, intracellular parasitisms. Resistance to infection should certainly no longer be regarded as either humoral or cellular, but rather as an exceedingly complex, and as yet imperfectly understood, integration of both humoral and cellular mechanisms.

EPILOGUE

Infectious diseases are situations in which a microparasite and the cells and tissues of a host are simultaneously struggling for survival in the same ecosystem. Through some fifty years I have been studying the factors comprised in these struggles with the hope of understanding, and hence of ameliorating, their consequences for the human hosts. A medical man could scarcely do less. And yet there is another side to the coin, and this is a very significant side, indeed (109, 110).

Emily and I in our travels about the world have had occasion to observe the quality of life in the slums of Puerto Rico and of Colombia, of Manila, Hong Kong, Seoul, and Calcutta. Indeed, we need not leave Philadelphia to observe the consequences of the dispossessed rural poor from the South who flood into our northern cities. Subhuman qualities of life are obviously the result of many factors, but one factor of cardinal importance has been the sudden application of preventive medicine to populations with no countervailing attempts to make corresponding adjustments in population growth. In India, the decline in annual deaths per thousand population in the years 1960–1964 as compared with 1940–1944 was 52.9 per cent; in Ceylon, the per cent of decline was 56.9. There was no corresponding adjustment of the birth rates (111).

We have only recently returned from a trip to India and Ceylon and I should like to offer, as it were, a cross-section through the populations as we saw them.

In Darjeeling, India, our hotel room looked out at the precipitous mountain slopes of the Himalayan foothills, and in the distance to Kantenjunga and a majestic range of the high Himalayas. The province of Darjeeling used to be very sparsely inhabited, but the British Colonials discovered it as a hill resort—a refuge from the heat of India. Below our window was a large convent in which hundreds of little girls in uniform assembled each morning. If the precepts of *Humanae Vitae* are inculcated into these girls (and tens of thousands more like them throughout India), it is not difficult to imagine how soon the population will exceed the means of subsistence even in that rugged mountainous region, as in other parts of India.

In Ceylon, one need not imagine. Determined and successful public health efforts some ten years before had minimized malaria and other diseases, and the population growth had increased enormously. One drove along the roads between double streams of bullock carts, bicycles, and pedestrians, mostly children approaching child-bearing age. Already that beautiful and fertile island does not raise enough rice to feed its population. What of the future?

In Calcutta, one sees degradation of life which is almost beyond belief. Men, women, and children eating, sleeping, living, and dying on the pavements, shared with the sacred cows and the dogs rummaging in the garbage. Facilities for acceptable human life are simply swamped by the excess of population.

As physicians we must strive for the preservation of life; but must we not likewise strive for the preservation of the quality of life, and support in every possible way the emerging programs of foundations, institutions, and governments to bring populations and resources into states of viable equilibrium?

LITERATURE CITED

1. Clifton, C. E., *Ann. Rev. Microbiol.*, **20**, 1–12 (1966)
2. van Niel, C. B., *Ann. Rev. Microbiol.*, **21**, 1–30 (1967)
3. Taliaferro, W. H., *Ann. Rev. Microbiol.*, **22**, 1–14 (1968)
4. "Perhaps even those things at some future time it will be fun to remember"—Virgil, *Aeneid*, **1**, 203
5. Clopton, M. B., *J. Missouri State Med. Assoc.*, **30**, 499–501 (1933)
6. Memorial Service, Dr. Harvey Gilmer Mudd, *J. Missouri State Med. Assoc.*, **31**, 168–70 (1934)
7. Shapleigh, J. B., *Surgery, Gynecol. Obstet.*, 372–74 (Sept. 1925)
8. Mudd, H. G., *Surgery, Gynecol. Obstet.*, 579–81 (Apr. 1926)
9. Reagan, F. P., MacMorland, E. E., Mudd, S., *Anat. Record*, **12**, 265–80 (1917)
10. Burnet, Sir M., The Newer Immunology; An Evolutionary Approach. In *Infectious Agents and Host Reactions*. (Mudd, S., Ed., W. B. Saunders Co., Philadelphia, Pa., in press, 1969)
11. Mackaness, G. B., Blanden, R. V., Cellular Immunity. In *Infectious Agents and Host Reactions*. (Ibid.)
12. Mackaness, G. B., The Mechanism of Macrophage Activation. In *Infectious Agents and Host Reactions*. (Ibid.)
13. Mudd, S., *Anat. Record*, **14**, 55-56 (1918)
14. Mudd, S., Goldman, A., Grant, S. B., *J. Exptl. Med.*, **34**, 11–45 (1921)
15. Mudd, S., Grant, S. B., Goldman, A., *Ann. Otol., Rhinol. Laryngol.*, **30**, 1–73 (1921)
16. Wells, W. F., *Airborne Contagion and Air Hygiene*. (Harvard Univ. Press, Cambridge, Mass., 421 pp., 1955)
17. Mudd, S., *J. Bacteriol.*, **8**, 459–78 (1923)
18. Warren, S., Mudd, S., *J. Bacteriol.*, **9**, 143–49 (1924)
19. Mudd, S., Mudd, E. B. H., *J. Bacteriol.*, **9**, 151–67 (1924)
20. Mudd, S., Mudd, E. B. H., *J. Exptl. Med.*, **40**, 633–45, 647–60 (1924)
21. Mudd, S., Filters and Filtration. In *Filterable Viruses*, 55–94. (Rivers, T. M., Ed., Williams & Wilkins Co., Baltimore, Md. 1928)
22. Mudd, S., *J. Gen. Physiol.*, **7**, 389–413 (1925); 73–79 (1925); **9**, 361–73 (1926)
23. Lurie, M. B., *Resistance to Tuberculosis; Experimental Studies in Native and Acquired Defensive Mechanisms*. (Harvard Univ. Press, Cambridge, Mass., 391 pp., 1964)
24. Mackaness, G. B., Blanden, R. V., *Progr. Allergy*, **11**, 89–140 (1967)
25. Mackaness, G. B., *J. Exptl. Med.*, **120**, 105–20 (1964)
26. Mackaness, G. B., *J. Exptl. Med.*, **116**, 381–406 (1962)
27. Mitsuhashi, S., Sato, I., Tanaka, T., *J. Bacteriol.*, **81**, 863–68 (1961)
28. Blanden, R. V., Mackaness, G. B., Collins, F. M., *J. Exptl. Med.*, **124**, 585–600 (1966)
29. Mudd, S., Mudd, E. B. H., *J. Exptl. Med.*, **43**, 127–42 (1926)
30. Mudd, S., Fürth, J., *J. Immunol.*, **13**, 369–83 (1927)
31. Mudd, S., Mudd, E. B. H., *J. Exptl. Med.*, **46**, 167–72, 173–95 (1927)
32. Mudd, S., Lucké, B., McCutcheon, M., Strumia, M., *Colloid Symposium Monograph*, **6**, 131–38 (1927)
33. Mudd, S., Lucké, B., McCutcheon, M., Strumia, M., *J. Exptl. Med.*, **49**, 779–95 (1929)
34. McCutcheon, M., Strumia, M., Mudd, S., Mudd, E. B. H., Lucké, B., *J. Exptl. Med.*, **49**, 815–31 (1929)
35. Strumia, M., Mudd, S., Mudd, E. B. H., Lucké, B., McCutcheon, M., *J. Exptl. Med.*, **52**, 299–312 (1930)
36. Mudd, S., Lucké, B., McCutcheon, M., Strumia, M., *J. Exptl. Med.*, **52**, 313–29 (1930)
37. Mudd, S., Mudd, E. B. H., *J. Gen. Physiol.*, **14**, 733–51 (1931)
38. Mudd, E. B. H., Mudd, S., *J. Gen. Physiol.*, **16**, 625–36 (1933)
39. Lucké, B., Strumia, M., Mudd, S., McCutcheon, M., Mudd, E. B. H., *J. Immunol.*, **24**, 455–90 (1933)
40. Mudd, S., *J. Immunol.*, **26**, 447–54 (1934)
41. Mudd, S., McCutcheon, M., Lucké, B., *Physiol. Rev.*, **14**, 210–75 (1934)
42. Joffe, E. W., Mudd, S., *J. Gen. Physiol.*, **18**, 599–613 (1935)
43. Goodman, J. R., Moore, R. E., Baker,

R. F., *J. Bacteriol.*, **72**, 736–45 (1956)

44. Zucker-Franklin, D., Hirsch, J. G., *J. Exptl. Med.*, **120**, 569–76 (1964)

44a. Mudd, S., Mudd, E. B. H., *J. Immunol.*, **17**, 39–52 (1929)

45. Report of Proceedings, *Intern. Congr. Microbiol., 2nd London, 1936*, 502–10

46. Colebrook, L., *Intern. Congr. Microbiol., 2nd, London, 1936*, 134–35, Report of Proceedings

47. Austrian, R., *J. Clin. Pathol.*, **21**, Suppl. 2, 93–97 (1968)

48. Breinl, F., Haurowitz, F., *Z. Physiol. Chem.*, **192**, 45–57 (1930)

49. Mudd, S., *J. Immunol.*, **23**, 423–27 (1932)

50. Flosdorf, E. W., Mudd, S., *J. Immunol.*, **29**, 389–425 (1935)

51. Mudd, S., Flosdorf, E. W., Eagle, H., Stokes, J., Jr., McGuinness, A. C., *J. Am. Med. Assoc.*, 956–59 (1936)

52. McGuinness, A. C., Stokes, J., Jr., Mudd, S., *J. Clin. Invest.*, **16**, 185–96 (1937)

53. Flosdorf, E. W., Mudd, S., *J. Immunol.*, **34**, 469–90 (1938)

54. Hughes, J., Mudd, S., Strecker, E. A., *Arch. Neurol. Psychiat.*, **39**, 1277–84 (1938)

55. Flosdorf, E. W., Stokes, F. J., Mudd, S., *J. Am. Med. Assoc.*, **15**, 1095–97 (1940)

56. Mudd, S., Flosdorf, E. W., *New Engl. J. Med.*, **225**, 868–70 (1941)

57. Flosdorf, E. W., Hull, L. W., Mudd, S., *J. Immunol.*, **50**, 21–54 (1945)

58. Mudd, S., Thalhimer, W., Eds., *Blood Substitutes and Blood Transfusion.* (Charles C. Thomas, Springfield, Ill., 407 pp., 1942)

59. Wells, W. F., Wells, M. W., Mudd, S., *Am. J. Public Health*, **29**, 863–80 (1939)

60. Wells, W. F., Wells, M. W., Mudd, S., Lurie, M. B., Henle, W., Air-Borne Infection and Experimental Air-Borne Disease. In *Aerobiology*, Publ. 17, 284–89. (Am. Assoc. Advan. Sci., 1942)

61. Mudd, S., *Am. P. Public Health*, **34**, 578–86 (1944)

62. Mudd, S., *Med. Clin. N. Am.*, Philadelphia No., 1293–1319 (1944)

63. Mudd, S., *Bull. N. Y. Acad. Med.*, **21**, 393–418 (1945)

64. Wells, W. F., *Airborne Contagion and Air Hygiene.* (Harvard Univ. Press, Cambridge, Mass., 423 pp., 1955)

65. Hillier, J., Mudd, S., Smith, A. G., Beutner, E. H., *J. Bacteriol.*, **60**, 641–54 (1950)

66. Mudd, S., Smith, A. G., *Optik*, **7**, 319 (1950)

67. von Borries, B., Glaser, W., *Kolloid-Z.*, **106**, 123–28 (1944)

68. Winkler, A., König, H., *Zentr. Bakteriol. Parasitenk.*, Abt. I, *Orig.*, **153**, 9–15 (1949)

69. Marton, L., *J. Bacteriol.*, **41**, 397–413 (1941)

69a. Mudd, S., Lackman, D. B., *J. Bacteriol.*, **41**, 415–20 (1941)

70. Mudd, S., Anderson, T. F., *J. Am. Med. Assoc.*, **126**, 561–71 (1944)

71. Mudd, S., *J. Am. Med. Assoc.*, **126**, 632–39 (1944)

72. Anderson, T. F., Stanley, W. M., *J. Biol. Chem.*, **139**, 339–44 (1941)

73. Mudd, S., Anderson, T. F., *J. Immunol.*, **42**, 251–66 (1941)

74. Goto, S., Homma, J. Y., Mudd, S., *J. Bacteriol.*, **94**, 751–58 (1967)

75. Sall, T., Mudd, S., Davis, J. C., *Arch. Biochem. Biophys.*, **60**, 130–46 (1956)

76. Mudd, S., Yoshida, A., Koike, M., *J. Bacteriol.*, **75**, 224–35 (1958)

77. Sall, T., Mudd, S., Takagi, A., *J. Bacteriol.*, **76**, 640–45 (1958)

78. Mudd, S., Winterscheid, L. C., Brodie, A. F., Mitteil. deut. Gesell. Elektronen-Mikroskopie. In *Physikal. Blätter.*, **4**, 79–80 (1951)

79. Mudd, S., Brodie, A. E., Winterscheid, L. C., Hartman, P. E., Beutner, E. H., McLean, R. A., *J. Bacteriol.*, **62**, 729–39 (1951)

80. van Iterson, W., Symposium on the Fine Structure and Replication of Bacteria and their Parts. II. Bacterial Cytoplasm. *Bacteriol. Rev.*, **29**, 299–325 (1965)

81. Vanderwinkel, E., Murray, R. G. E., *J. Ultrastruct. Res.*, **7**, 185–99 (1962)

82. Mudd, S., Hartman, P. E., Payne, J. I., Phillips, A., In *Symp. Japan. Med. Congr., 14th*, 675–81. (Medical Journals Publ. Assoc., Tokyo, Japan, 1955)

83. Mudd, S., Takeya, K., Henderson, H. J., *J. Bacteriol.*, **72**, 767–83 (1956)

84. Kawata, T., Sall, T., Mudd, S., *J. Bacteriol.*, **79**, 459 (1960)

85. Mudd, S., Kawata, T., Payne, J. I.,

Sall, T., Takagi, A., *Nature*, **189**, 79–80 (1961)

86. Lenhart, N. A., Mudd, S., Yoshida, A., Li, I. W., *J. Immunol.*, **91**, 771–76 (1963)

87. Yoshida, A., Mudd, S., Lenhart, N. A., *J. Immunol.*, **91**, 777–82 (1963)

88. Hisatsune, K., De Courcy, S. J., Jr., Mudd, S., *Biochemistry*, **6**, 586–94 (1967)

89. Hisatsune, K., De Courcy, S. J., Jr., Mudd, S., *Biochemistry*, **6**, 595–603 (1967)

90. Staphylococcal Symposium, *Am. J. Public Health*, **48**, 277–318 (1958)

91. Conference on Staphylococcic Infections, *J. Am. Med. Assoc.*, **166**, 1177–1203 (1958)

92. *Proceedings National Conference on Hospital-Acquired Staphylococcal Disease, 1958.* (U.S. Dept. Health, Education, Welfare, Public Health Service, Communicable Disease Center, Atlanta, Georgia)

93. *Proceedings Conference on Relations of the Environment to Hospital-Acquired Staphylococcal Disease, 1958.* (U.S. Dept. Health, Education, Welfare, Communicable Disease Center, Atlanta, Georgia)

94. Gladstone, G. P., Mudd, S., Hochstein, H. D., Lenhart, N. A., *Brit. J. Exptl. Pathol.*, **43**, 295–312 (1962)

95. Mudd, S., Gladstone, G. P., Lenhart, N. A., *Brit. J. Exptl. Pathol.*, **46**, 455–72 (1965)

96. Lenhart, N. A., Mudd, S., Yoshida, A., Li, I. W., *J. Immunol.*, **91**, 771–76 (1963)

97. Li, I. W., Mudd, S., *J. Immunol.*, **97**, 41–45 (1966)

98. Shayegani, M. G., Mudd, S., *J. Bacteriol.*, **91**, 1393–98 (1966)

98a. Good, R. A., Bulwarks of the bodily defense. Lessons from disturbances of ontogenic development of the lymphoid and other immunologic apparatuses. In *Infectious Agents and Host Reactions.* (Mudd, S., Ed., W. B. Saunders Co., Philadelphia, in press, 1969)

99. Taubler, J. H., *J. Immunol.*, **101**, 546–49 (1968)

100. Taubler, J. H., Mudd, S., *J. Immunol.*, **101**, 550–55 (1968)

100a. Mudd, S., Zappasodi, P., Taubler, J. H., *Bacteriol. Proc.*, M16 (1969)

101. Baker, A. G., Taubler, J. H., Mudd, S., *Bacteriol. Proc.*, M200, (1968)

102. Mudd, S., Host-Parasite Interaction in Staphylococcal Infection: An Examination of a Successful Parasite. In *Topics in Medicinal Chemistry.* (Rabinowitz, J. L., Myerson, R. M., Eds., Intersci. Publs., New York, 360 pp., 1968)

103. Mudd, S., A Successful Parasite: Parasite-Host Interaction in Infection by *Staphylococcus aureus.* In *Infectious Agents and Host Reactions.* (Mudd, S., Ed., W. B. Saunders Co., Philadelphia, in press, 1969)

104. Sevag, M. G., Ashton, B., *Nature*, **203**, 1323–26 (1964)

105. De Courcy, S. J., Jr., Sevag, M. G., *Nature*, **209**, 373–76 (1966)

106. De Courcy, S. J., Jr., Sevag, M. G., *Antimicrobial Agents Chemotherapy — 1966*, 235–44 (1967)

107. Sharda, D. C., Cornfeld, D., Michie, A. J., *Arch. Disease Childhood*, **41**, 400–1 (1966)

108. Horwitz, M. R., Eshelman, J. L., Sevag, M. G., De Courcy, S. J., Jr., Mudd,, S., Blakemore, W. S., *Surg. Forum*, **19**, 532–34 (1968)

109. *World Academy of Art and Science, 2. The Population Crisis and the Use of World Resources.* (Mudd, S., Ed., Dr. W. Junk, Publ., The Hague, Netherlands, 562 pp., 1964; American Ed., Indiana Univ. Press, Bloomington, Ind., 562 pp., 1964)

110. *World Academy of Art and Science. The Population Crisis. Implications and Plans for Action.* (Ng, L. K. Y., Mudd, S., Eds., Indiana Univ. Press (Paper), 364 pp., 1965)

111. Population Reference Bureau, Inc., Washington, D.C., *Annual Report, 1967*

Ann. Rev. Biochem., Vol. 41

BIOCHEMISTRY AS PART OF MY LIFE

DAVID NACHMANSOHN

David Nachmansohn

BIOCHEMISTRY AS PART OF MY LIFE 782

DAVID NACHMANSOHN

Departments of Neurology and Biochemistry
College of Physicians and Surgeons, Columbia University
New York, NY

Character, emotions, literary and artistic experience, philosophy, and political involvements form an integral part of a personality. Since scientists are human, all these factors determine their reactions, their way of thinking, and must be essential elements in the formation of scientific ideas and views, motives, and attitudes. Knowledge alone of a special scientific field, however solid and profound, provides only the tools. What is achieved with these tools depends to a very large extent on the complex factors of a personality. When I was asked to write the prefatory chapter, I was thrilled. It seemed to me not only a great honor but a great challenge, an opportunity to describe my general views, my philosophy, the people who inspired me, the factors which influenced my scientific thinking, the background of my scientific theory. I wanted to omit as much as possible autobiographical data. But soon it became apparent to me, just as to some of my illustrious predecessors of these prefatory chapters, that a purely philosophical-historical essay would lose its special character and flavor and resemble some scientific abstracts without any experimental data. It would lose its attraction and vitality. It would fit into philosophical and historical books rather than into these chapters, where readers may like to know how a scientist who proposed a new concept found his particular way and what specific factors influenced his development and his theories. If one accepts the premise that science forms only a part of a personality, personal experience and background should be included.

An important factor in the development of each person is family background, cultural environment, and early education, experience, and inspiration. Coming from a Jewish liberal and democratic intellectual family, I was early imbued with the importance of cultural, spiritual, and moral values. My parents spoke several languages. My mother was a passionate piano player. She took the children rather early to concerts, operas, and theaters; she loved social life and we had frequent interesting visitors in the house. Since my parents wanted me to have the best possible education, I was sent to a "humanistic" gymnasium, at that time considered to be the most superior system for a really good education, rather than a preparation for a profession or other career. Latin and Greek, literature, and history were the dominating features in the program of our school. The standard was very high. One

1

THE EXCITEMENT OF SCIENCE

8243-2601/78/1127-0453$01.00 © 1978 ARI   453

teacher of Greek was particularly inspiring, especially when we read in the last school years Homer, Plato, etc. But in the whole curriculum we were never exposed to science. We had excellent mathematical courses, but only a perfunctory course of a few months in physics, never any chemistry, never any biology. In my early teens I started to become an avid reader of literature and history. The dramas of Schiller and Kleist, the novels of Balzac, Zola, Tolstoy, and Dostojewsky were among my favorites.

However, the writings which had the most profound influence on my thinking and outlook on life were those of Goethe. At the age of 17 I read the second part of *Faust* which gives the essence of Goethe's evaluation of man's life, much of his philosophy and of his great wisdom. Faust, having enjoyed all that life has to offer, love, wealth, honor, power, all kinds of pleasures, etc, realizes at the end that the greatest value in life, that which gives life real meaning and the deepest satisfaction, is creative achievement. When he has the vision of transforming by his activity a desert into a blooming fertile land where men, women, and children would live and work, he feels that such an accomplishment would make him happy.

> *Zum Augenblicke duerft ich sagen,*
> *Verweile doch, du bist so schoen!*
> *Es kann die Spur von meinem Erdenleben*
> *Nicht in Aeonen untergehn.*

Having this vision, Faust feels deep happiness; he loses, therefore, his pact with the devil and dies. Reading this drama time and again was a stirring experience, and the notion that creative work is more valuable and counts more than anything else in life remained the guideline of my life. Many other writings of Goethe, the beauty of his poems, his enthusiasm for life, the nobility of his thinking were a continuous source of inspiration.

In 1918 I graduated from the gymnasium. In view of my diversified interests I was undecided what I should study at the University. My parents insisted on medicine in view of the bleak economic and political outlook in Germany. This seemed to offer at least a certain economic security. The system in German universities permitted, in fact encouraged, attendance at a variety of lectures outside the special professional field, especially during the first year or two. A broad cultural background was felt to be desirable in addition to the special professional training. Having never been exposed to biology or science, but being deeply involved in the humanities, I attended lectures on Plato by Ulrich von Wilamovitz-Moellendorf, lectures on philosophy and history by E. Cassirer, F. Meinecke, and others. In the summer semester of 1920 in Heidelberg I attended lectures of Jaspers, A. Weber, H. Oncken, and a seminar on Kant's *Kritik der reinen Vernunft*. On the other hand, my first experience in the medical school was very discouraging. I found the lectures on anatomy, which at that time still dominated the preclinical training, extremely dull and boring. In the fall of 1920 I seriously considered giving up medicine to study philosophy, literature, and history. My parents and friends advised me to continue medicine. One reason was the disastrous situation of Germany at that time. The country was in the grip of terrible economic, political, and social turmoil; violence from extremist left- and right-wing groups was common. But the crucial argument was the violent and frightening antisemitism of students and even of some professors, especially in smaller universities. The Jews were claimed to be responsible for the terrible situation of Germany, for the humiliation by the Versailles Treaty, and for many other evils. Although brought up in a liberal and completely unorthodox family, I was very proud to be a Jew, proud of my Jewish heritage. I knew well Jewish history, I was deeply stirred by many parts of the Bible. Under the influence of Kurt Blumenfeld, one of the leaders of German Zionists, I became an enthusiastic and active supporter of the movement to create a Jewish home in Palestine where Jews would be free to develop their creative forces. The liberation of that country at the end of World War I from the centuries of Turkish rule, which had reduced the country to a desert, and the Balfour declaration seemed to open a new chapter. Certainly the country would need physicians. I decided to get a medical degree.

By the time I passed the medical state board in the winter 1923–1924, my whole outlook had drastically changed. The study of biology and medicine had opened to me an entirely new and exciting world full of challenging and interesting problems. I had read much about the history of science, about Galilei and Newton, Lavoisier, Pasteur, Claude Bernard, Helmholtz, etc. Biomedical research seemed to offer superb opportunities for creative work. Biochemistry, although a relatively young field in medicine, appeared to me the most promising and appealing one. I joined the laboratory of Peter Rona, Director of a biochemical laboratory in the Pathology Department of Berlin University. It offered an opportunity to get a good background in chemistry, physical chemistry, and especially in enzyme chemistry. It attracted many people who had graduated in medicine but intended to go into biochemical research. Rona was a superb teacher; in his lectures he discussed the latest progress in enzyme chemistry, especially the work of Neuberg, Warburg, and Meyerhof. Quite a few people who later contributed much to biochemistry were trained in this laboratory, among others F. Lipmann, H. A. Krebs, R. Schoenheimer, E. B. Chain, and Karl Meyer. Most fortunate for my future scientific development was my association with Hans H. Weber. He was a superb teacher. Under his guidance I worked on the problem of the independence of protein hydration from protein ionization. During that year, in 1925–1926, I acquired not only a more solid background in physical and protein chemistry, but also a critical approach to the experimental analysis of a problem. Since both Weber and I were passionately interested in history and politics, literature and philosophy, we had endless discussions which often lasted until midnight. Frequently we missed the last bus and had to walk from the center of Berlin to the far-away Western section where we both lived. A close friendship developed between us which has grown steadily over the decades. His daughter Annemarie spent five years in my laboratory in the

1950s. One day, Weber suggested to me to work with Otto Meyerhof as the most outstanding leader in the kind of problems I was interested in. When I went to see Meyerhof at the Kaiser-Wihelm-Institutes (KWI) in Dahlem, Meyerhof asked me about my training. When he heard five years medicine and three years with Rona he said, "Usually I don't accept beginners," but after an hour-long discussion about muscular contraction, about phosphocreatine, just newly discovered, and other problems, he accepted me. Thus started a period of my scientific formation which had the most profound impact on all my future scientific work.

The KWI were located in Dahlem in a fashionable suburb of Berlin, built on lovely grounds and surrounded by lawns and gardens. They were purely research institutes founded in 1910. Large endowment funds had been provided by industrialists and bankers who realized that Germany's wealth was based on the rapid development of basic science. Chemical, pharmaceutical, electronic, optical, and many other science-based industries had transformed Germany in a half-century from one of the poorest to one of the wealthiest countries. Germany's economic wealth and power would greatly benefit by strong support of scientific research. The members of the Institute had no teaching obligation. Therefore, in contrast to the Universities, no special fields had to be represented. The selection of a director of an Institute—or a subdivision—was based exclusively on his scientific stature, competence, and the excellence of his achievements. This accounts for the extraordinary collection of brilliant scientists in a small area in about six Institutes. In the late 1920s the KWI in Dahlem were one of the foremost and most outstanding scientific centers of Europe.

Meyerhof was even in his high school years deeply interested in philosophy and was a close friend of the philosopher Leonard Nelson. Both belonged to the school of Kant in the version of Fries, a philosopher of the early nineteenth century ("neo-Kantianer"). His deep involvement in philosophy is documented by the fact that for more than two decades he published articles in the *Abhandlungen der Fries'schen Schule*. For several years he was one of the two editors of these transactions. Following his graduation in medicine in Heidelberg in 1909 he entered the medical clinic of L. Krehl. There he met, in 1910, Otto Warburg. Warburg worked on the mechanism of respiration; his chief interest was the molecular mechanism of biological oxidation. The meeting and collaboration of these two brilliant men had a profound influence on their own development and in fact on biochemistry in general. It determined Meyerhof's interest in the field of biology, which was to become his life work. But he always retained his interest in philosophy. His philosophical approach to the study of life processes is apparent in his 1913 article, *Zur Energetik der Zellvorgaenge*. The article also reveals the depth of his knowledge and his remarkable ability to integrate a variety of physical, chemical, and biological phenomena.

The main problem which became the focal point of interest of his research was the question of how the potential energy of foodstuffs is made available to the cell. Between the initial energy input and its final dissipation as heat there must occur a series of energy transformations which may serve the living organism and provide the energy for its function and maintenance. Meyerhof chose the muscle as experimental material because it offered the most convenient and promising opportunity, with the methods then available, to correlate chemical transformations with both heat production and mechanical work. He started these investigations in 1918 and within a few years emerged some of the most fundamental notions and concepts of modern biology. A few of these contributions may be mentioned as an illustration. He introduced the basic notion of chemical and energetic coupling of cell reactions. From 1925 on he isolated the muscle enzymes which permitted the complete elucidation of the sequence of reactions during glycolysis, the "Meyerhof-Embden pathway." He showed that only one fifth to one sixth of the lactic acid formed during contraction is oxidized, while the rest is resynthesized to glycogen, the famous "Meyerhof quotient." This was the beginning of the notion of the cyclic character of cellular reactions. Detailed descriptions of Meyerhof's impact on biology may be found elsewhere (e.g. 1–5).

Meyerhof had only a few laboratories in the KWI of Biology, which housed six different sections. Only a small group of young scientists, usually only five or six, worked in his laboratory in those years. Somehow, Meyerhof's personality attracted a special type of people. Among my contemporaries were Severo Ochoa, Fritz Lipmann, Hermann Blaschko, Karl Lohmann, Ken Iwasaki, and Dean Burk. The smallness of the group facilitated close associations and friendships between us, and we had continuously lively and fruitful discussions. The quiet atmosphere was very conducive to creative work. Whenever we liked we went to see Meyerhof, on the average every two or three weeks, to discuss with him our data and ideas. These discussions were always inspiring. It was, however, not only our own particular experimental work, but all the various aspects of the problem of muscular contraction and beyond that of biology which were discussed and which influenced our thinking and approach.

Two flights upstairs in the same building were the laboratories of Otto Warburg. An extraordinary feature of his personality was the concentration of all his energy and will power exclusively on his research. He was already superbly trained in chemistry and physics at an early age, apparently under the influence of his father, physicist Emil Warburg. His brilliantly conceived experiments, the elegance and precision of their execution, evoked general admiration. An unforgettable experience was his lecture in 1928 on the spectrophotometric analysis of the heme. When Willstaetter criticized the application of spectrophotometry to such complex structures as cells, Warburg's reply was characteristic; if one finds appropriate reactions specific for the cell component which one wants to analyze, the rest of the cell is part of the test tube. In contrast to Meyerhof, philosophy, literature, and art played no role in his life. It was perhaps his deep devotion to his creative work and the strength of his personality which fascinated Meyerhof and attracted him to enter this field. Although the two men were different in many respects, they both felt, in spite of occasional frictions, a great admiration and respect for each other. In 1949 Warburg spent a summer in Woods Hole. One evening, Warburg and I were sitting for several hours in Meyerhof's garden. Meyerhof was in a happy mood and in particularly good shape. He discussed modern trends in biology and some of the recent remarkable achievements, frequently adding some interesting and stimulating philosophical comments. Warburg was visibly fascinated. On the way home he told me "You know, he is the greatest personality of all of us," quite a statement from a man who certainly did not suffer from an inferiority complex. Although he disliked visitors who came out of curiosity to meet the famous man, he was friendly and accessible to us when we came to see him with serious scientific questions. Warburg had even fewer academic collaborators than Meyerhof; H. A. Krebs and H. Gaffron were among my contemporaries. But the pioneering character of his experimental work and its impact on biochemistry is widely recognized. On the occasion of his eightieth birthday, 80 leading biochemists payed tribute to him in a Festschrift.

At a walking distance of a few minutes was the Institute of Carl Neuberg, the third giant of biochemistry in the KWI. He had an encyclopedic knowledge of chemistry and biochemistry. Fermentation was one of his main interests, and he proposed several schemes of intermediary steps which later had to be modified but which had, at the time, a great stimulating effect on biochemical thinking. Among the three laboratories existed many problems of common interest. Neuberg was always pleased when we visited him. In discussions with him, we often learned more in an hour than by reading in the library for a week.

Three such outstanding laboratories would have offered ample inspiration to any young biochemist. But one of the remarkable features of the KWI were the deliberate efforts, under the leadership of Fritz Haber, to break down the barriers between physics, chemistry, and biology. Haber's great fame is based on his success in producing ammonia from the atmospheric nitrogen and hydrogen, which he achieved by performing the reaction under high pressure in the presence of a uranium catalyst, and then transforming the ammonia into nitrates. This was a great contribution to agriculture and industry and was hailed at the occasion of the Nobel Prize award as a "triumph in the service of his country and of all mankind." But although these procedures may have been his most famous contribution, they are only one small fraction of his many scientific achievements of a fundamental character which deeply influenced almost all aspects of physical chemistry and chemistry. In addition to the basic aspects of his work, he had also the rare ability to visualize immediately how the new scientific knowledge might be applied for the benefit of industry or agriculture. Although imbued with Fichte's idea that the intrinsic and paramount purpose of science was the promotion of its own development, the aim of getting an increasingly deeper insight into the physical world, *rerum cognoscere causas,* he strongly felt that the ultimate aim of science should be to help the life of man and to benefit society. It is not surprising, in view of the wide range of his knowledge, his manifold interests, and his dynamic and forceful personality, that he organized seminars, the famous "Haber Colloquia," which became a great attraction and played an important role in the activities of all the KWI. In these seminars, which took place every second week, one met not only members of Haber's institute (Ladenburg, Freundlich, Polanyi, Bonhoeffer, etc) but also many others, among them Hahn and Lise Meitner from the chemistry, von Laue, and members of the three biochemistry groups and of the other biology departments (Mangold, Goldschmidt, Correns, Hartmann) and their associates. In these seminars, Haber's extraordinarily brilliant mind was at its best. He stimulated vigorous and exciting discussions. His unusual ability to recognize and grasp the essential aspects of the many different topics presented, even if they were not in his own field, to discover weaknesses and to raise pertinent questions, led to lively exchanges and sometimes to vigorous fights among the many outstanding people present. All these factors helped create a unique and exciting scientific atmosphere and made these seminars an unforgettable experience. One of the impressive features, especially for us younger people, was the fact that these outstanding scientists, including Haber, had never the slightest hesitation to admit mistakes or serious errors with the greatest ease, to withdraw their claims or objections in the face of a good argument. I was recently reminded of the refreshing and challenging atmosphere in these seminars, of these determined efforts to explore in common important and challenging problems without any vanity, prestige, or prejudice, when I read the recently published exchange of letters between Albert Einstein and Max Born (6). In a letter to Born, Einstein writes: "I have made a monumental blunder ....., but one has to console oneself. Only death prevents one from making blunders." Born too admits without the slightest hesitation serious errors as natural and unavoidable parts of scientific endeavors and activities. The fame of the Haber seminars attracted visitors not only from various parts of Germany, but even from abroad: scientists came to Berlin just to attend a Haber Colloquium.

The KWI in Dahlem had an extraordinary impact on science in the twentieth century. In the field of physical sciences it may suffice to quote as an illustration the discoveries of Hahn and Strassman: they opened the way to the use of atomic energy which Rutherford only a few years earlier had thought to be impossible. The question of whether atomic energy will be used for nearly unlimited potential benefits or for the total destruction of mankind is a problem which surpasses the frame of this article, but the discoveries probably mark a decisive turning point in history and have opened a new age. In the biological sciences the most profound and perhaps unique contribution of the Dahlem era was the systematic and most successful efforts of a group of brilliant scientists aimed at the understanding of living organisms in terms of physics and chemistry, i.e., the analysis of biological and especially cellular mechanisms on a molecular level. The inspiration to biologists derived from the genuine interest of outstanding physicists and chemists in their problems and the continuous mutual exchange of ideas between the two groups—the Haber Colloquia formed only one, though the most visible expression—were essential factors. By their revolutionary thinking and concepts and by their superb achievements the Dahlem group and their pupils, who especially in the Nazi era were dispersed all over the world, were instrumental in the development of a new era in biological sciences; they were pioneers in laying the foundation of an approach frequently referred to as molecular biology, which is increasingly penetrating all fields of biological sciences.

For a full appreciation of the atmosphere of Berlin, of which the KWI represented only a small aspect, one has to consider it in the context of the cultural life of the Weimar Republic in general. In spite, or perhaps as a result, of the disastrous war and the economic, political, and social convulsions of the postwar period, the cultural forces in Germany flourished as never before. The creative forces in the Weimar Republic were at a pinnacle. Berlin became a center of nearly magic attraction to intellectuals in all fields, to artists, writers, philosophers, and scientists. It had 30 to 40 theaters of world fame under leaders such as Max Reinhardt, Barnowski, Erwin Piscator, Leopold Jessner, and many others. The superb quality of the performances by some of the greatest living actors could hardly be surpassed. The musical life was traditionally very strong. There were four superior operas in Berlin playing during the whole year. Many other cities in Germany, big and small, had their own opera houses. The number of poets and writers was astonishing: Rainer Maria Rilke, Stefan George, Thomas and Heinrich Mann, Franz Werfel, Ernst Toller, Walter Hasenclever, Wolfskehl, Stefan and Arnold Zweig, Bertold Brecht, Fritz von Unruh, Ricarda Huch, to mention just a few of the better-known names. Painting and sculpture were thriving: Kaete Kollwitz, Barlach, George Grosz, Zille, Liebermann, again to quote a few of the best-known names as illustration. In architecture, the Dessauer Bauhaus began to attract worldwide attention. It is well known how some of its members, such as Gropius and Mies van der Rohe, have influenced modern architecture all over the world. Although the universities had suffered severe and grievous losses during the war and the early postwar period and were beset by all kinds of difficulties, there were many outstanding and world famous scholars both in science and humanities. It is true that many of these developments, which seemed to explode in the Weimar Republic, had already started early in the century, i.e., in the prewar period. But the deeply stirring experiences of the war and the tremendous revolutions during the postwar period were apparently an extraordinarily powerful catalyst, a source which seemed to have inspired and electrified deep human emotions, which came to the surface in the Weimar Republic especially in the later 1920s. No visitor from Europe or the United States could fail to notice the fascination, the vitality, the dynamism of Berlin. The scientific atmosphere of the KWI must be viewed in the perspective of the general cultural life of this era. An analysis of the political weakness of the Weimar Republic and the reason for its seemingly sudden collapse would by far surpass the frame of this article.

By 1930 it had become apparent that the days of the Weimar Republic were numbered, in view of the catastrophic economic deterioration and the concomitant rise of the Nazi Party. A few weeks after the Nazi regime came into power, in January 1933, I left Germany with my family. I went first to Palestine, but there was not the slightest possibility of continuing my research. Weizmann personally strongly urged me to wait a few years and to continue my research wherever the conditions appeared favorable. He had great plans for the future of science in Palestine; he was particularly anxious to develop biochemistry there and asked for my active help. In fact, in 1936 he invited H. A. Krebs and myself to Israel to discuss with us detailed plans. Soon afterwards, however, the riots began and the rapidly deteriorating situation forced the postponement of all these plans until long after World War II.

Among several offers I chose Paris. One of the main reasons was the opportunity of complete independence in my research. The years 1933 to 1939, which I spent at the Faculté de Sciences in Paris, had a deep influence on my life. I experienced the fundamental and amazing difference between knowing something about a civilization from the outside, by studying books, and the exposure to it from the inside. When I arrived in Paris, my French was poor and my first immediate effort was to learn French. While I had read much French literature in translation, it was a revelation to realize the unbelievable difference between reading Anatole France or Balzac, Hugo or Verlaine, in their original language or in translations. The charm and the monuments of Paris and of the French countryside, Provence, Bourgogne, Bretagne, cannot be grasped and appreciated without the knowledge of the language, the literature, and the history. It really takes years to get the true flavor of the extraordinary richness and charm of French civilization. It is hard to describe in a few words the new world which was opened to me and hard to find a proper expression of how greatly these years in France enriched my life and left a deep and permanent imprint.

Although the life of the French scientific community in the period between the two wars was not comparable in intensity to that in Dahlem, I was fortunate to meet several outstanding scientists who had an inspiring influence on my scientific thinking. I must mention in particular René Wurmser and his group. I also had an opportunity to meet personally some of the famous scientists of the rue Pierre Curie and the Collège de France, among others Jean Perrin, Langevin, Mme. Curie, Joliot-Curie and his wife Irène, Bauer, and many others.

Before turning to my scientific activities in Paris, I would like to mention an event which had important consequences. I visited the United States in September 1937 and gave a lecture at Yale University in John F. Fulton's department. He invited me to join his laboratory, which I did in the summer of 1939 shortly before the outbreak of World War II. Meyerhof and his wife Hedwig visited the United States, incidentally, at about the same time and we spent much time together. Meyerhof was quite depressed; his position in Germany had become extremely difficult. He had hoped to find a place in the United States, but nothing happened during this visit except an offer of a minor and unsatisfactory position in an industrial concern. Knowing the respect and admiration for Meyerhof in Paris, I proposed to inquire about the possibility of bringing him to Paris. On my return I contacted immediately René Wurmser, Henri Laugier, and Jean Perrin. I got their enthusiastic support and he was offered a research professorship. Since an open correspondence was dangerous, we had arranged in New York a special code. He arrived in Paris in September 1938. This period marked the beginning of an extremely close and intimate personal friendship. We had many common interests and friends. We made excursions to Chartres, the chateaux de la Loire, etc. His amazing knowledge of art and history, combined with an extraordinary memory, made every one of these excursions a delightful experience. After the collapse of France he came in 1940 to this country as a research professor at the University of Pennsylvania in Philadelphia. From the time he fled from Paris until his arrival here, he and his wife passed through a trying period of great physical hardships. Meyerhof spent most of his summers in Woods Hole, and our friendship grew more intensive and remained very close until his death on October 6, 1951. Severo and Carmen Ochoa also spent most of their summers in Woods Hole, and the three couples spent many happy days together. Meyerhof loved Woods Hole and was in an unusually relaxed mood there. Discussions on many topics and on many occasions revealed the richness of his personality and his great wisdom.

Quite a few of my scientific friends from Germany had gone in 1933 to England. I was anxious to keep my personal associations with them and each year regularly attended a few meetings of the Physiological Society, especially since there were always stimulating discussions. After all, it was a few hours trip from Paris to London. I never anticipated that by attending these meetings my scientific interest would take a new and unexpected turn and that the new direction would determine my scientific life work for a period of over 35 years.

One of the main controversial topics of these meetings of the English Physiological Society in the 1930s was the role of acetylcholine (AcCh) in nerve activity. Dale and his associates had proposed that AcCh acts as a mediator of nerve impulses across junctions (synapses) between nerve and nerve, and nerve and muscle, in contrast to the electrical currents that propagate impulses along nerve and muscle fibers. Two major facts seemed to support this idea in a most impressive way. First was the powerful action of

AcCh on some synaptic junctions, in striking contrast to its complete failure to affect conduction of nerve or muscle fibers. Secondly, on stimulation of nerve fibers, AcCh appeared in the perfusion fluid of synaptic junctions, although this appearance was observed exclusively in the presence of eserine, an extremely potent inhibitor of acetylcholinesterase, the enzyme rapidly hydrolyzing AcCh.

The idea of neurohumoral transmission met vigorous opposition from many prominent neurobiologists, from the school of Sherrington in England, and from physiologists in this country and in continental Europe. The experimental observations were not questioned, but their interpretation was. The assumption that two fundamentally different mechanisms propagated impulses along axons and across junctions seemed difficult to reconcile with the evidence based on electrical signs, which indicated many basic similarities in the properties of the membranes in fibers and in junctions, in spite of some marked differences.

The lively and passionate discussions for and against neurohumoral transmission were fascinating to experience. Coming from biochemical laboratories in which the role of enzymes and proteins was in the center of all notions of cellular function, I was struck by the complete absence of any biochemical approach to such a controversial and important question. Knowledge of the enzymes hydrolyzing and forming AcCh, the bioenergetics of the processes involved, and their integration into general metabolism seemed to me imperative and indispensable for finding satisfactory explanations of the many open problems concerning the physiological function of AcCh.

There are many different types of ester-splitting enzymes in the organism, but virtually nothing was known about the enzyme which has the specific function of hydrolyzing AcCh. There were immediately a number of interesting and challenging questions which required answers: what was the distribution of this enzyme in different tissues of a given species or in various species throughout the animal kingdom; what was its concentration; did the enzyme have special characteristic features; when did it appear during growth and development in relation to function, etc. I started this work in 1936. One of the first interesting findings was the presence of the enzyme in high concentration in a great variety of different types of excitable fibers of nerve and muscle, in brain, and in vertebrates and invertebrates, whereas it was hardly detectable in organs such as kidney or liver. The concentration appeared to be a few times higher at the level of junctions than in the fibers. On the basis of some erroneous assumptions I estimated that the concentration of the enzyme at the junctions must be extremely high compared to that in the fibers.

While reading the literature about the neuromuscular junction, I came across an article of Lindhard (7) in which he describes electric organs of fish as a kind of modified set of muscle fibers comparable to motor endplates, but in which the muscular elements are either absent or exist only in rudimentary form. Obviously, it would be of greatest interest to test the acetylcholinesterase activity in this tissue. A few weeks later I saw by chance for the first time electric fish (*Torpedo marmorata*) at an exhibit at the World's Fair in Paris in the summer of 1937. Dr. Catherine Veil from the Faculté de Sciences was in charge of the fish and I asked her for a few specimens. When a young chemistry student working with me at that time, Mlle. Arnette Marnay, determined the enzyme concentration, the result was simply stunning: 1 kg of electric tissue (fresh weight) hydrolyzed 3–4 kg of AcCh per hr, although the tissue is 92% water and only 3% protein.

This astonishing concentration of the enzyme in an organ so highly specialized in its function and so poor in proteins made a deep impression on me. In my lectures to students I like to quote the words of August Krogh: "Nature has created quite a few animals with the special purpose to help biologists to solve their problems." For a biochemist interested in the analysis of the proteins and enzymes associated with the function of AcCh, it was immediately obvious that here was not only an extraordinary, but probably a unique material for the type of biochemical research I had in mind. Information about the proteins associated with the function of AcCh and the bioenergetics of the reactions might well provide insight about the controversial role of this ester and help to overcome the impasse in which this problem found itself as a result of the vigorously opposing view of physiologists.

The use of electric tissue in the more than 30 years which have passed since this discovery was instrumental in the isolation, identification, and characterization of the proteins directly associated with the function of AcCh. It was of paramount importance in correlating many chemical and physical events in bioelectrogenesis. Willstaetter (8) writes that a scientist needs for success four Gs (Geduld, Geschick, Geld, und Glueck). It was certainly great luck that I came at an early stage of my investigations across such an extraordinary material, nearly tailor-made for my aims. But while this material was an invaluable factor in the progress, it seems to me imperative to emphasize two other factors which were essential for the advances accomplished in the last three decades. The first factor was the inspiration which I received from the Dahlem atmosphere and, in particular, by my association with Meyerhof as to the fundamental notions and concepts for analyzing the chemical basis of cellular functions such as the early emphasis that cellular reactions are chemically and energetically coupled. The importance of bioenergetics and the sequence of energy transformations for the analysis of cellular mechanisms have already been mentioned. I was in Meyerhof's laboratory when he and Suranyi discovered, in 1927, that phosphocreatine breakdown, in contrast to that of ordinary phosphate esters, was associated with a large enthalpy. This was the birth of the notion that some phosphate derivatives decompose with a large change of free energy, and in the following years several other such phosphate derivatives were found, in particular ATP, which is used by the cell to trap the energy of oxidation. Meyerhof (9) summarized these discoveries which had such a great influence on the development of modern biochemistry in a review article in 1937. These ideas were later further elaborated by Lipmann (10) and Kalckar (11); their historical development has been summarized by Kalckar (12). Another exciting experience for all of us in the laboratory were the advances following successful extraction of the glycolytic enzymes from muscle, the systematic analysis of the reactions during glycolysis now usually referred to as the Meyerhof-Embden pathway.

The second crucial factor was the spectacular progress of biochemistry in general during the last three decades, due to a large extent to improvements in instrumentation and to highly refined methods provided by progress in physics and chemistry. These opened unforeseen possibilities to explore cellular mechanisms at the cellular, subcellular, and molecular levels; they permitted application of the previously mentioned notions and of the way of thinking prevailing in the KWI in Dahlem to the analysis of biological functions, which at that time, though only a few decades past, seemed to be in many areas beyond the reach of experiment. For my particular problem, the analysis of the proteins associated with AcCh and their function in bioelectrogenesis, two special developments were essential: the information obtained during the last decade about biomembranes in general; and the investigations of the protein properties culminating in the exploration of the three-dimensional structure of by now probably more than a dozen enzymes and proteins.

As a result of the biochemical approach a modification of the original theory became necessary. AcCh is not a neurohumoral transmitter between two cells; it is never released from the cell. Its appearance in the outside fluid is an artifact. Its release and action are intracellular, taking place *within* the excitable membrane. It is the trigger which initiates and controls the permeability changes permitting the ion movements across the excitable membranes during electrical activity. The picture, as it has evolved during the last three decades, is as follows: excitation apparently leads to the release of AcCh present in the membrane in a bound form. It acts as a signal recognized within the membrane by a specific AcCh-receptor protein. The reaction induces a conformational change of the protein, thereby possibly releasing, by allosteric action, Ca^{2+} ions bound to the protein. The release of Ca^{2+} ions may induce further conformational changes of phospholipids and other polyelectrolytes in a membrane element referred to by Changeux as the ionophore. The end result of these chemical reactions is a change in the permeability to ions, permitting the movements of many thousands of ions, possibly as many as 20,000 to 40,000 in each direction, per molecule of AcCh released. These reactions thus act as typical amplifiers of the signal given by AcCh. A new circuit has been generated which stimulates the adjacent point. There the same processes are repeated and thus the impulses are propagated along the fibers. Acetylcholinesterase rapidly hydrolyzes the ester, in microseconds, thereby permitting the return of the receptor protein to its original conformation and the reestablishment of the barrier for the ion movements. AcCh in its bound form and the two proteins reacting directly with the ester (receptor and esterase) are presumably linked together structurally as well as functionally and form a protein assembly in the excitable membrane in a way comparable with other well-known multienzyme systems such as the electron transfer system in mitochondrial membranes, the fatty acid synthetase, etc. The structural organization of the system may account for the efficiency, precision, and high speed of the events in the membrane during electrical activity. While an essential role of Ca^{2+} ions in the permeability changes of excitable membranes appears likely, their release requires a specific control mechanism. Among the cell components only proteins have the ability of recognizing specific ligands and of thereby providing the proper control for the initiation or termination of specific cellular function. These views are supported by a vast and, in the last few years, rapidly increasing number of experimental data.

The function of a prefatory chapter is not to present the work of the author, even in a nutshell. The reader seriously interested in the chemical basis of nerve activity is referred to the literature for details. The early work is described in a monograph (13); the progress of the last decade is summarized in several review articles (14–21). However, a few milestones in the development may be briefly mentioned as an illustration of how the combination of various factors, which formed the basis of the work, led to a series of developments resulting in the new concept. In particular, some of the ideas which stimulated the work on the three proteins may have some general interest.

Acetylcholinesterase.—In view of the wealth of information resulting from the isolation of the glycolytic enzymes, the discovery of the extraordinary acetylcholinesterase concentration in electric tissue immediately suggested to me the use of this material for the isolation of the enzyme. In 1938 a highly active enzyme solution was obtained. This was the first acetylcholinesterase preparation obtained in solution. Until that time serum esterase was considered to be the enzyme which had the physiological function of hydrolyzing AcCh and was widely used as a test enzyme to study the reaction with potent and competitive inhibitors. In the early 1940s AcCh esterase was purified from electric tissue about 500-fold. Procurement of electric tissue in those years (World War II) was extremely difficult. The amounts of highly purified protein were extremely small, but the purified enzyme preparations were adequate for kinetic studies. The topography of the active site was analyzed and many of the reaction mechanisms of enzyme inhibitors used in biology and medicine were explained. Of particular interest was the interpretation of the mechanism of action of organophosphates, widely used as insecticides and potentially powerful chemical warfare agents. These organophosphates react covalently with the serine oxygen in the active site of AcCh esterase and form a phosphorylated enzyme. In contrast to the physiologically formed acetylated enzyme, which reacts in microseconds to form acetate and restored enzyme, the phosphorylated form is relatively stable and, because of the vital role of the enzyme for the organism, the action of organophosphates is fatal. The most likely way to restore the enzyme activity appeared to be an attack by nucleophilic agents which would compete with the oxygen of the serine residue and thereby remove the phosphoryl group in a displacement reaction. A potent nucleophilic agent was developed, pyridine-2-aldoxime methiodide (PAM), and was found at very low concentrations to reactivate rapidly the phosphorylated enzyme. When the compound was tested on animals, it proved to be a most efficient antidote against insecticide poisoning which acted by repairing the biochemical lesion. When injected into animals in combination with atropine, a compound which protects the receptor protein against excess AcCh, the antidote protects against doses of insecticides more than twenty times higher than those lethal to 100% of unprotected animals. PAM is today widely used in many countries as antidote in organophosphate insecticide poisonings.

Beginning in 1939 biochemical evidence began to accumulate in our laboratory that the enzyme is membrane bound. But these data were of necessity of indirect nature. Only during the last decade has it been shown by electronmicroscopy in combination with histochemical staining methods that the enzyme is indeed exclusively located in excitable membranes. This has been demonstrated with a great variety of different fiber types from animals throughout the animal kingdom. Of special interest is this exclusive localization of the enzyme in the excitable membrane of the electroplax: since the membrane forms 10^{-4} or less of the cell mass, the observations indicate that 1 g of excitable membrane of the electroplax may hydrolyze 30 kg of AcCh (or more) per hr.

The exciting developments in the 1960s which led to the exploration of the three-dimensional structure of several proteins and enzymes made it desirable to work out a large-scale preparation of acetylcholinesterase which led, in 1967, to the crystallization of the enzyme as hexagonal prisms. The enzyme has a molecular weight of 260,000 and is formed by two different polypeptide chains. The number of subunits and active sites is still under investigation.

Choline-O-acetyltransferase (Choline acetylase).—The isolation, purification, and crystallization of acetylcholinesterase was essentially the outcome of the discovery of a uniquely favorable source for extracting this enzyme. The discovery of the enzyme that acetylates choline was based on general ideas of biochemistry; but in this particular case the application of bioenergetic considerations and the notion of the coupling of chemical reactions in cell metabolism provided essential clues which led to the success of these investigations. When observations began to indicate that AcCh is associated with bioelectrogenesis in general and is not a "transmitter" the question arose: how is its formation and hydrolysis integrated into the sequence of chemical reactions associated with electrical activity? The ionic concentration gradients apparently provide the energy required for the ion movements during activity following the concentration gradients; AcCh was proposed to act as the trigger making this potential source of energy effective. But how was the energy spent during activity restored; what reactions provided the energy for the resynthesis of acetylcholine? I had worked in Meyerhof's laboratory for several years on the role of phosphocreatine in muscular contraction. I was surprised to find that its concentration in electric tissue of the eel, in spite of the extremely low protein content of this tissue, is as high or higher than in striated muscle. ATP is nearly but not quite as high as in muscle. In view of the essential role of these two compounds in the energy transformations in muscular contraction, their conspicuously high concentrations in the highly specialized electric tissue suggested to me the possibility that they may also be essential in providing the energy requirements for genesis of bioelectricity although, for many reasons, through different mechanisms. A strong breakdown of phosphocreatine was indeed found to coincide roughly with electrical activity (22). It appeared safe to assume that phosphocreatine was used for the resynthesis of ATP used during activity. Even if a large part of the energy of ATP hydrolysis were used for the restoration of the ionic concentration gradient, as was later shown to be the case, the data suggested that part of the energy of ATP hydrolysis may be used for the acetylation of choline.

When ATP was added to electric tissue extracts, acetylation of choline was disappointingly small. This could be due to the rapid hydrolysis of ATP by ATPase. Indeed, when ATP was added in the presence of sodium fluoride, a strong AcCh formation was obtained (23). This was the first enzymic acetylation obtained in a soluble system in which the free energy of ATP hydrolysis was used. The finding was so unexpected that three journals refused its publication (*Science, J. Biol. Chem., Proc. Soc. Exp. Biol. Med.*). John F. Fulton accepted it for the *Journal of Neurophysiology*. Acetylphosphate, discovered by Lipmann (24), was not active. Lipmann refused to accept my findings for two years. But at the end of 1944 he came to my laboratory with a box containing a number of phosphate derivatives to test their action in my system. In the beginning he had hoped that acetylphosphate would work in my test system, but for some reason it did not work in his. After a number of experiments he realized that only ATP was effective of all the compounds tested, and he then accepted my view of its role as the energy source for acetylation. The mechanism of acetylation remained obscure, however, until the coenzyme of acetylation (CoA) was discovered (25–27) and its chemical nature elucidated (28). ATP, CoA, and acetate form acetyl-CoA; acetyl-CoA transfers the acetyl group to acetyl acceptors, including choline. The mechanism of this reaction was finally elucidated by Berg (29) who found that the donor of the acetyl group of CoA is acetyl adenylate formed by the reaction of ATP and acetate with liberation of pyrophosphate. It was a deep satisfaction that Meyerhof, in his opening address at the first *Symposium on Phosphorus Metabolism* (30), shortly before his death, mentioned my findings among the three important contributions to the biological role of ATP.

Acetylcholine receptor protein.—Although the notion of receptors as cell components reacting specifically to highly active chemicals affecting the cell was postulated early in the century (Paul Ehrlich, Langley, etc), no methods were available to characterize, isolate, or identify them. This applied also to the action of AcCh. A turning point was the development of the isolated monocellular electroplax of *Electrophorus* in 1956 (31). The many unique and extraordinary features of this preparation have been repeatedly described. The truly outstanding characteristic is the predominance of the excitable membrane extremely rich in acetylcholinesterase and receptor protein in contrast to the nearly negligible metabolic activity of the rest of the cell. In the following decade it was demonstrated that the receptor was a protein, as I had postulated in my Harvey lecture in 1953, and many of the molecular properties of this protein were analyzed. They finally permitted the isolation and characterization of this receptor protein. Applying α toxin of *Naja*, Changeux and his associates, using electric tissue, were recently able to separate unequivocally this protein from the enzyme. Thus, the hopes raised in 1937 by the discovery of the high acetylcholinesterase concentration in electric organs, namely the possibility of characterizing the properties of the proteins and enzymes associated with the function of AcCh, have been borne out by the progress of the last three decades.

Information about the three proteins associated with the action of AcCh, no matter how pertinent, does not by itself provide evidence for their proposed role. For example, neither the high concentration nor the localization of the enzyme or of the receptor in excitable membranes provide conclusive evidence for the postulated function. The theory is based on many additional and different types of data; some of them may be briefly mentioned as an illustration. 1. Potent, competitive, and reversible inhibitors of the enzyme, such as physostigmine, reversibly affect axonal conduction, especially when suitable preparations are used. Applied, e.g., to a single frog sciatic nerve fiber, the depressing effect of the inhibitor on conduction is as powerful and as rapid as on the synaptic junction: it takes place within seconds and in low concentrations. 2. Organophosphates, irreversible inhibitors of the enzyme, block conduction irreversibly. The contradictory results reported in early studies on these compounds have been explained by factors by which the behavior of enzymes in solution may strikingly differ from that in intact cell structures. 3. The reported disappearance of the enzyme activity prior to the block of electrical activity on exposure to organophosphates turned out to be the result of the lack of methods capable of quantitative determinations of the total enzyme activity in tissues, even in normal ones; the available methods permit quantitative evaluations only in solution (20). 4. In several preparations conduction irreversibly blocked by organophosphates may be restored by exposure to PAM which, as mentioned before, repairs the specific biochemical lesion postulated to be responsible for the irreversible block. 5. The essential role of the AcCh-receptor protein for the electrical activity in conducting as well as in pre- and postsynaptic membranes has been demonstrated by a variety of physicochemical data. For instance, structural analogs of AcCh (so-called "local anesthetics") are receptor inhibitors; since their effect is due to the competition with AcCh for the AcCh receptor, their block of electrical activity in conducting as well as synaptic parts of the excitable membrane demonstrates the essential role of the receptor protein in these two parts. In a series of compounds the molecular groups of AcCh were systematically substituted and tested on the electroplax. By successive modifications of the chemical structures it was shown how AcCh is gradually transformed from a receptor activator, acting on junctions only, into a group of receptor inhibitors capable of penetrating the barriers protecting the excitable membranes in the conducting parts and reacting with the receptor. 6. The failure of AcCh and related compounds to act on conduction, i.e., the limitation of the effects to the junction, has been explained in a variety of ways as being due to the presence of structural barriers surrounding the excitable conducting membranes and impervious to quaternary ammonium derivatives. For instance, curare, a potent receptor inhibitor long believed to act on junctions only, reversibly blocks conduction in isolated frog sciatic nerve fibers; at the nodes of Ranvier the compound may reach the excitable membrane, elsewhere protected by the myelin. 7. AcCh and curare do not affect conduction even in unmyelinated axons. In experiments on the squid giant axon the Schwann cell, about 4000 Å thick, formed a barrier for certain quaternary ammonium derivatives; they do not penetrate into the interior, in contrast to their tertiary analogs. After a brief exposure to snake venom in low concentrations (phospholipase A is the active principle), AcCh and curare do penetrate into the interior and do affect conduction. 8. A direct action of AcCh and related compounds on conduction was found on some axons, e.g., on those of the walking leg of lobster in which the excitable membrane seems to be incompletely protected. 9. The Hodgkin-Huxley theory, which dominated the thinking of many electrophysiologists for two decades, assumed that ion movements during electrical activity are a simple diffusion process. This assumption is contradicted by a variety of data. For instance, drastic changes of ion concentrations on the inside and outside of the axonal membrane have little effect on the electrical parameters in contrast to the predictions of the theory.—There is a strong heat production during the rising phase of the action current and a strong heat absorption during its falling phase. There is no alternative to the assumption of chemical reactions being responsible for the heat produced and absorbed.—The theory was based on the Planck-Nernst equations which are applicable only to systems in equilibrium, whereas events in cell membranes require the use of nonequilibrium thermodynamics. For detailed accounts of these and many other recent developments the reader is referred to the literature quoted (especially 17–21).

Conduction along axons and transmission across junctions.—The special features of synaptic junctions cannot be discussed here. It may, however, be briefly mentioned that a vast amount of evidence has accumulated during the last decade that AcCh initiates the permeability changes in the pre- and post-synaptic membranes in a way similar to that in the membranes of nerve and muscle fibers (32). It acts in the junctional membranes as a signal amplified by a series of chemical reactions: per 1000 molecules of AcCh released in the nerve terminal membrane by an impulse, many millions of potassium ions cross the nonconducting gap and induce a release of AcCh in the postsynaptic membrane, initiating the same processes there. The appearance of AcCh in the perfusion fluid of junctions is an artifact due to its incomplete hydrolysis in the presence of physostigmine, the potent inhibitor of acetylcholinesterase; in its absence, no trace of AcCh appears. Otto Loewi's experiments on the frog heart, in which no physostigmine was used, are not reproducible. The great diversity of pharmacological and physiological phenomena on junctions, compared to those in axons, are due to the great variety of differences in structure, shape, organization, and environment, but not to the basically different mode of action of AcCh. The proposed concept integrates the biochemical data with those on which the idea of neurohumoral transmission was built and with the views of many prominent neurophysiologists who assumed a basically similar mechanism for conduction along axons and transmission across junctions.

I was fortunate to have been able to attract to my laboratory, especially in the last twenty years, quite a few competent, qualified, and devoted collaborators. The results of the joint efforts are described in the review articles. These associations were pleasant and inspiring and have been of great benefit for the work.

Although the combination of the three factors outlined above provided the basis which permitted the accumulation of information on the proteins associated with AcCh and their function in excitable membranes, progress in science obviously depends on additional elements which are more difficult to define, but are nevertheless real and essential. One may refer to them as imagination, fantasy, or intuition: ideas that permit one suddenly to see connections between seemingly independent facts, interpretations of data which appeared unexplainable, the ability of integrating a variety of loosely or unconnected observations. In his recent work Heisenberg (33) writes: "It is a common mistake to think that all that matters in science is logic and the understanding and application of well-established laws. Actually, imagination plays a decisive role in science, especially natural science. For even though we can hope to obtain the facts only after many sober and careful experiments, we can fit the facts themselves together only if we can feel rather than think our way into the phenomena." Meaningful experiments are frequently, if not usually, the results of ideas without which most of the experiments would never have been performed. However, new ideas frequently clash with long-established beliefs and are therefore often resisted. This is only natural and not altogether undesirable; actually, constructive and justified criticism stimulates the design of new and more convincing experiments.

However, the problem of the resistance to new scientific ideas touches an important topic with wide implications; it has very deep roots and is as persistent as that encountered in most other fields of human life and activities; it is only less widely known and recognized. Changes of political, social, economic conditions, of religious or ideological systems have been, as is well known, fought throughout history; in contrast, the notion of the open-mindedness of scientists is believed not only by laymen but frequently even emphasized by scientists themselves. De facto, however, there is a nearly infinite number of examples where outstanding and competent scientists strongly resisted the introduction of new concepts whether they were limited in scope and confined to a special problem or had revolutionary implications for the whole field of science, for our notions of the physical world, and beyond that for the thinking of man, for philosophy, religion, etc. One may recall, for instance, the fierce resistance of brilliant contemporary astronomers to the heliocentric theory proposed by Copernicus.

It may be valuable to illustrate not only the correctness but also the profound implications of this phenomenon for science as well as for scientists. Young scientists should be taught in their most formative years that the knowledge presented by their teachers or in their textbooks may look very different within a decade or two. This is particularly true at present in the biological sciences which only in this century have begun to join the "exact" sciences. It may be useful to recall the situation which prevailed in physics in 1890 before the revolution which has taken place since the turn of the century. Doubt and the problematic character of "established" facts are the basic ingredients which stimulate imagination and new ideas in creative minds. When Max Planck proposed in his doctor's thesis in 1879 some new ideas on the second law of thermodynamics and on the notion of entropy, he was shocked by the lack of interest and understanding even among the leading physicists most closely connected with the topic, e.g. Helmholtz or Kirchhoff. Only much later did he find out that Gibbs had proposed similar ideas years before him, but with the same lack of success. One of the most famous examples in history is, of course, the fight of Lavoisier for explaining the increase of the weight of metals on heating by oxidation. Stahl's phlogiston theory had penetrated so deeply the thinking of the eighteenth century that such brilliant scientists as Cavendish, Scheele, and many others died still believing in phlogiston, including Priestley, the discoverer of oxygen, who died in 1804. Any history book on science, physics, biology, or medicine provides a great number of equally striking examples. In his article on *Theoretical Concepts in Biological Sciences*, Krebs (34) quotes several interesting examples of the rejection of new concepts, the hostility, vituperation, and vitriolic comments with which scientists met new scientific ideas (see also 35). This response may reflect a natural human reaction, a defense mechanism against the overthrow of seemingly well-established and cherished views. It reminds one of the story in the novel by Anatole France: *L'Ile des*

Pingouins. A historian who has a new concept of writing history consults a member of the Academy. As advice he gets a strong warning to abandon his plan: "Si vous avez une vue nouvelle, une idée originale . . . vous surprendrez le lecteur. Et le lecteur n'aime pas à être surpris. Il ne cherche jamais dans une histoire que les sottises qu'il sait déjà. Un historien original est l'objet de la défiance, du mépris et du dégoût universelle."

However, one of the most remarkable conceptual fights in physics in this century may be briefly discussed for several reasons: (*a*) because of its great revolutionary implication on contemporary thinking; (*b*) because some of the greatest scientists in the history of mankind were involved; and (*c*) because it illuminates the many factors involved in such fights, such as philosophy, basic notions and thinking, personality, conviction bordering on religious belief, and many others. In Newton's views, no assumptions in his physics are made which are not necessitated by experimental data. Space, time, and the law of causality are absolute notions. Were these conceptions correct, this theory would never have required modification. The philosophy of Kant, presented in his *Kritik der reinen Vernunft,* is founded on Newton's physics. He distinguishes between knowledge induced from experience, therefore referring to it as "empirical" knowledge, and knowledge that is a priori: a judgment which has complete generality, i.e., which makes it impossible to imagine any exception, is always a priori. In regard to physics, Kant applied the latter term to space, time, and the law of causality.

Einstein revolutionized modern epistemology when he emphasized, in contrast to Newton, that the physical scientist arrives at his theory only by speculative means. His theory of relativity necessitated a new precise formulation of space and time, which could no longer use the scientific thinking of Newton's mechanics. But when Niels Bohr (36) introduced in 1927 the notion of complementarity, another of the great revolutionary scientific and philosophical concepts of the century, Einstein remained not only unconvinced but hostile. A crucial factor underlying Bohr's views is the impossibility of a sharp separation between the behavior of atomic objects and the interaction with the measuring instruments which serve to define the conditions under which the phenomena appear. Therefore, different experimental conditions may provide different pictures which are considered as complementary in the sense that only the totality of the phenomena observed exhausts the possible information about the object. Every perception refers to an observational situation that must be specified in accounting for the result. A deduction thus achieved is complete for this particular observational situation but incomplete for another observational situation. In classical physics as in Kant's philosophy, the law of causality is taken for granted. It cannot be proved or disproved and is part of the category referred to by Kant as a priori. In quantum phenomena the law of causality does not apply. As elaborated by several physicists, in particular by Heisenberg (37), the concepts of indeterminacy, statistical description, and probabilistic distribution are inherent aspects of the phenomena described. However, the new concepts recognize that no matter how far quantum phenomena transcend the scope of classical physical explanation, the account of all evidence must be expressed in classical terms, including space, time, and causality. But what Kant had not foreseen was the fact that these concepts can be the condition for science and at the same time have only a limited range of applicability. The exchange of letters between Einstein and Max Born (6) is an impressive document of the fight of Einstein against the new concepts of quantum theory. Born's sustained efforts, which stretched over decades, to convince Einstein of the new concepts of quantum theory remained unsuccessful. The famous statement of Einstein that "God does not play dice" expresses his belief in the law of causality and his rejection of the new concepts of quantum theory.

The revolution in modern physics and the resulting conceptual and philosophical changes should and will have a profound influence on the attitude and thinking of all those interested in biological sciences. The brilliant accomplishments in molecular biology in the last few decades by applying physics and chemistry have created an intellectual excitement in this field comparable to that prevailing in the early part of this century in physics. But they have at the same time revealed the extraordinary complexity of living matter which increases in a fantastic way when we move from molecules to organized structures, to whole cells, to organs, and finally to the whole organism. Biological sciences are still at the very beginning of the road. We are still very far from understanding the secrets of life. In his lecture series *What is Life* Schrödinger (38) raises the question whether it will one day be possible to explain what happens in a living organism at a given time in a given space in terms of physics and chemistry. He answers this question in the affirmative. Few biologists will at present take exception to this attitude provided one limits the answer to the specific question raised. Progress in biology is the result of applying physics and chemistry to living organisms, and there is no place for assuming "vital forces." But there are aspects of life which belong to different categories: ethical values, intelligence, consciousness, free will, emotions and psychological phenomena, etc. Whether these aspects will be eventually explainable in terms of physics and chemistry is an open question. Niels Bohr, in his many discussions of this problem, has seriously questioned such a possibility and has expressed the view that it will be necessary to apply the notion of complementarity to dichotomies such as body and mind, physical necessity and causality and free will, or to the understanding of psychology, ethics, etc. Since even in the much simpler systems of physics different answers may result from different observational situations, one should expect similar difficulties in the analysis of the much more complex problems of biology and life. Hence the necessity of different types of approaches which would not contradict, but supplement each other. Similar views are expressed by Heisenberg (37): there can be no doubt that the brain acts as a physicochemical mechanism when treated as such. But complete understanding of psychological phenomena, physics, chemistry, evolution, etc will, in Heisenberg's view, not be sufficient in describing them. In the light of contemporary philosophical concepts evolved from quantum theory we must keep our mind open for the possibility, or even probability, that an understanding of certain categories of life will become possible only by introducing new revolutionary conceptual notions and new dimensions into our thinking, just as complementarity and indeterminacy were indispensable for understanding quantum phenomena when physics moved into the atomic field.[1]

My scientific work has formed a central and integral part of my life. When John F. Fulton invited me to join his laboratory at Yale University, one of the important factors which determined me to accept this invitation was the impressive strength and vitality of science in the United States. I had many strong attachments to Paris and it was a difficult decision. I had spent my decisive formative years in Europe and I knew I would miss many aspects of European life by moving to the States. But the political situation was rapidly and ominously deteriorating. I personally was convinced of the inevitability of a devastating and catastrophic war. Moreover, the rapid growth of science in the United States was an extremely tempting attraction, and the research possibilities offered to scientists seemed to surpass by far all those available in Europe, especially in that period of a depressing political situation, and these factors strongly influenced my decision.

I spent three most stimulating and pleasant years at Yale University in Fulton's laboratory. Fulton received me with great warmth and cordiality. He was the first prominent neurobiologist to endorse fully and publicly my views, and he remained a resolute and firm supporter of my ideas until his death in 1960. However, in one respect these years were extremely depressing due to the events of World War II, especially for those who had strong personal and emotional ties to Europe. In 1942 I accepted an offer from Columbia University, which attracted me as a great scientific and intellectual center. In the Medical School basic research was strongly emphasized, particularly by the forceful leadership of H. T. Clarke and R. F. Loeb. Clarke's liberal attitude and his remarkable and generous personality had attracted in the 1930s a number of prominent investigators, such as R. Schoenheimer, D. Rittenberg, E. Chargaff, K. Bloch, E. Brand, D. Shemin, D. E. Green, K. Meyer, D. Sprinson, to mention only a few. It was a dynamic and inspiring biochemistry group with many able collaborators and frequent distinguished guests. There was also a strong neurobiology group, initiated by T. Putnam and continued by H. H. Merritt, and in addition several other departments which strongly supported basic sciences. The Downtown Campus is a great intellectual center in humanities and the Departments of Physics and Chemistry belong to

[1] In my first conversation with Einstein, in 1926 in Berlin, we discussed the mind-body problem. When I mentioned a lecture on this problem which I had just attended and which the speaker closed with DuBois Reymond's words: *ignoramus, ignorabimus*, Einstein vigorously objected to this remark as an "unscientific attitude."

THE EXCITEMENT OF SCIENCE   475

the world's leading centers in their fields. Most of my life work has been carried out at Columbia University. The inspiration which I received in this atmosphere and by the many contacts with friends and colleagues in other great centers of the New York area was invaluable.

Soon after I joined Columbia, an unexpected event greatly helped my research. The Defense Department asked me to help them in exploring the effect of organophosphates on cholinesterase. It had become known that the Nazi Army had apparently prepared large stores of these potentially dangerous chemical warfare agents. I was assured complete freedom in my work, although it was secret as long as the war lasted. It was immediately apparent to me that these compounds, independently of all practical aspects, offered scientifically a most challenging problem and raised many questions of a close and direct relationship to my main scientific interests. An additional attraction was the willingness of the Army to accede to my requests and to help me to get a large number of electric eels from the Amazon River.[2] Organophosphates turned out to be an extremely useful tool in the studies of acetylcholinesterase and have provided pertinent insight into its role in excitable membranes.

My different approach to the problem resulted, however, in sharp conflicts with the then-prevailing views of many pharmacologists and electrophysiologists, as discussed in the reviews quoted (14–21). News of the death of my theory was frequently announced, but was "greatly exaggerated," to use Mark Twain's remark in the version of Sigmund Freud. My deep conviction of the right direction and value of my approch has never been shaken in the slightest degree and helped me to face and overcome difficult situations. Actually, the objections raised spurred me to greater efforts for meeting the challenges. I must gratefully acknowledge the unflinching and, under these circumstances, invaluable financial and moral support of the National Science Foundation and the National Institutes of Health, which permitted me to continue my research. In the last few years, partly due to the general developments referred to earlier, the necessity of applying biochemical notions and methods to studies of the mechanism of conduction and especially the value of electric tissue for analyzing the chemical aspects of excitable membranes are increasingly recognized.

In spite of my enthusiasm, and my genuine and deep devotion to my scientific research, science was and is only part of my life. My love for literature, art, and archaeology, for music, opera, and theater, remained for me an integral part of my life. I have always been happy that I had a broad and intensive training in humanities before I turned to science, although it seems to me in retrospect that the educational program of my high school should have been more balanced. If one has spent many years with intensive studies

[2] The procurement officer in the Pentagon in discussing the necessary arrangements told me: "You know, doc, we had many crazy requests during this war, but that is the craziest I ever heard. Electric eels from the Amazon River for the war effort!"

of Greek and Roman civilization, if one is familiar with the glory and the tragedies of Athens and Rome, then visits to the places where these historical events took place, the monuments of the past still showing the splendor and the creative manifestations of the epoch, have a deep and stirring meaning. These sights revived many of the exciting impressions of my high school years. The visits to the Acropolis and Delphi, the Greek islands, the temples of Segesta or Agrigento, the Forum Romanum and Palatinum, meant obviously much more to me than the pleasure and enjoyment of beautiful architecture or what remained of it. The history of a great era of the past becomes suddenly a reality and arouses deep emotions, bringing back the marvels of the Periclean age, the greatness of Greek sculpture, drama, and philosophy, the great achievements of Rome in administration and law in the admirable building and planning of a great city. These civilizations laid the foundations for the development of the Western world. Many of these visits were made with my friends Severo and Carmen Ochoa; we shared the same deep feelings and emotions looking at these magnificent and meaningful monuments of the past.

During my many trips all over Europe attending meetings or giving lectures, I always made a special effort to save a maximum of time to visit museums and see cathedrals, monuments of the Renaissance or of the Gothic, of the Middle Ages or of the nineteenth century. To be in the Uffici or in the Rjiks Museum, in the Prado or in the Hermitage, have always been precious and unforgettable highlights.

Human life manifests itself in an almost infinite variety of exciting forms and activities. One must only have the drive and the ability to see and enjoy them, as Goethe expressed it in *Faust:*

> *Greift nur hinein ins volle Menschenleben*
> *Ein jeder lebts, nicht vielen ist's bekannt*
> *Und wo man's packt, da ist es interessant.*

Throughout my life Israel played a special role: the country is not just a place for rescuing Jews from persecution by violent antisemitism. It is the country of the Bible and the Prophets, in which the great ethical and spiritual values of the Jewish people originated and to which they remained firmly attached through the 2000 years of their history after the physical destruction of the Jewish State. The Zionist movement found in this century a forceful leader in Chaim Weizmann: with his almost magic personality he exerted a remarkable influence on many leading statesmen of this era as well as on many prominent scientists and intellectuals. In the 1920s I had the privilege to discuss with him on several occasions the role of science in Israel. For him building up modern science in Israel on the highest level was not a luxury, but an absolute necessity for the establishment and survival of a viable society built on modern and sound principles and adjusted to the requirements of our age.

My deep personal attachment is based not only on ideological grounds.

Part of my family lives there and many of my closest friends and colleagues moved from Germany to Israel during the Nazi period. On my frequent visits to Israel, especially during the last two decades, I have followed the breath-taking development of the country in all fields of human endeavors: trans-forming a barren desert into a blooming country, building modern cities and social institutions, and creating great centers of science and an intensive artis-tic life on levels comparable to those in the highly developed centers in Eu-rope and the United States. One of the great impressions on a visit to Israel is the extraordinary type of their young people: their enthusiasm and vitality, their feeling of having a full, creative, and meaningful life, and their devotion to the rebuilding of the country.

But the unique character of the country is due to the blending of this modern life with the setting and within the framework of ancient history. Everywhere one finds oneself in places familiar from biblical times, one sees ancient monuments all over the country recalling the events of the Bible and of the later history. When in 1936 Weizmann took Hans Krebs and myself on a tour from Jerusalem to Rehovot, he showed us a number of sites of historical events ranging from biblical times up to World War I, thus convey-ing the feeling of the continuity of history, of the many close bonds between past and present. For obvious reasons, I have many close personal and scien-tific connections with the Hebrew University and the Weizmann Institute. I have been active for both institutions for many years. A great satisfaction is to have been able to train in my laboratory several very talented Israeli scien-tists who are now active in the various scientific centers there.

This century has seen almost miraculous achievements of science and technology in many areas: in medicine and biology, in nuclear physics and electronics, communication and chemistry, etc. The vision of Archimedes of the power of science expressed 2000 years ago seems to become reality:

$$\Delta \grave{o} \varsigma\ \mu o\iota\ \pi o \tilde{\upsilon}\ \sigma \tau \tilde{\omega}\ \varkappa \alpha \grave{\iota}\ \varkappa \acute{o} \sigma \mu o \nu\ \varkappa \iota \nu \acute{\eta} \sigma \omega^{3}$$

These advances are inseparably associated with changes in men's lives at a speed and range unprecedented in history. Largely due to this progress wide sections of the world's population, although unfortunately still much too small ones, live healthier, longer, happier, and more comfortable lives than ever before. But almost unavoidably, many of these advances produced unde-sirable and dangerous side effects, created many serious problems and un-foreseen difficulties for society. Frequently, this was due, not to the advances of science and technology, but to their misuse by incompetent handling and lack of leadership in their application and administration. Wide sections of the population have reacted in recent years with hostility, and blame science and technology for many evils. Scientists must be in the forefront of the fight against these emotional, irrational, and dangerous tendencies; they must

[3] Give me an appropriate place and I will move the universe.

make serious efforts to enlighten their fellow citizens. For the scientist the main attraction of science may be the intellectual excitement, the deep satisfaction provided by creative work, the enthusiasm felt when he is able to unfold the secrets of nature. But for society it offers the best, if not the only hope for finding better and more satisfactory ways of using the power of science for a deeper understanding of man and the physical world in which he lives, and for building a better world for future generations.

LITERATURE CITED

1. Nachmansohn, D. 1950. *Metab. Funct.*, 1–3
2. Nachmansohn, D., Ochoa, S., Lipmann, F. A. 1952. *Science* 115: 365–69
3. Nachmansohn, D. 1952. *C. R. Congr. Int. Biochem.*, 34–37
4. Peters, R. 1954. *Obit. Roy. Soc.* 9: 175–200
5. Weber, H. H. In *Molekulare Bioenergetic und makromolekulare Biochemie*, ed. H. H. Weber. Heidelberg: Springer. In press
6. Einstein, A., Born, M. 1969. *Briefwechsel 1916–1955.* Muenchen: Nymphenburger Verlagsanst. 330 pp.
7. Lindhard, J. 1931. *Ergeb. Physiol.* 33:337–57
8. Willstaetter, P. 1949. *Aus meinem Leben.* Weinheim: Verlag Chemie. 463 pp.
9. Meyerhof, O. 1937. *Ergeb. Physiol.* 39:10–75
10. Lipmann, F. 1941. *Advan. Enzymol.* 1:99–121
11. Kalckar, H. M. 1942. *Biol. Rev. Cambridge Phil. Soc.* 17:28–45
12. Kalckar, H. M. 1969. *Biological Phosphorylations. Development of Concepts.* Englewood Cliffs, NJ: Prentice Hall. 735 pp.
13. Nachmansohn, D. 1959. *Chemical and Molecular Basis of Nerve Activity.* New York: Academic. 235 pp.
14. Nachmansohn, D. 1963. In *Cholinesterases and Anticholinesterase Agents. Handb. Exp. Pharmakol. Ergw.*, ed. G. Koelle, Bd. 15, 40–45, 701–40. Heidelberg: Springer. 1220 pp.
15. Nachmansohn, D. 1964. *New Perspect. Biol. Proc. Symp.* 4:176–204
16. Nachmansohn, D. 1966. *Ann. NY Acad. Sci.* 137:877–900
17. Nachmansohn, D. 1968. *Proc. Nat. Acad. Sci. USA* 61:1034–41
18. Nachmansohn, D. 1969. *J. Gen. Physiol.* 54 S:187–224
19. Nachmansohn, D. 1970. *Science* 168:1059–66
20. Nachmansohn, D. 1971. In *Handbook of Sensory Physiology, Vol. I: Principles of Receptor Physiology*, ed. W. R. Loewenstein, 18–102. Heidelberg: Springer. 600 pp.
21. Nachmansohn, D. In *The Structure and Function of Muscle*, ed. G. Bourne. New York: Academic. In press
22. Nachmansohn, D., Cox, R. T., Coates, C. W., Machado, A. L. 1943. *J. Neurophysiol.* 6:383–96
23. Nachmansohn, D., Machado, A. L. 1943. *J. Neurophysiol.* 6:397–404
24. Lipmann, F. 1940. *J. Biol. Chem.* 134:463–64
25. Nachmansohn, D., John, H. M., Waelsch, H. 1943. *J. Biol. Chem.* 150:485–86
26. Nachmansohn, D., Berman, M. 1946. *J. Biol. Chem.* 165:551–63
27. Lipmann, F., Kaplan, N. O. 1946. *J. Biol. Chem.* 162:743–44
28. Lipmann, F., Kaplan, N. O., Novelli, G. D., Tuttle, L. C., Guirard, B. M. 1947. *J. Biol. Chem.* 167:869–70
29. Berg, P. 1956. *J. Biol. Chem.* 222: 991–1013, 1015–23
30. Meyerhof, O. 1951. In *Phosphorus Metabolism*, Vol. I, ed. W. D. McElroy, B. H. Glass, 3–10. Baltimore: Johns Hopkins Univ. Press. 762 pp.
31. Schoffeniels, E., Nachmansohn, D. 1957. *Biochim. Biophys. Acta* 26:1–15; Schoffeniels, E. 1957. *Biochim. Biophys. Acta* 26:585–96

32. Nachmansohn, D. 1971. *Proc. Nat. Acad. Sci. USA* 68:3170–74
33. Heisenberg, W. 1971. *Physics and Beyond. World Perspectives,* ed. R. N. Anschen. New York: Harper & Row. 247 pp.
34. Krebs, H. A. 1966. *Curr. Aspects Biochem. Energ.,* 83–95
35. Barber, B. 1961. *Science* 134:596–602
36. Bohr, N. 1958. *Atomic Physics and Human Knowledge.* New York: Wiley. 101 pp.; Bohr, N. 1949. *Albert Einstein: Philosopher-Scientist,* ed. P. A. Schilpp, 99–241. Evanston, Ill: Library of Living Philosophers. 781 pp.
37. Heisenberg, W. 1952. *Physics and Philosophy. The Revolution in Modern Science. World Perspectives,* ed. R. N. Anschen. New York: Harper. 206 pp.
38. Schrödinger, E. 1945. *What is Life?* Cambridge, Mass.: Cambridge Univ. Press. 91 pp.

Ann. Rev. Phys. Chem., Vol. 20

FIFTY YEARS OF PHYSICAL CHEMISTRY
IN GREAT BRITAIN

By R. G. W. Norrish

R.G.W. Norrish.

FIFTY YEARS OF PHYSICAL CHEMISTRY
IN GREAT BRITAIN

By R. G. W. Norrish

Department of Chemistry, Cambridge University, Cambridge, England

In reviewing the contributions made to physical chemistry by British scientists in the past 50 years, I do not propose to be limited by artificial boundaries which, largely for historical reasons and administrative purposes, have tended in our time to restrict the public conception of our discipline. It is true that the recognition of physical chemistry as a discipline in its own right may be said to have originated in Germany and Sweden and Holland subsequent to 1880 by the fortunate integration of the brilliant work of Ostwald, Arrhenius, Van't Hoff and their schools of work and thought, but this does not make Davy or Faraday or, for that matter, Boyle any less of a physical chemist within the meaning of the narrowest interpretation of the term. Accordingly, I propose to view physical chemistry as a field of human endeavour which includes all those works and discoveries which depend for their success on the simultaneous application of principles which have hitherto been described as "physical" and "chemical." Science indeed knows no boundaries: it is therefore important for future progress that we should not allow traditional divisions to inhibit the unity of its grand philosophy. I am aware that in describing the physical chemistry of Britain of the past 50 years I may be accused of presumption by those who label themselves as physicists or chemists or biologists for daring to claim a right to their "preserves"; but I am not troubled. Nuclear physics is no less part of physical chemistry because it now depends upon the application of engineering principles; its roots lie deep in the newer atomic theory which owes so much to the pioneering work of Rutherford, Bohr, Moseley, Aston, and Einstein; the understanding of its recently discovered kinetic processes was foreshadowed by the chemist's study of combustion reactions and, in particular, by the treatment of branched chain reactions and degenerate explosions by N. Semenov. I shall not, however, consider the development of nuclear chemistry and physics beyond the pioneering works of Cockcroft and Walton. So far as British physical chemistry in this field is concerned, its great contribution would appear to end with the death of Rutherford and the onset of the second world war. Any truly further British developments are beyond my ken, but much may have been done within the four walls of Harwell.

The period we have under review starts at the end of the first world war in 1918. At the beginning of 1919 the writer returned from the war in France to commence his studies in chemistry, physics, and botany as a scholar of

1

Emmanuel College, Cambridge. During the previous 10 months as prisoner of war he had been able to prepare himself by reading for the career which was to keep him at Cambridge for 50 years—first as student, later as Fellow of his college and lecturer in Physical Chemistry, and finally as Professor of Physical Chemistry in the University from 1937 to 1965. Thus, including his years of retirement, he may be said to have had a comprehensive view of the growth of chemistry and physics in one of the most vital phases of their development in Great Britain. But though much of what follows is coloured by the Cambridge scene, he does not contemplate any charge of partisanship, for it may be justly claimed that Cambridge during a considerable part of this time was a centre of British scientific growth second to none, standing indeed as one of the great beacons of natural learning and science in the world.

To establish the background of the development of physical chemistry in Great Britain from the end of the first world war in 1918, it is necessary to make clear the main growing points of thought bequeathed by earlier workers both at home and abroad. The discovery of the rare gases of the atmosphere by W. Ramsay and Lord Rayleigh had put a new complexion on the periodic table of the elements. The work of W. H. and W. L. Bragg following Von Laue had opened up the whole field of chemical crystallography to exact measurement by X rays and focussed attention upon the study of the solid state. The researches of J. J. Thomson in Cambridge into the discharge of electricity through gases had not only emphasised the particulate character of the electron by determination of the ratio of its charge to mass (e/m) but had established it as one of the fundamental constituents of the atom. His characterisation of canal rays—or positive rays, as they came to be called—as streams of positive ions, and his detection by their means of the isotopes of neon, bequeathed to F. W. Aston the principle of the mass spectrograph.

In 1895 there came to the Cavendish Laboratory a young research student from New Zealand, Earnest Rutherford, whose physical work was destined to revolutionise the fundamental basis of chemistry. The phenomenon of radioactivity had just been discovered by Becquerel, and the radioactive elements polonium and radium isolated by Pierre and Marie Curie. Though Rutherford was engaged on the study of Hertzian waves, the wide discussion of the new and strange facts was not lost upon him, particularly the suggestion of Thomson that radioactivity may be attributed to a regrouping of the constituents of the atom, giving rise to electrical effects similar to those produced in the ionisation of a gas. At McGill University in Montreal, as professor of physics in the years before 1907, Rutherford seized upon the discovery of radioactivity to make its further study very much his own. Like Faraday, Rutherford was endowed with an intuitive mind and a power of devising simple yet direct experiment which enabled him not only to characterise the α- β- and γ-radiations emitted by radioactive elements, but also with F. Soddy to propose the theory of radioactive disintegration, later to be related to the periodic classification of the elements. By his study of the scattering of α- radiation, he laid the foundation of the modern theory of the nuclear structure of the atom and opened the way for N. Bohr and E. C. Stoner later in 1924 to develop the theory of atom building in terms of the Pauli principle, which was to form the basis of atomic and molecular spectroscopy.

Rayleigh received the Nobel prize for physics and Ramsay the prize for chemistry in 1904; Thomson received the prize for physics in 1906 and Rutherford and Soddy the prize for chemistry in 1908 and 1921 respectively. W. H. and W. L. Bragg shared the prize for Physics in 1915. These men laid the foundation for a great growth of physical chemistry in the first years of our time. There were others, such as H. G. J. Moseley at Cambridge, whose work (following C. G. Barkla in London) established the linear sequence of the characteristic X radiation as we pass from one element to the next in the periodic classification, and so made clear the reality of atomic numbers. They would have taken their places beside the leaders had they survived the carnage of the "great" war.

While these dramatic events revealing the basic pattern of the atom were taking place, early work by W. A. Bone in Leeds and later London on gaseous combustion, and by T. M. Lowry in London on dynamic isomerism, was quietly emerging and due to develop after the war into vigorous lines of research in reaction kinetics; at the same time F. G. Donnan in Liverpool and London was creating new and fundamental thought about membrane equilibria, in terms of the classical theories of Arrhenius and Van't Hoff, which was destined to prove of basic importance to biology.

We shall commence our review with the work of F. W. Aston. By subjecting a beam of positive rays to an electric and magnetic field coincident in position and direction, Thomson had shown that the constituent charged particles undergo a displacement in two directions at right angles to each other, such that all particles having the same value of e/m fall in a parabolic trace on a photographic screen placed perpendicular to their path. By this means it was possible to distinguish separate parabolic traces as representing ions of identical elementary charge and differing mass, and in this way the isotopes of neon were first detected. This was the first conception of a mass spectrograph, but it is to Aston, who joined Thomson from the brewing industry comparatively late in life, to whom credit for the perfection of the instrument is due. By altering the directions of the electric and magnetic fields so that they were at right angles to each other and to the direction of flight of the positive rays defined by two collimating slits, he was able to focus all particles of the same e/m onto a photographic plate and record them as a sharp spectral line. In this way Aston established the widespread existence of isotopes throughout the periodic table and so extended the work of Soddy, who with Fajans and Russell had been responsible for the displacement law and for the discovery of isotopes in the radioactive series. Aston also discovered the "whole number" law for atomic weights by mass spectrographic analysis, and thereby re-established the early hypothesis of W. Prout by showing that fractional atomic weights are due to the presence of more than one isotope of integral weights. By further refinements, however, he was able to detect small variations of atomic weight from whole numbers in certain cases—particularly in hydrogen—and so to establish the fundamental principle of mass defect, involving the conversion of mass into energy; when, for example, 4 hydrogen nuclei are combined to form a helium nucleus with loss of mass, the equivalent generation of energy amounts to 27 million electron volts. Such mass defects were the first indication of the enormous stores of energy available if nuclear transformation could be achieved.

Aston received the Nobel prize for chemistry in 1922. Since his time his apparatus has been improved and refined, but there has been no change in its basic construction. It has passed into general use as an analytical tool in chemistry for the study of organic reactions and chemical kinetics, and has been used extensively for the determination of ionisation potentials of atoms and molecules.

In 1908 William Henry Bragg returned to England from Adelaide, Australia, as professor of physics at Leeds University. Originally a pure mathematician, he slowly gravitated to the experimental study of radioactivity. His view that X rays and γ rays are of a corpuscular nature consisting of "an electron which has assumed a cloak of darkness in the form of sufficient positive electricity to neutralise the charge" could not be sustained, for with the discovery of X-ray diffraction by Von Laue, Friedrich, and Knipping in 1912, the wave character of X rays was established. Bragg, in association with his son Lawrence, was not slow to realise the importance of the crystal lattice as a three-dimensional diffraction grating and, with a directness characteristic of his experimental approach, to accept the paradox presented by the coexistence of particle and wave properties.

Making use of the regularly spaced planes of the crystal lattice as "reflecting surfaces," it was shown that for reflection to occur, cooperation is required among a number of such planes parallel to one another. The favoured angles of incidence and reflection were proved to depend quantitatively on the wave length of the X rays and the distance between the reflecting crystal planes by a simple relationship now well known as Bragg's law. Thus in 1913 and 1914 was born the X-ray spectrometer which in the hands of workers in all branches of chemistry over the last 50 years has done probably more than any other instrument to elucidate the nature of the solid state, the dimensions of molecules, and the inner secrets of the life process itself.

From 1923 onward W. H. Bragg at the Royal Institution, in association with A. Muller and G. Shearer, perfected the X-ray spectrometer, using both the ionisation chamber and the photographic plate as the detector of the reflected ray and building instruments of ever greater X-ray power and precision.

At the same time he gathered around him workers who were destined profoundly to affect the course of physical chemistry in many laboratories. His own research interest was mainly directed toward the crystal structure and dimensions of long chain organic compounds such as fatty acids and the theory of space groups. In this work he was assisted by Müller, Shearer, and Miss Yardley (now Dame Kathleen Lonsdale), who established in 1928 the geometric structure of the benzene ring for the first time and later pioneered the study of thermal vibrations in crystals and the diffuse scattering of X rays both theoretically and experimentally.

Another brilliant colleague was J. D. Bernal, who in 1923 studied the structure of quartz and graphite by X-ray spectroscopy and later with R. H. Fowler established the tetrahedral structure of water based on the X-ray examination of ice and of water itself. Bernal emerges as an inspiring influence in British physical chemistry. He became a leader in the application of X rays to the elucidation of the structure of highly complex organic molecules such as the sterols; his crystallographic data made possible the recognition of the polycyclic framework of cholesterol, ergosterol, and calciferol. With students and associates he studied the structure of the simpler proteins such as pepsin and insulin, and laid the foundation for preliminary work on the structures of haemoglobin and chymotrypsin. The greatness of his inspiration is acknowledged by all who have worked with him and is marked by the success of many of his associates, including Dorothy Crowfoot Hodgkin, who later received the Nobel prize for chemistry (1965) for her work in elucidating the structure of vitamin B_{12}, penicillin, and sterol derivatives.

Meanwhile, W. L. Bragg, first at Manchester, later at Cambridge, and finally at the Royal Institution, followed a parallel course, concentrating with his school mainly on the structure of inorganic ionic crystals, diamond, metals and alloys, and order-disorder phenomena including phase changes. They conceived X-ray diffraction as a branch of optics and so introduced practical methods of interpreting data based on the analogy of diffraction of visible light. By measuring the intensity of X-ray diffraction from oscillating or rotating crystals, they developed methods of plotting electron density maps which finally made possible the solution of the structure of highly complex polymeric substances such as proteins and their derivatives. In the development of this technique, J. M. Robertson, later at Glasgow, played an outstanding part. M. F. Perutz and J. C. Kendrew determined the structure of haemoglobin, myoglobin, and other proteins, and J. D. Watson and F. H. C. Crick at Cambridge, using the X-ray data of Miss R. Franklin and M. H. F. Wilkins, who were working at King's College, London, finally elucidated the structure of DNA. Kendrew and Perutz received the Nobel prize for chemistry in 1962, and Crick, Wilkins and Watson received the prize for medicine in 1962.

Rutherford returned to England from Montreal in 1907, when he accepted the Chair of Physics at Manchester University, and in a short time attracted many workers from all parts of the world. Assisted in the direction of general research by L. W. Geiger, he continued with unbounded energy his radioactive researches. Having proved the α particles to be doubly positively charged helium atoms, he studied with Geiger and E. Marsden their scattering in passing through gases and with great intuition arrived at his conception of the nuclear atom, upon which Bohr in 1913, who had spent some time in the Manchester laboratory, built his theory of the hydrogen atom. Equally penetrating was his work with H. Robinson and W. F. Rawlinson on the β-ray spectra of radium B & C, which confirmed and extended the work of Hahn and Meitner that β rays appeared to be emitted in groups of defined energy and that secondary β rays, excited in heavy metals by γ-radiation, are similarly constituted, leading later to a conception of quantised levels within the nucleus. The further extension of this work by C. D. Ellis in the Cavendish Laboratory, however, showed that the discrete β-ray spectrum is a secondary phenomenon produced by the emission of γ rays as a discrete spectrum. The primary β rays of radioactive substances were found to be emitted in a continuous spectrum of velocities. This puzzling phenomenon constituted an apparent anomaly in a world of quantised relationships until it was solved by Pauli in 1933 with the recognition of the neutrino.

In 1919 Rutherford published a result obtained in the intervals between war work, which was pregnant with promise for the future—the first achievement of the deliberate transmutation of matter. By bombarding nitrogen with α particles he obtained long-range particles which were identified as hydrogen nuclei and proved that they came from the cracking of the nitrogen nucleus. The impressive growth of Rutherford's genius went on with ever increasing brilliance as Cavendish professor of physics at Cambridge. Continuing his experiments on α-particle bombardment with the collaboration of J. Chadwick, it was soon shown that most of the lighter elements could be disintegrated with the emission of long-range protons, finally leading to the view that the α particle penetrates the nucleus through the surrounding potential barrier with the emission of a proton. This was later demonstrated by P. M. S. Blackett, who photographed the disintegration of the nitrogen atom in the Wilson expansion chamber. Further consideration of nuclear reactions led Rutherford to the certainty of the existence of the neutron as a constituent of the nucleus, but the many efforts to detect it experimentally were unrewarded, until finally Chadwick in 1932 identified it as the penetrating radiation derived from beryllium when it is bombarded by α particles. Thus was found a nuclear building stone which was to change the early conception of nuclear structure as a complex of protons and electrons to one composed of protons and neutrons; the origin of β rays was recognised as the transformation of the neutron by nuclear reaction into a proton and an electron.

A natural development of the discovery of "nuclear chemistry" was the project of artificial disintegration first announced by Rutherford in 1927 at his presidential address to the Royal Society. To this end he encouraged T. E. Alibone to design high voltage tubes for the acceleration of protons, charged atoms, and electrons. Based on this work J. D. Cockcroft and E.T.S. Walton in 1932 were ultimately successful in accelerating protons to 600,000 volts and by bombarding a target of lithium, to detect the emission of α particles resulting from the break-up of the lithium nucleus. Thus, at the end of his life Rutherford saw the dream of transmuting the elements come true. He pursued the theme with vigour, discovering with M.L.E. Oliphant evidence for the existence of helium mass 3, and of the hydrogen isotope, tritium. The technique of artificial transmutation was greatly developed, but with Rutherford's death in 1937 and the outbreak of war in 1939, his great work came to an end, departing forever from the Cavendish Laboratory but burning bright and in more and more sinister fashion in places around the world as the autocatalytic chain reactions of nuclear fission and the processes of nuclear fusion came to be discovered.

Of Rutherford's many brilliant colleagues in the Cavendish Laboratory, six were awarded Nobel prizes: namely, C.T.R. Wilson (1927), J. Chadwick (1935), E. V. Appleton (1947), P.M.S. Blackett (1948), J. D. Cockcroft (1951), E. T. S. Walton (1951), as well as F. W. Aston, already mentioned.

After the first world war, a great development of interest in chemical reactivity became manifest, especially in relation to the kinetic mechanisms of homogeneous and heterogeneous reactions in gaseous and liquid phases. The conception of activation bequeathed to us by Arrhenius as an explanation of the profound effect of temperature on reaction velocity was examined in particular by F. A. Lindemann and C. N. Hinshelwood at Oxford. From their early work grew the idea of kinetic activation depending both for unimolecular and bimolecular reactions upon the attainment of a critical molecular energy for reaction, in accordance with the statistical principles of Maxwell. The comparatively simple methods of experimentation had, however, their limitations, and it eventually became clear that the few examples of homogeneous unimolecular and bimolecular processes conforming to the theory of kinetic activation are in effect misleading and more complex than at first supposed, being the integrated effect of two or more component reactions, involving transient intermediates. This fact became more and more clear as the reality of the chain reaction and the branched chain reaction, first clearly expressed for thermal reactions by Christiansen and Kramers in 1923, was realised. Hinshelwood and his colleagues prosecuted their study of chemical reactivity with great vigour and, simultaneously with other workers (notably Semenov in U.S.S.R., M. G. Evans and M. Polanyi in Manchester, and Eyring in the United States), were responsible for the great growth of the kinetic theory of reaction. Hinshelwood's view of kinetic activation was modified by Rice and Ramsperger in the United States and later improved by N. B. Slater in England, while the theory of the transition state, owing partly to the work of Evans and Polanyi in 1935, pointed a new approach to the calculation of absolute reaction rates which was complementary rather than competitive to the theory of kinetic activation. The application of both theories to reactions in solution has been studied by E. A. Moelwyn-Hughes in relation to collision frequency and the influence of the nature of the solvent, and by R. P. Bell, who emphasised the importance of the quantum mechanical tunnel effect, especially for electron and proton transfer reactions. C. N. Hinshelwood and N. N. Semenov shared the Nobel prize for chemistry in 1956 for their contributions to the understanding of the problems of chemical reaction kinetics.

In 1920 T. M. Lowry was appointed professor of physical chemistry at Cambridge, the first chair to be specifically so named in Great Britain. The significant contribution to physical chemistry made by Lowry is not always fully realised. His early work in London with H. E. Armstrong before the war had established his reputation as a highly skilled experimental chemist with a rare insight into the problems of dynamic isomerism and mutarotation, two terms which in fact he coined. His further studies into these problems at Cambridge brought an entirely new understanding to the field of acid-base catalysis and contributed greatly to the Bronsted-Lowry theory of acids and bases conceived independently and simultaneously by the two workers. Their view of acids and bases as proton donors and proton acceptors respectively, and the realisation of the amphoteric quality of water in this respect, was brilliantly applied by Lowry to the study of mutarotation in general and of tetramethyl glucose in particular. These reactions he considered to be dependent on simultaneous addition and removal of a proton at different parts of the molecule under the influence of water, or alternatively by a mixture of acid and basic catalysts such as cresol and pyridine in anhydrous solution. Lowry's extensive study of optical rotatory dispersion as a function of wavelength in terms of the Drude equation, his extension of the Cotton effect to rotatory dispersion, and his study of anomalous rotatory dispersion gave not only new understanding, but also a body of accurate and reliable data which has provided valuable working material for many physical chemists who have followed him, and which indeed is becoming again a live interest at the present day.

Further important work on acid-base catalysis in 1950 to 1956 was contributed by R. P. Bell and his students by the study of such reactions as the hydration of acetaldehyde and the halogenation of acetone. His results gave support to the Bronsted-Lowry theory, which owes much to his clear exposition in his treatise on *Acid-Base Catalysis*, published in 1941.

Following World War I, A. J. Allmand, who had been active as a pupil and assistant to Donnan, continued with his researches in electrochemistry, including studies of passivity and irreversible electrode phenomena with marked success; but after the great contribution of Debye and Huckel in Germany to the theory of electrolytes in 1923, a contribution which had been anticipated by the work of R. S. Milner at Sheffield in 1913, the subject tended to become "somewhat arid" from the point of view of Allmand, and progress with the theory of concentrated solutions which he would have liked to advance became slow; work on the subject moved strongly in the direction of thermodynamic treatment which was not to his liking, and his main interest now developed towards photochemistry, in which he and his collaborators made significant contributions, particularly to studies of the photolysis of ozone and the synthesis of hydrogen chloride.

This early period was also marked by the stimulating work of J. A. V. Butler, whose kinetic treatment of electrode reactions and equilibria provided a valuable model on which to base further experimental projects. Butler's later treatise on electrochemistry in 1939, together with the work of S. Glasstone and A. Hickling on electrolytic oxidation, provided a link through the tenuous period of the second world war with the present generation of electrochemical research.

Electrochemistry in Great Britain continued to develop from Allmand's work on passivity to the study of overpotential of the hydrogen and oxygen electrodes by E. K. Rideal, F. P. Bowden, and J. N. Agar, in the light of the earlier understanding of Tafel in Germany. While part of the overpotential of irreversible electrodes can be ascribed to changes in concentration of ions near the electrode, the greater part is due to the requirement of activation for processes such as the penetration by the relevant ions of the energy barrier between the solution and the metal electrode, the formation of atomic or molecular hydrogen or oxygen, and the liberation of the gas from the electrode. The work of the British authors, especially when using the very reproducible mercury electrode, was significant in the quantitative treatment of the problem, though it may be admitted that there is still uncertainty in many cases as to which of the above steps is the rate determining process leading to overpotential.

Such studies of passivity and overpotential led to the pioneering work of U. R. Evans and G. D. Bengough between the wars on the corrosion of metals. They elucidated the importance of the effects of passivity and depolarising reactions in bimetallic electrolytic systems, and emphasised the factor of differential aeration in the corrosion of single metals exposed to a suitable electrolytic environment.

The spirited controversy which arose between Evans and Bengough as to the detail and relevant importance of these factors greatly enlivened the study of electrochemistry in Britain and elsewhere at the time; it was largely due to misunderstanding but was responsible for much new work on the part of the respective protagonists until it was finally amicably resolved by the publication of a joint paper in 1938.

Other contributions by British authors to electrolytic studies which arose out of this earlier work have been made by T. P. Hoar on the oxygen electrode and the formation of oxide films on platinum and by F. T. Bacon, whose work on the hydrogen-oxygen fuel cell has developed through his invention of the biporous electrode into a viable source of energy for space travel. Work on the intractable problem of the fuel cell in general is still proceeding in Britain, as in many other places abroad.

Since 1950, new lines of research have been developed by W. F. K. Wynne-Jones and his colleagues at Newcastle, who, with the aid of advanced techniques, have greatly increased our knowledge of the anodic formation of solid films on electrodes and the cathodic deposition of metals. J. E. B. Randles at Birmingham has made significant contributions to the study of electrode reactions by a newly devised A/C technique, and R. Parsons at Bristol has contributed to experimental work on electrocapillarity and to the theory of the electrical double layer.

With newly developing electronic techniques, there is still a wide range of investigation open to the electrochemist, especially with reference to the recording of rapid electrode reactions. We may mention in particular the study of the effect of high powered light flashes and intense continuous irradiation applied to electrode reactions initiated by the present writer in collaboration with S. Paszyc and U. Heyrovsky.

In 1920 E. K. Rideal joined the department of physical chemistry in Cambridge. His lively mind caused him to range widely over the domain of physical chemistry, but his major achievements were in the field of surface chemistry and the chemistry of colloids, and also in the study of adsorption and surface catalysis. In the 1920's the pioneering work of Langmuir in the U.S.A. on unimolecular films of insoluble fatty acids spread on the surface of water had created wide interest, and the work of N. K. Adam at University College, London, and Rideal at Cambridge went far to establish the equations of state and the thermodynamics of spreading of insoluble long chain hydrophilic compounds on water. Simultaneously, W. M. McBain at Bristol was making fundamental contributions in elucidating the micellar character of soap solutions and the emulsification of insoluble oils in water by surface active additives such as the salts of fatty acids. In the 1930's Rideal established the Colloid Science department at Cambridge. Here, with J. H. Schulman and A. E. Alexander, he studied interactions of dissolved substances with insoluble monolayers, developing the surface potential technique and penetrating into the field of biological membranes. Here also J. K. Roberts studied adsorption by specially cleaned tungsten surfaces, demonstrating the rapidity of chemisorption and the decrease of adsorption heat with surface coverage. Rideal, together with G. H. Twigg and D. D. Eley, incorporated these results into new mechanisms of catalysis and introduced the use of evaporated metal films to study the interaction between "van der Waals" and chemisorbed layers of hydrogen by Eley, and later by B. M. W. Trapnell and C. A. Kemball. Eley and his students at Nottingham interpreted their studies of catalysis on palladium-gold alloys in terms of a covalent model for chemisorption bonds, and J. W. Linnett developed a method for investigating the recombination of hydrogen atoms and oxygen atoms at insulator, semiconductor, and metallic surfaces. It is true to say that the British school over the years has laid much stress on the chemistry and physics of surfaces, including work of importance on colloidal sols such as S. Levine's theoretical studies and A. S. C. Lawrence's studies on streaming birefringence.

At the present time, work continues in many centres in Britain directed to the testing of the modern theory of colloidal stability based on the electrical repulsion of double layers due to Verwey and Overbeck (Holland) and Derjaguin (U.S.S.R.). Rideal's influence on a large school was lasting and widespread: he has done much for the healthy development of physical chemistry in this country.

In 1930 or thereabouts, Lowry interested F. P. Bowden, who had been a pupil of Rideal, in the early work of William Hardy on sliding friction. This initiated Bowden's fundamental contribution of proving that sliding friction is basically the result of adhesion—that is, the force required to shear the strong "welded junctions" formed at the regions of real contact. As part of this work, he showed that the area of real contact between surfaces is vastly smaller than the apparent geometric area. The frictional energy is liberated at these very small regions so that the true temperature at the actual points of contact is very much larger than the average macroscopic temperature rise. These frictional hot spots play an important part in the polishing process and in the surface melting responsible for the low friction of ice. Bowden also demonstrated the importance of surface films in reducing friction and adhesion. If these are removed in high vacuum, the surfaces adhere catastrophically. Oxide chloride and sulphide films formed by attacking surfaces with the appropriate vapour are very effective in preventing this and allow lubrication of ceramic and metal surfaces at temperatures up to 700° C. In exploring the mechanism of so-called boundary lubrication, he showed that fatty acids are good lubricants only if they can react *in situ* to form the corresponding metallic soaps.

Finally Bowden turned his attention to the direct measurement of the forces between solid surfaces. Using the cleavage face of mica, he was able to work down to a separation of a few angstroms and to establish a transition from normal to "retarded" van der Waals' forces at a separation of about 200 Å. Bowden, who died in 1968, created a distinguished school of the physics and chemistry of friction and lubrication which, now located in the Cavendish Laboratory, continues under the direction of his former colleague, D. Tabor, with the collaboration of some of his past research associates.

The great advance in chromatography achieved by A. J. P. Martin and R. L. M. Synge in 1941 as a result of work originally carried out at Cambridge provided a technique of such precision and simplicity that it has virtually created a revolution in chemistry, spanning the wide field of organic and biological research. Arguing that the partition isotherm of a solute between two liquid solvents is more generally linear than for its distribution between a liquid and a solid, Martin and Synge invented the technique of paper chromatography, using paper strips as a support for the stationary liquid phase and gravity to elute the moving solution. There can be no doubt that paper chromatography has provided a supremely powerful method of separation and analysis. Martin and Synge early recognised that the flowing liquid could be replaced by a gas, and within 10 years A. T. James and Martin brought to successful development the technique of gas-liquid chromotography in which the gaseous mixture is eluted over a stationary liquid, supported in a column on an inert solid. The development of

this powerful technique, both for analysis and for plant control, in the succeeding years is too well known to require any further description here; it
has been fully treated by H. Purnell in his excellent work on *Gas Chromatography* (1962). Suffice to say that gas-liquid chromatography by its great
simplicity and speed of operation has probably facilitated pure reseacrh
and large scale chemical operation more than any other discovery of comparable character. For this most significant contribution, Martin and Synge
were jointly awarded the Nobel prize for chemistry in 1954.

Concurrently with the development of chromatography, G.B.B.M.
Sutherland in Cambridge and H. W. Thompson in Oxford simultaneously
developed the modern technique of infrared analysis and applied it in the
first case to the analysis of the complex mixtures of the hydrocarbons in
petroleum spirit. The methods pioneered by them at this time formed the
basis for all subsequent infrared analyses now so widely used in organic research as a guide for the characterisation of molecules and key chemical
groups and the determination of molecular structure. In this later work,
Thompson's study of the intensities, as well as the position of infrared bands,
has been very important. The use of polarised infrared radiation in the examination of oriented polymers by Thompson, and in the investigation of
proteins and nucleic acids by Sutherland, extended the use of infrared
spectrometers to many fields of great importance in industry and biophysics.

Further contributions have been made to chemical spectroscopy, particularly by W. C. Price in 1935 to 1938 in his pioneering researches on the
determination of ionisation potentials from ultraviolet absorption. His work
with the diatomic molecules O_2 and HI revealed electronic spectra composed
of a series of peaks which converged to a continuum and whose succession
could be represented by an expression involving the Rydberg constant. Such
spectra, which are excited by an electric discharge at low pressure and involve
nonbonding electrons, measure the ionisation potential at the point of convergence which gives the energy for the process $X_2 + h\nu \rightarrow X_2^+ + \theta$. Price
extended the method to a wide variety of polyatomic molecules, including
CO_2, CS_2, N_2O, H_2S, H_2O, and a number of organic vapours such as methyl
halides, formic acid, and butadiene. Valuable work was also contributed in
1954 by C. A. McDowell, who determined ionisation potentials by the
method of electron impact in the mass spectrometer. The highly original
work of A. D. Walsh at Dundee on the interpretation of the spectra of polyatomic molecules has led to the prediction of the shapes of free radicals and
electronically excited molecular states in terms of orbital theory and electronic structure of importance both to photochemistry and chemical reactivity. Reference must also be made to the distinguished spectroscopic
work of A. G. Gaydon in London and R. F. Barrow in Oxford, particularly
with reference to diatomic molecules and radicals. The latter has notably
established the "anomalous" dissociation energy of the fluorine molecule.

More recently, B. Bleaney at Oxford was one of the first to develop the
field of microwave spectra, initially by the analysis of the microwave spectrum of ammonia. By measuring the pressure broadening of the spectral

lines, he was able to determine the molecular collision diameters involved. Bleaney also made important contributions to the field of paramagnetic resonance; he investigated the spectra of inorganic salts, particularly at low temperatures, and discovered and analysed the hyperfine structure of tungsten, manganese, and cobalt salts. His work on the paramagnetic resonance spectra of transition metal ions in crystal lattices has facilitated the understanding of the electronic structure of complex ions. Bleaney's work on paramagnetic resonance applied to chemical problems has been continued and extended by R. Richards, also at Oxford, particularly with reference to structural problems.

In the years immediately before the first world war, the theoretical work of Stark in Germany—followed by the experimental studies of Bodenstein, Warburg, and Nernst—had put a new complexion on the face of photochemistry. By applying the quantum theory to photochemical kinetics, the conception of the primary and secondary photochemical reactions emerged which made possible the explanation of the whole variety of photochemical phenomena—photolysis, photosynthesis, fluorescence, phosphorescence and photosensitisation. The reality of the chain reaction became clear with the study of the hydrogen-chlorine reaction and led to the recognition of the participation of atoms and later free radicals in chemical reactions.

The only photochemical work of note at this time in Great Britain was the classical study of the hydrogen-chlorine reaction and its induction period by D. L. Chapman and his associates at Oxford, and also by Allmand; this was continued after the war with very significant results.

This was the state of the subject when the present author at Cambridge in 1922 took up the study of photochemical reaction in the light of the new knowledge. One of the first results was the discovery of the fluorescence and the photolysis of nitrogen dioxide and the reciprocal variation of fluorescence with quantum yield of reaction through the spectrum. This reaction furnished one of the best examples of the fundamental principle of Stark referred to above. Between the two wars, work was continued with collaborators on the hydrogen-chlorine reaction, amplifying the results of Bodenstein and Chapman, and also in particular with C. H. Bamford and other collaborators on the photolysis of carbonyl compounds, carboxylic acids, and esters. The detailed study of these latter reactions in gas phase and solution indicated two independent types of photolysis which were designated type I and type II. The former involves the production of free radicals—e.g.,

$$\begin{matrix} R_1 \diagdown \\ \qquad CO + h\nu \rightarrow R_1CO + R_2 \\ R_2 \diagup \end{matrix} \qquad\qquad \text{Type I}$$

where R_2 could be hydrogen, alkyl, or alkoxy; the latter involves an intramolecular rearrangement, without the formation of free radicals, e.g.:

$$\begin{matrix} CH_3CH_2CH_2 \diagdown \\ \qquad\quad CO + h\nu \rightarrow CH_2{:}CH_2 + CH_3COR_2 \\ R_2 \diagup \end{matrix} \qquad\qquad \text{Type II}$$

THE EXCITEMENT OF SCIENCE 495

For type II to occur, at least one of the radicals, R_1 or R_2 must be an alkyl radical and contain at least three carbon atoms, the break in the chain always occurring at a point $\alpha - \beta$ to the carbonyl group. Studied in solution, these reactions gave the first experimental verification of the Franck-Rabinowitsch principle, type I being inhibited and type II unaffected. This work later formed the starting point of a wide range of organic photochemical studies and at the same time provided a proof of the participation of free radicals in photolytic reactions and a ready source of free radicals for the study of pyrolysis, combustion, and polymerisation.

During the same period, much interest developed on the combustion of hydrocarbons in the gaseous and liquid phases, particularly in Oxford, Cambridge, and London. It became clear that in the gas phase such reactions are branched chain processes, initiated either thermally or photochemically by free radicals and presenting all the features of degenerate explosive processes in accordance with the mechanism of Semenov. However, much difference of opinion, especially between the Oxford and Cambridge schools of thought, developed as to the nature of the propagating radicals and the reactions involving branching. The Oxford workers with Hinshelwood supported a chain propagated by alternate links involving peroxide radicals and alkyl radicals, with the alkyl hydroperoxide as the intermediate branching agent, while the Cambridge school with the present writer believed chain propagation proceeds by way of hydroxyl radicals and alkyl radicals with the corresponding aldehyde as the branching agent. Other workers—for example, W. A. Bone and his colleagues in London—preferred to ignore the evidence for chain propagation and were satisfied to explain the overall stoichiometric results by the earlier theory of hydroxylation. This is no place to enter further into the controversy which became very prominent at the time; suffice to say that it had the effect of stimulating a great deal of new kinetic studies of hydrocarbon combustion, with corresponding improvements in the experimental techniques and the accumulation of valuable data. Concurrently, the study of polymerisation, which had been shown by Staudinger in Germany to follow in certain cases the kinetics of a chain reaction, became extremely widespread, particularly in the case of vinyl polymerisation. The fundamental reactions of copolymerisation cross-linking and chain branching were early studied at Cambridge, particularly by the present author and by H. W. Melville and co-workers, using techniques involving thermal photochemical and catalytic processes for the control of the reactions of initiation, degradation, propagation, and branching. Further reference to work on polymerisation is made below.

Simultaneously, E. J. Bowen and his colleagues at Oxford in the 1920's and 1930's had taken up the study of photochemistry from a different standpoint. After early studies of the quantum yields of the photolyses of simple gases such as ozone and the oxides of chlorine and photoxidation of aliphatic compounds, they turned their attention to the fluorescence, phosphorescence, and luminescence of aromatic polycyclic hydrocarbons particularly anthracene in solution. New principles were discovered relative to the relationship between fluorescence, photosensitisation, and photooxidation; measurement of the relative efficiencies of different molecules for quenching gave early information about the interactions involved in energy transfer.

In later years, Bowen has extended his studies significantly to the fluorescence of solids, showing, for example, that in solid solutions of anthracene in napththacene the excitation of one component may be transferred over considerable distances in the solid before exciting fluorescence in the other.

Theoretical work bearing on Bowen's researches and destined to have widespread influence on our understanding of the mechanism of fluorescence and photoconductivity of solids, as well as the nature and origin of the photographic latent image, was contributed by N. F. Mott, first at Bristol, where he was appointed Professor of Physics in 1933, and later at Cambridge as Cavendish Professor in 1954. It was at Bristol in the early thirties that a beginning was made on the study of certain aspects of the solid state by J. E. Lennard-Jones, W. E. Garner, and Mott, involving such topics as lattice energies, decomposition and catalytic reactions, and electronic processes in ionic crystals; this work gained considerable momentum and early secured for the Bristol school a high reputation for research into the physics and chemistry of the solid state. Mott's contributions with H. Jones were first to the application of quantum mechanical principles to the electronic theory of metals and alloys, which already owed much to the pioneering work of W. W. Hume-Rothery at Oxford. From this he moved through his work on semiconductors and insulators to what from the physical chemical point of view is of major significance—his study of the behaviour of ionic crystals under the stimulus of light. This culminated in collaboration with R. W. Gurney in the characterisation of the process of formation of the latent image in photographic emulsions. They visualised the primary process as the photolytic liberation of an electron from a bromine ion of the lattice into the conducting band of the crystal, by which means it can travel considerable distances, eventually to become trapped by essential impurities such as silver sulphide or silver atoms. The traps, becoming highly negatively charged, attract and neutralise mobile interlattice silver ions and those left unbalanced by the photolysis of the bromine ions. In this way, nuclei of silver collect, sometimes in the body but more often at the surface of the grains, and form centres for subsequent development by mild reducing agents. This explanation, which is based on a knowledge of the photoconductivity and electrolytic conductivity of ionic crystals, has been found fully adequate to explain the manifold and varied properties of the photographic plate and has been of great service to the technical development of photography in general.

The work of Garner and his colleagues, T. J. Gray and F. S. Stone, on chemisorption by semiconductor oxides is similarly related to lattice defects in the form of interstitial cations or cation vacancies. Chemisorption occurs at the surface in the form of ions by the gain or release of the electrons by the lattice, and in some cases the chemisorbed species (e.g., O^{2-} ions) may travel into the lattice forming cation vacancies. The addition of small quantities of oxides with cations of different valency to those of the parent lattice may promote adsorption by increasing the semiconductivity of the oxide. These effects may further be related to catalytic activity of semiconducting oxides by the electron transfer theories of Wolkenstein (USSR) and D. A. Dowden.

Arising further from the work of Garner and Mott were Garner's studies of the slow and explosive decomposition of crystalline solids such as salt hydrates and metallic azides, extended more recently by the work of F. C. Tompkins. Based on the theory of nucleation, the mechanism is envisaged as similar to the formation of the latent image, whereby metallic aggregations form as centres of catalytic activity. The process depends upon the presence of lattice defects and trace impurities. However, as has been demonstrated by Garner and by Bowden and A. Joffe following Rideal, the mechanism of explosive decomposition for highly exothermic processes depends also on other factors such as the production of "hot spots" by adventitious initiation processes, from which the explosive reaction is propagated adiabatically.

Other significant studies in the 1940's relating to the solid state have been made in Great Britain by A. R. Ubbelohde, who sought to create links between the accurate determination of crystal structure and solid state thermodynamics. This led, for example, to the first direct demonstration of zero-point energy in crystals in relation to isotope effects on replacing hydrogen by deuterium in "metallic" solution. Studies of premelting in crystal lattices and of lambda transformations in single crystals have shown the role of hysteresis to be intrinsic. More recently, Ubbelohde further established the various mechanisms of melting for different kinds of structures such as molecular, ionic, or metallic crystals, and demonstrated the existence of several types of liquids with diverse structure and properties.

Another development in the study of the solid state, stemming from 1945, was the discovery of the so-called clathrate compounds by H. M. Powell at Oxford. Such compounds are in reality solid solutions consisting of volatile solutes such as SO_2 or inert gases trapped within the structure of solid hydrogen-bonded solvents such as quinol and ice. The stoichiometric composition of such "compounds" is variable, and the matrix lattice of the host becomes thermodynamically unstable upon the removal of the volatile constituent. Clathrate compounds are to be distinguished from the stable adsorption systems formed in the open structure of zeolites, which can be decomposed reversibly by heating. Since the war, R. M. Barrer and his collaborators in London have made an extensive study of the adsorption of gases and solutes by natural and artificial zeolites, whose open structure contains channels of considerable length and sufficient width to permit the passage of vapours and cations, a property which has led to their development as molecular sieves.

In 1921 there arose at Oxford a man of penetrating intellect who was destined in the next 20 years to exercise a profound integrating influence over a wide field of chemistry, bringing order and cohesion to the problems of chemical structure and combination. Though he lived alone for 50 years in the same rooms at Lincoln College, N. V. Sidgwick was probably better known than any other chemist at the time. As an experimenter or director of research he had no significant success, but as a tutor and teacher of chemistry he was well liked by his pupils from the first days of his appointment in 1905. Sidgwick was a scholar, as was evidenced by the publication of his classic treatise on *The Organic Chemistry of Nitrogen* in 1910, but his love of travel and contact with other minds, especially Rutherford's at home and abroad eventually stimulated his enthusiasm for the main work of his life—the electronic theory of valency, and in its light, the description of the chemical elements and their compounds. His interest began in 1921 when he became impressed by the deficiencies of the Lewis-Langmuir octet theory, which gave no clear explanation of the formation and stability of the complex inorganic coordination compounds discovered by Werner, and was not in harmony with the physical principles of atom building laid down by Bohr and Stoner. Of his publication of *The Electronic Theory of Valency* in 1927, it was rightly said in 1937 that "for the first time the most diverse structural phenomena covering the whole field of chemistry were rationally systematized and that he had more widespread influence on the views of chemists in this country (Great Britain) than any other of this generation." Sidgwick was a true physical chemist. In 1931 he said in a public lecture at Cornell University:

> The ultimate object of the chemist is to express his conclusions in physical terms, but he must remember, if he tries to do this, that these terms have already a very elaborate and precise connotation; every concept which he uses involves a series of definitely established properties. That in fact is why it is important to be able to use them. But it is essential to use them rightly............ The chemist must resist the temptation to make his own physics; if he does it will be bad physics just as the physicist has sometimes been tempted to make his own chemistry and then it was bad chemistry.

By 1937 the development of wave mechanics was giving a new interpretation to Sidgwick's pattern of structural chemistry, but nothing basic had to be unsaid, and there emerged in addition the conception of resonance, which while objecting to the term, he characterised as the most important development that structural chemistry has had since it was extended to three dimensions by Van't Hoff. Sidgwick remained a scholar to the end, and completed his work by the publication of his magnum opus, *The Chemical Elements and Their Compounds*.

One of the foremost to become interested in the application of mathematical concepts to Sidgwick's theories was J. E. Lennard-Jones. He early became interested in gas kinetic problems and in 1924 published an expression for the viscosity of a gas which derived from a consideration of both attractive and repulsive forces between molecules. This was before the introduction of quantum mechanics. The formulation permitted the determination of molecular diameters and the minima in the interaction potentials from the measurements of viscosity and its temperature dependence. His model still survives as the principal source of information about elastic collisions in gases, and was early used in collaboration with A. E. Ingham for the calculation of the lattice energy of various crystal structures. First at Bristol and later at Cambridge this formed the starting point for his application of quantum mechanics to the problem of molecular structure and the solid state, and also in the light of wave mechanical theory, it gave rise to the theory of molecular orbitals which he, together with Hund and Mulliken, later founded. From 1932 at Cambridge until the outbreak of war, and later from 1946 to his premature death in 1954, Lennard-Jones built up a school of theoretical chemistry which became world famous. His researches, which largely depended on the development of molecular orbital theory and its application to all kinds of chemical structure and molecular properties, were carried out with such collaborators as A. F. Devonshire, C. A. Coulson, G. G. Hall, and J. A. Pople, all of whom have become leaders in this field of theoretical chemistry. During his sojourn at Cambridge as professor of theoretical chemistry, Lennard-Jones was a source of inspiration to all and sundry with whom he discussed chemical problems; his insight into the fundamentals of physical theory rarely failed to give help to what was often a sorely troubled mind.

More recently, J. W. Linnett has proposed a simplified model for molecular building consisting of two tetrahedrally disposed quartets rather than four pairs of electrons. He shows that wave functions based on this arrangement have considerable value in calculating molecular properties.

The outbreak of the second world war brought most of the academic research work on reaction kinetics in Great Britain to an end, and it was not until 1945 that it could be resumed. There were, however, certain technical investigations that occurred in the interim period which were of importance to further research when peace was restored. One of these was the suppression of gunflash and exhaust flaming, in which the present author was involved with others. By the application of combustion theory it was found possible to suppress ignition of one combustible fuel by the addition of a second, and so to achieve the aim with a considerable measure of success.

Another was the development by B. Bleaney of clystron oscillators giving 3 and 1 cm waves which contributed to his subsequent development of microwave spectroscopy as a new and profitable field of research.

There must have been many other promising ideas derived from technical research during the war years, such as the lead sulphide photoelectric cell for use with infrared radiation and new developments in the field of solid state science, but they were not published so it is not possible to speak of them further.

When academic research on reaction kinetics was ultimately resumed, attention was directed to seeking new ways of elucidating the role of free radicals and atoms in flames, in the pyrolysis and oxidation of hydrocarbons and hydrides in general, and in photochemical reactions, including processes involving energy transfer. Most of the methods employed involved a spectroscopic approach and endeavoured where possible to emphasise the quantitive aspect of reactivity.

We shall first refer to a technique of wide applicability to gaseous and liquid systems, the method of flash photolysis and kinetic spectroscopy. From the previous studies of photochemical, thermal, and combustion reactions the role of free radicals had become abundantly clear, but the part played by these short-lived transients could only be deduced from the circumstantial evidence of reaction kinetics and spectroscopy, since they could neither be observed nor isolated by classical means.

It therefore became important, if further progress was to be made, to endeavour to obtain objective evidence of the participation of atoms and free radicals both in thermal and photochemical reactions. Using continuous sources of the highest attainable intensity, the present author and his co-workers attempted to obtain evidence by spectroscopic means of a stationary concentration of intermediates in photolytic and photo-oxidation reactions of suitable reactants such as ketene, but without success. It soon became apparent that their reactivity was so great that no sufficient stationary concentration for detection by the means then available could be achieved.

It was the realisation that enormously greater 'instantaneous' light intensities could be obtained from a powerful light flash than from a conventional light source that led G. Porter and the author to examine the effect of applying such flashes to suitably responsive photochemical systems. Using an electric discharge from a condenser bank through inert gas dissipating about 10,000 joules, it was found that the resulting light flashes of about 2 to 3 milliseconds duration were able to create large measures of photo-decomposition in reactants such as nitrogen dioxide, chlorine, ketene, acetone, and diacetyl, amounting to 100 per cent in some cases. It was obvious that momentarily there must be very high concentrations of free radicals or atoms in such reacting systems which by suitable means should be detectable by absorption spectroscopy. This was first achieved by Porter, who by using a second less powerful flash triggered mechanically by the method of Oldenberg at specific short intervals after the first was able to observe the complete dissociation of chlorine by the disappearance of the Cl_2 absorption spectrum and its return over a period of milliseconds as the atoms recombined.

The modern method of flash photolysis and kinetic spectroscopy uses an electronic technique by which the first flash of about 2000 joules and 10 to 15 μsec duration is caused to trigger the second flash of $ca.$100 J and 2 μsec duration, at specific intervals measured in microseconds or milliseconds after the first. By this means a series of spectroscopic absorption photographs taken at increasing intervals of time through the length of the cylindrical reaction vessel reveal the growth and decay of transients during the course of the reaction and often contribute highly relevant data to the elucidation of the kinetic mechanism of the process.

It early became apparent, however, that even the dissipation of only 0.05

per cent of the total output of energy by the main flash (2000 J) by say 150 ml of gas at 1 mm pressure will raise the temperature of the reactant by about 5000° C. Thus, under such conditions, the results of flash decomposition are more properly to be regarded as flash pyrolysis than flash photolysis; but if steps are taken to neutralise this temperature rise by the addition of a large excess of inert gas (some 100- to 500-fold), the process can be kept isothermal to within 10° C; and with reactants in solution, there is no problem. Thus there are two ways in which we can employ the technique of flash spectroscopy the adiabatic and the isothermal. In the first, advantage is taken of flash heating in the undiluted system to administer an adiabatic shock, which makes possible the spectroscopic detection of the growth and decay of transient intermediates generated in pyrolysis and explosive processes. This arises from the fact that by flash heating the whole system is instantaneously and nearly homogeneously heated to high temperatures, so that the flame front is virtually as thick as the length of the reaction tube. It then became possible for the first time to observe the reaction of unexcited transients leading to and participating in explosive processes, as well as the electronically excited species to which we were previously limited.

In the isothermal method we may observe the growth and decay of transient species formed by photolysis and other photochemical processes and detect their reactions. From their spectra we can study their structure, and since many are produced in a high state of vibrational excitation, it is possible to follow the process of energy transfer on collision. In cases where the extinction coefficient of the relevant atom or free radical can be derived, measurements of reaction velocity and energy transfer may be made quantitative. In solution, reactions of polymerisation and decomposition of inorganic and organic reactants can be kinetically analysed and especially the generation of the triplet state and its relation to the reactivity of a wide range of organic compounds. By kinetic spectroscopy, reactions taking place in times as short as a few microseconds can now be effectively kinetically analysed; by further improvement of the time resolution through development of light flashes of shorter duration for a given energy, still more precisely detailed results may be expected. For the development and application of flash spectroscopy, R. G. W. Norrish and G. Porter shared a part of the Nobel prize for chemistry in 1967. The wide variety of studies which have been carried out by their collaborators, B. A. Thrush, A. B. Callear, M. W. Windsor, N. Basco, K. Erhard, D. Husain, W. D. McGrath, J. A. Smith, M. I. Christie, and many others, some of whom are now directing research schools of their own, assures that flash spectroscopy will long remain a powerful method for the investigation of chemical reactions.

Simultaneously with the development of flash photolysis in the nineteen-fifties, the study of radiation chemistry took shape mainly under the direction of J. H. Baxendale (Manchester), F. S. Dainton (Leeds), and J. J. Weiss (Newcastle), by which the early pioneering work of C. S. Lind (U.S.A.) in 1921 was renewed. Interest centered on the effects of α- β- and γ-radiation and X rays on water, on aqueous solutions, and on polymerisation. The importance of the ractions of the free radicals H and OH in aqueous systems was recognised early, and evidence for the existence of short-lived hydrated electrons had been accumulated by 1960. The absorption spectrum of the hydrated electron was first observed by Boag and Hart (U.S.A.) in 1962, using the technique of pulse radiolysis, analogous to flash photolysis. This method has been used extensively since that time to measure rate constants for reactions of the solvated electron H, OH, and other free radicals, to study transitory inorganic ions of unusual valency, and to obtain information about excited states and energy transfer in organic systems. Parallel studies of the same problems have been made using the complementary technique of matrix isolation.

During the past 20 years, the study of atomic, radical, and ionic reactions occurring in flames of hydrocarbon oxidation proceeded apace by various spectroscopic techniques. A. G. Gaydon, by examining the chemiluminescence of flames, has characterised many of the elementary reactions taking place, including energy transfer processes giving rise to electronically excited species such as OH^*, CO_2^*, CH^*, C_2^*, SO^*, HCO^* and NO_2^*. His studies and those of D. T. A. Townend of "cool flames," observed in the slow combustion of hydrocarbons, have established excited formaldehyde as the luminescent agent. His work with shock tubes has further extended our knowledge of the pyrolysis and reactions of simple molecules in terms of the electronically excited products. With H. G. Wolfhard, he has made progress in determining "instantaneous temperatures" of excited transients in their various degrees of freedom.

Using the technique of microwave attenuation, including also the use of resonant cavities, T. M. Sugden and his collaborators have been able to measure the change in concentration of electrons in oxy-hydrogen flames produced by the addition of alkali metal halides. The ionisation potentials of the reactants of the pure hydrogen flame either in air or oxygen are very high; the conductivity is consequently very low and the electron concentration negligible. The ionisation potentials of the alkali metals, however, are low enough to give appreciable ionisation equilibria which are modified by the formation of hydroxyl ions by reaction of electrons with the free hydroxyl of the flames. Stable alkali hydroxide molecules are also formed. The extent to which these reactions cause departure from the Saha equation for the ionisation equilibrium of the alkali metal makes possible the study of equilibria in these and similarly treated flames in a manner which is somewhat analogous to the treatment of ionisation in solution.

Parallel studies of chemiluminescent emission from oxy-hydrogen flames containing metal atoms have also detected high energy reactions between atoms and free radicals leading to electronic excitation of the metallic atoms or their derivatives, e.g., Fe^* CuH^* and $MnOH^*$. These excited products are formed in amounts exceeding those corresponding to thermal equilibrium; they lead quantitatively to estimates of the high energy liberated by the elementary reactions of atoms and radicals involved.

Other work on energy transfer was contributed by Mc. F. Smith, in collaboration with the present writer, by measurement of the quenching of sodium resonance radiation by a wide range of gases and vapours. Big differences were observed. The Noble gases and saturated hydrocarbons are ineffective or weak in action, while unsaturated molecules show high efficiency. The results parallel the quenching efficiencies for the quenching of mercury resonance radiation, though the mechanism is different. Further light on the quenching and photochemistry of mercury radiation has been contributed by A. B. Callear by means of flash spectroscopy.

Gaydon first showed in 1944 that oxygen atoms react with nitric oxide with the emission of a characteristic yellow-green spectrum attributed to the reaction $O + NO \rightarrow NO_2^*$. This result forms the basis of the measurement of atomic oxygen concentration in flow systems developed by K. Kaufman while in Cambridge by the so-called method of gaseous titration. This and the luminiscent reaction $N + NO = N_2 + O$, discovered by Kaufman and Kistiakowski, has been applied extensively by B. A. Thrush and his collaborators at Cambridge to study a wide range of atomic and free radical reactions, including intercombination involving hydrogen, oxygen and nitrogen atoms, hydroxyl, etc. Used spectroscopically, the method yields quantitative results of particular value to the kinetics of atomic reactions and of chemiluminescence.

In the 1950's, work was also resumed on the catalysis and inhibition of chain reactions in which S. G. Foord and F. S. Dainton had previously been engaged in collaboration with the present writer. By the combination of photochemical and thermal studies of the effect of NO_2 and $NOCl$, etc., on the velocity and explosion limits of the $2H_2 + O_2$ reaction, they had been able to establish the importance of the chain-thermal mechanism of explosive processes. The resumption of this work by P. G. Ashmore, and later with his students, threw new light on these and analogous reactions, emphasising the role of nitric oxide in the kinetics of sensitisation. His study and analysis of the thermal $H_2 - Cl_2$ reaction in the presence of sensitisers has further contributed valuable support to the chain-thermal theory of explosion in gases.

Subsequent to 1950, much study has been made of the kinetics of the pyrolysis and combustion of hydrocarbons. J. H. Purnell and his collaborators at Cambridge and Swansea have carried out detailed studies of the early stages of the pyrolyses of C_2-C_5 paraffins which have established the general—and in many cases the particular—features of the complex radical chain processes occurring. This work has also been significantly extended to the pyrolysis of Silane. The numerous studies of combustion by A. D. Walsh, A. F. Trotman-Dickinson, J. H. Knox, C. F. H. Tipper, R. R. Baldwin, and C. F. Cullis have focussed attention and thrown light on many aspects of the kinetics, including the effect of the treatment of the surface of the reactor, competitive combustion, and general problems of mechanism. With the introduction of gas chromatography, it has been possible to achieve in detail the product analysis of pyrolysis and oxidation. The present writer and his collaborators have studied the mechanism of oxidation of complex hydro-carbons such as benzene, isobutene, hexane, and neopentane by chemical methods, with special reference to intermediate and final products. Further enlightenment has been contributed by kinetic spectroscopy by which it has been established that the OH radical is a reactant common to all hydrocarbon oxidations examined and also to that of hydrides in general. This has led to the proposal by the writer of a generalised chain mechanism of oxidation of hydrides involving propagation by OH and hydride radicals with branching dependent on the formation of an unstable oxygenated intermediate in the propagation reaction.

Studies of kinetics relating to polymerisation have received considerable attention in Great Britain, particularly concerning addition polymerisation involving substituted ethylenes and related monomers; significant progress has been made by several workers along original lines. W. T. Astbury and C. W. Bunn were the first to characterise by X-ray analysis the structure of vinyl polymers in relation to the distribution of amorphous to crystalline forms and to the orientation effects of stressing and deformation; by cold drawing of fibres they were able to measure the dimensions of the constituent units of the oriented polymer chain.

G. Gee in Manchester with his collaborators contributed much original work to the thermodynamic aspects of polymer solutions; in this connection he made an extensive study of the swelling and elasticity of vulcanised rubbers and cross-linked polymer networks in general.

H. W. Melville in Aberdeen and Birmingham was early engaged in the kinetic study of vinyl polymerisation. His work combining photochemical and thermal techniques went far to establish the role played by free radicals in initiation; in particular, he and his colleagues were able to determine the short life span of the propagating polymer radicals and also to devise methods for synthesising graft polymers. Melville's work has been continued and extended by J. C. Bevington at Lancaster, by G. M. Burnett at Aberdeen, and their students.

New reactions of initiation have been discovered and investigated by C. H. Bamford and M. J. S. Dewar and Bamford and R. P. Wayne et al. involving photosensitisation by dyestuffs and by certain metal carbonyls of Group VII in the presence of carbon tetrachloride, respectively. F. S. Dainton has studied initiation by photochemical electron transfer processes and redox reactions. In this connection he has notably applied his studies with E. Collinson of the radiolysis of aqueous solutions to initiation by free radicals. The present author and his students have variously contributed studies of copolymerisation, cross-linking and branching of polymer chains, and have elucidated the physical conditions involving increase of viscosity leading to the slowing of termination and acceleration of polymerisation as the reaction proceeds in poor solvents. Independently, they and Bamford have also discovered and investigated the trapping and immobilisation of growing polymer radicals in insoluble products such as polyvinylchloride and polyacrylonitrile and the acceleration of reaction to which it gives rise.

These studies and those of other British authors have contributed original thought and progress to the wide field of polymerisation reactions. The subject has grown as a whole from the early work of Staudinger, with contributions from workers widely separated in different parts of the world. In this growth the contribution of British physical chemists has played a significant part.

As I close these pages, familiar faces I once knew so well fade slowly in the wake of time. Others take their place, moving along the trail they blazed. I count myself fortunate to have known them. This account of their 50 years and more of work is incomplete, limited by my own ignorance of much that should have been recorded, but I have enjoyed writing it such as it is, and I hope it may be accepted by the reader with this understanding.

Science is international; none of the work described here would have happened but for work of others in countries round the world, for nature reveals herself to those with eyes to see and minds to comprehend, and such are not limited by geographical or political frontiers. Only by working together in pursuit of knowledge can we hope to achieve that mutual understanding upon which the well-being of our world depends.

ACKNOWLEDGEMENTS

It is a pleasure to thank those several colleagues who have materially assisted me with information and advice in the compiling of this review.

Ann. Rev. Astron. Astrophys. 1977. 15 : 1–17

ABOUT DOGMA IN SCIENCE, AND OTHER RECOLLECTIONS OF AN ASTRONOMER

E. J. Öpik

Ann. Rev. Astron. Astrophys. 1977. 15 : 1–17

ABOUT DOGMA IN SCIENCE,   ×2105
AND OTHER RECOLLECTIONS
OF AN ASTRONOMER

E. J. Öpik

Armagh Observatory, Armagh BT61 9DG, Northern Ireland and University of Maryland, College Park, Maryland 20742

DEFINITIONS

The notion of dogma, or the unquestionable acceptance of certain propositions— the doctrines—is usually associated in our minds with religious (or antireligious) beliefs. Yet it has a much wider application. However alien to science, and not widespread there, it *de facto* sometimes infiltrates the realm of research. Usually based on some recognized authority and accepted in a group or "mini-establishment" of true believers, scientific dogma lacks the punitive aspects of its religious counterpart and therefore is open either to ultimate destruction when it is proven wrong, or to logically justified acceptance when finally it is vindicated by facts. Dogma differs from a hypothesis by the refusal of its adherents even to consider the aspects of its validity. Legitimate disagreement or controversy creates dogma when arguments are no longer listened to. Although usually belonging to the realm of theoretical models where direct experiment (or observation) is not possible, dogmatism may sometimes induce its followers to misquotation or misrepresentation of the most undisputable facts, even of the statements made in print by their opponents : if the statements themselves may be subject to doubt or erroneous, the fact of the printed word is indisputable. As shown in the following, my astronomical experience has met several examples of such a "prejudiced blindness." In any case, these misquotations are not intentional but seem to be caused by a specific "dogmatic" superficiality, something like knowing in advance what the other fellow would say and therefore not listening to him, or not reading properly his work.

A UNIVERSITY TREK TO CENTRAL ASIA

Nevertheless, sometimes the dogmatic counterpart may listen, and even may see the light. In the following, a dramatic episode of this kind is described.

In the beginning of 1919, when Bolshevik rule was firmly established in most of Russia, with some fighting still continuing on the fringes, the new rulers decided to found a university in Tashkent, the capital of the just reconquered Russian possessions of Central Asia (so called, although they are rather in the West of the continent). Over 100 professors and other teaching staff with their families volunteered to leave starving Moscow and to start a new life in the food-rich but otherwise risky Asiatic surroundings. As the only astronomer in the group, I was to be Chairman of astronomy and put new life into Tashkent Observatory. A former military geodetic observatory, it had been reorganized by V . V . Stratonoff, but after three decades of respectable research activity it had somehow fallen into disarray during the revolution, and it was now to be made an integral part of the new Turkestan University.

Rail communications in Russia were at that time in a state of disorganization, and so it was no wonder that our legendary trek of 3000 km from Moscow to Tashkent took 70 days, from end of January to beginning of April, 1919. The obstacles were chiefly on the first half of the journey; after entering Asia behind the river Ural the sponsorship of the Tashkent authorities helped us to make the second half of the journey in only three days. There was no coal or fuel storage, so we had to saw and split the raw timber for the locomotive ourselves. Beyond the Volga, there are oak forests, and, to our surprise, freshly felled raw oak turned out to be excellent fuel despite its wet appearance! In March, at a small town on the border of Europe and Asia, we were prevented by the local Soviet from proceeding further for a whole fortnight: it was an act of sheer ingratitude. When we arrived in the town, the Soviet, aware of the presence of so many learned persons, asked us to give a lecture and perhaps a variety show to the populace of the isolated township, and we readily agreed. There had just been a splendid display of aurora, with streamers reaching to the zenith, and secret rumours were spreading that these were the artillery flashes of the cavalry of Dutov, advancing from the north to wipe out the Communists (who were not too much liked by the population). So I gave a lecture on aurorae, geomagnetism and solar activity, and then a concert followed. There was quite a good tenor among the professors; I accompanied him at the piano, and other items of entertainment followed.

As a result of our success, the next day we got a letter from the Soviet requesting a repeat of our performance, with a threat that they would not let us proceed unless we complied. We did not bow to the threat and, in protest, refused to deal with them; instead we sat out in our carriages until a strong order from Tashkent forced the local Soviet to lift its embargo. At the next station we again lost three days: our engine with its fuel—the fruit of many hours of our hard work—was stolen by a trainload of Red Marines while we were asleep. After having obtained another engine and prepared the fuel again, we put out sentries overnight (I was one of them), and not in vain: another migrant group approached us at night, intending to take over the fruits of our labour, but they retreated without a fight. The next morning we went on without further adventures—changing engines and preparing fuel, no longer of regular logs but from ragged, thorny, yet very dry saksaoul shrub of the Kirghiz semidesert steppes; "goblins, not fuel" as some of us resentfully remarked. But it burned well and carried us in three days through all Central Asia to our final destination. Before departure, on the meter-thick snow cover, frozen hard on the surface under the intermittent action of the springtime sun and the frost at night, we celebrated in jubilation and I performed a wild improvised dance in long leaps—I called it the dance of the polar bear: my only solo dance in my life.

A CONFRONTATION RESOLVED

After arrival in April, 1919, I served for two years on the faculty of the newly organized Turkestan University, with my main concern being the revitalization of the Tashkent Observatory. Placed on the outskirts of Tashkent at an elevation of only 440 m above sea level, but near the Central Asian plateaus and the highest mountain ranges of the world, the Observatory enjoyed a most favorable astronomical climate. Besides astronomy and astrophysics, it comprised meteorology and seismology as well—the latter being of especial local importance in view of the frequent earthquakes (one night I was almost thrown out of my shaking bed, but the tremor stopped, and I—quite lightmindedly—did not run out into the safety of open space, as the other inhabitants did). The observatory grounds occupied a vast area, and the single detached structures—laboratories (I had a living room alone in the Astrophysics Laboratory), telescope domes, and seismic and meteorological instruments as well as living quarters—were widely spread over the area without a proper view of each other. Most of Russia was already under the Bolshevik (Communist) rule, after the second or Red revolution of October 1917, and this included also Central Asia. In the transition time, a local physics teacher, V. N. Milovanoff, consented to take over the Directorship of the Observatory, and after my arrival he stayed on and became mainly concerned with administration, while I became Vice-Director and looked after the scientific side. The changing circumstances, especially the takeover by the new rulers while law and order were far from being guaranteed over the vast, sparsely populated expanses of Central Asia, created many administrative problems.

Seismology was represented by one person, G. V. Popoff, an athletically-built bearded religious fanatic who lived with his middle-aged housekeeper Maria Abramovna. She looked also after me, in so far as laundry and cooking of lamb chop was concerned; I bought the provisions myself (Tashkent was at that time a cornucopia for food, the envy of starving Russia). Popoff volunteered to meet the visitors, much to our relief, and hundreds flocked to his popular lectures, which were held on open ground. What he was lecturing about, we did not know—he was competent enough to deal with all the aspects of astronomy, meteorology, seismology. Then suddenly we got a shock: a letter came from the local Commissariat of Education informing us that Popoff was preaching religion under the cover of scientific popularizations, and that the people were flocking to his lectures because of their religious superstitions. The letter insisted that this was contrary to the proclaimed antireligious principles of the ruling party and was also nonscientific. Therefore, the letter concluded, Popoff was not worthy of keeping a scientific post and must be immediately dismissed.

We were deeply worried by this. Not only because we believed in humanitarian considerations and the freedom of speech and thought, but also because Popoff was a very much wanted member of the staff whom we did not wish to lose, Milovanoff and I decided to plead for him at the Commissariat. This was indeed a dangerous enterprise, in view of the political situation and the sensitivity of the new rulers to violations of the Marxist dogma, but we did not mind the risk, and it went off quite happily. The Commissar, a young man of about my age (26) named Dvolaitsky, received us in the entrance hall of his office. We pointed out that Popoff was a good scientist and the only seismologist available, and that, while supplies of photographic registering paper were cut off by the revolution, he managed to resuscitate the almost forgotten mechanical method of using smoked paper scratched by the seismograph needle (with levers in between) and thus continued a makeshift uninterrupted registration and recording of Earth tremors. "Well," said the Commissar, addressing me, "you are an astronomer and you should know that astronomy has proven that there is no God." The naïveté of this sentence was obviously genuine—not just an officially imposed party attitude—and seemed to invite discussion. I burst into laughter and explained to the perplexed Commissar that, while belief or disbelief in God is a matter of inner feeling and personal freedom, science is powerless to prove or disprove His existence. We had quite a long talk, and after listening to me attentively Dvolaitsky then decreed: all right, Popoff may remain in office, but he must limit his activity to purely scientific professional functions and he shall not be allowed to lecture to the public any further.

THE AFTERMATH

After a couple of years—I had already left Tashkent at that time for my homeland Estonia—I learned that Popoff was not at all grateful and almost strangled Milovanoff, the Director. Popoff apparently did not appreciate what was done for him and considered Milovanoff the culprit who terminated his lecturing. One day, when Milovanoff descended underground to have a glance at the seismic laboratory, he was accused of "spying" and attacked by Popoff. A passerby heard the screams and rescued the Director.

Many years later, during Stalin's purges of his colleagues, among the lists of comrades—Bolsheviks put to death by their overlord—I recognized the name of Dvolaitsky. The name is uncommon—I never have seen it before or afterwards. It was almost certainly the name of my Tashkent Commissar of Education, a top Communist who listened to reason against the party dogma. Could this have been the cause of his "liquidation" by Stalin?

A WARNING

In 1940/41, at Tartu Observatory, during the first Communist occupation of Estonia, I had another experience along similar lines, although not so dramatic. I gave a course of popular lectures to the public at Tartu Observatory, with philosophical digression into the mysteries of existence and the meaning or purposefulness of the Universe. People flocked to my lectures and demonstrations in about the same manner as to Popoff's in Tashkent, and when I finished my course, I was specially thanked by representatives of the audience for idealistically lifting them up from Earth, nearer to Heaven. Although religion was never explicitly mentioned in my lectures, somebody apparently reported on my idealistic approach, and one evening, while I was outdoors explaining the constellations to my audience, an important-looking young man approached and warned me that I was on a slippery path and that I must avoid the themes of God and religion; otherwise the lectures would have to be stopped. "You know," he said, "many years ago there was an old professor at Tartu who said that when searching the heavens his telescope had never shown him either angels or God." This time I did not burst into laughter— a more sinister threat was behind this naïve sentence. Neither was the course of my lectures interrupted.

DOGMATISM IN SCIENCE

While religious—or antireligious—abstract dogma stands beyond the reach of science, in science itself the dogmatic approach has played, and is still playing, a conspicuous role, but with a difference: the preconceived ideas in science are subject to verification or rejection through further research. They remain "dogma" only as their adherents refuse to consider alternatives to their doctrines, rejecting criticism beforehand. A group of scientists proclaiming such a dogma would thus form an "establishment" or "mini-establishment" (depending on the extent of the group), an extreme case being a single person persistently building on a certain unproven assumption.

Scientific dogma as here defined may prove correct, and even when ultimately disproved, it may serve a useful purpose, as a stimulant to research and the accumulation of facts. Often it may also be harmful, as an obstacle to freedom of research, especially when influencing editors and reviewers of scientific magazines or directors of institutions.

The discovery of America by Christopher Columbus, who underestimated the size of our globe and until his death firmly believed that he had reached the outskirts of Asia, is a case to this effect. It has been often said that he would not have embarked on the voyage had he known the actual distance to be covered. Yet— and the analogy may also hold in research—Columbus had other indications, such as old stories, washed-up twigs of unfamiliar trees, etc., which implied that a continent must exist not too far away in the West. Even knowing the true dimensions of the Earth, he could have concluded that land was not too far away across the Atlantic, and he could still have gone out in search of it, Asia or not.

Scientific dogma or "mini-dogma" is still usually based on some recognized authority whose pronouncements are unconditionally accepted by his followers or, at least, by himself. Newton's laws of mechanics and gravitation are a splendid example of dogma justified by centuries of research and still basically valid, despite the *corrigenda* introduced by the Theory of Relativity. On the contrary, the dogma that lunar craters are volcanoes, which was maintained for so long by the mini-establishment of (chiefly) amateur lunar observers, has been proven wrong, although this does not detract from the value of their observations.

In the following are briefly described some cases of astronomical dogma that are part of my personal experience. Of course, by a kind of "natural selection," only examples of misjudgment or misinterpretation are pointed out, while those instances that proved correct are not mentioned, simply because it is difficult to distinguish between the roles of dogma and of critical research in such cases.

STELLAR STRUCTURE

While Emden produced purely hydrostatic models of gaseous spheres, the genius of Eddington put life into them by introducing the concepts of energy transfer, chiefly by radiation, and of energy generation. Eddington's merits in deciphering the internal structure of stars are incomparable and his work, especially in laying out the physical principles, still serves as a basis for continuing research, despite his failure in explaining the structure of giants. And the failure itself was not in the lack of physical or mathematical methods of approach; everything was already contained in Eddington's own papers and equations. Yet Eddington's models were conveniently assumed to be of uniform chemical composition, which became a doctrine for himself and for the mini-establishment of his followers and which, as we know now, cannot produce "inflated" or giant stellar structures. Against the well-known fact that the recognized nuclear energy source—the conversion of hydrogen into helium—creates a concentration of matter with heavier molecular weight around the stellar core, Eddington produced the von Zeipel concept of forced rotational convection in operation in the hydrodynamically stable radiative-equilibrium layers (1). However, he failed to put numbers into his equations, and when I did it (2), it turned out that the time scale of would-be mixing for the Sun is of the order of 10^{14}–10^{15} years, 10,000 to 100,000 times longer than the age of the solar system or the time scale of nuclear reactions. This refers to most slowly or moderately rotating stars, especially to giants, and the assumption that the helium produced from hydrogen in the hot central regions would somehow get mixed into the entire volume of the star is untenable, and by a large margin. It should be emphasized that in this respect I did not add a single bit to Eddington's admirable creation; I only performed the calculation according to his prescribed formulae. Eddington's failure to pursue the consequences of his own theory can only be explained in terms of a "blind spot," a dogmatic refusal to abandon his model of convenience—that of uniform chemical composition. My next step was the numerical integration of "composite" nonuniform models, which was much more complicated than the application of Eddington-Emden's homology formulae. However, it was realistic and brought the reward of explaining the structure of giant stars, with high central temperatures and densities that are adequate for advanced nuclear reactions, but with large radii and low mean densities (3).

MIXING LENGTH

A model of convection, initiated by Schmidt and Prandtl, pictures the vertical transport of excess heat in a gaseous or liquid medium through the symbol of a "mixing length," L_m, such that a hotter element ("a bubble"), while rising over this length, does not exchange heat with the colder surroundings and delivers all its excess at the top. It is matched by a similar cold bubble descending from the top and absorbing its prescribed share of excess heat only when arriving at the bottom. Such a lateral isolation of the moving "bubbles" or streams could be achieved only by a miracle, and I devised a realistic model of convective transport (4, 5), taking into account lateral exchange, which agrees with laboratory experiments within $\pm 20\%$, while the mixing-length model predicts a transport by $+3000\%$ in excess of the true value. In a model of cellular convection, the rising current is, of course, always warmer than the descending current *at the same potential level*, but, because of lateral exchange, this is only one tenth of the total adiabatic temperature excess between the extreme levels; the rising current is gradually precooled through lateral contact with its gradually preheated descending counterpart. The stream velocity, proportional to the square root of the equipotential temperature difference, is reduced to one third, and the real convective transport is thus reduced in a ratio of $\frac{1}{10} \times \frac{1}{3} = \frac{1}{30}$th of the mixing-length prediction, while dimensionally the transport equation remains unaltered.

The mixing-length symbolism, probably meant only as a simile, was grasped by a school of astrophysicists in its literal sense and used for numerical applications. In the deep stellar interiors, with their high temperatures and densities, the superefficient mixing-length model would require deviations from the adiabatic temperature gradient of about one part in three million, while the realistic model would require about one part in one hundred thousand—both small enough not to be reckoned with in the calculations of stellar models. However, in the outer layers near the stellar surface the difference is enormous. In this context, the dogma of the mixing-length has become an obstacle to progress, as can be seen from the following incident.

In 1970, D. J. Mullan—a pupil of mine who has now risen to prominence with numerous researches, especially in the physics of stellar atmospheres, offered for publication in *Astronomy and Astrophysics* (the European Journal) a paper on "Cellular Convection in Stellar Envelopes." By applying my theory, he explained a score of various spectral traits, created by the bottleneck of inefficient convection in stellar atmospheres, which could not be accounted for by the mixing-length doctrine, with its over-efficient transport. On the report of a referee who disagreed with my theory, the paper was rejected. The fact of rejection on such grounds, even if disputable, is in itself very ominous. Although the story had a happy end— the paper was then published without much delay by the Royal Astronomical Society (6) and represents undoubtedly a gem of a contribution to the knowledge of stellar atmospheres—the attitude of the referee (a staunch believer in the mixing-length, yet officially anonymous) was characteristically dogmatic. By misrepresenting —apparently from unwittingly misreading the texts but perhaps true to his creed— the unwanted alternative to the mixing length, it is a remarkable example of wishful thinking. Here are a few citations from the referee's report, which was communicated to the author. He (or she) writes (exact excerpts are in quotation marks, with my comments following):

1. In Öpik's investigation, "The largest part of the convective upward heat transport is assumed to be transported down again since the matter presumably

cannot get rid of its surplus energy." Quite contrarily, I show that the contact heat transport is so powerful that the excess heat escapes laterally into the downward current before reaching its ultimate destination and that matter gets rid of its surplus energy much too readily and sooner than in the mixing-length analogue. Further, convective transport is not to be equated to the total heat content of matter moving upwards, but is only the net difference delivered at the top, and this, once delivered, cannot "be transported down again," a physically meaningless suggestion and something I nowhere had intimated.

2. The referee then continues, "This then leads to the conclusion that a hot gas stream ($\Delta T > 0$, $\Delta \rho < 0$) will sink in the atmosphere, which seems impossible to the referee." The naive term "hot" betrays the root of his misunderstanding: there is nothing like absolutely "hot" or "cold," nor is there any absolute definition of the temperature excess, ΔT. What matters is the difference between the warmer rising current and the colder descending current at the same potential level, and for the "sinking" stream ΔT is always less than 0 or colder, although the difference is decreased ten times compared to the mixing-length figment, while, compared to the bottom level, both rising and descending currents are colder because of adiabatic expansion.

This would suffice, though there is more to it. The quotations as cited above are typically similar to the dogmatic criticism of misrepresented tenets of another faith, often as wholehearted as it is prejudiced; unwillingness to get acquainted with the actual pronouncements of the other side is common to both. Such an attitude, though alien to science, is nevertheless there as the consequence of human weakness, and we have to reckon with it as a fact.

As an outstanding example of dogmatism in science, the mixing-length syndrome still persists and comprises an influential mini-establishment, though it is harmless as long as it is confined to stellar interiors and keeps clear of the surface layers, or of planetary atmospheres.

CRATERING

Impact cratering is important in the process of shaping the surfaces of planets, especially of the Moon and Mars, as well as in cosmogonic processes of building the planets from aggregates of smaller stray bodies. Until lately experiments at cosmic velocities were not available, and I developed a theory based on first principles which now, when compared with the experimental data, turns out to allow the calculation of crater volume, diameter, depth, and ellipticity of craters with an unexpected accuracy of better than 20% in linear measure and at all velocities (7). No empirical parameters are used. The crater volume is essentially proportional to the *translational momentum* of the projectile, with a *corrigendum* for vapor formation and a secondary shock caused by it, while the cohesive strength of the target enters as an independent variable determining the coefficients of proportionality. The success of this a priori theory, now empirically verified over a range of velocities from 2 to 20,000 m sec^{-1} without using ad hoc coefficients of proportionality, is its main justification. Consideration of the *radial momentum*, created in the target by the intruding projectile, is the main feature of the theory.

Regrettably, experimenters in hypervelocity impact have used and are still using kinetic energy as the argument for interpolation, even without proper regard for the cohesive strength of the target. While this procedure may be practically satisfactory for representing experiments with the same materials within a limited range of velocity and a more or less constant mass of the projectile, extrapolation by kinetic energy beyond the experimental range may lead to errors of many orders of magnitude: mv^2 can never be made to correspond uniquely to mv. Actually, because of vapor formation and second shock, the velocity exponent for crater volume may be about $\frac{4}{3}$, while at constant velocity the crater volume must be proportional to the mass of the projectile. An intermediate formula for cosmic velocities, something like $mv^{4/3}$, could be suggested, instead of mv. My theory actually allows for this, although without the mathematical oversimplification. The empiricists, however, having discovered the lesser power of velocity but insisting on kinetic energy as the argument, would put the crater volume proportional to $(mv^2)^{2/3} = m^{2/3} v^{4/3}$, with the unnatural $\frac{2}{3}$ power for mass (instead of 1). There are more details to it, to be looked for in the relevant publications.

At present, however, a mini-establishment exists around the doctrine of kinetic energy as the impact argument. This is an impediment to progress and, until the successful correct theory (especially with regard to cohesive strength) is applied, extrapolations and speculations on the cosmogonic role of cratering are subject to major pitfalls.

Estimates of the mass of projectiles that produced meteor craters on the Earth and Moon offer a relevant example. For the Arizona crater, estimates based on the doctrine of kinetic energy were about 40 times lower than the mass corresponding to the criterion of translational momentum, and independently confirmed by the depth of penetration. If such were the efficiency of meteorite impacts (which actually waste most of their energy on the inelastic radial shock and heating of the target material), the number of craters in lunar maria would be by almost two orders of magnitude higher than observed, amounting to saturation cratering and equal to that on the continentes. It has been shown (8, 7) that, with the observed population of stray bodies in the solar system and my theory of cratering, the frequency of small and medium-sized craters in the maria is closely accounted for by impacts during the past 4,500 million years, while larger craters show an excess, accountable by survival of premare craters through the event of mare lava flooding (itself a result of a major impact). This in itself is a most impressive confirmation of the cratering theory, obtained well before experiments with hypervelocity (and non-hypervelocity) impacts were made on Earth.

By arguing ad absurdum, we could say that, if energy were directly relevant to the size of a crater, a bonfire lit on a rock surface should lead to "progressive cratering" because heat is also kinetic energy. Cratering is the result of action of forces, and action is in direct relationship to momentum. In a kind of transfiguration of momentum, the translational momentum of the projectile creates radial momentum of the displaced target (rock) material in a constant ratio of from two to five (depending on velocity as determining the secondary shock from vaporization); the total momentum of the symmetric radial shock is, of course, zero, while the translational momentum of the projectile is absorbed by the main body (planet).

THE EXCITEMENT OF SCIENCE 517

In the target, the inelastic shock conserves its momentum separately in each radial direction as long as destruction of the solid material and hydrodynamic flow takes place. The velocity, U, of the radial motion thus decreases as the volume and mass involved increase, until the hydrodynamic resistance ρU^2 (ρ = density) becomes equal to the crushing strength of the material, s. This determines the limit of destruction and the volume of the crater.

Agreement with experiment and with observation (frequency of lunar craters) completely supports this theory (which, of course, is more complicated than could be sketched here). It can only be wished that the dogmatic eclipse of the realistic cratering theory by the kinetic energy scaling would be lifted and that the overlooked and neglected *perfect* theory be raised to its proper place, for realistic dealings with cosmic or cosmogonic events. (Years ago it was found that the theory correctly predicts the armour-blasting properties of artillery shells, a "practical" confirmation of, unfortunately, too sombre associations).

LUNAR AND MARTIAN VOLCANOES

Despite all the eloquent statistical arguments of interplanetary astronomy, the thesis that lunar craters are presumably of volcanic origin was quite widespread, chiefly among amateur astronomers and professional geologists. Decades ago some of them even tried to deny a meteoritic origin of the Arizona crater. The dogma was very strong and was proclaimed by a considerable group or establishment of true believers. At present it has been completely destroyed, at least as concerns the Moon, by direct space exploration and landings, and need not interfere with scientific progress any further. Of course, Mars has now become the refuge of planetary volcanoes, although no longer in a dogmatic sense. It is conceded that most of the Martian craters also originated from impact, because their frequency (surface density) corresponds to statistical expectations for the fringe of the asteroidal belt. But a few large structures are still called "calderas," perhaps wishfully implying their volcanic origin. From the total evidence available, I still prefer to consider them impact craters, surviving an immense lava flooding in the Martian northern hemisphere from an impact of a large asteroid "in the beginning" (9). Since they are similar to the larger surviving lunar craters, and since they are placed on elevated ground well above the average level, such survival in the midst of a lava sea is quite plausible. The matter, however, cannot yet be considered as finally settled. Besides, from the slowness of erosion on Mars—which on Earth is a necessary link in mountain building—Mars cannot yet have entered the phase of volcanism that may be billions of years ahead.

ANCIENT MARTIAN "RIVERS"

The identification of some gigantic meandering cracks as the beds of ancient rivers (of water or lava) on Mars is in danger of becoming a mini-dogma, misleading and perhaps impeding progress. The only reason for such an identification is their meandering shape and formation of systems of succursals closely reminiscent of terrestrial river systems. I have pointed out that exactly similar meandering and branching systems of cracks are omnipresent on asphalt or concrete sidewalks, e.g. on all university campuses I have visited (10, 9). These are caused by the pressure of encroaching vehicles without any relation to fluid flow, and the cracks or clefts on the Martian surface are most probably of similar origin, caused by the pressure of readjustments of the deeper crust on which the top layer rests. Water rivers are definitely excluded—with large amounts of water, cloud and snow formation (at present solar luminosity, though it was *lower* in the past) would have depressed the mean global temperature on Mars from the present low temperature of $-42°C$ to $-62°C$, equal to the coldest Siberian midwinter. Water would everywhere be completely frozen under such circumstances, unable to flow and create rivers. Although lava flow, dubious as it is, cannot be excluded by such an argument, cracks on the surface of a thermally evolving planet must inevitably arise, and before looking for the "rivers," let us look for the few real cracks: they are there, relegating the "rivers" to the realm of fantasy until better confirmation is available. The lunar "rilles," which also were regarded as traces of liquid flow, are now more definitely identified as cracks or rifts (10), and this may serve as an analogy for Mars.

ORIGIN OF METEORITES

The physical and mineralogical structure of meteorites implies that they are collision fragments of asteroidal or sublunar sized bodies. Consideration of encounter probabilities ensures that collisions between members of the asteroid belt do happen; hence there is a possibility that meteorites actually arrive from the asteroid belt, their orbits being changed by the velocity imparted at collision and by subsequent planetary perturbations. The newly determined density of the largest asteroid, Ceres, indeed confirms the hypothesis that asteroids are compact stony bodies and not fluffy objects like cometary nuclei. This falls now in line with the hypothesis of the asteroidal origin of meteorites, which is now seemingly becoming a mini-dogma, accepted without further doubts. Yet it has been shown (11) that contemporary collisions and perturbations cannot account either for the yield or for the orbits of meteorites, which resemble those of short-period comets brought inside Jupiter's orbit by nongravitational forces. The orbits of the so-called Apollo class of "asteroids" belong to the same type, which suggests that they are extinct remnants of disintegrated gigantic cometary nuclei. The apparent conclusion, to be substituted for the fruitless dogma, would relegate their origin to collisions among quite another class of primeval asteroids and to the dawn of the solar system. The original fragments would then have become incorporated into the ices of accreting comet nuclei, and been ejected by planetary (Jupiter) perturbations to the outskirts of the solar system (Oort's sphere of comets), where they would be stabilized by stellar perturbations and sent back to the inner regions of the solar system by similar perturbations. After being captured by Jupiter into short-period orbits, the "rocket-effect" of evaporating gases would cause some of the orbits to shrink (namely those with retrograde rotation of the nuclei) and thus to escape Jupiter's dangerous vicinity. With evaporation of the volatiles, the meteoritic fragments or the Apollo "asteroids" are released into our interplanetary surroundings, to be ultimately removed by planetary encounters within a lifetime of the order of 100 million years. This model is also in harmony with the cosmic-ray ages of meteorites, which represent the time since they were released from shielding inside the cometary nuclei, and not the time of their collisional break-up. If this were so, their relevance to the origins of the solar system would be greatly enhanced.

TIDAL ORIGIN OF THE SOLAR SYSTEM

The hypothesis by Chamberlain and Moulton, so diligently pursued by Jeans, that a close tidal encounter of the Sun with another star led to the formation of the solar system, has enjoyed widespread (though not universal) acceptance and is still on the books, despite its improbability bordering on impossibility. It offers an example of dogmatic attitude with formation of its peculiar establishment group, which is especially strong in popular writings. Such an encounter, of course, is quite possible but extremely improbable. Further, as pointed out by Russell, the hot gases ejected from the Sun could not condense into planets, especially not during the short time of the stellar passage, but would instead disperse into space. Their angular momentum (which raises the cardinal challenge to all cosmogonic theories) at ejection would be short of the requested value by a factor of the order of 20. This difficulty, and then putting the planets into circular, regularly spaced orbits, created formidable problems that Jeans attempted to answer through appropriate perturbations by the passing star, a gigantic mathematical task never convincingly concluded. The problem is in itself interesting and it was worthwhile to treat it, but physically the outcome of the encounter would have been the ejection of uncondensed gaseous matter, which would at first have formed a nebula. Any wisps of gas, ejected into various intersecting or interpenetrating orbits, would then through collision settle into a circularly rotating aggregate with conservation of the sum of angular momentum, i.e. a primeval nebula from which later the planets could have condensed. Yet this returns us to the nebular hypothesis, and there is no way whereby we can distinguish between a nebula directly condensed from interstellar space, and one formed in the tidal encounter—except for the criterion of probability. And in this respect the solar system gives an answer. The systems of the satellites of the outer planets show the same kind of regularity exemplified by the mother system itself: a coplanar succession of near-circular orbits with a more or less regular spacing (Titius-Bode Law). Instead of the extremely low probability of stellar encounters, the formation of solar or planetary satellite systems appears to be the rule rather than an exception. The nebular hypothesis is thus able to account for everything. While the improbability of a tidal encounter can be partly brushed away by assuming that it happened when the stars were much closer together (this, however, would require an improbably high age for the solar system), the ensuing regular spacing of the planets (or the satellites) requires the intervention of another improbable configuration, so that the idea of a tidal encounter can hardly be maintained in this context. Of course, there could have been tidal encounters during the early phases of evolution of our stellar universe, and the theoretical work done in this respect is not quite in vain. Yet the outcome of such encounters could be very different from the formation of something resembling the solar system.

PUBLICATION BIAS AND EXCHANGE

When, in 1938, my papers (3) on "Stellar Structure and Stellar Evolution" (with calculations of unmixed stellar models and those of giant stars) were printed in the *Publications of the Tartu (Dorpat) Observatory* in Estonia, I soon afterwards received a letter from George Gamow, underlining the importance of my work but reproaching me for publishing in such an "obscure" place, wherefore—in his opinion— progress in the study of stellar structure must have been unnecessarily delayed.

The view that, by all means, publication must be achieved in the internationally recognized "important" journals pervaded and still pervades the astronomical establishment and especially the young generation; the latter of course for obvious practical reasons. Yet the fact that Gamow—within a year—got hold of my papers, and that others soon continued on these lines (sometimes referring, sometimes not, to my work), is the best answer. Tartu Observatory, in the centuries-old astronomical tradition of exchange of publications (possibly explained by the fact that we are all dealing with the same cosmic laboratory called the Universe) exchanged its publications with all astronomical institutes of the world, so that the work did not remain unknown to those who cared (physics and most other branches of science do not adhere to such a tradition). And, as to the economical side, the publication costs in Estonia were very low (as they are also at Armagh where the same tradition of exchange is continued). Although, for instance, the editorial setup of the *Astrophysical Journal* was friendly toward my aspirations, the page charges and reprint costs for some 200 pages of mathematical and tabular material would have been absolutely prohibitive. Also, it was certain that full-length publication of such extended papers would not be possible. I had already had previous experience with editors and reviewers requiring great reductions in size, to the detriment of detail which is so essential in pioneering work.

I wish here to emphasize further the importance of the traditional exchange of publications. Not all observatories (especially the smaller ones) are in a position to subscribe to all the "important" journals. Also, a search in libraries for the relevant articles a scientist may need would involve unnecessary psychological effort and waste of time. Thus, because of human weakness, communication between scientists would considerably suffer unless, as in the astronomical tradition, reprints and independent publications of an observatory are systematically numbered and kept in one place. In such a case it is easy and even rather tempting to look among the systematized publications of another astronomical institution for the collected printed papers of a colleague who is known to work on a definite subject. Theoretically, the convenience may appear to be irrelevant, but practically it is of utmost value.

The institutions that do not follow the tradition of exchange usually send out Lists of Reprints, available on demand. This procedure disregards the fact that any research of value is not meant to satisfy the interests of individuals of today, but should address itself also to the future. It cannot be known in advance which work will be of relevance within decades (or even centuries) to come. A selection made to satisfy the temporary interests of today may and certainly will miss the works of relevance for tomorrow. A library always contains more works than would be ever needed or read: but it is impossible to foretell the interests of the future, and the collection must necessarily be very much more complete than the actual needs that may arise.

In the matter of exchange, the giving hand should take the initiative. As his moral and vocational obligation, a creative spirit must make his results known, at least where similar work is, or could be done. It is like seeding. Few of the seeds may fall on fertile soil, yet nobody can predict for certain which of them will grow. And, where there is no seeding, there will be no growth.

THE CRITICS: EDITORIAL REFEREEING AND CENSORSHIP

Another reason for not always publishing in the accepted "important" journals is a danger of being rejected, either because of a sincere failure of the editorial apparatus to appreciate new developments (often because of the fear of appearing ridiculous), or because of dogmatic and personal prejudices. These last two should not enter into the editorial judgment, which should be as impartial as possible, yet actually they sometimes do.

As an example of editorial changes that do not infringe on impartiality, I would cite a case with *The Irish Astronomical Journal,* where I am Editor, and where, in an article on the "Lunar Surface" offered by Patrick Moore (12), he voiced his support for the volcanic theory of lunar craters. Although completely disagreeing, it was not my business to interfere except for one minor change: the two groups with opposing views were called "authorities" by Patrick Moore—one, favoring the volcanic origin, consisting chiefly of amateur astronomers, the other, siding with the impact origin, consisting of professional scientists (one of them a Nobel laureate); I had the word "authorities" changed to "authors," with the author's consent.

In my experience, however, editors have not always been so impartial, although usually they are. An article entitled "The Optical Oblateness of Mars" and based on my microphotometric measurements of Mount Wilson photographs during the Opposition of 1958 was accepted by *Icarus* for publication, on the condition that I omit two concluding pages and a figure, actually containing my chief results. The photometrically measured diameters showed a consistent variation with areographic latitude, closely similar in the two colors—the blue and the yellow—and were interpreted as revealing climatic zones of atmospheric circulation similar to those on Earth. I maintained namely that what we measure as the limb is the top of a dust layer that is mainly responsible for the reflectivity of the atmosphere, and not its gas (this is now confirmed by the Viking 1976 landings on Mars). An upward current would lift the dust up, a downward current would carry it down. The measurements showed an equatorial uplift, a subtropical depression as for the anticyclone trade-winds, again an increased diameter in the zone of middle latitudes corresponding to the terrestrial temperate zone of westerlies, and again a subpolar—apparently anticyclonic—depression; only the polar diameters (all relative to an equipotential surface) were increased, contrary to the expectation of an anticyclonic depression, but this clearly seemed to be caused by snow or ice crystals of high albedo replacing the yellow dusty haze of the other latitudes. Now, on this *observational* evidence (whatever its interpretation, right or wrong), the Editor of *Icarus* and my professed friend had put his veto! Another journal was then ready to accept publication, but on the evidence of previous rejection the Editor changed his mind. The article was ultimately published in The Irish Astronomical Journal (13).

From my long experience, both with my own papers, and with those sent to me for refereeing, I have a feeling that the "recognized" journals usually accept without difficulty papers with a middle-of-the-road content, useful contributions to research which already has established itself or accumulations of additional new material. Papers that make little sense are mostly rejected, but some of them are slipping through. As to pioneering work, papers of this kind often are running the risk of rejection or of excessive curtailment.

The practice of anonymous referees is much to blame for editorial malfunctioning. A referee for a scientific journal is a scientist, morally committed to seek and openly proclaim the truth as he feels it, and he should never hide behind the screen of anonymity. As referee, I always send the author an exact carbon copy of my letter to the Editor with all my comments. Among about 150 of such reviews, I have received many letters oɪ thanks (for my suggestions) and only one of what practically amounted to abuse; in most cases, however, there was no reaction. There were a few cases when I was asked to be arbiter in an unfavorable referee report, and I succeeded in rehabilitating some authors from unfair criticism by anonymous colleagues who appeared to think that they alone were entitled to write about a certain subject. Anonymity in refereeing is like kicking somebody in the dark, without a chance of response; it "protects" the reviewer but not the author. The sooner this scourge of anonymity is abandoned, the better for the honest pursuit of research. If fewer referees can be found when there is no anonymity, it will be only to an advantage: those who consent will be a more qualified selection for the job of critics.

We may ask here how many geniuses have been crushed by the unsympathetic and prejudiced attitude of editors and critics in the sciences as well as in the arts, and remained unknown forever? The late discovery of forgotten geniuses testifies to the existence of a graveyard of misunderstood or mishandled originalities which did not fit into the "establishment" of the critics. The sad record of George Bizet, who died in desperation witnessing the failure of his "Carmen" in the eyes of the critics and the Paris public influenced by them, serves as a reminder: "Carmen" has become unquestionably the most popular opera of all times. By independently printing in the "obscure" Tartu or Armagh publications, my work has ultimately made itself known. Would I have been forced to limit myself to the "recognized" big journals, much of it—possibly some most original contributions—could have remained buried forever.

THE EXCITEMENT OF SCIENCE   523

THE BIRTH OF A MYTH

A remarkable article by Leighton (14) supplied with artistically rendered humorous illustrations almost true to life, emphasizes the fact that scientists seldom listen to others and, if they are not dozing during lectures, they may either be preoccupied with their own thoughts or enter into private discussions. Although partly explained by acoustical difficulties, this attitude may not be limited to conference lectures.

The following example describes a case to the point. The printed word is an indubitable fact, irrespective of whether or not its statements are subject to debate. Yet here we have a critic who first completely misrepresents a published work, and then—rightly—sets out to destroy this figment of his own imagination which, incidentally, is just the opposite of what the author of the criticized work was saying.

In the universally recognized international journal *Science*, R. K. Ulrich (15) refers unfavorably to my work (16) on stellar structure and variations of solar luminosity. He purports to describe my model and calls it "physically untenable," but the described model is not mine—only the critic's invention and, so to speak, the very opposite of what I had proposed. In my reply (17) I point out that, while I consider inward diffusion of *hydrogen* into the core depleted by hydrogen burning, he objects to inward diffusion of the *heavy elements* which in my model are not diffusing at all. In my model, turbulent mixing suddenly transports more hydrogen to the core, triggering thus an increase in the nuclear energy output, this being the most important point in my theory of variability. Yet Ulrich never mentions "hydrogen" by name or "nuclear energy generation" in this context. While I trace the heavy-element content in the Sun to interstellar *diffuse* matter (dust) during the process of star formation by accretion, thus predating all the development stages of the future Sun, the critic insists that I am putting them into the core by internal diffusion inside the Sun! etc. etc. Possibly, the words "diffuse" matter and negligent reading, with a mind concerned with gas diffusion, may have led him to this gross misrepresentation.

Now, as often happens, other authors could rely on the second-hand information of such a source, and a ready-made legend or myth, perhaps a new dogma could emerge, something of the sort "Öpik wants diffusion of the heavy elements in the Sun to be responsible for its variability, which of course is too slow on the time scale of stellar evolution." Note that a similar, perhaps not so extreme misstatement about "hot bubbles going down" has been mentioned above in connection with the ill-conceived notion of the "mixing length."

It is not a question of whether I am right or wrong in my theory, but only of what I had actually said in print, thus of the complete distortion of an undisputable fact. In this case—as probably in many others—the editorial reviewer system has goofed, while the critical author, instead of a straightforward apology and admission of fault, in a "reply" (18) just vaguely expresses some of his own views on solar models and restricts himself to considering the (irrelevant and practically nonexistent) diffusion of the heavy elements "relevant to hydrogen," instead of considering the diffusion of hydrogen itself into the depleted core.

FOR A POSTSCRIPT

These nonsystematic recollections, based on personal experience, are concerned chiefly but not exclusively with preconceived notions and dogmatism. The research topics mentioned above as examples are of necessity those close to the author's scientific activity; he considers the points raised in their connection of great importance in the study of the Universe, but by no means implies that he is always right. He sincerely wishes that his words may not completely remain a lonely cry in the wilderness, but may perhaps at some time help someone in the impartial search for truth.

Literature Cited

1. Eddington, A. S. 1929. *MNRAS* 90:54
2. Öpik, E. J. 1951. *MNRAS* 111:278
3. Öpik, E. J. 1938. *Tartu Obs. Publ.* 30, No. 3 (118 pp.), No. 4 (48 pp.)
4. Öpik, E. J. 1950. *MNRAS* 110:559
5. Öpik, E. J. 1953. *Geophys. Bull. (Dublin),* No. 8, 14 pp.
6. Mullan, D. J. 1971. *MNRAS* 154:467
7. Öpik, E. J. 1969. *Ann. Rev. Astron. Astrophys.* 7:473
8. Öpik, E. J. 1960. *MNRAS* 120:404
9. Öpik, E. J. 1973. *Irish Astron. J.* 11:85
10. Öpik, E. J. 1969. *Irish Astron J.* 9:79
11. Öpik, E. J. 1968. *Irish Astron. J.* 8:185
12. Moore, P. 1965. *Irish Astron. J.* 7:106
13. Öpik, E. J. 1973. *Irish Astron J.* 11:1
14. Leighton, R. B. 1971. *Phys. Today* 24: No. 4, 30
15. Ulrich, R. K. 1975. *Science* 190:619
16. Öpik, E. J. 1965. *Icarus* 4:289
17. Öpik, E. J. 1976. *Science* 191:1292
18. Ulrich, R. K. 1976. *Science* 191:1293

THE EXCITEMENT OF SCIENCE   525

Ann. Rev. Physiol. 1978. 40:1–17
Copyright © 1978 by Annual Reviews Inc. All rights reserved

RHAPSODY IN SCIENCE

P. F. Scholander

Author with rare gourmet mushroom (*Sparassis*)—just another childhood hobby.

Ann. Rev. Physiol. 1978. 40:1–17

RHAPSODY IN SCIENCE ♦1183

P. F. Scholander

5508 N.E. 180th Street, Seattle, Washington 98155

Dear readers, I find it quite shocking to have reached a stage in life when I am supposed to be retrospective. Like a vulgar streaker I shall expose myself in the limelight. I frequently ask myself how has it been possible to live such an exciting and happy life being such an irregular and irresponsible person?

My Norwegian mother was an accomplished pianist. I have just noticed in her copy of Speemans' encyclopedia, *Goldenes Buch der Musik* (1909), that she underlined the following advice to prospective musicians (which in my opinion applies to scientists as well): "Nur dem kann ich raten, Künstler zu werden, der aus reiner Liebe zur Kunst selbst, aus innerem Drang und Bedürfnis—unbekümmert um glänzenden Erfolg—ihr sein Leben weihen will und darin sein Genügen und sein Glück findet."[1]

GROWING UP IN SWEDEN

Emotion has certainly been a strong ingredient in my life, whether in music or science. As a small boy I crawled underneath my mother's Steinway when she practiced, smothered by the waves of emotion in the music of Bach, Grieg, Sinding.... Once when I was about two years old I walked on the ice of a little pond, and I can still see in detail the reeds I passed, namely *Scirpus lacustre,* then twice as high as I, now only half my size. Music comes early, and I believe also fondness for nature.

As an adolescent I was exclusively interested in natural history. I loved zoology, Brehm's *Tierleben,* Seton Thompson, botany, physics, and as-

[1]Only such a person can I advise to become an artist, who from pure love of art itself, through internal urge and need—unconcerned about spectacular successes—desires to devote his life to it and therein to find his fulfillment and his happiness."

tronomy (Arrhenius, Jeans, Eddington, etc). While in school I flunked history, but an unusual faculty let me make up for it by merits in the sciences. I might have been stopped dead right there, but my teachers *protected* me. There is one thing I realized very early. I had a curiosity that *craved* research of any kind, and could not think of anything else. I was deadly afraid of winding up as a school teacher and decided that the surest way of avoiding that would be to go into medicine.

MEDICAL SCHOOL IN OSLO, WITH ARCTIC EXPEDITIONS (1924–1939)

During the first years in the medical school in Oslo, reading Latin, philosophy, anatomy, I felt dreadfully wanting in active research, so one winter day I threw my fetters and went out to look at some little plants growing on the bark of trees. Consulting a book on lower plants, I recognized that these were lichens, which I could easily identify. In a spruce forest I found little yellow tufted lichens, which turned out to be the very common *Cetraria pinastri.* It so happened that one specimen had little round fruiting bodies along its powdery margin, and this was stated to be *extremely rare.* I was struck as if by lightning, sped across town to the Botanical Museum, and there met the famous lichenologist Professor Bernt Lynge. I showed him my insignificant treasure and he said: "This is most exciting, and I will give you working space right here. If you stick with this hobby, I'll send you to Greenland. I cannot go because of my rheumatism."

This was a fantastic welcome, the like of which I had never experienced. It changed my whole life. Lynge certainly knew his fishing, and he hooked me like a young trout on a "Professor" fly.

As a consequence I spent three summers (1931–1933) on arctic expeditions to NE Greenland, SE Greenland, and Spitzbergen. The latter was a glaciological expedition and came about because the camp of Andrée, the famous Swedish polar balloonist, had just been found on White Island. The great Swedish glaciologist, Hans Ahlmann, saw his chance to finance a Swedish-Norwegian expedition, ostensibly to put up a commemorative plaque in honor of Andrée. His main purpose, of course, was to map and study the inland ice of Northeast Land. I was asked to serve as botanist and medical man. This gave me my first extensive contact with outstanding men in the geophysical sciences.

A Generous Gift of Ph.D. in Botany

On these botanical summer expeditions going through the pack ice I saw a great many seals, polar bears, and diving birds; intrigued by Starling's textbook in physiology, I decided I would take up diving as a research project if and when I ever got through medical school. My textbooks have always been full of derogatory marginal notes about explanations I did not believe. This somewhat irreverent attitude toward the accepted doctrines made my grades mediocre to say the least.

Finished at last with medical school, where I came out at just about the bottom of my class, I moved up to the Botanical Museum and through the help of my outstanding friend, curator Johannes Lid, I completed my project with vascular plants and wrote two regional floras, one on Greenland and one on Spitzbergen. With Lynge and Dahl I coauthored two lichen floras on Greenland.

To my utter surprise I was approached by Professor Jens Holmboe, head of the Botany department, who told me that my plant work would qualify me for a Ph.D. in Botany if I would take exams in plant histology and physiology. Reading the plant physiology I became fascinated by the old problem of sap rising in trees. At that time it was accounted for by the cohesion theory of Dixon, which predicted that the hydrostatic pressure in the sap would be negative by easily −30 to −40 atm. It seemed to me that this would be utterly impossible to accept without clear experimental demonstration, which was not at hand. (But see below.)

Two years after barely squeezing through my medical exams, I subjected myself to the quite formal Ph.D. ritual. This involved two public lectures, one proposed by the faculty and the other by myself. The latter was on lichens and their striking symbiosis with specific algae. As is the case with many other symbiotic conditions, in Tridacnas, corals, and even rumen, it is still unclear how the nonsymbiotic spawn of the host gets infected with the appropriate clone of the symbiont.

Studies in Diving Animals

Having a Ph.D. made things very much easier for me. I got a developmental grant and a big laboratory in the physiology department of the medical school, originally planned for lab courses but until then never used. Harald Erikson (a medical student) and I got hold of several seals and, using Haldane and van Slyke apparatuses and a seal in harness in a bathtub, we got some preliminary information on the physiology of diving. It became immediately clear that in order to follow this in detail we needed recording instrumentation. This took us about a half year to construct and gave us a running record of all essential respiratory data. In addition, arterial samples and EKGs were taken. After a couple of years we had a good picture of what happens during diving: the tremendous bradycardia, the great change in blood distribution, etc.

ROCKEFELLER FELLOW WITH LARRY IRVING (1939–1943)

On his own initiative, my friend and mentor Bernt Lynge went down to Copenhagen and contacted the great physiologist, August Krogh, who invited me to his lab and arranged with Larry Irving for the Rockefeller Fellowship that took me to Swarthmore. Larry had already done distinguished work on diving physiology, and together we greatly expanded the field at Swarthmore. The Second World War approached and Larry went into the Air Force. I followed later, after having spent some time at the Fatigue Lab with Professor Jack Roughton, working out a microtechnique for the measurement of blood- and respiratory-gases—always in collaboration with the great glassblower and friend, Jim Graham. One day I asked Jack, "If I were to fake the oxygen content of water by adding oxyhemoglobin to it, what chance would you give me that the oxygen transport through the solution would be increased?" He said, "Two per cent." I took his reply with a grain of salt and filed the problem for a later time.

While working out the carbon monoxide micromethod, I was all the time in the grip of the Arctic allure. I thought I might influence Larry in that direction, and suggested that we study the old problem of monoxide poisoning in tents and snow-houses. (It had been suggested that Andrée's party very likely died from monoxide poisoning.) We made an expedition to Mt. Washington and found that the flame of a kerosene stove in a poorly ventilated tent or snowhouse will go out at some 3% CO_2, and even at 1.5% a match or candle will not burn, but no monoxide is formed. However, the minute you cook on the flame monoxide pours out and rises rapidly in the blood.

Little did I know then how susceptible Larry was to the Arctic lure; as everybody now knows, he wound up with great honors in the Alaska Hall of Fame after founding the Arctic Institute for Biology at the University at Fairbanks.

IN THE UNITED STATES AIR FORCE (1944–1947)

Covered Life Rafts in the Aleutians

Commissioned as Captain in the Air Force on Larry's research team, I spent some time looking at the difficulties our pilots had in the Aleutians. Their main problem was exposure, and so we devised a covered, four-man life raft, which we took to Attu in March. A folded tent was sealed along the sides. It could be erected on a paddle front and aft and could be closed at the top. Its occupants sat on an air mattress, dry and warm. Tied by a floating line several hundred feet behind a standby vessel in the mouth of a southern fjord in Attu, we tested two rafts for four days and nights. Half of the time we spent in storms with 80 mph gusts and mountainous waves. In my tent were Sir Hubert Wilkins, the arctic explorer, and Dr. George Sutton, the well-known ornithologist, plus a young Navy man. Everything went fine and we all enjoyed Sir Hubert's long tales of his great admiration for his old friend Frederick Cook, whose negative celebrity was undeserved.[2]

Altar Wine and Rescue from the Sky

Shortly after this successful demonstration, a telegram notified Sir Hubert that a B-47 had crashed near Cold Bay at the tip of the Alaskan peninsula. Would he please lend his arctic experience to the rescue efforts? We flew in and passed over the crash site, which was at the edge of a 3000 foot mountain plateau. One survivor had been seen. Tractors and dog teams had tried in vain to reach the site. There was general turmoil at headquarters. Clearly somebody had to jump down there. I had plenty of experience in snow and ice, and called for a pilot. Somebody told me that jumping had been vetoed by the camp commander, to which I snarled "Go to bloody hell!" Risking his career, a young pilot called Estes volunteered, as did a medical doctor, John Weston, and a priest whose name I cannot recall. We located some emergency chutes, raced to the B-17, and in ten minutes were flying over the site. None of us had ever jumped before. Once down, I was dragged at a good clip by the wind. Pulling in the shrouds, I stopped but had some trouble getting my leg straps unbuckled. It occurred to me that my friends might be in danger. I raced downwind and caught one after the other as they were helplessly dragged along by the wind. Throwing myself on top of them, I got them loose from their harnesses.

It took me two hours to walk up to the crashed plane. It was upsidedown and contained three survivors. Two men had perished. The doc administered medical aid, while I took on the cooking. I was greatly helped by the fact that the plane was filled with cases of altar wine, for this was a supply plane for Passover, or "Hangover" as we called it. Appropriately, our priest was the bartender, and I hope all that sanctified altar wine will give me a sorely needed Brownie point for the hereafter. Finally, a bush plane from Anchorage got us all out of there.

Our covered raft fell into oblivion, but was reinvented in a brilliant edition some ten years later by the British and was universally adopted.

Meanwhile, the war had ended. It turned out that our short stay in the Aleutians counted as foreign service; I had been there two days (!) more than required for United States citizenship, which I was gallantly offered.

[2]See: Mowatt, F. 1968. *Polar Passion.* Boston: Little Brown.

Accepting it involved an oath of allegiance, and I was deathly afraid of being asked to recite the second verse of "The Star Spangled Banner," which I simply could not learn.

POINT BARROW AND STUDIES IN COLD

Larry had now really been taken by the polar passion and had started a laboratory at Point Barrow where the Navy was engaged in petroleum research. He asked me to give him a hand, which of course I did. With splendid help from Otto Hebel of Swarthmore we got a quonset hut well insulated and heated for exacting physiological work. Before winter freeze-up we marked suitable habitats with flags, locating animals and plants under the snow cover.

Mammals and Birds

Our work on warm-blooded animals concentrated on measurements of insulation, metabolic rate, and body surface temperatures. It turned out that mammals larger than the fox could sleep with basal oxygen consumption at temperatures as low as $-30°$ to $-40°C$, which were the coldest we could produce. The smaller ones, like weasels or lemmings, started to shiver at $+15°C$; below that, they increased their heat production essentially proportional to the deviation from their internal temperature of $+37°C$, as would be expected from Newton's law of cooling. By comparison, tropical animals in Panama had a critical temperature only a few degrees below the body temperature and also followed Newton's basic law. Arctic and tropical birds did the same. It goes without saying that the animals adjusted their insulation by raising or lowering furs or feathers, curling up with the nose under the tail, etc. In the appendages of aquatic animals (seals, whales, water fowl), arteriovenous countercurrent systems are common heat savers.

Cold-blooded Animals and Plants

In contrast, the cold-blooded animals and plants showed a rather regular Q_{10} sequence that went from 2 to 4 between $0°$ and $10°C$ and then decreased with increasing temperature. Our arctic material badly needed tropical material for comparison. I secured it in Cuba, and it turned out that findings from these specimens were practically identical to those from the arctic material. This line of inquiry was promptly discontinued.

Freezing the Living Engine

However, another aspect was much more attractive. Going out to the little flags that marked our desired pools, we cut out vertical slices of the ice. When we melted their surfaces with our warm hands we found that quite a few insect larvae (chironomids, i.e. gnats) were frozen there. Within several hours of thawing, they started to crawl around and feed. While thawing, they changed from an opaque yellow to a transparent red, so they had obviously been completely frozen. We therefore ran a series of experiments to discover to what extent the metabolic rate would vary with the ice-content in the larvae and the lichens. This required a flotation technique where the specimen initially sank to the bottom of a vial containing, say, $-5°C$ kerosene. Heavy bromo-benzene of exactly the same temperature was added until neutral buoyancy was attained. The density of the mixture was then equal to that of the object and was easily determined by simple pycnometry. The procedure was repeated at $0°C$. The difference in density gave the ice content, provided the sample was gas-free, which was checked by melting it under a microscope. It turned out that at $-5°C$ some 90% of the water was already frozen in most material.

Putting the frozen samples in the smallest possible amount of air, either singly or packed together in an airtight glass vial, one could wait until enough change had taken place in the gas composition to tell the story. It turned out that below freezing temperature the Q_{10} was between 20 and 50. At $-20°C$ one had to wait 1880 hr to get a significant gas change. If we now consider that these animals or plants readily survived in a melted state for a month or ten days at $0°C$, one may extrapolate that at $-40°C$ they would survive for several million years. Whether or not such extrapolation holds is another story.

Gas Penetration Through Ice

In securing our frozen chironomid larvae our attention was attracted by the fact that right under the ice surface there regularly appeared a great many gas bubbles. Puncturing these with a micropipette through a little pool of cold mercury, we could micro-analyze a few mm^3 of the gas. It turned out to contain large amounts of CO_2, and not a trace of oxygen; the rest was nitrogen and undoubtedly methane. This gas had been sitting for 7–8 months under a sheet of ice only a fraction of a millimeter thick. Its retention suggested a very high impermeability of ice to gases.

A series of experiments was made to check this. Ice sheets 0.1 mm thick were cast between flat hydrophobic surfaces and floated on cold mercury. A known volume of a gas, e.g. oxygen, was introduced under the sheet and its surface area measured. We could now go in with a curved micropipette after a certain period and sample the bubble for microanalysis. We were never able to detect any penetration of oxygen or nitrogen; as far as CO_2 was concerned, at $-10°C$ it penetrated ice at least 70,000–80,000 times more slowly than water. Later more accurate checks performed by Hemmingsen gave figures for CO_2 ten times slower than mine. This right away gave us the idea that the air trapped in glacier ice when the snow compacts (visible as tiny bubbles) carries a permanent record of the composition of the atmosphere, provided that no melting has occurred.

WOODS HOLE OCEANOGRAPHIC INSTITUTE (1949–55)

Supercooling in Fish

The productive period at Pt. Barrow resulted in a huge volume of material to be written up. I moved to the famous library at the Marine Biological Laboratory (MBL), and was soon invited by Dr. Alfred Redfield to join the Oceanographic Institute.

Dr. David Nutt of Dartmouth had told me that the Hebron fjord in Labrador maintained its winter temperature (–1.9°C) all year at the bottom, even though the temperature of the surface layer in the summer would regularly reach +5° to +10°C. Now, an oceanic fish usually has a freezing point depression of about –0.5° to –0.8°C, somewhat like ours. The question then is, what happens in fishes living at or near the freezing point of sea water? Do they raise their osmolality? Do they supercool, or what? Through several expeditions, summer and winter, we found that at the bottom of the Hebron fjord the fishes retain their normal freezing point depression of about –0.8°C all year round, and are supercooled by about 1°C at all times. At the surface they are supercooled only in the winter.

At Woods Hole we found that several species of local fishes could be supercooled to –3°C and survive perfectly well. But, if the supercooled water was *seeded with ice* the fishes froze and died almost instantly. Even in sea water at –1.5°C, i.e. warmer than the freezing point of sea water, a fish touched by a piece of ice would freeze and die. In other words it is the absence of seeding that protects the Labrador fish from freezing at the bottom. A summer fish there with its freezing point at only –0.8°C will freeze when touching ice in a vat of sea water cooled to its freezing point. Shallow water fish swimming about in the winter ice at Hebron double their freezing point depression and then are not seeded by touching the ice. In contrast to marine fish, marine invertebrates are always nearly isosmotic with the sea water and therefore do not freeze unless they are exposed to severe winter air by low tide (and even then some survive). Recently these early data on arctic fish and invertebrates have been expanded qualitatively as well as quantitatively, mainly by the beautiful work of Dr. Arthur DeVries and collaborators. They have shown that in essence the antifreeze in antarctic fish is provided by mucoproteins that stabilize against nucleation and ice propagation.

Sap Rising in Grapevines as Part of Ocean Circulation

Ever since my preparations for the Ph.D. in botany in Oslo I had been bugged by the problem of sap rising in trees. No direct measurements were available to confirm Dixon's brilliant cohesion theory of sap rising, which essentially said that water evaporating from the leaves pulled sap in continuous strings under tension up from the ground. Strasburger had shown that trees 20 m tall that had been sawed off at the base under water would transport poisons like copper sulfate or picric acid up into their leaves. Obviously no living pumps could be involved.

While in Woods Hole at the Oceanographic Institute I noticed that when native grapevine twigs were broken off before the leaves were out, sap dripped from the plant. In other words, at that time the sap pressure was surely positive. A micromanometer was constructed that could be clamped airtight to a branch, and it measured the positive pressure. By removing it, carefully sealing the puncture, and clamping it at various heights, one obtained very accurately a perfectly normal hydrostatic gradient of 0.1 atm per 1 m elevation, regardless of the general pressure, which might easily be up to + 5 atm. In other words, there was no mysterious lifting force in the xylem, no "wiggling hairs" or "matric potential." With the leaves out, measurement became impossible; the pressures were clearly negative. For instance, if the base of a vine were cut off under water and connected airtight with a burette, the vine drew up water. Even if the burette was filled to the brim and closed vacuum tight, the drinking rate remained the same, indicating high negative pressure. It could be demonstrated that no air leaked into the burette from the cut.

My colleague John Kanwisher and I spent much spare time working on the grapevines. To the rest of the faculty this hardly seemed a fitting job for an Oceanographic Institute. But the day was saved when Alfred Redfield pointed out that evaporation from the ocean falls as rain and is drawn up by the vines. The rising of sap is, indeed, a neglected part of this grand circulation.

The reader will discover that we return several times to this intriguing old problem below.

UNIVERSITY OF OSLO (1955–1958)

Hunting Reindeer in the Cold

We were intrigued by our experience with cold acclimation in arctic and tropical warm-blooded animals, and it was natural to take a look at how humans manage it. Our first task was to define a reasonable standard for cold stress, avoiding such extremes as putting naked people in a barrel of ice water. We were surprised to find that lying naked on a sofa at +20°C gives humans a very uncomfortable night of shivering. The cold is hardly bearable after four o'clock in the morning. We found that a similar stress could be obtained by sleeping naked in a single-blanket sleeping bag at 0°C. That these standards were about the same was determined by checking oxygen consumption (the sleeper's head being enveloped in a ventilated hood connected with a spirometer), and by measuring rectal and skin temperatures during the entire night. The advantage in using 0°C as the outside temperature is that it is relatively easy to find in nature.

Using the bag apparatus in the mountains of Norway, we subjected a group of students dressed in summer clothing to autumn temperatures around 0°C (and gave them reindeer hunting as a reward). They spent their nights sleeping naked in the single-blanket bag under an open rain shelter, no fires permitted. After five or six days they had no trouble; when checked in the lab test, they slept warm all night, with bouts of shivering that raised their basal metabolic rate to 150%.

Silly Bugger White Man

We were now prepared to tackle the aborigines of Australia. We were greatly aided in this by the great pioneer in the field, Sir Stanton Hicks, of Adelaide University, who some 20 years earlier had led several expeditions to central Australia for this purpose. Members of our group were Hugh Le Messurier (Adelaide), Ted Hammel (Yale), Sandy Hart (Ottawa), John Steen (Oslo), and myself. Our work essentially supported and extended Sir Stanton Hicks' findings.

We arrived at our destination (150 miles west of Alice Springs in central Australia) in their midwinter. Frost formed on the ground at night. We found that the natives slept naked on the sand between two little brush fires. As the sun rose, the aborigines gathered around a big fire to chat and warm up. We had an excellent interpreter with us, and arranged the test as a competition between the natives and the "silly bugger white man." We began by lying down next to them naked on the sand behind their little brush windbreak and between two little fires. Sleeping was very difficult for us. While they stoked their little fires a dozen times during the night, we had to do it thirty or forty times. They got a kick out of our awkwardness.

After that we introduced them to the bag test. The outside temperature was just about 0°C. They accepted the hood and a series of skin thermocouples with no trouble, but how Hammel ever talked these young spear-throwing warriors into accepting the intimacy of a rectal thermocouple I shall never understand. Again we placed one of ourselves next to the subject, identically equipped. We suffered through the night shivering and chilling, raising our basal rate to 150%. In contrast, the aborigines slept through the bag test with a normal basal rate, but with leg temperatures going down to around 10°C. In other words, they had learned to adapt the economical way, by accepting peripheral cooling.

Eskimos and Lapps reacted in the test as we do; by virtue of their warm clothing and bed gear they, like us, live in a tropical microclimate. But Hammel's similar studies of the few surviving Indians in Tierra Del Fuégo and of cold-hardy tribes in the Kalahari desert, found essentially the same phenomenon we found in the Australian aborigine. It is interesting that when Hammel tested the latter in scorching hot midsummer in a 0°C cold van at Darwin, they retained their cold-hardy, economical winter reaction.

Search for Ancient Atmosphere in Norwegian Glaciers

Our studies at Point Barrow of the frozen aquatic larvae had given us data on the virtually complete gas-tightness of ice, even in sheets only a tenth of a millimeter thick. This suggested the possibility of finding ancient atmospheres preserved in the gas bubbles so characteristic of glacier ice. The technique was worked out in collaboration with the famous Danish isotope glaciologist Willy Dansgaard, on a glacier in the Norwegian mountains. We mined large pieces of ice and transported them by snowmobile. One ^{14}C dating required vacuum distillation of 20–30 tons of ice. The CO_2 was collected in KOH. With the technique in good order and the project recommended by such outstanding colleagues as Hans Ahlmann, Carl Rossby, and Harald Sverdrup, we were funded through the Arctic Institute of North America to go after the Greenland ice. (All Norwegian ice samples had been contaminated by melting of the glacier. The composition of the gas samples, instead of reflecting that of the atmosphere, reflected the gross differences in solubility coefficients of CO_2, argon, and nitrogen.)

Expédition to Greenland Glaciers (1958)

The expedition was organized with a laboratory built on the deck of the sealer "Rundöy." Senior scientists were Willy Dansgaard (Copenhagen), David Nutt and Larry Coachman (Dartmouth), and myself. We made our first attempt to obtain ice from a desirable iceberg by boarding it with instrumentation. In less than an hour it capsized. Dansgaard and his two assistants were nearly lost. We never boarded an iceberg again, but stayed with the ship at a safe distance. We towed ton-sized pieces, which had fallen off the berg, to our ship where they were split (by a narrow, long loop of steampipe) into pieces that could be hoisted onboard. On deck they were split further to fit into our two 150-liter vacuum boilers. Laboring day and night, we could process a single sample of 0.2 gr ^{14}C as CO_2 in two to three days. The technique worked fine; but as it turned out, there were slight differences in the gas compositions from adjacent 2–3 kg pieces. Even in

Greenland, every locality displayed contamination of the gases by melting. That melting does occur was later confirmed by the United States Air Force. Landing on the Greenland Ice Cap in the summertime, observers frequently found pools of meltwater at almost any altitude and latitude. This, however, would not invalidate our ^{14}C and oxygen-isotope data. From these Dansgaard could estimate whence our samples came. Our oldest iceberg was caught at latitude 73°N. It was about 3000 yrs old and had formed at about 3000 m altitude, which placed it some 300 km inland. It became quite clear from our findings that if one wishes to try for a sample of the old atmosphere trapped in a glacier, one must seek it in the coldest part of our globe, namely in the antarctic ice cap.

SCRIPPS INSTITUTION OF OCEANOGRAPHY (1958–   )

Facilitated Oxygen Transport with Baird Hastings

Since my lab at Scripps was not yet ready when I joined the Institution in 1958, I had a great time with Baird in his roomy lab suite at Scripps Clinic and Research Foundation in La Jolla. I had already spent a most fruitful time in his lab at Harvard Medical School, where I had worked out an ultrasensitive respirometer for single cell divisions. Now I was eagerly getting into facilitated O_2 transport through hemoglobin solutions, a phenomenon I and my students had discovered in Oslo five years earlier with an instrument designed in collaboration with Jim Graham (see above). Using essentially the original technique, I did a comprehensive series demonstrating enhanced diffusion of oxygen through hemo- and myoglobin, along with some other heme pigments (also in vitro).

As was to be expected, my sloppy handling of mercury demanded plastic replacements of the brass fixtures in half a dozen of the sinks. However, I was redeemed when my boss tripped aboard one of Scripps' ships, spilling all the mercury out of a van Slyke apparatus! Nobody knows what that did to the engine room.

Reverse osmosis in mangroves

Speculation about rising sap as a case of reverse osmosis naturally focused attention on mangroves. The interesting thing about mangroves is that they grow in sea water; thus the substrate is known, at least if they are not growing in estuaries. Therefore, if the cohesion theory were right, the trees would require at least −25 atm sap pressure in order to pull fresh water out of the sea. In the daytime when the process would be driven by evaporation, at least −40 to −50 atm would be expected. We found that mangrove sap is indeed very close to being fresh water; some carries a little salt, which is excreted by the leaves.

Finally on a Baja California expedition the problem of measuring the negative pressure was solved. Using a small portable lathe aboard ship, I made a little pressure chamber of about 100 cc capacity. The chamber could hold 140 atm nitrogen pressure and it had an airtight seal in the lid for letting through the stem of a mangrove seedling. The idea was to try a reverse osmosis experiment with a seedling replanted in sand and sea water in a beaker that could be placed inside my "bomb."

I was called away to Scripps for a few days. While I was away Hammel and Hemmingsen got the idea of putting a mangrove twig in the pressure chamber with the cut end sticking out. When they applied nitrogen pressure, the sap suddenly appeared at the cut end. The balancing pressure could be accurately determined; it was between 30 and 50 atm, and was characteristic for each different bush. Next the technique was expanded to produce pressure-volume curves: in steps, more and more sap was extruded and measured together with the new balancing pressure. It turned out that the extruded sap was almost completely saltfree, often containing less than 0.01% sodium chloride, even in salt-secreting mangroves that normally had 0.1% salt in the ascending xylem sap. Later on, after shooting down high twigs from Douglas fir and Redwoods (from triangulated heights), we found within ± 10% that there was a normal gravitational gradient in the negative hydrostatic pressure. As with the grapevines, there was no active pumping.

THE RESEARCH VESSEL *ALPHA HELIX* (1966–)

My contacts with a great many marine-oriented scientists here as well as abroad, along with my own experience, convinced me that there did not exist a vessel that made even the slightest attempt to cater to the needs of modern experimental biology. It became quite clear to me that something ought to be done about it. I was introduced by Larry Irving to marine biology as a general and fundamental tool in the Marine Biological Laboratory at Woods Hole. Its history of numerous contributions to fundamental biological and medical knowledge is impressive. Modern advances in biomedical science depend more than ever upon access to precisely the kind of material that exhibits most clearly what we are after, regardless of what plant, animal, or microbe is involved. It seemed to me highly appropriate to propose that a specially built ship be acquired to serve as a transportable biological laboratory for modern experimental research.

In support of this proposal was the attractive possibility of generating on these expeditionary efforts an interdisciplinary and international cooperation, where each member was isolated from his routine academic chores, and where all shared the enormous privilege of having uninterrupted time with selected colleagues otherwise not available. Not least of all, I felt such expeditions would be a legacy passed on from the senior to the young—an opportunity like those I had myself so richly experienced.

A national advisory board headed by Baird Hastings was constituted in 1963 with a dozen highly distinguished senior members. As a result, the *Alpha Helix* was funded and built. To myself and hundreds of colleagues from various disciplines and countries who have had the privilege of using this facility, the research vessel *Alpha Helix* stands as a proud landmark of *nonpolitical U.S. generosity,* as a gift to international friendship and scientific cooperation. During its ten years of operation it has visited many parts of the globe, from the tropics to the arctic and antarctic. It is presently (for the second time) a guest in the Amazon, joining in most fertile cooperation with scientists from our great South American hosts. Its scientific productiveness has been prodigious.

Some of Our Work on the Alpha Helix

The first place we took the *Alpha Helix* (1966) was to the Great Barrier Reef, into sheltered waters between the Flinders Islands north of Cairns, in Queensland. One of our programs was a continuation of inquiry into water relations in mangroves, a study in reverse osmosis. It showed that seedlings transplanted in sand and sea water, when given ample time to heal the damaged root, did indeed respond to pressure in the "bomb" by filtering out essentially fresh water. Another thing of fundamental interest, found by our renowned colleague Dr. Stanley Miller in wilted (i.e. turgor free) twigs of the mangrove *Lumnizera,* was that when such a twig was put in an accurately thermostated pressure bomb the balancing pressure *varied with temperature.* Going down from 27°C to 3°C and then back up to 37°C, using seven steps in all, he found that within a few per cent plus or minus the balancing pressure was indeed proportional to the absolute temperature —i.e. was of kinetic origin. This was later superbly confirmed by Hammel on a conifer and was the last information from plants necessary to convince us that the purely thermodynamic description of osmosis (such as that pioneered by G. N. Lewis) based on the *activity* construct could be transcribed into an easily visualized *kinetic mechanism* accessible to direct experimental verification.

Negative Interstitial Fluid Pressure in Animals

In April, 1967, while we were working on supercooling in fish in the Bering Sea, the *Alpha Helix* got stuck in the ice for a week. I used the time to pay attention to a simple physical problem. I had long wanted to examine the physical meaning, if any, of what plant physiologists and others call "matric potential;" to determine, in other words, whether there is any anomaly in the hydrostatic gradient in a vertical matrix through which water can move. I let starch (from the galley) sediment in a cylinder. By stuffing a cotton tuft in one end of a fine plastic tube and a standard glass capillary in the other, I made a wick probe. The wick, pervious only to water, was placed at the thixotropic bottom sediment and the capillary was mounted vertically. An identical wick probe was placed in the clear water surface. Of course the meniscuses in the identical capillaries stayed level, in accordance with the Second Law. When starch was added so that it pushed against the water surface, capillarity turned the water pressure negative and both wicks registered this. Again, of course, the depressed meniscuses stood at identical levels—i.e. connecting the capillaries could not maintain a perpetual motion. In a submerged matrix (homogeneous or not) the inside hydrostatic pressure of motile water at equilibrium must, level by level, be exactly the same as that of the outside water.

In preparing for our Amazon expedition (1967) at Scripps, a colleague happened to use our laboratory to do some work on iguana lizards. It struck me that the sharp skin folds so characteristic of these animals indicated that there must be a negative fluid pressure in the subcutaneous tissue to hold these folds together; otherwise edema would develop. Why should one not check this with the simple wick method which worked so well in all sorts of nonliving matrixes? At this time I was not aware of Dr. Guyton's pioneer work, claiming negative tissue fluid pressure, and the strange controversy raised by this simple concept. Anyhow, a wick stuck under the skin of an iguana sure enough registered a few cm negative water pressure. We worked out the proper technique for this simple approach and took it with us on the expedition.

Down in the Amazon we confirmed our finding in a number of organs and animals. Together with Stanley Miller, student Alan Hargens and I had a great deal of fun comparing a big water snake (anaconda) with a big land snake (boa constrictor). Putting these three-meter-long snakes on a tilt table, with one subcutaneous wick in the neck and one in the tail, we found that the anaconda compensated beautifully against hydrostatic pressure changes when tilted head up or head down—an ability, one may respectfully point out, that is not vital in a submerged animal. Head down, the boa also compensated beautifully against positive changes; but head up, as when climbing a tree, he developed edema (positive pressure) in his tail. It appears the Powers That Be must have gotten their wires crossed.

THE SOLVENT TENSION THEORY OF OSMOSIS

Dixon's cohesion theory of the ascent of sap is closely related to a proposed kinetic explanation of the osmotic process. The "solvent tension theory" has been proposed very clearly several times, starting with Hulett (1901), who accounted for all colligative properties by solvent tension. Later it was stated by Herzfeld (1935), Mysels (1959), and others. This kinetic concept lent itself to conclusive experimental verification, as first demonstrated by the brilliant work of Perrin on the Brownian motion (1909).

If solute molecules like sugar are deposited on the bottom of a beaker of water they will rise by thermal motion; if stopped by a semipermeable membrane fastened across the beaker they are reflected, exerting a *pressure on the membrane.* If the membrane is removed they proceed, putting drag on the water until they are stopped by another barrier, namely the free surface; and as Professor Theorell in Stockholm commented, "They are blind and don't know what the hell they hit!" Being reflected this time by the surface, they exert their *pressure on the surface* and hence leave the solvent under tension. Based on equipartition of energy, this is the simple reason for the lowered vapor pressure over a solution, expressed in Poynting's relation.

This is a *kinetic,* easily comprehended *physical explanation,* which takes into consideration everything known about osmotic processes. As already pointed out by Hulett, and lately in great detail by Hammel, solvent tension is the direct cause of all colligative properties. Using magnetic solutes one can superimpose a known magnetic force upon the thermal force; the two forces are exactly additive with respect to osmotic pressure. The same additive effect can be obtained by the buoyancy effect of heavy and light solute molecules in a reversing osmometer.

In contrast we have Lewis' thermodynamic activity concept. Useful as a mathematical construct to keep the numerical aspects of the colligative properties in order, it was never designed to give insight into the physical mechanism. Nevertheless, it has bred an array of free-for-all concepts like water concentration, molfraction component of hydrostatic gradient, the abolition of gravitational gradients by chemical potential, etc. No wonder then that one can point to thermodynamic inconsistencies in the presently popular "water concentration theory." Take a rigid semipermeable cylinder submerged in a jar of water. At equilibrium the *solution* in the cylinder registers say +24 atm on a gauge that accepts solute and solvent. It is now assumed that the *water* pressure in the solution is +24 atm as registered on the gauge. If that were so, the water in the solution would be compressed and heavier than the outside water and would maintain a perpetual motion by sinking through the solution. On the contrary, the solvent pressure (and its vapor pressure) in such a system is identical layer by layer inside and outside.

COSMOLOGICAL VISTAS

When I first read about it I became fascinated by Mach's postulation of the relativity of mass. Of course I had already been thrilled by Cavendish's elegant measurement of the universal gravitational constant, where in a sense he overpowered locally the gravity of the earth. It so happened at this time that our institution needed a simple ultracentrifuge; a standard Beckman L with a little window in the lid was ordered. As an extra item we ordered a 4 kg rotor with a flat upper surface upon which z-bent capillaries could be fastened, à la Briggs, for studying cohesiveness of fluids by submitting them to G forces. The window was above the rotor edge.

With a brilliant young student, Bent Schmidt Nielsen, I often discussed the thrill of doing a relativistic experiment focused on Mach's principle. Imagine a battery-driven centrifuge in absolutely empty space. In such a circumstance it would be meaningless to ask which rotates around the other, the rotor or the housing. This seemed to open a possibility that a large spinning mass might induce an ever-so-slight centrifugal force in a closely adjacent smaller mass. Right or wrong the notion filled us with surges of cosmological elation.

So, returning from our exciting expedition to the Great Barrier Reef in Australia (1966), we fitted a half-meter-long brass tube, vacuum tight, through the window fixture in the centrifuge lid. The tube was closed at the bottom with a thin brass plate that could be lowered to within a fraction of a millimeter of the spinning rotor. Inside the tube a 200 gr brass weight was suspended as a pendulum with a flat base almost touching the bottom plate. The pendulum was damped by a dashpot and the optical readout by a galvanometer mirror was easily sensitive to 0.01 mg. We elevated the cylinder and started the rotor. Once full speed was attained, we lowered the cylinder to within a fraction of a millimeter of the rotor spinning in high vacuum. The nearness was made audible by the scraping of a thin steel string (a violin's E string) attached near the bottom plate. The first time this was tried we stood on a ladder rather than in the direct firing line, in case the rotor should blow out as it did for Svedberg in Uppsala, where it shot a big hole in the wall. Everything went well. The stability of the system was admirable and the effect of the rotating mass upon the pendulum was absolutely and completely *nil,* luckily for us.

Ten years later I had the pleasure of showing the remnants of this my greatest experiment to my colleague, the renowned astrophysicist Dr. William Thompson of the University of California at San Diego. He calculated that the theoretical displacement of our pendulum should have been on the order of $2.5 \cdot 10^{-11}$Å!! The experiment was a resounding fiasco, but so what? My reward? A wonderful *Appassionato* with the *Cosmos!*

Ann. Rev. Biochem., Vol. 37

REFLECTIONS FROM BOTH SIDES OF THE COUNTER

By Arne Tiselius

Ann. Rev. Biochem., Vol. 37

REFLECTIONS FROM BOTH SIDES OF THE COUNTER[1]   659

By Arne Tiselius

Institute of Biochemistry, Uppsala University, Uppsala, Sweden

Contents

Like many of my colleagues in different parts of the world, I have spent a large part of my life both as an active research worker—which, as everybody knows, also involves finding money to support one's work—and as a member of organizations which distribute funds or awards, or by other means attempt to stimulate and support scientific progress. Working in a field in which basic research frequently leads to practical applications in medicine or in industry, the question of balance between "pure" and "applied" often turns up. It has happened to me several times that these double functions have given rise to a dialogue between the two halves of a Janus face—if this rather wild parable may be permitted here.

Take for example an after-dinner speech in presence of authorities and other distinguished people who (hopefully) listen attentively to a scientist describing research as the key to the well-being of mankind. When he returns to his laboratory the following morning and takes a look at his experiments and his notes, a certain hesitation may develop (which is not necessarily a kind of hangover): "That well-being of mankind which I promised yesterday may, alas, have to wait several more years still." It is, however, fortunate, although sometimes a little embarrassing, that people nowadays understand that research, and especially fundamental research, will have to take its time before it pays dividends. Governments and private foundations are prepared to spend funds for research which must be regarded as enormous, compared to what was usual only a few decennia ago. We still have to remind them, in

---

[1] This chapter was written during a stay at Villa Serbelloni, Bellagio, in October 1967. I wish to express my gratitude and appreciation to the Rockefeller Foundation and to the hosts of the Villa, Mr. and Mrs. Marshall, for the hospitality and excellent facilities which I enjoyed.

1

THE EXCITEMENT OF SCIENCE

most cases, that the share given to long-range fundamental work always tends to lag behind.

Being aware of the fact that an increasing amount of the national income is being spent on research in most developed countries, those who are responsible for the distribution of such funds (many of them active scientists themselves) may often feel that the background for wise decisions is rather insufficient. I shall not enter upon a discussion of this much-debated problem here but will limit myself to some personal reflections in this connection, especially concerning the promotion of basic scientific research.

This is perhaps the most difficult problem in the field. We know far too little about the optimal conditions for scientific productivity, chiefly because the decisive factor is the individual. We may try to trace the development of a great discovery by reading papers. This is done, of course, and "research about research" is becoming an active and useful discipline. Nobody believes however, that this or similar activities will result in a publication *How to Make Great Discoveries* similar to the well-known *How to Win Friends and Influence People*. To stimulate basic research, even if we have access to many more facts than at present, is not a "push-button" affair. Neither could we hope to develop a flourishing period in the arts, like the Renaissance, by some ever so well-organized campaign.

A scientific paper, however, is usually of rather limited value if we want to know how things really happened. The facts and the conclusions may be presented in a perfectly logical order and with admirable elegance, but, did it really happen in this way? The author writes his paper when he has arrived at the final results and then first realizes which way he should have gone. Probably the way he actually took was much more complicated, involving sidesteps, mistakes, disappointments, wishful thinking, and still more of that kind. Thus, a scientific paper mostly involves a certain kind of after-rationalization, naturally perfectly understandable and permissible. Scientific work is objective and its results should be devoid of the personality of the author. In a way this is a pity, since I believe that in science the human subjective factor is involved in the act of creation almost as much as in art, literature, and music.

In his book *The Sleepwalkers*, Arthur Koestler (1) has presented and commented upon several examples of how scientific progress has evolved through the ages along much more irregular paths than generally realized. One can also find in works on the history of science many stories—some of them well-known—about what I would like to call the "triggering" mechanism which releases a scientific discovery (Newton's apple, Kekule's busride, etc.). And a Danish philosopher, Ludwig Feilberg, who towards the end of the last century wrote a very interesting book on such phenomena, stated that his most productive period for new ideas was when he was brushing his teeth. Probably he then felt no obligations to the rest of the world and his mind could play around freely.

We cannot go deeper here into a discussion of these psychological phenomena. Neither can we recommend that the authorities encourage scientists to sit under apple trees, ride on buses, or brush their teeth. But it is well to keep in mind how essential an anti-stress atmosphere is for any creative effort, including the sciences.

More essential and easier to grasp is the situation described as "the prepared mind." The release mechanism must act upon something, and I believe that an analysis of the prepared mind, with all factors involved, provides a good deal of the kind of information we are looking for. It seems to me that one of the particularly attractive and interesting aspects of the Prefatory Chapters in this Annual Review is that they may provide excellent first-hand information in this respect, which is otherwise not easily accessible. I am proud to have been invited to join the group of distinguished authors to these chapters, and my ambition has been to try to present some experiences which might be of interest to some readers from this particular point of view, in line with what I have said above.

There is, however, always the difficulty that any kind of introspection during the creative process probably will disturb and perhaps even spoil everything. Too much observation interferes and it is even tempting to generalize the Heisenberg principle to many phenomena other than those in the world of atoms: rare birds and plants, beautiful landscapes and the genuine traditions of their inhabitants being transformed into tourist attractions, the object of interest being obscured or scared and losing its very character by too many indiscrete onlookers. Even "confessions" of a scientist about how things really happened may also involve a kind of after-rationalization. This has been emphasized to me by my friends in the field of History of Learning when I have tried to persuade them to pay attention also to today's events in science fields, for example, by personal interviews with scientists while they are still alive and active.

EARLY WORK IN ELECTROPHORESIS

To most research workers the decisive factor in preparing their minds in a general way is obviously their impressions and experiences during their university years, particularly if they have the good fortune of having a great scientist as their teacher. This was so in my case and it should be obvious to all those familiar with the work in physical and biochemistry in Sweden how much I owe to The Svedberg—a great personality and a good friend. Last year I was asked to write a prefatory chapter on "50 years of Physical Chemistry in Sweden" for the *Annual Review of Physical Chemistry* and decided to have this chapter deal with The Svedberg and his work (with his successor Stig Claesson as co-author) (2). I became research assistant to Svedberg in 1925—a very productive period at the Institute, when the ultracentrifuges were developed and this new technique started to give such significant results in the protein field. Svedberg is a fascinating mixture of a physicist and a biologist, but perhaps not too much of a chemist (as he once confessed to me). To a certain extent I believe he saw the ultracentrifuge also as an instrument to classify the animal and plant kingdoms according to the physico-chemical properties of their macromolecular components. I remember the excitement when it was found that pure proteins sedimented as strictly homogeneous substances: a surprise to a colloid chemist but not unexpected to a chemist who regarded them as ordinary chemical substances but of very large molecular weight. I managed to make some minor contribution to the ultracentrifugation technique, introducing (with Ole Lamm) refractive index observation methods and also a theoretical treatment of the influence of charge and of electrolytes on the sedimentation, which at that time was in-completely understood. My main interest, however, became electrophoresis and my first crude experiments were a continuation of earlier work done in Madison by Svedberg and Scott. It was very stimulating to do this work against the background of the development of the ultracentrifuge. Elec-trophoresis appeared to me at first much less fundamental, but it turned out to have some interesting characteristics of its own which fascinated me. Svedberg was very encouraging but was, of course, too much absorbed in his own work with the centrifuge to give me much of his time. And electrophore-sis appeared so much simpler from a technical standpoint, suitable for a young man to play around with more or less on his own. Moreover, the excellent instrument workshop greatly facilitated the testing of new ideas.

In spare time I read some biochemistry (which at that time was not in-cluded in the chemistry curriculum in Uppsala). I remember being fascinated by the enormous variability and above all the specificity of biochemical substances, so new and so strange to a physical chemist. My daily worries in the electrophoresis work were connected with impure or badly defined ma-terials. Even those substances that had the blessing of the ultracentrifuge as being homogeneous did not always behave well in my apparatus. This was particularly true with the serum proteins. Gradually I became convinced that the definition and the purification were all-important problems not only for the substances in my hands, but for the whole of biochemistry. Thus, separation became the key problem and I became convinced that this would require a number of alternative methods, considering the multitude of sub-stances one would have to deal with. Not only ultracentrifugation, not only electrophoresis, but other methods would also have to be explored, preferably those which depended on physicochemical phenomena, as these are more likely to be gentle. When working with biological materials I had learned to have the deepest respect for their sensibility to drastic treatments. I remem-ber speculating much about further development of chromatographic and adsorption methods but, fortunately for myself, decided to pursue the ex-ploration of electrophoresis technique first. This work led to my doctoral dissertation (3), which was published and defended in late 1930. Although it was very well received by the Faculty and by The Svedberg himself and led to appointment as "docent," I remember very vividly that I felt disap-pointed. The method was an improvement, no doubt, but it led me just to the point where I could see indications of very interesting results without being able to prove anything definite. I can still remember this as an almost physical suffering when looking at some of the electrophoresis photographs, especially of serum proteins. I decided to take up an entirely different problem, but a scar was left in my mind which some years later would prove to be significant.

DIFFUSION AND SORPTION IN ZEOLITE CRYSTALS—A DEVIATION

I had read about the unique capacity of certain zeolite minerals to exchange their water of crystallization for other substances such as ethyl alcohol, bromine, mercury, etc., the crystal structure remaining intact even if the crystal was evacuated to remove water. The "empty" crystal served as a kind of molecular sieve. Much work on the diffusion of proteins in solution had been going on for some time at the institute, making use of optical methods for the observation. I collected some particularly high-quality specimens of zeolites during a visit to the Färöe Islands, experiencing the joy of field work. Sometimes I was left alone for a whole day among freshly weathered rock on waterfront shelves at the foot of very steep mountains, sea puffins my only company until fishermen came with their boat to fetch me in the evening. I remember well asking them what would happen if the sea became too rough for their boat. They replied, pointing upwards: "We will take you that way with a rope," an experience which fortunately did not prove necessary.

I doubt if my work on zeolites has left any traces in the scientific literature and I am not sure to what extent it was inspired from my interest in separation, thus being a forerunner of work on molecular sieving done at the Institute of Biochemistry 30 years later. I knew, however, that I had very little chance of getting a permanent position in my country in those fields which attracted my interest: biochemistry and biophysics. There were simply no chairs available, but there was one in inorganic chemistry to be available in 1938, and might not zeolites be counted as inorganic chemistry?

Anyhow, I enjoyed this work, particularly when I could follow the diffusion by observing the crystal in a polarizing microscope in a specially constructed vacuum chamber (4). The anisotropy of diffusion came out beautifully. I believe that the most important outcome of this work was that it prompted me to go to Princeton to work with H. S. Taylor on adsorption phenomena. I was fortunate enough to be granted a Rockefeller Foundation fellowship for this purpose and thus spent September 1934 to August 1935 in the U.S.A. This turned out to be a most stimulating year of decisive influence on my career. The atmosphere in the Frick Chemical Laboratory was very inspiring; I remember particularly Henry Eyring's seminars and the frequent discussions by groups of people from many different fields, with a frankness and informality which I had not experienced before. Of equal, or perhaps even greater, importance for my future were the frequent contacts with the Rockefeller Institute, both in Princeton and in New York, which led to lasting and inspiring friendship with J. Northrop, W. Stanley, M. Anson, and many others, as well as opportunities of meeting men like Landsteiner, Michaelis, and (later) Heidelberger. They all expressed to me a strong belief in the future of "experimental biology." They knew my work in electrophoresis and when I told them about my difficulties they nevertheless greatly encouraged me to carry on. In discussions with them I found that many times their problems needed something which had been in my mind for years, but which, so far, I had failed to realize. Thus, while still in the U.S.A. I started to make plans for a systematic investigation of disturbances and sources of error in electrophoresis; even though this would take several years, I was now convinced that it was worthwhile.

A New Attempt at Electrophoresis; the Resolution of the Main Protein Components of Serum

Actually this now took less time then I had expected, although it involved both experimental work and theoretical calculations. I learned something which I later tried to teach others, namely that if difficulties arise one should not look for an alternative procedure too quickly, but should try to go deeper into the subject first. One of my most enjoyable experiences during this work was to learn how useful it was that Nature has arranged the density maximum of water at $+4°C$, just where I needed it. The greatest difficulty in electrophoresis of solutions of high conductivity (e.g. serum) is the risk of convections caused by the heat produced by the electric current. By choosing a temperature where the density varies very little with temperature this risk is largely eliminated, and one can apply much stronger currents, giving much better resolution. By using alternating current instead of direct current heat convection could be studied separately.

The introduction of a refractive index method of observation, the schlieren method, also meant a great advantage as compared to the ultraviolet absorption used earlier in both ultracentrifugation and electrophoresis. It was soon considerably improved by Longsworth, Philpot, and Svensson and was also refined to a high degree by Lamm in his "scale" method.

Now I was convinced that my new apparatus would work and I was impatient to demonstrate this. Thus, instead of trying a sample of a reasonably homogeneous protein, I picked out a sample of serum from the refrigerator, dialyzed it against a buffer solution and put it into the machine. If it worked with serum, it should work with almost anything else. After about two hours I observed four distinct schlieren bands, indicating the migration of albumin and three globulin components which were named α, β, and γ. This was a great surprise to me, although there had been some indications of this in my earlier work. I wanted to share my enthusiasm with somebody, and I brought my old friend K.O. Pedersen into the room, who after a thorough inspection muttered: "Maybe there is something there." (He has always been a very critical and careful investigator.)

This work soon led to a publication describing in some detail the new apparatus and the background for its construction, as well as the discovery of the main serum components. To my surprise it was not accepted in a biochemical publication to which it was sent first, being too "physical." Thus, in 1937 it appeared in the *Transactions of the Faraday Society* (5). The reaction was immediate and extremely positive; as my American friends had predicted, I was flooded with letters and requests for reprints and even a telegraphic order (from Edwin S. Cohn) for an apparatus to be built in our workshop. I remember particularly a very kind letter from Dr. Harvey Cushing, asking me for a reprint for his library of the classics of medicine, a very memorable event, indeed, for a young scientist. Svedberg, who was just about to leave for a lecture tour in the U.S.A., took with him some of my results, which no doubt contributed to the rapidly developing interest in the new method. I was particularly happy when I saw a paper by Landsteiner (who had been so encouraging and kind to me) which demonstrated the electrophoretic differentiation of duck and hen egg albumin in the new apparatus. The Rockefeller Institute was engaged very early in the new technique under the very able guidance of Longsworth, who introduced some important improvements, above all the automatic scanning of the schlieren diagrams and the theoretical interpretation of the results. A close contact between our groups was established involving also a highly appreciated and lasting friendship. Frank Horsfall, Jr. came to Uppsala for a year to study the method, and we had great fun together in "crossing snails" as we put it. We demonstrated that the reversible dissociation-association of hemocyanins from *Helix pomatia* and *Helix nemoralis*, achieved by shifting the pH, led to the formation of hybrid molecules, whereas the same experiment with *Helix pomatia* and the more distant species *Litorina litorea* appeared to give no or very little hybridization (6). In 1939 I received an invitation from Dr. Walter Bauer to spend a year at the Rockefeller Institute Hospital to continue our collaboration. Unfortunately, my stay was limited to only two months because of the war. The fruitful contacts and the exchange of information (even of unpublished results) has, however, continued ever since. I wish to mention here particularly also the names of Moore and Stein, and Kunkel, who spent a year in Uppsala from 1950 to 1951.

Today, when exchange of scientists between countries (even those who are rather distinctly apart politically and geographically) is promoted by governments and academies by different forms of bilateral cultural agreement, it is well to remember that an absolute reciprocity in the number of scholars who are sent from each country is not essential. A scientist coming to a laboratory to learn something will usually contribute experience and many new ideas which more than compensates for the expense of the host institution or the host country. This has, at least, been our experience and I believe it is generally so, at least with countries of similar scientific standards.

The Research Professorship: Continued Work on Electrophoresis

In 1937, mainly on the initiative of Svedberg, a research professorship was established at the University of Uppsala by a generous donation of Major Herbert Jacobsson and Mrs. Karin Jacobsson of the well-known Gothenburg shipping family Broström. I was the first to be appointed to this chair, intended "for research and teaching in those fields of chemistry and physics, which are of importance for the processes of life." Thus, being in a secure position and with the promise of support from the Rockefeller and Wallenberg Foundations, I could plan for the future. Eight years later, in 1946, a department of biochemistry in the Faculty of Science was officially established with its own budget and personnel, and in 1952 we moved into a new building, having had until then only a few rooms in the Institute of Physical Chemistry at our disposal. In the meantime, however, many things had happened.

Looking back, I believe there has been a tendency in my mind not to stick to a subject when convinced that it is in good hands with other people. This applies to my scientific work as well as to administrative functions of different kinds. I may wish to continue by giving advice and stimulation as much as I can, also perhaps criticism, without really doing anything with my own hands. I have often felt the desire to try something new, as I have already indicated above in connection with my early speculations about the significance of separation in general—not only by electrophoresis.

People have sometimes asked me why we did not study at an early stage the many clinical applications which electrophoretic analysis of pathological sera seemed to promise. It is true that we did some work, for example, the demonstration (with Kabat) of an antibody as a separate electrophoretic component in the gamma globulin group. But I felt that with our particular background and our experience it would be better to concentrate upon a further improvement of the method in different directions and on applications to problems which were closer to our field of interest. I must admit, however, that I was a bit skeptical about the great expectations raised by my medical colleagues. I remember, however, how one of my many good friends among them once mentioned that some people already had investigated sera from mental patients by electrophoresis. He added: "Tiselius, what I particularly like about your new method is that so many patients will have to be released, but still more, that quite a number will have to be taken in."

Even though early work in Uppsala and, above all, elsewhere pointed the way for the clinical application of electrophoretic analysis in certain pathological conditions, the real breakthrough in this field came with the introduction of the much simpler method of filter paper electrophoresis. In 1927 I made some experiments in this direction and separated red phycoerythrin from blue phycocyanin by electrophoresis in a slab of gelatin, obtaining beautiful narrow migrating zones. I did not pursue this work and never published anything. I was above all interested in the quantitative aspects and disliked introducing an ill-defined medium such as gelatin. This showed a lack of foresight on my part; in many cases, the separation is the all important issue, even if one has to abstain from the quantitative aspects and thus the possibility of predicting what is going to happen. Much later (in the early sixties) Hjertén demonstrated that in weak agar gels there is hardly any influence of the medium, and before that Smithies in his beautiful experiments in concentrated starch gels obtained extremely good resolution, to which the medium itself contributes essentially by a sieving mechanism.

I have already mentioned above that right from the beginning of this period I had the good fortune to have very able collaborators to whom I could entrust the fulfillment of the electrophoresis work. Thus, Harry Svensson considerably improved the optical system of the moving boundary method and made fundamental contributions, with experimental verifications, to the theory of electrophoretic migration. This resulted in the first dissertation from the Institute [in 1946 (7)]. Svensson became docent of biochemistry and also later, after having left us, continued such work in Stockholm, in the U.S.A., and later as professor of physical chemistry in Gothenburg. A number of other collaborators working with serum proteins and enzymes provided us with many questions about methods and gave us a solid ground of practical applications to use in our planning for the continued methodological work. I firmly believe (as I have often stated) that in such work it is very essential to have a constant exchange of ideas and experiences between those who develop methods and those whose prime interest is to apply them—if possible under the same roof. The latter do not always seem to have the patience to go deeper into methods if there is trouble, and the former run the risk of becoming perfectionists or gadgeteers if they forget that methods should work not only in model experiments. The best proof of the usefulness of a new method is that it is applied by others without any persuasion.

THE WAR YEARS: CHROMATOGRAPHY OF COLOURLESS SUBSTANCES

As Sweden was fortunately not involved in the World War of 1939 to 1945, work in the Institute went on almost as usual, although with reduced personnel (many were called up for military service). I served for a time as a member of the Board of Defence Research. We tried to make ourselves useful by contributing some work of immediate importance, as our country was suffering from shortage of many essential commodities. The feeling that we were for a time being almost completely cut off from a large part of the world with which we always had enjoyed friendly relations, was depressing. The great tragedy which was going on all around us and the fear that we ourselves might become involved made pure scientific work for one's own pleasure sometimes appear out of place. During this time, however, I became strongly interested in chromatography and studied some of the earlier literature on the subject. I felt that electrophoresis was hardly specific enough for separation of the multitude of substances occurring in materials of biological origin. I was surprised to find that very little use had been made of such methods, except for carotenoids and some other coloured substances. And not even Willstätter, who made such extensive use of adsorption in the purification of enzymes, and who applied chromatography to some of his other problems, came to use this method to any considerable extent.

I decided to attempt "chromatography of colourless substances" or, as I preferred to call it, "adsorption analysis" by observing the separation, not on the column but in the eluate. Influenced by my previous experiences, I decided to use optical methods for continous determination of the concentration of the separating substances as they were leaving the column. With the very able help of Stig Claesson (8) (who later became Svedberg's successor as professor of physical chemistry) I built a microrefractometer for this purpose, and started extensive investigations on the adsorption analysis of sugars, amino acids, peptides, and many other colourless substances. I learned a great deal from this work, becoming aquainted with the phenomena behind the separation. I distinguished between frontal analysis, elution analysis, and displacement analysis, and, as far as I know, this had not been done before (9). Especially the displacement phenomena must have been observed many times before with coloured substances, as one can see it when a mixture is applied to a column, even before any eluant has been added. I could account for this phenomenon theoretically and this work gave me a great deal of satisfaction. The key problem in chromatography by elution is to eliminate the "tailing" of the migrating zones, which are due to the fact that the adsorption isotherms are curved, giving rise to a stronger binding at low concentrations than at high. In displacement analysis, the zones form a procession where any tailing material from one particular zone will be displaced to where it belongs by the following zone. The drawback of the method is, however, that the zones emerge in immediate contact with each other. Thus the elution analysis, with well-separated zones, is superior in most cases (with the possible exception of large scale preparative work, e.g., in the separation of rare earths on ion exchangers).

In most of this work we used active carbon as adsorbent and we tried many methods to rectify the adsorption isotherms, e.g. by pretreatment of the carbon with small amounts of strongly adsorbed materials. Some of this work led to useful results, and the methods came into practical application. In general, however, I must say that it never led as far as I had hoped. The reason was simply that these problems were solved in a superior way by others: above all by Martin & Synge in their partition chromatography and by Moore & Stein by introducing fraction collectors and ionic exchange resins for amino acid analysis. I still believe, however, that this early work of ours came to play a role in much of the subsequent and more successful work in our laboratory. We became aware of the key problems and our wishful thinking was influenced accordingly. This was also a great period for research workers in separation. It was now generally realized that separation is not only essential for substances, but it is also of the utmost importance as a tool in determining the structure of large molecules, demonstrated in a most striking way by Sanger in his insulin work and by many who followed in his footsteps.

THE DEXTRAN STORY

My account of what we were doing during the war years and immediately afterwards would be rather incomplete if I did not mention dextran. Not only did this substance come to play an important role in our activities (even up to this day), but the dextran story may also be of interest as an example of how things may happen in research.

We had been asked to do some work on the freeze-drying of plasma for use in military medicine, where our experience in that field with our own protein materials was considered to be of value. We had also been approached by the Swedish Sugar Manufacturers Corporation, who were having trouble with some of their beet extracts becoming contaminated with slimy substances which obstructed the filters used in the raffination. The problem was essential as there was a shortage of coal and fuel in general. We found or rather confirmed that the slime was due to dextran produced by bacterial infection (*Leuconostoc mesenteroides*).

We decided to look for a sensitive and specific reaction for dextran and hence, tried to produce an antiserum by injecting dextran into rabbits. To our surprise no reaction was observed and the animals seemed to be quite indifferent to even large doses. The two young biochemists who were active in both the plasma and the dextran projects, being aware of the importance of finding a substitute for plasma, now asked themselves: why not try dextran? [Grönwall (10) and Ingelman (10, 11)]. Earlier substitutes (gum arabic, etc.) had proved unsatisfactory and even dangerous, as they produced reactions or accumulated in the kidneys or in the liver. Dextran of a suitable molecular weight could be obtained, it did not appear to give rise to reactions, and was likely to be gradually broken down in the body. They tried (and successfully) to introduce large quantities of dextran, first in rabbits, then in dogs, and finally in human patients. The pharmaceutical company, Pharmacia, which had just moved from Stockholm to Uppsala, became interested and started to produce dextran ("Macrodex") which is today one of their most important products. Thus, a close collaboration started between Pharmacia and the Institute of Biochemistry which I believe has been of great mutual benefit. There is an open exchange of information in seminars and private discussions and some research workers in Pharmacia had their earlier training with us. It was a particular pleasure to be able to tell the Sugar Manufacturers that instead of trying to prevent the formation of dextran in their extracts they should now build a factory in order to produce it in large quantities and sell it to Pharmacia for further treatment.

This story has always seemed to me to be interesting and significant since it demonstrates how a prepared mind (in this case, the awareness of a need) and a cross-fertilization of ideas may lead to useful results in a very unexpected direction. And many, perhaps most, significant results are unexpected.

The Nobel Prize

In 1948 I was awarded the Nobel prize in chemistry for work on electrophoretic and adsorption analysis, especially for the discovery of the heterogenous nature of serum proteins. One year earlier, at the 1947 Nobel banquet,

I had the privilege of speaking to the Nobel laureates of that year. I will quote from this speech (12):

Nobel Laureates:

In science and in medicine your work in search for the truth has disclosed new laws of Nature and opened up new and vast fields for research of the utmost importance for the welfare of mankind. And, in literature, new truths of a different kind have been brought to light in a way of which only art in its highest and most subtle forms is capable. When a new thought is born, or when one of the deep secrets of Nature yields to the searching scientist—in this very act of creation—there is a pure and primitive happiness deeper than anything of this kind which can ever be granted a human being to experience.

This view has been formulated in one simple sentence by the great Swedish chemist Carl Vilhelm Scheele, who some 150 years ago wrote 'Det är ju sanningen vi vilja veta, och vad är det väl icke för en ljuvlighet att få tag på den.' (It is the truth we are searching for, and what a delight it is to find it.)

Nobel Laureates: Like all creative work, your achievements must have given you many moments of that sublime happiness which Scheele had in mind when he wrote these words. Probably your work has also many times, perhaps even more frequently, involved disappointments. In any case, we do not believe that to you even the highest awards and the most whole-hearted recognition can be more than a faint reflection of the deep satisfaction you must have experienced in your work. We do believe, however, that the Nobel prizes afford us all a suitable way of expressing our indebtedness—an indebtedness from the whole civilized world—to those pioneers in science, medicine, and literature who have by their deeds enriched this civilization and pointed the way for its further development. We would like you to consider your awards for what they should be and what they are: an expression of the gratitude of mankind.

And then, in 1948, again at the Nobel banquet, I expressed my feelings of gratitude on being myself awarded the Nobel prize in a speech from which I quote the following (13):

Alfred Nobel was a great idealist, who thought in international terms. The Foundation which bears his name was created with the intention of furthering achievements which he considered to be of great benefit to mankind. That he awarded prizes for these achievements shows that he believed individual endeavour to be an important part of progress in all cultural fields.

This banquet, which is dedicated to the memory of Alfred Nobel and serves as the ceremonial background for the distribution of the prizes, may well prompt one to ask: To what extent is progress in literature and science linked with the personality of poets and scientists? That this is indeed the case is beyond all doubt, but there is a difference between these two forms of endeavour—one cannot judge them by the same standards. Poets, like all creative artists, impress the stamp of their own personality on their work far more than scientists can. The work of the poet bears his individual mark. Thus, as an individual, he is indispensable for the development of culture and civilization. The scientists, however, seeks objective truth, which is and must be completely free from all traces of his individual personality.

It can be said with certainty of all scientific discoveries that if they are not made by one scientist, they will sooner or later be made by another. Naturally there is a strong emphasis on the 'sooner or later.' The enormous progress made in science and in medicine seems today to demonstrate a process of organic growth which proceeds according to its own laws, and in which it is frequently difficult to distinguish between individual achievements. It not infrequently occurs that the same discovery is made in different parts of the world at more or less the same time, with perhaps a few days or weeks between them. Rapid communication of all new discoveries and intensive correspondence between scientists in all parts of the world has contributed to the advancement of science in a spirit of team work in numerous fields. This can only be of benefit to science and thus to civilization as a whole. These thoughts should serve as a reminder for the individual scientist when considering his own role in assisting development, and should deter him from any false conception of his own importance. These were just a few ideas which occurred to me on being myself the recipient of the greatest honour which can be awarded to a scientist.

There is more to be said about Nobel awards and the Nobel Foundation, but I will return to this below.

AN EXPANDING PROGRAM: NOT ONLY SEPARATION

The Nobel prize had some consequences. One was that the plans for a building for the Institute of Biochemistry were accelerated. I remember paying a visit to the Undersecretary for Education and Research a few weeks after the announcement of the award, asking him if now the outlook was not much brighter. He agreed, but added: "If you have done all these nice things we read about in the papers with only three rooms at your disposal, I have some difficulties in understanding why you need a whole new building." One of the rooms was originally intended to be a pantry, and a Stockholm newspaper one day showed a caricature of myself as the man who won a Nobel prize working in a kitchen. Such things may be helpful nowadays, and we got the building and moved to the new premises in 1952.

Biochemistry now attracted an increasing number of students and more advanced research workers. We could organize a more diversified program often based upon the interest shown by particularly able people who gradually became leaders of groups: Thus Porath (14) constructed columns for zone electrophoresis, also for preparative use, and applied them in the study of certain pituitary hormones. He also continued work on serum fractionation, later continued by Bennich, who recently discovered a new γ-globulin of considerable interest. Porath inspired and successfully led many other investigations and gradually became my second in command, now director of the Institute. Malmström brought with him a strong interest in metal-combining enzymes and after some years published a dissertation on enolase. He organized an "enzyme group" which attracted some very able students. He carries on this work as professor of biochemistry in Gothenburg. Boman introduced molecular biology in the institute after a very profitable year with Lipmann at the Rockefeller Institute. He is now professor of microbiology in the University of Umeå in north Sweden. Roos (15) made a very penetrating study of great clinical interest on the purification of certain pituitary hormones. His dissertation was published in October 1967. Weibull studied the purification and some of the properties of bacterial flagellae (16). His continued work on the lysis of bacteria led to remarkable results and he is now professor of microbiology in the University of Lund.

Our orientation towards microbiology made us feel very hampered by the fact that there was no chair for general microbiology in Sweden. We decided to work for this, with much support from colleagues in related fields. We started a teaching course in 1957 and found in von Hofsten a very enthusiastic and devoted teacher. The authorities were gradually convinced that this subject needed support and today there are chairs in the faculties of science in Uppsala, Lund, and Umeå, and more are coming.

It would lead too far to go into further detail about all these new activities, even if I would have liked to mention many more names, also among our foreign guests who have contributed so much. In the literature references at the end of this chapter the reader will find a list of dissertations in biochemistry from this Institute up to the end of 1967. A review of recent work at the Institute on separation is found in (17). It seems appropriate in this chapter that I limit myself to work where I was more directly involved. I just want to add that in most of the cases our experience in separation and our equipment for such work provided a platform for new ventures of a different kind. It is mostly not easy to start something new, especially not today when international competition is more severe than ever.

Except for the above-mentioned work by Porath and his collaborators on preparative-zone electrophoresis in columns packed with different kinds of inert stabilizing materials (especially modified cellulose), extensive investigations on different types of zone electrophoresis in gels were now performed by Hjertén & Jerstedt. Hjertén was the first to discover the advantages of using gels of agarose (that is, agar from which the agaropectin has been removed) and worked out convenient methods for preparing it in bead form, easy to pack into columns. Such suspension columns (also of polyacrylamide) are often convenient in electrophoretic analysis also of large molecules or even particles which do not migrate easily in a compact gel. Hjertén's main contribution was, however, the development of the "free-zone electrophoresis" where the zones separate in a horizontal quartz tube, slowly revolving around its long axis. Convections due to gravity are thus eliminated and no stabilizing medium is required. This micro method gives high resolution and accurate values of mobilities. A complete account of this method is given in Hjertén's dissertation (18).

In separation work, one's ambition is mostly to obtain substances as pure and homogeneous as possible. It is a common experience, however, that certain materials will resist purification beyond a certain level, even if different methods are attempted. This may be disappointing, but such a result may be extremely valuable from an entirely different point of view. An "impurity" which sticks tenaciously to the substance we wish to isolate may do so because in the original biological material it belongs to this substance in a way which is significant from the structural or functional point of view. Thus, an "impure" substance may represent a very valuable piece of information. It is tempting in such cases to formulate the distinction between the organic chemist and the biochemist by saying that the latter should also devote himself to the study of impure substances.

Submicroscopic particles or more generally fragments of a biological structure are of course "impure" from a strictly chemical point of view but represent just that kind of information. It is therefore natural that they are being investigated both from the point of view of structure and of function. I have often speculated about the possibility of a systematic study of fragments, which should be obtained by some kind of successive dispersion of some biological material (19). Given methods for the separation of such fragments according to size and some other properties, it ought to be possible to reconstruct on paper the original structure in a way somewhat analogous to sequence determination of amino acids in a protein molecule by analysis of fragments obtained by hydrolysis. Such structure analysis by successive dispersion might be useful in conjunction with electron microscopy.

Nothing like this has been achieved so far, but considerations of this kind have stimulated our interest in the study of the separation of particles, submicroscopic and even microscopic. In addition to Hjertén's work already referred to above, the very interesting partition method of Albertsson, described in his dissertation (20), should be mentioned. It can be demonstrated that the larger the molecular or particle size, the more closely similar propperties are required for two phases if a defined and not too one-sided distribution of the molecules between these two phases is to be possible. Moreover, with biochemical material, the phases should consist mainly of water. Albertsson demonstrated that it was possible to obtain such two-phase systems by dissolving small quantities of certain high polymers, such as polyethylene glycol or dextran in water. One will then observe a sharp boundary separating two solutions, one containing mainly dilute polyethylene glycol, the other dilute dextran, and both containing 95 to 98 per cent water. Low molecular weight substances will distribute equally between such phases, but as the molecular weight goes up the material will go preferentially into one or the other phase, or sometimes will form an interphase. The method is not sensitive to temperature changes (as with the lower glycols used by Martin & Synge in their partition chromatography of proteins). By changing such parameters as the electrolyte composition or the molecular weight of the (commercially available) phase-forming polymers, it is often possible to change the partition in one direction or the other. This interesting method (which also can be operated easily on a very large scale) has proved very useful, for example, in virus purification. Albertsson, who is now professor of biochemistry in the university of Umeå, has stepped up the separation efficiency by constructing a counter-current apparatus adapted for this type of work.

FURTHER ADVANCE IN CHROMATOGRAPHY: GEL FILTRATION

Adsorption on calcium phosphate gels has long been used for purification of enzymes. We felt the need of a greater range of specificity in chromatography of proteins, nucleic acids, etc., than was available with the commonly used materials for separation of such substances. When trying columns of calcium phosphate we ran into great trouble, which was difficult to account for (work with Hjertén and Levin). It was gradually found that the phosphate changed its properties during the experiment and we decided to try the most stable modification of calcium phosphate, namely hydroxyl apatite, which is easy to prepare. This has worked well and is now often used in various connections. Application to nucleic acids and nucleotides appears particularly interesting.

During many years the desirability of working out a chromatographic method based upon differences in molecular weight or size had been the subject of much discussion at the Institute. I made some attempts at a kind of zone ultrafiltration—a logical extension of zone electrophoresis and zone ultracentrifugation. I used gel columns but they always cracked when I applied high pressure to get the liquid through. Synge and I discussed these experiments and concluded that the only way of getting the liquid through a compact gel column was to apply electroosmosis. Synge and Mould continued such experiments and could demonstrate the separation of uncharged substances on strips of nitrocellulose when an electric current was sent through.

A great step forward was when Porath & Flodin (21) demonstrated that columns of gel particles could act as a kind of molecular sieve, the smaller molecules being retarded by penetrating more or less into the gel, the larger being less affected and therefore appearing first in the filtrate. It was again our old friend the dextran which came to our help. I have been told that all this started by an unexpected observation. Research workers at the Institute tried to use particles of dextran gel (made by cross-linking of dextran) as a filling material in electrophoresis columns. They forgot to turn on the current but observed nevertheless a beautiful separation when buffer solution ran through the column. A study of this phenomenon was taken up by Porath & Flodin and gradually led to the almost explosive development of the field. At an early stage we sent some dextran gel to Stanford Moore at the Rockefeller University, together with an account of our first results. We got a most enthusiastic reply. Flodin was invited to a Gordon conference in the U.S.A. and the new method spread almost epidemically.

It is true that similar attempts were made in other laboratories at the same time or somewhat earlier, but it seems that dextran gels were so superior that the method came into general practical application first when this material was introduced.

A very fruitful collaboration with the Pharmacia company was again established. They now produce dextran gels of various degrees of cross-linkage under the commercial name of Sephadex, recently used also for work in certain organic solvents.

Gel filtration seems to be essentially an exclusion process with no affinities involved (except in certain cases). The absence of "tailing" is remarkable even at fairly high concentrations and also in many cases where some adsorption affinities appear to be involved. It seems that our dream of chromatography with linear isotherms has come true at last; the physico-chemical background is not yet quite clear, however, especially not in the behaviour of larger molecules. Other gels have now also come into use for the same purpose (agarose, polyacrylamide) largely through the work of Hjertén. The most recent development in this field [by Porath and his group (22)] involves the introduction of specific groups or even proteins on the gel matrix which makes it possible to use highly specific columns which are of great interest in, for example, immunochemistry and enzymology.

EXPERIENCES FROM THE OTHER SIDE OF THE COUNTER, 1946 TO 1967

World War II demonstrated in a very convincing manner what scientific research can contribute and what great potentialities may be involved not only in the military field but also—and this now became imminent—in peaceful activities for the restoration of human welfare and for our continued existence as a whole. Governments approached scientists for advice. Some of them preferred to remain in a corner of the laboratory, others were more willing to contribute, even if they were somewhat embarrassed by this sudden benevolence and startled by the limelight of public attention.

In 1945 I was asked by the Government to become a member of a committee which was to recommend measures for improving conditions for research in the field of sciences, especially basic research. I accepted although I had no previous experience of committee work. I found to my surprise that people listened to me and often followed my advice. The committee was quite successful and practically all its recommendations were approved by the Parliament—something very unusual in our early experience. Thus among other things, a research council for sciences was established (similar councils for technological and medical research had already been organized during the war period). I was asked to become the chairman of the new council. This was rather unexpected because for obvious reasons one usually prefers to have in the chair a person who is not actively engaged in the field and may be suspected of representing local interests. I learned later that my name had been submitted by the rectors of the Universities of Stockholm and Lund who evidently felt confident that I would not speak unduly for Uppsala. Anyhow, I accepted a four years' appointment, being flattered and feeling that I might be able to help the new council fulfill the hopes of my fellow scientists and also to gain the confidence and even the appreciation of the authorities and of the general public—all this is very essential for its future existence and expansion. Many people had expressed doubt that it was wise to give a large sum of money (at that time 1 million Swedish crowns, today 25 million Swedish crowns per year) to a group of scientists to play around with, with very little or no government control.

I remember with particular pleasure how the council soon found that speaking too much for local interests was considered unstylish and the support of promising research, wherever it was done, was the all-important goal we were aiming at. I remember also the excellent collaboration with our secretary general, Dr. Funke, who still has this function today, and is also secretary of our Atom Energy Council and head of the Administrative Council of CERN.

I had hardly left the council (in 1951) when I was asked to join a committee for cancer research, which was to plan for an organization where activities were to be financed mainly by raising private contributions. The Government offered to pay a sum equal to what we could collect. This committee worked very slowly and I was often disappointed as it appeared so difficult to come to an agreement. Differences between representatives of clinical and basic research was the main source of disagreement. When finally the "National Society for Cancer Research" was formed, I was asked to be chairman of the research committee which was to deal with applications and recommend grants. Surprisingly enough there were few disagreements. Conflicts between "pure" and "applied" faded into the background as there were comparatively few applications for clinical work. This was not the first time that I had seen prestige arguments loose some of their weight when it came to actual practice.

In 1947 the first International Congress of Chemistry after the war was held in London under the auspicies of the International Union of Pure and Applied Chemistry (IUPAC). I had been asked to give one of the plenary lectures about our work on electrophoresis and separation in general. It was a very memorable reunion of scientists from many countries. Our English hosts did their utmost, despite all the hardships and sufferings they had just come through, and were very successful indeed. There were many discussions about future international collaboration in a spirit of general optimism. I was elected as one of the vice presidents in charge of the Section for Biological Chemistry. Four years later, at the IUPAC Congress and Conference in Washington in 1951 I was elected president of the Union for a four year period, succeeding H. R. Kruyt.

My work in the Union involved a great deal of travelling (perhaps too much) but brought me into contact with international affairs and personalities. At this time IUPAC was run almost as a family affair; Raymond Delaby as General Secretary and Leslie Lampitt as Honorary Treasurer were the dominant figures. Both were very forceful personalities and highly devoted to the Union, but they sometimes disagreed. I had to be the mediator and was usually successful. I believe the International Unions through their Commissions and Conferences do very useful and necessary work and act as clearance centers for much valuable information and many interesting personal contacts. Some of the work, for example questions of nomenclature or standardization, may not appear very fascinating but has to be done. I often hoped that IUPAC could play a more active rôle in organizing international research projects in chemistry, but perhaps time was not yet ripe for this. I believe, however, that such collaboration will become necessary in many fields. With the present expansion of scientific research, we will soon find that no single country can afford to do everything which appears possible or even promising. There tends to be a waste of effort by unnecessary duplication, by secrecy, and competition based upon prestige. It is to be hoped that the international Scientific Unions will play their part in guiding the development in organized collaboration, such as we have seen in the International Geophysical Year, and which seems to be coming in the International Biological Program now being launched.

A difficult problem during my presidency of IUPAC was that many biochemists wanted to form a Union of their own instead of remaining in the Chemistry Union as a Section for Biological Chemistry. I was surprised to find that there existed in some countries (especially in Britain) conflicts and disagreements between biochemists and chemists, and that these feelings came to expression in our discussions. I remember saying once that any biochemist who goes sufficiently deep into his problem runs the risk of becoming a chemist. But enforced collaboration never works. The Biochemistry Union was started and I believe that a very good and increasing collaboration has been established. The tendency will probably be towards an increasing number of international unions as science continues to become more and more differentiated. Just because of this, it becomes increasingly important that a closer collaboration between the unions be established directly or through the International Council of Scientific Unions.

Two of my later functions "on the other side of the counter," namely as one of the initiators and members of the Science Advisory Council to the Swedish Government and as a member of the board of the Wallenberg Foundation—Sweden's most important private foundation for the support of science and culture—are both too recent to be viewed in retrospect. I shall, therefore, not deal with them here. In the latter case, I have been given an opportunity to practice some of the viewpoints I have expressed in Warren Weaver's recently published book on U.S. Philanthropic Foundations (23). It has been a great satisfaction in both cases to be able to do service to those who have given me support and encouragement throughout my scientific career.

In my experience, I have come to this conclusion (also expressed by many others in similar activities): in the support of fundamental research, the individual research worker is more essential than the research project, when judging priorities. In a small country like my own this may cause the research front to advance in a somewhat uneven manner. There may be strong and active schools in certain fields with a fine tradition, and none at all in other no-less-essential domains. I believe this has to be accepted. I have even been a spokesman for giving priorities to fields where we are particularly strong ("You mean of course biochemistry" our Prime Minister Erlander once remarked). The main argument would be that in such fields we are likely to have a number of young research workers of great promise. Any investment in them is likely to prove fruitful. In applied research, where one has to pay more attention to immediate needs of the society, of industry, public health, agriculture, etc., priorities must be given on a somewhat different basis and one may have to accept a certain amount of government direction. Even so, personalities mean a great deal also here; but if for some reason there is no "nucleus," it has to be created, for example, by sending some young people abroad to study and to do research under some leading personality in the field.

And last, I would like to add a few words about my activites in the Nobel Foundation, with which I have been connected in various functions since 1947 when I became member of the Nobel Committee for Chemistry. To act as a member of jury which has to choose those who have contributed most to human progress is very difficult and hardly enviable. No wonder that hesitation was expressed by some members of those institutions which were entrusted in Alfred Nobel's will with the responsibility of awarding the prizes which were to carry his name. The tremendous growth of science and other human activities since the will became known (in 1897) has not made this task easier. However, we have been encouraged in the mostly positive reaction in the world, especially as regards the scientific prizes. Another kind of moral support comes from the experience that there seems to be fairly evident international opinion, in that certain candidates are nominated year after year. This despite the fact that nominations (by invitation only) come from many different countries and many different university professors and other leading personalities, invited according to a plan which aims at considerable variation and a world-wide coverage over a period of several years. Often the problem is to select one or a few among a group of let us say five to ten obviously highly worthy candidates (the total number of candidates proposed being perhaps 50 to 100). Here one often faces the problem of comparing the incomparable, and in order to reach a decision one may have to resort to a natural desire to pay attention also to a fair distribution among different important regions within each prize field.

Obviously the members of the prize-awarding institutions can not fulfill the task entrusted to them if they are not themselves in close contact with the international development in the five Nobel prize fields of human endeavour. Some of them should be actively engaged in such work. This requires a high overall level of activity in these fields. The statutes of the Nobel Foundation provide the possibility of organizing Nobel Institutes for this purpose, also by investigating results which have been proposed for an award. This somewhat unrealistic idea has to my knowledge never been realized, even though the Nobel Institutes indirectly have been of great importance as offering a possibility of providing especially deserving personalities a secure position and good facilities. It would, of course, be far beyond the resources of the Nobel Foundation to assume even a partial responsibility for the high level of research in, for example, physics, chemistry, or medicine in our country. As a matter of fact, the Nobel Institute of Medicine is today largely financed by the State and the Nobel Institute for Physics and Chemistry was taken over completely by the State a few years ago. I suggested to the Foundation that we ought to find other means of serving the original purpose of those institutes, namely to aid the work of the prize-awarding institutions by arrangements which should facilitate international contacts. Thus, we have recently been able to start some new and—as I believe— promising activities, such as inviting Nobel guest professors and Nobel guest lecturers, and organizing Nobel Symposia. The latter are a kind of informal roundtable conferences, in limited areas of great actual interest, to which leading personalities from different parts of the world may be invited. We believe that these new initiatives may be of great value also to the Nobel countries (Sweden and Norway) as a whole. After all, they are situated in a corner of the world. The symposia were made possible by a generous grant from the recently established Bank of Sweden 300th Anniversary Fund.

The president of the Nobel Foundation has to address the distinguished audience assembled in the Stockholm Concert Hall for the prize-awarding ceremony on December 10th every year. I found this interesting and a good platform to express some personal views of the distinctions in general. Naturally, I felt somewhat hampered in my attempts to hail the laureates by being a laureate myself. Thus I came to stress the significance of the prize more as a challenge than as a personal distinction. I believe this is justified, as it seems to agree with Alfred Nobel's own intentions. In the light of recent discussions of the sometimes embarrassing consequences of a Nobel prize for the recipient (24), I would like to quote here from my last presidential address (1964) the following (25):

> While it is true that the Nobel prize is today primarily regarded as a personal award and distinction, it is nevertheless dubious whether this was Alfred Nobel's main intention. He was himself, to say the least, indifferent to worldly honours. It is also said that he wished his prizes to help and support unworldly pioneers and dreamers who were not shrewd enough to profit financially from the results of their work. He wanted to make it possible for such people to continue their work without financial worry. This conception of the prizes has of course, for obvious reasons, increasingly faded into the background. But a Nobel prize is still regarded as having a purpose and value beyond the immediate personal distinction it confers.
>
> It is obviously impossible to distinguish between the work and the man behind the work in this connection. But we can refuse to let the one overshadow the other. The honouring of some outstanding achievement can often bring real support in the completion of a valuable piece of work, not merely by helping the prizewinner himself, but also by encouraging those who will follow in his footsteps. There are areas within the domains of scientific research and literature, not to mention work for peace, where such support for a good cause can be of the greatest value. The institutions responsible for awarding the Nobel prizes have, I believe, often been guided by considerations of this kind when they have taken decisions designed to realize Alfred Nobel's intentions. There are fundamental discoveries which, because of their theoretical character for example, do not attract the attention they deserve if long-term developments are to be taken into account. It is primarily immediate practical results which are looked for by the general public, the authorities, and others on whose support writers and scientists depend. And the subdued

kinds of literature, not least when they have what Nobel called an "idealistic" orientation, may require help and support to make their voices heard.

Among the Nobel prize-winners there are many, of course, who do not need this extra limelight, whose personalities and work are already well known and widely appreciated. But they may even so be willing to use their fame and achievement to draw attention to others to whom this can mean a great deal. And this is not only the case with the unworldly dreamers in whom Alfred Nobel believed and whose work he wished to forward. It is, after all, the prize-winners who make the prize and it is their achievements which ultimately form the basis for the prestige of the Nobel prize in the world.

I believe that all this could be condensed in a kind of "code of conduct" once formulated by a very prominent colleague of mine who in a discussion of such questions exclaimed: "I have my prestige to spend it." And this applies not only to Nobel laureates but to scientists in responsible positions in general, on either or on both sides of the counter.

While this is being written, the plans for a new building and a new organization of the Institute of Biochemistry in Uppsala are well on their way. They will involve an intensified collaboration above all with other biomedical research, a special center for research and service in the field of separation, and arrangements for an extensive cooperation with industrial research. I have tried to prepare the ground for all of this and I believe that the future development is in good hands.

LITERATURE CITED

1. Koestler, A., *The Sleepwalkers* (Hutchinson & Co. Ltd., London, 1959)
2. Tiselius, A., Claesson, S., *Ann. Rev. Phys. Chem.*, **18**, 1–8 (1967)
3. Tiselius, A., *The Moving Boundary Method of Studying the Electrophoresis of Proteins* (Dissertation, Uppsala, 1930)
4. Tiselius, A., *Nature*, **133**, 212 (1934)
5. Tiselius, A., *Trans. Faraday Soc.*, **33**, 524 (1937)
6. Tiselius, A., Horsfall, F., Jr., *J. Exptl. Med.*, **69**, 83 (1939)
7. Svensson, H., *Electrophoresis by the Moving Boundary Method. A Theoretical and Experimental Study* (Dissertation, Uppsala, 1946)
8. Claesson, S., *Studies on Adsorption and Adsorption Analysis with Special Reference to Homologous Series* (Dissertation, Uppsala, 1946)
9. Tiselius, A., *Arkiv Kemi*, **14B**, 31 (1941)
10. Grönwall, S., Ingelman, B., *Acta Physiol. Scand.*, **9**, 1 (1945)
11. Ingelman, B., *Investigations on Dextran and its Application as a Plasma Substitute* (Dissertation, Uppsala, 1949)
12. Tiselius, A., in *Les Prix Nobel 1947*, 52 (Norstedt and Sons, Stockholm, 1948)
13. Tiselius, A., in *Lex Prix Nobel 1948*, 60–61 (Norstedt and Sons, Stockholm, 1949)
14. Porath, J., *Zone Electrophoresis in Columns and Adsorption Chromatography on Ionic Cellulose Derivatives as Methods for Peptide and Protein Fractionations* (Dissertation, Uppsala, 1957)
15. Roos, P., *Human Follicle-Stimulating Hormone* (Dissertation, Uppsala, 1967)
16. Weibull, C., *Investigations on Bacterial Flagella* (Dissertation, Uppsala, 1950)
17. Tiselius, A., Porath, J., Albertsson, P. Å., *Science*, **141**, 13–20 (1963)
18. Hjertén, S., *Free Zone Electrophoresis* (Dissertation, Uppsala, 1967)
19. Tiselius, A., *Arkiv Kemi*, **15**, 171 (1960)
20. Albertsson, P. Å., *Partition of Cell Particles and Macromolecules* (Dissertation, Uppsala, 1960)
21. Porath, J., Flodin, P., *Nature*, **183**, 1657 (1959)
22. Axén, R., Porath, J., Ernback, S., *Nature*, **214**, 1302 (1967)
23. Tiselius, A., in *U. S. Philanthropic Foundations*, 249–51 (Weaver, W.,
   Ed., Harper and Row, New York, 491 pp., 1967)
24. Zuckerman, H., *Am. Sociol. Rev.*, 391–403 (1967)
25. Tiselius, A., in *Lex Prix Nobel 1964*, 11–18 (Norstedt and Sons, Stockholm, 1965)

Dissertations from the Institute of Biochemistry, Uppsala

In addition to those dissertations referred to above (7, 8, 11, 14, 15, 16, 18, 20), the following dissertations have been published up to January 1st, 1968:

Danielsson, C. E., *Investigations on the Seed Proteins of the Gramineae and Leguminosae* (Dissertation, Uppsala, 1952)
Malmgren, H., *Enzymatic Breakdown of Polymetaphosphate* (Dissertation, Uppsala, 1952)
Frick, G., *Studies on Desoxyribonucleoprotein and Desoxyribonucleic Acid* (Dissertation, Uppsala, 1954)
Gelotte, B., *Activin. A Low-Molecular-Weight Substance in the Contractile Element of Muscle* (Dissertation, Uppsala 1954)
Brattsten, I., *Continuous Zone Electrophoresis by Cross Velocity Fields in a Supporting Medium* (Dissertation, Uppsala, 1955)
Drake, B., *Chromatography Combined with Automatic Recording of Electrolytic Conductivity* (Dissertation, Uppsala, 1955)
Leyon, H., *The Structure of Chloroplasts* (Dissertation, Uppsala, 1955)
Malmström, B. G., *The Mechanism of Metal-Ion Activation of Enzymes. Studies on Enolase* (Dissertation, Uppsala, 1956)
Boman, H. G., *Ion Exchange Chromatography of Proteins and Some Applications to the Study of Different Phosphoesterases* (Dissertation, Uppsala, 1958)
Rosenberg, A., *The Role of Metal Ions in the Catalytic Action of Peptidases* (Dissertation, Uppsala, 1960)
Flodin, P., *Dextran Gels and Their Applications in Gel Filtration* (Dissertation, Uppsala, 1962)
v. Hofsten, B., *Some Aspects of the Growth and Enzyme Formation* (Dissertation, Uppsala, 1962)
Levin, Ö., *Electron Microscope Studies on Protein Molecules with Special Reference to the High Molecular Weight Respiratory Proteins* (Dissertation, Uppsala, 1963)

THE EXCITEMENT OF SCIENCE   571

Broman, L., *Chromatographic and Magnetic Studies on Human Ceruloplasmin* (Dissertation, Uppsala, 1964)

Heilbronn-Wikström, E., *Phosphorylated Cholinesterases. Their Formation, Reactions and Induced Hydrolysis* (Dissertation, Uppsala, 1965)

Björk, W., *Studies of the Purification and Properties of Snake-Venom Phosphodiesterase and 5'-nucleotidase, and the Use of These Enzymes in the Studies of Some Other Enzymes Which Also Attack Mono- and Polynucleotides* (Dissertation, Uppsala, 1967)

Ann. Rev. Microbiol., Vol. 21

THE EDUCATION OF A MICROBIOLOGIST; SOME REFLECTIONS

By C. B. van Niel

Ann. Rev. Microbiol., Vol. 21

THE EDUCATION OF A MICROBIOLOGIST; SOME REFLECTIONS

By C. B. van Niel

Hopkins Marine Station of Stanford University, Pacific Grove, California

CONTENTS

The newly emerging individual can attain some degree of stability and eventually become inured to the burdens and strains of an autonomous existence only when he is offered abundant opportunities for self-assertion or self-realization. He needs an environment in which achievement, acquisition, sheer action, or the development of his capacities and talents seems within easy reach. It is only thus that he can acquire the self-confidence and self-esteem that make an individual existence bearable or even exhilarating.   Eric Hoffer, *The Ordeal of Change.*

PRE-DAWN

EARLY EXPERIENCES

My father died early in 1905, when I was a little over seven years old. He had been the junior partner in a furniture business which, at the time of of his death, had done well enough to provide his widow with an adequate income. My mother's father and brothers, as also those on my father's side, were all self-made businessmen; none had received a professional education. They all lived in Haarlem, The Netherlands, and took it upon themselves to guide and supervise the affairs of the small household consisting of my

1

mother and her two children. I seriously doubt that the aptitudes and peculiarities of my sister and myself were ever taken into account in the family councils; the decisions reached were simply imposed by the "Olympians," the term by which Kenneth Grahame, in "The Golden Age," so aptly describes the grownups as seen by a child, at the beginning of the 20th century.

It seemed a foregone conclusion that eventually I should step into my father's place in the furniture establishment. Accordingly, at the end of the usual six years in grammar school I was sent to a secondary school where the curriculum of the final two years was designed to prepare the pupils for a commercial career. Even if I had been consulted as to my own preferences, I would have been at a loss to suggest alternatives; my experience was strictly limited to what went on in the closely-knit family circle, and there a business occupation seemed to be considered as the only decent way of earning a living.

But a drastic change occurred in 1913, at the end of my first three years in high school, when my mother, sister, and I spent several weeks of the summer vacation as guests of a friend of the family. He resided in Drenthe, then one of the least developed provinces of The Netherlands, where he had acquired what impressed me as a rather large estate. Besides the mansion, stables, and garden it consisted mostly of forest and moorland, much of which had been converted into agricultural land. A sizable portion had been set aside for use as a private experiment station; here, various soil treatments were being compared as to their effectiveness for the cultivation of different kinds of crops.

During those weeks I regularly accompanied our host to this fascinating tract and listened spellbound as he explained the reasons for and the significance of the experimental plots. It was this experience which kindled in me a profound appreciation of what I later came to recognize as the scientific approach to the solution of a problem. The very fact that one could raise a question and obtain a more or less definitive answer to it as the result of an experiment was a revelation that deeply impressed me, particularly because I had grown up in a milieu where any kind of question was invariably answered by the stereotyped reply: "Because somebody (usually a member of the family) said so," or simply, "Just because." Though this sort of reasoning had often seemed quite unsatisfactory to me, I had not previously been aware that a totally different kind of answer could be made possible by conducting a carefully devised test.

My interest in the "experiment station" and its broad significance must have been sufficiently apparent to my new mentor, for he pleaded with my mother to provide me with a college education. This would necessitate my transfer to a different high school where academic subjects were emphasized, and it would set me back a whole year. As usually, mother could not make up her mind without further consultations at home, and the trouble was that no one in the family knew exactly where a college education might lead. But one of the uncles was acquainted with somebody who had prospered financially as an industrial chemist, and hence it was finally decided that I was to study chemistry at the Technological University in Delft.

In the new high school I had the privilege of being exposed to chemistry by Dr. H. van Erp. He was a splendid teacher, though rather dreaded by his pupils because he insisted that, during the daily quizzes, we should "express ourselves accurately." Many were the occasions when the poor student, groping for an adequate answer in front of the blackboard, was told to go back to his seat and that he had failed again. The customary remonstrance: "But, Sir, I meant . . ." was cut short by the rejoinder: "Oh, I know what you mean, all right; but you must learn to say what you mean!" Apart from thus teaching us the proper use of language, van Erp, who was a member of the Deutsche Chemische Gesellschaft and at least once a year contributed an original paper to its "Berichte," also presented the basic principles of chemistry in so forceful and exciting a manner that I became more and more absorbed in the subject and soon began to assemble various pieces of equipment and an assortment of chemicals so that I could have a laboratory of my own —in my bedroom at home. The experiments I performed there were mostly repetitions of those done at school; but it was a new and thrilling experience when I managed to identify the ingredients of a trade-marked chemical fertilizer used in our vegetable garden.

During my final year in high school, van Erp exposed us to the principles of organic chemistry, and it was at this time that I began to toy with an intriguing prospect. We had learned that the composition of sugar is exactly expressed by the term carbohydrate: a combination of carbon and water; and I clearly remember the day when, standing in front of the coal bin, it occurred to me that it should therefore be possible to synthesize sugar from these cheap raw materials.

In the Army

In 1916 I graduated from high school with a record that let me enroll as a student in the chemistry division of the Technological University without having to take an entrance examination. On account of the lower expenses it had been determined that I was to commute; the one-hour train trip twice a day could be used for study. As was expected of me, I worked diligently and did particularly well in analytical chemistry. The laboratory course in qualitative analysis was completed in three instead of the regular ten months, mainly because I could draw upon the very thorough training in van Erp's courses and the additional experience gained in my home laboratory. So it looked very promising; however, after the Christmas vacation I did not return to Delft. On the third of January, 1917, I was inducted into the Dutch army in which I served till the end of December, 1918.

It was a shattering experience. In the protective and strongly Calvinistic environment of the large family, my mental development had barely reached that of a ten-year old. I had no personal opinions on anything and was utterly unaware of the many problems to which man is exposed and with which he must learn to cope. Though I could speak, read, and write tolerably well in four languages—the result of compulsory training in French (eight years), German (five years), and English (four years)—and had "learned" many other subjects, the knowledge acquired did not extend beyond facts, without any apparent interconnections or comprehension of their potential meaning. Now, for the first time on my own and away from the family, I should have been totally lost had it not been for a former high school classmate, Jacques de Kadt, who had been inducted at the same time and place and took me under his wing. He was an "intellectual" who had read much with considerable comprehension and developed personal opinions on many matters that had never even faintly suggested themselves as problems to my mind. He opened completely new vistas to me for which I shall always be grateful. And I am happy that during a visit to Holland in 1964 I could express my gratitude to him in person.

After the first few days in the extremely primitive military camp outside the city limits of Amersfoort, he proposed that we rent a room in the city where we could spend our evenings comfortably. We were joined by a friend of Jacques', the now famous Dutch sculptor, Mari Andriessen. We found a room and spent our evenings and weekends talking about many things; and while listening to their discussions I began vaguely to perceive that the attitudes and views to which I had hitherto been exposed at home and in the family circle were woefully limited. For the first time I became aware of the existence of art, literature, a social system; and I found to my amazement that, even at the age of 19, one could have opinions on such matters quite different from those I had passively imbibed.

Our room also seemed to make it possible for me to continue at least some of the studies begun three months earlier in Delft. However, the time between the end of the evening meal, about 7 p.m., the walk into town and back, and the curfew at 10 p.m. was too short to make much headway. On the advice of my roommates I wrote to the Professor of Analytical Chemistry, H. ter Meulen, asking him for a letter in support of my application to the Company Commanding Officer for a permit to stay out later. After examining my application and Professor ter Meulen's supporting letter, the C.O. grinned and gleefully remarked, "Denied"; he evidently had no use for soldiers who intended to become "intellectuals." This experience may have been one of the causes of my general dislike of the military mind.

The three of us did not stay together for long. As high school graduates, de Kadt and I were automatically sent to an officer's training school in the northern part of the country where, again, we rented a room to spend our evenings and weekends. Here, Jacques made what seemed to me a terrible mistake. As a form of homework we had been assigned to write an essay on military discipline. While I dutifully regurgitated the pronouncements of the officer-teacher, my friend wrote a sharply critical appraisal, quite at variance with the instructor's views, and supported it by quotations from the writings of many ancient and modern philosophers! I feared that he would be severely punished, but to my surprise he was not. Instead, he was transferred to the postal service. After having expressed his regrets that he would have to leave me, he prepared a list of books which he advised me to read and study and I remember that it included works by Zola, Anatole France, Ibsen, Strindberg, Shaw, and Nietzsche. At first they disturbed me deeply because they treated with human situations in a manner that cast renewed doubts upon the validity of the Calvinist tradition. Nevertheless, they had the effect of making me recognize that much of what I had thus far taken for granted was debatable and rested on a shaky foundation. And this initiated the rebellious phase of my life which lasted for several years.

Promotion from private to corporal and sergeant came in due time. But my growing preoccupation with the problems of society, with morals and ethics, had more and more disillusioned me with the military establishment. Hence, I deliberately failed the examinations prerequisite for promotion to lieutenant, and I was returned to regular army duty. Some months later I was, however, sent to another officer's training school where I made new friends who fanned my rebellion and added greatly to my education. This is true in particular of H. S. Keuls, who has become known as the "nestor of Dutch poetry." In December, 1918, the prospective officers were given the choice of finishing their training or returning to civilian life; it need hardly be mentioned that I chose the latter.

College

Now, after an interruption of exactly two years, I was ready to resume my studies in Delft. However, the situation had changed considerably. Two months earlier I had come of age and was thus entitled to my inheritance. This made me financially independent. Meanwhile, my mental outlook had undergone radical changes and the reading done during the intervening years had impressed upon me the great significance of literature so that I had seriously begun to consider abandoning the study of chemistry and becoming a writer instead. This became even more appealing when I learned that the requirements for a Chem. E. degree included many subjects that did not interest me in the least, so that I reasoned that the time spent on them would be rather wasted and could much more profitably and excitingly be devoted to reading and writing about the new ideas that had begun to germinate.

It was impossible to discuss my problems rationally with any member of the family, by whom I was now treated as the proverbial black sheep, except for an aunt by marriage who, on account of her unconventional attitudes and behavior, had also fallen into disgrace and in whose judgement I had great confidence. She strongly advised me to continue my chemical studies even though this implied that I should have to spend much time on subjects I did not find important. She pointed out that the reasons I had advanced for my inclination to change might not be altogether honest and could in part have been the result of a desire to shirk unattractive tasks.

Did I realize that I would inevitably have to face the same kind of difficulties in any other occupation? How could I know whether I would be successful as a writer? Would it not be better first to show myself that, when faced with obstacles, I had the ability and will power to overcome them? And this could be determined by finishing my studies for the Chem. E. degree. If thereafter I still wanted to devote my life to writing, I could easily do so, with the added advantage that I would not be financially dependent on literary productions because I could earn a living as a chemist. The arguments impressed me and I promised to return to Delft where I soon found pleasant living quarters.

But the mental turmoil did not make it easy to take up the thread that had been temporarily dropped two years before. Instead of regularly attending lectures, working in laboratories, and studying at home, I spent much of the time delving deeply into the great products of French, English, Scandinavian, and Russian literature of the 19th century. The inevitable result was that in June, 1919, I was totally unprepared to pass the examinations customarily taken by the chemistry students at the end of their first academic year. This did not pose a serious problem; I simply did not sign up for examinations at that time. By postponing them for a year I might not even lose anything because it might be possible during that time to master not only the first-year subjects but also those covered in the second year.[1] It worked and, in spite of the fact that I did not settle down to really intensive study until November, 1919, I passed all of these exams in June, 1920.

Back in the swing and fortunate enough to be able to learn easily and rapidly, I now found time to take several courses that promised to be particularly interesting to me in addition to those required by the chemistry division for graduation. The most appealing were van Iterson's courses in plant anatomy, the microchemistry of plant products, the chemistry and technology of wood and of plant and animal fibers, and genetics; and Beijerinck's courses in microbiology which, owing to his impending retirement at the end of the 1920–21 academic year, were the last ones he taught. These elective courses had once more exposed me to biology, and in a manner far more exciting than the high school courses had done.

In November, 1921, I had fulfilled all of the requirements for graduation as a Chem. E., with the exception of the final year's work in a specialized area selected by the candidate himself. At this point I was strongly inclined to choose either van Iterson's Laboratory of Technical Botany or the Laboratory of Microbiology, both of which were incorporated in the chemistry division for this purpose. The following passage from La Rivière's essay, "Kluyver as seen by his students," admirably expresses my reasons:

[1] Much can be said in favor of the system which permits university students to take their examinations when they consider themselves well enough prepared to pass them. It is contrary to the practice in United States colleges, where the examinations must be taken immediately at the end of a course, with the possible exception of examinations in foreign languages required for an advanced degree.

The students at the Technological University in Delft who have qualified as candidates for the Chem. E. degree and elect to specialize during their final year in microbiology have had a rigorous training in all branches of chemistry, in physics, and in mathematics; but they have no more than a faint inkling of the properties of animate matter. There can be no doubt that their choice has usually been influenced by a more or less conscious disillusionment with the inflexible behavior of molecules on the one hand, and on the other by the intriguing opportunity to study a new subject that will expose them to the alluring mysteries of life itself [(1), p. 49].

Now, Beijerinck's courses had certainly whetted my appetite for microbiology and mainly, I believe, because most of what he had taught had been "way over my head." But the famous man had retired and, though A. J. Kluyver had been appointed as his successor and was already in residence, he had not yet begun to teach. Moreover, none of my friends knew anything about the new professor except that he had been in the then Netherlands' East Indies for several years and that he was quite young. His regular duties were to start after the delivery of his inaugural address, a formal and public lecture scheduled for 18 January, 1922. Ordinarily, such an inaugural lecture contained a general outline of the program the new appointee intended to develop, and this, I reasoned, might provide me with a good opportunity to "size up" the professor, and guide me in reaching a final decision as to the subject in which I wanted to specialize. Considering my still very undeveloped mind and lack of knowledge of the discipline, it was a rather inexcusable and, surely, a cocky attitude. However that may be, I attended the lecture and was very favorably impressed with what I saw and heard. This pertained less to the program which I was not in a position to evaluate than to the manner in which it was presented, and some remarks that seemed to my rebellious spirit unconventional enough to be promising tipped the balance. The next morning I sought an audience with Professor Kluyver, expressed my desire to specialize in microbiology, and began working in his laboratory then and there.

After having acquired some proficiency in the more essential techniques, a special problem had to be chosen for intensive study. Kluyver suggested an experimental investigation of a problem that had been posed by Max Rubner's treatise (2) on the physiology of yeast. The purpose was to determine the longevity of a yeast cell under conditions that did not permit multiplication but are otherwise adequate to sustain normal metabolic activities. It was anticipated that a medium with a reasonably high concentration of sugar and a low concentration of an appropriate nitrogen source would answer the requirements for this study. The occurrence or nonoccurrence of growth was to be ascertained by cell counts and nitrogen determinations. When I was ready to embark on this project the summer vacation had begun, and this I spent in my home laboratory where I practised the micro-Kjeldahl technique and learned to incinerate yeast so that the residual ash could be used to provide a complete supply of minerals. On the side, I made a successful attempt at isolating *Bacillus calfactor*, the thermophilic bacterium which Miehe had implicated as the causal agent of the self-heating of moist hay, and when a pure culture had been obtained I tried but failed to render its flagella visible by some of the recommended staining procedures.

In September, back in Delft, I began the studies on the yeast problem in earnest; they occupied me for the next nine months. But the experimental results were rather discouraging for, in all media even in those with the lowest concentration of nitrogen compounds, the yeast cells multiplied. During that time I also continued my attempts to stain flagella, now under conditions much more favorable than in my home laboratory where the ingredients needed for a particular staining method had not always been available. Nevertheless, though I used every procedure described in the literature, the attempts remained fruitless for several months. When finally I succeeded in preparing a satisfactory slide, I found that I could reproduce this result with any and all of the different methods. However, I am still at a loss to explain the earlier failures. These studies resulted in my first publication (3). At the end of the academic year I received my Chem. E. degree and spent a few days celebrating the occasion with my fiancée and family.

THE BREAK OF DAWN

Assistant

In the course of the three and one half years preceding this event, I had spent my inheritance, much of it on books. During my last year as a student I should, of course, have considered the need to secure a future position. But I had not made any move in this direction. The work in Kluyver's laboratory had been utterly absorbing and I had temporarily abandoned the notion of becoming a writer, hoping that somehow it might be possible to devote my life to the study of microbes. At the time of my graduation I had only 80 florins left; yet, instead of belatedly starting to look for a job, I decided that the small balance would permit me to spend a few more weeks in the laboratory. Thereafter, well . . . .

It was during the second week that Kluyver called me in for a brief conference. He told me that one of his two assistants, both inherited from Beijerinck, was going to leave and asked whether I would be willing to fill the vacancy. Willing? What more could I have wished for? When would I be expected to start? Tomorrow? Wonderful!

It was hard to believe. More than 90 per cent of the fresh crop of Chem. E.'s who had made serious efforts throughout the year were still desperately trying to find a position, any position. And without the slightest exertion on my part I was offered the most attractive one I could imagine.

Tomorrow came, and the new assistant was told of the duties he was expected to perform: taking charge of the extensive pure culture collection, helping to look after the still very small number of students, and preparing the demonstration material for Kluyver's two lecture courses, both running through the full academic year. One of them, the general microbiology course, was repeated each year; the other was intended to cover various aspects of applied microbiology. This was too large and diversified a topic to be adequately treated in a single year, hence, the course was spread out over a three-year period. As the special subject for the 1923–24 course, Kluyver had selected the microbiology of water and sewage. He now mentioned that two groups of microbes, the iron and the sulfur bacteria, were supposed to play an important role in the economy of bodies of water, but that both he and his other assistant were totally unfamiliar with these organisms. From reports in the literature it appeared, however, that they were hard to handle and grew very slowly, so that it could be anticipated that the preparation of appropriate demonstration material would prove to be difficult and time-consuming. Kluyver therefore suggested that I should begin right away to acquaint myself with these creatures so that at the appointed time I would be in a position to supply the desired demonstrations.

It was an exciting challenge. For the first few weeks I delved into the available literature, familiarizing myself in particular with the different methods that had been used for the cultivation of iron and sulfur bacteria; thereafter I spent most of my time setting up numerous experiments from which I learned much about the organisms. And when Kluyver had arrived at the point where his lectures on these bacteria fitted in, I was able to supplement them with a respectable display of various cultures of representative types. This was, of course gratifying. But during my exploratory work I had also become aware of some fundamental problems associated with the behavior of these microbes, for which a number of conflicting explanations had been advanced, notably by Winogradsky, Engelmann, and Molisch. This "made it tempting," as the Master would say, to study them in greater detail in the hope that additional information might aid in straightening out the confused situation.

THE SULFUR BACTERIA

It is a moot question whether the earlier envisaged possibility of manufacturing sugar from coal and water had any influence on the further developments of this phase of my work. The simple fact is that I had become enamored with the aesthetically attractive purple sulfur bacteria. An added impetus to investigate them more closely was provided by the publication of Bavendamm's monograph (4) in 1924, in which he proposed a presumably novel interpretation of the metabolism of these organisms which, on closer examination, proved to be essentially identical with the one that had previously been advanced by Buder (5). Despite the fact that this interpretation aimed at reconciling the contradictory views of the earlier students of the purple bacteria, it nevertheless seemed to leave some crucial questions unanswered.

In view of the arduous assistant's duties, it was impossible to spend much time on this project, and progress was accordingly slow. However, I soon found that purple sulfur bacteria could be grown in bottles completely filled with a mineral medium supplemented with sulfide and bicarbonate and exposed to light. This method of cultivation proved far more convenient and permitted a more rigorous control over environmental conditions than the method used by Bavendamm. To be sure, growth was far from rapid; four to six weeks were required for the population in such bottle cultures to reach its maximum. However, this was not altogether a disadvantage, for it meant that my cultures could be perpetuated by subculturing them at six-week intervals. In this manner I eventually succeeded in isolating pure cultures of a small *Chromatium* species and of *Thiosarcina rosea*, the latter from sewage and composed of beautifully regular, motile cell packets. The pure cultures were subsequently used for experiments in which the influence of different sulfide concentrations on growth was tested. The results, though not of a dependably quantitative nature, clearly indicated that the final cell yield was proportional to the initial sulfide concentration of the medium, and vastly in excess of that obtainable with colorless sulfur bacteria in media of comparable composition.

This simple fact suggested an interpretation of their metabolism that seemed to eliminate the unsatisfactory aspects of the Buder-Bavendamm hypothesis. It was a logical extension into the realm of photosynthesis of the general concept, then being developed by Kluyver and his co-workers, that fermentative as well as oxidative metabolic processes can be considered as composites of more or less elaborate series of consecutive and chemically intelligible step reactions, each one of which represented an inter- or intra-molecular transfer of hydrogen atoms from a donor to an acceptor molecule or site of a molecule. The results I had obtained by 1926 had shown that the purple sulfur bacteria, or Thiorhodaceae, can grow in strictly mineral media but only when exposed to light. This meant that they had to be considered as photosynthetic organisms. On the other hand, the requirement for H_2S and their failure to produce O_2 could now be interpreted to mean that they use H_2S as the specific H-donor for the reduction, or assimilation, of CO_2. The gist of this idea was incorporated by Kluyver & Donker in their epoch-making treatise on "The unity in biochemistry" (6).

Interspersed with the experiments with the purple sulfur bacteria were some studies originating from chance observations of a few other organisms. The careful examination of a yeast culture sent by a Japanese scientist to the pure culture collection of the Delft laboratory had shown it to be contaminated with another yeast which exhibited a curious property. When inoculated on an agar plate and incubated in an upside-down position, it produced mirror images of the colonies in the lid of the Petri dish. A study of this phenomenon showed that the yeast formed spore-like bodies, now designated as "ballistospores," and that these were violently discharged from the mother cell by a mechanism which Kluyver, whose knowledge of the microbiological literature was phenomenal, immediately recognized as similar to that causing the dispersal of the basidiospores of higher fungi. A few other yeasts with the same property were soon found, and it was suggested that this group, collected in a new genus, *Sporobolomyces*, should be considered as primitive or reduced members of the Basidiomycetes (7,8). Another study dealt with the striking morphological features of a fortuitously isolated spore-forming bacterium which, in liquid cultures, grew as tightly coiled, many-stranded ropes. As far as I know, this peculiar organism, described as *Bacillus funicularius* (9), has never been reported since.

However, these were clearly side issues. My main concern remained the study of the purple bacteria. In 1926, the earlier observations were amplified when a number of pure cultures of nonsulfur purple bacteria, or Athiorhoda-ceae, had been isolated and it was found that they could develop in one and the same medium either anaerobically, but only if illuminated, or aero-bically in complete darkness, so that for these organisms light and oxygen appeared to be equivalent. This result had obvious implications for an inter-pretation of photosynthesis, and I fully expected that a promising basis had been prepared for what eventually could be expanded into a Doctor's dis-sertation on the purple bacteria. But this was not to be.

THE PROPIONIC ACID BACTERIA

The Chemistry Division of the Technological University in Delft ad-ministered a fund whose income was used to award annually a "Gijsberti Hodenpijl Prize" to the person who had submitted the best essay on a sub-ject proposed by the faculty. In 1924, the topic had been selected by Kluy-ver, and the announcement read in part:

In the J. Ind. Eng. Chem., *15*, 729, 1923, Whittier and Sherman comment on the significance which propionic acid might attain for the synthesis of various tech-nically important products if it could be manufactured on an industrial scale at low cost.

Because the studies so far published on the propionic acid fermentation, particu-larly on its bacteriological aspects, still leave important problems to be resolved, a more fundamental investigation would be of great significance.

The notice aroused some interest in Kluyver's institute. However, the deadline for submission of the essays was only a year away, and none but Kluyver's first assistant, H. J. L. Donker, was sufficiently far advanced and had enough time to consider seriously competing for the prize. Donker ac-tually spent a few months trying to isolate cultures of propionic acid bacteria but dropped this work when the attempts proved unsuccessful. Shortly after-wards, a chemist with some microbiological training came to the laboratory to try his hand at the problem; he, too, gave up after he failed to isolate cultures during the first few months. At this point there was clearly not enough time left to conduct an adequate study within the prescribed limit. Nevertheless, Kluyver was anxious to initiate a comprehensive study of the propionic acid bacteria, and assigned this subject consecutively to an unem-ployed Chem. E. who wanted to broaden his experience in microbiology, and to a new graduate student. The former secured a position before he had accomplished more than preparing some elective cultures; the latter, who could start off with these cultures, soon decided that he was more interested in music than in science and left to become a professional musician. These four successive failures made Kluyver remark that there seemed to be a curse on the propionic acid bacteria, and the subject was abandoned.

One Sunday morning I noticed some reddish spots in a piece of cheese on our breakfast table. My supervisory association with the previous abortive attempts to isolate propionic acid bacteria had served to familiarize me with the literature pertaining to this group, and thus I knew that such spots were sometimes associated with propionic acid bacteria. So I went to the lab, prepared a batch of yeast autolysate-lactate medium which Orla-Jensen had used in his early studies on the organisms, and inoculated a series of bottles filled with the medium with some of the reddish spots. A week later these cultures looked promising; all of the bottles showed copious growth and contained propionic acid. Streak plates were prepared and the predominant colony types that appeared on the plates were subsequently replated for pure culture isolation. Rather to my surprise, none of the isolates grew in bottles with the yeast autolysate-lactate medium. Of course, the result was disappointing, but it did not much disturb me. I was not particularly interested in the propionic acid bacteria, and my efforts to isolate them had been little more than a brief excursion into an area in which I had not previously worked myself.

About a month later another piece of cheese with red spots turned up. The earlier experiment was repeated, but this time I exercised particular care to isolate *all* of the different, rather than just the predominant, colony types that developed on the plate cultures. In spite of this the final result was the same as before: none of the isolates grew in bottles with yeast extract-lactate. At this point I did become disturbed. Since 1907, when von Freudenreich and Orla-Jensen had first isolated and characterized propionic acid bacteria, at least four other microbiologists had succeeded in doing so, and this now raised the question whether I should not be able to accomplish it if I really put my mind to it. The time to test this was propitious for the summer vacation had just begun and, apart from two technicians who washed dishes, sterilized glassware, and prepared media, I was the only one working in the laboratory. With their help I devoted myself exclusively to the problem and when Kluyver returned from his vacation I could show him a tray with stab cultures of eight or ten strains of propionic acid bacteria. The Master was delighted and announced that here was the beginning of my Doctor's thesis. My reaction was different; the experiments had been intended to convince myself that I, too, could isolate these somewhat elusive creatures, and this having been accomplished, I now was ready to continue the experiments with the purple bacteria. Kluyver lectured me for more than an hour and the gist of his argument was that, in view of the slow growth of these organisms, it might take me ten years or more before that project could be brought to a reasonable conclusion, whereas, with the propionic acid bacteria cultures already available and many others on the way, it should be possible to complete a comprehensive study of this group in much less time. Afterward I could resume the study of the purple bacteria. The argument did not really convince me for I was not in a hurry to get a Doctor's degree, especially since I had meanwhile been promoted to "Conservator" of the institute, a position which was secure, and which had permitted me to marry after an engagement that had stretched out over six and a half years. Nevertheless, in the end I let myself be persuaded and spent the next two years studying the propionic acid bacteria and their fermentation. The result was a monograph published in 1928 (10). It was, I believe, a sound piece of work, though not truly a "labor of love." Even so, one of the consequences was rather exciting; this was the identification of the substance largely responsible for the characteristic flavor of high-grade butter. The starting point was the observation that some strains of propionic acid bacteria, grown anaerobically on yeast extract-glucose-calcium carbonate-agar plates, emitted an odor strongly reminiscent of high-quality butter. This feature proved to be correlated with the ability of these strains to produce acetylmethyl carbinol, which led to the identification of diacetyl as the major component of the typical butter flavor (11). When Kluyver informed the president of the "Netherlands' Yeast and Alcohol Manufacturing Co." of this development, I was offered a position as research microbiologist at this large industrial establishment under what seemed at first sight to be quite attractive conditions. But when I was informed that I would not be allowed to discuss my work with others, I declined the offer.

SUNRISE

A Visitor from California

During the latter part of 1927, Professor L. G. M. Baas Becking, on sabbatical leave of absence from Stanford University, visited the Microbiological Laboratory in Delft and delivered a lecture on the sulfur bacteria based on studies he had recently published (12). While working on the propionic acid bacteria I had also maintained my cultures of the purple bacteria, and these provided the basis for a lengthy and animated discussion with Professor Becking. Some months later he again visited the laboratory, on a Saturday morning when Kluyver was attending a meeting of the Royal Netherlands' Academy of Sciences in Amsterdam. On this occasion he talked a great deal about a new addition to Stanford University's Hopkins Marine Station, then abuilding, which was to be devoted to experimental investigations of problems in the field of marine biology, and we discussed many microbiological and biochemical matters of mutual interest. The discussions were begun in the laboratory and continued during and after lunch in our home. On a beautiful sunny morning in the early spring of 1928, when Kluyver and I were preparing a list of purchases for the coming academic year, Becking paid his third and last visit to the laboratory. After the preliminary greetings had been exchanged, he abruptly asked me whether I would accept a position at the Hopkins Marine Station if Stanford University were to offer me one. The question was so totally unexpected that I was stunned for awhile, and finally I asked when he would like to have my answer. He seemed surprised; couldn't I make up my mind right now? This was impossible for I needed time to think about the proposition and to discuss it with Kluyver and my wife. Becking urged me to decide soon and let him know within a week, the last one he would be in Holland prior to his return to California. He quickly jotted down his address and left.

Evidently Kluyver had not been taken by surprise. While I was wistfully looking out of the window at an almond tree that had just come into bloom, thinking also of the intense happiness I had experienced during the more than six years in his institute, Kluyver admitted that he had anticipated this development, though he realized that to me it was quite unforeseen. He further expressed his opinion that for my progress it might be advantageous to be on my own, and finally assured me that, in case I should not like the situation in California, there would always be a place for me in the Delft laboratory. When I came home for lunch and related the morning's events, my wife also claimed that, on the occasion of Becking's visit during which he had so eloquently presented the potentialities for work at the Jacques Loeb Laboratory, the name of the new addition to the Hopkins Marine Station, she had strongly suspected that he thereby hoped to arouse my interest in the place sufficiently to make me willing to join its staff. But she, too, knew that I had been blissfully unaware of this possibility.

During our long engagement and the first two years of our marriage we had agreed that there were two parts of the world in which we did not care to live: the colonies in the East Indies and the United States. We detested the preoccupation with material possessions that seemed to be so prevalent in those countries where wealth and comfort appeared to be more important than an intellectually satisfying life. But Becking's personality had impressed us very differently, and his prognosis of the future of the Hopkins Marine Station had been so idealistic that it did not take us very long to discard our previous misgivings. Accordingly, I wrote the letter to Becking in which, at Kluyver's insistence, I emphasized that I would feel free to work on whatever seemed important to me and that I intended first of all to continue the study of the purple bacteria.

For several months there was no response; but in September I received a cablegram: "Appointment secured. Letter follows." The letter contained the information that I was expected to start my work at the Hopkins Marine Station in January, 1929. We booked passage on the MS "Axel Johnson" which was scheduled to leave from Antwerp early in November. During the previous five years there had never been time for more than a few days' vacation and the sea voyage promised to provide ample opportunity to take a much needed rest, while at the same time eliminating the troubling prospect of having to cross the American continent with a 15-months old daughter in the close confinement of a railroad carriage.

The voyage took six weeks. After a severe storm during the first week out, the skies cleared and the weather turned beautiful, so that we could thoroughly enjoy the magnificent scenery around the Panama Canal and the West coast of Central America. And in the last few days I experienced two sights that have left a profound and lasting impression.

As a student I had developed the habit of studying till late in the night, a regime to which I have adhered until a few years ago. Because I slept very soundly it was always necessary to have someone wake me up in time to begin the next day's work at 9 a.m. I was therefore greatly surprised to find myself wide awake, without any obvious cause, on the morning of the 24th of December while it was still totally dark outside. On coming out on deck the Eastern sky had just begun to turn light; after awhile an impressive mountain range became visible and I witnessed the sunrise over the San Diego mountains—a glorious sight! The ship stayed in the San Diego harbor for two days and a wire was sent to Becking, announcing our scheduled arrival in San Francisco early on December 28th. On the morning of that day I again awoke by myself before dawn and watched another sunrise, this time right over the Golden Gate. In view of later events, I have sometimes been inclined to look upon these extraordinary occurrences as favorable omens.

The Hopkins Marine Station

We were met at the pier by Becking, traveled by train to Palo Alto for a brief visit with Ray Lyman Wilbur, then President of Stanford University, and Carl Lucas Alsberg, head of the Food Research Institute. Later that day we continued our trip by train to Monterey and by car to Carmel, where we spent the first few days as guests of the Beckings, who had already rented a house for us in their neighborhood.

Our first impressions of the new environment were more than favorable. At the end of December we could walk in brilliant sunshine without an overcoat; during an hour's walk on our first Sunday we found no fewer than 20 different plants in bloom; the setting of the small village, then largely an artist's colony, was exquisite. And the laboratory, built on a small promontory at the southern tip of Monterey Bay, was equally enchanting. Here, I was provided with a large room overlooking part of the bay and it seemed difficult to imagine that in this place it would be possible to concentrate my attention on science. But after six weeks of relative inactivity it was nonetheless exciting to get back to work. During the first few days a large number of experiments of different sorts were initiated and, although I have never been unaware of the scenic beauty of the surroundings and ever been grateful for the opportunity to carry on my work in this lovely spot, it did not interfere with my studies which soon occupied all of my time.

Soon after my arrival, Dr. Walter K. Fisher, Director of the Hopkins Marine Station, inquired about the research program I intended to develop. When I had answered his question in rather general terms, he asked whether I had any plans to emphasize marine microbiology. I explained that my interests covered a broad spectrum, comprising the morphology, taxonomy, ecology, physiology, and biochemistry of bacteria, fungi, algae, and protozoa. And, because the fundamental problems pertained to fresh-water and soil organisms as much as to marine forms, it seemed to me unwise to restrict my attention to the latter, whose only distinguishing feature was their occurrence in salt water. Thus, I intended to choose problems for special investigation without regard to the habitat of the organisms that could be used advantageously for my work. The response was gratifyingly simple and direct: "You go right ahead, Doctor." On this basis I have conducted my "program" at the Hopkins Marine Station.

When I began my association with the new Jacques Loeb Laboratory its staff was composed of a group of specialists: a general physiologist, a marine ecologist, a biophysicist, an organic chemist, a physical chemist, an analytical chemist, a geologist, and myself, a general microbiologist; there was only one graduate student who was almost immediately assigned to me. Becking had been responsible for assembling the greater part of this group and expected that all of us would collaborate on a comprehensive study of the geophysico-chemical and biological features of the ocean and various kinds of salt lakes and their interrelations. It sounded like a grand program, but the collaboration did not work out well. From the beginning each of the specialists got involved with problems of his own, and this left little time for cooperative ventures. My primary concern was a continuation of the studies on the purple bacteria and I hoped to benefit from the experience of the organic chemist when trying to characterize their pigments, and from that of the biophysicist in designing a simple apparatus for determining action spectra of the photosynthetic and other manifestations of these organisms. But the former displayed no interest in bacterial pigments and was totally absorbed in his attempts to synthesize C_3N_4, the equivalent of CO_2 in the "nitrogen system" of organic chemistry, while the latter, who was studying the shifts in absorption spectra of marine algae induced by various treatments, advanced so many arguments to show that the sort of experiments I had in mind were too primitive to yield conclusive results that I soon gave up consulting him. He seemed to be unaware of the great significance of G. N. Lewis' trenchant statement that "the strength of science lies in its naiveté" (13). Hence, I realized that the logical path for me to follow was to plan my work so that it would not have to depend on the cooperation of others. This experience also taught me that the attack on a problem which requires the joint efforts of diverse specialists is likely to be successful only if it develops through the gradual accretion of a group whose members have already evinced a desire to work on specific aspects of that problem. For this reason I have consistently tried to leave my co-workers free in the choice of subject rather than urging them to apply their special knowledge to further the researches of others. In a congenial group this often happens voluntarily, and then true collaboration will be achieved.

During my first few months at the Hopkins Marine Station, studies on several different topics were started. Soon, however, the photosynthetic bacteria began to require undivided attention, expecially after it was found that their growth was greatly accelerated by continuous illumination. Under such conditions, fully grown cultures developed in as many days as they did in weeks when exposed to daylight. The isolation of pure cultures and experiments designed to determine the best conditions for their growth proceeded apace, and I had also begun to learn much about the cultivation of green sulfur bacteria, so frequently encountered in association with the purple bacteria. All of these experiments could be performed without the help of others.

YELLOWSTONE PARK

Although the staff members at the Hopkins Marine Station were expected to teach courses during the spring and summer quarters, I was relieved of this duty during my first year in Pacific Grove. I do not remember the reason for this dispensation. It allowed me, however, to spend a few weeks of the summer of 1929 in Yellowstone Park for the purpose of making a preliminary study of the microorganisms found in hot springs. The invitation had been extended to Becking, but he urged me to go in his place. It was an interesting experience for, under the expert guidance of Dr. Allen of the Carnegie Institution of Washington, I was confronted with a wide diversity of environments in which many kinds of microbes were at work. Particularly instructive was the first day, and it taught me another lesson.

We had just arrived at the Mammoth Hot Springs area where the massive travertine deposits with hot, H_2S-containing spring water running down the slopes exhibited a striking color display. Near the top were streaks of reddish mucoid material; below this was a zone of ochre-colored matter, and lower down, as well as in the rivulets, an abundance of blue-green patches could be observed. After having contemplated the situation for awhile, I explained it to Dr. Allen as follows.

The reddish masses obviously consisted of purple sulfur bacteria, growing at the expense of light and the H_2S in the water. The ochre bands were just as easily accounted for; they clearly represented mass developments of iron bacteria, organisms which presumably use the oxidation of ferrous ions as energy source. They can usually be found in places where seepage water, after passage through an iron-containing and reducing substratum, emerges and becomes exposed to air. In the hot spring effluent, the reduction of iron compounds had, of course, been caused by the H_2S; this having been removed by the activity of the sulfur bacteria at the higher level, both the composition and the lower temperature of the effluent below that zone would favor the development of iron bacteria. Finally, after the iron had also been converted to its highest oxidation level, the effluent was suitable only for the growth of algae. The Mammoth Hot Springs thus provided a conspicuous illustration of naturally occurring ecological niches for three groups of microbes.

The "explanation" seemed to surprise Dr. Allen who told me that biologists previously had reported only blue-green algae in and on the travertine deposits; purple and iron bacteria had never been mentioned. I suspect that I dismissed this information by contending that those investigators were probably not familiar with such bacteria and that the picture was immediately apparent to a general microbiologist.

After dinner I went out to collect samples from the variously colored areas which I later examined in my cabin under the microscope. Right from the start, the results made me uneasy for there was no evidence for the presence of either purple or iron bacteria in any of the samples. In spite of their different colorations, the only organisms that could be detected seemed to be blue-green algae. Therefore I was forced to admit that I had jumped to conclusions, and the next morning I apologized to Dr. Allen, retracting my previous "explanation."

In the course of the following days I did encounter some mass developments of purple sulfur bacteria in the area, though only in narrowly confined localities. A fairly large *Chromatium* was found swimming in a tiny pool on top of the Mammoth Falls where the thermometer registered a temperature of about 90° C. During my stay in Yellowstone Park there was no opportunity to conduct experiments but I collected samples of spring water which were shipped to the Hopkins Marine Station. Upon my return they served to provide a tentative explanation for the almost exclusive presence of blue-green algae in and around the Mammoth Falls. Aliquots of the samples were distributed over a number of culture flasks, inoculated with local soil, and exposed to light. Only blue-green algae developed in these cultures. But, in corresponding flasks in which traces of nitrate or ammonium salts had been added to the water, green algae and diatoms appeared as the predominant organisms. Thus, it could be concluded that the virtual absence of nitrogen compounds rendered the hot spring water unfit for any but blue-green algae, whose ability to use atmospheric N_2 as their sole source of nitrogen had been announced by Beijerinck as early as 1901 (14). At that time, the occasional occurrence of purple sulfur bacteria in the hot springs should have alerted me to the possibility that they, too, might be able to fix N_2. However, this did not occur to me, and not until 20 years later did Kamen & Gest (15) establish this property.

BACTERIAL PHOTOSYNTHESIS

Following the Yellowstone Park interlude, the work with the purple and green sulfur bacteria was continued, and at the end of 1929 the results obtained over a period of more than six years were presented before the Western Society of Naturalists which held its winter meeting in Pacific Grove (16). The recent experiments had corroborated and extended the studies begun in the Delft laboratory by rigorous quantitative determinations of cell yields as a function of the amount of available H_2S and of the oxidation products formed; they had also shown that the green sulfur bacteria follow the same pattern of behavior. The accumulated data supported the view that photosynthesis can be considered as a light-dependent reaction in which different substances, specific for different kinds of photosynthetic organisms, serve as H-donors for the reduction of CO_2. This could be expressed by the generalized equation:

$$2H_2A + CO_2 \xrightarrow{\text{light}} 2A + H_2O + (CH_2O),$$

from which some important consequences could be deduced. First, the equation implies that photosynthesizing plants use H_2O as the exclusive H-donor and hence that the O_2 they produce should be derived entirely from the dehydrogenation of this substrate and not in whole or in part from CO_2. Second, it indicates that the green and purple sulfur bacteria do not use H_2O but H_2S and other reduced sulfur compounds as H-donors, which explains both the fact that their photosynthetic activity is strictly dependent on the presence of such substances and the inability of the organisms to produce O_2. And, third, it could readily account for the properties of the Athiorhodaceae by postulating that they can use only organic substrates as H-donors. It was later found that the Thiorhodaceae can also use organic substrates, an aspect which was studied in some detail by Muller (17).

It now seemed attractive to undertake a comparative study of the various organisms that perform different kinds of photosynthesis in the hope of discovering correlations that could provide a key to a better understanding of the photosynthetic mechanism. And because of the immediately apparent differences in the pigmentation of the higher plants and algae, the green sulfur bacteria, the Thiorhodaceae, and the Athiorhodaceae, a chemical investigation of the pigment systems suggested itself as an obvious starting point. An opportunity to initiate such a study was provided during the spring quarters of 1933 and 1934, when I taught a lecture course at Stanford University, and Dr. H. A. Spoehr, Director of the Laboratory of Plant Biology of the Carnegie Institution of Washington, made the facilities of his laboratory, located on the Stanford campus, available to me. Here I began a study of the carotenoid pigments of *Rhodospirillum rubrum*, one of the Athiorhodaceae, in which I benefited greatly from the experience and cooperation of Dr. J. H. C. Smith, one of Spoehr's associates who had worked in the area of carotenoid chemistry. In 1936, a Rockefeller Fellowship permitted me to acquire some familiarity with chlorophyll chemistry in the research institute of Professor Arthur Stoll in Basle, Switzerland, under the competent guidance of Dr. E. Wiedemann.

But these studies did little to advance our knowledge of the photosynthetic mechanism. Besides, a critical look at the generalized equation revealed that, even though it rested on sound experimental evidence, it seemed to lead to consequences that were manifestly at variance with certain well-established phenomena. In its simple form the equation implied that radiant energy should be needed either for the oxidation of the specific H-donors or for the reduction of CO_2. The first alternative was difficult to reconcile with the existence of a group of colorless sulfur bacteria which can oxidize H_2S

and other reduced S-compounds in complete darkness, as also with the fact that a large variety of organisms can similarly oxidize organic substrates; and the second alternative conflicted with the firmly established ability of the chemoautotrophic microbes, such as the thiobacilli, the hydrogen-oxidizing, and the nitrifying bacteria, to synthesize their cell material from CO_2 as the sole carbon source without the benefit of radiant energy. Thus, it became necessary to search for an interpretation of the photosynthetic mechanism that would circumvent these difficulties. It gradually evolved during the period 1935–49.

During the examination of the numerous cultures of green and purple sulfur bacteria I had often been struck by the wide range of cell shapes and sizes found among the population of even a single pure culture whose microscopic appearance suggested that it was composed of a mixture of different representatives of the group of Thiorhodaceae, some of which resembled, more or less closely various genera and species established in 1888 by Winogradsky (18), on the basis of his observations of slide cultures periodically supplied with H_2S. Winogradsky had not worked with pure cultures, however, and when these had become available, appropriate experiments clearly showed that the morphological characteristics of a strain can vary considerably, particularly in response to differences in the sulfide concentration and pH of the medium. This not only rendered an unambiguous identification of my isolates impossible, it even made me suspect that some of Winogradsky's taxa might actually represent growth forms of a relatively small number of species. Thus, it was clear that a critical evaluation of Winogradsky's classification of the Thiorhodaceae required elaborate comparative studies with a wide diversity of isolates, and I hoped that such studies would provide the basis for a taxonomic treatment of the group in which morphological variability and differential physiological characteristics were taken into account.

That this was not accomplished was mainly because of my failure to isolate any of the morphologically most distinctive purple sulfur bacteria. Even though they were sometimes present as the majority organisms in the inocculum used to start new cultures, the latter invariably yielded only a few types of small Thiorhodaceae which could be distinguished because their growth rates were differentially affected by sulfide concentration and pH of the medium, and thus could be isolated at will by the use of media of specified composition for the primary elective cultures. However, the taxonomic problem had to be shelved until a more representative collection of pure cultures, which should include at least members of most of Winogradsky's genera, was available.

This problem was discussed at some length in my paper on the green and purple sulfur bacteria published in 1931 (19). It was reviewed by E. G. Pringsheim (20), who took issue with my contention that purple sulfur bacteria were pleomorphic and supported his criticism by including a photomicrograph of a sample taken from a natural mass development of Thiorhodaceae in which unmistakably identifiable representatives of several of Winogradsky's genera appear side by side. It may here be remarked that my reservations as to the validity of Winogradsky's classification had never extended so far as to deny the existence of clearly definable taxonomic entities within the group; obviously, I could not have doubted the reality of, e.g., *Thiosarcina*, a pure culture of which I had isolated in Delft, or of *Thiospirillum*, which I had often encountered in material collected from natural sources. My reservations were based on the variability exhibited by pure cultures of the small types, suggesting that other members of the group might likewise respond to differences in environmental factors; and this could be resolved only by studies with pure cultures of such organisms which, unfortunately, were not obtained.

Why had I been unable to isolate any of the more conspicuous purple sulfur bacteria? This question has been answered by the recent studies of Schlegel and Pfennig (21, 22) which have shown that many Thiorhodaceae require vitamin B_{12} for growth. To be sure, I had not altogether overlooked the possibility that representatives of the group might need special growth factors. On many occasions, when material was available in which one or more easily distinguishable types were particularly abundant, I had used media supplemented with yeast autolysate, both for elective and direct isolation cultures. But this had not altered the results in any way: in none of such cultures did the organisms multiply, and only the small ones, which grew equally well in nonsupplemented media, developed. The use of yeast autolysate was predicated on the assumption that it should contain all of the growth factors needed by saprophytic microorganisms; it had often proved superior in this respect to most other sources of growth factors for such organisms and been found eminently suitable for the cultivation of all the strains of Athiorhodaceae in my collection. Nevertheless, in retrospect this seems but a poor excuse for not having tested other supplements, especially extracts of mud with which mass developments of Thiorhodaceae are usually associated in nature. This was the approach followed by Schlegel and Pfennig, leading to the discovery of B_{12} as an essential growth factor. It has now made many representatives of this group, as well as some kinds of green bacteria previously observed only in natural habitats, accessible for further experimentation.

BRIGHT DAY

COMPARATIVE BIOCHEMISTRY OF PHOTOSYNTHESIS

As mentioned earlier, the dilemma posed by the generalized equation of photosynthesis occupied me for many years. The first advance toward a more satisfactory interpretation grew out of the recognition that the evolution of O_2, the unique feature of green plant photosynthesis, could provide a clue to the solution of the problem (23). If, as implied by the equation, O_2 originated from the dehydrogenation of H_2O, then the energy required for this partial process would be essentially the same as that for the overall photosynthetic reaction. This means that the radiant energy would have to be used *in toto* for the splitting of H_2O and that the reduction of CO_2 should occur by way of light-independent ("dark") reactions. The comparative biochemical approach to photosynthesis suggested that this might apply equally to the bacterial photosyntheses. As far as the CO_2 assimilation is concerned, this did not present a problem; it was, however, difficult to accept the apparent consequence that H_2A instead of H_2O should here participate directly in the photochemical step. But was it necessary to invoke such a reaction? Clearly not; the very fact that for the Athiorhodaceae light and O_2 had been shown to be equivalent, now suggested that these organisms—and by extrapolation all other photosynthetic bacteria—performed a photochemical reaction in which H_2O is converted to a substance, comparable to but not identical with O_2, which can act as H-acceptor for the dehydrogenation of H_2A in a dark reaction. Then the common denominator of all photosyntheses would be a photodecomposition (photolysis) of H_2O to a reducing and an oxidizing moiety, the former serving for the reduction of CO_2, the latter for the production of O_2 in green plants and for the oxidation of H_2A in other photosyntheses. This concept can be represented by the equations:

$$
\begin{array}{ll}
4H_2O \longrightarrow 4[H] + 4[OH] & \text{light reaction} \\
4[H] + CO_2 \longrightarrow (CH_2O) + H_2O & \left.\vphantom{\begin{array}{c}a\\a\end{array}}\right\} \text{dark reactions} \\
4[OH] + 2H_2A \longrightarrow 4H_2O + 2A &
\end{array}
$$

$$
\overline{4H_2O + CO_2 + 2H_2A \longrightarrow (CH_2O) + 5H_2O + 2A}
$$

It may be noted that the "addition of extra H_2O" to the old de Saussure equation for green plant photosynthesis: $H_2O + CO_2 \rightarrow (CH_2O) + O_2$, had been instrumental in developing the first generalized equation which implicated H_2O as the source of O_2, a view that was subsequently supported by the studies of Winogradov & Teis (24) and of Ruben et al. (25) with O^{18}-labeled CO_2 or H_2O. Similarly, the further addition of H_2O to the generalized equation sufficed to eliminate the difficulties encountered earlier.

These ideas were tested after I had returned from my sabbatical leave of absence at the end of 1936. In numerous experiments with various cultures of Athiorhodaceae, it was established that the rate at which these bacteria oxidize a particular organic substrate is identical during anaerobic incubation in sufficiently strong light and in darkness in the presence of air. At light intensities below saturation, the rate of oxidation under anaerobic conditions was lower, but it could be increased to the maximum level by supplying O_2 at the same time (26, 27). These results were consistent with the conclusion that the oxidation of H_2A proceeds by way of dark reactions. A search for the nature of the oxidant formed in the photochemical reaction was, however, unsuccessful. Illumination of cell suspensions in the absence of an oxidizable substrate did not lead to the accumulation of an oxidant. Hence, it was concluded that the latter might be the oxidized form of an enzyme present in too small an amount to be detectable by the methods used (28, 29). The photosynthetic mechanism could then be represented by the diagram:

This, in essence, is the extent of my contribution to a better understanding of the photosynthetic mechanism. Considering that it took me about 25 years to reach this point, and in view of the numerous and important advances made more recently by others, the result seems rather meager, and the fact that it has received so much recognition has often been a source of embarrassment. On the other hand, it has been gratifying to witness the important role which the photosynthetic bacteria have played in these new developments.

FIRST STUDENTS

During the first few years in Pacific Grove I soon realized that neither the Beijerinckian elective culture methods, a deliberate application of the principle of natural selection (30–32), nor the potentialities of Kluyver's "comparative biochemistry" developed during the seven years I was privileged to spend in his laboratory (33), were receiving much attention in this country. This induced me to organize a general microbiology course in which these aspects were emphasized and integrated.

The first course, given in the spring of 1930, was attended by a single student, R. E. Hungate, who became sufficiently interested to make him choose a microbiological problem, the decomposition of cellulose by the intestinal inhabitants of termites, for his Ph.D. dissertation. Immediately thereafter, H. A. Barker enrolled in the summer course and he, too, decided to specialize in microbiology after first augmenting his chemical training by taking a Ph.D. degree in chemistry. In the next few years the course was taught to somewhat larger groups, and in the late 1930's I was forced to limit the enrollment in order not to overburden my capable and faithful technical assistant, Mrs. Pearl Murray Wagner, who prepared the large amounts of glassware and media needed for the experiments.

The number of graduate students working on a full-time basis was very small. The first, Lewis A. Thayer, who was studying diatom blooms and their potential relation to petroleum formation, I had "inherited" upon my arrival. Hungate came next, though he spent most of the year at Stanford where he had a teaching position, and worked at the Hopkins Marine Station only during the summer months. In September 1930, F. M. Muller arrived from Holland, supported by a special fellowship, and began a study of the metabolism of organic compounds by Thiorhodaceae (17). Soon afterward, R. M. Bond, wanting to test the validity of Pütter's theory that marine animals had to satisfy their food requirements at least in part by using dissolved organic substances in the sea, embarked on an investigation of the nutrition of the brine shrimp, *Artemia salina* (34). In 1933, H. A. Barker, now a Ph.D. who had been awarded a National Research Council fellowship, started his microbiological career with studies on marine dinoflagellates and diatoms (35, 36), on the colorless green alga, *Prototheca zopfii* which led to the concept of oxidative assimilation (37, 38), and on the anaerobic decomposition of amino acids by bacteria. He stayed for two years during which the two of us were the only occupants of the Jacques Loeb laboratory; the large staff present at the end of 1928 had dispersed.

GROWTH OF A FAMILY

Following my return from Europe the situation began to change rapidly. From early 1937 until 1942, a sizeable group of co-workers assembled. It included Drs. W. A. Arnold, Hans Gaffron, J. W. Foster, and A. L. Cohen as postdoctoral fellows, and S. F. Carson, E. H. Anderson, M. Doudoroff, R. Y. Stanier, J. O. Thomas, and H. Bliss as graduate students, most of whom stayed for two or more years. Many new problems were tackled and in 1939 a new and promising approach to biochemical problems was initiated. This owed its inception to Dr. Arnold who, during a year's study in Copenhagen, had become impressed with the potentialities that isotopically labeled compounds offered for the study of biochemical reactions. He acted as mediator in establishing a close cooperation with S. Ruben and M. D. Kamen at the University of California in Berkeley, in whose laboratories some experiments with $C^{11}O_2$ were conducted for the elucidation of special aspects of the propionic acid fermentation and of the anaerobic metabolism of *Tetrahymena geleii*, then the only ciliate that could be grown in pure culture (39–44). It was a period of great and joyous activity, and I am profoundly grateful to this wonderful community from whose association I have learned so much.

The United States entrance into the second world war brought this association to an end, and in 1942 the future prospects for microbiology at the Hopkins Marine Station began to appear dim. Dr. M. P. Starr was the only newcomer; he remained from September, 1944 till January, 1946, studying the nutritional and taxonomic aspects of plant pathogenic bacteria and the pigment production by one of them. In January, 1946, I started my second sabbatical leave of absence and it was during that year that I became more and more reconciled to the idea that I should not expect any future collaborators. Nevertheless, to my surprise and delight, at the beginning of 1947 I found that a new group was forming. During the next three years it included Drs. S. R. Elsden and B. E. Volcani who left after the first year's summer course; Drs. G. Fåhraeus, I. Tittler, Jane Pinsent (now Mrs. Q. Gibson), A. Schatz, and Raul Trucco, who stayed for a year; and Drs. J. L. Stokes and M. B. Allen, who remained for three or more years, as did the new graduate students, Wolf Vishniac, Barbara Wright, Helge Larsen, Barbara Bachmann and F. G. Lara. It was as wonderful a group as the one that had assembled a decade earlier, though now truly international in character, comprising British, Israeli, Norwegian, Swedish, Brazilian, Argentinian, and American scientists. They worked independently, yet with much intercommunication, on many problems, ranging from thermophilic bacteria and blue-green algae to myxobacteria, bacterial fermentations, denitrification, pigment formation, green sulfur bacteria, thiobacilli, hydrogen oxidizing bacteria, photosynthesis, and protozoan metabolism.

In 1950, a new influx raised the size of the microbiological family to fifteen, the result of a totally unexpected development. Early that year I had received letters from about ten persons who had applied for postdoctoral fellowships and expressed their desire to work at the Hopkins Marine Station. Anticipating that at most only a very few would actually be awarded fellowships, I had agreed to accept them. To my surprise, all but one of the applications were approved, and consequently the family was enlarged by the arrival of Drs. R. K. Clayton, J. Silliker, Max Eisenberg, Helen Whiteley, Alma Whiffen, Harry Katznelson, Adelaide Brokaw, and Marian Cramer, Kjell and Kjellrun Baalsrud, and some new graduate students. This was the period of maximum activity at the laboratory and even though the facilities were quite limited, which often necessitated planning ahead for the use of special instruments, the spirit was as delightful as ever while a good deal of sound work was accomplished.

SUNSET

After the 1953 summer course, the population had dwindled to a single postdoctoral fellow, Dr. M. Winfield from Australia, and one graduate student. In spite of this, the new year proved to be quite a hectic one. As President of the Society of American Bacteriologists, now the American Society for Microbiology, I was confronted with some very touchy problems facing the society. In addition, I had accepted an invitation to deliver, jointly with A. J. Kluyver whom I had not seen since 1936, the Prather Lectures at Harvard University just prior to, and the Messenger Lectures at Cornell University immediately after the S.A.B. meeting in Pittsburgh. The preparations for this meeting had required a vast amount of correspondence, occupying every weekend, the planning of the lectures had taken up all the rest of my time, and when I returned to the West coast I was very tired. Nevertheless, the opportunity to take a rest had not yet come for the 1954 summer course would begin shortly and a new postdoctoral fellow, Dr. W. Verhoeven from Kluyver's laboratory, arrived for a four months' sojourn.

At the end of that summer quarter I began to make serious plans for another sabbatical leave of absence which, with the support of a Guggenheim Fellowship, I intended to spend in Europe. It was to begin in England with the delivery of the second Marjory Stephenson Memorial Lecture before the British Society for General Microbiology at its meeting in London in April, 1955 (31). Because I had not enjoyed even a brief vacation for several years, it was enticing once again to travel by ship through the Panama Canal; this would also provide an opportunity to prepare the lecture under, for me, ideal circumstances.

Early in March we embarked on a freighter of the Holland America Line which left from Oakland late in the afternoon. The sky was clear and deep blue, the weather calm; and as the ship slowly moved out of San Francisco Bay, I now watched a superb sunset just as we passed under the Golden Gate bridge. With retirement only nine years away, it seemed like another and appropriate omen.

After the London meeting we traveled through England, Scotland, and Norway, where I visited many laboratories and learned much. The culmination of the Norwegian trip came in Trondheim where I could celebrate the appointment, just announced, of my former student, Helge Larsen, to the Chair for Biochemistry at Norway's Technological University. Next I spent some months in Holland, particularly in Delft, where Kluyver and I put the finishing touches on our joint Prather Lectures which the Harvard University Press wanted to publish; *The Microbe's Contribution to Biology* finally appeared in 1956 (45).

Scientifically most exciting were the three months I was allowed to work at the Pasteur Institue in André Lwoff's department. Here, I could familiarize myself with the new developments which, led by Lwoff, Monod, Jacob, Wollman, and G. Cohen, were opening new vistas in the area of microbial biochemistry and genetics. They were followed by a month's sojourn in Israel, mostly at the Weizmann Institute; visits to several Italian and Swiss laboratories; and a ten days' stay in Strasbourg where I learned to handle L forms of bacteria under the guidance of Tulasne and Minx. The last few weeks were spent in Holland, trying to consolidate the many new experiences gathered during the year and catching up with my correspondence which had been rather neglected. Two days before our return to California we paid a final visit to the Delft laboratory and its Master. When we left I sorrowfully realized that it would be the last time I was to see him, and the wireless message announcing his death, which I received a week later while crossing the Atlantic Ocean, did not come as a surprise.

Back at the Hopkins Marine Station in time to prepare for the 1956 summer course, it soon became evident that the sunset watched a year earlier had not presaged a rapidly approaching period of obscure existence. In September, another new group of co-workers assembled: Drs. Hans Veldkamp from Holland, June Lascelles from England, Sanatkumar Vias from India, who were joined by Drs. Carlton Bovell and Sydney Rittenberg, the only "true Americans." Clayton, who had meanwhile become a staff member of the U.S. Naval Postgraduate School in nearby Monterey, kept in close touch with us and participated regularly in our seminars. The following year, Kjell Eimhjellen, a pupil of Helge Larsen's came for a two years' stay, the second of which was spent in the company of Drs. J. W. M. La Rivière, the last of Kluyver's doctoral students whose work and attitude I had learned to appreciate greatly during the visits in Delft in 1955–1956, and Stanley Scher from Rutgers University.

The studies of this group dealt chiefly with the anaerobic metabolism of *Cytophaga* species, free-living spirochaetes, and hydrogen-oxidizing bacteria; the cultivation of *Spirillum volutans* and of the colorless sulfur bacterium, *Thiovulum*; and the utilizability of aromatic compounds as substrates for photosynthesis by Athiorhodaceae. La Rivière was also engaged in formulating his concepts on the interrelations between life, thermodynamics, and information theory. This involved many lengthy and stimulating discussions which greatly clarified my own ideas on the significance of patterns of organization in relation to the properties of complex systems and on the origin of life on earth.

As before, the spirit in the laboratory was stimulating. But the work of the following year was uninspired for the two new postdoctoral fellows were plagued by personal problems which rendered their activities in the laboratory largely ineffective. Once again it looked as if the microbiological investigations at the Hopkins Marine Station were coming to an end. Nevertheless, a period of renewed and joyous activity was ushered in soon afterward. During the next few years, Jack London began his enzymological studies with thiobacilli, Dr. George Kidder spent a sabbatical leave of absence working on some aspects of pteridine metabolism of protozoa, discovered the need for sulfur compounds for the germination of spores of a mold species and enlivened the small community by his endearing personality and the splendid series of seminars he gave on protozoan nutrition. Drs. J. Bennett, Ercole Canale-Parola, Norbert Pfennig, Lynn Miller, Edward Leadbetter, Roger Whittenbury, and Jeanne Stove Poindexter tackled aspects of the taxonomic status and metabolism of *Lineola longa*, the production of pulcherrimin by yeasts and *Bacillus* species, the isolation of pure cultures of various purple and green sulfur bacteria, the fermentation of galactose and lactose by yeasts, the induction of fruiting body formation in myxobacteria, the efficiency of photosynthesis by green sulfur bacteria at the expense of thiosulfate and H_2, respectively, and the biology of the *Caulobacter* group of bacteria. One other guest, Dr. D. I. Arnon, spent a year preparing the foundations for a new book on photosynthesis and contributed much to the daily discussions of many and diverse topics.

During that period I also found time to collaborate with Dr. L. R. Blinks, then Director of the Hopkins Marine Station, on a problem that had suggested itself while I was writing a paper for the *Annual Review of Plant Physiology* (46). Emerson and co-workers had discovered that in green plant photosynthesis two different pigments can cooperate in the effective utilization of radiant energy, which had led to speculations as to the specific reactions mediated by the two pigment systems, and it seemed that similar studies with photosynthetic bacteria might aid in pinpointing the nature of the reactions in which they are involved. Our experiments showed convincingly that *Rhodospirillum rubrum* does not exhibit the Emerson ("enhancement") effect; when illuminated with two light beams of different wavelengths below saturating intensity, the rate of photosynthesis is equal to the sum of those measured with each of the beams separately (47). This suggested that the enhancement effect in green plants is connected with the evolution of O_2, an inference strongly supported by the experiments done by Gaffron (48) with green algae performing a bacterial type of photosynthesis known as photoreduction, in which O_2 is not produced. These experiments, too, failed to disclose an enhancement effect.

After 1956, a direct participation in experimental studies had become more and more difficult. The daily consultations with co-workers and the increasing correspondence, particularly the writing of letters of recommendation, had come to occupy much of my time. In addition, the general microbiology course had to be kept up to date and, owing to the ever-expanding use of microorganisms, beginning with the late 1940's, for the study of particular biochemical processes, spectacular advances in this area were being made at an almost exponential rate which required constant revision and augmentation of the material to be covered. Meanwhile, the number of participants in the course had grown from the earlier maximum of 8 to well over 20, about half of them auditors, and the caliber of the "classes" had also undergone a marked change. Many prominent physicists and biochemists, recognizing the great advantages stemming from the use of different kinds of microbes for biochemical and biophysical investigations, now attended the summer course. I was often greatly embarrassed when I had to develop the basic principles of microbiology with the aid of very simple experiments before so illustrious an audience, though it was always a pleasant relief to discover that at least part of the material presented was new to the group.

Finally, I had become involved in another kind of activity. Upon my return from Europe in May, 1956, I received many urgent requests for obituaries of A. J. Kluyver, only six of which I could honor before the end of the year. In September I learned that a group of Kluyver's admirers, mostly former students residing in the Netherlands, were planning a book in memory of the Master. It was deemed desirable that it be published in English, but many of the articles selected and solicited for inclusion had been or were being written in the Dutch language, and I offered to prepare the English translations. I had also accepted an invitation to write an evaluation of Kluyver's contributions to microbiology and biochemistry for the book. Altogether, I spent more than two years on this undertaking. It was a labor of love and a partial redemption of the great debt of gratitude for the incalculable role Kluyver had played in guiding my education and development in much more than just a scientific sense. The book was published at the end of 1959 (49).

No sooner had this task been accomplished than another and somewhat similar one was broached: the editing of a book honoring E. G. Pringsheim on the occasion of his 80th birthday in 1961. It was to contain a selection from among the large number of important contributions he had made in the course of an amazingly productive life, which even now is happily continuing. Some of these had originally been published in German and were to be republished in English translations. Unfortunately, several events conspired to delay the date of publication and the *Selected Papers of Ernst Georg Pringsheim* (50), printed with the aid of a generous subsidy from the "Foundation of Microbiology" established by S. A. Waksman, appeared two years late.

Were these activities, and those concerned with the preparation of special lectures and review articles, a means of satisfying the old urge to become a writer? I don't know. But this I do know: in the course of time, writing has become more and more difficult for me. Even after an apparently sound framework has been worked out for the presentation of a subject and a preliminary version has been drafted, it requires endless reconstructions and revisions before an acceptable product begins to emerge, and this takes much time and effort. Nevertheless, I have had my fling at it, as well as at science, whose fundamental significance, power, and limitations I have learned increasingly to comprehend and appreciate.

The result has been a full and truly exhilarating life, and I am well aware that this has been made possible because many persons—former teachers, colleagues, associates, and students; officials of universities and foundations who have supported my work; and, last but not least, my wife and children who have had to put up with a husband and father most of whose time has been spent in the laboratory and study, and perhaps too little with his family —have unstintingly provided me with the opportunities whereof Hoffer speaks in the passage quoted at the beginning of this paper. Here I want to express my profound gratitude to them all.

LITERATURE CITED

1. La Rivière, J. W. M., in *Albert Jan Kluyver, His Life and Work*, 49–67. (Kamp, A. F., La Rivière, J. W. M., Verhoeven, W., Eds., North-Holland Publ. Co., Amsterdam, 567 pp., 1959)
2. Rubner, M., *Die Ernährungsphysiologie der Hefezelle bei alkoholischer Gährung.* (Veit, Leipzig, 396 pp., 1913)
3. van Niel, C. B., *Zentr. Bakteriol. Parasitenk., Abt. II.,* **60,** 289–98 (1923)
4. Bavendamm, W., *Die farblosen und roten Schewefelbakterien.* (G. Fisher, Jena, 156 pp., 1924)
5. Buder, J., *Jahrb. Wiss. Botan.,* **58,** 525–628 (1919)
6. Kluyver, A. J., Donker, H. J. L., *Chem. Zelle Gew.,* **13,** 134–90 (1926)
7. Kluyver, A. J., van Niel, C. B., *Zentr. Bakteriol. Parasitenk., Abt. II,* **63,** 1–20 (1924)
8. Kluyver, A. J., van Niel, C. B., *Ann. Mycol.,* **25,** 389–94 (1927)
9. Kluyver, A. J., van Niel, C. B., *Planta,* **2,** 507–26 (1926)
10. van Niel, C. B., *The Propionic Acid Bacteria.* (Boissevain and Co., Haarlem, 187 pp., 1928)
11. van Niel, C. B., Kluyver, A. J., Derx,

H. G., *Biochem. Z.*, **210**, 234–51 (1929)

12. Baas Becking, L. G. M., *Ann Botany (London)*, **39**, 613–50 (1925)
13. Lewis, G. N., *The Anatomy of Science.* (Yale Univ. Press, New Haven, 219 pp., 1926)
14. Beijerinck, M. W., *Zentr. Bakteriol. Parasitenk., Abt. II*, **7**, 561–82 (1901)
15. Kamen, M. D., Gest, H., *Science*, **109**, 560 (1949)
16. van Niel, C. B., in *Contributions to Marine Biology*, 161–69. (Stanford Univ. Press, 277 pp., 1930)
17. Muller, F. M., *Arch. Mikrobiol.*, **4**, 131–66 (1933)
18. Winogradsky, S., *Beiträge zur Morphologie und Physiologie der Bakterien. Heft 1: Zur Morphologie und Physiologie der Schwefelbakterien.* (Felix, Leipzig, 120 pp., 1888)
19. van Niel, C. B., *Arch. Mikrobiol.*, **3**, 1–112 (1931)
20. Pringsheim, E. G., *Naturwissenschaften*, **20**, 479–83 (1932)
21. Schlegel, H. G., Pfennig, N., *Arch. Mikrobiol.*, **38**, 1–39 (1961)
22. Pfennig, N., *Zentr. Bakteriol. Parasitenk., Abt. I*, Suppl. **1**, 179–89; 503–4 (1965)
23. van Niel, C. B., *Bull. Assoc. Diplomes Microbiol. Fac. Pharm. Nancy*, **13**, 3–18 (1936)
24. Winogradov, A. P., Teis, R. V., *Compt. Rend. Acad. Sci. USSR*, **33**, 490–93 (1941)
25. Ruben, S., Randall, M., Kamen, M. D., Hyde, J. L., *J. Am. Chem. Soc.*, **63**, 877–78 (1941)
26. van Niel, C. B., *Advan. Enzymol.*, **1**, 263–329 (1941)
27. van Niel, C. B., in *Photosynthesis in Plants*, 437–95. (Franck, J., Loomis, W. E., Eds., Iowa State Coll. Press, Ames, Iowa, 495 pp., 1949)
28. van Niel, C. B., *Am Scientist*, **37**, 371–83 (1949)
29. van Niel, C. B., in *The Microbe's Contribution to Biology*, 73–92. (1956) (see Ref. 45)
30. van Niel, C. B., *Bacteriol. Rev.*, **13**, 161–74 (1949)
31. van Niel, C. B., *J. Gen. Microbiol.*, **13**, 201–17 (1955)
32. van Niel, C. B., *Quart. Rev. Biol.*, **41**, 105–12 (1966)
33. Kluyver, A. J., *The Chemical Activities*

of *Microorganisms.* (Univ. of London Press, London, 109 pp., 1931)
34. Bond, R. M., *A contribution to the study of the natural food cycle in aquatic environments. With particular consideration of micro-organisms and dissolved organic matter.* (Bull. Bingham Oceanog. Collection **4**, Pt. 4, 89 pp. 1933)
35. Barker, H. A., *Arch. Mikrobiol.*, **6**, 141–56 (1935)
36. Barker, H. A., *ibid.*, **6**, 157–81 (1935)
37. Barker, H. A., *J. Cellular Comp. Physiol.*, **7**, 73–93 (1935)
38. Barker, H. A., *ibid.*, **8**, 231–50 (1936)
39. Carson, S. F., Ruben, S., *Proc. Natl. Acad. Sci. U.S.*, **26**, 422–26 (1940)
40. Carson, S. F., Foster, J. W., Ruben, S., Barker, H. A., *ibid.*, **27**, 229–35 (1941)
41. Carson, S. F., Ruben, S., Kamen, M. D., Foster, J. W., *ibid.*, **27**, 475–80 (1941)
42. Foster, J. W., Carson, S. F., Ruben, S., Kamen, M. D., *ibid.*, **27**, 590–96 (1941)
43. van Niel, C. B., Ruben, S., Carson, S. F., Kamen, M. D., Foster, J. W., *ibid.*, **28**, 8–15 (1942)
44. van Niel, C. B., Thomas, J. O., Ruben, S., Kamen, M. D., *ibid.*, **28**, 157–61 (1942)
45. Kluyver, A. J., van Niel, C. B., *The Microbe's Contribution to Biology.* (Harvard Univ. Press, Cambridge, Mass., 182 pp., 1956)
46. van Niel, C. B., *Ann Rev. Plant Physiol.*, **13**, 1–26 (1962)
47. Blinks, L. R., van Niel, C. B., in *Studies on Microalgae and Photosynthetic Bacteria*, 297–307. (Japanese Soc. Plant Physiol., Ed., Univ. of Tokyo Press, 363 pp., 1963)
48. Gaffron, H., in *Bacterial Photosynthesis*, 3–14. (Gest, H., San Pietro, A., Vernon, L. P., Eds., Antioch Press, Yellow Springs, Ohio, 523 pp., 1963)
49. *Albert Jan Kluyver, His Life and Work.* (Kamp, A. F., La Rivière, J. W. M., Verhoeven, W., Eds., North-Holland Publ. Co., Amsterdam, 567 pp., 1959)
50. *Selected Papers of Ernst Georg Pringsheim.* (van Niel, C. B., Ed., Inst. Microbiol., Rutgers Univ., New Brunswick, N.J., 331 pp., 1963)

Ann. Rev. Plant Physiol. 1972. 23:1–28

A CHEMIST AMONG PLANTS

Hubert Bradford Vickery

Ann. Rev. Plant Physiol. 1972. 23:1–28

A CHEMIST AMONG PLANTS 7520

Hubert Bradford Vickery

Biochemist Emeritus, Department of Biochemistry, Connecticut Agricultural Experiment Station, New Haven, Conn.

CONTENTS

When one looks back upon more than 50 years of activity in chemistry, a natural question is how did this all start? I have a clear memory of the incident. Among the books given me by my father for Christmas when I was in the seventh grade in school in Yarmouth, Nova Scotia, was a copy of Steele's *Fourteen Weeks in Chemistry*. This was an elementary text at about high school level, but I read it with increasing interest, learned the symbols for the elements, and carried out such experiments as the preparation of carbon dioxide from baking soda and vinegar. During the subsequent 4 years in high school, I took every course that was offered in chemistry, physics, and botany, and by the time I was graduated at age 17, I knew that I wanted to become a chemist. There followed a year of teaching in a two-department country school, but I spent many evenings in the laboratory at the high school working through a text book on qualitative analysis.

In the fall of 1912, I entered Dalhousie University in Halifax, Nova Scotia, with credit for most of the courses required of freshmen. This was possible at that time because the fourth year of high school, then an appallingly rigorous course, was accepted by the colleges in Nova Scotia as the equivalent of the freshman year. But not the chemistry! Being young and very naive, I applied for entry into the honors course in chemistry. Professor MacKay questioned me thoroughly on what I had read and what laboratory experience

1

I had had, and then said that I might take the lectures and laboratory of the second-year course in chemistry, but that I must also do the laboratory work of the freshman year and write the final examinations of the first-year course. He indicated that if I was successful, they would consider admitting me into the honors course the following year. At Dalhousie, then as now, the willing horse got the heaviest load.

The honors course at Dalhousie allowed one to specialize in his chosen subject. He could substitute advanced courses in his field for the otherwise required courses for the Bachelor degree. In my case, this meant that I omitted Latin, history, economics, philosophy, and so forth, but took all of the lecture and laboratory courses offered in chemistry and as much of the advanced work in physics as could be crowded in. English, French, and German were of course required. They also had another interesting little trick. If one took a regular course and turned in a final examination paper that was practically perfect, he got a simple pass mark. To obtain anything better than this, he applied to the professor early in the year and was given a list of extra reading. Then at the final examination he was required either to write a second paper on this reading or to answer the first group of questions on the examination paper and then go on to a second group on the theory that if he was any good he could answer the regular questions very quickly and could spend the greater part of the 3-hour period on the really tough ones on the extra reading at the end. It was a demanding system. The attitude of the professors was, simply, here are the pearls, do as you like with them. If you successfully picked them up you got honors, otherwise you went home. Certainly, no one asked you embarrassing questions about why you failed the quiz last week.

In addition, to obtain honors in chemistry, one took all of the courses available, worked in the laboratory all day and every day, wrote the regular examinations, and then at the end of the senior year wrote a second set of four papers on everything one had been taught or had read about during the entire 4 years. There was also a requirement for a brief research, which in my case was a revision of one of the methods for the qualitative analysis of the alkaline earths. I was able to determine the limits within which each earth could be detected in the presence of excess of the others.

After graduation with honors in chemistry and chemical physics in the spring of 1915, there followed 2 years of teaching science in one of the high schools in Halifax, but the third year was interrupted on December 6th, 1917, by the explosion of a munitions ship in Halifax harbor which destroyed the entire north end of the city, killed more than 2000 people, and injured many thousands more. Luckily the school in which I taught was well over the brow of the hill above the point at which the ship exploded so that, although every window was broken and there was much structural damage, no one in the building was killed although most of us were cut by flying glass. If that morning I had gone up to my room instead of stopping to talk a few minutes with one of the other teachers, I should have been found later under the bricks of a chimney that had been overturned and had broken through the roof directly above my desk and chair.

It was many weeks before the school system of the city could be reorganized, and meanwhile I learned that the Imperial Oil Company had a position for a chemist at the refinery that had just been built across the harbor in the town of Dartmouth. However, 18 months as an analytical chemist in the oil industry convinced me that teaching offers a much more pleasant life, and I applied for and obtained the post of science teacher at the Provincial Normal School in Truro, Nova Scotia. Meanwhile I had completed the requirements for a Master of Science degree at Dalhousie by an extensive course of reading and a research project in which I determined the approximate position of the isochlor lines for the natural waters of the Province (1), a problem of some importance in sanitary surveys and one in which Professor MacKay was much interested.

My career as a teacher of prospective teachers was, however, a short one. In the early spring of 1920, I was summoned to Dalhousie to undertake the lectures and supervision of the laboratory classes for all of the courses in chemistry currently taught. Professor MacKay had died in the fall, and Professor Nickerson, who had attempted to carry on alone until further assistance could be found, had become ill and had collapsed in class. Looking back, I think that those two months when I tried to do the work of two men were about the toughest of my entire life. It was, of course, an impossible task, but the students were considerate, they also worked hard, and most of them passed their examinations successfully. And the university was not only kind but generous. They paid me well, and at the end rewarded me with an appointment to an 1851 Exhibition scholarship.

This scholarship, funded by the profits made from the great Exhibition of 1851 in London under the shrewd direction of Prince Albert, the husband of Queen Victoria, is not as well known as the later established Rhodes scholarship which was patterned after it. The Rhodes has no restrictions, but the 1851 is restricted to British citizens among whom it carries prestige equal to the Rhodes, and is further restricted to science. Today it is a moderately valuable scholarship, but in 1920 it was only 200 pounds, and the pound had dropped seriously in value just after the war. It was not much, but my wife and I decided that it was enough. Acting on the advice of Dr. Howard Bronson, the professor of physics at Dalhousie, a Yale man, the fall of 1920 found us in New Haven registered in the Yale Graduate School. During the next 2 years my wife earned the degree of PHT while I earned the PhD.

For the second time an element of pure chance directed the course of my scientific career. It happened that one of Dr. Thomas B. Osborne's assistants at the Connecticut Agricultural Experiment Station resigned early in 1921 while in the midst of a research problem in which Osborne was greatly interested. Chancing one day to meet Professor Treat B. Johnson, with whom I was studying at Yale, he asked if there was a young man in Johnson's group who would be willing to come out to the Station and do his research there.

Johnson had put me to work on a problem supported by government funds on the nitration of benzene, and it occurred to him that, come summer vacation, the man on this problem might be required to continue the work in one of the government military laboratories. Although security matters were not so gravely considered in those days as they were in later years, it was obvious that a Canadian citizen might not be welcome under the circumstances. Accordingly, a few days later, Johnson walked out to the Station laboratory with me and introduced me to Osborne, who asked if I would be willing to do my research under his direction. To a young and impecunious student, the chance to work in a large, well-equipped, and comfortable laboratory under a man of international reputation, with no fees or charges for reagents or apparatus, offered a choice to which there was only one answer. A few days later, after some intensive reading on fundamental protein chemistry, I began work in the laboratory where I have spent the rest of my professional life.

THE CONNECTICUT AGRICULTURAL EXPERIMENT STATION

Osborne was an extraordinary man. He was an acknowledged world authority of the time on the chemistry of proteins. A charming although sometimes crusty gentleman of independent wealth, for more than 30 years he had directed a laboratory which had demonstrated that the proteins of plant seeds are substances of definite and ascertainable chemical properties, that each seed protein differs in some respect from all others and is thus a specific substance, and together with Professor Lafayette B. Mendel of Yale, had, during the previous 10 years, shown that proteins may differ from each other in their nutritive effects. They had demonstrated that lysine and tryptophan are essential in the nutrition of rats, and had independently discovered vitamin A. Currently the work of the laboratory had to do with the occurrence of vitamin B in such food materials as fruits and leaf tissues as well as in yeast, and especially with the effect of carefully controlled diets of known composition on the phenomena of growth. Osborne himself was engaged in a study of the proteins of green leaves and in the fractionation of extracts from leaves in the hope of obtaining some evidence concerning the chemical nature of vitamin B. He set me to work on a study of the stability to acid hydrolysis of the form of nitrogen in proteins which is liberated as ammonia, the so-called amide nitrogen. At that time it was purely speculative that this nitrogen is combined in the protein as the amide groups of asparagine and glutamine.

The problem developed into a study of the rate at which the protein gliadin from wheat is hydrolyzed, with respect to both amide and peptide bonds, by boiling acid or alkali as a function of the concentration of the hydrolyzing reagent. A year later I was able to present to the authorities at Yale a dissertation which, to my intense relief, they accepted (2). The oral examination was more or less of an anticlimax. With the exception of Johnson himself, there was no one at that time on the chemical faculty at Yale who knew enough about protein chemistry to ask any but the most elementary questions, and I survived without too much damage.

Osborne then offered me an assistant's position in the laboratory, which at that time was mainly supported by annual grants from the Carnegie Institution of Washington. The salary was only a little more than I had been earning as a teacher in Nova Scotia, for Osborne was extremely careful with his budget, and he pointed out that under annual grants there was no assurance that the position would last for more than the current year. But the opportunity to continue a life of research in a laboratory directed by such an outstanding scientist was far too good to be missed, and I have never regretted my decision.

During the next 2 years, the laboratory was enlivened by the presence of A. C. Chibnall, who came to us from Schryver's laboratory at Imperial College in London. He had been busy for some time with efforts to prepare proteins from cabbage leaves after plasmolysis of the cells by treatment with ether water. Osborne and Wakeman's publication in 1920 of a paper describing a protein preparation from spinach leaves, which was the first on the proteins of leaves to be published for more than a century, stimulated him to apply for a postdoctoral fellowship which brought him to the only other laboratory in the world where work in this field was in progress. He arrived almost unannounced before Osborne got back from his prolonged summer vacation—which ended only after the partridge season opened—so I arranged for Chibnall to join Professor Mendel's group. When Osborne returned a few weeks later, I told him of this Englishman who wanted to join us. Osborne was very busy that morning, and my tale was greeted with a succession of grunts and finally, "Well, is he any good?" Assured on that point, he finally said, "All right, bring him out."

Chibnall succeeded in charming Osborne within the hour, and I was instructed to install him in the laboratory at once. As soon as Osborne saw his command of technique, his original approach, and his industry, he happily turned over all of the work on leaf proteins to him, and during the next 2 years Chibnall established methods for the isolation of preparations of leaf proteins that served him as well as others for the next decade or more to carry out distinguished research. Meanwhile we formed a friendship that, in spite of innumerable arguments, has lasted to the present day.

THE BASIC SUBSTANCES IN ALFALFA

Osborne invited me to share with him and Dr. Alfred J. Wakeman in the work on the search for evidence of the presence of vitamin B in fractions prepared from the juice of the alfalfa plant. At that time, the only methods available for the separation of nitrogenous organic substances involved the use of salts of heavy metals as precipitants. Silver, mercury, and lead salts, and phosphotungstic acid were the main reagents. Hydrogen sulfide was used to decompose the precipitated organic compounds, and barium hydroxide to decompose the phosphotungstic acid precipitates. All reagents must be removable by some simple technique. Wakeman had been using mercuric acetate and sodium carbonate, added alternately until an excess was present in a faintly alkaline solution, to precipitate amino nitrogen from extracts of the alfalfa plant, and nutrition tests for the presence of the vitamin in such fractions were in progress, a slow matter at best. It seemed to me that one must work on as large a scale as possible if one were to obtain chemical evidence for the identity of the substances in such fractions, for isolation of crystalline products and ultimate analysis were the only acceptable proofs of identity. The simple chromatographic methods on trace amounts of material used today were then many years in the future. The progress of the fractionation was to be followed by determinations of total, amino, and amide nitrogen.

Accordingly, I undertook the analysis of an extract prepared by grinding a large quantity of freshly cut alfalfa and expressing the juice with the hydraulic press. The juice was treated with an equal volume of alcohol which precipitated any protein present together with much inorganic material. The filtrate was then concentrated and "clarified," as it was called, by the addition of an excess of basic lead acetate, and lead was removed from the filtrate. Next was treatment with an excess of mercuric acetate and sodium carbonate, a procedure advocated years before by Neuberg and Kerb as a general method to precipitate amino acids. Removal of reagents, followed by precipitation of basic substances by phosphotungstic acid and treatment of both precipitate and filtrate with barium hydroxide to liberate the organic bases and remove the reagent, completed the process. This was done on solutions from both the mercuric acetate precipitate and the filtrate from it. One finally obtained fractions that were concentrated for fractional crystallization. The base fractions were treated with silver sulfate according to the Kossel method for the separation of the basic amino acids, and arginine and lysine were sought for before and after severe acid hydrolysis.

The outcome of this and several subsequent elaborate fractionations was most disappointing. After more than 3 years of study, Charles Leavenworth and I were able to account for only about 55 percent of the nitrogen and 30 percent of the solids in the alfalfa extract as identified substances weighed as well-characterized products in crystalline form. The main components were asparagine, a few amino acids separated by the Fischer esterification process and including serine, which was isolated from a plant for the first time, together with arginine, lysine, stachydrine, choline and adenine, and trace amounts of a few other basic substances (3). In 1958 in Canada, while studying stachydrine synthesis in alfalfa with the use of modern chromatographic methods, Marion found that homostachydrine, the betaine of pipecolic acid, is also present and is a contaminant of samples of stachydrine obtained by the classical procedure. He was kind enough to examine several of my preparations and, to my delight, found that one of them was in fact pure.

Many modifications of the procedure were made. For example, basic lead acetate was given up as a clarification reagent as the result of the work of a student, Carl Vinson, who joined us for awhile and isolated adenine, arginine, lysine, stachydrine, aspartic acid, and tyrosine, in small amounts to be sure, from a carefully washed basic lead acetate precipitate obtained from an extract from alfalfa. Obviously it is impossible to wash such voluminous precipitates adequately, since these substances were among the main components of the original extract, and only aspartic acid should form an insoluble lead compound. Inasmuch as barium hydroxide must be used frequently in later stages of the analysis, we subsequently clarified the extracts by adding an excess of barium hydroxide and alcohol at the start of the operation. This threw out a huge precipitate of barium salts of organic acids, mainly malic and malonic acid, together with barium compounds of much carbohydrate material. It also clarified the solutions as well as the lead reagent and precipitated little nitrogenous material other than any aspartic acid present.

The identification of adenine provided us with a puzzle for several months (4). While I was on vacation one summer, Leavenworth had observed that the substance we had isolated as a picrate, which crystallized in fine hair-like needles that decomposed at 298°, is precipitated from the extract by silver salts at acid reaction. This is a general property of purines, and we accordingly tried all known tests for purines on it. It did not give the murexide test, a fact which in the light of later knowledge should have been strong evidence that the substance was adenine, but any suspicions of this we entertained were suppressed by the descriptions of adenine picrate in the literature. This substance was alleged to separate in small rhombic crystals which decompose in the vicinity of 281°. One or two papers mentioned hair-like needles, but where this appearance was noted no decomposition point was given. Furthermore, in previous years Leavenworth had repeatedly isolated adenine of the behavior described in the literature from liver. Fortunately, Professor Mendel still had one of these old preparations. When heated under the conditions we were using in determing the decomposition point of our preparation from alfalfa, it decomposed at 290°. To settle the matter, we isolated adenine by the conventional process from liver, obtaining a preparation of the picrate that decomposed at 295°, converted this to the beautifully crystallized sulfate, and then back to the picrate. This time it crystallized as long pale yellow hair-like needles that decomposed sharply at 298° and was obviously identical with our preparations from alfalfa. This taught us a great deal. We no longer accepted descriptions in the literature at face value, and we learned that to assure ourselves of the purity of a preparation, it is necessary to crystallize the substance as at least two different derivatives or salts. Only in this way may major impurities be eliminated. The adenine isolated from alfalfa apparently separated as its picrate with little or no trace amounts of other purines as contaminants. It was too pure to be easily recognized!

The nutrition tests for the presence of vitamin B in the fractions from alfalfa were unsatisfactory, so I turned to a study of brewers yeast, an especially rich source of the vitamin. In the course of the fractionation I obtained a specimen of nicotinic acid, a substance which had previously been obtained from yeast by Funk and years later was recognized as one of the water-soluble B vitamins. In the 1920s, however, the observations that adenine is present in alfalfa and nicotinic acid in yeast were simply moderately interesting facts. The biochemical relationships of these substances were not recognized until many years later. In spite of many frustrations, however, we had learned a great deal about the composition of protein-free extracts from plants. Amino acids are present in appreciable amounts, asparagine often being the dominant one. They were also present in combinations which we took to be peptides since acid hydrolysis increased the amounts of amino nitrogen. The bases arginine and lysine occurred not only free but also in peptide combination. Furthermore purines, especially adenine, were present together with methylated bases, frequently in large amounts, the major one in alfalfa being stachydrine, in tobacco nicotine, in yeast choline. The greatest difficulty with the methods of analysis then available was, however, the interpretation of what had come to be called the basic nitrogen, that is, the total nitrogen precipitated by phosphotungstic acid. To be sure, much of this nitrogen did indeed belong to known basic substances, but considerable amounts could be shown to be present as peptides which contained monoamino acids. In addition, the barium phosphotungstate obtained when decomposing the alleged base precipitates contained much nitrogen which stubbornly resisted extraction even by boiling hydrochloric acid. The acid did indeed extract an assortment of monoamino acids from the precipitate, but never all of the nitrogen. Today we know that extracts from leaves may contain 50 or more soluble nitrogenous substances demonstrable by paper chromatography, but the identity of the peptide-like material that we encountered is still unknown, and in no case has anything approaching a quantitative accounting of the soluble nitrogen of extracts of leaves been obtained.

BASIC AMINO ACIDS

By 1927 it had become clear that the current methods for the analysis of extracts of plant tissues were hopelessly inadequate, and I accordingly turned to a study of the methods for the determination of the basic amino acids of proteins, a far simpler problem. Leavenworth and I improved the method for the separation of the silver compounds of histidine and arginine as a result of the observation that histidine silver is completely precipitated at pH 7.0 and that arginine silver does not begin to precipitate until the reaction becomes considerably more alkaline (5). Clearly the silver compounds form most completely at or near the isoelectric points of the respective bases. This led to a number of analyses of the bases yielded by several proteins with results which we believed were more nearly accurate than any previously published (6), and which were of material assistance to Professor Edwin Cohn at Harvard, with whom we collaborated in his attempts to account for the acid-binding capacity of proteins. We developed gravimetric methods to determine arginine and histidine, using flavianic acid for arginine and 3,4-dichlorobenzenesulfonic acid, a reagent suggested by Max Bergmann, for histidine, and learned to prepare both bases in quantity by fairly simple procedures. We also had the satisfaction of preparing lysine as a free base in crystalline form for the first time (7). A student, Charles A. Cook, who worked with us in 1931, prepared crystalline ornithine also for the first time.

Tobacco Leaf Chemistry

Meanwhile the work on the chemistry of leaf tissue also went on. At a conference in 1928, after Osborne had retired, and in which Chibnall, who had come over for a visit, joined Osborne and me, we discussed the direction in which the work should go. It was obvious that the complete analysis of the nitrogen compounds present in a leaf was impossible with available methods. Alfalfa was not a good choice of a leaf tissue to study since the plant has small leaves and is cut with a scythe; the tissue worked with was therefore largely stem. If we were to learn anything about the metabolism of the leaf we needed a large leaf from a plant easily grown in quantity in Connecticut, and especially one of known genetic relationships that would be available from year to year. The obvious choice was the tobacco plant, a major crop in the state. It is an amusing instance of my continued naiveté that I at no time considered how this decision would be regarded by the administration of the Station. When I told Director Slate of our plans, I was met with an enthusiastic response and the promise of greatly increased financial support. As a result, Dr. George W. Pucher joined us, and during the next 20 years the team of Vickery and Pucher, together with their associates, published no less than 74 papers and Station Bulletins mostly dealing with the chemistry of the tobacco plant.

Pucher was the most skilled and thorough laboratory worker that I have ever met. His approach to a problem was always original, difficulties were patiently surmounted, and in his work on the development of analytical methods, no possible source of interference was overlooked. When he had finished a project to his own satisfaction, he would prepare a voluminous report supported by table after table of data, and then, if I did not interfere, he would put notebooks and reports away in the vault and go on to the next project. Once he had found out how to do something, he lost all interest in the detail of how the results had been accomplished. It devolved upon me to prepare such reports for publication. In doing this, I recall no instance when I had failed to understand some point and questioned him that he could not find in his notebook some experiment or test that settled the matter at once.

From the start, Pucher was intrigued by the analytical problems presented by the tobacco plant. Nicotine is just volatile enough in hot alkaline solution to interfere with the determination of ammonia, a matter of fundamental importance in the study of amide metabolism. He solved that one, first by the use of Permutite, a cation exchange resin, to absorb the ammonia, filtering it and then distilling the ammonia from the Permutite (8). Later he devised equipment for distillation in vacuo with the use of alkaline buffers, a rapid, convenient, and accurate method now widely used (9). However, he soon settled down to the study of the organic acids of the plant. It was clear from the composition of the barium hydroxide and alcohol precipitates we were obtaining when clarifying leaf extracts that organic acids make up a substantial part of the organic solids present, sometimes even exceeding the amount of protein in the leaf. Furthermore, the tobacco technologists were interested in organic acids since the prized "graininess" of the surface of finished cigar wrapper tobacco was alleged to arise from crystals of salts of organic acids. In 1930, the only method known for the separation and determination of the common plant organic acids was the distillation of their esters in vacuo. Reliable specific methods were available for very few except for oxalic acid, although there was a polarimetric method to determine malic acid and a gravimetric pentabromoacetone method to determine citric acid, both subject to various uncertainties. Pucher greatly improved the technique of the ester distillation method to identify and determine the organic acids of the tobacco leaf (10). A few years later, his mastery of the method enabled him to identify the so-called "crassulacean malic acid" of the *Bryophyllum* leaf as isocitric acid (11), a fact which led in turn to an extensive study of the organic acid metabolism of this and related plants, as well as to the development of methods to prepare this almost unknown but very important acid in quantity and to improve its synthesis (12, 13). Both of these problems were carried further by me after Pucher's untimely death in 1947.

What was probably Pucher's most significant contribution was his development of a succession of methods to determine citric acid. When citric acid in solution in the correct concentration of sulfuric acid is treated with potassium bromide and potassium permanganate, it is converted quantitatively into pentabromoacetone. The earlier method described in the literature involved weighing this substance and, with proper corrections for solubility, gave excellent results. It occurred to Pucher that it should be possible to decompose this compound and titrate the bromide ion produced. Obviously this could be done satisfactorily on a smaller amount of citric acid than was necessary for the gravimetric method, and held out hope that a semi-micro method could be developed. The reagent chosen for the debromination was sodium sulfide, and when the appropriate conditions were found, and especially when the beautiful titration method of Sendroy was applied to determine the bromide, an excellent method accurate in the 1 to 16 mg range became available (14, 15). But Pucher did not stop there. He had noted that the debromination was accompanied by the appearance of a yellow to red color, depending on the amount of citric acid taken for the test, when the petroleum ether extract containing the pentabromoacetone was treated with sodium sulfide. This led to the development of an accurate colorimetric method to determine quantities of citric acid of the order of 1 mg or less (16), a method that subsequently has been modified by many workers, although without notable improvement in either accuracy or convenience, but which served essentially in its original form to enable Krebs to develop the tricarboxylic acid cycle hypothesis as the explanation of respiration in pigeon breast muscle, for which he later received a Nobel prize.

A valuable by-product of the study of the method to determine citric acid was the observation that any malic acid present in the mixture of organic acids subjected to oxidative bromination is converted quantitatively into a substance which is volatile with steam and which forms a very insoluble compound with dinitrophenylhydrazine. This product, when dissolved in hot pyridine and made alkaline, gave a strong blue color in every way suitable for the colorimetric estimation of malic acid (17). The method was undeniably tricky, many precautions were essential, and rigid adherence to the technique was necessary. However, in skilled and experienced hands it gave accurate recoveries of malic acid and served us for years until the development of a simple chromatographic method by Palmer in our laboratory in 1954. I shall never forget the delighted expression on Otto Folin's face when he visited us one day soon after Pucher had worked out the details of the method and demonstrated the magnificent blue color of the final solution to him. Folin did not understand the chemistry of the reaction, nor for that matter did we, but he fully appreciated the significance of that color. It was years later that the volatile oxidation product of malic acid was identified as glyoxal.

In our studies of the forms of nitrogen present in extracts from leaves, we had frequently encountered evidence for the presence of a substance which is rather easily decomposed with the formation of ammonia. H. E. Clark, who worked with us in the mid-thirties, found a great deal of this in the stems of tomato plants raised upon ammonium salts as the source of nitrogen. The behavior was that to be expected of glutamine, and Clark accordingly undertook the isolation of glutamine by the classical procedure of Schulze. He was rewarded with several grams of beautifully crystallized glutamine, then one of the rarest of the amino acids. This observation led to many further studies of amide metabolism as well as to the modernization of the Schulze procedure for preparing glutamine in quantity from beets. A micro method to determine it was worked out in collaboration with Chibnall, and, within a few years, glutamine (of various degrees of purity to be sure) became a regular item in chemical catalogs although an expensive one.

HISTORY OF CHEMISTRY

In 1927, Osborne was asked by the editors of *Physiological Reviews* to prepare a review of the literature dealing with the many hypotheses that had been advanced to account for the structure of proteins. Osborne himself was too busy in this his last year as head of the laboratory and asked me to write the paper. For months, I read the early and more recent literature and wrote my account of it. We had an excellent library and collection of reprints going back to the earliest days of the subject, and I rarely had to go down to the Yale library and then only for obscure papers. Each morning, when Mendel came out for his daily conference with Osborne on the nutrition work, I was asked to read what I had prepared. I was then subjected to the criticism of two of the keenest minds in American biochemistry. No infelicity of English expression escaped their attention, and I was showered with helpful suggestions for improvement and extension of what I had done. It was the most thorough course in technical writing I have ever experienced, as well as a revelation of what two masters of their own fields knew about their subjects. The outcome was a manuscript of which I was, I hope pardonably, proud

(18). Then the question of authorship came up. I maintained that since Osborne had been asked to write it and had shared in every paragraph, he should be the author. He replied that since I had written it, I was the author and he had had little to do with it. The debate grew quite heated, but I finally closed it with the question, "Are you ashamed to have your name on this paper?" Osborne thought for a moment and then said, "All right, but you are the senior author." For the second time in his life Osborne took the second position in authorship; the first was a paper with Chittenden in 1892!

In 1924, Mendel had asked me to give a course of ten lectures on protein chemistry in the department of Physiological Chemistry, an appointment which was continued annually until my retirement in 1963. Second-year graduate students were required to take it, and it served for a long time as a means to eliminate students who were unwilling or unable to take things seriously. Although this requirement was modified as time went on, the course was well attended. In the early years it was the only formal instruction in protein chemistry given at Yale. Later, as protein chemistry became more and more to be treated in the regular courses, I turned to fundamental amino acid chemistry as my subject with an occasional course on the history of protein chemistry. In 1930, I took occasion to work up lectures on the discovery of each of the protein amino acids. That fall, when visiting Professor C. L. A. Schmidt in Berkeley for a few days, I found that he had also worked up and delivered a similar course. We discussed the matter of preparing a joint paper for one of the review journals and sat down at once and divided up the list of amino acids for our individual treatment. Later we exchanged manuscripts as they were written, edited each other's material, and finally offered the whole to *Chemical Reviews*. To my surprise it was at once accepted, and in the printed form took 149 pages of the journal, nearly a whole issue (19). I suspect that that paper has repeatedly been the subject of seminar discussions in biochemical departments. The stock of reprints was exhausted in a few months, and I still get an occasional request for one, an evidence that the history of biochemistry is a matter of increasing consideration even to the present generation of students to most of whom a paper 10 years old is hopelessly outdated and of no further account.

Recently, at the request of the editors of *Advances in Protein Chemistry,* I have brought the material in that now 40-year-old review up to date. During this period at least 16 new amino acids have been detected in hydrolysates of various proteins, mainly of animal origin. Most of them are derivatives of the 20 amino acids that make up the list of commonly found products of the hydrolysis of proteins, and are the result of highly specific interactions of enzyme systems with already assembled polypeptide chains whereby various methylation or oxidation or halogenation reactions occur. This is an active field of study at the present time, and there is little doubt that the list of authentic protein amino acids will ultimately be at least twice as long as that recognized in 1931.

One of the most interesting bits of historical research I have been concerned with is the matter of the origin of the word protein. Since it first appeared in his papers, I had always assumed that the term was invented by Mulder in the late 1830s as a designation for the radical which he thought combines with phosphorus and sulfur in the so-called albuminous substances as they occur in nature. In 1948, Sir Harold Hartley gave an address in Stockholm on the occasion of the centenary of the death of Berzelius. He pointed out the great service Berzelius had made to chemistry in the matter of nomenclature, and in listing terms suggested by Berzelius mentioned the word protein. My friend Chibnall picked up this point in the published account of Sir Harold's lecture and wrote a concerned letter to me on the mistake I had made in a published paper on proteins. Correspondence, and later a delightful personal interview with Sir Harold, led to the preparation of a brief note on the origin of the word to be published simultaneously with one of his own (20). He had not appreciated the fact that Berzelius' suggestion had been totally overlooked by protein chemists.

It appears that Mulder was accustomed to send his analytical data to Berzelius for discussion—the correspondence between the two men runs to hundreds of pages in Soderbaum's magnificent edition of the Berzelius letters— and in a letter early in 1838 he set forth his idea that the albuminous materials in their native condition are compounds of "the animal substances lacking sulfur and phosphorus" with sulfur and phosphorus. This was an awkward expression and it seemed to Berzelius that a new technical term was required. In a letter to Mulder written on July 10, 1838, he suggested the word protein, derived from the Greek for "to be in the first place," as a suitable one on the grounds that these substances are clearly in the first place in human and animal nutrition. Mulder had not asked for a suggestion, and Berzelius obviously felt no strong proprietary rights in this one, for there are many words derived from the Greek *protos* meaning first. At all events, Mulder gladly accepted the suggestion and used the new word freely in a series of remarkable publications in which he expounded his ideas. Later, as a result of the long and bitter polemic with Liebig, who soon showed that Mulder's view was inadequate, the meaning of the word was expanded and later was universally adopted as a far better term than the German *Eiweiss* or English *albumin* with their implied restriction to the proteins of eggs.

REVISION OF THE NOMENCLATURE OF AMINO ACIDS

At a meeting of the editorial board of the *Journal of Biological Chemistry* in 1945, the current confusion in the nomenclature of the amino acids was discussed. For example, two papers on the metabolism of tryptophan had been published recently in the journal in one of which the amino acid was named $l(+)$-tryptophan, in the other $l(-)$-tryptophan although the same substance was meant. The difficulty arises from the fact that the specific rotation of six of the common protein amino acids is dextro if observed in acid solution but levo if observed in water. A further difficulty had arisen with the name of threonine. This had been called $d(-)$-threonine by W. C. Rose, its discoverer, but since its α-carbon atom has the configuration of the l-family (the β-carbon is d), it was frequently referred to in the literature as $l(-)$- threonine or even simply as l-threonine.

The suggestion made by H. B. Lewis and W. M. Clark, that the whole problem could be resolved if the small capital letters L and D were used as prefixes, was enthusiastically accepted. These prefixes were intended to desig- nate the configurational family to which the α carbon belonged. I was asked to prepare a proposal to be made to the nomenclature committee of the American Chemical Society, and also to similar committees in Britain, for the complete reform of the nomenclature of amino acids. The outcome was a set of rules of nomenclature, developed after many revisions with the aid and criticism of committees in both this country and abroad, which was accepted as official by the American Chemical Society in 1947 and at once came into general use. The International Union of Pure and Applied Chemistry ap- proved them in 1952 (21).

However, the rules as then worked out failed to deal satisfactorily with the difficult problem of naming the isomers of amino acids that have more than one center of asymmetry. Our committee had one version, our British colleagues another, and a number of insoluble problems had begun to accu- mulate. In 1961, I attempted to resolve this difficulty by appropriating the prefixes of carbohydrate nomenclature to the names of the more complex amino acids so as to define the configurational relationship. This general method of naming these substances had already been used by a few workers, but there was no consensus regarding precise details. Having had consider- able experience with the behavior of committees on nomenclature, I ap- proached the top committee, that of the International Union of Biochemistry, simultaneously with my approach to the national committees, and supported my suggestion with letters of approval from several leading amino acid and carbohydrate authorities. It worked; and in spite of my flagrant breach of protocol, the rule was officially approved within only 2 years (22). It serves as at least a temporary solution of the difficulty we had had 15 years before, and the new rule is now in common use.

Biochemistry of Leaves

Pucher's sudden death in the fall of 1947 from a heart attack left the laboratory in an extremely awkward position. Wakeman had retired a few years before and had been succeeded by Marjorie Abrahams, who had be- come an expert in the methods to determine the organic acids. Pucher was in the midst of his studies of the organic acid metabolism of the *Bryophyllum* leaf and with the fundamental chemistry of isocitric acid. Leavenworth and I were busy with the basic amino acids and with glutamine, work in which Pucher had also shared, and there were two long reports with accompanying notebooks in the vault, one on the determination of starch in leaves (23), and a second on an elaborate experiment that Pucher had carried out at my suggestion in which a number of organic acids had been fed in culture solution to tobacco leaves (24). These had to be prepared for publication without the occasional conferences which had made the writing of previous papers a much simpler and far more pleasant task. Leavenworth's death on the first anniversary of Pucher's deprived me of a valued collaborator and friend with whom I had worked for more than 25 years, and it became clear that the program of the laboratory must be thoroughly revised. Hitherto I had spent most of my time on the basic amino acids and on amide metabolism in leaves. Pucher had worked with the organic acids and carbohydrates, and we worked together on many experiments in which detached tobacco leaves were studied during culture under various conditions and during the technical operation of curing.

In all of our culture work, we had depended upon the selection of a series of samples of leaves that we hoped would be initially identical with each other at the start of the treatment. One or two samples would be prepared for analysis at once to furnish a base line for the detection and measurement of the chemical changes which subsequently occurred during the treatment. We tried many methods. Random selection from a large pile of leaves, adjustment of the samples to initial equal weight, selection of leaves of the same size, and so forth. No method we had been able to devise gave a set of eight or ten samples in which the total nitrogen content, for example, had a coefficient of variation of less than about 5 percent. Thus our conclusions regarding any chemical changes that had occurred during the treatment were always subject to a sampling error of at least this magnitude, and we were constrained to accept as real only changes of the order of 10 percent or more. This was clearly not good enough.

STATISTICAL SAMPLING OF LEAVES

I enlisted the aid of Dr. Chester Bliss, the statistician on the station staff, took him out to the greenhouse, and asked him how one could pick a set of ten samples of initial identical composition from a row of large handsome tobacco plants. He examined the plants for a few minutes and then pointed out that there are three sources of variation to be considered. One arises from differences between individual plants, a second from the position of the leaf on the plant and a third, which he called a random component of variation, from such matters as slight differences in the composition of the soil in which each plant grew, slight differences in exposure to light, and such accidents as minor damage from insects and so forth. If one picked the leaves in such a way that each plant and each position of the leaf on the plant were equally represented in each sample, one should be able to minimize the variation between samples, and thereby reduce to an acceptable level the coefficient of variation of any component which was not altered by the treatment. In a few minutes, he worked out a method of picking the leaves, based on a systematic Latin square, and suggested that we try it.

Our response was to pick sets of samples from a group of *Bryophyllum calycinum* plants according to three systems. In one system reliance was placed on the assumption that opposite leaflets on the same pentafoliate leaf should be nearly identical in composition, in another we assumed that leaves of the same size should be identical, and in the third we picked the leaflets according to the systematized Latin square so that each plant and each leaflet position was equally represented in each sample. All samples were dried for analysis, and solids, ash, total nitrogen, protein nitrogen, and starch were determined. Bliss then subjected the data to an elaborate analysis which showed the great superiority of the statistical method (25). To obtain an accuracy equal to that given by this method, from four to five times as many samples would need to be taken by either of the other methods for each point examined. In other words, for equal accuracy the statistical method is about five times as efficient in terms of the analytical work required. Subsequently, we always obtained sets of samples of tobacco or *Bryophyllum* leaves which gave coefficients of variation of the nitrogen or ash content and sometimes even of the fresh weight of less than 2 percent. Since the analytical error for most components was itself of the order of 1 percent, it is clear that a considerable improvement in the reliability of our work had been made. There is a great deal of satisfaction, when one comes to plot the data from such a set of samples, in having the figures for some component that was not affected by the treatment yield a horizontal straight line. Admittedly, there are certain difficulties. To obtain samples for some of the experiments with *Bryophyllum* leaves, my whole crew had to turn up in the greenhouse well before sunrise and put in an hour or so of concentrated hard work. One's popularity on a bright summer morning is not improved by demands of this kind, but all later admitted, when the data were ready, that it had been worthwhile.

There were three lines of investigation that seemed important; the continued study of Crassulacean metabolism, an examination of the effects of culture of tobacco leaves in solutions of organic acids, and renewed study of the chemistry of isocitric acid, for our laboratory was probably the only one in the world where this hitherto extremely rare substance was available in substantial quantities.

CRASSULACEAN METABOLISM

Plants of the family Crassulaceae exhibit the phenomenon of diurnal variation of acidity to an extreme degree, for from 15 to 20 percent of the organic solids are often involved in the diurnal changes in composition that occur. During the night the concentration of organic acids increases at the expense of starch, while during the day it decreases again and starch accumulates. Malic acid, and to a lesser extent citric acid, are the chief acid reactants.

Because of an unfortunate systematic error in calculating the protein content of *Bryophyllum* leaves exposed to diurnal variation of light, a few experiments that Pucher had recently carried out seemed to show that protein varies in amount in a manner correlated with the variation in organic acid content. Luckily the error was found and corrected in an unhappy paper that we published shortly before Pucher died, but it was necessary to check a number of points and to use the new sampling method so as to improve our general accuracy. Accordingly, a long series of culture experiments was planned in which we hoped to confirm and follow up many of the earlier observations. One that caused us considerable trouble was an observation that starch may accumulate in small amounts in *Bryophyllum* leaves excised at daybreak and cultured in darkness, that is to say exposed to an artificially prolonged night. This was surprising as one normally assumes that starch is a product of photosynthesis. How could it accumulate in the dark? The outcome was that the starch in such leaves did indeed continue to diminish for the first few hours after the leaves were picked as malic acid continued to increase, but after the malic acid reached its maximum and began to diminish, as it invariably does in prolonged darkness, starch increased moderately for a few hours. Thus, if malic acid is at a high enough level, starch is formed independently of the illumination (26).

We also repeatedly confirmed the observation that isocitric acid in *Bryophyllum calycinum* leaves undergoes little if any change in amount during culture of the leaves either in light or darkness, this in spite of the later observation of Attila Klein, who worked with us in 1962 and made use of a specimen of isocitric acid labeled with ^{14}C that I had prepared. He found that isocitric does indeed play an important part in the general organic acid metabolism of the *Bryophyllum* leaf (27). To be sure, we had seen an occasional set of samples, particularly one which was unusually low in starch at daybreak, in which the isocitric acid was drawn into the metabolism for a short time after the leaves were maintained in culture in darkness, but as soon as the malic acid began to diminish, the drain upon the isocitric acid ceased. It became clear that much is still to be learned about the metabolism of isocitric acid in this species, the most important point probably being why and by what means it accumulates in Crassulacean plants in so phenomenal a way. And if, as seems likely, most of that present is segregated into a pool which is remote from the part which is exposed to the activity of the enzyme systems, especially the aconitase of the cells, how is this segregation accomplished?

We studied the effect of prolonged exposure of *Bryophyllum* leaves to both light (28) and darkness (29) and the capacity of the leaves to recover from stresses thereby brought about (30). Prolonged exposure to light appeared to damage the enzyme systems concerned with the synthesis of malic acid when the stressed leaves were placed in darkness. This effect was presumably associated with the small loss of protein that was observed. Other experiments dealt with the effect of temperature on the fundamental transformation of starch to malic acid which occurs in darkness, and with the effect of light upon the citric acid of leaves collected at sunrise (31). Citric acid behaves in a manner much like that of malic acid although on a lower scale of quantity, but the increases seen in the first few hours of exposure to light, when malic acid is still at a high level although diminishing in amount, led us to conclude that citric acid arises from malic acid by a series of reactions for which there are many analogies in the behavior of these two acids in the tobacco leaf. Thus the level attained by citric acid and the variation in quantity present are functions of the concentration of malic acid. The level attained by malic acid is in turn a function of that of the starch.

METABOLISM OF ORGANIC ACIDS IN TOBACCO LEAVES

Our interest in the organic acids of tobacco leaves, aside from the fact that they are major components of extracts of the tissue, was greatly stimulated by the observation made in the early 1930s that citric acid increases, sometimes to a phenomenal extent when the leaves are cured. This also occurred when leaves were cultured in water in darkness, and evidence gradually accumulated indicating that, far from being merely a buffer system provided to stabilize the hydrogen ion activity of the cells, the organic acids are extremely reactive metabolites in the cell system. The preliminary experiment in which Pucher had fed a wide assortment of acids to leaves cultured in darkness gave convincing evidence of this. Plans were accordingly laid to embark upon a study of the effects upon the organic acids of culture of tobacco leaves in solutions of a long series of organic acids. For this we needed improvements in our analytical methods, and Chester Hargreaves, who came to us as a postdoctoral fellow in 1950, was asked to examine the method for citric acid and to improve the method to determine isocitric acid based upon the use of aconitase from beef heart to transform it to citric acid. This he successfully accomplished (15). The whole problem of the analytical determination of the organic acids was placed upon a new footing, however, by the brilliant work of James K. Palmer, who joined us in 1954. He was asked to see what could be done by the chromatographic methods which were then coming into use in many fields. The outcome was a technical improvement of methods originally described by Busch, Hurlbert, and Potter in which the acids are separated on Dowex-1 by elution with a gradually increasing strength of formic acid (32). The apparatus was simple, the results were accurate, and the whole procedure was rapid. Moreover, by suitable modification of the reagents, the method could be adapted to the determination of a wide range of organic acids. But we did not discard Pucher's method for citric acid. On the contrary, it became even more useful. Citric acid, isocitric acid, and a third component that we ultimately identified as phosphoric acid are eluted together from Dowex-1 as a single peak on the plot of the titration of the fractions. Accordingly, these fractions were combined, citric acid and phosphoric acid were determined in the mixture, and when working with *Bryophyllum* leaves, Hargreaves' aconitase technique was used to determine isocitric acid. Later, when the beautiful enzymatic method of Grafflin and Ochoa to determine isocitric acid became available it was used exclusively.

Thus, with improved analytical methods and assisted in many of the more recent experiments by the availability of organic acids labeled with ^{14}C, we undertook to examine the behavior of most of the components of the tricarboxylic acid cycle as well as a number of other acids when fed to tobacco leaves cultured in darkness.

There are two approaches to such problems. With radioactive substrates, only very dilute solutions of the acid are necessary because of the great sensitivity of the analytical methods. Furthermore, small samples of leaf tissue, even single small disks, are adequate for many kinds of experiments. But if one is interested in the behavior of the carbohydrates, and the proteins, as well as in the loss or gain of organic material from respiration or photosynthesis, large samples are necessary and, to detect changes by ordinary analytical methods of which one can be certain, it is necessary to, as it were, flood the leaf tissue with substrate. Thus one uses 0.2 M solutions of the substrate and prolongs the period of treatment. To be sure, thoroughly abnormal conditions are established, but one obtains unequivocal evidence of the enzymatic behavior one is seeking to establish.

Perhaps a word about the fundamental technique of these experiments may be helpful. With the improvement in accuracy from the use of statistical sampling, it became possible to consider quantitative relationships between the amounts of the substrate taken up and the amounts of its metabolic products, and, as the work progressed and our confidence increased, we were sometimes in a position to suggest definite enzymatic reactions to account for the observations. The general method depends upon the assumption that tobacco leaves placed with their bases in a 0.2 M solution of the potassium salt of an organic acid will take up a quantity of the acid which can be determined in several ways. The increase in potassium, and the increase in the alkalinity of the ash usually gave closely agreeing results. With aspartic or glutamic acid, the increase in nitrogen also served, and with oxalic or phosphoric acid and, as we found, tartaric acid, the increase in the administered acid furnished an additional measure, since neither oxalic nor tartaric acid is metabolized detectably in the tobacco leaf but merely accumulates. The respiration of leaves cultured in darkness was measured by the loss of organic solids, the net photosynthesis in light by their increase. Most of the work was done upon leaves cultured in darkness. The sets of samples, usually ten in number and with tobacco each consisting of 20 leaves, were arranged in V-shaped troughs in an air-conditioned room at constant temperature and humidity. A sample was dried for analysis at the start, and samples were removed at intervals or at the termination of the experiment, usually 24 or 48 hours. After being dried and weighed, the leaves were equilibrated in the constant temperature room until they came to constant weight, and the tissue was then ground and preserved in closed bottles in the same room. The bottles were removed only for the short time required to weigh out samples for analysis. All data, obtained as percentages of this weight, were then calculated to the amount in grams or milliequivalents that would have been found if each of the samples of fresh leaves had weighed initially one kilogram. Thus, any difference between the amount of a component in the treated sample and the initial sample was a direct measure of the effect of the treatment. In all experiments, control samples were cultured in water and in 0.2 M potassium succinate to afford data for comparison with the effect of the acid under study.

The outstanding reaction seen in tobacco leaves cultured in water in the dark is the transformation of malic acid into citric acid. There was also a moderate loss of organic solids due to respiration and evidence for a little proteolysis. Both of these latter reactions were substantially increased by culture on salts of organic acids. As data accumulated, it became clear that two moles of malic acid are used for each mole of citric acid produced. This reaction was greatly promoted by culture on malate (33, 34), and also by succinate. It was reversed by culture on citrate (35). Succinic acid is normally present in tobacco leaves in only minute quantities and, when fed to the leaves, about 90 percent of that taken up disappears, being converted mainly into malic and citric acids.

Only about one-half of the fumaric acid taken up by tobacco leaves is converted into other acids, citric acid being the major product. In contrast, only about one-third of the maleic acid fed at pH 6 underwent chemical change. At pH 5, this fraction was one-fifth. There was a minor increase in citric acid, but the ultimate fate of most of the maleic acid that underwent change was not clearly evident (36). The major effect was an inhibition of several of the enzyme systems of the leaves, in particular the systems involved in the formation of citric acid and the utilization of malic acid, and also the proteolytic system. It could be inferred that the presence of active thiol groups in the structure of the enzymes involved is essential.

From culture solutions of glycolic acid adjusted to pH 5 and 6, nearly 90 percent of the anion taken up by the leaves disappeared, and glycolic acid is thus nearly as active a metabolite as succinic acid in this system (37). The major product is citric acid which accumulated to nearly the same extent as it does when citric acid itself is administered. In addition, reactions occurred which led to the accumulation of moderate amounts of succinic acid and of substantial amounts of oxalic acid, the first instance we have encountered in which an important effect upon oxalic acid occurred. Oxalic acid is usually regarded as an end product of organic acid metabolism, and the present evidence suggests that one of its precursors is glycolic acid. Malic acid behaved in a manner which suggested that it serves as an intermediate in the transformation of the glycolic acid to citric acid. It increased moderately during the first 24 hours of culture, but then diminished substantially.

Malonic acid, in contrast, underwent transformation to only a small extent. Traces of this acid are normally present in the tobacco leaf, and it is a substantial component of the organic acids in leaves of leguminous plants such as beans and alfalfa, but its position in the general metabolic scheme is by no means clear. However, when present in tobacco leaves at a concentration that approached 0.1M at the end of 48 hours of culture, it led to a marked accumulation of succinic acid, an evidence of the inhibition of succinic dehydrogenase (38, 39). The loss of malic acid was substantial, but there was little if any change in the amount of citric acid present until the second

24-hour period of the experiment: at pH 5 there was a small increase and at pH 6 this was greater. The evidence suggested that malonic acid contributed to some extent to the formation of citric acid. Interpretation of the general results in terms of the tricarboxylic cycle is manifestly difficult.

Several experiments in which (+)-tartaric acid was fed to tobacco leaves led to the then unexpected result that this substance is not at all metabolized; it merely accumulates and has no effect whatever upon the series of reactions whereby malic acid is converted into citric acid even when the total amount of tartaric acid in the system finally exceeded the total amount of malic acid present. This observation enabled us to look more closely into the validity of our assumption that the uptake of an organic acid at any reasonable pH of the culture solution can be accurately calculated from the increase in the alkalinity of the ash and the percentage of the acid neutralized at the pH of the solution as deduced from the dissociation constants. Sodium tartrate was fed to a series of samples in solutions in the range pH 2.3 to 6.2 in which the acid was neutralized from 23 to 99 percent. The uptake was calculated from the increase in the alkalinity of the ash and from the tartaric acid found. The mean result of eight observations was that the calculated uptake was 100.6 ± 8.3 percent of the amount of tartaric acid found in the leaves after 48 hours of culture, and the conclusion could be drawn that our calculations were valid within these limits (40, 41).

A further outcome of this experiment was the evidence for the substantially increased respiratory loss of organic solids when salts of organic acids are fed to tobacco leaves, presumably because of the increased demand for the energy required to convey the acid from the vascular system into the cells, and especially the evidence for a stimulation of decarboxylation reactions when the culture solution is adjusted to a pH reaction lower than that of an extract of the normal leaf. This can be interpreted as a protective response to the advent of acid from the culture solution; there is only minor loss of carboxyl groups when the reaction of the culture solution is at or above the normal reaction of extracts of the leaves.

The analogy in the behavior of (+)-tartaric acid and oxalic acid when fed to tobacco leaves raised the question, why is (+)-tartaric not metabolized? The literature indicates that in other than the grape vine, few higher plants are positively known to contain any tartaric acid whatever; it is not a common plant organic acid. However, in the leaves and fruit of the grape it is the dominant acid, and the situation is one that invites speculation. The demonstration in 1951 by Bijvoet and his associates that both asymmetric carbon atoms of naturally occurring (+)-tartaric acid are configurationally related to D-glyceraldehyde, and are thus of the same configuration as carbon atoms 2 and 3 of glucose, led us to suggest that tartaric acid arises from glucose by a sequence of oxidation reactions whereby carbon atoms 5 and 6 are removed and the endstanding carbon atoms of the remaining substance are oxidized to carboxyl groups. Apparently there is no enzyme system in the grape plant or in the tobacco leaf that attacks the resulting (+)-tartaric acid and it therefore accumulates. It is thus also an end product of organic acid metabolism although to be sure it is not commonly seen. The subsequent demonstration by Vennesland and her associates that *meso*-tartaric acid, in which one of the carbon atoms certainly belongs to the L-family, is metabolized in leaf tissues lent some color to this view and there is more recent evidence in its favor.

To obtain some light on the so-called dark fixation of carbon dioxide, tobacco leaves were cultured in the dark in 0.1 and 0.2 M solutions of potassium bicarbonate for 24 and for 48 hours, and the effects upon the organic acid metabolism were examined (42). Control samples were cultured in potassium fumarate, succinate, sulfate, and secondary phosphate. The data indicated that from one-fifth to one-quarter of the bicarbonate ion taken up entered into the organic acid metabolism, and that a stimulation of the formation of citric acid occurred of the same order as that observed when succinate or fumarate is fed. The molar relationship between the amounts of precursor used to that of the citric acid produced is essentially constant whether bicarbonate, succinate, fumarate, or L-malate is made available to the enzyme systems of the leaf, and it could be suggested that the bicarbonate ion is metabolized with the production of a four-carbon organic acid, possibly oxaloacetic acid, which was subsequently converted into citric acid.

The examination of the composition of the control sample cultured in secondary phosphate, the purpose of which had been merely to provide a control at an alkaline reaction not far from that of the bicarbonate solutions used, led to the unexpected observation that phosphoric acid is eluted by formic acid together with citric acid from the Dowex-1 column, and thus in turn to the identification of the puzzling unknown acid we had invariably found in this fraction. Even more interesting was the observation that citric acid accumulated to a considerably greater extent than it did in the water or potassium sulfate control. Analysis of the data made it quite clear that the alkaline solution of phosphate had taken up carbon dioxide from the air of the culture room, and that this culture solution was also, in effect, one of bicarbonate. However, more detailed study was obviously required, and an experiment was carried out in which the effect of phosphate solutions adjusted to reactions from pH 4.4 to 8.8 was examined (43). There was no marked effect upon the behavior of the malic and citric acids save in the sample cultured at pH 8.8. In this, citric acid accumlated to twice the amount found in the potassium sulfate control, confirming the previous observation. The only specific effect of the phosphate detected was the accumulation of minor amounts of water-soluble substances which contained organically bound phosphate.

The repeated observation that the major effect of the culture of tobacco leaves on solutions of the common organic acids is an increase of citric acid led to a test of the effect of pyruvic acid (44). Citric acid did indeed accumulate in substantial amounts, and there was only a small decrease in malic acid; pyruvic acid was clearly extensively involved in the transformation.

There was a substantial uptake, although little pyruvic acid survived the operations of drying the leaves and evaporation of the formic acid eluates from the column. Proof that pyruvic acid became involved in the reactions that occurred was obtained in a parallel test in which a single tobacco leaf was cultured in a solution that contained pyruvic acid labeled with radioactive carbon. After 48 hours, the citric acid present had approximately the same specific activity as the labeled pyruvic acid, and the trace amounts of succinic acid and fumaric acid in the leaf were also highly labeled. Malic acid was only moderately labeled, but a consideration of the large amount present from the start showed that about as much radioactive carbon was present in the malic acid at the end as had found its way into the citric acid. The evidence pointed to the presence in the leaves of a condensing enzyme system similar in function to that of Stern and Ochoa.

Analysis of the data suggested that citric acid is present in the tobacco leaf in two compartments or pools. One of these receives and stores newly formed citric acid, the other and probably much smaller pool is the scene of rapid enzymatic reactions which may well follow the plan of the tricarboxylic acid cycle. However, the evidence that this scheme of reactions is responsible for the entire respiration of the leaf is still far from convincing.

The availability of substantial amounts of the potassium salt of isocitric acid in our laboratory naturally led to a study of the metabolism in the tobacco leaf of this rare but extremely important substance. One of Pucher's last researches was the development of a method to prepare isocitric acid in quantity from the *Bryophyllum* plant. After much study, he had hit upon the dimethyl ester of isocitric lactone as a substance of suitable properties and had accumulated several hundred grams of this compound (13). When D.G. Wilson joined us for a 2-year postdoctoral tour in the mid-1950s, I suggested that he might carry out a preparation by Pucher's method and then go on to see if a chromatographic method could not be worked out, for we knew that isocitric lactone could be separated on the Dowex-1 column from both citric acid and isocitric acid which are eluted together. He succeeded in doing this, but had difficulty in obtaining a satisfactory preparation of the lactone. It occurred to him one day that monopotassium tartrate is an unusually insoluble salt, and that, since isocitric lactone is another dicarboxylic acid, its monopotassium salt might also be reasonably insoluble. The literature suggested that the first carboxyl group of isocitric lactone would be maximally neutralized at about pH 3.25, so he adjusted a concentrated solution of the lactone to this reaction and added alcohol. A beautifully crystallized salt promptly separated in nearly quantitative yield (45). Samples of this salt and the corresponding rubidium salt supplied to Dr. A. L. Patterson of the Institute for Cancer Research in Philadelphia enabled him to work out the crystal structure in detail and later to establish that the asymmetric α-carbon atom belongs to the D family in confirmation of the extraordinary work of Katsura and his associates in Japan in 1960.

When Wilson left us to return to Canada, there remained a collection of many solutions that represented ether extracts of the acids from *Bryophyllum* leaves, and it occurred to me to try to isolate his potassium salt directly from them. This was possible, although the yield was small and the material rather impure, but one day a small crop of crystals which were considerably more insoluble than the potassium salt of the lactone separated from an aqueous mother liquor. Analysis showed that it was the monopotassium salt of isocitric acid itself. The outcome of this observation was a method to prepare isocitric acid as its potassium salt which is simple enough to serve as an excellent student preparation, and within a few weeks we had hundreds of grams of it on hand (46). Until recently when it became available commercially, we were the source of supply of this unique compound to colleagues all over the world who were interested in the metabolism of isocitric acid.

When isocitric acid is fed to tobacco leaves, it behaves quite differently from its behavior in the leaves of Crassulacean plants (47). From a culture solution at pH 5, about two-thirds of the substantial amount taken up disappears as such, and about 40 percent of this quantity is apparently transformed into citric acid. Malic acid increased slightly, and in a parallel single leaf experiment in which synthetic isocitric acid labeled with ^{14}C in carbon atoms 3 and 4 was used, malic and succinic acids were extensively labeled as well as the citric acid. It could be concluded that, although a part of the citric acid may have arisen through the condensation reactions of the tricarboxylic acid cycle, much of it also arose through direct transformation by an aconitase.

An experiment in which L-glutamic acid was fed to tobacco leaves in darkness showed that this substance ranks in metabolic activity with succinic and glycolic acids in that about 90 percent of that taken up is transformed into other substances (48). The nitrogen appeared mostly as asparagine, together with an appreciable amount of aspartic acid and ammonia and a little glutamine amide nitrogen. More than four-fifths of the acquired nitrogen could be thus accounted for. Citric acid increased strikingly, and succinic acid increased to about one-half the extent that it did in the control sample cultured on succinate. The small increase in the so-called minor acids, which are eluted from the Dowex-1 column in advance of succinic acid, could be quantitatively accounted for by the glutamic and aspartic acids in this fraction. In the parallel single-leaf experiment in which radioactive glutamic acid was administered, both aspartic and succinic acids attained a specific activity equal to that of the glutamic acid supplied, and malic and citric acids were highly labeled, the glutamine somewhat less so. The trace of D-glutamic acid in the culture solution obviously was not attacked at all. The general picture suggested that three moles of glutamic acid yielded one mole each of citric acid, asparagine, and ammonia, and succinic acid is obviously an important intermediate in the reactions.

Among the many puzzles encountered in our studies of the chemical changes that occur when tobacco leaves are subjected to the commercial operation of curing was the apparent complete stability of asparagine under circumstances where all other amino acids save possibly glutamine are completely oxidized with the liberation of ammonia (49). Tobacco leaves to be used for cigar wrappers, after being picked from the plants as the leaves successively became technically ripe, are strung on cords attached at each end to wooden laths and hung in the curing shed where they remain for many weeks. The leaves soon turn yellow and then brown, most of the water they contain evaporates, and about one-half of the protein disappears. The outstanding chemical event is the formation of asparagine in an amount equivalent to about 70 percent of the nitrogen of the protein which disappeared with subsequent oxidation of the amino acids produced. Amino acids other than asparagine do not accumulate save in negligible traces. In an attempt to throw some light upon this behavior, tobacco leaves were cultured in darkness on aspartic acid adjusted to pH 5 and 6 and also upon asparagine (50). In a parallel single-leaf experiment, radioactive aspartate at pH 5.5 was used. About three-quarters of the aspartic acid taken up was metabolized, asparagine and ammonia being the major products together with a little glutamine. There was a small but detectable increase in protein during the first 24 hours, a most unusual observation. Citric acid increased in substantial amounts, but there was only a minor increase in malic acid. Succinic acid also increased appreciably, an evidence that it must have served as one of the important intermediates in the reactions that occurred. In the single-leaf experiment in which radioactive aspartic acid was fed, succinic acid attained the same specific activity as that of the L-aspartic acid in the aspartic acid taken up, while the small amount of aspartic acid present at the end had twice this specific activity, an evidence that the trace of radioactive D-aspartic acid in the substrate was not metabolized at all.

The leaves cultured on asparagine after 48 hours contained appreciably more asparagine than was taken up in spite of the fact that about one-quarter of the uptake had been metabolized. The main nitrogenous products were ammonia and glutamine, and the organic acids behaved in a manner very much like that in the leaves cultured on aspartic acid. The general picture was that of a system capable of metabolizing aspartic acid and asparagine efficiently in the presence of another system capable of synthesizing asparagine. After 48 hours of culture, the effects of the second system predominated, thus leading to a false impression of the unusual stability of asparagine in tobacco leaves whether cultured in solutions or subjected to the technical operation of curing. Aspartic acid and glutamic acids are obviously as readily metabolized as most of the other organic acids we have studied, but aspartic acid is unique in that tobacco leaves, and probably many other plant tissues, make use of the ammonia produced by the oxidation of amino acids for the formation of asparagine which accumulates. Observations of this accumulation led Prianishnikov many years ago to draw an analogy between the behavior of asparagine in plants and urea in the animal. He considered both substances to be produced as a response to the presence of ammonia. This old "detoxication" hypothesis is no longer insisted upon today, for there is much evidence that ammonia can accumulate in leaves of certain species, particularly rhubarb, without harm. However, it is a fact that the formation of asparagine is the most common outcome in any situation where plant tissues—whether seeds, shoots, or leaves—encounter conditions where the protein is hydrolyzed at a rate greater than that at which it is synthesized.

In looking back upon these busy years when we tried to throw a little light upon the behavior of organic acids in one narrowly restricted phase of plant metabolism, one sometimes wonders what it all amounted to. One conclusion is clear. At the beginning of our work very little was known about the organic acids of leaves save that they form a substantial part of the organic substances in this extraordinary tissue. Why they are there and what they do was mostly a matter for speculation. Today the organic acids are recognized to be the central metabolites of the systems involved in carbohydrate and protein chemistry, in the phenomena of photosynthesis and respiration, and in many more functions of the life process in plants.

The present-day appreciation of the significance of the organic acids is mainly the result of the work of biochemists who have concerned themselves with enzymes. Although there is a long history of the detection of this or that enzyme reaction in plant tissues, it was not until the fundamental techniques of protein fractionation at low temperatures began to be widely applied some 40 years ago that various systems of coordinated enzyme reactions were discovered and offered as explanations of the chemical events observed in living cells. Pucher and I were both trained as organic chemists with a bent for analytical chemistry. I was chiefly interested in nitrogenous substances, he in non-nitrogenous: we were not physiologists at all, but as we learned more about the composition of leaf tissues and how it changed as the result of treatment, the physiological, or rather the biochemical aspects of the observations impressed themselves more and more upon us. Back in the early thirties there were very few investigators who concerned themselves with the chemical composition of plants, in fact astonishingly little was known about it, and one of the greatest and most pleasant surprises of my professional life was the award in 1933 of the Hales prize by the American Society of Plant Physiologists. I had had no inkling that our work with leaves had been noticed by colleagues in this discipline. For many years we were chiefly busy with the development of analytical methods suitable for application to extracts of plants. Accuracy was the main consideration and revision of the techniques was continuous. Our concern was with substances and how they behaved. Thus I have been content to leave the development of detailed biochemical mechanisms to those whose special training equipped them to deal with this difficult field. In my view accurate measurement of the chemical changes that occur must precede attempts to account for them in terms of enzyme-catalyzed reactions.

The development of chromatographic methods in recent years has enormously broadened the field of analytical attack upon physiological problems and has diminished the scale upon which one may work to what an organic chemist quite properly regards as traces of material. Yet I am still old-fashioned enough to prefer a crop of crystals that one can weigh to the measurement of the location of a spot of color on a strip of filter paper. Thus as I approach 80 years of age I am happy to turn over to my younger and more broadly trained colleagues the responsibility for further progress in the study of the organic acids of plants.

I cannot close, however, without a word of appreciation for the devoted help of the members of my staff over the years. To Israel Zelitch, my successor as head of the laboratory, I am indebted for innumerable helpful discussions as well as for direct aid in many experiments. And to K. R. Hanson, J. K. Palmer, A. Klein, A. N. Meiss, C. A. Hargreaves II, C. C. Levy, and especially in the early years to G. W. Pucher, A. J. Wakeman, C. S. Leavenworth, Abraham White, H. E. Clark, and Emil Smith I owe an equal debt. For more than 40 years Laurence Nolan gave practical assistance in preparing the innumerable samples for analysis and in the invention of apparatus (51) to simplify the work. Marjorie Abrahams and Katherine Clark have given devoted and skilled technical assistance. I have unblushingly picked the brains and used the ideas and abilities of all, and if there is any merit in what we have done, by far the greater part of it belongs to these devoted colleagues.

LITERATURE CITED

1. Vickery, H. B. 1918. *Trans. Nova Scotia Inst. Sci.* 14:355
2. Vickery, H. B. 1922. *J. Biol. Chem.* 53:495
3. Vickery, H. B. 1924. *J. Biol. Chem.* 60:647; 61:117. Ibid 1925. 65:81, 91, 657
4. Vickery, H. B., Leavenworth, C. S. 1925. *J. Biol. Chem.* 63:579
5. Vickery, H. B., Leavenworth, C. S. 1927. *J. Biol. Chem.* 72:403
6. Vickery, H. B., Leavenworth, C. S. 1928. *J. Biol. Chem.* 76:707; 79:377
7. Vickery, H. B., Leavenworth, C. S. 1927. *J. Biol. Chem.* 75:115. Ibid 1928. 76:437, 701; 78:627
8. Vickery, H. B., Pucher, G. W. 1929. *J. Biol. Chem.* 83:1
9. Pucher, G. W., Vickery, H. B., Leavenworth, C. S. 1935. *Ind. Eng. Chem. Anal. Ed.* 7:152
10. Vickery, H. B., Pucher, G. W. 1931. *Conn. Agr. Exp. Sta. Bull.* 323
11. Pucher, G. W. 1942. *J. Biol. Chem.* 145:511
12. Pucher, G. W., Vickery, H. B. 1942. *J. Biol. Chem.* 145:525. 1946. 163:169
13. Pucher, G. W., Abrahams, M. D., Vickery, H. B. 1948. *J. Biol. Chem.* 172:579
14. Pucher, G. W., Vickery, H. B., Leavenworth, C. S. 1934. *Ind. Eng. Chem. Anal. Ed.* 6:190
15. Hargreaves, C. A. II, Abrahams, M. D., Vickery, H. B. 1951. *Anal. Chem.* 23:467
16. Pucher, G. W., Sherman, C. C., Vickery, H. B. 1936. *J. Biol. Chem.* 113:235
17. Pucher, G. W., Vickery, H. B., Wakeman, A. J. 1934. *Ind. Eng. Chem. Anal. Ed.* 6:288
18. Vickery, H. B., Osborne, T. B. 1928. *Physiol. Rev.* 8:393
19. Vickery, H. B., Schmidt, C. L. A. 1931. *Chem. Rev.* 9:169
20. Vickery, H. B. 1950. *Yale J. Biol. Med.* 22:387
21. Vickery, H. B. 1947. *J. Biol. Chem.* 169:237
21a. Vickery, H. B. 1952. *Chem. Eng. News* 30:4522
22. Vickery, H. B. 1963. *J. Org. Chem.* 28:291
23. Pucher, G. W., Leavenworth, C. S., Vickery, H. B. 1948. *Anal. Chem.* 20:850

24. Pucher, G. W., Vickery, H. B. 1949. *J. Biol. Chem.* 178:557
25. Vickery, H. B., Leavenworth, C. S., Bliss, C. I. 1949. *Plant Physiol.* 24:335
26. Vickery, H. B. 1952. *Plant Physiol.* 27:231. 1957. 32:220
27. Klein, A. O. 1964. *Plant Physiol.* 39:290
28. Vickery, H. B. 1953. *J. Biol. Chem.* 205:369
29. Vickery, H. B. 1954. *Plant Physiol.* 29:520
30. Vickery, H. B. 1956. *Plant Physiol.* 31:455
31. Vickery, H. B. 1959. *Plant Physiol.* 34:418
32. Palmer, J. K. 1955. *Conn. Agr. Exp. Sta. Bull.* 589
33. Vickery, H. B., Hargreaves, C. A. II 1952. *J. Biol. Chem.* 197:121
34. Vickery, H. B. 1955. *J. Biol. Chem.* 214:323
35. Vickery, H. B. 1952. *J. Biol. Chem.* 196:409
36. Vickery, H. B., Palmer, J. K. 1956. *J. Biol. Chem.* 218:225
37. Vickery, H. B., Palmer, J. K. 1956. *J. Biol. Chem.* 221:79
38. Vickery, H. B., Palmer, J. K. 1957. *J. Biol. Chem.* 225:629
39. Vickery, H. B. 1959. *J. Biol. Chem.* 234:1363
40. Vickery, H. B., Palmer, J. K. 1954. *J. Biol. Chem.* 207:275
41. Vickery, H. B. 1957. *J. Biol. Chem.* 227:943
42. Vickery, H. B., Palmer, J. K. 1957. *J. Biol. Chem.* 227:69
43. Vickery, H. B., Levy, C. C. 1958. *J. Biol. Chem.* 233:1304
44. Vickery, H. B., Zelitch, I. 1960. *J. Biol. Chem.* 235:1871
45. Wilson, D. G. 1963. *Can. J. Biochem. Physiol.* 41:1571
46. Vickery, H. B., Wilson, D. G. 1958. *J. Biol. Chem.* 233:14
47. Vickery, H. B., Hanson, K. R. 1961. *J. Biol. Chem.* 236:2370
48. Vickery, H. B. 1963. *J. Biol. Chem.* 238:2453
49. Vickery, H. B., Meiss, A. N. 1953. *Conn. Agr. Exp. Sta. Bull.* 569
50. Vickery, H. B. 1963. *J. Biol. Chem.* 238:3700
51. Nolan, L. S. 1949. *Anal. Chem.* 21:1116

Ann. Rev. Physiol., Vol. 31

A HALF CENTURY IN SCIENCE AND SOCIETY

By Maurice B. Visscher

Maurice B. Visscher

A HALF CENTURY IN SCIENCE AND SOCIETY

By Maurice B. Visscher

Department of Physiology, University of Minnesota Medical School
Minneapolis, Minnesota

Writing for one's colleagues, young and old, about a half-century of experiences in science and society is an opportunity which challenges one to be pertinent and constructive in what one chooses to treat. Anyone who has worked in large institutions for a lifetime and lived through the societal explosions and turmoil of the twentieth century has seen enough to provide material for volumes of anecdote and commentary. In this brief account my intention is to recount a few of my own experiences as a scientist and as a citizen.

It may not be inappropriate to start out by saying that I was born into a Calvinist Dutch family, a grandson of immigrants who left the Netherlands in the eighteen-forties to escape the then-current wave of religious intolerance and to seek their better economic fortunes in the New World. These immigrants began building a college before their own houses were completed. Their motive was to provide clergymen and teachers. My father and his three brothers and two sisters graduated from that college, as did I. My own tendency to participate in the actualities of academic, social, and political life of our times has at least a partial explanation in the intellectual milieu in which I was raised. Despite the authoritarian nature of Calvinist doctrine, more than lip service was paid to devotion to evidence and logic. Hope College, which the pioneers established, very early appointed professionally trained scientists to its staff and encouraged serious study of the sciences. So successful was it, in fact, that in an analysis of the origins of Doctors of Philosophy in the natural sciences in the United States, Hope College was found to be among the first ten institutions in the country as to the fraction of its baccalaureates obtaining such degrees in the first half of this century. The college did also, however, provide a very large number of clergymen and teachers. Its "success" in the sciences was not what its founders envisioned as its mission, and I know that some of its recent governing board members have not been happy with the fact that, as in my own case, many of its graduates have strayed away from Calvinist theology.

There is a certain irony in the fact that the very rigidity of the Calvinist position promotes a sense of urgency in those individuals who for logical reasons have had to reject its theology, to be candid about their rejection of it, and even to be more "evangelistic" about their newfound intellectual outlook than they would probably have been if they had not had such strict upbringing. I doubt that if I had been raised in a humanistic Unitarian fam-

THE EXCITEMENT OF SCIENCE

8243-2601/78/1127-0637$01.00 © 1978 ARI   637

ily, I would have been as active in the promotion of humanistic religious organizations as I have been. My father took a relaxed view of religious matters. He had prepared himself for teaching and did teach in elementary and high schools for part of his life, turning to part-time and full-time farming later.

In the latter connection, I should like to mention that my own experiences in participation in the hard work of a family earning a living from the soil have made me very skeptical about the idea that there is some soul-saving grace in youngsters being obligated to spend much of their out-of-school hours in gainful occupation. For me, at least, it consumed so much time and energy that only a fortunate set of genes allowed me to get a reasonably good education in spite of the necessity to participate in earning a living. Perhaps I am wrong, but I think that my father's interest in intellectual matters and in nature study had much more to do with my interest in scholarly things and in my energetic pursuit of them than did the long out-of-school work hours imposed on me by the economic status of my immediate family. There may be poison in economic affluence, but it probably stems from the perversion of family interest from pursuit of knowledge to pursuit of pleasure in some wealthy families, rather than from the lack of necessity for gainful occupation on the part of the children.

My boyhood, adolescent, and college years were undoubtedly fortunate for me because they opened my eyes to the great range of opportunities which were open to reasonably intelligent and diligent persons. Before I graduated from college, I had literally no fixed idea as to what I wanted to do with my life. I saw so many attractive things to do that it was hard for me to make a choice. Two exceptionally stimulating biology teachers piqued my interest in the biological sciences. One was Miss Lydya Rogers in my high school, and the other, Dr. Frank N. Paterson, in my college years. I took courses in college in botany, zoology, embryology, bacteriology, physiology, comparative anatomy, and genetics, as well as inorganic, qualitative and quantitative analytical and physiological chemistry. My physiology was a skimpy one-year, and my mathematical background in college was also inadequate. But I enjoyed literature, history, political science, philosophy and psychology, and some Greek and Latin which have been valuable to me in vocabulary and language structure.

My family circle included a number of lawyers, several ministers and missionaries, one physician, several teachers, farmers and businessmen. One of my older brothers was earning his doctorate in zoology at Johns Hopkins while I was a college student. My physician uncle was actually my favorite. He had reduced fractures of my arm twice as well as having sewed up a partly severed finger, and his gentle, confident way of giving professional help impressed me with the usefulness of his calling. He was such a kindly person that even when he left my blood aunt for another woman, I couldn't but think that my aunt was in large measure at fault. A divorce in a Calvinist family is not a conventional or approved procedure, and I kept my counsel pretty largely to myself on this matter.

Family attitudes undoubtedly have great impact on the career choices of young people, and in my case it was a virtual certainty that I would enter some so-called learned profession. Even my father, who had abandoned teaching himself, appeared to take it for granted that I would.

I ended up deciding to go into some medical science. I applied for scholarships for study in medicine and any one of the three sciences, bacteriology, biochemistry, or physiology. Bacteriology was on the list because I had registered for a "Problems" elective in my senior year, and had made a rather serious study of the colon bacillus count, the organic matter, and other chemical characteristics of the water in the river and lake system into which the city of Holland, Michigan, in which Hope College is situated, dumped its raw sewage at that time. In connection with that study, I read what I could about sanitary engineering and found it to be interesting.

At this point came the decisive turn toward the medical research field, and almost accidentally it happened that the dean of the Medical School of the University of Minnesota, Elias Potter Lyon, had come from Hillsdale College, another of the small private colleges in Michigan, and made it a practice to look into small liberal arts colleges for likely candidates for graduate and medical education at Minnesota. He had sent announcements of scholarship opportunities at Minnesota to my biology professor, who in turn advised me to apply. Dean Lyon offered me a modest stipend as a Teaching Fellow to study at Minnesota. I accepted gladly. It was the circumstance of the scholarship offer in a physiology department that undoubtedly decided for me that my career would be in physiology rather than biochemistry or microbiology.

The fate that brought me to the University of Minnesota for my graduate and medical education was a fortunate one for my subsequent career for many reasons. I put first among the advantages the fact that my graduate advisor, the late professor Frederick Hughes Scott, encouraged me to use the field of physical chemistry as the minor field for my Ph.D. program. The background of physical chemistry permitted me to approach the problems of material transport and other basic problems in ways that would have been impossible without it.

Another advantage that accrued from my acceptance as a student at Minnesota derived from the very permissive attitude of the faculty of the Medical School with regard to curriculum choices for medical students. The faculty had set up an "honors program" for specially selected students which eliminated entirely the lock-step curriculum for such students. It permitted satisfying the requirements of any department by private study and comprehensive examination, in lieu of registration for its regular courses. It resulted in my being able to satisfy the requirements for both the Ph.D. and the M.D. degrees in four calendar years of registration in the University but with considerable independent study. I spent one summer delivering babies and another in a substitute internship in internal medicine under the supervision of a superb clinical teacher, Dr. George E. Fahr, during which time I learned enough about the problems of diagnosis and treatment of disease in patients to give me a healthy respect for the clinical investigator and for the competent physician as well.

The fact that virtually all fixed formalities, aside from examinations and a modest amount of clinical experience, were eliminated from the M.D. requirements at Minnesota for honors students permitted and encouraged me to gain a broader and more useful background for both teaching and research than I would otherwise have obtained. It may be pertinent to note that at the present time, when reform in medical education is in the air and the emphasis is being placed in many institutions upon integrating, streamlining, and shortening the so-called core curriculum, one is actually making the process even more rigidly lock-step than it was before. The elaborately planned "integration" of material, supposedly for easier comprehension by the student, contributes to the rigidity. It is to be questioned whether serious students need such spoon-feeding, and furthermore it is doubtful that medicine as a learned profession will be improved by the imposition of greater rigidities in the core curriculum. I regret the fact that students at Minnesota will in the immediate future have less freedom of action than I myself had forty-seven years ago. It seems to me to be odd that in this age of emphasis upon flexibility in education, medical schools should be moving in the direction of discouraging independence of action.

Major factors in my own education as a scientist were my experiences as a postdoctoral National Research Council fellow in two great centers of physiological research, University College, London, and the University of Chicago. Working at the laboratory bench and the operating table with the late Ernest Starling, and discussing problems with A. V. Hill, E. B. Verney, and a dozen other staff members and students at University College, and doing the same with Anton J. Carlson, Arno B. Luckhardt, George E. Burget, and others at the University of Chicago, were priceless opportunities to broaden my knowledge and sharpen my scientific wits. At the University of Chicago I also met A. Baird Hastings and imbibed from him the points of view of the Van Slyke school. In fact Baird Hastings taught me practicable micromethods for measuring the $[H^+]$ in blood, which allowed George Burget and me to discover the pH dependence of epinephrine in its action on blood vessels, a phenomenon which has been found to be a general one for most catecholamine actions upon receptors.

I consider that my most creative work has been done in the field of transport of materials. It may be of interest that my work in this whole field began with a request in 1930 from a post-doctoral student, Dr. W. R. Pendleton, that I help him investigate the possibility of using the entire intact intestine as a dialyzing mechanism in acute renal failure. I suggested that the first thing to do would be to study the transport of urea and other crystalloids across the intestinal mucosa at various levels of the gut. So began a dozen years of work because I became intrigued with the complexities of the processes involved. Univalent ion impoverishment by the gut proved to occur against electrochemical gradients, and water moved at rates not proportional to water activity ratios. At about this time I had the good fortune to move to an institution where my colleagues in physics were making radio isotopes with a van de Graaf machine and the late Professor John H. Williams provided me with isotopes of chloride and sodium, permitting me to study directly the bidirectional movements of those ions. Professor A. O. C. Nier had just perfected his mass spectrometer, and his generous cooperation allowed us to study mass isotopes of various elements as well.

The point I am trying to make is that I owe to students and colleagues the greatest of debts. A University—a community of scholars—from the beginner in the search for knowledge to the most sophisticated scholar, is an environment in which creativity can and does flourish. Our society has not produced any other mechanism which is as generally successful in the advancement of basic knowledge. It is certainly true that many basic discoveries have been made in research institutes, divorced from educational activities, but historically until the present most basic discoveries have come from the universities and other similar teaching institutions.

There are voices today suggesting that this era is passing, because teaching must be the primary function of educational institutions and for them additions to knowledge are secondary goals. However, I would point out that graduate education is impossible except in the context of creative scholarship. Insofar as an educational institution even pretends to carry on graduate instruction it must of necessity support research as a major function. Unless it does, its graduate programs are a fraud, both upon its students and upon society at large. The proportionate emphasis upon teaching and research activities that may be optimal may be arguable, but again let it be noted well that graduate schools are the training grounds for almost all investigators. An environment in which mediocre research is done is not optimal for the training of superior investigators. Thus, if only to have superior investigators to work in industrial and other nonteaching research institutes, one must support universities liberally enough to provide them with adequate quotas of the most competent researchers. To do otherwise would be to kill the goose that lays the golden egg.

Scientific Communication

The last half-century has seen the most rapid changes in the support of the scientific enterprise that have ever occurred. Over that time the annual budget of the Medical School with which I am asssociated has increased by more than two orders of magnitude. When I began work as a graduate student I could have read all of the pages published in every physiological journal in the world and still have had much time for laboratory research and teaching. Today it has been calculated that to read or rather to scan the physiological periodical literature appearing in a single year, allowing one minute per page, would require the full time of a scholar for more than three years.

We in scientific work must recognize that the very success of science in transforming society in ways agreeable to its populations has brought about a demand for further advances and a willingness to spend, even an insistence upon providing, the additional funds to multiply the magnitude of the enterprise in one lifetime. In the biological sciences practical exploitation of knowledge of genetics, pest control, and plant and animal nutrition has greatly affected all of agriculture and animal husbandry. In the biomedical field we have seen the virtual disappearance in many countries of numerous diseases, and the therapeutic control of many others. Virus attenuation for vaccination, the chemotherapeutic and antibiotic agents, advances in diagnoses and management of surgical diseases, and many other developments based upon scientific discoveries have so pleased the public at large that it has come to believe that all that is necessary to solve other health problems is to direct the allocation of money to employ people to solve them. There is a real danger, first, that uninformed persons in positions of influence and power may waste vast sums in attempting to achieve practical results and, second, that after failing to get the desired practical results, they may bring disillusionment to a public that has come to expect the impossible, with consequent loss of support for the scientific enterprise. The public at large does not realize that, although creativity and originality can be encouraged by appropriate incentives, fresh ideas cannot simply be ordered and produced. Furthermore, successful practical solutions of particular problems require that basic background knowledge be available. Open-heart surgery, for example, could not have been achieved without a great deal of background knowledge of myocardial metabolism and other aspects of cardiac physiology, as well as of respiration, blood coagulation, and other physiologic phenomena. Practical solutions to organ transplantation problems will not be achieved until the mechanisms of the immune response are better understood. Approaches to prevention of arteriosclerosis and hypertension are very unlikely to succeed until facts now unknown are discovered.

The huge expansion of the scientific effort has brought severe problems in the field of scientist-to-scientist communication. As I have already noted, it has become impossible for any scientist even to scan the literature in any broad field. The research scientist now must rely upon others to bring information to his attention. He must even rely upon the judgment of others as to what is important for him to know. As yet we have very imperfect tools to classify, store, and retrieve information. The explosion of knowledge is itself the major cause of our distress in this regard.

My own activities in attempting to help solve this problem have been directed toward improving the effectiveness of abstracting and indexing and of critical compendiums and reviews. In the first category I undertook, as President of the Board of Directors of Biological Abstracts, Inc., to obtain financial subsidy from federal sources to allow it to continue its operations, after it had incurred large deficits which had prevented publication of indexes for a number of years. *Biological Abstracts* was put back on its financial feet and has since become an important element in the field of information source identification and retrieval.

In the second category I served on the Board of Publication Trustees of the American Physiological Society, part of the time as its Chairman, and had to do with the initiation and the first ten years of production of the *Handbook of Physiology* series as Chairman of its Editorial Committee. In this series, of which thirteen volumes have appeared, we have attempted to have summarized by the most competent experts the present state of knowledge in large fields of physiology. The entire field has not yet been covered, but already a second treatment has been begun for the first section, neurophysiology, which was published ten years ago. This effort is pointed at providing a solid base for graduate students and teachers, as well as investigators, to start from in beginning serious study of any problem about which they may have little first-hand knowledge of the primary literature.

As to the third category, my main interest has been in helping, as a member of its Board, in the development of the program of *Annual Reviews*. Annual Reviews, Inc. now publishes critical review volumes yearly in each of fourteen major areas of science. It is expanding into other fields. It aims to bring new knowledge into focus. It also provides a fairly convenient and very inexpensive indexing mechanism for searches by scientists in their private libraries. As a nonprofit organization it is able, largely because of the generous unpaid service of all of its distinguished invited authors and the token payments to its Editors and Editorial Committees, to provide these services at a fraction of the cost of comparable volumes published commercially.

My interests in these activities are related to my belief that scientists must concern themselves with the control of the information exchange process, if the element of usefulness is to be given first priority. I am also concerned that the costs of information retrieval mechanisms be held to a minimum.

The mechanisms we are now using for secondary publication are certainly not optimal for all aspects of information storage and retrieval. They are fragmentary in their coverage of the huge volume of primary publication and they are time consuming for exhaustive searches. However, there is no substitute for expert judgment in any critical review and I venture to predict that such periodic reviews will be useful for a very long time to come. Machine methods may soon lighten the loads of the review authors, by locating for them all the pertinent papers in a limited field which they should see. Criticism, integration, and interpretation will, I expect, remain the job of the scientific scholar.

Scientist as Citizen

Two women whom I wish I had never encountered have played roles in my life as a citizen. The first was Irene Castle McLaughlin, a famous dancer of the World War I period, and the second was Christine Gesell Stevens, a very prominent New York and Washington socialite. The former was the leader of the antivivisectionist movement in Chicago in the thirties and the latter is the most influential proponent of regulatory and restrictive legislation to control animal experimentation today on the Washington scene. Both of them are quite obviously more interested in protecting pets than people. Irene Castle McLaughlin operated the Orphans of the Storm, a pet refuge near Chicago. Christine Gesell Stevens operates the Animal Welfare Society in New York, which provides a forum for her views.

Irene Castle McLaughlin had much to do with inducing William Randolph Hearst to order the editors of his chain of newspapers to initiate and maintain a barrage of scurrilous and fraudulent attacks on animal experimentation in the thirties. It will be remembered that the Great Depression was causing newspapers to disappear in those years. One of my friends, Llewellyn Jones, had been the Literary Editor of a major Chicago daily, which went bankrupt, and was pounding the pavements looking for a job. He was approaching the absolute end of his financial rope before he took a feature-writing job with the Hearst papers in Chicago as a last resort to feed his family. His first assignment was to assemble more material on the "brutality" and "futility" of animal experimentation. His first call was to my office in the Department of Physiology of the University of Illinois. He told me what his predicament was and asked whether I would help him write a constructive story on the great service that animals, especially unwanted pound animals, were making to human welfare through research. He also intended to point out how additional financial support could improve the housing and care of such animals. It turned out that his hope that his editor would accept a reasoned factual account and a constructive proposal was a vain one, and his writings were rejected. A month later he came back to visit me saying, "I have just received my certificate of sanity. I have been fired by the Hearst people!"

The scientific community at first believed that the orders given by W. R. Hearst, Sr. to his editors to attack "vivisection" might have been based on genuine misinformation and that he might be amenable to persuasion. A large group of Chicago biologists, headed by the then President of the National Academy of Sciences, Frank R. Lillie, and including of course Anton J. Carlson, addressed a long telegram to Mr. Hearst at his San Simeon ranch, detailing the reasons for supporting rather than attacking the use of animals in scientific study. Our naiveté was apparent to us a few days later when we learned that Irene Castle McLaughlin herself was at that very time a house guest at his ranch. No change in the Hearst policy occurred until a number of important drug and cosmetic advertisers in the Hearst magazine chain withdrew their accounts because of the attacks. The reaction then was quite prompt.

Mrs. McLaughlin attempted in 1934 to get the Council of the City of Chicago to rescind an ordinance which authorized the poundmaster to release to approved scientific institutions unclaimed impounded animals which would otherwise be put to death. She appeared before the Aldermen in public hearings and attempted to gain the support of various civic groups. In one of her appearances before the City Council she was asked somewhat facetiously by an Alderman as to why she was so much interested in dogs. Her reply was, "In my lifetime a lot of men have, but no dog has ever let me down." In her zeal to stop animal experimentation she challenged scientists to debate the justification for "vivisection" with her before women's groups, thinking she would have sympathetic audiences. She had exhausted the patience of such hardy debaters in Chicago as Anton Carlson and Andrew Ivy who then headed the Departments of Physiology at the University of Chicago and Northwestern University. Since I was their very junior colleague as Head of the Department of Physiology at the Chicago branches of the University of Illinois, they insisted that it would be my turn to meet the lady in her challenge to debate before a thousand or so of the members of the Chicago Women's City Club in the huge ballroom of the Michigan Boulevard Blackstone Hotel, and also later before a little smaller group in the Oak Park Arms hostelry in the swank west suburban area of Chicago. I was forewarned that Mrs. McLaughlin made it a practice in such debates to be the second speaker and always to time her arrival so as to be able to make an entrance interrupting the first, dressed in arresting attire and leading a couple of beautiful dogs on leash to the platform. So, anticipating such behavior, I arranged that the wife of a colleague would sit in the front row with a four-year-old son who would, on the proper signal, come up to the platform and sit on my lap if "Irene" came in with her canine stage props. I also arranged to speak entirely from lantern slides, with the ballroom entirely darkened, and spent much of my allotted time on the benefits that "vivisection" had brought to child health and welfare.

Mrs. McLaughlin behaved according to prediction and made an entrance to the stage—in pitch darkness—with two sleek white Russian wolfhounds on leash. She was sitting quietly when I asked for the lights as I finished. After she was "introduced", I waited a decent amount of time to let her get well into her tirade against the sadistic scientists before I motioned for my own stage stealer to cross the platform. When he did and climbed on my knees, a loud titter went over the audience of women. Mrs. McLaughlin faltered, flustered, turned to me and said, "If I had known that you would use such unfair tactics, I would have brought a half-a-dozen children myself." Which, as a matter of fact, she did two weeks later at the Oak Park Arms Hotel. A "vivisectionist" certainly could hardly be expected to give undue quarter to a superannuated misanthropic danseuse who put pets before people. It would be simpler if the 1968–1969 version of opponents of animal experimentation were as naive as Irene Castle McLaughlin, but some of them are not.

Irene Castle McLaughlin was a straightforward antivivisectionist and was recognized by the saner sections of the public as a crank. She never succeeded in putting her ideas across. Christine Gesell Stevens is a person of much greater plausibility and asks, not for abolition, but rather for rigid regulation and restriction of the use of animals in research. I have labeled her and her ilk as the neoantivivisectionists, since they maintain that they are not opposed to all use of animals in scientific study, but only oppose what are in their opinions unnecessary and improper uses.

Mrs. Stevens is the daughter of the late Dr. Robert Gesell who was Professor of Physiology at the University of Michigan. A few years before his death Robert Gesell shocked his physiological colleagues by beginning a crusade for the regulation and restriction of the use of living animals in scientific study. He appeared before a business meeting of the American Physiological Society and proposed a resolution which would have placed that Society in the position of requesting the Congress of the United States to enact legislation similar to the British Cruelty to Animals Act of 1876. He charged that cruelties were being perpetrated upon animals in many research laboratories and that millions of animals were sacrificed annually in useless and painful experiments. He asserted that tissue cultures could be substituted for whole animals in large segments of medical research, as for example in testing anticancer and other drugs. As might be anticipated, his ideas did not receive a very warm reception from his scientific colleagues. His motion was referred to a committee of three past-presidents of the Society, of which I happened to be one, who met for several hours with him to try to ascertain what precisely was in his mind. We attempted to learn what specific incidents Gesell could describe, but he persisted in generalizations such as, "you know as well as I do that many investigators are careless" and "millions of mice are uselessly made to suffer in cancer therapy studies."

After his death his daughter, who had married Mr. Roger Stevens, a very wealthy and powerful behind-the-scenes politician, has carried on the vendetta her father began in his dotage against former colleagues.

The drive on the part of the antivivisectionists of various shades to put major impediments in the way of animal experimentation made it necessary for the medical and related biological science community to organize an educational arm, the National Society for Medical Research, to counteract the propaganda for special legal measures to regulate and harass scientists and scientific institutions in their employment of animals. The late Professor

Anton Carlson persuaded the Association of American Medical Colleges to spearhead the organization of a hundred or more scientific and professional societies to carry on this work. He himself was the President of the NSMR until shortly before his death and was followed by Dr. Lester Dragstedt, Dr. Hiram Essex, and four years ago by myself. I had been a Board member since the inception of NSMR and its Vice President for a number of years, and became convinced of the pressing need to expand the educational work of the NSMR when important political figures on the national scene were enlisted by Mrs. Stevens on the side of the neoantivivisectionists.

The former legal counsel for the Stevens' financial empire, now Associate Justice of the U.S. Supreme Court, Mr. Abe Fortas, wrote for her the draft of a bill which is the U.S. counterpart of the British Cruelty to Animals Act of 1876. On account of her husband's position of political influence, having been Chairman of the Finance Committee of the Democrats and a friend of Presidents and of many Senators and Congressmen, she was able to induce a dozen top leaders in the Congress to sponsor her proposed legislation. Many of her one-time supporters have since backed away from her camp, but in this era in which Federal regulation of industry, business and trade is regularly accepted practice, it does not seem improper to some Congressmen to extend regulation into the scientific enterprise, especially since the Federal government is the major supporter of such work. The idea that general anti-cruelty laws and professional self-regulation might constitute superior control mechanisms does not seem to appeal to those members of Congress who have listened to Mrs. Stevens as a personal friend and a political power by matrimony.

The biological science community is not finished with the problems of legislation in the animal welfare field. Further regulatory legislation threatens at the present time and the animal welfarists have access to mass circulation magazines with inflammatory and misleading, even false statements about the use of animals in scientific research. The general public is still inadequately informed about the realities of the situation and it is my personal view that it would condone legislation which would be harmful to its own interests. Furthermore, many scientists, fortunately not a majority, are so frightened by the "little old ladies of both sexes" in the antivivisectionist camp that they advocate accepting bad legislation in order to avoid something worse, as Darwin and Huxley did in Britain in 1876.

LOYALTY-SECURITY PROBLEMS AND SCIENCE

The key role of science in modern military matters has made scientists the objects of suspicion and distrust as well as the instruments of innovation in weaponry and defense. The discoveries that led to the development of atomic fission and fusion bombs were obviously not entirely fortunate for scientists or for society. The atomic bomb has made life insecure for the human race as a whole, and the drive for secrecy about its science and technology spread the poison of suspicion as to the loyalty and reliability of scientists over the whole community, not simply in the military establishment. The fear that scientists might reveal the "secret" of the bomb to Russians, Chinese, or even Frenchmen, set the stage for the Oppenheimer and Condon persecutions and provided the atmosphere in which the then President Truman felt obliged to issue the government-wide Executive Order in 1947 requiring loyalty-security clearance for every Federal employee in all categories. That Executive Order in turn provided the base upon which the Joe McCarthy communists-in-government scare was built. Truman undoubtedly thought he was inaugurating a controllable system of clearance procedures but as it turned out, he began the McCarthy-Nixon-HUAC era.

This era had major implications for science. Instead of encouraging loyal service for government agencies it discouraged any service at all by many scientists who saw in the system the beginnings of a police-state with "thought-control", comparable in many respects to the practices of the authoritarian regimes that World War II was fought to eliminate as threats to individual freedom. As the representative of the American Association of Scientific Workers, I introduced a resolution into the Council of the American Association for the Advancement of Science calling for a committee study of the effects of the loyalty-security order on science and scientists. The resolution was adopted and I was asked to chair a Special Committee on Civil Liberties of Scientists for the AAAS. It was fortunate that one Committee member from the field of political science was already engaged in a Foundation-supported study of the entire loyalty-security problem and therefore highly competent legal and other staff talent was available for the study. We made a detailed documented report to the Executive Committee of the AAAS and submitted a digest of it for publication in *Science*. It was eventually published there, but only after a battle lasting several months with some timid souls, and also, it must be admitted, a few apparent supporters of Joseph McCarthy, on the governing body of the AAAS.

My own loyalty-security reliability was brought up for questioning after this activity on my part. I was a member of the first Study Section of the National Heart Institute and my dossier was not automatically cleared by the Agencies Loyalty-Security Board. I was presented with an interrogatory document which asked me not only to provide precise information about my travels abroad with dates of visits to each city, about all of my organizational connections, about persons I had met or known or corresponded with in my contacts at home and abroad, but also asked me what I knew about the opinions and organizational affiliations of a score or more of particular named persons, several of whom I had never met. I debated whether to comply with the demand for such information, but I decided that I would go ahead in order to be able to speak freely as a "cleared" person in the future rather than be impeded by a cloud of suspicion. I resolved, however, to resign my appointment to the Study Section of the NIH once I had fought

the thing through. Actually I knew nothing about any possibly "communist-
or communist-front" connections of any of the people they named and I
myself had done nothing much worse from the loyalty board's viewpoint
than support Norman Thomas' Democratic Socialist Party a few times and
Franklin D. Roosevelt's campaigns a few more times, aside from having been
an officer of the American Association of Scientific Workers and having been
instrumental in developing organized efforts to publicize the menace of the
entire loyalty-security system to the future of civil liberties, especially in the
scientific and academic communities.

In due time, but only after I was informed by the chief security officer at
the University of Minnesota that my telephone was being tapped and that
he was under orders to open private office files in loyalty-security investigation cases, my clearance came through. A person with slightly more paranoid
tendencies might have broken under such strain, but I confined myself to
carrying out my previous resolve to have nothing to do in the future with
any government work which required the filing of documents detailing under
oath anything more than my affirmation of loyalty to my country. This has
resulted in declinations on my part to serve in several capacities, the most
recent one being as a Special Consultant to the World Health Organization.
The United States and the USSR as well as a few other countries refuse to
allow their nationals to serve the United Nations Specialized Agencies without full and fresh security-loyalty clearance. I have continued to serve in
capacities where such clearance is not required.

An amusing, if it were not so tragic, incident in this connection occurred
in relation to service as an official U.S. Delegate to the General Assembly of
the International Union of Physiological Sciences in 1956. The U.S. was
entitled to five Delegates. The five were elected by the U.S. National Committee for the IUPS and their names forwarded to the Department of State
five months before the Assembly meeting for official designation, a practice
since discontinued, in part because of this episode. The Department cleared
two of the delegates very promptly, and two more after a short delay. The
fifth person was never cleared, despite the fact that he was at the time a high
official in another Federal agency and had within the immediately preceding
six months been "cleared" by that agency after a full investigation. I refrain from mentioning names because the Joe McCarthy spirit is still alive
in the land, albeit somewhat discredited but not muted, as the existence of
the John Birch Society attests.

INTERNATIONAL COOPERATION IN THE SCIENTIFIC ENTERPRISE

Physiologists have pioneered in the development of scientist-to-scientist
communication. Except for periods of world war, International Congresses
have been held triennially since 1889. Initially the Congresses were small
gatherings, because physiologists were few. Arrangements were relatively
simple and a Permanent Committee for the International Physiological

Congresses, with one representative from each major country from which participants came, constituted the only continuing body to provide contacts between the groups of scientists in various parts of the world. The Secretary of that Committee was the agent of continuity. After World War II it became apparent to many persons that occasional large international meetings, although useful, could not serve as the sole mechanism of international direct communication and cooperation in physiology. In several other fields of science international unions had been developed, and were cooperating through the International Council of Scientific Unions (ICSU) in promoting both international and interdisciplinary projects. When the United Nations Educational, Scientific, and Cultural Organization (UNESCO) was established it undertook to assist the work of ICSU by substantial subventions which supported substantive projects rather than the organizational expenses of the Unions.

Since UNESCO is supported as an intergovernmental agency, its programs and budgets are controlled by governmental representatives. Initially the "S" in UNESCO was quite liberally supported, but it began to lose its place of relative importance, particularly because the United States Department of State became more interested in developing the general communications field. This was not out of harmony with the stated purposes of UNESCO which stress the importance of building "the defenses of peace in the minds of men", but combined with the reluctance of governments to spend adequate sums of money on the U.N. Specialized Agencies, it has put a crimp in the programs of international cooperation in science.

At the 1950 Physiological Congress in Copenhagen the decision was made by the Permanent Committee to explore the possibility of setting up an International Union of Physiological Sciences to become an adhering part of ICSU. The question was to be considered further and the organizing meeting for the Union was planned to occur in 1953 in conjunction with the next International Congress in Montreal. As the representative of the United States and the Secretary of the Committee I had major responsibility in the matter. The IUPS was organized in its present form at that time. There was much skepticism as to whether an International Union would actually provide the physiologists of the world with much additional opportunity to carry on cooperative projects.

The events of the intervening years have justified the Union concept but it must be admitted that the IUPS has not yet accomplished as much as its more enthusiastic promoters, among whom I was one, had hoped. As one of its officers, its General Secretary for the first six years of its operation, I can hardly criticize too freely. We set up a number of Travelling Lecture-Conference Teams which were supported by special grants. A Newsletter was established by IUPS and opportunities for special smaller symposia were provided. The organization automatically provided for a more acceptable and more democratic method of control of all international activities in matters physiological. The IUPS failed, however, in one of its initial objectives, namely to provide an umbrella organization for related groups such as biophysics, nutrition, biochemisty, and pharmacology. Although the latter group was initially a part of the Union, and became a Sub-Union, it has now split off completely. The others never were part of the IUPS.

The IUPS also joined the organization now known as the Council of International Organizations of Medical Sciences (CIOMS), supported in part then by UNESCO and the World Health Organization (WHO) by regular subventions, but now supported regularly only by WHO and dues from its members. CIOMS has assisted in providing travel grants for younger investigators attending scientific meetings and in providing modest funds to assist in publications related to Congresses and Conferences. It has organized Conferences and Symposia of importance to the progress of physiology. CIOMS has taken the lead, among international nongovernmental organizations (NGOs), in considering the problems of ethics and legality in the use of human subjects in scientific study. In sum the CIOMS has performed useful services.

None of the NGOs has been as useful in solving problems of international cooperation in science until now as one might have hoped they would be. But international tensions, jealousies, and suspicions, not to mention financial niggardliness, have been the main inhibiting factors. At the scientist-to-scientist level accomplishments have been real. I do not doubt that, if human society survives its present state of international anarchy on the military side, the scientific community will be partly responsible for having built bridges of understanding and cooperation in ventures of human importance.

As an American of great loyalty to my country—I mean its people rather than any particular party in power that may misrepresent them—I cannot refrain from mentioning another episode which disturbed me greatly in the area of international relations in science. As I noted before, I was the General Secretary of the IUPS from 1953–1959. In that capacity I received the applications of physiological societies in various parts of the world for memberships in the IUPS. About a month before the meeting of the General Assembly of the IUPS in Buenos Aires in 1959, I received an application from the Secretary of the Physiological Society group in Taipai, Taiwan for its admission as the "Chinese Physiological Society". It was a regulation of the IUPS that the statutes of adhering organizations be provided for evaluation as to their consistency with IUPS principles, and that information be given as to the geographic area within which resided the physiologists who were members of the organization applying for membership. Neither had been supplied and I wrote asking for them. I suspected that a political problem was being thrown at the IUPS but I did not realize how hot the problem would turn out to be. No action was possible before the Council met in Buenos Aires.

I arrived in Buenos Aires by air at about 3:00 a.m. on the morning before the Council was to meet and reached my hotel at 4:30 to go to bed for some sleep. At 8:00 a.m. I was awakened by a telephone call from the American Embassy asking me to call on an official there about some unspecified urgent matter before the Council meeeting. I was not left in doubt long about the nature of the urgent matter because I had no sooner turned over to get another wink of sleep than another call from the Embassy of the Republic of China aroused me again. The Ambassador very much wanted to see me that day to reinforce the application for membership of the Chinese Physiological Society in IUPS. I invited him to have tea or sherry with me at my hotel that afternoon. By that time I surmised that the application from Taipai was not simply one from a group of scientists but I still did not realize how much importance was attached to it by the U. S. Department of State.

When I called later in the morning at the United States Embassy office, housed in the upper floors of a building in downtown Buenos Aires whose lower floors were occupied by a major private American bank, the first Secretary informed me that the Department of State was greatly interested in having the Republic of China recognized as the spokesman for Chinese in "cultural" affairs and read to me a long cablegram from Washington outlining particularly the importance attached to having the Republic of China become the Chinese participant in the Olympic Games Committee and he indicated that the same interest applied to international scientific organizations. I told him quite candidly that the Council of the IUPS was not likely to approve of the application in its present form and that the U.S. delegation to the Assembly had discussed the matter and was not disposed to vote for approval of the application in the form submitted, because the Nationalist Chinese on Taiwan did not in fact represent the major Chinese physiological community.

At this juncture the Embassy official asked me whether the American Physiological Society did not receive and use Federal funds, and whether it might not be embarrassed if such funds were to be withheld. I replied that it did carry on some work under Federal subsidy but that it did so only to serve the national interest and that I doubted that the U.S. National Committee for the IUPS would be willing to subordinate scientific organizational principles to international power politics.

I brought the substance of this message from the U.S. Embassy to the Delegates from the U.S. to the General Assembly and there was unanimous agreement that, although we would welcome the Taiwanese physiologists into the IUPS, we would not admit their organization under false pretenses as to the geographic area and population group that it represented.

The Ambassador from the Republic of China proved to be a very charming and intelligent person. I explained what I thought would be the difficulty in accepting the request for admission to the IUPS of the Taiwanese as the "Chinese Physiological Society" and suggested that with some more accurate limitation of its scope the group would be very welcome as a member organization. At our first meeting he was not averse to some kind of alteration in the name of the organization. Later, however, he apparently received instructions from Taipai to press for the name implying representation for the whole Chinese community, and changed his stand on this score.

The Council of the IUPS objected, as I thought it would, to the proposal to admit the Taiwanese physiologists under the guise of representing all Chinese physiologists. The General Assembly accepted the decision of the Council. We thought the matter would rest at that point, at least for three years, but we were wrong. A few days after the actions of the Council and Assembly were made known to the U.S. Embassy, we were called to the Embassy to be told that the Department of State wanted us to request an extraordinary meeting of the General Assembly to reconsider the matter. The five members of the U.S. delegation to the Assembly decided reluctantly to make a pro forma request to the President of IUPS to consider calling such a meeting, making it clear to him that we ourselves were not changing our position. Professor C. Heymans, the President, declined to call an extraordinary meeting. The tragedy was that we were forced to make known to our foreign colleagues that the United States Department of State was attempting to intrude its power and influence into international scientific union business. With the change in the administration in Washington the next year, this policy of the Department of State was abandoned and attempts at coercion upon scientific unions have ceased, hopefully permanently.

It would not leave a correct impression to refer only to troubles that physiologists have had with the Department of State. Actually in all matters in which major "cold war" policies were not involved, the Department has been helpful. In connection with arrangements for the 1968 International Congress it has made every effort to avoid problems in obtaining visas for foreign participants. It is unquestionably aware of the valuable role that international communication and cooperation can play in the progress of science and its utilization.

A POSTLUDE

I have purposely dealt primarily with the nontechnical aspects of my experiences because I hope they may illuminate some of the social and political problems with which scientists have had to deal in the last half-century. Physiologists are going to be confronted with attempts to restrict and regulate the use of animals for many years to come. The era of almost unquestioned acceptance of the virtue and the value of scientific progress is already at an end. The United States Congress is moving into a phase of interest in early practical results, rather than continuing to have confidence that scientists themselves will take advantage of new basic knowledge to make practical advances. Without doubt larger fractions of the scientific enterprise will be devoted to planned and organized attacks on specific applied problems in the near future than was true in the past. What effect this will have on long-term progress, even in applied science, is by no means certain. It could result in stultification of the enterprise if it brings about discouragement to basic scientists. It has yet to be shown that basic science progress can be a byproduct of applied research. The opposite we know frequently happens.

The growth in magnitude of the scientific enterprise in the twentieth century forces scientists themselves to take an active interest in the social and political forces that actually control their lives and their work. The best hope that we can have is that the public spokesmen for science may have good judgment and that they may be successful in influencing the public and the political leaders in society to adopt policies that will not in the end be self-defeating. In a democratic society one must be optimistic about the ultimate power of reason.

Reprinted from
ANNUAL REVIEW OF PHYSIOLOGY
Volume 36, 1974

SOME BIOPHYSICAL EXPERIMENTS FROM FIFTY YEARS AGO

Georg von Békésy

Georg von Békésy

Reprinted from
ANNUAL REVIEW OF PHYSIOLOGY
Volume 36, 1974

SOME BIOPHYSICAL EXPERIMENTS FROM FIFTY YEARS AGO

❖1103

Georg von Békésy
Laboratory of Sensory Sciences, 1993 East-West Road, Honolulu, Hawaii

... was man ist, das blieb man andern schulding.

Torquato Tasso, Act 1, Scene 1
Johann Wolfgang Goethe

Editor's Note

Georg von Békésy died June 13, 1972. In early March of that year, the Editors and Editorial Committee members of the Annual Review of Physiology invited him to write the prefatory chapter for this 1974 volume. He accepted promptly and added "I am sure I will have it finished long before your deadline of June 1973." He mailed us his completed manuscript three weeks later, early in April 1972, two months before his death.

Professor von Békésy received the Nobel Prize for Physiology and Medicine in 1961. In the presentation speech, Professor C. G. Bernhard of the Karolinska Institute said

The field of Physiological acoustics has a noble ancestry. von Békésy's distinction is to have recorded events in a fragile biological miniature system. Authorities in this field evaluate the elaborate technique which he developed as being worthy of a genius. By microdissection, he reaches anatomical structures difficult of access, uses advanced tele-technique for stimulation and recording, and employs high magnification stroboscopic microscopy for making apparent complex membrane movements, the amplitudes of which are measured in thousandths of a millimeter.

Professor von Békésy, your outstanding genius has given us an intimate knowledge of the elementary hearing process. As a whole this is a unique contribution. The main reasons for the award are, however, your fundamental discoveries concerning the dynamics of the inner ear. With reference to Nobel's intentions it is also a great satisfaction to be able to award the prize for outstanding discoveries which are entirely the result of one single scientist's work.

At the time of his death, Georg von Békésy was Professor of Sensory Sciences, University of Hawaii.

1

THE EXCITEMENT OF SCIENCE

8243-2601/78/1127-0657$01.00 © 1978 ARI   657

Introduction

I like to read the history of science, especially of medicine. It describes the fight between man and nature, which consists of a continuing series of successes and failures. Unfortunately, the majority of present day scientific papers are written in a cold, impersonal style. In earlier times many case histories were presented and by reading them it was possible to make a connection between the conclusion and the actual situation. Today case histories are no longer published and reviews sometimes sound almost like autopsy reports.

Some researchers are of the opinion that it is no longer necessary to study the history of science because science has progressed so fast and always linearly, that earlier experiences are of no value. I do not believe this, but believe, like many of my otological colleagues, that scientific progress is more like a spiral, as in the cochlea of the ear, which always progresses in one direction but with many ups and downs and with many repetitions of earlier forms. I was therefore very pleased to receive from the Editors of this series an invitation to write a free-form article that includes whatever autobiographical, anecdotal, or philosophical comments I wish to make. The invitation encouraged inclusion of personal reminiscences and presentation of my views of any aspects of physiological science of special personal significance. In writing such an article, the pattern of ups and downs, advances and setbacks in my research and other endeavors may be revealed. The Editors believe that this and other prefatory chapters, taken together, will provide a valuable historical perspective of physiology in the twentieth century.

I quote these instructions from the Editor's letter in an effort to eliminate some misunderstandings that can occur when reviewing earlier periods. One possible mistake is that we do not like the time just past. In art it is very well known that the style of twenty to forty years ago is generally disliked today; you can find the best art objects of this period in junk shops and attics. This held also for art Nouveaux until some seventy years had passed, after which the art of that period became collector's items. Such aversion to art of the recent past probably results from our rejection of the opinions of our parents and older people and the desire to change, to do something new and better. In not succeeding so well we start to dislike the earlier period.

Another mistake is exactly the opposite of the first—sometimes we tend to make the past more attractive than it really was, omitting the unpleasant parts. And a third problem, ever present in an autobiography, is making your life more important than it really was. I tried to avoid this because it is misleading for the younger generation and can foster mistrust and dislike of autobiographies. I hope I have been successful in avoiding these errors.

One reason my biography is a little different is that my parents were not immigrants, as were the parents of many scientists in this country. I come from a well known Hungarian family of means and therefore our circumstances in general became worse and worse with time. Certainly there were not such improvements as with families who came to the United States and started life fresh.

South Germany and Switzerland Before World War I

I received my basic education in Munich and Switzerland. Munich then belonged to the kingdom of Bavaria and had a certain freedom from the North German atmosphere. At that time South Germany and Switzerland were among the leading countries in science. It was Munich that had the first automatic telephone and it was Munich where garbage was collected in three different cans: one for paper, one for spoilable material, and one for the rest.

In Munich there was much interest in paintings and sculpture, and it was there that I, as a schoolboy at the age of about 8, first saw a well known sculptor, who lived in our neighborhood in the Königinen Strasse, work on a live model. The cooperation between the model and the sculptor was something that I will never forget. Both of them worked on the opposite end of the problem, but they helped each other and produced an exceptional work.

Munich had a museum for the history of science, unique at that time, and many museums for fine arts. The people knew how to live and how to let other people live and how to work. Formally it was a kingdom, but in behavior it was perhaps the best democracy I have even seen.

Switzerland was more commercially oriented. The schools were excellent and life was well organized. There were many refugees from Russia and other countries, mostly revolutionaries like Lenin. They were free to do and say what they wanted. But it is extremely difficult to make a revolution in Switzerland. The intended revolutionaries would listen to the speeches but at noon when the church bells started to ring all over the town, most would remember that they had steaming soup on their table at home and leave. Revolution just does not work after a hot meal.

Life was comfortable and some groups in Berne, the capital, were accused of not doing any work at all. The Imperial and Royal Embassy of Austria and Hungary belonged in this category and it was said that the chancellery of this Embassy had written a sign on the door, "Office hours from 12 to 1."

But despite the apparent beautiful and friendly atmosphere there was tremendous tension in 1913 because everybody expected a war. Even Switzerland had a large military buildup to protect her borders, and the students were drafted periodically because the front between France and Germany could easily be extended into Switzerland if there were no proper defense lines.

World War I

The impossibility of avoiding a war in the next years influenced every boy's thinking. At that time it seemed to make no sense to study mathematics or theoretical physics when you could be drafted a year later into the war. It was known even in 1912 that the Siberian Russian Army had moved partially to the border of Germany and Austria-Hungary. Such masses of soldiers and material were involved that there was no way to stop it. My colleagues in Germany, those who were not Swiss citizens, tried to emigrate from Europe. One of my best friends committed suicide as he simply could see no other way out. Perhaps the worst feeling about the future was the fear that after World War I had ended there would be nothing left of Europe.

Given some of the preparations on both sides, France and Germany, it was obvious that the war would last very, very long, in spite of some of the military who felt they could occupy each other's capital in a week.

Only a few students favored the war. I had one friend from Prague who wanted to serve there so he could participate in the liberation of the Czechs from Austria-Hungary. He volunteered for military service in Prague but unfortunately the regiment to which he was assigned revolted. The liberation force was immediately crushed by the Austrian Army and every tenth member of the regiment was shot. My friend was one of them.

After this and similar experiences, I became an outspoken coward. I did not want to live from day to day waiting for a tragedy, so I made a program for my future life. Once you do so, all your behavior is directed against the disturbing factors that you want to avoid and in this way I completely lost my enthusiasm for military achievements.

Another disturbing factor was the question of a career in music. My piano teacher was like an income tax accountant—very precise, checking everything constantly. His technique was superb. He envied me to a certain degree because the span between my thumb and my small finger, as a boy at that time, was already larger than his, and because I had "good bones" in my hand. I could get from a concert piano about the maximum it could give.

But the real issue was that he played Chopin, for instance, with the precision of a Swiss watch. He realized that my way of playing, with a little Hungarian twist, was much more interesting for the Swiss public. There were things I could not explain to him. For instance, some of the very fast runs across the whole keyboard impressed him very much and I could not tell him that these are always somebody's own manufacture and have nothing to do with the notes which are written on paper. It was Liszt who used them extremely successfully, and knowing about four to five variations, could ornament a piano piece exactly so that it fit the group of listeners. My teacher told me in a very clear way that a concert pianist needed a repertoire of about six pieces and no more, and could travel all through Europe by having only that small knowledge. It would take about two or three years, he said, until I would play them so that I would be a top performer for two or three pieces.

My opinion was that I did not know how to play the piano at all but I did know what the public wanted. However, on this basis I would not start a life career. Another point which for me was very disturbing was the fact that music stuck with me. On hearing a good musical melody, I had to hum it for days and weeks. A painting or a sculpture never stuck with me. I could look at it and draw it and after a few minutes it somehow faded out; but music was different, it occupied my brain and that handicapped good logical thinking on my part. So for this reason, I gave up the playing of music even as a sideline.

Education in Switzerland

Education was developed mainly by Pestalozzi, a simple, practical teacher who worked out new methods of transmitting information from an older man to a young student. I had the good fortune to be educated in a private school, the Institute

Minerva in Zurich. The great advantage of that gymnasium was that it used the so-called mobile class system, in which there were four to six classes in each subject but on different levels of progress. It was therefore often necessary to have six or more professors simultaneously teaching the same thing. A student had free choice to select the class on a level that fit him the best. This type of impedance matching, as we would call it today in electronics, was very useful and generally avoided inferiority complexes in the students and frustration in the teacher. If somebody failed in one subject, he could study the same subject for another year and not be disturbed in his progress in other subjects. In this way I finished my course in physics in about two years so I was able to do experiments at home in my own style and pace. Therefore when I did attend the university I had a definitely better education than many others. I was very poor in English and even more so in German. Unfortunately, Hungarian is a language which has few similarities to other languages, except Finnish, so knowing it did not help in learning other languages.

The University in Berne is financed by the citizens of the Kanton of Berne, a very small group of people. The capital of Switzerland, Berne, had only 50,000 inhabitants at the time. Therefore it was taken as an honor to be able to study at the University. When I first shook hands with the Rector Magnificus he made it clear that to be a chemist and to learn at the University cost the University, as far as I remember, about 5000 Swiss francs per year, although the education of a lawyer there cost the University only around 600 Swiss francs per year. This made it obvious to students that the few chemistry students were using a large amount of the citizens' money for their study and, therefore, the citizens of Berne expected the chemists to produce something of importance.

As a student I had the impression that it was most important to study mathematics. I found out that at the University, just as in most universities, there are very good courses in elementary and in highly advanced mathematics but nothing in-between. This made it extremely difficult to learn the subject. I ended up learning four dimension tensor analysis, which at that time seemed to be important in the treatment of the relativity theory. But finally I had to admit to myself that this was a very poor choice. Mathematics, and so geometry, makes a few assumptions and builds an empire based on these assumptions; it is a sort of closed circuit performance. It does not teach anything about what and how to observe.

From mathematics I went for a short time into astronomy, in which I wanted to stay but unfortunately the nights in the cool observatories produced a trememdous physical strain on me. Later on I went into chemistry. Unfortunately, of all the chemistry professors at the University, only one was really original. He was a colloid chemist who was disliked by everybody and who hated to teach. He gave only one lecture every week on colloid chemistry. But looking back at the things that I learned at the University of Berne, I have to say that he did teach me more than any other professor. The reason for this was that he was teaching his own experiences. For his lectures he had one book which he used as a framework, but everything else was his own, and these personal experiences stuck much better than the logically built up performances in the textbooks. Probably one of his most valuable pieces of advice was given to me when I asked him for help on a problem I could not solve: he listened for a while and then told me simply that the library was on the second floor and walked away. I was shocked that first day, but having learned in the military that if you want to complain about somebody you should do so after a night's sleep, I slept and the next morning realized that he had given me very valuable advice. Since I did not know the answers and he also did not know the answers, the solution of the problem could be in the library.

At the University of Berne, the student had complete freedom to select lectures. There was not even a minimum number of lectures he had to attend. I had, therefore, a chance to compare the different professors. Besides the colloid chemist, an anatomist was the most interesting. He was able to draw with both hands simultaneously. His drawings of the nervous system and circulatory system were almost as beautiful as the world famous drawings of Leonardo da Vinci.

The more I stayed at the University, the clearer it became that the subject a person chooses practically decides his future. I did not have a chance to become the assistant of a Nobel Prize winner who could show me the road on which he walked so it would have been easy to continue. My most important decision became how to select something of good quality. I asked practically everybody questions on that subject and I found that the art dealers are probably the best advisers in this field.

I asked an art dealer how can you learn which is an original art object and which is a fake. He smiled and told me there is only one solution—to constantly compare. You should never buy the things you like but buy one type, and then buy many different pieces of the same type. Then you will be able to decide on the first look which is genuine and which is fake without really being able to give a reason for the decision. I had real success with this method when comparing bronze statues. I bought several bronze statues from the Byzantine period. The bronzes in Anatolia have a very well defined patina, but in an antique shop it was never possible to tell if that patina is the same that fits into the Byzantine period. Having borrowed the piece and putting it together with all the other Byzantine pieces, there was no difficulty at all in deciding if that patina was a Byzantine type patina or not. If it had the same patina it was sure to be a Byzantine; if it did not, it still could be a genuine Byzantine piece but it may have been cleaned or somthing else could have happened. I learned to leave out such pieces even when they were extremely attractive.

To me this method of comparing seems to almost guarantee success over a longer period of time. But I paid a very high price in working hours because comparing involves studying the ideas of several people, not just one; a great number of people whose work I studied and learned to know were simply dropped later, along with all of the work I had done in their interest, simply because the comparison revealed their work to be of lesser importance. But I think the process is worthwhile since, as any archeological excavation shows, it is the quality which determines if something remains or is lost.

As a student I was very unjust toward my professors. At the end of every semester I made a sort of inventory of what I learned and what of the learned material could be useful in later years. Such an inventory showed an unbelieveable lack of efficiency. There were lectures at which I spent three or four hours per week without taking anything useful home. One reason for this was that they were very hard to memorize. The more I studied what is memorized and what is forgotten, I came to the conclusion that the Arab way of teaching by telling anecdotes, used around 1200 to 1400 A.D., was a very good method. It cannot be used in chemistry or mathematics but certainly it can be used much more than is done today. Today we mistrust anecdotes because somehow they do not represent a statistical mean value, but in general they do represent certain principles. I still remember very well the fairy tales my mother told me, scientific diaries such as of Faraday's, and some of the books on the beginning of electrophysiology. The anecdotes seemed to be successful because they rounded up in small, meaningful units what the memory can use and keep.

Listening to illuminated films, let us say on the method of doing certain surgery, never taught me anything useful because I could not remember the details. If I watched the actual surgery I went from landmark to landmark with the surgeon and could remember the landmarks. In most movies on surgery the landmarks are never even mentioned so they were, in spite of the huge amount of work invested in them, of very little practical use.

After the revolution in 1918, my family lost practically everything. I could have stayed in Switzerland and continued my studies, since the Swiss were very nice and they even offered me a job to sustain myself, but I had the feeling that I should somehow help to reconstruct Hungary. That is the optimism of a young man of 20 years. I misjudged tremendously the speed with which a country is able to rebuild and I misjudged also the new situation in Europe, which was more and more approaching financial chaos. There were no experts on economy in Hungary, only excellent people in science.

I received my PhD from the University of Hungary; my thesis dealt with a method to determine the diffusion coefficient of fluids in a very short time (sometimes less than three minutes). From the diffusion coefficient the molecular weight could be determined. It was a method I should not have given up but should have developed much further.

My general feeling at the University was that I wasted my time, especially the years which were the most valuable for a young man, namely when memory is still good and judgment becomes more and more objective. This feeling was further increased when it was impossible in Hungary for a PhD in physics to get a job. I visited factories and laboratories one after another and I was always told the same thing—what do you want us to do with a physicist. This was probably the most difficult time of my life and it was my mother who kept me going.

After a certain time I decided to look around systematically and find out which was the best equipped laboratory in Budapest. I found that it was the laboratory of the government controlling the research in long distance telephones, telegraphy, and radio stations. Hungary was in the middle of Europe and therefore communication was a very important feature. The government was forced by peace treaties to spend a certain amount of money to keep the transmission lines in good shape. To do so, they constructed a laboratory and gave the laboratory a certain amount of money to buy the necessary equipment. It was this financial support which started my research. There was a fixed income to the laboratory with no questions asked as to how it was spent. I still think that this is the basis of every big discovery.

The laboratory gave me a salary that was the lowest of all of the 80 people in the laboratory group; it was less than the salary of a carpenter in those days. But I tremendously enjoyed the possibility of learning new things. Every day was a new experience. On one day the telephone line between Prague and Belgrade would be down, on another day the radio station would have some problems. So I had the chance to study large fields of very different background. Sometimes chemistry became important because the cables, with lead mantle, corroded. Or there were stray currents in the ground and they produced trouble. This was the field where I learned to pick up and make conclusions from stray ground current about a sort of large scale encephalography. In some cases I was very successful; for instance, at that time it was necessary to check the condition of international transmission lines. To do so from Budapest a loop was made to London and from London back to Budapest, and the input of the voltage transmitted to London and back was measured in the loop arriving at Budapest. From this, conclusions were made about the state of the transmission lines. In general this measurement took about 15 or 20 minutes. Since there was a great number of telephone lines of this type, there was much excitement every morning in the control room while the lines were being checked. I developed a new method by which I could check the telephone line in a loop in one second. My new method consisted of not using sinusoidal tones, in general use at that time, but by using the transients. By looking up the transients it was immediately possible to see not only the amplitude distortions in the telephone line but also their phase distortions, and the phase distortions were much more sensitive and gave a much better control of the stability of the telephone line. With such small tricks I was able to escape much of the routine work, and I later applied them in the field of hearing.

How I Became Interested in Hearing

Of all the developments after the war, communication became one of the most progressive. Hungary was constantly forced to build cables of international quality. They tried to standardize them in endless international meetings, but they were just as unsuccessful as the peace treaties.

In trying to fulfill the often very strict requirements, the government always had to ask for several bids. I was very much surprised to learn in reviewing some of these bids that those submitted by different companies for cables to connect the same two cities differed by less than 1%. At first everyone thought there was a secret agreement among the companies since most of them were controlled by financial groups outside Hungary. However, after having reviewed all their calculations, beginning with the price of copper, paper, and lead, from which the cable is made, I was sure that the mathematics of cable construction was so well developed that it was amazingly precise.

A more difficult point resulted from the fact that a communication line consists of three parts: the telephone apparatus, the cables, and the central switchboard. The cost of the switchboard did not play a very important role in the calculations of cost, but the price of the telephone sets, because of the large number of them in a city, was comparable in importance to the price of the cables. Therefore, the question was, if we wanted to improve the quality of a telephone transmission, where should we invest the money—in telephone sets or in cable? This was purely a question of economics, but I had the feeling that only the ear could supply the answer. It was a bioeconomical question.

It was possible to calculate the cost of improving the cable compared to that of the telephone set. Unknown, however, was which improvement the ear would most appreciate. My first experiment was to show that the ordinary telephone membrane vibrates in a much more distorted way than does the eardrum; to have a perfect transmission system, the telephone membrane should vibrate in such a way that the quality and the damping is comparable to that of the eardrum. These observations were again made with transients and they gave a clear answer as to where to invest further improvements—international cables or the local telephone system. After settling this question, I received all the financial support I needed to investigate the mechanical properties of the ear and to match the earphone to the membrane of the ear in such a way that sound transmission would be optimal.

My first conflict was with the institute of anatomy where I wanted ears so that I could measure the mechanical properties of the middle ear. In general it was said that the physicist, even if he does not have a job as a physicist, should not get involved in anatomy. Especially as I was employed by an engineering laboratory, they did not want me present at autopsies. Unfortunately I had no alternative and so used the simple fact that the anatomical institute had two doors: one front door with a beautiful stairway used by professors and a back door where I was able to walk in and out with a few temporal bones. Naturally, there were difficulties because removing parts of human bodies (I had been taking them outside the anatomical institute) was improper conduct. The institute made it clear to me several times that if the police became involved I would have difficulty proving I used the bones only for scientific purposes. But in time everything quieted down and I was able to extend my research further to live anesthetized animals. I think this was real biophysics.

It is an extremely peculiar situation that a completely exposed nerve trunk (the chorda tympani) runs across the middle ear. That seemed to be one of the best places to measure the velocity of electrical transmission in the nerve fibers. The professor of physiology at the University of Budapest, Dr. Beznak, heard of my interest in the chorda tympani and immediately brought an anesthetized cat to my lab so we could do the measurements. Measurements at that time were quite difficult because we recorded with a Siemens loop oscillograph. Any time there was an overload on the loop it simply burned out this very expensive equipment; but we found that by putting a glow lamp parallel to the loop with a transformer, we could do peak clipping. I do not know who discovered peak clipping but we used it extensively before it was described in the literature. It made the oscilloscope and the Edelman galvanometer foolproof. Today, I think the biggest discovery is the oscilloscope because it has the advantage that it never burns out. You can use any voltage without damage to electronic beams. It has one disadvantage in that anyone, even if he knows nothing about electronics, can suddenly become an expert. It was the beginning of the age of bioelectronics.

Experiments with the first cat were a shock to the professor of physiology and to me. The professor said the chorda tympani serves mainly to stimulate the salivary glands with electric discharges. Therefore he put a tube in the salivary gland duct, and under a microscope it was possible to see that every time a small condenser discharge was transmitted through the chorda tympani the fluid in the capillary tube moved along one or two millimeters. It was a beautiful, clear experiment, very impressive to me because it showed that the secretion in a gland is just as precise as the reading in a voltmeter.

Unfortunately, my idea was that the chorda tympani picks up the stimulation of the taste nerves, and the transmission of the electric line is in the opposite direction. We could not agree and finally we gave up the experiment. Seemingly it turned out that the chorda tympani does both things and in different cats the distribution is very different. The chorda tympani is thus not the best object of research.

I was very lucky that I had this failure because it focused, for all my life, my interest on the importance of the material used for experiments. There are animals on which certain experiments cannot be done. That became clear in these conditions and therefore the selection of the right animal, just as it happened later with the squid or the limulus eye, is just as important as is the development of new methods in research.

The Elasticity of the Membranes in the Cochlea

I had the impression that the rapid development of electroacoustical and telephone engineering methods would make it possible to retest the different fields of hearing and biophysics, and I decided, therefore, to direct my attention to the theory of hearing. At that time, around 1930, there were about five different theories of hearing, just as we have today. Also at that time one of the main questions concerned the form of the vibration pattern of the basilar membrane for a pure tone. Since this is a purely physical question, I felt that this could be solved with the modern method.

Helmholz had looked at the basilar membrane 150 years earlier, as had Corti. But they prepared the basilar membrane by chipping off the bones. Since the cochlea is embedded in the hardest bone of the human body, the basilar membrane was generally displaced during preparation and this prevented making precise measurements. In Corti's preparations, the whole cochlea dried out almost completely by the end of the dissection and this resulted in distortion of its structures. With Helmholz it was probably the same way. To avoid this drying, it is best to do the entire anatomical dissection under water or physiological solution. Therefore I used a square bath and let the fluid flow in one side and out the other. The fluid stream was kept constant in the whole cross section and by using a drill instead of scissors for dissecting it was possible to slowly peel off thin layers of the bone. Any time the drill was used, a formation of bone dust clouded the water, but the streaming water washed it away in a few seconds and the whole field of view was again clear and

ready for new dissections. This method of underwater dissection was very convenient. If there was a membrane to be lifted, it was picked up with forceps and, by
using an underwater (plankton) microscope with a magnification of 180 or even a
little greater, the membrane could be pulled off carefully. By opening the forceps
the membrane piece which was pulled off flowed away with the water. It was a
pleasure to do dissection that way. It had the advantage too that there was no danger
in dissecting an infected ear. With the drill it was very easy to expose one full turn
on the tip of the cochlea allowing good observations of the basilar membrane.

The next question was whether we could make an opening in the cochlea without
disturbing the vibrations of the basilar membrane. I spent too much time in developing an underwater glue which could fasten a window over the opening that I made
in the cochlea. Later it turned out that the best way to fix a window on the cochlea
opening is to use a highly viscous fluid, perhaps physiological solution with gelatin.
Being highly viscous it is still moveable for DC displacement but not so for frequencies above 30 cycles per second, when it becomes almost completely rigid. Under
stroboscopic illumination this could easily be checked.

In almost no time it was possible to show that there are traveling waves on the
basilar membrane going from the stapes to the softer parts of the membrane. It is
quite interesting that traveling waves were not readily accepted mainly because of
the mathematics of the whole problem. Even today, most of the mathematical
treatments of this problem have so many omissions and simplifications that they do
not describe the movement of the basilar membrane properly. Even for me the
traveling wave looked a little strange at the beginning, but as time went on it was
obvious that whenever a system changes its mechanical properties continuously
there is always a traveling wave. It is the only wave form by which energy is
transmitted in systems with a lateral extension. Therefore, the traveling wave is the
natural transmission form for the cochlea. After a while the theory and the whole
measuring process were simplified so that by the simple means of testing the deformation of the basilar membrane under the DC pressure of a needle tip, it was
possible to determine that the basilar membrane in the human cochlea should have
vibrations corresponding to a traveling wave and not to resonance or other type of
vibrations.

After having done measurements on the temporal bone of human cochleas, the
question was whether these measurements were reliable. To answer this question it
was necessary to make the vibration amplitude of the basilar membrane so large that
we could see it with a magnification of 200 under a stereoscopic microscope with
stroboscopic illumination. Once this was accomplished, the additional question was
raised as to whether these vibrations would be the same as those in the basilar
membrane if the amplitude was about 100,000 times smaller. This is the question
of nonlinearity and it is quite clear that if we listen to a very weak tone at 1000
cycles, we can increase it almost 100,000 times and still the tone is unchanged. In
all my measurements I never went to a higher amplitude of vibration than the
vibrations of a pure tone. This can easily be checked if we keep the middle ear intact
because the moment the amplitude goes higher we will have a tickling in our ear
and a change in the vibration pattern in the middle ear. In general I used a sound

producer attached to a T tube. On one side was the preparation and on the other side was my own ear. Since in my own ear I heard a pure tone for that amplitude, there was no good physical reason to assume that anything would be different in the preparation.

Another question was, is the elasticity of a living basilar membrane different from that of a basilar membrane without a blood supply? As far as I could see there was no real difference. I developed a very peculiar and very sensitive method of stroboscopic illumination which again did not measure the amplitude but rather the phase of the vibration. It could be shown that a vibration pattern measured in a live lightly anesthetized guinea pig did not change when the animal was killed by an overdose of pentobarbital or by inhalation of nitrogen. It was stated many times in the literature that the tissues change their physical properties in ten to twenty minutes after death, but I never could prove that. There is some change, for instance in the eardrum, produced by the stopping of the blood supply, but since the eardrum consists of three thin layers, when blood flow stops the humidity in the ear channel is immediately decreased and the eardrum starts to dry out.

If a patient is taken to an operating room, he is usually rolled into the room receiving a continuous intravenous perfusion of physiological solution. In animal surgery, this is seemingly seldom done and therefore in all animal experiments dehydration is a problem. The smaller the animal, the faster it dehydrates because the surface relative to the weight is increased. It is incredible how many severely dehydrated cats I have seen, with almost brittle tongues, on which records from the cortex and the inner ear were made.

In time, I came to the conclusion that the dehydrated cats and the application of Fourier analysis to hearing problems became more and more a handicap for research in hearing. Therefore, my interest went more into the psychological questions. I am very thankful to one otologist who, in cases in which the labyrinth had to be taken out because of disturbance in the vestibular organ, gave me a few minutes before the operation to test the mobility and the difference between DC displacement and vibration transmission in the middle ear on patients who were under anesthesia but whose blood supply, and therefore humidity, were normal.

At this point I had a well developed and productive laboratory. Unfortunately all my work was interrupted by World War II.

World War II and After

At the start of World War II our impression in Hungary was that a scientific laboratory would never be bombed by American airplanes. But it was not so: on the second day we found out that the American airplanes really did not hit a specific target at all. Instead, they used the tactic of carpet bombing, in which the leading airplane makes a circle and the following airplanes throw all their bombs inside that circle. The bombing was extremely inefficient; it killed people who had very little to do with the war and, in many cases, were definitely opposed to it.

A few days later a building near my lab and most of my equipment and writings were completely destroyed. It is interesting to note that the largest destruction was not done by the bomb itself but by the air suction produced by the explosion wave.

It pulled out the windows and everything in the cabinets and built it up into a mess which could not be separated.

The Russian attack came on the ground. They fought man against man. Toward the end, it seemed to be one German against eight Russians. Everything was depressing because there was no visible reason or logic in the whole behavior. The Russian Army worked like a machine; every morning at 7 o'clock they started with the Haubitzen to shoot and destroy one section after another very systematically. At 5 o'clock they stopped, had their supper, and the next morning they moved ahead again and went sometimes a few miles, sometimes only a few hundred feet. At the end the whole section under attack was destroyed, including the section where I lived, near the Danube. The highest wall left was about one meter high. I had many friends living in that section so I visited them before I decided to leave. I shouted their names under the blue sky but nobody came out so I went from one opening to another as I knew they had to come out for water. In some of these openings you could see the tragic history of Budapest. On the upper level was the modern type of buildings built in the nineteenth century, something of a modern empire style. One layer deeper you could see a clear empire style and going deeper in the layers there was the Baroque style in the staircases and cuttings of the stones on which the buildings and the cellar was based. If you went even deeper, there was a definite Gothic cutting of the stones and later a Romanesque style. It showed that Budapest had its history beginning from Rome up to modern times and it was destroyed several times during the last 2000 years. But every time it was built up again on the same place.

I have been asked several times why Hungarians are relatively successful compared with other people, especially in science. I have the impression that this sticking to one place and to one aim is the main reason why in the long run Hungary still produces important contributions to the culture of this world.

Since it was obvious that I would not be able to continue my scientific work, I decided to leave Hungary. Professor G. Holmgreen from Stockholm invited me to Sweden and from Sweden I went to Harvard where Professor S. S. Stevens took me into his laboratory.

Unfortunately, while at Harvard a great tragedy for my research occurred when the tower of Memorial Hall burned down. My entire working place in the basement of Memorial Hall was flooded and I lost the most cherished writings and old books, that I collected after I left Hungary.

I have found, on numerous occasions in my life, that it is impossible to rewrite the same idea the second time with the same freshness and logic as it was done the first time. After the fire in Memorial Hall some of my writings lost their precision because of the destruction of most of my data. To write a paper takes me, in general, one or two years. It takes almost one year to formulate the question that I would like to answer. It takes a half year to carry out the experiments and a half year to put it in the correct shape. I work on several problems simultaneously and the older I got the more problems I had to work on; this takes away the freshness of the logical buildup. Lately, the ordering of equipment, working on the financial aspects, getting an award from the funding agency, and receiving and setting up equipment in the laboratory can take two or three years, even in the case of a simple experiment. This has completely changed my way of building up an experiment and my method of working, which was not to collect data but to collect different methods, and not to do every observation many times but to do the same observation with different equipment and different methods. This approach now has become almost impossible to carry out. In earlier times I had no difficulty in measuring, for instance, the elasticity of the basilar membrane with three or four different methods. In general I published one method which I figured out was the most simple and the most reliable one. It always amused me when later somebody picked from all the possible methods exactly that method I had found most unsatisfactory and introduced it as a totally new method.

Life in the United States of America

I really had very little information about the situation in the United States of America when I lived in Europe and a large part of the information was distorted. For instance, American films of good quality were so expensive that they did not reach Hungarian theaters. Those we saw were cowboy films and gave a relatively distorted picture compared with the German films which were excellent and of philosophical value. When I arrived in Sweden I was sure I would get good information about the US and I found the booklet given by the United States Army to their soldiers when they went to France. It contained instructions on how to behave when contacting French people and on the difference between a Frenchman and an American. The interesting thing was that at the end of this booklet was a chapter on how to survive in the United States. My first clear information came from that booklet. Unfortunately this information was also useless to me, being rules for survival of an American, not for a European who just arrived in the United States.

I could read English very well at that time, especially the technical language, but I could not speak the language at all and there were many incidents which were quite amusing, at least in looking back. The entry for a Hungarian into the United States was at that time quite difficult. Having arrived at La Guardia Airport after a long, long flight on a two-engine airplane with a long stopover at Laborador, I approached the officer of the Health Department. He looked at my passport and then he asked me if I was healthy and I told him no. He asked me again if I was healthy and again I told him no. Then he just put his hand on my shoulder, pushed me across the line, and I was in the United States. I was very much disturbed that already in the first seconds in the USA, I had probably made a very big mistake. My English-German dictionary soon made clear that I had mistaken the word healthy for wealthy. Since I had only $100 in my pocket and was thinking about the Rockefeller fortune, I was not able to tell him yes to his misunderstood question.

This small incident right at the beginning gave me an inferiority complex for which even today I am not really able to compensate. At Harvard University my colleagues were very friendly and helped me with my English. I learned from them good classical English. I still cannot write a good paper but I can judge if a paper is good or not. The secretary at the laboratory told me several times that I should read Churchill's book on the English speaking people[1] because it is written in beautiful English. I bought the book and I agree the English is just magnificent, but I did not like it for two reasons. One was the fact that it did not correspond exactly to the situations I had first-hand knowledge of and the other was that Churchill is a very poor painter. My general theory is that if a person is really bright and is good in his own field, he is bright in his amateur field also. It was a peculiar experience that a year later I was told to forget Churchill. I had an inkling that Churchill published at that time his controversial volume on the American revolution.

The American attitude toward science was very different from that to which I was accustomed in Europe. In Europe there was a certain pessimism about the degree of progress scientists can achieve, and there was an important difference in the role which financial support plays in new discoveries. In Europe you were born to be either an artist, a scientist, or a banker. You had to inherit the specific qualities which made you important in that field. In America everybody could learn to draw, everybody could make a million or lose it, and everybody could make excellent new discoveries. It was quite difficult to switch from one style to another style, especially after a certain age. My opinion was that both of these assumptions were extreme. The consequence of my optimism was incredible. Biophysics, which had been more biomechanics, became suddenly molecular physics, an absolutely new field with a tremendous potential for new ideas. Some old telephone experiments were developed into information theory. It was quite clear that in this field the optimism really paid off. But sometimes optimism went too far: a fund raiser told me that if he succeeded in raising $9 million, he would solve the cancer problem—that was about twenty years ago.

In the beginning the most interesting events of my life in the United States were trips to meetings. I met people whom I had known only from books and papers, and I saw the country and the large museums which contained collections I had never seen before. I soon enlarged my field of view and became very interested in the West Coast because it was so different from Boston. Eventually I decided to live in Hawaii and build a laboratory there. The decision was a good one and in the last six years we have made a few measurements which I think are new. Hawaii is basically just as different from the mainland as Europe is from the United States. The United States is very impressive for anyone who comes from Europe because it is so large and varied. In the United States you can have almost anything that you want, it is only a question of money. It was not so in Europe. In Hawaii you cannot have anything you want, but you can have things that you never expected before. Living in paradise, as it turns out, is not a very simple thing. It is so beautiful that there is really too much beauty and too much color. It is well known that the nervous system is more sensitive to variations than to continuous stimuli because of the role of adaptation. The same holds true for the life circumstances.

The University of Hawaii is a new university and therefore tradition does not play any role. This can produce many differences. For instance when I first came to

[1]Churchill, Winston Leonard Spencer. *A History of the English-Speaking Peoples,* Vol. 3: *The Age of Revolution.* New York: Dodd, Mead.

THE EXCITEMENT OF SCIENCE   671

Hawaii I was very much surprised by the fact that between lectures the students walked on the grass of the campus and not on the beautifully planned roads. On the Harvard campus all the students walked on the roads, never on the grass. It took me a long time to explain this phenomenon. Obviously, the architect who designed the roads of the Hawaii campus did not know the different doors to the lecture halls, so the students had to make shortcuts. When I went back to Harvard I was told that the roads were there 200 years ago. Originally all the different houses on the campus were homes of professors and every professor owned a cow. These roads were selected by the cows when they went to and from the pasture. The cows had very good sense in making shortcuts, much better than the modern architects; and I have a very high respect for Harvard in that if something is good, they keep it, completely independent of who made it first.

Lately it seems that the different places in the United States are becoming more and more similar to each other, which makes life less and less interesting, and I have come to the conclusion that the most fascinating things today can be found mainly in museums. That is the one thing I enjoyed so much on the East Coast. The museums taught me in many ways how to look at the great diversity that human genius has produced during the past milleniums.

Ann. Rev. Pharmacol., Vol. 11

PIECES IN THE PUZZLE

U. S. VON EULER

Ann. Rev. Pharmacol., Vol. 11

PIECES IN THE PUZZLE 6500

U. S. von Euler

Karolinska Institutet, Stockholm, Sweden

I suppose the main reason for the Editors to honor me by asking me to write a Prefatory Chapter for the Annual Review of Pharmacology is that my activities seem to have dealt more with pharmacology than my official title as professor of Physiology might indicate. If so, this is no doubt due to the fact that some of my most prominent teachers have been professors of pharmacology or served as directors of pharmacological laboratories—G. Liljestrand, H. H. Dale, and C. Heymans—although I would still regard them as essentially physiologists. However, their knowledge of pharmacology was profound and it is not surprising that their familiarity with pharmacological tools has been to some extent transferred to their pupils. Dale introduced the term "autopharmacology," signifying the actions of certain substances occurring naturally in the body ("Körpereigene" in German) which in some respects resemble pharmacologically active drugs. At least the description and the analysis of their action often followed patterns similar to those applicable to various drugs used for pharmacotherapeutical purposes or as tools in the attempts to classify and understand the action of other drugs.

After the end of World War I experimental physiology and pharmacology expanded rapidly. O. Loewi and W. B. Cannon gave the long wanted definitive proof of chemical neurotransmission, which was to have such far-reaching consequences. Acetylcholine, chemically synthesized many years previously and studied by Reid Hunt and Taveau in 1906, had caught the interest of many research workers including H. H. Dale, who soon noticed the potential importance of this compound for biology. Histamine also came into the picture, and was even implicated in the cause of wound shock during the 1914–1918 war. Several other "biogenic amines" were isolated during this period and their actions studied.

The conditions for scientific work in the medical field in Scandinavia after World War I could hardly be characterized as very favorable as regards localities, laboratory equipment, personnel, and funds. Many of those who still found research work attractive, in spite of the few positions available and the modest salaries, were undoubtedly beset by the same peculiar urge to find out what the mechanisms were behind the biological phenomena with

1

THE EXCITEMENT OF SCIENCE

8243-2601/78/1127-0675$01.00 © 1978 ARI 675

which they had been vaguely acquainted in the preclinical studies. Intellectual curiosity and a desire to "explain" various phenomena are probably still the most effective forces driving the young student into an uncertain future in research. The motivation may of course be more complex, and in addition to the hope of making useful discoveries, a pleasant feeling of becoming a member of what appeared to be a distinguished fraternity might also enter. An incentive of a rather special kind was sometimes offered by the early discovery that some explanations in the textbooks did not appear to be too convincing. This was almost an invitation to provide the correct answer as a result of clever experiments and new approaches which presumably had not entered the mind of the textbook writer. This included no doubt a touch of emulation which may not be considered as very distinguished in research, but nevertheless can have an activating effect. In spite of certain hardships, those who started research work in the twenties were in a privileged positions, or at least this was my own feeling. The research climate was improving and we were aware of an upsurging interest which to a large extent was to be credited to a local group of young scientists of Karolinska Institutet.

The genetic factor must of course not be overlooked. In my own case such factors may have been inherited both from my father, who was a biochemist, and my mother who began with botany and specialized in diatomes, following up the work of her father, who was also professor of inorganic chemistry in Uppsala and discovered the elements Scandium and Thulium. It is probably quite common that a son of scientists at an early stage becomes engaged in some kind of research work. Through my father's wise suggestion I received an excellent introduction into research by joining the group around Robin Fåhreaus in 1925. Fåhreaus, who had then already won considerable fame by his discovery of the sedimentation reaction in the blood, had a singular gift of enthusing his disciples, spurring their curiosity and interest. Every observation was "important" and could lead to results of great biological significance. Fåhreaus had all the charm of a gifted scientist who could provide the "ignition" impulse to start one's mind in the research direction. Equally helpful was my teacher Göran Liljestrand, who set off the second stage, in the form of a Rockefeller Fellowship, allowing the happy holder to study abroad with outstanding scientists. One can only wish that today's young research workers could experience the same happiness as befell one when the letter of acceptance arrived. I have since learned that many of my colleagues agree with me that these Fellowships have been truly instrumental in creating the solid basis for a research career, including a certain status.

Also here the choice suggested by my teacher was excellent; I was to study for half a year at The National Institute for Medical Research in London under H. H. Dale. At this time, 1930, rapid progress was made in experimental pharmacology and physiology, and Dale was one of the most renowned and successful leaders in this field. Dale combined precision, careful experimenting, and critical evaluation of the results with thorough knowledge of the literature and scientific phantasy. His laboratory was an ideal place for a young scientist with open eyes and a willingness to learn.

Acetylcholine was in the center of interest in many laboratories and Dale must have felt that it played a central role in physiology. The intense action of this drug on the motility of the gut suggested that it might serve as an intestinal motility hormone, or "Hormon der Darmbewegung," as previously ascribed to choline by le Heux, who based this concept on the results obtained with drugs on the isolated intestine as used in R. Magnus' laboratory in Utrecht.

My first task in Dale's laboratory was to try to demonstrate the presence of such a hormone in the effluent of a perfused intestine of the turtle and rabbit after nerve stimulation. This failing, I resorted to the simpler task of searching for it in extracts of rabbit intestine. The result appeared most encouraging when the extracts were tested on an isolated piece of rabbit jejunum, since this contracted very nicely on addition of the extract to the bath. However, addition of atropine to the bath fluid did not suppress this effect, which seemed then to exclude choline and acetylcholine, and with youthful enthusiasm I declared that a new biologically active substance had been discovered! This was of course not immediately accepted, but at Dale's suggestion the effects observed became subject to further study, in which I had the privilege of working with John H. Gaddum, then first assistant in Dale's laboratory. After some months of hard work and valuable advice from the experienced chemists at the Hampstead Institute it became reasonably certain that the active factor was a new active principle, which was simply called "P" (for our standard Preparation) and later "Substance P" which it still is called. A few years afterwards I found that it could be salted out and behaved like a polypeptide. Later, B. Pernow in our laboratory found a simple method of obtaining the substance in a high degree of purification.

Thus almost the first turn of the spade brought up a new biologically active factor which certainly was to a large part due to luck. Whatever the cause, it had a strongly encouraging effect on the young scientist. Clearly, the demonstration of a new compound with certain actions only constituted the introductory step, and it remained to show what possible physiological function it could have. Definite conclusions as regards the physiological role of Substance P have still not been reached, but its occurrence in the gut, its high biological activity, and its atropine resistance would make it a strong candidate for a motility hormone of the gut, rather than acetylcholine.

These early experiments not only whetted the appetite for finding more active substances in biological material but also provided the necessary "know-how" for making such attempts successful. Competition was strong, however, and this was the time for new unidentified biologically active substances to appear in large numbers. Some of these later proved to be mixtures of known substances or the result of misinterpreted effects.

On my return to the Pharmacological Department of the Karolinska Institute in the early thirties, further studies of the biological action of tissue extracts led to the observation of what appeared to be adrenaline in the prostate gland. J. B. Collip had made similar observations and carried the purification of the active sympathomimetic factor further. He arrived finally at the conclusion that it might be tyramine. Considering that this amine releases noradrenaline, which is indeed the active factor in the vesicular gland, the conclusion was not far from the truth.

There was then only a short step to testing seminal fluid in 1934. The lowering effect of a small volume of the native material on the urethane-treated rabbit's blood pressure was truly startling, and again suggested a new active principle. Learning that M. W. Goldblatt in England had published a note of some similar results in a little known Journal the year before was encouraging and provided at the same time a challenge to continue these studies. A systematic study of prostate and vesicular glands from various animals gave the surprising result that, except for human material, only the sheep vesicular gland contained the new factor in large amounts.

In the course of the purification work it became clear that the active factor was of lipidic character, and it could soon be characterized as an unsaturated lipid soluble acid which was named prostaglandin. This was definitely proven with Professor Hugo Theorell's ingenious electrophoresis apparatus, which made it possible to follow the mobility of the active principle. The oily material fortunately yielded a watersoluble barium salt on addition of barium hydroxide which precipitated large amounts of impurities, and on desiccation it gave a dry amorphous powder, suitable for storage. From a large batch of vesicular glands from sheep collected by the helpful Icelandic Slaughter Company enough material could be obtained to serve as starting material for further purification work and biological tests. It was also natural to approach the lipid specialist S. Bergström, who soon became interested in the purification problem. After long and systematic work he not only succeeded in isolating several members of the prostaglandin family but also tackled the intricate structural problem, a masterpiece of chemical knowledge and skill.

I believe these events illustrate in a convincing way both the value of early training in an outstanding laboratory like H. H. Dale's, and the great advantage of belonging to a school fostering scientists like Theorell and Bergström. Under less fortunate circumstances the basic observations might not have been made, and the subsequent development not achieved in our research groups.

At the time of its discovery in 1934 and in the next few years afterwards the prostaglandin was mainly regarded as a curiosity, and in the interval passing until its isolation by Bergström and his group in 1960, other interests came into the foreground. The excretion of amines in the urine either as free amines or as conjugates had been demonstrated by several laboratories and it seemed tempting to look further into this field. Before long the biological tests revealed a nicotine-like substance in urine. I rejected the suggestion of my friend Irvine Page that is was just nicotine, since it occurred not only in the urine of my two boys, then 8 and 10 years of age, but also in bovine urine. This stimulated my interest and the following year, 1944, the active compound was isolated and identified—it was piperidine. This might have started off a systematic study of its occurrence in the body and its formation, but some other results claimed preferential interest since they appeared to be of special significance.

The story of the different "sympathins," told by Cannon and Rosenblueth in the early thirties raised great interest as a continuation of the discovery of chemical transmitters by Loewi and Cannon. The situation became increasingly more intriguing, however, by some observations made by other research workers. Bacq's hypothesis in 1933 that "Sympathin E" might be identical with noradrenaline was not accepted by Cannon and Rosenblueth, however. As late as 1939 Cannon concluded that tissue sympathin actually was adrenaline, as generally held at that time.

This was evidently a field offering some chances of obtaining more information by analysis of tissue or nerve extracts. Such extracts in our experiments clearly showed the presence of an adrenaline-like substance, but it became gradually clear that the activity pattern did not wholly agree with that of adrenaline. Could it be noradrenaline? This was a real challenge, sweeping away other seemingly promising research projects. It must also be remembered that at this time research funds were very limited and teaching took a good deal of the available time. It was necessary to select one field and not split the resources on several topics.

Once the suspicion had been raised that the nerve transmitter was different from adrenaline it became easier to design the right kind of experiments. It then turned out that the results fitted in very nicely with the assumption of noradrenaline as the active catecholamine in adrenergically innervated organs.

An attempt to check the possibility of noradrenaline occurring in the adrenal medulla offered an instructive experience. For this purpose I used an extract of rabbit adrenals which were easily available in the laboratory. The result was perfectly clear: the active amine was adrenaline. Not knowing that I had picked the only mammal with practically only adrenaline in its suprarenals, I had a good chance to consider the dangers of generalizations in the following year, when Peter Holtz' work on noradrenaline in cat adrenals was published.

The hypothesis of noradrenaline as adrenergic neurotransmitter did not in the beginning meet with great credence, probably because of the authority of those who regarded adrenaline as transmitter. This was perhaps to be expected and saved me from what might have been a disappointment. A latency period of a couple of years before the findings and the implications would be generally accepted would seem almost unavoidable. As usual some scientists at once saw the significance, like B. A. Houssay, while others only reluctantly were prepared to accept the fairly obvious evidence on which the concept was based or continued to use the term "Sympathin." On the other hand it is, I think, a common observation that some new theories, results, and concepts which are, at least to some people, clearly doubt ful or erroneous, may readily find their way into textbooks. However, interest grew rapidly, particularly in the U.S.A. and in England, where my friend Gordon Wolstenholme arranged several Ciba Symposia on various branches of the new area. The differential analysis of adrenaline and noradrenaline in urine, and the development of useful biological and chemical methods of assay no doubt stimulated the growth of this field.

If the neurotransmitter was present in the adrenergic axons, as we knew it was, how could it survive there in constant amounts? This was a question that had to be answered. The electron microscopic findings of two pioneers in adrenergic nerve biochemistry (H. Blaschko) and morphology (N. Å. Hillarp) seemed to provide the answer. Why should not subcellular particles rich in noradrenaline occur in adrenergic nerve axons if they could occur in the homologous chromaffin cells? Hillarp at once responded to my proposal of looking for "granules" in adrenergic nerves that were rich in noradrenaline, and before long we had the necessary data to express the view that the adrenergic neurotransmitter was stored in a protected form in the axons, and consequently in all organs supplied with adrenergic nerves.

The isolated nerve or organ granules have subsequently served as a readily available material for studies of their properties in several laboratories including our own, and their multi-faceted reaction pattern to various drugs has gradually helped to elucidate several types of drug-action on adrenergic activity. The enormous growth of the number of studies in this field, regarding both the CNS and the periphery, makes it increasingly difficult to maintain a clear view of the field. Still, the mechanism behind the release of the adrenergic neurotransmitter is a challenging problem and we stand here before events that take place at a level close to molecular biology. Although knowledge about autonomic neurotransmission has advanced conspicuously in a brief period of some 25 years I feel personally that some day, perhaps in the next 10 years, an increased insight into molecular and membrane biology will allow us to see more clearly how the transmitter is liberated from the axon. A salient point is still: If acetylcholine is implicated, where is its place?

Only rarely the research worker is able to make contributions that are so rounded off that they seem to form a closed chapter. Perhaps this is more true for physiology than for other disciplines. Very often there is a system of cross-connections that seem to extend in all directions. As a finishing illustration I shall only mention one such cross-connection that appears to bind together two fields in which I have been specially interested. I am referring to the release of prostaglandins on adrenergic nerve stimulation and the action of the PGs on the effect of such stimulation. The recent finding in our laboratory (P. Hedqvist) that prostaglandin in minute concentrations can under certain conditions block adrenergic nerve transmission represents, I think, an example of an unexpected crosslink between two systems. Thus research impulses leading to new and useful combinations clearly depend on the environment. Variations in the "programming" of the scientists would seem to be a prerequisite for ensuring a healthy variability of the research.

Sometimes even apparently lackluster and uninspired, tedious work may give good information. Certainly it appeared a dull preoccupation to determine the noradrenaline content of a large number of organs and tissues in the body, but it did reveal some facts of importance such as the presence of the neurotransmitter in the brain, and the occurrence of remarkably high amounts in the male accessory glands. Finally it offered a means of determining the relative supply of adrenergic nerve fibers to different organs, as confirmed by the later developed histochemical fluorescence technique.

Even if testing extracts of various tissues and organs on a battery of pharmacological test objects may not be characterized as sophisticated science, it can lead to valuable findings, often not predictable. It is perhaps worth recalling that research in Portuguese is called "pezquisas," which literally means fishing. For those who have a feeling for where to fish and recognize a good fish when they see it, this method has its merits. On the other hand the intellectual pleasure behind an intelligently planned experiment giving the desired answer is great and well earned.

Which advice, if any, should be given to young scientists as regards research procedures in a wider meaning? Considering the large number of people actively engaged in research at the present time one can be reasonably sure of two things. One is that a large proportion of these will develop into good and useful conventional scientists, using available techniques on generally accepted problems, and applying good statistics to the required number of experiments. The results of these efforts occupy a large part of the space in the growing number of journals. Much of this work is of confirmatory type and would be useful but for the often too large number of pages. A tendency in papers of this type to expand on long and tedious discussions is noticeable.

For a young scientist it is of great concern that he should be recognized by those who decide the distribution of grants and handle the applications for positions. A board of first rate scientists generally chooses the best of the young generation for grants and positions, thereby tending to perpetuate itself, which, alas, may be said also of a board of less distinguished members. In a large country with many universities and research institutions the situation is never serious, but in a small country the swing in either direction may be very marked. In fact a few authoritative persons of the convincing type may on questionable grounds promote certain types of research and hold back others. The young scientist therefore may have to adjust his work, or at least his program, so that it fits in the prevailing pattern. It is thus tempting for a young research worker to fall in line with a reasonably profitable line of research which has reached a state of general acknowledgement. It is safe; and since work is going on in the field there are good chances to be quoted, often in connection with a recognized colleague. The population of research workers has sometimes been compared to a heap of soapbubbles. When a small one collides with a big one it just unifies with the large one and makes this increase.

The advice given to young scientists, approaching the entrance of the temple of science, had formerly often an idealistic touch, and to a previous generation, more accustomed to sermons, the wise men were expected to preach on chosen occasions. No doubt the present generation of young scientists is less amenable to listen to the big words; they look at their job in a more practical way and their primary thoughts are often more concerned with salary than research philosophy. They know that in order to proceed along the career road they must have publications in sufficient number. Consequently they are tempted to prefer investigations within an established field using techniques that are acknowledged, and concepts that are accepted by the Granting Boards. Such studies will often lead to the desired position without too original thoughts or lucky findings. Unfortunately the products will often bear a slight mark of safe banality stamped upon them although they serve their purpose in other respects. The harsh competition in fact appears to foster a new kind of research worker who is more like an engineer who produces numerous data with the aid of sophisticated machines.

A recurring dilemma for any scientist is to know when to go on with a problem and when to leave it. This also implies that those who are responsible for directing or financing other people's work must judge the probability of success or failure. When a scientist builds up a hypothesis on the basis of his own data and those of others, in pharmacology as in other biological sciences, the process mostly rests on the more or less solid ground of analogies. To decide whether a new hypothesis is likely or not is a formidable task that requires judicious weighing of a large number of factors, which can hardly be assessed on a precise basis. There is little doubt, however, that some prominent scientists have a great ability to separate the gold nuggets from the uninteresting material and have a "feeling" for the probability of a hypothesis or theory, and for its significance.

Obviously some scientists have the gift of selecting topics and problems better than others. When working in Sir Henry Dale's laboratory in London in the thirties we used to say that "a guess by Sir Henry is more likely to be right than many so-called established facts from other sources." In other words, we had accepted the use of probability without evidence. This way of proceeding has dangers but as long as there are limits to man hours and funds this method has its virtues and experience seems to verify this. Of the numerous suggestions and theories put forward only those will be selected for closer study which appear to be likely. It is of course inevitable that this occasionally leads to overlooking single projects that should be followed up, but on the whole this procedure has many advantages.

The fast growth of the volume of scientific achievements has been shown to follow an exponential function, doubling every 10 to 15 years. This applies to the number of individuals engaged in the field, as well as to the number of journals and other parameters, and varies only to a small degree between different subjects and different countries. The rapid growth of pharmacology, like other sciences, has some interesting corollaries. Not only does the exponential increase mean that about one-half of all scientists who have ever existed on our planet have worked for less than a dozen years, but—on a quantitative basis—one-half of our accumulated total knowledge or, let us say the number of data, is the result of research work done in the last decade. No wonder it is hard to keep track of them. Considering this menacing avalanche of data it becomes a "must" to present them in as "pure" form as possible, freed from all unnecessary outgrowths. Some journals adhere successfully to this principle for the reader's benefit, but much remains to be done. Overdocumentation and verbosity are still too frequently encountered. For the young scientist it is often a disappointing fact that he is rarely quoted unless by himself, and his papers are mostly rapidly buried in an ever increasing stream of publications. On the other hand it is increasingly evident that a small number of papers are quoted over and over again.

To observe the events and trends in a research field is sometimes fascinating, often interesting and even intriguing. At times one is reminded of a flock of starlings rapidly switching from one direction to another, following some directional signals unseen to the watcher. Sometimes an idea is "fashionable" and will rapidly ensure the adherence of a large group. It can be the correct one for that matter, but the opposite may happen. To turn the tide in such cases is often a slow and tiresome process. Controversial opinions—even with the wrong one at the top—are not necessarily an unwanted state of affairs since they may challenge the ingenuity and experimental skill of the proponents and lead to a quicker solution of the problem. Tenacity in maintaining fixed positions and unwillingness to accept evidence from the counterpart are often characteristic features in this game, however.

The diversity of problems and difficulty of following the work done in other sectors tend to limit the outlook for the experimentalist. Many attempts have therefore been made to integrate the findings in different or at least adjacent fields to a more complete picture. To take an example: in pharmacology as in physiology one of the most pertinent tasks is to solve the problems of interaction between agonists, antagonists, and receptors. Protein and lipoprotein chemistry, combined with morphological studies on membranes and electrophysiological experiments revealing ion currents are here likely to give useful information. The increasing necessity to consider events at a subcellular or molecular level is apt to bring new and perhaps unforeseen difficulties of conception. We have still a very incomplete idea how the molecules move and shift in their microworld; the consistency and geography of the cytoplasm is still imperfectly known and so are the molecular pathways within the cell. The time scale is another complicating factor. We know that on depolarization of the axon membrane at the neuro-muscular junction a momentary shift in calcium ion distribution occurs, followed by a release of the transmitter acetylcholine, which on the other side of the synaptic "cleft" reacts with receptors on the muscle in the endplate region of the muscle cell. All this occurs within a millisecond, as shown by the beautiful studies of Katz and others. Increased knowledge of the kinetics of fast reactions in organic media are here a prerequisite for better understanding of the events.

Of the many new research fields associated with recent developments in the neurotransmission area perhaps the central actions, including psychopharmacology, have the most far-reaching consequences. This was predicted by H. H. Dale who wrote (about acetylcholine) in his Nobel Lecture in 1936: "The possible importance of such an extension, even for practical medicine and therapeutics, could hardly be overestimated." The discovery by B. B. Brodie that after administration of reserpine amines were no longer stored in the brain in the normal way and that these disturbances were associated with psychic alterations was a breakthrough in this field. We are witnessing a steady progress in psychopharmacology, and one might predict that by the judicial use of drugs it will be possible to adjust the mental instrument to considerable extent. There is no need to emphasize what can be achieved by improper use of such tools.

The steadily increasing number of scientists in the biological as in other fields might perhaps suggest that they should be increasingly powerful as a body, particularly considering the importance of the biological sciences for human welfare. Recognizing their own potential value, biologists sometimes express the view that they should be called upon to a greater extent to take an active part in the handling of the city's, the nation's, or the world's affairs.

It would not be difficult to find a large number of highly competent, honest, and hardworking scientists, willing to assume such a role as advisers to the population in toto, and yet the number serving in such capacities is very limited to say the least. The reason is of course that those in power—regardless of the nomenclature of the system—are for the most part only moderately interested in aspects of science other than those that can serve their political aims. In such cases a suitable expert can always be found, serving both as a proof of recognition of research as such and as a scientific support for a specific purpose. It may be well to recognize that this situation is not likely to change and that the ambitions of the scientist should profitably be linked also in other directions.

Although research work is a job like many other kinds of work it is often so engaging that it may indeed sucessfully compete with most hobbies. My revered master and friend Bernardo Houssay, on the third day of his holidays at the seaside wrote a postcard to the laboratory: "I envy you who can stay in the lab and do research." This love for the work makes it possible to find individuals who are willing to do high quality work at a modest remuneration.

Research requires many co-factors in addition to hard work and some luck in order to be fruitful. Perhaps one might use a travesty: "En recherchant, il faut être dégagé de toute autre préoccupation." Everyone engaged in research recognizes the importance of a good research climate. To define this is not easy but it does include a certain amount of freedom of time and freedom from want, a reasonable supply of equipment and means to continue along unexpected lines, relatively loose ties as regards research program and so forth. These requirements may seem too liberal for many rationalists and authorized research planners and for those who allocate the money, yet there is no guarantee that even under these favorable conditions anything really worth while will turn up for a long time. Fortunately, there are almost always some products of work that can be accepted as a reasonable return for the money spent.

Looking back, it is fairly obvious that the past 50 years have been a favorable time for research, reasonably unhampered by such interference as might check the freedom and choice of subjects. It is not equally clear that this will be the case in the future. An increasing tendency to control the activities can be noticed and this may be felt as a tether that does not encourage those to enter the field for which a certain freedom is a prerequisite for successful work. It is to be hoped that even in the future a certain liberty will be allowed in this respect.

Most of the reflections laid down in this article have been said before and probably better, but the experience of those who have been in research for a long time and who have met scientists of all kinds and calibers may still have an interest for others who are in the beginning of a research career. After all, homeostatic mechanisms operate also in the human mind, and certain basic rules are likely to remain unaltered, even if a variety of external factors may produce temporary changes in the outlook.

The Excitement and Fascination
of Science: Volume One (published 1965)

CONTENTS